THE PHYSICS OF ELEMENTARY PARTICLES

by

H. MUIRHEAD

NUCLEAR PHYSICS RESEARCH LABORATORY
UNIVERSITY OF LIVERPOOL

PERGAMON PRESS

OXFORD · LONDON · EDINBURGH · NEW YORK
PARIS · FRANKFURT

Pergamon Press Ltd., Headington Hill Hall, Oxford
4 & 5 Fitzroy Square, London W. 1
Pergamon Press (Scotland) Ltd., 2 & 3 Teviot Place, Edinburgh 1
Pergamon Press Inc., 122 East 55th St., New York 22, N.Y.
Pergamon Press GmbH, Kaiserstrasse 75, Frankfurt-am-Main

First Edition 1965

Library of Congress Catalog Card No. 64–15737

CONTENTS

PREFACE

THE object of this book is to discuss the physical principles which are used to describe the properties and behaviour of the elementary particles. The book is intended for graduate students working in the subject of elementary particle physics. The emphasis is, however, on explaining theoretical ideas to the experimentalist, and so the book should be also useful to more senior workers in this field.

The book is intended to give its reader a working knowledge of the symmetry laws and of the methods of discussing interactions with the aid of graph techniques and dispersion relations (together with their limitations). The theoretical basis of these methods has been discussed, but it is not claimed that the treatments are necessarily rigorous.

In the book I have tried to keep the mathematics as simple as possible. In places this simplicity has meant some sacrifice of mathematical elegance, but I believe this is justifiable since the experimentalist is not as familiar with mathematical techniques as the theorist. I also believe some stress should be laid on the theoretical foundations of the subject, for example the implications of special relativity have been examined in some detail. A working knowledge of quantum mechanics has been assumed.

In writing this book I have had help and advice from many colleagues. I am indebted in particular to Professor A. W. Merrison, Drs. G. R. Allcock, L. Carroll, J. Eades, A. N. Kamal, D. N. Olson, G. Rickayzen, J. C. Willmott, T. Woodhead, J. R. Wormald and J. N. Woulds for reading parts of the manuscript.

I am also indebted to the editors and publishers of the following books and journals for permission to reproduce certain figures appearing in the text.

The Aix-en-Provence International Conference on elementary particles, Figures 14.11, 14.17, 14.18, 14.22, 14.35.
Annual Reviews Inc., Figures 12.3, 12.4, 14.13, 14.14, 14.15.
CERN Information Service, Figures 14.1, 14.20, 14.21, 14.29, 14.30.
Il Nuovo Cimento, Figure 11.5.
The Physical Review, Figures 11.11, 11.27, 11.28, 14.3, 14.5, 14.9, 14.26, 14.27.
Physical Review Letters, Figures 9.11, 10.25, 10.26, 10.27, 11.25, 11.26, 12.22, 12.27, 13.4, 13.5, 13.6, 13.8, 13.9, 14.12, 14.33, 14.34, 14.36.
Proceedings of the Physical Society, Figure 13.7.
Reviews of Modern Physics, Figures 13.10, 13.11.
John Wiley and Sons, Inc., Figure 6.1.

H. MUIRHEAD

CHAPTER 1

THE DISCOVERY AND CLASSIFICATION OF THE ELEMENTARY PARTICLES

1.1. EARLY WORK WITH THE COSMIC RADIATION

1.1 (a) *Electrons to pions*

The discovery of many of the elementary particles has been associated with work on the constitution of the cosmic radiation. For example, examination of the electronic showers observed in cloud chambers led to the identification of the positron (Anderson, 1932; Blackett and Occhialini, 1933).

Some particles were well known by this period; the electron and proton had been identified in the classical work of J. J. Thomson and others on gaseous discharges at the turn of the century. The study of the radiation from a black body had led to the realisation that electromagnetic waves were quantised; the quanta are now called photons (Einstein, 1905). The neutron had been found as a result of the early investigations into nuclear reactions (Chadwick, 1932).

The discovery of the positron was especially important, however, for two reasons. First, because it marked the starting-point of a series of discoveries of new particles in the cosmic radiation which were to take place over the next twenty years. Secondly, because it provided the first important experimental proof of a prediction arising from relativistic quantum theory.

The discovery of the positron demonstrated the correctness of Dirac's relativistic wave equation for the electron (1928). One of the properties of this theory was that it required the electrons to exist in states of negative energy as well as the more familiar positive energy states. This was at first believed to be a fundamental difficulty in the theory. Dirac believed such states could exist and proposed the 'hole' hypothesis as an explanation (1929). He suggested that the particles occupying states of negative energy could not be observed directly, but that a few of the states were unoccupied – these he called *holes*. Dirac showed that the holes should behave like particles of positive charge, and suggested that they might be protons. It was pointed out by Oppenheimer (1930), however, that if the proton was associated with the electron hole, then the e^-p^+ system should annihilate within a period of $\sim 10^{-10}$ sec. Later it was shown by Weyl (1931) that the hole should have the same mass as the particle, and so a particle of positive charge and electronic mass should exist.

Blackett and Occhialini identified the positron with the hole corresponding to the electron in Dirac's theory; they also pointed out that the falling of a negative electron into a hole would appear as the collision of a positron with an electron and would result in their mutual annihilation. The energy liberated was assumed to appear as two γ-rays. Experimental confirmation of this process was obtained by Thibaud (1933).

The hole theory is not limited to electrons and positrons. Any particle satisfying Dirac's equation should possess an equivalent hole which behaves as an antiparticle. Further proof of Dirac's theory has been given by the discoveries of the antiproton (Chamberlain, Ségrè and Ypsilantis, 1955), the antineutron (Cork *et al.*, 1956) and the anti-Λ hyperon (Prowse and Baldo-Ceolin, 1958).

In addition to the positron, the examination of the processes associated with the development of the electronic showers in the cosmic radiation also led to the discovery of the meson. An apparent anomaly was observed in the absorption of these showers in various materials. Some of the particles penetrated greater thicknesses of matter than was feasible if they were either electrons or protons (protons would have lost energy too quickly by ionisation and electrons by radiation). Furthermore the particles appeared to possess either positive or negative electric charge. The analysis of the experimental data strongly indicated the existence of a particle of mass intermediate between that of the proton and electron, and probably in the region of 100–200 electron masses (m_e). This suspicion was verified by Neddermeyer and Anderson (1938) who photographed a particle, with a mass estimated to be $\sim 240\ m_e$, stopping in a cloud chamber.

The existence of a particle with about this mass value had been predicted by Yukawa (1935). In his paper Yukawa suggested that the strong, short range forces, which bound *nucleons* (neutrons and protons) together in a nucleus, arose from the mutual exchange of massive quanta or mesons which were strongly coupled to nucleons. The range of this force was of order $1/m_\pi \, (\equiv \hbar / m_\pi c)$, where m_π denotes the mass of the mesonic quantum.

Yukawa made a further suggestion about the properties of his hypothetical particle. In order to account simultaneously for nuclear β-decay and for the fact that the meson had not (at that time) been observed, Yukawa suggested that it decayed spontaneously into an electron and a neutrino in a time which he estimated to be about 10^{-7} sec. Apparent verification of this property was obtained by Williams and Roberts (1940). These workers observed the β-decay of a particle of mass $\sim 250\ m_e$ in a cloud chamber.

An identification of the meson observed in the cosmic radiation with Yukawa's particle was made at this period. Apparent support for the identification was given by the measurements of the lifetime of the mesons by Rasetti (1941) and by many others. The determinations gave an average value of $\sim 2 \times 10^{-6}$ sec. The hypothesis was proved to be untenable, however, by the experiments of Conversi, Pancini and Piccioni (1947). These workers found that when negatively charged mesons were brought to rest in carbon virtually all decayed, but when they were arrested in iron no decay electron appeared. This result was surprising, since it had been expected (Tomanaga and Araki, 1940) that when negatively charged mesons were brought to rest in matter they would undergo rapid nuclear absorption because of Yukawa's postulated strong interaction between nucleons and mesons. Thus no decay electrons were expected.

Subsequent work showed that the negative mesons underwent either decay or capture when brought to rest, and that the capture rate λ_c increased as the atomic number Z of the stopping element increased. It was found that roughly half the mesons decayed and half were captured in the region of $Z \sim 12$. Thus we find

$$\lambda_c \sim 10^6 \text{ sec}^{-1}$$

for $Z \sim 12$, since the decay rate is given from experiment by $\lambda_d \sim 10^6 \text{ sec}^{-1}$. When brought to rest in matter, a negative meson falls into a Bohr orbit about a nucleus; this orbit is roughly 200 times smaller than that for the electron, since the mass of the meson is $\sim 200\,\text{m}_e$. A simple calculation shows that the meson would then spend about 10^{-3} of its time inside a nucleus for $Z \sim 12$. Thus if the interaction had been of the strong type predicted by Yukawa, the expected capture rate would have been about

$$\lambda_{\text{strong}} \sim 10^{23} \times 10^{-3} \sim 10^{20} \text{ sec}^{-1}$$

where 10^{23} represents the inverse of the characteristic nuclear time† $[(1/c)(\hbar/m_\pi c)]$. Thus the discrepancy was enormous; the ratio

$$\frac{\lambda_c}{\lambda_{\text{strong}}} \sim \frac{10^6}{10^{20}} \sim 10^{-14} \tag{1.1}$$

provides a number which is characteristic of the weak nuclear interactions, for example β-decay (compare § 8.7(c)).

Thus the meson of mass $\sim 200\,\text{m}_e$ and lifetime 2×10^{-6} sec could not be of the type predicted by Yukawa, and it was suggested by Bethe and Marshak (1947) that other types of meson might exist in the cosmic radiation.

Evidence for the existence of two types of meson was presented by Lattes, Muirhead, Occhialini and Powell (1947). These workers obtained photographs in nuclear emulsions of the decay of one meson into another. Subsequent work (Lattes, Occhialini and Powell, 1947; Goldschmidt-Clermont *et al.*, 1948) showed that the parent particle in the decay process was of mass $\sim 270\,\text{m}_e$, and that it interacted strongly with nuclei and was produced copiously in high energy nuclear interactions. The product particle was found to have mass $\sim 200\,\text{m}_e$ and weak nuclear interactions; it was therefore identified with the weakly interacting particles which had been observed earlier by other techniques.

The parent meson was called a π-meson, and its product a μ-meson. Nowadays it is customary to reserve the word meson for the strongly interacting

† The nuclear size is of order $\hbar/m_\pi c$; thus the value of $[(1/c)(\hbar/m_\pi c)]$ gives an order of magnitude for the time taken by a particle to cross the nucleus.

particles of mass intermediate between an electron and proton. The π and μ particles are now also called *pions* and *muons* respectively.

The observations with the nuclear emulsions showed that the muons arising from the stopped pions possessed a unique kinetic energy, thus indicating that the pion decayed into two particles. The kinematics of the decay process indicated that the unobserved particle accompanying the muon was of low mass. It is now known to be the neutrino† so that the pion decay process is

$$\pi \rightarrow \mu + \nu .$$

Subsequent work showed that the lifetime for this process was $\sim 10^{-8}$ sec (Camerini *et al.*, 1948; Richardson, 1948).

The pion was thus identified with the strongly interacting heavy quantum of Yukawa's theory. An important extension of Yukawa's work was made by Kemmer (1938). Experiments in low energy physics had shown by that time that proton–proton and neutron–proton forces were approximately equal. Kemmer pointed out that in order to explain this result it was necessary to assume the existence of neutral mesons as well as charged ones. Evidence for the existence of neutral pions was obtained by Carlson, Hooper and King (1950), and independently by Bjorkland *et al.* (1950). The latter experiment is of interest since it was performed with a machine rather than with the cosmic radiation. The former experiment was more informative, however. The presence of the mesons was identified through the appearance of electron–positron pairs close to high energy nuclear disintegrations in the emulsions. Carlson, Hooper and King suggested that the decay process was

$$\pi^0 \rightarrow \gamma + \gamma$$

and that the photons materialised as electron–positron pairs. They estimated that the neutral meson possessed a lifetime of $\sim 10^{-14}$ sec and a mass of $\sim 300\ m_e$. The near equivalence of the mass value to that of charged pion suggested that the new particle was its neutral counterpart.

1.1 (b). *Kaons and hyperons*

In 1947 Rochester and Butler (1947) obtained evidence for the existence of massive unstable particles in the cosmic radiation. Whilst examining penetrating (mixed muon and electron) showers with the aid of a cloud chamber they obtained two classic photographs. One picture showed the decay of a neutral particle into two charged ones and the other the decay of a charged particle. Rochester and Butler were able to assign lower limits of $770 \pm 200\ m_e$ and

† The mass of the neutral particle is ~ 0 and the obvious choices are that it is a neutrino or a photon. A search was made by O'Ceallaigh (1950) for electron pairs materialising from photons along the possible flight paths of the neutral particles in nuclear emulsions. No events were found, and the data indicated that the probability for the process $\pi \rightarrow \mu + \gamma$ was less than 4×10^{-3}.

980 ± 150 m$_e$ for the masses of the unstable neutral and charged particles respectively.

Further evidence for the existence of heavy unstable particles was provided by Brown *et al.* (1949), who reported on the observation in a nuclear emulsion of the decay of a charged particle into three charged mesons. One of the mesons appeared to be a π-particle, whilst the other two were either pions or muons. The parent particle was called a τ-meson, and its mass was estimated to be ~ 1000 m$_e$. Subsequent photographs obtained by the Bristol group, and by other workers, showed that the decay process yielded three pions

$$\tau^+ \rightarrow \pi^+ + \pi^+ + \pi^-.$$

In what follows, the present-day nomenclature for the heavy unstable particles will be used. Those which possess strangeness and whose final decay products include a nucleon are called collectively *hyperons*, those with mass ~ 965 m$_e$ are called *K*-mesons or *kaons*. In addition the hyperons and nucleons are given the generic title of *baryons*. It is also customary to indicate the nature and number of the decay products by subscripts; thus the τ-meson is called a $K_{\pi 3}$.

Following the early reports considerable research was carried out on the heavy unstable particles during the early 1950's. The work revealed a rich variety of phenomena, and attention was focused on the classification of the particles according to masses, lifetimes and decay schemes.

Thus Armenteros *et al.* (1951) were able to show that at least two types of neutral particles existed, one, which is now called the Λ^0 hyperon, decayed according to the scheme

$$\Lambda^0 \rightarrow p + \pi^-$$

and the other probably decayed as follows

$$K^0 \rightarrow \pi^+ + \pi^-.$$

(The K^0 particle was called a θ^0 at the time of its discovery.)

Many decay modes were found for the kaons, although at first it was not realised that they represented alternative decay modes of the same particle. Thus O'Ceallaigh (1951) found a particle of mass 1125 ± 140 m$_e$ which decayed to a muon and at least two neutral particles; the parent particle could therefore be labelled as $K_{\mu 3}$. By way of contrast the Paris group, using a double cloud chamber, established the existence of a kaon which decayed into a muon with unique range (Gregory *et al.*, 1954). The kaon could thus be labelled as $K_{\mu 2}$. The energy of the muon was consistent with the decay scheme

$$K \rightarrow \mu + \nu.$$

This scheme was chosen rather than the process $K \rightarrow \mu + \gamma$ because no electron-positron pairs were observed in close association with the point of decay.

Further decay modes involving the conversion of kaons into pions were also found. Crussard *et al.*, (1954) showed that an alternative decay mode of the τ-meson ($K_{\pi 3}$) existed

$$\tau' \equiv K_{\pi 3}^{+} \rightarrow \pi^{+} + ? + ?.$$

The question marks are now known to be π^0 mesons. An alternative decay mode of the kaon was also found by Hodson *et al.* (1954)

$$K^{+} \rightarrow \pi^{0} + \begin{pmatrix} \pi^{+} \\ \mu^{+} \end{pmatrix}.$$

They pointed out that the mass value for the kaon gave a more consistent fit with other data if the decay scheme

$$K^{+} \rightarrow \pi^{+} + \pi^{0}$$

was chosen. This conclusion received later confirmation.

Further types of hyperon were also found in the cosmic radiation. Thus York, Leighton and Bjornerund (1953) obtained a photograph in a cloud chamber which was tentatively identified as

$$\Sigma^{+} \rightarrow p + \pi^{0}.$$

The Σ-hyperon was more massive than the Λ^0 hyperon. The workers quoted above estimated its mass to be $\sim 2200\ m_e$. Confirmation of the existence of the Σ-hyperon was obtained by Bonetti *et al.* (1953), who also identified the alternative decay mode

$$\Sigma^{+} \rightarrow n + \pi^{+}.$$

Later the negative counterpart was observed in a diffusion chamber exposed to the negative pion beam of the Brookhaven cosmotron (Fowler *et al.*, 1954).

A picture of unusual interest was obtained by Cowan (1954). It revealed the existence of a 'cascade' hyperon Ξ^- of mass $\sim 2600\ m_e$ which decayed to yield the Λ^0 hyperon. The presence of the latter particle was identified by its decay in the same photograph

$$\begin{array}{l} \Xi^{-} \rightarrow \Lambda^{0} + \pi^{-} \\ \qquad\ \ \downarrow \\ \qquad\ p + \pi^{-} \end{array}$$

It is noteworthy that no positively charged counterpart of the Ξ^- particle has been observed. The neutral variety will be discussed in § 1.4(b).

This completes the account of the early discoveries of the elementary particles. During the period 1954–6 the emphasis shifted from work using the cosmic radiation to work on the large accelerators. The general classification of the known particles into various groups was completed by this period, however. The classification is shown in Table 1.1 (more detailed tables of properties of the particles are given in A. 9 Appendixes, p. 715). In this table masses are given in m_e units.

By 1955 it was known that the kaons possessed masses which lay close to 965 m_e, and the masses of the hyperons were also reasonably well determined. Rough lifetimes had also been set upon the particles from observations with cloud chambers. The cloud chamber has an effective lifetime τ for the observation of particles given by

$$\tau \sim \frac{d}{c} \tag{1.2}$$

where d is the diameter of the cloud chamber and c is the velocity of light. If d is about 30 cm, then τ becomes $\sim 10^{-9}$ sec. Thus a study of the distribution of the

TABLE 1.1

Bosons	Fermions	
	Leptons	Baryons
Photon γ 0	Neutrino ν 0	Nucleon $N \sim 1840$
Pion $\pi \sim 270$	Electron e 1	Hyperons $\Lambda \sim 2200$
Kaon $K \sim 965$	Muon $\mu \sim 207$	Hyperons $\Sigma \sim 2350$
		Hyperons $\Xi \sim 2600$

decays inside the cloud chamber provides a measure of lifetime of the unstable particles (see, for example, Bartlett, 1953). Lifetimes of $\sim 10^{-10}$ sec were found for the hyperons and neutral kaons; those for the charged kaons were shown to be in excess of 10^{-9} sec, and probably in the region of 10^{-8} sec. The work with cloud chambers which had magnetic fields had also shown that both K^+ and K^- mesons existed.

From time to time events have been reported which do not fit into the general classifications given above. The reports may arise either from errors associated with the evaluation of the event, or from the presence of genuine particles which do not readily reveal themselves by present experimental techniques. The latter point will be discussed in § 1.5.

1.2. THE NEUTRINO

The experiments on the energy spread of the electrons from β-decay, made in the 1920's, showed an apparent failure of the principle of conservation of energy. Later work indicated that the conservation of angular momentum was also violated. Pauli (1933a) suggested that the conservation laws could be saved by postulating the existence of a particle of zero (or near zero) mass, no electrical charge and spin $\frac{1}{2}$ (in units of $\hbar$). This particle is now called the neutrino. Fermi (1934) assumed the existence of the particle when he developed his theory of β-decay. The theory has proved to be extremely successful.

The neutrino is a particle which can be described by the Dirac equation, and so it should have a corresponding antiparticle. The neutrino ν and antineutrino $\bar{\nu}$ are defined by the processes

$$n \to p + e^- + \bar{\nu} \tag{1.3}$$

$$p \to n + e^+ + \nu .$$

The β-decay of the free proton is not observed since it is energetically forbidden. The decay can occur when the proton is a member of a complex nucleus (A and Z denote atomic weight and number respectively)

$$A, Z \to (A, Z - 1) + e^+ + \nu .$$

Many attempts have been made to detect the neutrinos directly (for early references see Allen, 1958). A successful experiment was carried out by Cowan *et al.* (1956). Antineutrinos resulting from β-decay in a nuclear reactor were detected in an inverse decay process

$$\bar{\nu} + p \to n + e^+ .$$

By using arguments associated with the principle of detailed balance (§§ 2.4(b) and 9.2(a)), it is possible to work out the cross-section for this reaction from known data on β-decay lifetimes. The cross-section is energy dependent (§ 12.12), and for neutrinos coming from reactors an appropriate cross-section was calculated to be $\sim 10 \times 10^{-44}$ cm^2.

In order to detect this small cross-section, very large amounts of target material and intense beams of neutrinos were required. The neutrinos were detected in a large volume ($\sim$1400 l) of organic liquid scintillator containing cadmium. The basic principle used was the following:

$$\begin{array}{lll} & & e^+ + e^- \to 2\gamma \\ & & \uparrow \\ \bar{\nu} + p \to & n + & e^+ \\ & \downarrow & \\ & n + Cd \to Cd + \gamma . & \end{array}$$

The neutrino could collide with a proton in a hydrogen atom in the scintillator. If a reaction occurred, the positron annihilated with an electron yielding γ-rays in a period of $\sim 10^{-9}$ sec. The neutron has a flight time of several microseconds before being captured by the cadmium; capture γ-rays were then produced. In order to identify the appropriate signal from the scintillator, the energies of the pulses from the two sets of γ-rays and the time intervals between them were measured as a function of the power of the reactor. A cross-section of $(12\,^{+7}_{-4})\ 10^{-44}$ cm^2 was obtained in a first experiment; an improved version yielded $(11{\cdot}0 \pm 2{\cdot}6)\ 10^{-44}$ cm^2 (Reines and Cowan, 1959). These figures compare favourably with the theoretical figure of $\sim 10 \times 10^{-44}$ cm^2.

The same need to conserve energy, momentum and angular momentum arises in processes other than nuclear β-decay, and so neutrinos have been associated with them, for example

$$K \to \mu + \nu$$

$$\pi \to \mu + \nu$$

$$\mu \to e + \nu + \bar{\nu}.$$

A question of considerable importance is whether the neutrinos associated with muons are necessarily the same as those appearing in nuclear β-decay (this problem is discussed in §§ 12.11 and 12.12). Recent experimental evidence has shown that they are different (Danby *et al.*, 1962). A high energy neutrino beam from the decay processes $\pi \to \mu + \nu$ and $K \to \mu + \nu$ was passed through a spark chamber weighing 10 tons. Although the neutrino flux was small compared to that from the reactor in the Cowan and Reines experiment, the loss in intensity was compensated for by a rise in the cross-section to about 10^{-38} cm^2 (§ 12.12). Virtually all the observed events in which a single charged particle were produced were consistent with the reaction

$$\nu + N \to N + \mu$$

and none with the process

$$\nu + N \to N + e$$

where N indicates a nucleon in the target nucleus, and we have used the symbol ν generically for both neutrinos and antineutrinos. Thus one may conclude that $\nu_e \neq \nu_\mu$, where the subscripts indicate neutrinos associated with electrons and muons respectively.

1.3. CONSERVATION LAWS AND ELEMENTARY PARTICLES

1.3 (a). *The 'strong' conservation laws*

The conservation laws (or symmetry properties) of physics are of great importance in establishing the properties of the elementary particles. Some of the laws are well known from classical physics, others are quantum mechanical concepts. In addition some of the laws appear to hold absolutely, whilst others are obeyed under certain conditions. The former category include:

(1) conservation of charge,
(2) conservation of mass–energy,
(3) conservation of linear momentum,
(4) conservation of angular momentum,
(5) conservation of baryons,
(6) conservation of leptons,
(7) time reversal invariance.

Those which hold under certain conditions are:

(8) conservation of parity,
(9) charge conjugation invariance,
(10) isopin invariance,
(11) conservation of strangeness.

Other invariance laws can exist; some are combinations of the above laws.

The laws (1–3) are used to established the charge and rest mass of particles. Conservation of angular momentum is used to determine the spin.

The conservation of angular momentum refers to the total angular momentum, that is the (vectorial) sum of the orbital and the intrinic (or spin) angular momenta. The orbital angular momentum possesses only integral values of $\hbar$, whilst the spin can have integral or half-integral values. One important consequence of this statement is that total angular momentum can only be conserved if particles with half-integral spin appear in pairs in any process, for example

$$\pi^- + p \rightarrow n + \gamma$$

$$\text{spin} \quad 0 \quad \tfrac{1}{2} \quad \tfrac{1}{2} \quad 1.$$

Particles with integral spin obey Bose–Einstein statistics and are called *bosons*, whilst particles with half integral spin obey Fermi–Dirac statistics and are called *fermions*. This relationship will be discussed in more detail in § 9.2.

This completes our consideration for the time being of the strong conservation laws. The remaining laws mentioned at the beginning of this section will be examined later.

1.3(b). *The conservation of parity*

The concept of parity is associated with the spatial properties of a system. The principle is one which arises naturally in quantum mechanics, but does not appear to have a classical analogue. Consider a non-degenerate quantum mechanical system which is represented by a one-dimensional wave function $\Psi(x)$. If an operator P causes the coordinates of the system to be reflected, we may write

$$P\Psi(x) = \Psi(-x) = \xi_P \Psi(x) \tag{1.4}$$

and if P is applied a second time

$$P^2\Psi(x) = P\Psi(-x) = \xi_P^2 \Psi(x). \tag{1.5}$$

We have now returned the system to its original condition, and so we expect that

$$\xi_P^2 = 1, \qquad \xi_P = \pm 1. \tag{1.6}$$

The eigenvalues of P then lead to the following definitions for even and odd parity

$$\begin{aligned} \xi_P &= +1 \quad \text{even parity} \\ \xi_P &= -1 \quad \text{odd parity.} \end{aligned} \tag{1.7}$$

The argument can be extended in an obvious manner to a three dimensional system†; if the system possesses orbital angular momentum a simple relation exists between the value of the orbital angular momentum quantum number l and the parity of the system, namely

$$\begin{array}{lll} l & \text{even} & \text{even parity} \\ l & \text{odd} & \text{odd parity.} \end{array} \tag{1.8}$$

This relationship will be proved in § 5.4(b).

Apart from the parity of the spatial part of a wave function, it is possible to consider the intrinsic parity of a single particle, if that particle is a boson. Thus experiment indicates that the pion has odd parity (§ 9.5). The concept of intrinsic parity has no meaning for a single fermion, however, since fermions are either conserved in reactions or created and destroyed in pairs (§ 1.3(a)). It is possible, however, to discuss the relative parity of a pair of fermions. For example in § 5.4(d) we shall show that the parity of a fermion is opposite to that of its antiparticle.

Experiment has shown that the conservation of parity between the initial and final states of an interacting system only applies to certain classes of interaction. The interaction of the elementary particles with each other can be separated into three main classes, each with its own coupling strength. They are the following:

(1) Electromagnetic interactions, for example Compton scattering

$$\gamma + e^- \to \gamma + e^-.$$

The common parameter appearing in the electromagnetic processes is the fine structure constant

$$\alpha = \frac{e^2}{4\pi} \equiv \frac{e^2}{4\pi\hbar c} \sim \frac{1}{137}.$$

Photons are always present as real or virtual particles in electromagnetic interactions.

(2) Strong interactions, for example pion–nucleon scattering

$$\pi + p \to \pi + p.$$

The strength of this interaction is characterised by the dimensionless coupling term

$$\frac{g^2}{4\pi} \equiv \frac{g^2}{4\pi\hbar c} \sim 15.$$

The strong interactions are associated with mesons, nucleons and hyperons. The kaon couplings are somewhat weaker than those involving pions.

† In certain situations ξ_p^2 can equal -1. This point will be discussed in § 5.4.

(3) Weak interactions, for example β-decay

$$n \to p + e^- + \bar{\nu}.$$

The Fermi coupling constant G, which determines the rate of β-decay, is $\sim 10^{-49}$ erg cm^3, and can be put in a dimensionless form only if a length is specified. A convenient length is the Compton wavelength of the pion $1/m_\pi$ [$\equiv \hbar/m_\pi c$]. The rate of β-decay is proportional to G^2, and hence we may obtain a dimensionless number by writing

$$G^2 \sim 10^{-98} \sim 10^{-14}\, \hbar^2 c^2 \left(\frac{\hbar}{m_\pi c}\right)^4.$$

We will therefore take the figure 10^{-14} as a reasonable dimensionless measure of the strength of weak interactions. The weak interactions are associated with electrons, muons and neutrinos (collectively called *leptons*) and with certain decay processes for the mesons and hyperons.

Experiments indicate that parity is conserved to a high degree in strong and electromagnetic interactions (§ 9.2), but is violated in weak interactions.

1.3(c). *The failure of parity conservation*

The realisation that parity might not be conserved in certain types of weak interactions arose from the θ–τ puzzle. By 1956 it was becoming increasingly apparent that the masses and lifetimes of all the K-particles were identical, and so the various processes of kaon decay were probably different decay modes of the same particle. This feature can be seen in Table 1.2 which was compiled by Lee and Yang (1957b).

TABLE 1.2

Type	Abundance (%)	Mass of K (m_e) from		Lifetime (10^{-8} sec)
		Primary particle	Decay products	
$K_{\pi 3}$ (τ)	5·56 ± 0·41	966·3 ± 2·1	966·1 ± 0·7	1·19 ± 0·05
$K_{\pi 3}$ (τ')	2·15 ± 0·47	967·7 ± 4·0		
$K_{\mu 2}$	58·20 ± 3·00	967·2 ± 2·2	965·8 ± 2·4	1·24 ± 0·02
$K_{\pi 2}$ (θ)	28·90 ± 2·70	966·7 ± 2·0	962·8 ± 1·8	1·21 ± 0·02
$K_{\mu 3}$	2·83 ± 0·95	969·0 ± 5·0		0·88 ± 0·23
K_{e3}	3·23 ± 1·30			1·44 ± 0·46

Now consider the two decay modes

$$\tau \equiv K_{\pi 3} \to \pi + \pi + \pi$$

$$\theta \equiv K_{\pi 2} \to \pi + \pi.$$

If parity is conserved the intrinsic parity of the kaon will be determined by the net parity of the pion system. If we assume that the kaon has zero spin†, then its intrinsic parity will be given by $(\xi_{P\pi})^n$, where $\xi_{P\pi}$ is the intrinsic parity of the pion and n the number of pions in the daughter state. But $\xi_{P\pi} = -1$ (§ 9.5) and so the parity of the τ-meson wave function is given by $(-1)^3$ and that of the θ-meson by $(-1)^2$.

Thus the existence of the two decay modes pointed to the fact that τ and θ were different particles with parity -1 and $+1$ respectively, whilst the evidence from their masses and lifetimes suggested that they were identical. The problem was thus one of reconciling two conflicting pieces of evidence, and the solution was provided by Lee and Yang (1956a; see also Yang, 1957). They pointed out that the analysis of the data on the decay of the kaons had been carried out on the assumption that parity is preserved in the decay process. Now this decay process is a weak interaction, since the lifetime for kaon decay is $\sim 10^{-8}$ sec and the characteristic nuclear time is $\sim 10^{-23}$ sec, and so

$$\frac{\lambda_K}{\lambda_{\text{strong}}} \sim \frac{10^8}{10^{23}} \sim 10^{-15}. \tag{1.9}$$

Lee and Yang showed that whilst good experimental evidence existed for the preservation of parity in strong and electromagnetic processes, no data existed for the weak interactions. They listed certain consequences of the failure of parity conservation which could be subjected to experimental test. In particular they pointed out that there should be asymmetries in the emission of electrons about the direction of spin of the parent system in β and μ-decay.

Both predictions were rapidly confirmed. Wu *et al.* (1957) observed an electron asymmetry in the β-decay of partially orientated Co^{60} nuclei. Garwin, Lederman and Weinrich (1957) and Friedman and Telegdi (1957) examined the positrons emitted in the decay of positive muons and found asymmetry about the direction of muon spin. Later work showed the nonconservation of parity in the decay of kaons and hyperons.

1.3(d). *The conservation of isospin*

The concept of the conservation of isotopic spin (isospin) is associated with the experimental evidence for the principle of the *charge independence* of nuclear forces. This principle states that, at identical energies, the forces between any of the pairs of nucleons n–n, n–p and p–p depend only on the total angular momentum and parity of the pair and not upon their charge state. The equality of the n–p and p–p forces for free nucleon scattering in the singlet $^1S_0$ state has been established to good experimental limits. The property of charge independence is revealed also in the positions and properties of the energy levels in certain light nuclei; the example of the triplet B^{12}, C^{12}, N^{12} is shown in Fig. 9.5 (p. 377).

† There is good, but not conclusive, evidence to support the assumption of zero spin for the kaon; a more extended discussion of the θ–τ puzzle will be given in § 9.2(e).

Cassen and Condon (1936) showed that the principle of charge independence could be elegantly expressed by the concept of *isotopic spin* or *isospin*. The isospin of a system is formally similar to angular momentum but is linked to the charge states of the system. If a group of nuclei or particles exist in n charge multiplets, then the isospin number T for the group is given by

$$2T + 1 = n. \tag{1.10}$$

The charge state of a particle or nucleus in the multiplet is related to the third (or z) component of an isospin operator†

$$Z = \frac{Q}{e} = \left(T_3 + \frac{1}{2}A\right) \quad \text{for nucleons and nuclei}$$

$$\frac{Q}{e} = T_3 \quad \text{for pions} \tag{1.11}$$

where Z is the atomic number, Q the total charge and A the atomic weight of the system. Thus the neutron and proton can be considered as two substates of a nucleon doublet with isospin $T = \frac{1}{2}$; if χ_p and χ_n denote the isospin functions for the proton and neutron respectively then T_3 has eigenvalues of $\frac{1}{2}$ and $-\frac{1}{2}$ respectively

$$T_3\chi_p = \tfrac{1}{2}\chi_p, \quad T_3\chi_n = -\tfrac{1}{2}\chi_n. \tag{1.12}$$

The association of mesons with charge independence was first made by Kemmer (1938), and led to his proposal for the existence of a neutral meson (§ 1.1(a)).

The total isospin of a system is conserved in strong interactions. In some classes of weak interactions a failure of isospin conservation occurs. Electromagnetic interactions are manifestly charge dependent, and so the concept of isospin cannot be usefully applied to them.

1.4. STRANGE PARTICLES

1.4(a). *Associated production*

The early work on kaons and hyperons showed that the cross-section for the production of these particles was probably a few per cent of that for pions (Fowler *et al.*, 1951; Armenteros *et al.*, 1951). This result indicated that strong interactions existed between the heavy unstable particles and the nucleons and pions. In contrast to this conclusion the lifetimes for the particles to decay into nucleons and or pions were relatively long, thus indicating that the particles were coupled to nucleons and pions by weak interactions only.

† In the present chapter no formal distinction will be made between eigenvalues and operators.

Consider, for example, the following hypothetical case (Gell-Mann and Pais, 1955a). If the production of a Λ^0 hyperon occurred in the following reaction:

$$\pi^- + p \rightarrow \Lambda^0 + \pi^0$$

then its decay could go by way of the virtual process

$$\Lambda^0 \rightarrow \pi^0 + p + \pi^- \rightarrow p + \pi^- .$$

The first step should be a strong interaction by virtue of the process $\pi^- + p \rightarrow \Lambda^0 + \pi^0$, and the second step involves the absorption of a pion by a proton and is also strong. Thus the characteristic nuclear time of $\sim 10^{-23}$ sec might have been expected for the decay of the hyperon. Instead, the process is slower by a factor

$$\frac{\lambda_\Lambda}{\lambda_{\text{strong}}} \sim \frac{10^{10}}{10^{23}} \sim 10^{-13}. \tag{1.13}$$

This figure is again in the realm of weak interactions (compare (1.1)). The contrast between the production and decay rates for the kaons and hyperons thus indicated that totally different interactions must come into play in the two processes. The kaons and hyperons were given the collective appellation of *strange particles*.

The separate mechanisms of the decay and production processes caused Pais (1952) to propose the hypothesis of *associated production*. Pais suggested that at least two strange particles must be involved in the production process in order that a strong interaction could occur; on the other hand, a weak interaction occurs if only one strange particle is present, as in the decay process.

The hypothesis received experimental confirmation in the work of Fowler, *et al.* (1953, 1954, 1955). This work represented the first major contribution from an accelerator project to the physics of strange particles. The authors operated a hydrogen-filled diffusion chamber in a beam of pions of momentum 1·35 GeV/c and identified the following processes:

$$\pi^- + p \rightarrow K^0 + \Lambda^0 \tag{1.14}$$

$$K^0 + \Sigma^0 \tag{1.15}$$

$$K^+ + \Sigma^-. \tag{1.16}$$

Reaction (1.15) is of particular importance and we shall return to it later.

1.4(b). *Isospin and strange particles*

By 1953–4 it was apparent that the strange particles were produced in strong reactions. In addition, the work on the values for their masses had reached a sufficient degree of refinement to show that they existed in charge multiplets. Isospin quantum numbers were therefore assigned to the particles. The assignment was carried out independently by Gell-Mann (1953, 1956) and by Nakano and Nishijima (1953; see also Nishijima, 1955). The satisfactory nature of their scheme lay in the fact that it predicted the existence of two particles (the Σ^0 and Ξ^0 hyperons) which were later found by experiment. Basically the scheme assumed that the conservation laws for T and the component T_3 were preserved or broken in the manner shown in Table 1.3.

TABLE 1.3

Interaction	Conserved	Broken
Strong	T, T_3	
Electromagnetic	T_3	T
Weak		T, T_3

The preservation of T_3 in electromagnetic interactions is necessary since it is linearly related to the charge.

The Λ^0 hyperon offers a satisfactory point at which to start the isospin assignments. It exists as a charge singlet and so must correspond to $T = 0$. Thus both conservation of T and T_3 fails in the decay process

$$\begin{array}{cccc} & \Lambda^0 \rightarrow & p & + \pi^- \\ T & 0 & \frac{1}{2} & 1 \\ T_3 & 0 & +\frac{1}{2} & -1. \end{array}$$

The K^+ meson is next considered; it can be seen that a decay of the type $K \rightarrow 3\pi$ can be forbidden (as a strong interaction) if an isospin assignment $T = \frac{1}{2}$ is made. The neutral kaon K^0 can then form the second member of the charge doublet, so that

$$\frac{Q}{e} = (T_3 + \tfrac{1}{2}). \tag{1.17}$$

These assignments are then consistent with the facts that the decay processes

$$\begin{array}{ccccccccc} & K^+ \rightarrow & \pi^+ & + \pi^+ & + \pi^- & \qquad & K^0 \rightarrow & \pi^+ & + \pi^- \\ T & \frac{1}{2} & 1 & 1 & 1 & & \frac{1}{2} & 1 & 1 \\ T_3 & +\frac{1}{2} & +1 & +1 & -1 & & -\frac{1}{2} & +1 & -1 \end{array}$$

proceed at the rates expected for weak interactions. They also fit with the conservation of T and T_3 in the fast (that is strong) interactions observed in the diffusion chambers

$$\begin{array}{ccccc} & \pi^- + & p \rightarrow & \Lambda^0 + & K^0 \\ T & 1 & \frac{1}{2} & 0 & \frac{1}{2} \\ T_3 & -1 & +\frac{1}{2} & 0 & -\frac{1}{2}. \end{array}$$

The K^- particle can be interpreted as an antiparticle to the K^+ according to this scheme. Particle and antiparticle should have opposite values for the T_3 components (§ 9.4), and so the eigenvalue of T_3 should equal $-\frac{1}{2}$ for the negative kaon. The existence of a neutral particle $\overline{K}^0$ was postulated to make up the doublet, so that

$$\frac{Q}{e} = (T_3 - \tfrac{1}{2}) \tag{1.18}$$

The existence of a particle and antiparticle of zero charge possesses interesting consequences which will be discussed in § 12.9(d).

The remaining hyperons will now be considered. In order to describe the dissociation of the Σ^+ particles into a pion and a nucleon as a slow (weak) process which violates isospin conservation, it is necessary to assign integral isospin to them. The value $T = 1$ was chosen for the multiplet so that

$$\frac{Q}{e} = T_3. \tag{1.19}$$

This choice for T fits the observed processes for the production and decay of the hyperon

$$\begin{array}{lcccccccc} & \pi^- & + p \rightarrow & \Sigma^- & + K^+ & \qquad & \Sigma^- \rightarrow & n + & \pi^- \\ T & 1 & \frac{1}{2} & 1 & \frac{1}{2} & & 1 & \frac{1}{2} & 1 \\ T_3 & -1 & +\frac{1}{2} & -1 & +\frac{1}{2} & & -1 & -\frac{1}{2} & -1 \end{array} \tag{1.20}$$

and the fact that more than one charge type exists, but the choice $T = 1$ implies that the charge multiplet should have three members. The existence of a Σ^0 hyperon was therefore postulated.

One consequence of this scheme is that the weak decay of the Σ^0 hyperon must compete with a (fast) radiative decay

$$\begin{array}{lccc} & \Sigma^0 \rightarrow & \Lambda^0 & + \gamma \\ T & 1 & 0 & 0 \\ T_3 & 0 & 0 & 0. \end{array}$$

This process conserves T_3 and is therefore an allowed electromagnetic transition. It will occur less rapidly than a strong interaction, but will still be very rapid compared with a weak interaction. Taking the characteristic nuclear period of 10^{-23} sec, the lifetime for the Σ^0 hyperon is roughly given by the relation

$$\tau_{\Sigma^0} \sim 10^{-23} \times \frac{g^2}{e^2} \sim 10^{-20} \text{ sec.} \tag{1.21}$$

The discrepancy between this figure and that for a weak decay ($\sim 10^{-10}$ sec) is so great that the weak decay mode

$$\Sigma^0 \rightarrow p + \pi^0$$

has never been observed. On the other hand, Fowler *et al.* (1955) and Walker (1955) found events in hydrogen-filled diffusion chambers which were incompatible with the kinematics for the process

$$\pi^- + p \rightarrow \Lambda^0 + K^0$$

but satisfied the conditions for the process

$$\begin{array}{l} \pi^- + p \rightarrow \Sigma^0 + K^0 \\ \qquad\qquad\quad \downarrow \\ \qquad\qquad\quad \Lambda^0 + \gamma . \end{array}$$

Further support for the occurrence of the radiative decay process was provided by the observations of Alvarez *et al.* (1957). These workers examined the absorption processes for K^- mesons in hydrogen, and found Λ^0 hyperons whose kinetic energies were only compatible with the scheme

$$\begin{array}{l} K^- + p \rightarrow \Sigma^0 + \pi^0 \\ \qquad\qquad\quad \downarrow \\ \qquad\qquad\quad \Lambda^0 + \gamma . \end{array}$$

The Ξ^- hyperon was postulated to be part of a charge doublet by Gell-Mann. He proposed the existence of a neutral hyperon Ξ^0 as the second member of the doublet, so that the charge on the particles is given by

$$\frac{Q}{e} = (T_3 - \tfrac{1}{2}).$$

This scheme fitted with the observed decay of the Ξ^- hyperon

$$\begin{array}{lccc} & \Xi^- \rightarrow & \Lambda^0 + & \pi^- \\ T & \tfrac{1}{2} & 0 & 1 \\ T_3 & -\tfrac{1}{2} & 0 & -1 \end{array}$$

and was confirmed when the Ξ^0 particle was found by Alvarez *et al.* (1959).

1.4(c). *The strangeness quantum number*

It was pointed out by Gell-Mann (1956) and by Nishijima (1955) that a more elegant classification of the strongly interacting particles than that based on isospin alone could be made if a parameter S, called the *strangeness number*, was introduced. This term is defined by the relation

$$\frac{Q}{e} = T_3 + \frac{B}{2} + \frac{S}{2}. \tag{1.22}$$

Here B represents the *baryon number* (baryon is a generic name for nucleons and hyperons). Its inclusion represents the *conservation of baryonic charge*, namely that the number of baryons minus the number of antibaryons is conserved in any process (§ 9.3).

Using the relation (1.22) and the isospin assignments discussed in the previous section, the classification of the strongly interacting particles may be carried out as shown in Table 1.4.

TABLE 1.4

T	T_3					B	S
	$+1$	$+\frac{1}{2}$	0	$-\frac{1}{2}$	-1		
$\frac{1}{2}$		Ξ^0		Ξ^-		1	-2
1	Σ^+		Σ^0		Σ^-	1	-1
0			Λ^0			1	-1
$\frac{1}{2}$		p		n		1	0
$\frac{1}{2}$		K^+		K^0		0	$+1$
$\frac{1}{2}$		$\bar{K}^0$		K^-		0	-1
1	π^+		π^0		π^-	0	0

An inspection of equation (1.22) shows that the condition that T_3 is conserved in strong interactions is equivalent to conserving S. Similarly the change of T_3 in weak interactions is equivalent to changing S. Numerically the conditions are

$$\Delta T_3 = 0 \equiv \Delta S = 0 \quad \text{strong interactions} \tag{1.23}$$

$$\Delta T_3 \neq 0 \equiv \Delta S = \pm 1 \quad \text{weak interactions.} \tag{1.24}$$

Certain important consequences are associated with these rules. For example it is more difficult to create K^- than K^+ mesons, since $S = -1$ for negative kaons. Thus the threshold energy for the production of K^- particles is much higher than that for K^+, since a particle of strangeness $S = +1$ must be created with a negative kaon. For example the reactions

$$\pi^- + p \rightarrow K^+ + \Sigma^-$$

$$K^+ + K^- + n$$

have thresholds of 0·9 and 1·4 GeV respectively in the laboratory reference frame.

Another consequence of the rule (1.23) is that the cross-section for the interaction of K^- mesons with protons can be expected to be much larger than that for K^+ mesons at low kaon energies. The only process possible for the positive particle is elastic scattering

$$K^+ + p \rightarrow K^+ + p$$

whereas, in addition to elastic and charge exchange scattering

$$K^- + p \rightarrow K^- + p, \qquad K^- + p \rightarrow \bar{K}^0 + n$$

the following channels are open to the negative kaon:

$$\begin{aligned} K^- + p \rightarrow\ & \Lambda^0 + \pi^0 \\ & \Sigma^- + \pi^+ \\ & \Sigma^0 + \pi^0 \\ & \Sigma^+ + \pi^-. \end{aligned}$$

Kaon–nucleon scattering will be discussed in Chapter 14.

1.5. ON THE EXISTENCE OF FURTHER PARTICLES

The discoveries of new particles have occurred sometimes as a result of a theoretical impetus and sometimes by accident. The strange particles fall into the latter category for example.

Before proceeding further we will discuss what is meant by the word 'particle' in the present context. From our previous discussions we may conclude that the decay processes fall into three main categories with lifetimes in the following regions:

strong	10^{-23} sec
electromagnetic	10^{-16} to 10^{-20} sec
weak	10^{-10} sec.

The strong decays appear simply as resonant states—the $T = J = \frac{3}{2}$ resonant state of the pion–nucleon system is a well-known example. They can scarcely be considered to have a separate existence since they vanish near their point of production and so will be ignored in this section. They will be discussed, however, in Chapters 13 and 14. For the present we shall consider particles which decay by weak or electromagnetic processes.

There is no good theoretical reason why further particles should not exist—indeed, the classification scheme of Gell-Mann and Nishijima permits the existence of many more particles than are known at present. If we limit ourselves to particles with $T = 0$ and allow for the fact that all known particles have

$$\left|\frac{Q}{e}\right| \leqq 1 \quad \text{and} \quad |B| \leqq 1$$

then the particles listed in Table 1.5 could exist (together with their antiparticles).

TABLE 1.5

$\frac{Q}{e}$	B	T	S	
0	0	0	0	mesons
+1	0	0	+2	
1	1	0	1	baryons
−1	1	0	−3	

A baryon with strangeness −3 has been found by Barnes *et al.* (1964). Its predicted existence was crucial to the unitary symmetry scheme, which we shall discuss in§ 13.5(a). This theory predicts the masses of the particles together with certain quantum numbers.

Since the classification schemes for elementary particles are semi-empirical, there is no good reason why a new particle may not have the same quantum numbers as one we have already discussed, or it even may lie outside the present classification schemes altogether. Consider, for example, the hypothetical case of a meson M with charge states +1 and 0 and with $T = \frac{1}{2}$ and $S = 1$—the same quantum numbers as the K^+K^0 doublet. Let us assume that it has spin 1 and is slightly heavier than the kaon, so that it could decay radiatively

$$M \to K + \gamma.$$

The distance M would travel before decay would be of order $10^{-18}\,c = 10^{-8}$ cm for a radiative decay, and so its track would not be observable in a visual detection device. The other place at which it might be detected is at its point of production. Consider, for example, pions colliding with protons in a hydrogen bubble chamber, then we might have the scheme

$$\begin{array}{l} \pi^- + p \to M^+ + \Sigma^- \\ \qquad\qquad\;\; \downarrow \\ \qquad\qquad K^+ + \gamma \end{array}$$

and the observer would see the process as

$$\pi^- + p \to \Sigma^- + K^+$$

with apparent lack of energy–momentum balance. The lack of balance would reveal itself if the mass difference between M and K was large but not otherwise.

Furthermore, a hydrogen bubble chamber is the wrong medium in which to detect a γ-ray.

Thus it can be seen that a particle could easily miss detection in certain circumstances. We next consider the properties of a particle which make its detection by chance feasible with the apparatus existing at present. We consider first charged particles; the most likely medium for their chance discovery is probably the nuclear emulsion, since the emulsion permits the best identification of decay processes. The time range covered by an emulsion for the observation of a decay process is roughly 10^{-11} sec to several days. The lower limit has been set by requiring a track of ~ 1 cm upon which to make reasonable mass measurements, whilst the upper limit represents the time between exposure of the emulsion and its development. It can be seen that this time range covers weak but not electromagnetic decays, and, as we have shown in our hypothetical example above, particles which decay by the latter mode are also difficult to detect in production processes if their mass lies close to a known particle.

The time ranges available to undiscovered neutral particles are more elastic. The neutral particles are most easily found by their decay processes in bubble chambers or emulsions; each device has a characteristic detection time given by the time of flight of the particle before decay $\sim 1/c$ (dimensions of detector). Taking distances of 1–100 cm for bubble chambers and 1–100 microns for emulsions, we find times of roughly 10^{-10} to 10^{-8} sec and 10^{-14} to 10^{-12} sec respectively. Thus it can be seen that the time gaps are considerable.

The situation at present is, therefore, that new particles could remain to be discovered, especially outside the time ranges indicated above. The discovery of such particles could well affect the present classification schemes. On the other hand, new discoveries in the future may well occur through an initial theoretical requirement, as we have seen already in the case of the neutrinos and the Σ^0 and Ξ^0 hyperons.

CHAPTER 2

THE INTRINSIC PROPERTIES OF THE PARTICLES

2.1. INTRODUCTION

This chapter will be concerned with the experimental determination of the basic particle data—masses, lifetimes, spins and magnetic moments. This information, together with the physical constants, is required when the interaction of the particles is considered.

A description of the accurate evaluation of the basic physical constants is outside the scope of this book and will not be attempted. The reader may find excellent accounts of this work in the books by Cohen, Crowe and Dumond (1957) and by Sanders (1961). The constants of immediate interest are given in Table A.9.1† (Cohen and Dumond, 1958).

2.2 THE MEASUREMENT OF MASS

There are many methods of determining the masses of the particles. The techniques used may be summarised as follows.

(1) *Measurement of the trajectory of particles*. Most of the early determinations were made in this manner. The technique usually involves measurements of two quantities which are functions of mass and velocity; the latter quantity is then eliminated. The quantities measured are two of the following: momentum, ionisation, residual range, multiple Coulomb scattering. The most accurate determinations of mass by this method have been made by combining a measurement of the momentum p and the residual range R of a particle; a typical experiment is that of Barkas, Birnbaum and Smith (1956), which measured the masses of pions and muons. The momentum was found by measuring the curvature ϱ of the track of a charged particle in a magnetic field of strength H‡

$$p = H \frac{e}{c} \varrho . \tag{2.1}$$

† For the convenience of the reader the tables associated with this chapter are grouped in A.9 (Appendixes, p. 715).

‡ In this chapter e will be given in e.s.u. and H in e.m.u.; $c = 1$ units will not be used in equations involving electromagnetic quantities.

Now the energy loss, denoted by dE/dR, of a charged particle in matter is a function of velocity alone (see, for example, Ritson, 1961)

$$\frac{dE}{dR} = f(v) \tag{2.2}$$

hence

$$R = m\varphi(v). \tag{2.3}$$

The function $\varphi(v)$ is well known from both experiment and theory, and therefore v may be eliminated from (2.1) and (2.3). The experiment of Barkas, Birnbaum and Smith employed nuclear emulsions for determining R, and the magnetic field was that of the synchrocyclotron in which the pions were produced. The following results were obtained

$$\begin{aligned} \pi^+ &= 273{\cdot}3 \pm 0{\cdot}3\ \mathrm{m_e} \\ \pi^- &= 272{\cdot}8 \pm 0{\cdot}3\ \mathrm{m_e} \\ \mu^+ &= 206{\cdot}9 \pm 0{\cdot}2\ \mathrm{m_e}. \end{aligned} \tag{2.4}$$

(2) *Measurements based upon kinematics.* These methods involve a process in which all the masses except one of the particles are known. By measuring the kinematic quantities involved in the process the mass may be found. As an example we may take a determination of the mass of the Λ^0 hyperon (Bhowmik, Goyal and Yamdagni, 1961). The proton and pion resulting from the decay process

$$\Lambda^0 \to p + \pi^-$$

were both detected in nuclear emulsions, and their residual ranges were determined. Their kinetic energies, and hence their total energies and momenta, were then determined from known range–energy relationships.

Since the rest mass of a particle is a Lorentz invariant quantity (§ 3.2(g)), the mass of the Λ^0 hyperon could then be determined from the following relations ($c = 1$ units)

$$\begin{aligned} m_\Lambda^2 &= E_\Lambda^2 - \mathbf{p}_\Lambda^2 \\ &= (E_p + E_\pi)^2 - (\mathbf{p}_p + \mathbf{p}_\pi)^2 \end{aligned} \tag{2.5}$$

where m, E and $\mathbf{p}$ represent rest mass, total energy and momentum respectively. The mass of the Λ^0 hyperon was found to be 1115·46 ± 0·15 MeV.

(3) *Measurement of mesic X-rays.* When slow, negatively charged particles are captured into atomic orbits about nuclei, they fall through a series of atomic states and photons are emitted in the process. If relativistic corrections are ignored, the simple Bohr formula gives the energies of the photons as

$$\omega_{ab} = \tfrac{1}{2}\alpha^2 m Z^2 \left(\frac{1}{n_a^2} - \frac{1}{n_b^2}\right) \tag{2.6}$$

where m represents the mass of the captured particle, Z the nuclear charge and $\alpha \sim 1/137$. It can be seen that the energy of the photon can therefore be used to establish the value of the mass m. The method has been used to determine the mass of the pions and muons. The photons detected lie in the X-ray region, and their energies are determined by using known discontinuities and absorption coefficients in X-ray absorption spectra. As an example of this method, the paper by Devons *et al.* (1960) may be quoted. An accurate measurement was made of the energy of the photons emitted in the transition by negative muons from the $3D_{5/2}$ to $2P_{3/2}$ state about the phosphorus nucleus. The energy was found to be $88{,}017\,^{+15}_{-10}$ eV, and by using the relativistic equivalent of equation (2.6) a muon mass of $206{\cdot}78\,^{+0{\cdot}03}_{-0{\cdot}02}\,m_e$ was deduced.

(4) *Methods involving electromagnetic precession.* These are by far the most accurate, but can only be applied to particles with magnetic moments and therefore spin. They can be discussed more appropriately when we have described the measurement of magnetic moment in § 2.5.

Tables of the masses of the particles are given in A.9 (Appendixes, p. 716).

2.3. THE MEASUREMENT OF LIFETIME

Two direct methods are commonly employed – the time of flight technique and direct measurement by electronic methods. The former method is now used mainly for particles which have a short lifetime, although it was originally used in the late 1930's for measuring the lifetime of the muons in the cosmic radiation. It is based upon the fact that if a particle lives for t sec and moves at a velocity v during that time it will travel a distance $d = vt$. Hence a value for the mean life τ may be found by studying a measured distribution of the values of d and by making appropriate relativistic corrections for the dilation of the particle's time scale in the laboratory system (§ 3.2(c)). A recent example of the use of this technique may be found in the paper of Glasser, Seeman and Stiller (1961). A measurement of the lifetime of the π^0 meson yielded a value of $(1{\cdot}9 \pm 0{\cdot}5) \times 10^{-16}$ sec.

The measurement of decay rates by electronic methods is limited to lifetimes of $\sim 10^{-10}$ sec and greater because of the finite rise times encountered in photomultipliers and electronic apparatus. If strong beams of particles are available, great statistical accuracy can be achieved by this method, since large amounts of data may be handled electronically. For example, the experiment of Lundy (1962) made 7×10^6 measurements of the decay times of positive muons, and a mean lifetime $\tau = (2{\cdot}203 \pm 0{\cdot}004) \times 10^{-6}$ sec was obtained.

In addition to the techniques described above more indirect methods are possible for processes involving very short lifetimes. The ultra-short lifetimes, for example the resonant states (§ 13.1), may be measured by observing the width ΔE of the states and then applying the indeterminacy principle. The lifetime of the π^0 meson has also been estimated by examining apparently unrelated physical processes, in which the decay of the π^0 particle can occur as a virtual process. One method will be discussed in § 11.1(e).

2.4. THE MEASUREMENT OF SPIN

The intrinsic spin of a particle can be inferred from the conservation laws for angular momentum. The techniques used for elucidating the spin of each particle vary considerably.

2.4(a). *Data from atomic transitions*

The spins of the electron, photon and proton were fixed by examination of the processes listed below.

(1) Electron, spin $\frac{1}{2}$, from the existence of the alkali doublets and anomalous Zeeman splitting (Uhlenbeck and Goudsmit, 1925).

(2) Photon, spin 1, from the deduction of the selection rules, $\Delta l = \pm 1$, $\Delta j = 0, \pm 1$, $0 \to 0$ forbidden, in allowed atomic transitions.

(3) Proton, spin $\frac{1}{2}$, from the alternations in intensity of the lines of the band spectrum of hydrogen molecules (Hori, 1927; Kapuscinski and Eymers, 1929). Briefly, alternate lines of the rotational band spectra of diatomic homonuclear molecules vary in intensity in the ratio $(s + 1)/s$, where s represents the nuclear spin. A clear discussion of the reasons for this effect may be found in the book by Bethe and Morrison (1956). The assignment of spin $\frac{1}{2}$ for the proton was deduced independently by Dennison (1927) from a study of the anomalous behaviour of the specific heat of molecular hydrogen.

(4) A half integral spin was assigned to the neutron as soon as it was discovered, since the observations on the rotational band spectra, mentioned above, led to the condition that nuclei with even or odd atomic weights possessed even or odd spins respectively (Heitler and Herzberg, 1929; Rasetti, 1930). This result could be easily explained if nuclei consisted only of neutrons and protons†, and if the former possessed half integral spin.

Definite proof that the neutron possessed spin $\frac{1}{2}$ was obtained when measurements were made on the scattering of slow neutrons from *o*- and *p*-hydrogen (see, for example, Evans, 1955).

2.4(b). *The spin of the pion*

Following a suggestion by Marshak (unpublished), the spin of the meson was deduced from measurements of the cross-sections for the processes

$$p + p \to \pi^+ + d$$

$$\pi^+ + d \to p + p$$

† Prior to the discovery of the neutron, nuclei had been assumed to consist of protons and electrons. The assumption led to difficulties for the spin of N^{14} which was observed to be 1, but required 14 protons and 7 electrons according to the proton–electron hypothesis.

by applying the principle of detailed balance (§ 9.2(a)). Before stating this principle we will discuss the cross-sections for the two processes. We shall show in § 7.4(c) that for any process of the type

$$\underbrace{a + b}_{i} \rightarrow \underbrace{d + e}_{f}$$

it is possible to write the expression for the angular distribution in the centre of momentum (c-) system (§ 3.2(h)) in the following form (7.74):

$$\frac{d\sigma}{d\Omega} = \frac{1}{(2\pi E_c)^2} \frac{1}{(2s_a + 1)(2s_b + 1)} \frac{1}{n^2} \frac{p_c^f}{p_c^i} \sum_i \sum_\alpha |T_{fi}|^2$$

where p_c^i and p_c^f represent the momentum of the initial and final particles respectively in the c-system, s_a and s_b the spins of the particles in the initial state, $\sum_i$ and $\sum_\alpha$ indicate summations over initial and final spin states respectively and the factors $1/(2\pi E_c)^2 \times 1/n^2$ are constants which may be ignored for the present purpose, and so we will replace them by C. $|T_{fi}|^2$ represents the dynamic features of the reaction.

We will therefore represent the cross-section for the process $i \rightarrow f$ as

$$\frac{d\sigma_f}{d\Omega} = \frac{C}{(2s_a + 1)(2s_b + 1)} \frac{p_c^f}{p_c^i} \sum_i \sum_\alpha |T_{fi}|^2. \tag{2.7}$$

Now consider the reverse process $f \rightarrow i$. At the same total energy in the c-system the differential cross-section becomes

$$\frac{d\sigma_i}{d\Omega} = \frac{C}{(2s_d + 1)(2s_e + 1)} \frac{p_c^i}{p_c^f} \sum_\alpha \sum_i |T_{if}|^2. \tag{2.8}$$

But the principle of detailed balance states that under certain conditions

$$\sum_i \sum_\alpha |T_{fi}|^2 = \sum_\alpha \sum_i |T_{if}|^2$$

and we shall show in § 9.2(a) that these conditions are fulfilled in the experiments under discussion, hence

$$\frac{d\sigma_f}{d\Omega} = \left(\frac{p_c^f}{p_c^i}\right)^2 \frac{(2s_d + 1)(2s_e + 1)}{(2s_a + 1)(2s_b + 1)} \frac{d\sigma_i}{d\Omega} \tag{2.9}$$

and if we represent the $\pi^+ d$ and pp systems as f and i respectively

$$\frac{d\sigma_{\pi d}}{d\Omega} = \frac{p_\pi^2}{p_p^2} \frac{3}{4} (2s_\pi + 1) \frac{d\sigma_{pp}}{d\Omega} \tag{2.10}$$

where we have inserted the factor $(2s_e + 1) = 3$ for the deuteron and $(2s_a + 1) = 2$ for the proton. If this equation is integrated over all angles it yields

$$\sigma_T(p + p \to \pi^+ + d) = \frac{p_\pi^2}{p_p^2}\frac{3}{2}(2s_\pi + 1)\,\sigma_T(\pi^+ + d \to p + p) \qquad (2.11)$$

where the symbol σ_T represents the total cross-section. The factor 2 appears because the two protons are indistinguishable.

The total cross-section for the absorption of π^+ mesons by deuterons was measured by Clark, Roberts and Wilson (1951, 1952) and by Durbin, Loar and Steinberger (1951); they obtained values of $(4{\cdot}5 \pm 0{\cdot}8)\,10^{-27}\,\text{cm}^2$ and of $(3{\cdot}1 \pm 0{\cdot}3)\,10^{-27}\,\text{cm}^2$ respectively. The cross-section for the reverse process of pion production was measured by Cartwright *et al.* (1953). Upon inserting the factors given above they were able to estimate the cross-section for the process $\pi^+ + d \to p + p$; they found values of $(3{\cdot}0 \pm 1{\cdot}0)\,10^{-27}\,\text{cm}^2$ and $(1{\cdot}0 \pm 0{\cdot}3)\,10^{-27}\,\text{cm}^2$, depending on the assignment of spin 0 or 1 to the pion respectively. Although the errors were large, the data clearly suggested zero spin for the pion. A later compilation of data from many laboratories, by Cohen, Crowe and Dumond (1957), yielded a value $(2s_\pi + 1) = 1 \pm 0{\cdot}10$.

The assignment of spin 0 to the π^0 meson is consistent with the experimental data on single pion production and charge exchange scattering, but no single unambiguous measurement has been made. The fact that the π^0 meson decays to two γ-rays may be used, however, to exclude the possibility that it has spin 1. The argument will be given in § 9.5(b).

2.4(c). *The spin of the leptons*

We have indicated previously that the spin of the electron is $\frac{1}{2}$ (§ 2.4(a)). This fact, together with data from β-decay, may be used to deduce the spin of the neutrino. The decay rates of β-active nuclei may be classified as allowed, first forbidden, second forbidden, ..., depending on the orbital angular momentum quantum number l of the emitted electron. The allowed transitions correspond to $l = 0$, and can be recognised by their (relatively) rapid decay rate. The occurrence of allowed transitions with nuclear spin changes of $0 \to 0$ and $0 \to 1$, for example

$$O^{14} \to N^{14*} + e^+ + \nu$$

$$\text{spin} \quad 0 \to 0$$

$$B^{12} \to C^{12} + e^- + \bar{\nu}$$

$$\text{spin} \quad 1 \to 0$$

therefore imply that the spin of the neutrino must be $\frac{1}{2}$ if angular momentum is to be conserved.

The spin of the muon may be deduced to be half integral from many facts, for example

$$\begin{array}{llll} & \mu \rightarrow e + \nu + \bar{\nu} \\ \text{spin} & \tfrac{1}{2} \quad \tfrac{1}{2} \quad \tfrac{1}{2} \end{array}$$

Probably the clearest evidence that the muon spin is $\frac{1}{2}$ has been pointed out by Kabir (1961), who showed that an experiment by Hughes *et al.* (1960), which detected the formation of muonium (a bound state μ^+e^-), also demonstrated the fact that the muon possesses a spin of $\frac{1}{2}$. The muonium can be made to precess in a magnetic field of strength H with an angular velocity

$$\omega = g \frac{e}{2mc} H$$

where g is the Landé factor for the muonium; to a good approximation it is given by

$$g = \frac{1}{2s_\mu + 1} g_e \tag{2.12}$$

where s_μ represents the muon spin and g_e the Landé factor for the electron. The g factor in the experiment of Hughes and his co-workers was half that for the electron and hence s_μ must be $\frac{1}{2}$.

2.4(d). *The helicity of the neutrino*

It will be shown in § 3.3(l) that the spin of a fermion of mass zero should lie either parallel or antiparallel to its motion. This property is called *helicity*; particles whose spin states always lie parallel to their momentum are said to have positive helicity, and vice versa. The observation of the failure of parity conservation in β-decay led to the serious consideration of whether the neutrino possessed a definite helicity. This property was established by an experiment performed by Goldhaber, Grodzins and Sunyar (1958). The helicity was found to be negative.

The experiment combined an analysis of the circular polarisation and the resonant scattering of the γ-rays emitted following orbital electron capture. We have stated in Chapter 1 (1.3) that a neutrino is *defined* as the particle accompanying positrons in β-decay.

$$(A, Z) \rightarrow (A, Z - 1) + e^+ + \nu.$$

Now, we shall show later that the absorption of a particle and creation of an antiparticle are equivalent processes, hence the lepton emitted in electron capture processes should be a neutrino

$$e^- + (A, Z) \rightarrow (A, Z - 1) + \nu \tag{2.13}$$

The principle of the experiment is illustrated in Fig. 2.1. The process used was the following: electron capture in Eu^{152} yielded the isomeric state Sm^{152*} (Fig. 2.1 (a))

$$e^- + Eu^{151} \rightarrow Sm^{152*} + \nu$$
$$\downarrow \gamma$$
$$Sm^{152}$$

and the γ-rays were detected by resonant scattering in Sm^{152} (Fig. 2.1 (b))

$$\gamma + Sm^{152} \rightarrow Sm^{152*} \rightarrow Sm^{152} + \gamma.$$

The kinematic conditions for resonant scattering are best fulfilled for those γ-rays which are emitted whilst the Sm^{152*} nucleus is still recoiling and which emerge along the direction of recoil (Fig. 2.1 (c); light and wavy arrows indicate

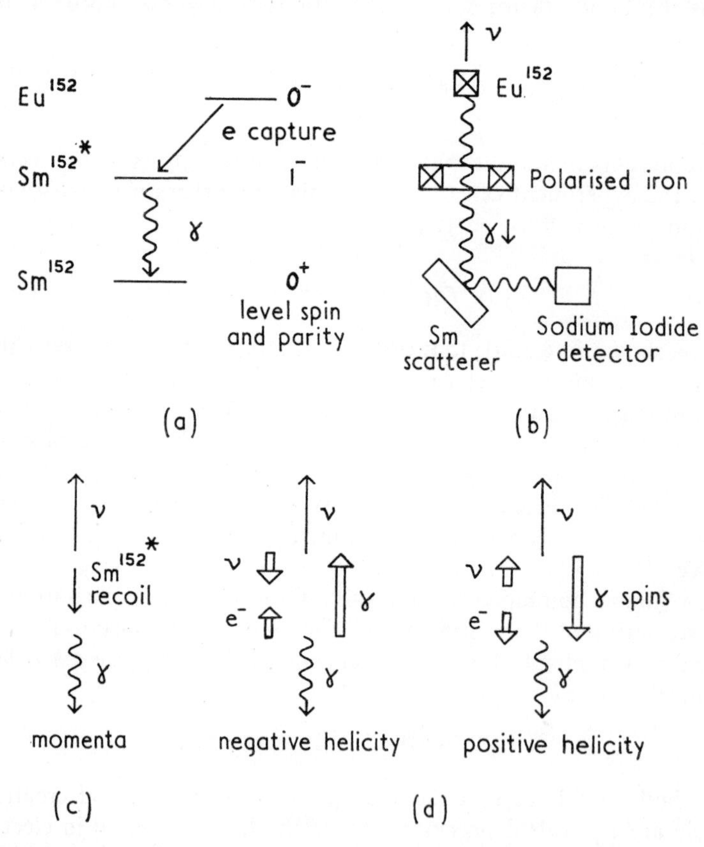

FIG. 2.1. Principle of experiment to determine the helicity of the neutrino associated with β-decay. Light and wavy arrows refer to directions of momenta; thick arrows indicate directions of the spins.

momentum in diagrams (c) and (d); heavy arrows show spin direction). The overall conservation of angular momentum then requires the spins of the particles to point in the directions shown in diagram (d) depending on the helicity of the neutrino. It can be seen that the helicity of the neutrino and the γ-ray must be the same. The latter property was detected by passing the photons through magnetised iron (Fig. 2.1 (b)). Different fractions of the γ-ray beam then reached the Sm^{152} scatterer depending on whether the direction of magnetisation was made parallel or antiparallel to the direction of motion of the γ-rays. From the experimental results it was deduced that the γ-rays, and hence the neutrinos, possessed negative helicity. The data was found to be compatible with 100 per cent polarisation for the neutrinos.

2.4(e). *The spins of the kaons and hyperons*

Processes of the type

$$K^+ \to \pi^+ + \pi^0$$

$$\Lambda^0 \to p + \pi^-$$

show that kaons and hyperons have integral and half integral spin respectively. There exists strong evidence that the spins are zero for the kaons and half for the hyperons.

The evidence for the kaon spin may be cited as follows.

(1) The observation of the decay mode $K^0 \to 2\pi^0$ (Plano *et al.*, 1957) shows that the spin of the kaon must be even, since the pions are identical particles and must be emitted in a state of even orbital angular momentum in order to satisfy Bose–Einstein statistics (§ 9.2(g)).

(2) If $s_K \neq 0$ the kaon could decay electromagnetically

$$K^+ \to \pi^+ + \gamma$$

and this decay mode would compete favourably with the weak decay modes, for example $K^+ \to 2\pi^+ + \pi^-$ (Dalitz, 1955). On the other hand, if $s_K = 0$ this decay mode is forbidden; the kaon has never been observed to decay in this manner.

(3) An analysis by Dalitz (1953; see § 9.5(c)) of the experimental data for the decay process $K^+ \to \pi^+ + \pi^+ + \pi^-$ has shown that they are consistent with $s_K = 0$.

(4) If $s_K \neq 0$ the processes of kaon production would sometimes lead to polarised kaon beams. No evidence for polarisation has been found.

Two main methods have been used for examining hyperon spins; they are due to arguments put forward by Adair (1955) and by Lee and Yang (1958).

The former method is based upon the assumption that the spin of the kaon is zero. Consider the process

$$\pi + N \to Y + K$$

where N represents a nucleon and Y a hyperon. Now if hyperons are examined which travel nearly parallel or antiparallel to the direction of the pion momentum (the z-axis of quantisation), then the component of hyperon spin along these directions must be $m_Y = \pm\frac{1}{2}$, since $m_l = 0$ (§ 6.1(a)) and the nucleon has spin $\frac{1}{2}$. Both states are equally populated if the nucleons are unpolarised. If the hyperon then decays by the process $Y \to N + \pi$, the angular distribution of the pions emitted at an angle θ from the z-axis in the hyperon rest frame is determined by the hyperon spin (s_Y) alone. The possible distributions are given in Table 2.1; the principle of the method used in the calculation becomes obvious if § 14.1(b) is examined.

TABLE 2.1

s_Y	Angular distribution
$\frac{1}{2}$	1
$\frac{3}{2}$	$1 + 3\cos^2\theta$
$\frac{5}{2}$	$1 - 2\cos^2\theta + 5\cos^4\theta$

The argument of Adair is independent of the orbital angular momentum $l = s_Y \pm \frac{1}{2}$ of the decay products $N + \pi$, and therefore of the problems associated with the violation of parity conservation in the hyperon decay. The experiments of Eisler *et al.* (1958) on the processes

$$\pi^- + p \to \Lambda^0 + K^0$$
$$\qquad\qquad \Sigma^- + K^+$$

have yielded results which are consistent with isotropy in both interactions, and therefore in contradiction with $s_Y = \frac{3}{2}$ or $\frac{5}{2}$.

An alternative approach to the problem of the hyperon spin was made by Lee and Yang. They pointed out that limits could be set on the magnitude of the spin by observing the magnitude of the asymmetry in the decay of polarised hyperons. Using arguments based on the conservation of angular momentum, they have shown that if θ represents the angle between the decay proton and the direction of hyperon polarisation in the latter's rest frame, then the average value of $\cos\theta$, measured for a large number of particles, should lie between the limits

$$\frac{-1}{2s_Y + 2} \leqq \langle\cos\theta\rangle \leqq \frac{1}{2s_Y + 2}. \tag{2.14}$$

The advantage of the method is that it requires no assumptions about the kaon spin. It has been used by Crawford *et al.* (1959a), who examined the decay of Λ^0 hyperons from the process $\pi^- + p \to \Lambda^0 + K$. They concluded that their data was consistent with a spin value of $\frac{1}{2}$ and that the assumption of spin $\frac{3}{2}$ failed to satisfy the data by three standard deviations.

2.5. GYROMAGNETIC RATIOS AND MAGNETIC MOMENTS

2.5(a). *Introduction*

The most accurately known properties of the particles are those which can be associated with their magnetic moments. The magnetic dipole moment $\boldsymbol{\mu}$ of a particle is given by the relation

$$\boldsymbol{\mu} = g \frac{e}{2mc} \mathbf{s} \equiv g \frac{e\hbar}{2mc} \mathbf{s} \tag{2.15}$$

where g represents the Landé factor and $\mathbf{s}$ the spin. The quantity

$$\mu_B = \frac{e}{2mc} \tag{2.16}$$

is normally used as a unit for the magnetic moment. It is called the *magneton* or *Bohr magneton*; sometimes the latter name is reserved for the electron.

In the presence of a magnetic field of strength and direction $\mathbf{H}$, a magnetic interaction occurs between the moment of the particle $\boldsymbol{\mu}$ and the field leading to $2s + 1$ substates. In practice only particles with spin $\frac{1}{2}$ are encountered and so only two states are formed, parallel and antiparallel to the field. The states are separated in energy by an amount

$$\Delta E = 2\mu H. \tag{2.17}$$

The system also undergoes *Larmor precession* about the direction of $\mathbf{H}$ with an angular velocity

$$\omega_L = g \frac{e}{2mc} H = \gamma H \tag{2.18}$$

where the term $\gamma = g\,(e/2mc)$ is often called the *gyromagnetic ratio* since

$$\gamma = \frac{\mu}{s}. \tag{2.19}$$

We note, finally, that if the particle is charged and in motion, then it will execute spiral orbits about $\mathbf{H}$. The orbits possess the characteristic *cyclotron frequency* ν_c

$$\omega_c = 2\pi\nu_c = \frac{e}{mc} H. \tag{2.20}$$

2.5(b). *The gyromagnetic ratio of the proton*

Techniques developed at Harvard (Bloch, 1946; Bloch, Hansen and Packard, 1946) and at Stanford (Purcell, Torrey and Pound, 1946) have enabled the gyromagnetic moment of the proton to be measured in an elegant and simple fashion.

If a magnetic field **H** is applied to a sample of hydrogeneous liquid (say water or a hydrocarbon) then the protons enter the two substates $m_s = \pm\frac{1}{2}$ with energy of separation $\Delta E = 2\mu_p H$. Now if a small oscillating field is applied at right angles to **H** (Fig. 2.2(a)) and if the frequency of the field is changed, then at a resonant frequency given by

$$\omega_R = 2\pi\nu_R = \Delta E = 2\mu_p H \tag{2.21}$$

transitions will be induced between the substates. Now the states are not populated equally – in thermal equilibrium there are slightly more protons in the

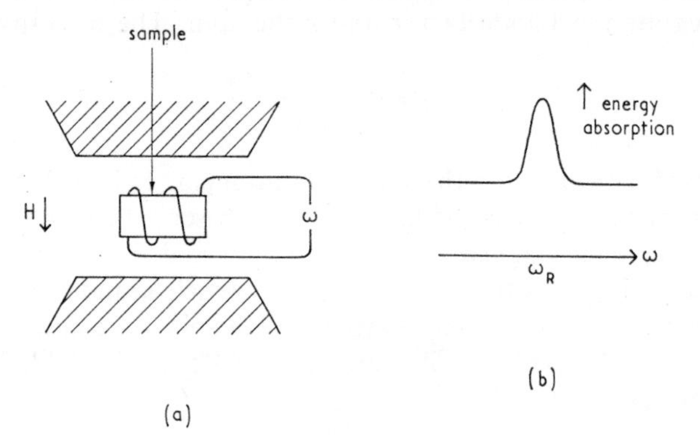

FIG. 2.2.

lower state than in the upper in the ratio $e^{2\mu_p H/kT}$. Thus, since the probabilities of transitions upwards and downwards are equal, the position of the resonant frequency may be observed by examining the net absorption of energy from the coil producing the alternating field (Fig. 2.2(b)).

Now we may write

$$\omega_R = 2\mu_p H = 2g \frac{e}{2m_p c} sH = \gamma_p H$$

hence

$$\gamma_p = \frac{\omega_R}{H}. \tag{2.22}$$

Thus an accurate measurement of ω_R and H yields γ_p. The errors associated with the technique have been steadily reduced, for example Driscoll and Bender (1958) have measured γ_p to an accuracy of 7·5 parts per million (p.p.m.). $\gamma_p = (2{\cdot}67520 \pm 0{\cdot}00002)\ 10^4$ rad sec^{-1} G^{-1}.

The technique described above is essentially very simple, and is often used in reverse to measure magnetic fields with great accuracy. Variants of the technique have been used to determine the gyromagnetic ratios of other particles – see, for example, Coffin *et al.* (1958), Bloch, Nicodemus and Staub (1948), and Beringer and Heald (1954). These papers describe measurements of γ for muons, neutrons and electrons respectively.

2.5(c). *The magnetic moment of the proton in nuclear magnetons*

Consider the ratio of equations (2.21) and (2.20)

$$\frac{\omega_R}{\omega_c} = \frac{2\mu_p H}{(eH/m_p c)} = \mu_p \frac{(2m_p c)}{e}. \tag{2.23}$$

Thus a measure of the ratio of the characteristic frequencies ω_R and ω_c enables a value for the magnetic moment of the proton to be ascertained in units of nuclear magnetons $(e/2m_p c)$.

Determinations of this nature have been made by Sommer, Thomas, and Hipple (1951) and by Collington *et al.* (1955). Small cyclotrons were constructed which could be placed between the poles of the same magnet used for supplying the constant field for the determination of ω_R, and simultaneous measurements of ω_R and ω_c were made. The first group of workers measured the frequency necessary to successfully accelerate the protons, whilst the second group used an inverse cyclotron principle. In this technique protons are fed into a cyclotron at high energy, and the frequency necessary to cause maximum deceleration was found. The experiments yield accuracies of about 20 p.p.m.; for example, that of Sommer *et al.* gave a value of $2{\cdot}792765 \pm 0{\cdot}00006$ nuclear magnetons.

2.5(d). *The proton magnetic moment in Bohr magnetons*

An important variant of the experiment described in the previous section is that in which the proton resonance frequency ω_R and the electron cyclotron frequency ω_{ce} are simultaneously measured. The ratio of these quantities then yields the proton magnetic moment in Bohr magneton units and other important data.

$$\frac{\omega_R}{\omega_{ce}} = \frac{2\mu_p H}{(e/m_e c)\,H} = \frac{\mu_p}{(e/2m_e c)} = \frac{\mu_p}{\mu_B}. \tag{2.24}$$

An experiment of this type, performed by Gardner and Purcell (1949; 1951), gave an accuracy of 13 p.p.m. Later work has improved the accuracy to 2 p.p.m., $\mu_p = (1{\cdot}521032 \pm 0{\cdot}000003)\,10^{-3}$ Bohr magnetons.

The data from experiments of this nature may be combined with that from the experiments quoted in the previous section to yield further important physical quantities, for example the ratio of the electron charge (in e.m.u.) to mass may be found by combining ω_{ce}/ω_R with γ_p, the proton gyromagnetic ratio

$$\frac{\omega_{ce}}{\omega_R} = \frac{e}{m_e c} H \frac{1}{\gamma_p H}$$

where (2.22) has been used, hence

$$\frac{e}{m_e c} = \frac{\omega_{ce}}{\omega_R}\,\gamma_p. \tag{2.25}$$

In a similar manner a combination of the results from the proton and electron cyclotron experiments yields the mass ratio of proton and electron

$$\frac{\omega_{ce}}{\omega_{cp}} = \frac{m_p}{m_e}. \tag{2.26}$$

2.5(e). *The g−2 values for the electrons and muon*

The observed g values for the electron and muon are slightly larger than 2. The Dirac theory for a relativistic electron predicts that this value should be exactly 2, whereas the more elaborate theories of quantum electrodynamics developed during the past two decades predict a value slightly larger than 2. Both muon and electron appear to have only weak and electromagnetic interactions, and so the comparison of theory and experiment for the g values of the electron and muon provide an important test of quantum electrodynamics since their weak interactions are of negligible strength compared with their electromagnetic interactions.

Let us assume that we have polarised charged leptons and we know their direction of polarisation. If they are then allowed to pass into a system with a magnetic field of strength H, they execute helical orbits about the direction of H and undergo Larmor precession with the following angular velocities:

$$\omega_c = \frac{e}{mc}H, \quad \omega_L = g\frac{e}{2mc}H.$$

Thus if $g = 2$ the direction of polarisation would remain fixed relative to the direction of motion of the particle; if $g \neq 2$ a phase angle opens up between the directions, and after a time t

$$\delta = (\omega_L - \omega_c)\,t = \frac{(g-2)}{2}\frac{e}{mc}Ht = a\frac{e}{mc}Ht \tag{2.27}$$

where

$$g = 2(1 + a).$$

Thus a measurement of the phase angle δ after a time t establishes the magnitude of the deviation of the g-value from 2. Measurements of $g-2$ have been made by Schupp, Pidd and Crane (1961) and by Farley and his co-workers (Charpak *et al.* 1961a; see also Charpak *et al.* 1962) for electrons and muons respectively. Both experiments require a knowledge of the direction of the particle spin in order to determine δ. In the electron experiment the spin direction was established with the aid of a double scattering experiment (compare § 6.2) in which the first and second scatterings were performed respectively before and after the passage of the electrons through the solenoid. Spin directions were determined in the muon experiment by observing the angular distribution of the decay electrons from the muons. Due to the violation of parity conservation these are emitted asymmetrically about the direction of muon spin (§ 12.3).

Values of

$$a_{\text{exp}} = (1160{\cdot}9 \pm 2{\cdot}4)\,10^{-6} = (1{\cdot}0011 \pm 0{\cdot}0020)\,a_{\text{th}} \quad \text{electrons}$$

$$a_{\text{exp}} = (1162 \pm 5)\,10^{-6} = (0{\cdot}9974 \pm 0{\cdot}0042)\,a_{\text{th}} \quad \text{muons}$$

were obtained in the experiments; $a_{\text{exp.}}$ and a_{th} refer to experimental and theoretical values respectively.

The last figure may be combined with accurate determinations of the muon gyromagnetic ratio (see, for example, Hutchinson *et al.*, 1961) to deduce a value for the muon mass. The comparison of the two experiments quoted yielded a value $m_\mu = (206{\cdot}768 \pm 0{\cdot}003)\,m_e$.

CHAPTER 3

PRELIMINARIES TO A QUANTISED FIELD THEORY

3.1. INTRODUCTION AND STATEMENT OF THE PROBLEM

A considerable section of the present theory of elementary particles is based on the assumption that for each type of particle there is associated a field for which the particles act as quanta. This theory is called the quantum theory of fields. The development of the theory has been mainly empirical. Its present form has been made to fit within the framework of

(i) relativity,
(ii) quantum mechanics,
(iii) classical field theory.

These topics give the theory a formal structure, but in order to make meaningful calculations and predictions, certain parameters, for example particle masses and coupling strengths, must be introduced into the theory in a phenomenological manner. The formal structure of the quantum theory of fields can then be used in association with these parameters to produce results which can be tested by experiment. The great strength of the quantum field theory lies in the fact that so many diverse pieces of experimental data can be explained by so few parameters.

Nevertheless, the theory is far from perfect; certain technical difficulties appear to be inherently associated with its formal structure, and the abandonment of field theory in its present form has been suggested by some workers. At the moment, however, no other completely satisfactory way of describing the properties of the elementary particles has been found.

In describing the behaviour of both classical and quantised fields it is often convenient to employ the Lagrangian notation. This system possesses certain advantages in that many of the conservation laws of physics may be readily demonstrated by considering the invariance of the Lagrangian under Lorentz transformations. The principle of relativistic invariance is an important one for field theory, since it imposes many restrictions on the possible forms for a field.

We commence this chapter, therefore, with a résumé of some of the main features of the special theory of relativity. This will be followed by an introduction to some relativistic wave equations. Finally, the Lagrangian for a classical field will be constructed, and some of the consequences of its invariance under Lorentz transformations will be examined.

3.2. RÉSUMÉ OF THE SPECIAL THEORY OF RELATIVITY

3.2(a). *Lorentz invariance*

In order to describe the processes taking place in nature, it is convenient to use a system or frame of reference; that is a system of coordinates against which we can locate the position of a particle in space and time. If a freely moving particle proceeds with constant velocity in a frame of reference, that is, it is not acted upon by external forces, that frame is said to be *inertial.*

The physics of elementary particles starts from the validity of the *principle of relativity*. In the context of our present knowledge concerning elementary particles the principle of relativity refers to special rather than general relativity. The principle is based on two axioms:

(1) Lorentz invariance,
(2) invariance of the velocity of light.

According to the first axiom, all the laws of nature are identical in all inertial frames of reference, that is the form of an equation describing a natural law is independent of its frame of reference. The second axiom implies that the velocity of light in a vacuum is the same in all reference frames. A necessary corollary of this statement is that no particle can travel with a velocity greater than that of light in a vacuum.

A further definition may be associated with the first statement — that of *covariance*. If an equation describing a physical process assumes a certain form in one inertial frame, and if, upon transformation into a new reference frame with new variables, the equation assumes the same form as the previous one, then that equation and its transformation is said to be *covariant*. It should be noted that the equation given in each reference frame is assumed to be a function only of the variables associated with that particular frame.

3.2(b). *Intervals*

The interval between two events a and b is defined to be

$$S_{ab} = [c^2(t_a - t_b)^2 - (x_a - x_b)^2 - (y_a - y_b)^2 - (z_a - z_b)^2]^{\frac{1}{2}} \qquad (3.1)$$

where t, x, y and z are the coordinates defining the position of the events in time and space, and c is the velocity of light.

For reasons of symmetry it is convenient to describe events and intervals in a *four-dimensional world* or *Minkowski space*. In this space we use the following coordinates:

$$x = x_1, \quad y = x_2, \quad z = x_3, \quad ict = x_4 \tag{3.2}$$

so that

$$S_{ab}^2 = -\sum_{\lambda=1}^{4} (x_a - x_b)_\lambda^2 = -(x_a - x_b)_\lambda^2 \tag{3.3}$$

and for points which are infinitesimally close together

$$\mathrm{d}S^2 = -\sum_{\lambda=1}^{4} \mathrm{d}x_\lambda^2 \equiv -\mathrm{d}x_\lambda^2 \tag{3.4}$$

where we have adopted the Einstein convention that a summation symbol is dropped when considering repeated suffices. Later on, for simplicity of writing, the suffix will also be dropped on certain summation terms. In general Greek symbols will be used for indices in four-dimensional space, and Latin indices for three-dimensional space.

3.2(c). *Invariance properties of the interval*

The terms S_{ab}^2 in (3.3) may be regarded as the square of a line element in Minkowski space. For convenience of discussion let us locate one event in space-time at $x_b = 0$, so that (3.3) can be written as

$$-S^2 = x_1^2 + x_2^2 + x_3^2 + x_4^2. \tag{3.5}$$

The mathematical properties of three- and four-dimensional systems are similar. Consider, for example, a vector $\mathbf{r}$ in ordinary (three-dimensional) space; its length is defined by the relation

$$\mathbf{r}^2 = x^2 + y^2 + z^2$$

where x, y and z are the lengths of its components along three rectangular axes. The length of this vector remains invariant under rotations of the axes – only its direction changes with respect to the axes

$$\mathbf{r}^2 = x'^2 + y'^2 + z'^2$$

where x', y' and z' represent the components of $\mathbf{r}$ after the rotation.

If it is postulated that space and time are homogeneous, one may similarly show that $-S^2$ is not changed by rotations of the coordinate system in a four-dimensional rectangular space, and so

$$-S^2 = x_1'^2 + x_2'^2 + x_3'^2 + x_4'^2.$$

A specific rotation will be examined in § 3.2(e).

It should be noted, however, that the axiom that the velocity of light is the same in all reference frames restricts the invariance property to the following form

$$-S^2 = x_1^2 + x_2^2 + x_3^2 - c^2t^2$$

$$= x_1'^2 + x_2'^2 + x_3'^2 - c^2t'^2. \tag{3.6}$$

We may apply this relation to a specific example – the birth and death of a meson. We will use the laboratory (L-) frame and the particle (c-) frame. In the L-frame an observer sees the two events as a creation of the meson at a space point $x_b(= x_b, y_b, z_b)$ and its death at a space point x_a at a time $t_a - t_b$ later. We can therefore write

$$t_a - t_b = \tau_L$$

$$\sum_{i=1}^{3} (x_a - x_b)_i^2 = v^2\tau_L^2$$

where v is the velocity and τ_L the lifetime of the meson as seen in the L-frame. In the c-frame the two events occur at the same space point so that $x_a' = x_b'$ for the x, y, z components, and the separation in time for the two events is given by

$$t_a' - t_b' = \tau_c.$$

The requirement of the invariance of the interval (3.6) then gives

$$-S^2 = \sum_{i=1}^{3} (x_a - x_b)_i^2 - c^2(t_a - t_b)^2$$

$$= v^2\tau_L^2 - c^2\tau_L^2 = -c^2\tau_c^2$$

or

$$\tau_L = \frac{\tau_c}{\sqrt{(1 - v^2/c^2)}}. \tag{3.7}$$

This is the time dilation relation of special relativity. It has been exploited in the measurement of the lifetime of the short lived π^0 meson (see, for example, Shwe, Smith and Barkas, 1962). The quantity $\tau_c = S/c$ is sometimes called the particle's *proper time*.

3.2(d). *Time and space-like intervals*

It can be seen that since the interval S_{ab} is given by the equation

$$S_{ab}^2 = c^2(t_a - t_b)^2 - (x_a - x_b)^2 - (y_a - y_b)^2 - (z_a - z_b)^2$$

the quantity S_{ab}^2 can be positive or negative. If $S_{ab}^2 > 0$ it is said to be *time-like* and if $S_{ab}^2 < 0$ it is said to be *space-like*. In the four-dimensional notation of equation (3.3) this statement is equivalent to saying

$$(x_a - x_b)^2_\lambda < 0 \quad \text{is time-like} \tag{3.8}$$

$$(x_a - x_b)^2_\lambda > 0 \quad \text{is space-like}. \tag{3.9}$$

Since no interaction can be propagated with a velocity greater than that of light, two events can only be related causally if the square of the interval between them is time-like.

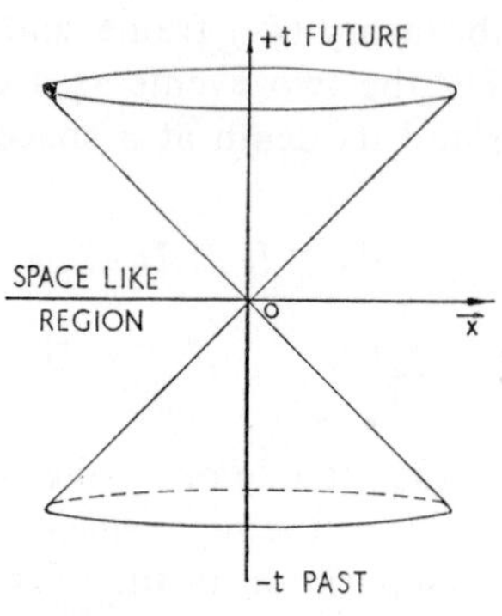

FIG. 3.1.

Consider an event located at the point $O = x_b$ in Fig. 3.1. Then any event related causally to that at O must lie inside a cone (*the light cone*) defined by the relation

$$c^2(t^2 - t_b^2) - \sum_{i=1}^{3} (x^2 - x_b^2)_i = c^2(t^2 - t_b^2) - (\mathbf{x}^2 - \mathbf{x}_b^2) = 0. \tag{3.10}$$

This equation in fact defines two cones; related events for which $t > t_b$ lie in the upper cone, that is in the future relative to t_b, whilst past events are located in the lower cone.

Events occurring outside the light cone are causally unrelated to that occurring at O. This region is frequently called the space-like (or remote) region.

3.2(e). *The transformation of coordinate systems*

The transformation of the coordinate system for an event from one inertial frame to another may be resolved into a series of parallel displacements and rotations of the coordinate system. The transformations lead to equations of great significance in physics. We will give a famous example. Each rotation in four-dimensional space can be resolved into six rotations in the planes

$$x_1x_2, \quad x_1x_3, \quad x_1x_4, \quad x_2x_3, \quad x_2x_4, \quad x_3x_4.$$

Consider a rotation in the x_1x_4 plane (Fig. 3.2), that is the y and z spatial coordinates remain unaltered.

The relation between the point P in the old and new coordinates is given by the equations

$$x_1 = x_1' \cos\theta - x_4' \sin\theta \qquad (3.11)$$

$$x_4 = x_1' \sin\theta + x_4' \cos\theta$$

where θ represents the angle of rotation. It is a simple matter to show that these relations satisfy the invariance equation (3.6).

Now consider an inertial frame K' which is moving relative to a frame K with a velocity βc along the x_1 axis ($0 < \beta < 1$). Only the spatial coordinate x_1 and time coordinate x_4 can be affected by this motion. Therefore any transformation between the coordinate frames can only link x_1 and x_4 with x_1' and x_4'. Thus if we consider the spatial origin of the K' system (the point $x_1' = 0$) from the K reference frame, we find

$$x_1 = -x_4' \sin\theta, \quad x_4 = x_4' \cos\theta$$

therefore

$$\frac{x_1}{x_4} = \frac{x_1}{ict} = -\tan\theta$$

but

$$\frac{x_1}{t} = \beta c$$

and therefore

$$\tan\theta = i\beta, \quad \cos\theta = \frac{1}{\sqrt{(1-\beta^2)}} = \gamma, \quad \sin\theta = i\beta\gamma. \qquad (3.12)$$

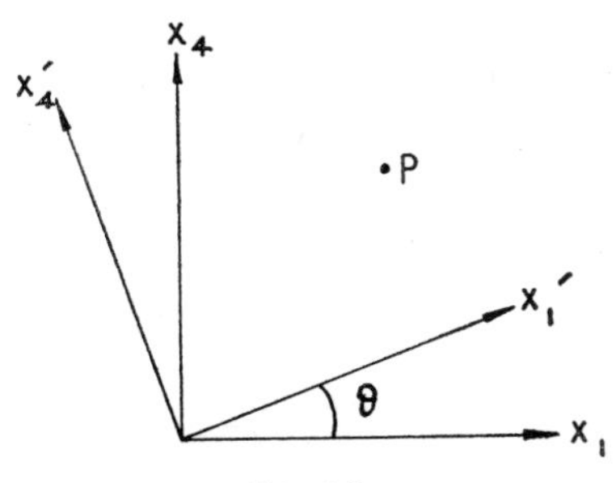

FIG. 3.2.

These equations give us the Lorentz transformation relations for two coordinate systems, which are in relative motion with relative velocity βc

$$x_1 = \gamma(x_1' - i\beta x_4') \qquad (3.13)$$

$$x_2 = x_2'$$

$$x_3 = x_3'$$

$$x_4 = \gamma(i\beta x_1' + x_4').$$

3.2(f). *Four-vectors*

The principles used in the derivation of equations (3.13) may be extended to four vectors. A *four-vector A* possesses components A_1, A_2, A_3, A_4 which satisfy the condition

$$\sum_{\lambda=1}^{4} A_\lambda^2 = \text{constant scalar quantity}$$

and which behave like x_1, x_2, x_3, x_4 for a transformation of the coordinate system

$$\begin{aligned} A_1 &= \gamma(A_1' - i\gamma A_4') \\ A_2 &= A_2' \\ A_3 &= A_3' \\ A_4 &= \gamma(i\beta A_1' + A_4'). \end{aligned} \tag{3.14}$$

Examples of commonly occurring four-vectors in the physics of elementary particles are the four-vectors for energy–momentum,

$$p_1 \equiv p_x, \quad p_2 \equiv p_y, \quad p_3 \equiv p_z, \quad p_4 \equiv \frac{iE}{c} \tag{3.15}$$

and current density

$$j_1 \equiv j_x, \quad j_2 \equiv j_y, \quad j_3 \equiv j_z, \quad j_4 \equiv ic\varrho \tag{3.16}$$

where the symbols E and ϱ refer to total energy and charge density respectively.

A consideration of equation (3.14) shows that the sum of the products of the equivalent components of two four-vectors, $A_\lambda B_\lambda$, is a scalar quantity and is the four-dimensional equivalent of the normal scalar product of two vectors. We shall frequently omit the subscripts on a scalar product of two four-vectors

$$AB \equiv A_\lambda B_\lambda;$$

for example, a quantity which is often encountered in the theory of elementary particles is the scalar product of two momentum four-vectors. Let the vectors be p and p', then the scalar product will be written as

$$pp' \equiv p_\lambda p_\lambda' = \mathbf{p} \cdot \mathbf{p}' - \frac{EE'}{c^2}. \tag{3.17}$$

Since we shall frequently work in units with $\hbar = c = 1$, this equation may be written as

$$pp' = \mathbf{p} \cdot \mathbf{p}' - EE'. \tag{3.18}$$

3.2(g). *The energy–momentum four-vector*

One of the most important four-vectors used in the physics of elementary particles is that for energy–momentum. First, we will re-write equation (3.4) as

$$dS^2 = -dx_\lambda^2 = c^2\,dt^2 - (dx_1^2 + dx_2^2 + dx_3^2)$$

$$= c^2\,dt^2\left[1 - \frac{(dx_1^2 + dx_2^2 + dx_3^2)}{c^2\,dt^2}\right]$$

$$= c^2\,dt^2(1 - \beta^2) = \frac{c^2}{\gamma^2}\,dt^2 \qquad (3.19)$$

where

$$\beta^2 = \frac{dx_1^2 + dx_2^2 + dx_3^2}{c^2\,dt^2}, \quad \gamma = \frac{1}{\sqrt{(1-\beta^2)}}.$$

Now dS is a Lorentz invariant quantity; thus if we introduce a dimensionless function

$$u_\lambda = \frac{dx_\lambda}{dS} = \frac{\gamma}{c}\frac{dx_\lambda}{dt} \qquad (3.20)$$

it will transform in the same manner as x_λ (see (3.13)), and furthermore since $dx_\lambda^2 = -dS^2$ the magnitude of $u^2\left(\equiv \sum_\lambda u_\lambda u_\lambda\right)$ is -1. Thus u_λ is a four-vector (§ 3.2(f)); it is called the *four velocity*, since in the nonrelativistic limit

$$\gamma \to 1 \qquad u_i \to \frac{1}{c}\frac{dx_i}{dt} \qquad (i = 1, 2, 3) \qquad (3.21)$$

and u is recognisable as the velocity of a particle or system divided by that of light.

Using this definition of four velocity, the components of four momentum for a particle of mass m can be constructed from the definition

$$p_\lambda = mcu_\lambda. \qquad (3.22)$$

The components of this equation are easily recognisable as the relativistic expressions for linear momentum and i/c (total energy)

$$p_i = \gamma m\frac{dx_i}{dt} = \frac{m}{\sqrt{(1-\beta^2)}}\frac{dx_i}{dt} \qquad (i = 1, 2, 3) \qquad (3.23)$$

$$p_4 = \gamma m\frac{dx_4}{dt} = \frac{i}{c}\frac{mc^2}{\sqrt{(1-\beta^2)}} = \frac{i}{c}E$$

and using the relation $u_\lambda^2 = -1$, we find

$$m^2c^2 = -p_\lambda^2 \qquad (3.24)$$

hence

$$m^2c^2 = \frac{E^2}{c^2} - p_i^2 = \frac{E^2}{c^2} - \mathbf{p}^2. \tag{3.25}$$

Since we shall work in units with $\hbar = c = 1$, this equation will normally be written as

$$m^2 = E^2 - \mathbf{p}^2.$$

An inspection of equations (3.1) and (3.25) shows that the relationship between the mass m and the energy–momentum four-vector is similar to that between the interval S and the coordinate four-vector. Thus m will be an invariant scalar quantity in all reference frames. This principle has been used to obtain the masses of the elementary particles, for example the mass of the Λ^0 hyperon was obtained by measuring the energy and momentum of its decay products in the transition

$$\Lambda^0 \to p + \pi^-$$

$$m_\Lambda^2 = -(p_p + p_\pi)_\lambda^2 \equiv -\sum_{\lambda=1}^{4} (p_p + p_\pi)_\lambda^2$$

$$= (E_p + E_\pi)^2 - (\mathbf{p}_p + \mathbf{p}_\pi)^2.$$

In § 3.2(d) it was stated that if for any two vectors x_a and x_b the relation

$$(x_a - x_b)_\lambda^2 < 0$$

was true, then the interval was time-like. Let the coordinates of the event a be located at $x_a = 0$. It is then customary to say that if the four-vector obeys the condition

$$x_\lambda^2 < 0 \tag{3.26}$$

it is a *time-like four-vector*. An inspection of equation (3.24) shows that $p_\lambda^2 < 0$ and hence p_λ is also time-like for a particle with real mass.

3.2(h). *Coordinate systems for the energy–momentum vector*

The four-vector for energy–momentum may be transformed between two coordinate systems, which are moving with uniform velocity with respect to each other, in the manner indicated in equation (3.14). Thus we find the following expressions

$$p_1 = \gamma(p_1' - i\beta p_4') \tag{3.27}$$

$$p_2 = p_2'$$

$$p_3 = p_3'$$

$$p_4 = \gamma(i\beta p_1' + p_4').$$

This principle is frequently used in considering the collision of two particles. In examining the behaviour of the particles after the collision, it is convenient to transform from the laboratory (L-) coordinate system to one in which the total momentum of the two particles is zero. This system is called the *centre of momentum* or *c-system*.

Let the energy and momentum of the particles before collision be E_a, E_b and $\mathbf{p}_a$, $\mathbf{p}_b$ respectively in the laboratory reference frame; the corresponding terms in the reference frame of the c-system will be denoted by dashes. Let

$$E = E_a + E_b$$

$$\mathbf{P} = \mathbf{p}_a + \mathbf{p}_b.$$

The x_1 and x_1' axes of the L and c-reference frames, respectively, will be chosen to be parallel to $\mathbf{P}$. Let the c-reference frame have a velocity β_c relative to the L-frame in $c = 1$ units. This term is often called the *velocity of the centre of mass;* the centre of mass system is, of course, the c-system. Since the y and z components of the momentum are zero before the collision, and remain unaltered in the transformation, our definition of the c-system as one in which

$$\mathbf{p}_a' = -\mathbf{p}_b' \tag{3.28}$$

implies that

$$p_{a1}' = -p_{b1}'.$$

We therefore find from equation (3.27) that

$$p_{a1} + p_{b1} = -i\beta_c\gamma_c(p_{a4}' + p_{b4}') = \mathrm{P} \tag{3.29}$$

$$p_{a4} + p_{b4} = \gamma_c(p_{a4}' + p_{b4}') = iE$$

where we are working in $c = 1$ units. These equations give

$$\beta_c = \frac{\mathbf{P}}{E} \tag{3.30}$$

and

$$E_c = \frac{1}{i}\,(p_{a4}' + p_{b4}') = \frac{E}{\gamma_c} = \sqrt{(E^2 - \mathbf{P}^2)} \tag{3.31}$$

where E_c is the total energy in the c-system.

The equations may also be derived by using the fact that the square of the resultant of two four-vectors must be a Lorentz invariant quantity

$$(p_a + p_b)_\lambda^2 = (p_a' + p_b')_\lambda^2.$$

Further relations between the L- and c-systems are given in A.8 (Appendixes, p. 713).

3.2 (i). *Four-tensors and their properties*

Equation (3.13) represents a special form of the Lorentz transformation. The most general form of the Lorentz transformation must involve both displacements and rotations of the coordinate axes. This transformation can be written as

$$x'_\lambda = a_\lambda + a_{\lambda\alpha}x_\alpha \tag{3.32}$$

and must be characterised by the invariance of the interval. The first of the terms on the right-hand side of equation (3.32) represents a change of origin for the space and time coordinates. For the present discussion we shall neglect it so that (3.32) reduces to

$$x'_\lambda = a_{\lambda\alpha}x_\alpha \equiv \sum_\alpha a_{\lambda\alpha}a_\alpha . \tag{3.33}$$

The terms $a_{\lambda\alpha}$ form a matrix; their values depend on the specific form of the transformation, for example the matrix for (3.13) is given by

$$a_{\lambda\alpha} = \begin{pmatrix} \gamma & 0 & 0 & -i\beta\gamma \\ 0 & 1 & 0 & 0 \\ 0 & 0 & 1 & 0 \\ i\beta\gamma & 0 & 0 & \gamma \end{pmatrix} \tag{3.34}$$

Since the interval remains invariant we may write

$$x'^2_\lambda = x^2_\alpha \tag{3.35}$$

that is

$$x'^2_\lambda = a_{\lambda\alpha}x_\alpha a_{\lambda\beta}x_\beta = x^2_\alpha \tag{3.36}$$

and so we find that

$$a_{\lambda\alpha}a_{\lambda\beta} = \delta_{\alpha\beta} . \tag{3.37}$$

This result immediately allows us to write the inverse of equation (3.33) as

$$x_\lambda = a_{\alpha\lambda}x'_\alpha \equiv a^{-1}x'. \tag{3.38}$$

This equation is obtained by writing (3.33) as

$$a_{\lambda\alpha}x'_\lambda = a_{\lambda\alpha}a_{\lambda\beta}x_\beta = \delta_{\alpha\beta}x_\beta = x_\alpha \tag{3.39}$$

and since the subscripts possess no special significance we may interchange λ and α, giving

$$x_\lambda = a_{\alpha\lambda}x'_\alpha .$$

Any linear transformation which possesses the property of equation (3.37) is called a *linear orthogonal transformation*. The matrix with components $a_{\lambda\alpha}$ is called the *transformation matrix*. It will be represented frequently by the symbol

, and the transformation written symbolically as

$$x' = ax \tag{3.40}$$

$$x = a^{-1}x'.$$

A further examination of the properties of the transformation matrix will be made in § 3.2(j).

It is obvious that the transformation (3.33) can be applied to any four-vector, and that the matrices involved in the transformation would have the property of equation (3.37). The equation (3.33) in fact represents one particular example of the transformation properties of a four-tensor. The definition of a tensor in four-dimensional space follows closely that normally used for three-dimensional space. For our purposes a *tensor* T of rank n may be defined as a quantity possessing 4^n components, $T_{\lambda\mu\varrho}\ldots$ (n indices), which behave in the following manner:

$$T'_{\lambda\mu\varrho}\cdots = a_{\lambda\alpha}a_{\mu\beta}a_{\varrho\gamma}\cdots T_{\alpha\beta\gamma}\cdots \tag{3.41}$$

during an orthogonal transformation of the coordinates.

Tensors in four-dimensional space are called *four-tensors*. Thus a four-tensor of rank zero will have one component and will act as a scalar quantity under orthogonal transformations. A four-tensor of rank one has four components, which transform as

$$T'_\lambda = a_{\lambda\alpha}T_\alpha \tag{3.42}$$

and thus it is completely equivalent to a four-vector.

A tensor of rank two transforms as

$$T'_{\lambda\mu} = a_{\lambda\alpha}a_{\mu\beta}T_{\alpha\beta}. \tag{3.43}$$

An example of a tensor of rank two is the derivative of a four-vector

$$\frac{\partial T_\lambda}{\partial x_\mu} = T_{\lambda\mu}$$

since, with the aid of (3.38)

$$T'_{\lambda\mu} = \frac{\partial T'_\lambda}{\partial x'_\mu} = \frac{\partial x_\beta}{\partial x'_\mu}\frac{\partial T'_\lambda}{\partial x_\beta}$$

$$= a_{\mu\beta}a_{\lambda\alpha}\frac{\partial T_\alpha}{\partial x_\beta} = a_{\lambda\alpha}a_{\mu\beta}T_{\alpha\beta}. \tag{3.44}$$

The tensor $T_{\alpha\beta}$ is said to be *symmetric* if $T_{\alpha\beta} = T_{\beta\alpha}$ and *antisymmetric* if $T_{\alpha\beta} = -T_{\beta\alpha}$. The diagonal components of an antisymmetric tensor must equal zero, since the latter quantity represents the only solution to equations of the type $T_{\alpha\alpha} = -T_{\alpha\alpha}$.

Tensors of higher rank, and certain aspects of the algebra of tensors, are considered in A.4 (Appendixes, p. 700).

3.2(j). *Classification of the Lorentz transformations*

The transformation matrix a in the Lorentz transformation (3.33)

$$x'_\lambda = a_{\lambda\alpha}a_\alpha$$

possesses certain properties which can be used to classify the Lorentz trans formations. Before doing this some of the mathematical properties of $a_{\lambda\alpha}$ ar worth noting. First, it was shown in (3.37) that

$$a_{\lambda\alpha}a_{\lambda\beta} = \delta_{\alpha\beta}.$$

This equation may be used to prove a second property, namely that the deter minant of a is ± 1. If we denote the transposed matrix of a by a^T, where a_λ $= a^T_{\alpha\lambda}$, equation (3.37) becomes

$$a^T_{\alpha\lambda}a_{\lambda\beta} = \delta_{\alpha\beta}. \qquad (3.45$$

This equation represents the usual rule for the multiplication of two matrice to yield a third one. It is obvious that the determinant of the third matrix i $+1$. If we use the symbol $\hat{1}$ to represent the unit matrix the equation can b written in matrix notation as

$$a^T a = \hat{1} \qquad (3.46$$

$$\det(a^T a) = \det a^T \det a = +1. \qquad (3.47$$

Now the determinant of a matrix is unaffected by the interchange of rows an columns and consequently

$$\det a^T = \det a. \qquad (3.48$$

Thus equation (3.47) shows that

$$\det a = \pm 1. \qquad (3.49$$

The transformation matrix given in (3.34) for the rotation of axes fulfils th conditions $\det a = +1$. On the other hand, a reflection of spatial coordinates o of the time component

$$a = \begin{pmatrix} -1 & 0 & 0 & 0 \\ 0 & -1 & 0 & 0 \\ 0 & 0 & -1 & 0 \\ 0 & 0 & 0 & 1 \end{pmatrix} \qquad \begin{aligned} x' &= -x \\ y' &= -y \\ z' &= -z \\ t' &= t \end{aligned} \qquad (3.50)$$

$$a = \begin{pmatrix} 1 & 0 & 0 & 0 \\ 0 & 1 & 0 & 0 \\ 0 & 0 & 1 & 0 \\ 0 & 0 & 0 & -1 \end{pmatrix} \qquad \begin{aligned} x' &= x \\ y' &= y \\ z' &= z \\ t' &= -t \end{aligned} \qquad (3.51)$$

obviously fulfils the condition $\det a = -1$.

The relation $\det a = \pm 1$ is one of the conditions used in the classification of he Lorentz transformations. They are divided as shown in Table 3.1.

TABLE 3.1

Class	1	2	3	4
$\det a$	$+1$	-1	-1	$+1$
a_{44}	>0	>0	<0	<0

n this table the classes have the following meanings:

Class 1 *proper Lorentz transformations* (also called the *restricted Lorentz group*)
Class 2 *space inversions*
Class 3 *time inversions*
Class 4 *space–time inversions.*

The proper Lorentz transformations involve those states which can be reached by an integration of a series of infinitesimal transformations. The rotation of coordinate frames (§ 3.2(e)) is in this category. The proper transformations are subdivided into *homogeneous* and *inhomogeneous transformations*, the former category involves rotations of the coordinate axes only, whilst the latter includes rotations plus displacements. Classes 2, 3 and 4 are called *improper Lorentz transformations*; they involve a discontinuity and cannot be reached by a sum of infinitesimal transformations. Classes 1 and 2 are called *orthochronous transformations* since they do not involve a reversal of time.

The orthochronous Lorentz transformations of coordinate frames lead to a classification of the behaviour of the observable quantities which normally occur in physics. If an observable quantity assumes a value O in one coordinate frame and O' in a second, then

$$O' = LO \tag{3.52}$$

where L is a function of the matrices $a_{\lambda\alpha}$. The behaviour of L determines the classification of O; the main categories are as follows:

(1) *Scalars*

$$S' = S, \quad L = +1 \tag{3.53}$$

thus L is a unit matrix.

(2) *Pseudoscalars*

$$P' = LP, \quad L = \det a \tag{3.54}$$

where $\det a = +1$ for proper Lorentz transformations

$= -1$ for reversal of spatial axes.

(3) *Vectors*; we have already shown that a four-vector transforms as

$$V'_\lambda = a_{\lambda\alpha} V_\alpha, \quad L = a_{\lambda\alpha}. \tag{3.55}$$

(4) *Axial (pseudo)-vectors*

$$A'_\lambda = LA_\alpha, \quad L = (\det a)\, a_{\lambda\alpha} \tag{3.56}$$

thus the space components of an axial vector fail to change sign upon reflection of coordinates.

(5) *Tensors*; these have already been discussed (§ 3.2(i))

$$T'_{\lambda\mu\varrho\ldots} = a_{\lambda\alpha} a_{\mu\beta} a_{\varrho\gamma\ldots}\, T_{\alpha\beta\gamma\ldots} \tag{3.57}$$

(6) *Pseudotensors*; the transformation matrix L is the same as for tensors multiplied by det a.

3.2(k). *Integration in four-dimensional space*

We note in this section some of the rules for integration in four-dimensional space. These are essentially the same as in ordinary space.

Integration over a hypersurface will first be considered – a *hypersurface* is the three-dimensional element of "area" for four-dimensional space. In this space

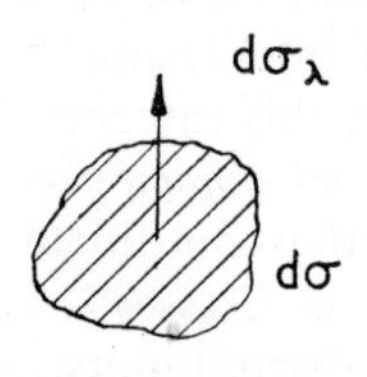

FIG. 3.3.

the hypersurface can be projected on to four hyperplanes given by $x_1x_2x_3$, $x_1x_2x_4$, $x_1x_3x_4$, $x_2x_3x_4$. Elements of "area" in the hyperplanes may be represented by the expression

$$d\sigma_\lambda = \frac{1}{i}\, n_\lambda\, dx_\alpha\, dx_\beta\, dx_\gamma = \frac{1}{i}\, n_\lambda\, d\sigma \tag{3.58}$$

$$\lambda \neq \alpha \neq \beta \neq \gamma$$

where n_λ is a unit vector normal to the element of area of the hypersurface. This definition is illustrated in Fig. 3.3. For example we may write

$$d\sigma_4 = \frac{1}{i}\, n_4\, dx_1\, dx_2\, dx_3 \equiv \frac{1}{i}\, n_4\, dx\, dy\, dz \equiv \frac{1}{i}\, n_4\, d\mathbf{x}. \tag{3.59}$$

An inspection of (3.58) shows that $d\sigma_\lambda$ is a four-vector; the presence of the $1/i$ term then ensures that the scalar product of $d\sigma_\lambda$ with another four-vector is a real scalar quantity, for example

$$\int d\sigma A \equiv \int d\sigma_\lambda A_\lambda \tag{3.60}$$

where the components of A can be written as $A_\lambda = n_\lambda A$ and $A_4 = in_4A_0$.

The element of four-dimensional volume is represented as

$$d^4x = \frac{1}{i}\, dx_1\, dx_2\, dx_3\, dx_4 = d\mathbf{x}\, dt. \tag{3.61}$$

The term $1/i$ again appears because of the definition $x_4 = ict$; if a coordinate scale with real fourth component is used the i can be dropped.

Analogous equations to Gauss' and Stokes' theorems

Gauss' theorem
$$\oint d\sigma_i A_i = \int d\mathbf{x}\, \frac{\partial A_i}{\partial x_i} \tag{3.62}$$

Stokes' theorem
$$\oint dx_i\, A_i = \int d\sigma_i (\nabla \times A)_i \tag{3.63}$$

$$i = 1, 2, 3, \quad d\sigma_i = n_i\, dx_j\, dx_k, \quad i \neq j \neq k$$

can be constructed for four-dimensional systems. For example, an integral over a closed hypersurface can be converted into an integral over the four-dimensional volume enclosed by it, by introducing the substitution

$$d\sigma_\lambda \equiv d^4x \frac{\partial}{\partial x_\lambda}. \tag{3.64}$$

Thus the four-dimensional equivalent of Gauss' theorem is

$$\oint d\sigma_\lambda A_\lambda = \int d^4x \frac{\partial A_\lambda}{\partial x_\lambda} \tag{3.65}$$

where A_λ is a four vector.

3.3 RELATIVISTIC WAVE EQUATIONS

In this section we shall consider equations which are inadequate to describe fully the properties of the elementary particles, but are instructive in that they include features which must be incorporated in any relativistic field theory. The equations are those of relativistic wave mechanics; their inadequacy lies in attempting to describe the properties of a set of elementary particles in terms of a fixed number of particles participating in any process. In practice provision must be made in any theory of particles for their possible annihilation or creation.

The equations considered are:

(1) The Klein–Gordon equation,
(2) The Dirac equation.

3.3(a). *The Klein–Gordon equation*

The basic equation of nonrelativistic wave mechanics uses the relation

$$E = \frac{\mathbf{p}^2}{2m} \tag{3.66}$$

of classical mechanics, and makes the substitution

$$E \to i\frac{\partial}{\partial t}, \quad \mathbf{p} \to -i\nabla \tag{3.67}$$

($\hbar = 1$ units), thus giving the Schrödinger wave equation

$$i\frac{\partial\psi}{\partial t} = \frac{-1}{2m}\nabla^2\psi. \tag{3.68}$$

A physical meaning can be attached to the wave function ψ by defining the probability density ϱ for finding a particle located at a point $\mathbf{x}$ at a time t as

$$\varrho = \psi^*(\mathbf{x}, t)\,\psi(\mathbf{x}, t) \tag{3.69}$$

and the probability current density as

$$\mathbf{j} = \frac{1}{2im}(\psi^*\nabla\psi - \psi\nabla\psi^*) \tag{3.70}$$

where ψ^* is the complex conjugate of ψ.

The probability density given by equation (3.69) is positive definite, and it is conserved since equation (3.68) may be used to show that the *continuity* or *conservation equation*

$$\frac{\partial\varrho}{\partial t} + \nabla\cdot\mathbf{j} = 0 \tag{3.71}$$

is satisfied.

Now the relativistic analogue of (3.66) is the equation

$$E^2 - \mathbf{p}^2 = m^2$$

and the substitution of the operators given in equation (3.67) then yields the Klein–Gordon equation

$$\left(\nabla^2 - \frac{\partial^2}{\partial t^2}\right)\varphi = m^2\varphi \tag{3.72}$$

which also may be written as

$$\left(\sum_\lambda \frac{\partial^2}{\partial x_\lambda^2} - m^2\right)\varphi \equiv \left(\frac{\partial^2}{\partial x_\lambda^2} - m^2\right)\varphi = (\Box^2 - m^2)\,\varphi = 0 \tag{3.73}$$

we shall use φ to represent the wave function of scalar particles obeying Bose–Einstein statistics).

If an attempt is made to set up a conservation equation of the form given in (3.71) using the Klein–Gordon equation, a problem arises. We first re-write equation (3.71) in covariant form by using the fact that the probability density is the fourth component of the current density four-vector (3.16)

$$\frac{\partial j_\lambda}{\partial x_\lambda} = 0. \tag{3.74}$$

A continuity equation obeying (3.74) can be constructed for the Klein–Gordon equation if j_λ is chosen to be of the form

$$j_\lambda = \frac{1}{2im}\left(\varphi^* \frac{\partial \varphi}{\partial x_\lambda} - \varphi \frac{\partial \varphi^*}{\partial x_\lambda}\right) \tag{3.75}$$

giving

$$\varrho = \frac{i}{2m}\left(\varphi^* \frac{\partial \varphi}{\partial t} - \varphi \frac{\partial \varphi^*}{\partial t}\right). \tag{3.76}$$

Thus the spatial components of j_λ are identical with those of $\mathbf{j}$ in (3.70) – apart from the trivial exchange of φ and ψ.

Now equation (3.76) can take both positive and negative values for ϱ since φ and $\partial\varphi/\partial t$ are arbitrary†. Therefore the probability density as defined by (3.76) is meaningless, since a negative probability is meaningless.

This unsatisfactory situation was resolved by Pauli and Weisskopf (1934) who pointed out that the term φ of the Klein–Gordon equation should not be treated as a wave function describing a single particle, but as an operator in a field equation, describing a field of relativistic particles of mass m. Furthermore, φ can be used to describe a field containing particles of both positive and negative electrical charge. Under these circumstances ϱ may be interpreted as the net charge density at any point. This feature will be discussed again in § 4.3(i.1).

The operator $(\Box^2 - m^2)$ is a Lorentz invariant, i.e. a scalar quantity. Thus if equation (3.73) is to be covariant, φ must be a scalar or pseudoscalar quantity. This implies that φ has no preferred direction, and so can be used to describe a particle of zero spin. This conclusion could have been reached alternatively by recalling that the wave function φ in equation (3.73) possesses only one component, and that in the nonrelativistic limit it reduces to equation (3.68), the Schrödinger equation without spin.

We may conclude, therefore, that the Klein–Gordon equation, or its equivalent in field theory, can be useful in describing the properties of spinless particles of finite mass, for example pions and kaons.

† Mathematically, the arbitrariness of ϱ with respect to sign may be demonstrated easily by writing

$$\varphi = \varphi_1 + i\varphi_2$$
$$\varphi^* = \varphi_1 - i\varphi_2.$$

3.3(b). *The Dirac equation*

The Klein–Gordon equation possesses symmetry in its four-dimensional coordinates $x_1, x_2, x_3, x_4 = ict$, by having all its derivatives in second order. A wave equation constructed in this manner, however, suffers from ambiguity in the interpretation of the probability density, ϱ, as shown in § 3.3(a).

An alternative approach to the construction of a relativistic wave equation was made by Dirac. He assumed that a particle could exist in several distinct states, for example orientations of spin, with the same momentum. Thus the wave function would have to possess more than one component, and could be represented by a column matrix

$$\psi = \begin{matrix}\psi_1 \\ \psi_2 \\ \psi_3 \\ \vdots\end{matrix} \equiv \psi_\alpha \quad (\alpha = 1, 2, 3, \ldots)$$

Dirac assumed that the probability density at any point was given by an expression similar in form to that appearing in nonrelativistic wave mechanics

$$\varrho = \psi^\dagger \psi \tag{3.77}$$

and satisfying the continuity equation (3.71)

$$\frac{\partial \varrho}{\partial t} + \nabla \cdot \mathbf{j} = 0.$$

In equation (3.77) $\psi^\dagger$ represents a row matrix

$$\psi^\dagger = \psi_1^* \psi_2^* \psi_3^* \ldots$$
$$\equiv \psi_\alpha^* \quad (\alpha = 1, 2, 3, \ldots)$$

so that

$$\varrho = \psi^\dagger \psi = \psi_\alpha^* \psi_\alpha . \tag{3.78}$$

If equations (3.71) and (3.78) are to be reconciled, then the wave equation must be a first order differential in both space and time coordinates. Dirac showed that a plausible equation, satisfying the condition of linearity in the derivatives of coordinates, was given by the expression

$$E\psi = (\boldsymbol{\alpha} \cdot \mathbf{p} + \beta m)\,\psi \tag{3.79}$$

leading to

$$\frac{\partial \psi}{\partial t} + \alpha_k \frac{\partial \psi}{\partial x_k} + im\beta\psi = 0 \tag{3.80}$$

where $k = 1, 2, 3$.

The adjoint function $\psi^\dagger$ (the Hermitian conjugate of ψ) satisfies the Hermitian onjugate to equation (3.80), namely

$$\frac{\partial\psi^\dagger}{\partial t} + \left(\alpha_k \frac{\partial\psi}{\partial x_k}\right)^\dagger - im(\beta\psi)^\dagger = \frac{\partial\psi^\dagger}{\partial t} + \frac{\partial\psi^\dagger}{\partial x_k}\alpha_k^\dagger - im\psi^\dagger\beta^\dagger = 0. \quad (3.81)$$

Now if equation (3.80) is multiplied on the left by $\psi^\dagger$, and (3.81) on the right by ψ, and the two equations added we find that

$$\frac{\partial}{\partial t}(\psi^\dagger\psi) + \left(\frac{\partial\psi^\dagger}{\partial x_k}\alpha_k^\dagger\psi + \psi^\dagger\alpha_k\frac{\partial\psi}{\partial x_k}\right) + im(\psi^\dagger\beta\psi - \psi^\dagger\beta^\dagger\psi) = 0. \quad (3.82)$$

f we wish to identify this equation with the continuity equation (3.71)

$$\frac{\partial\varrho}{\partial t} + \nabla\cdot\mathbf{j} = \frac{\partial}{\partial t}(\psi^\dagger\psi) + \nabla\cdot\mathbf{j} = 0$$

ve must choose α_k and β to be *Hermitian matrices*

$$\alpha_k^\dagger = \alpha_k, \quad \beta^\dagger = \beta$$

and the current takes the form

$$\mathbf{j} = \psi^\dagger\boldsymbol{\alpha}\psi \quad (3.83)$$

with components

$$j_k = \psi^\dagger\alpha_k\psi.$$

The matrices α and β in the Dirac equation must fulfil further conditions. If the Dirac wave function is to be truly relativistic, it must satisfy an equation of the Klein–Gordon type

$$(\Box^2 - m^2)\psi = 0.$$

An equation of this form can be constructed by operating on equation (3.80) from the left by

$$-\frac{\partial}{\partial t} + \alpha_l\frac{\partial}{\partial x_l} + im\beta.$$

This operation yields the equation

$$-\frac{\partial^2\psi}{\partial t^2} + \tfrac{1}{2}(\alpha_k\alpha_l + \alpha_l\alpha_k)\frac{\partial^2\psi}{\partial x_k\partial x_l} + im(\alpha_k\beta + \beta\alpha_k)\frac{\partial\psi}{\partial x_k} - m^2\beta^2\psi = 0. \quad (3.84)$$

Equation (3.84) will reduce to the form

$$(\Box^2 - m^2)\psi = 0$$

if the following conditions are fulfilled

$$(\alpha_k\alpha_l + \alpha_l\alpha_k) = 2\delta_{kl}, \quad \alpha_k\beta + \beta\alpha_k = 0, \quad \beta^2 = \hat{1}. \quad (3.85)$$

If $k = l$, the first of equations (3.85) implies that

$$\alpha_k^2 = \hat{1}. \tag{3.86}$$

3.3(c). *Covariant form of the Dirac equation*

Before considering the Dirac equation in greater detail, it is useful to put it in a form more suitable for ease of manipulation. The new form is called the covariant form, and we will use the four-dimensional Minkowski representation. Equation (3.80) may be multiplied by $-i\beta$ from the left, yielding

$$\beta \frac{1}{i} \frac{\partial \psi}{\partial t} - i\beta\alpha_k \frac{\partial \psi}{\partial x_k} + m\psi = 0 \tag{3.87}$$

where use has been made of the relation $\beta^2 = \hat{1}$ from equation (3.85)†. Upon making the substitutions

$$\gamma_k = -i\beta\alpha_k, \quad \gamma_4 = \beta \tag{3.88}$$

in equations (3.87), the covariant form of the Dirac equation is obtained

$$\gamma_\lambda \frac{\partial \psi}{\partial x_\lambda} + m\psi = 0. \tag{3.89}$$

The Hermitian conjugate of the Dirac equation was given in (3.81); we have since shown that $\alpha_k^\dagger = \alpha_k$ and $\beta^\dagger = \beta$ so that the equation can be written as

$$\frac{\partial \psi^\dagger}{\partial t} + \frac{\partial \psi^\dagger}{\partial x_k} \alpha_k - im\psi^\dagger \beta = 0.$$

This equation may be moulded into a similar from to that of equation (3.89), by multiplying with $-i\beta^2$ from the right. This operation yields

$$\frac{1}{i} \frac{\partial \psi^\dagger}{\partial t} \beta\beta - i \frac{\partial \psi^\dagger}{\partial x_k} \alpha_k\beta\beta - m\psi^\dagger\beta\beta^2$$

$$= \frac{1}{i} \frac{\partial \psi^\dagger}{\partial t} \beta\beta - i \frac{\partial \psi^\dagger}{\partial x_k} \beta\beta\alpha_k - m\psi^\dagger\beta = 0$$

where use has been made of equation (3.85). If we now make the substitutions for γ from equation (3.88) and introduce the function

$$\bar{\psi} = \psi^\dagger\beta = \psi^\dagger\gamma_4 \tag{3.90}$$

the adjoint form of the covariant equation is obtained

$$\frac{\partial \bar{\psi}}{\partial x_\lambda} \gamma_\lambda - m\bar{\psi} = 0. \tag{3.91}$$

† The unit operator has been omitted from (3.87) and from subsequent equations in order to avoid a clumsy notation.

In the covariant notation the separate terms for probability density (3.77) and current density (3.83) may be condensed into a single expression for a four-vector current density

$$j_\lambda = i\psi^\dagger \gamma_4 \gamma_\lambda \psi = i\bar{\psi}\gamma_\lambda \psi. \tag{3.92}$$

This expression satisfies the continuity equation (3.74).

3.3(d). *Some properties of the γ-matrices*

The relations given in (3.85)

$$(\alpha_k \alpha_l + \alpha_l \alpha_k) = 2\delta_{kl}, \quad \alpha_k \beta + \beta \alpha_k = 0, \quad \beta^2 = \hat{1}$$

may be written in a more compact form in terms of γ-matrices. Using the definitions given in equations (3.88), equations (3.85) become

$$\gamma_\mu \gamma_\lambda + \gamma_\lambda \gamma_\mu = 2\delta_{\mu\lambda}. \tag{3.93}$$

An anticommutation relation of this form can only be satisfied if the γ terms are matrices. This is not surprising since they were defined as products of the matrices α and β (equation (3.88)). For many physical problems it is not necessary to specify any mathematical property for the γ-matrices beyond that given in equation (3.93). It can be shown, however, that the γ terms can be represented by 4×4 matrices. This can be done by noting that, by virtue of equation (3.93), any product of an arbitrary number of γ's can be reduced to one of sixteen independent matrices

$$\Gamma_i = \begin{cases} \hat{1} & & (1) \\ \gamma_\lambda & & (4) \\ \gamma_\lambda \gamma_\mu & \lambda < \mu & (6) \\ \gamma_\lambda \gamma_\mu \gamma_\varrho & \lambda < \mu < \varrho & (4) \\ \gamma_1 \gamma_2 \gamma_3 \gamma_4 & & (1) \end{cases} \tag{3.94}$$

The number in brackets gives the number of matrices of each type; it can be seen that the sum is 16. Similarly a sum of products of γ's can be written as a sum of the sixteen basic terms with appropriate numerical coefficients (these can be real or complex). Now a 4×4 matrix has sixteen elements. It follows, therefore, that the sixteen basic terms and the operators γ_λ can be represented by 4×4 matrices.

The sixteen basic matrices of (3.94) may be simplified by introducing a new matrix

$$\gamma_5 = \gamma_1 \gamma_2 \gamma_3 \gamma_4. \tag{3.95}$$

It may easily be shown that the matrix γ_5 satisfies the same commutation rule as γ_λ

$$\gamma_5^2 = \hat{1} \tag{3.96}$$

$$\gamma_\lambda \gamma_5 + \gamma_5 \gamma_\lambda = 0, \quad \lambda \neq 5. \tag{3.97}$$

Two further general properties of the γ-matrices should be noted; they are concerned with the *traces* or *spurs of the matrices*

$$\text{tr } A = \text{sp } A = \sum_\lambda A_{\lambda\lambda}. \tag{3.98}$$

The trace represents the sum of the diagonal elements of the matrix, for example the trace of the unit (identity) 4×4 matrix is

$$\text{tr } \hat{1} = 4. \tag{3.99}$$

The properties, which we shall prove, are the following.

(1) The trace of the product of an odd number of γ-matrices is zero. To prove this we use the elementary property of a trace (see A.3 (Appendixes, p. 696)).

$$\text{tr } ABC = \text{tr } CAB. \tag{3.100}$$

Now from equations (3.96) and (3.97) we have

$$\gamma_5 \gamma_\lambda \gamma_5 = -\gamma_\lambda.$$

Thus, if we have a product of n (odd number) γ-matrices

$$\gamma_\lambda \gamma_\mu \gamma_\varrho \cdots \gamma_\varepsilon$$

then

$$\gamma_5 \gamma_\lambda \gamma_\mu \gamma_\varrho \cdots \gamma_\varepsilon \gamma_5 = -\gamma_\lambda \gamma_\mu \gamma_\varrho \cdots \gamma_\varepsilon$$

therefore

$$\text{tr } \gamma_5 \gamma_\lambda \gamma_\mu \gamma_\varrho \cdots \gamma_\varepsilon \gamma_5 = -\text{tr } \gamma_\lambda \gamma \; \gamma_\varrho \cdots \gamma_\varepsilon$$

but, from (3.100)

$$\text{tr } \gamma_5 \gamma_\lambda \gamma_\mu \; \gamma_\varrho \cdots \gamma_\varepsilon \gamma_5 = \text{tr } \gamma_5 \gamma_5 \gamma_\lambda \gamma_\mu \gamma_\varrho \cdots \gamma_\varepsilon$$

$$= \text{tr } \gamma_\lambda \gamma_\mu \gamma_\varrho \cdots \gamma_\varepsilon$$

We therefore have found that

$$\text{tr } \gamma_\lambda \gamma_\mu \gamma_\varrho \cdots \gamma_\varepsilon = -\text{tr } \gamma_\lambda \gamma_\mu \gamma_\varrho \cdots \gamma_\varepsilon$$

a condition which can be satisfied only if the trace is zero. Similarly it can be shown that

$$\text{tr } \gamma_5 = -\text{tr } \gamma_\lambda \gamma_5 \gamma_\lambda = -\text{tr } \gamma_\lambda \gamma_\lambda \gamma_5 = -\text{tr } \gamma_5 = 0. \tag{3.101}$$

(2) If the number of γ-matrices n in a product of matrices is even, the commutation rule may be used to reduce the expression to one involving $n - 2$ fac-

ors, for example since tr AB = tr BA then

$$\mathrm{tr}(\gamma_\lambda\gamma_\mu + \gamma_\mu\gamma_\lambda) = 2\,\mathrm{tr}\,\gamma_\lambda\gamma_\mu = 2\,\mathrm{tr}\,\delta_{\lambda\mu}$$

ence

$$\mathrm{tr}\,\gamma_\lambda\gamma_\mu = \mathrm{tr}\,\delta_{\lambda\mu} = 4\delta_{\lambda\mu}. \tag{3.102}$$

Further examples of the evaluation of traces are given in Appendices, A.5 (p. 701).

The properties given above complete the general description of the behaviour of the γ-matrices. Several representations are used for the matrices in scientific iterature; we shall use the representation given by Pauli, other forms are given n A.5 (Appendixes, p. 701).

The Pauli set, expressed in terms of 2 × 2 sub-matrices, are

$$\alpha_k = \begin{pmatrix} 0 & \sigma_k \\ \sigma_k & 0 \end{pmatrix}, \quad \beta = \begin{pmatrix} 1 & 0 \\ 0 & -1 \end{pmatrix} \tag{3.103}$$

$$\gamma_k = \begin{pmatrix} 0 & -i\sigma_k \\ i\sigma_k & 0 \end{pmatrix}, \quad \gamma_4 = \begin{pmatrix} 1 & 0 \\ 0 & -1 \end{pmatrix}, \quad \gamma_5 = \begin{pmatrix} 0 & -1 \\ -1 & 0 \end{pmatrix}$$

where the symbols 1 and 0 represent 2 × 2 unit and null matrices respectively, and the terms σ_k are the 2 × 2 Pauli matrices

$$\sigma_1 \equiv \sigma_x = \begin{pmatrix} 0 & 1 \\ 1 & 0 \end{pmatrix}, \quad \sigma_2 \equiv \sigma_y = \begin{pmatrix} 0 & -i \\ i & 0 \end{pmatrix}, \quad \sigma_3 \equiv \sigma_z = \begin{pmatrix} 1 & 0 \\ 0 & -1 \end{pmatrix}. \tag{3.104}$$

It can be seen that in this representation the γ-matrices are Hermitian

$$\gamma_\lambda^\dagger = \gamma_\lambda. \tag{3.105}$$

The sixteen basic matrices of equation (3.94) may be set out in this representation as

$$\Gamma_i = \begin{array}{lll} \hat{1} & & (1) \\ \gamma_\lambda & & (4) \\ i\gamma_\lambda\gamma_\mu & \lambda < \mu & (6) \\ i\gamma_\lambda\gamma_5 & & (4) \\ \gamma_5 & & (1) \end{array} \tag{3.106}$$

where γ_5 has also been used. The factors i have been included so that Γ_i is Hermitian

$$\Gamma_i^\dagger = \Gamma_i \tag{3.107}$$

and

$$\Gamma_i^2 = \hat{1}.$$

3.3(e). *The spin of a Dirac particle*

A particle whose wave function satisfies the Dirac equation may be readil demonstrated to have a spin of $\frac{1}{2}\hbar$. This may be done by making use of th Heisenberg condition, given in equation (4.49), that if an operator A commute with the total Hamiltonian H for any system, then that operator is a constant o the motion for the system. In other words, the expectation value for the operato A is a conserved quantity.

It is apparent from equations (3.79) and (3.88) that the total Hamiltonian fo a free Dirac particle is

$$H = \boldsymbol{\alpha} \cdot \mathbf{p} + \beta m = i\gamma_4 \boldsymbol{\gamma} \cdot \mathbf{p} + \gamma_4 m \tag{3.108}$$

where

$$\boldsymbol{\gamma} \cdot \mathbf{p} = \sum_k \gamma_k p_k, \quad k = 1, 2, 3.$$

Now, consider the commutation relation

$$[\mathbf{L}, H]$$

where $\mathbf{L}$ represents the orbital angular momentum operator. Writing $\mathbf{L}$ as

$$\mathbf{L} = \mathbf{r} \times \mathbf{p}$$

we find

$$\begin{aligned} [\mathbf{L}, H] &= [\mathbf{r} \times \mathbf{p}, i\gamma_4 \boldsymbol{\gamma} \cdot \mathbf{p} + \gamma_4 m] = [\mathbf{r} \times \mathbf{p}, i\gamma_4 \boldsymbol{\gamma} \cdot \mathbf{p}] \\ &= i\gamma_4 [\mathbf{r}, \boldsymbol{\gamma} \cdot \mathbf{p}] \times \mathbf{p} = -\gamma_4 \boldsymbol{\gamma} \times \mathbf{p} \end{aligned} \tag{3.109}$$

where use has been made of the commutation relation (compare (4.27))

$$[\mathbf{r}, \mathbf{p}] = i.$$

Equation (3.109) shows that $\mathbf{L}$ is not a constant of the motion for a Dira particle. A term with the properties of angular momentum may be added to $\mathbf{L}$, however, to produce an expression for total angular momentum which is a constant of the motion. This term (the Dirac spin operator) may be defined as

$$\hat{\boldsymbol{\sigma}} = -\frac{i}{2} \boldsymbol{\gamma} \times \boldsymbol{\gamma}$$

with components

$$\hat{\sigma}_{ij} = -\frac{i}{2}(\gamma_i \gamma_j - \gamma_j \gamma_i) = -\hat{\sigma}_{ji} \tag{3.110}$$

or

$$\hat{\sigma}_{ij} = \begin{pmatrix} \sigma_k & 0 \\ 0 & \sigma_k \end{pmatrix} = \hat{\sigma}_k$$

where i, j, k are treated cyclically, and the circumflex sign on $\hat{\sigma}$ is used to distinguish the 4×4 matrices from the 2×2 Pauli matrices.

It is apparent from equations (3.110) and (3.88) that

$$\hat{\sigma}_{ij} = -i\gamma_i\gamma_j = -i^3\beta\alpha_i\beta\alpha_j = -i\alpha_i\alpha_j, \quad i \neq j. \tag{3.111}$$

The component $\hat{\sigma}_{ij}$ is also frequently written as

$$\hat{\sigma}_{ij} \equiv \hat{\sigma}_k = i\gamma_4\gamma_5\gamma_k \tag{3.112}$$

where i, j, k are again taken cyclically.

The commutation of $\hat{\boldsymbol{\sigma}}$ with H yields the following expression:

$$[\hat{\boldsymbol{\sigma}}, H] = [\hat{\boldsymbol{\sigma}}, i\gamma_4\boldsymbol{\gamma} \cdot \mathbf{p} + \gamma_4 m] = [\hat{\boldsymbol{\sigma}}, i\gamma_4\boldsymbol{\gamma} \cdot \mathbf{p}] = 2\gamma_4\boldsymbol{\gamma} \times \mathbf{p}. \tag{3.113}$$

The commutation of the second term in H with $\hat{\boldsymbol{\sigma}}$ is obvious from commutation relations for γ-matrices (equation (3.93)). The result of commuting $\hat{\boldsymbol{\sigma}}$ with the first term in H is less obvious, but the answer may be readily verified by considering terms of the type

$$[\hat{\boldsymbol{\sigma}}, i\gamma_4\gamma_1 p_1].$$

It is apparent from equations (3.109) and (3.113) that the expression

$$\mathbf{J} = \mathbf{L} + \tfrac{1}{2}\hat{\boldsymbol{\sigma}}$$

is a suitable one for describing the total angular momentum operator of a Dirac particle. It commutes with H

$$[\mathbf{J}, H] = [\mathbf{L} + \tfrac{1}{2}\hat{\boldsymbol{\sigma}}, H] = 0 \tag{3.114}$$

and is therefore a constant of the motion. We may therefore identify the operator $\hat{\boldsymbol{\sigma}}$ as the intrinsic angular momentum (or spin) operator and conclude that a Dirac particle has a spin of $\frac{1}{2}\hbar$.†

Further properties of $\hat{\boldsymbol{\sigma}}$ may be readily deduced. First, $\hat{\boldsymbol{\sigma}} \cdot \mathbf{p}$ commutes with H

$$[\hat{\boldsymbol{\sigma}} \cdot \mathbf{p}, H] = 0 \tag{3.115}$$

so that $\hat{\boldsymbol{\sigma}} \cdot \mathbf{p}$ is a constant of the motion. Secondly, the properties of the components of $\hat{\boldsymbol{\sigma}}$ given in (3.110) imply that

$$(\hat{\boldsymbol{\sigma}} \cdot \mathbf{p})^2 = (\hat{\sigma}_i p_i)(\hat{\sigma}_j p_j) = \delta_{ij} p_i p_j = \mathbf{p}^2. \tag{3.116}$$

Thus the operator $\hat{\boldsymbol{\sigma}} \cdot \mathbf{p}$ must have eigenvalues $\pm |\mathbf{p}|$

$$\hat{\boldsymbol{\sigma}} \cdot \mathbf{p} = \pm |\mathbf{p}|. \tag{3.117}$$

Summarising the results of this section, we may conclude that the intrinsic angular momentum of a Dirac particle is of magnitude $\frac{1}{2}\hbar$ and that eigenfunctions of $\hat{\boldsymbol{\sigma}} \cdot \mathbf{p}$ represent systems in which the spin projections point either parallel or antiparallel to the momentum $\mathbf{p}$ of the particle.

† The spin operator for Dirac particles will be examined again in § 5.3(g).

3.3(f). *The magnetic moment of a Dirac particle*

The magnitude of the magnetic moment of a Dirac particle will be elucidated by considering its interaction with a magnetic field **H** which is constant in time. This field may be expressed in terms of a four-vector $A_\lambda(x)$ (§ 4.4(a)) with its fourth component $A_4 = 0$.

The effect of an electromagnetic field on a classical particle with electrical charge e is to change the four-momentum p_λ in the equation of motion for the free particle as

$$p_\lambda \to (p_\lambda - eA_\lambda). \tag{3.118}$$

Thus the Dirac equation (3.89) becomes

$$\gamma_\lambda\left(\frac{\partial}{\partial x_\lambda} - ieA_\lambda\right)\psi + m\psi = 0. \tag{3.119}$$

The i-term appears in this equation by virtue of the definitions given in equation (3.67). The equation (3.119) may be operated upon from the left by

$$\gamma_\mu\left(\frac{\partial}{\partial x_\mu} - ieA_\mu\right) - m$$

to yield a second order equation

$$\left(\frac{\partial}{\partial x_\lambda} - ieA_\lambda\right)^2\psi - \frac{ie}{2}\gamma_\lambda\gamma_\mu\left(\frac{\partial A_\mu}{\partial x_\lambda} - \frac{\partial A_\lambda}{\partial x_\mu}\right)\psi - m^2\psi = 0 \tag{3.120}$$

where once again use has been made of the commutation relation (3.93)

$$\gamma_\mu\gamma_\lambda + \gamma_\lambda\gamma_\mu = 2\delta_{\mu\lambda}.$$

Now the magnetic field **H** is given by

$$\mathbf{H} = \nabla \times \mathbf{A}$$

and therefore, since we are dealing with a static field, equation (3.120) becomes

$$\left(\frac{\partial}{\partial x_\lambda} - ieA_\lambda\right)^2\psi + e\hat{\boldsymbol{\sigma}}\cdot\mathbf{H}\psi - m^2\psi = 0 \tag{3.121}$$

where use has been made of equation (3.111). The expression given above can be re-written as

$$\left(\frac{\partial}{\partial x_4} - m\right)\left(\frac{\partial}{\partial x_4} + m\right)\psi = \left[-\left(\frac{\partial}{\partial x_k} - ieA_k\right)^2 - e\hat{\boldsymbol{\sigma}}\cdot\mathbf{H}\right]\psi \tag{3.122}$$

where

$$k = 1, 2, 3, \quad A_4 = 0.$$

Now in the nonrelativistic limit

$$\frac{\partial \psi}{\partial x_4} \rightarrow -m\psi$$

so that in this limit equation (3.122) becomes

$$-2m\left(\frac{\partial}{\partial x_4} + m\right)\psi = \left[-\left(\frac{\partial}{\partial x_k} - ieA_k\right)^2 - e\hat{\sigma}\cdot\mathbf{H}\right]\psi.$$

Therefore (compare (3.118))

$$i\frac{\partial \psi}{\partial t} = \left[m - \frac{1}{2m}\left(\frac{\partial}{\partial x_k} - ieA_k\right)^2 - \frac{e}{2m}\hat{\sigma}\cdot\mathbf{H}\right]\psi$$

$$\equiv \left[m + \frac{1}{2m}(\mathbf{p} - e\,\mathbf{A})^2 - \frac{e}{2m}\hat{\sigma}\cdot\mathbf{H}\right]\psi. \qquad (3.123)$$

It can be seen that the right-hand side of this equation contains the classical Hamiltonian term for the behaviour of a slow electron in a magnetic field plus

TABLE 3.2

Particle	Measured g-value
e^-	2·0023218 ± 0·0000048
μ^+	2·002324 ± 0·000010
p	5·58550 ± 0·00006

an additional potential energy term $(-e/2m\hat{\sigma}\cdot\mathbf{H})$. Since the potential energy of a magnet of moment $\boldsymbol{\mu}$ in a field of strength H is $-\boldsymbol{\mu}\cdot\mathbf{H}$, equation (3.123) shows that a Dirac particle with electric charge e should possess a magnetic moment

$$\boldsymbol{\mu} = \frac{e}{2m}\hat{\sigma} \equiv 2\frac{e}{2mc}(\tfrac{1}{2}\hbar\hat{\sigma}) = 2\frac{e}{2mc}\mathbf{s} \qquad (3.124)$$

where $\mathbf{s}$ represents the spin.

Now the g-value for a particle is defined by equation (2.15)

$$\boldsymbol{\mu} = g\frac{e}{2mc}\mathbf{s}$$

and so the Dirac equation may be used to describe particles with spin of $\frac{1}{2}$ and g-value 2. In Table 3.2 the measured g-values for particles of half integral spin are given (§ 2.5). It can be seen that the Dirac equation is fairly satisfactory for electrons and muons, but not for nucleons.

Pauli (1941) has shown that the Dirac equation can be modified to represent particles of arbitrary magnetic moment by adding a term of the form

$$-i\mu_a\gamma_\lambda\gamma_\nu\left(\frac{\partial A_\lambda}{\partial x_\nu}-\frac{\partial A_\nu}{\partial x_\lambda}\right) \tag{3.125}$$

to equation (3.119). The particle described by the resulting equation behaves as if it possesses an anomalous magnetic moment μ_a in addition to its normal moment. Anomalous magnetic moments will be considered again in §§ 11.3 and 11.4.

3.3(g). *Solutions of the Dirac equation*

As with the γ-matrices, a precise formulation is not required for the Dirac wave function in most problems. In addition it is obvious that if a solution is obtained for the Dirac equation, its particular form can depend upon that chosen for the γ-matrices. Nevertheless, specific solutions are worth investigating as they give additional physical insight into the properties of a Dirac particle.

In obtaining solutions for the Dirac equation, the form of γ-matrices given in equation (3.103) will be used. It was shown in § 3.3(b) that the Dirac wave function can satisfy an equation of the Klein–Gordon type

$$\left(\frac{\partial^2}{\partial x_\lambda^2}-m^2\right)\psi_\alpha=0$$

where the subscript α has been added to the wave function, since the Dirac wave function possesses more than one component.

It is apparent that the above equation can be satisfied by a wave function which includes the periodic term $e^{\pm ipx}$, where we use the notation

$$ipx \equiv ip_\lambda x_\lambda = i(\mathbf{p}\cdot\mathbf{x}-Et) \tag{3.126}$$

and the four-vector p_λ satisfies equation (3.24) $p_\lambda^2 = -m^2$.

Thus we will take as solutions of the Dirac equation

$$\psi_\alpha = u_{r\alpha}e^{ipx} \tag{3.127}$$

$$\psi_\alpha = v_{r\alpha}e^{-ipx}.$$

The terms $u_{r\alpha}$ and $v_{r\alpha}$ are called *spinors*. The term $+px$ refers to states of the Dirac particle with momentum $+\mathbf{p}$ and energy $+E$, and $-px$ to $-\mathbf{p}$ and $-E$ respectively. These states are sometimes called the positive and negative frequency parts respectively. The need for equations with negative energy states arise from the fact that equation (3.24)

$$p_\lambda^2 = -m^2$$

may be re-written as

$$E^2 = \mathbf{p}^2 + m^2.$$

This is a second order expression and so the first order terms must of necessity possess positive and negative roots

$$E = \pm(\mathbf{p}^2 + m^2)^{1/2}. \tag{3.128}$$

Upon substituting the solutions (3.127) into the Dirac equation one obtains

$$\begin{aligned}(i\gamma_\lambda p_\lambda + m)\, u_r &= 0 \\ (-i\gamma_\lambda p_\lambda + m)\, v_r &= 0.\end{aligned} \tag{3.129}$$

If the first of equations (3.129) is considered and the specific form of the γ-matrices given in equation (3.103) is inserted in it, we find

$$-\begin{pmatrix}0 & -\sigma_k \\ \sigma_k & 0\end{pmatrix} p_k u + i\begin{pmatrix}1 & 0 \\ 0 & -1\end{pmatrix} p_4 u + \begin{pmatrix}1 & 0 \\ 0 & 1\end{pmatrix} mu = 0. \tag{3.130}$$

We first consider a solution in the rest system of the particle

$$\mathbf{p} = 0, \quad p_4 = i\,|E| = im.$$

A suitable solution for u in this system is

$$u = \begin{pmatrix}1 \\ 0\end{pmatrix} N \tag{3.131}$$

where and 1 and 0 are 2×2 matrices and N is a normalisation term; similarly we may show that

$$v = \begin{pmatrix}0 \\ 1\end{pmatrix} N. \tag{3.132}$$

Solutions may now be sought for a coordinate system in which the particle is not at rest. In the limit $\mathbf{p} \to 0$ they must yield solutions (3.131) and (3.132). If u is written as

$$u = \begin{pmatrix}u_L \\ u_s\end{pmatrix} N \tag{3.133}$$

then upon substitution into equation (3.129) we find

$$-\begin{pmatrix}0 & -\sigma_k \\ \sigma_k & 0\end{pmatrix} p_k \begin{pmatrix}u_L \\ u_s\end{pmatrix} + i\begin{pmatrix}1 & 0 \\ 0 & -1\end{pmatrix} p_4 \begin{pmatrix}u_L \\ u_s\end{pmatrix} + m\begin{pmatrix}u_L \\ u_s\end{pmatrix} = 0$$

yielding

$$\begin{aligned}\boldsymbol{\sigma}\cdot\mathbf{p}u_s + ip_4u_L + mu_L &= 0 \\ -\boldsymbol{\sigma}\cdot\mathbf{p}u_L - ip_4u_s + mu_s &= 0.\end{aligned} \tag{3.134}$$

It can be seen that (3.134) does not yield unique solutions for u_L and u_s. If we wish u to take the form given in equation (3.131) in the limit $\mathbf{p} \to 0$, we may use the second of equations (3.134) and obtain

$$u_s = \frac{\boldsymbol{\sigma}\cdot\mathbf{p}}{-ip_4 + m}\, u_L = \frac{\boldsymbol{\sigma}\cdot\mathbf{p}}{|E| + m}\, u_L. \tag{3.135}$$

We will therefore write

$$u = \begin{pmatrix} u_L \\ u_s \end{pmatrix} N = \begin{pmatrix} 1 \\ \dfrac{\boldsymbol{\sigma}\cdot\mathbf{p}}{|E| + m} \end{pmatrix} N \tag{3.136}$$

It is customary to refer to u_L and u_s as the *large and small components of the Dirac spinors* respectively. An equivalent solution can be obtained for v

$$v = \begin{pmatrix} v_s \\ v_L \end{pmatrix} N, \quad v_s = \frac{\boldsymbol{\sigma}\cdot\mathbf{p}}{|E| + m}\, v_L$$

$$v = \begin{pmatrix} \dfrac{\boldsymbol{\sigma}\cdot\mathbf{p}}{|E| + m} \\ 1 \end{pmatrix} N. \tag{3.137}$$

Since the terms 1 and $\boldsymbol{\sigma}\cdot\mathbf{p}$ are 2×2 matrices, equations (3.136) and (3.137) may be expanded yielding the solutions shown in Table 3.3.

The table shows that u and v have their large and small components interchanged, for example

$$u_1 = \begin{pmatrix} u_{L1} \\ u_{s1} \end{pmatrix} N = \begin{pmatrix} v_{L1} \\ v_{s1} \end{pmatrix} N. \tag{3.138}$$

TABLE 3.3

	$u_{r\alpha}/N$		$v_{r\alpha}/N$	
α \ r	1	2	1	2
1	1	0	$\dfrac{p_z}{\lvert E\rvert + m}$	$\dfrac{p_x - ip_y}{\lvert E\rvert + m}$
2	0	1	$\dfrac{p_x + ip_y}{\lvert E\rvert + m}$	$\dfrac{-p_z}{\lvert E\rvert + m}$
3	$\dfrac{p_z}{\lvert E\rvert + m}$	$\dfrac{p_x - ip_y}{\lvert E\rvert + m}$	1	0
4	$\dfrac{p_x + ip_y}{\lvert E\rvert + m}$	$\dfrac{-p_z}{\lvert E\rvert + m}$	0	1
	$E > 0$		$E < 0$	

The terms u_1, u_2, v_1 and v_2 possess a particularly simple physical significance if we choose the z-coordinate axis parallel to the momentum

$$\mathbf{p} = (0, 0, p).$$

We noted previously (3.117) that the operator $(\hat{\sigma} \cdot \mathbf{p})$ possesses eigenfunctions with eigenvalues $\pm |\mathbf{p}|$. Although the terms u_1, u_2, v_1 and v_2 are not in general eigenfunctions of $(\hat{\sigma} \cdot \mathbf{p})$, if we let $\mathbf{p}$ lie parallel to the z-axis we may write

$$\hat{\sigma} \cdot \mathbf{p} \to \hat{\sigma}_3 p = \begin{pmatrix} \sigma_3 & 0 \\ 0 & \sigma_3 \end{pmatrix} p = \begin{pmatrix} p & 0 & 0 & 0 \\ 0 & -p & 0 & 0 \\ 0 & 0 & p & 0 \\ 0 & 0 & 0 & -p \end{pmatrix}$$

when we find that

$$(\hat{\sigma} \cdot \mathbf{p})\, u_1 = \begin{pmatrix} p & 0 & 0 & 0 \\ 0 & -p & 0 & 0 \\ 0 & 0 & p & 0 \\ 0 & 0 & 0 & -p \end{pmatrix} \begin{pmatrix} 1 \\ 0 \\ \dfrac{p}{|E| + m} \\ 0 \end{pmatrix} N = |\mathbf{p}| u_1 \tag{3.139}$$

$$(\hat{\sigma} \cdot \mathbf{p})\, u_2 = \begin{pmatrix} p & 0 & 0 & 0 \\ 0 & -p & 0 & 0 \\ 0 & 0 & p & 0 \\ 0 & 0 & 0 & -p \end{pmatrix} \begin{pmatrix} 0 \\ 1 \\ 0 \\ \dfrac{-p}{|E| + m} \end{pmatrix} N = -|\mathbf{p}| u_2 .$$

Thus the terms u_1 and u_2 can be associated with the spin directions of the positive energy states. They are called *positive energy spinors*.

Equations (3.139) can be re-written as

$$\frac{\hat{\sigma} \cdot \mathbf{p}}{|\mathbf{p}|} u_1 = u_1, \quad \frac{\hat{\sigma} \cdot \mathbf{p}}{|\mathbf{p}|} u_2 = -u_2 .$$

The term $\hat{\sigma} \cdot \mathbf{p}/|\mathbf{p}|$ is often called the *chirality operator*, and is denoted by the symbol σ_p

$$\frac{\hat{\sigma} \cdot \mathbf{p}}{|\mathbf{p}|} = \sigma_p . \tag{3.140}$$

A similar treatment to that in (3.139) can be carried out for v_1 and v_2 yielding

$$\sigma_p v_1 = v_1, \quad \sigma_p v_2 = -v_2 . \tag{3.141}$$

The terms v_1 and v_2 are called *negative energy spinors*. The physical content of equations (3.139) and (3.141) for the spinors u_1, u_2, v_1 and v_2 in the special coordinate system can be summarised with the aid of Fig. 3.4.†

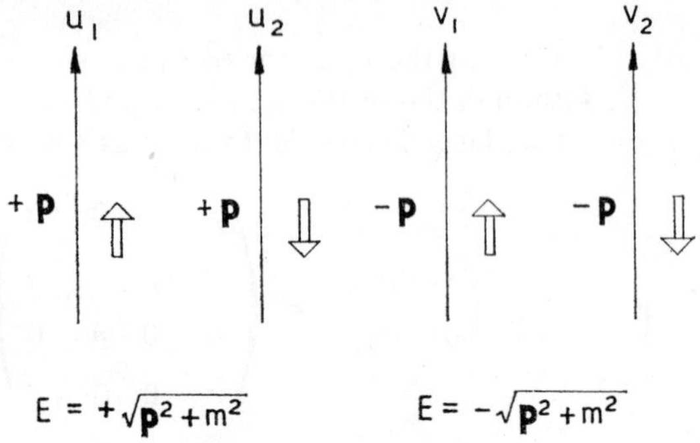

FIG. 3.4. Representation of Dirac spinors. The long arrows associated with the negative energy states indicate the direction of motion of the corresponding antiparticle.

It should be noted that in the nonrelativistic limit $\mathbf{p} \to 0$, the spinors u and v reduce to

$$u_1 = \begin{pmatrix} 1 \\ 0 \\ 0 \\ 0 \end{pmatrix}, \quad u_2 = \begin{pmatrix} 0 \\ 1 \\ 0 \\ 0 \end{pmatrix}, \quad v_1 = \begin{pmatrix} 0 \\ 0 \\ 1 \\ 0 \end{pmatrix}, \quad v_2 = \begin{pmatrix} 0 \\ 0 \\ 0 \\ 1 \end{pmatrix}.$$

(We shall show in the next section that if $\mathbf{p} \to 0$, $N \to 1$.) The relation of the spinors to the direction of the momentum therefore vanishes and we can write

$$\hat{\sigma}_z u_1 = +u_1, \quad \hat{\sigma}_z v_1 = +v_1$$
$$\hat{\sigma}_z u_2 = -u_2, \quad \hat{\sigma}_z v_2 = -v_2$$

where z can be defined in any convenient direction.

Furthermore, an inspection of the above relations reveals that

$$\begin{matrix} \hat{\sigma}_z u_1 \searrow \\ \hat{\sigma}_z v_1 \nearrow \end{matrix} \sigma_z \omega_1 = \omega_1, \qquad \begin{matrix} \hat{\sigma}_z u_2 \searrow \\ \hat{\sigma}_z v_2 \nearrow \end{matrix} \sigma_z \omega_2 = -\omega_2 \tag{3.142}$$

where σ_z is the z-component of the Pauli 2×2 spin operator, and the two-dimensional terms

$$\omega_1 = \begin{pmatrix} 1 \\ 0 \end{pmatrix} = \chi_{1/2}^{1/2}, \quad \omega_2 = \begin{pmatrix} 0 \\ 1 \end{pmatrix} = \chi_{1/2}^{-1/2} \tag{3.143}$$

are called *Pauli spinors*. The second notation is for later convenience.

† Spinors which are eigenfunctions of the chirality operator in any coordinate frame can of course be constructed. They are complicated expressions and are rarely used in practice.

Thus in the nonrelativistic limit we can replace the 4×4 spin operator by an operator with 2×2 components. Furthermore, the Dirac wave function becomes equivalent to the nonrelativistic Pauli wave function for particles with spin.

3.3(h). *The negative energy states of a Dirac particle*

In the previous section we have shown that four solutions are possible for the Dirac equation, and that two of these represent negative energy states. Some physical interpretation must be given for these states. If they were allowed to represent particles of negative mass, physical problems would arise in that particles with negative inertial energy could never be stopped by matter, since losses of energy by collision would imply that they move faster and faster. Furthermore, if they existed it can be shown that ordinary matter would rapidly decay to negative energy states.

To avoid this situation Dirac suggested that normally all the negative energy states were filled. Thus, since Dirac particles obey the Pauli exclusion principle, no further particles could enter the states of negative energy. He postulated that an unoccupied negative energy state, that is a *hole* in the sea of negative energy states, with energy $-E$ and momentum $-\mathbf{p}$, could be interpreted as a particle of energy $+E$ and momentum $+\mathbf{p}$, but with electrical charge opposite in sign to the particles occupying the energy sea. It is now customary to regard the unoccupied negative energy state as an antiparticle or a charge conjugate particle (relative to the particle in the positive energy state). We shall show later that the concepts of particle and antiparticle arise naturally in field theory.

3.3(i). *Normalisation of the Dirac equation*

Before examining the normalisation of the Dirac equation it is convenient to derive the following equation:

$$\bar{u}\gamma_\lambda u = \frac{p_\lambda}{im}\,\bar{u}u. \tag{3.144}$$

The insertion of the terms

$$\psi = u\,\mathrm{e}^{ipx}, \quad \overline{\psi} = \bar{u}\,\mathrm{e}^{-ipx} \tag{3.145}$$

into the equations (3.89) and (3.91) yields the following expressions:

$$(i\gamma p + m)\,u = 0 \tag{3.146}$$

$$\bar{u}(i\gamma p + m) = 0.$$

The first equation may be multiplied on the left by $\bar{u}\gamma_\lambda$ and the second on the right by $\gamma_\lambda\, u$ and the resulting equations added to yield

$$i\bar{u}(\gamma_\lambda\gamma p + \gamma p\gamma_\lambda)\,u = -2m\bar{u}\gamma_\lambda u.$$

Using the anticommutation properties of the γ-matrices (3.93) this expression reduces to

$$2i\bar{u}p_\lambda u = -2m\bar{u}\gamma_\lambda u$$

which can be re-written as in (3.144)

$$\bar{u}\gamma_\lambda u = \frac{p_\lambda}{im}\,\bar{u}u.$$

Equation (3.144) takes a particularly simple form when $\gamma_\lambda = \gamma_4$

$$\bar{u}\gamma_4 u = u^\dagger u = \frac{i\,|E|}{im}\,\bar{u}u = \frac{|E|}{m}\,\bar{u}u. \tag{3.147}$$

We commence the normalisation of the Dirac equation by adopting the convention that

$$\bar{u}_r u_s = \delta_{rs}. \tag{3.148}$$

This choice is not a unique one, some authors choose $u^\dagger u = 1$ as the point of normalisation. However, if we choose $\bar{u}_r u_s = \delta_{rs}$ equation (3.147) then implies that we should set

$$u_r^\dagger u_s = \frac{|E|}{m}\,\delta_{rs}. \tag{3.149}$$

In equation (3.49) the probability density for a Dirac particle was defined as

$$\varrho = \psi^\dagger\psi$$

thus the probability of finding a Dirac particle in a box of volume V is

$$\varrho V = \psi^\dagger\psi V = u^\dagger u V = \frac{|E|}{m}\,V_0\,\frac{m}{|E|} \tag{3.150}$$

where V_0 represents the volume of the box when it is in the same coordinate system as the particle. Thus the probability of finding the particle in the box is a Lorentz invariant quantity – a physically satisfactory situation.†

The equations (3.136) and (3.137) contained a normalisation term N; this quantity may now be evaluated. Consider, for example, u_1 in Table 3.3

$$u_1^\dagger u_1 = \frac{|E|}{m} = \begin{pmatrix} 1 & 0 & \dfrac{p_z}{|E|+m} & \dfrac{p_x - ip_y}{|E|+m} \end{pmatrix} \begin{pmatrix} 1 \\ 0 \\ \dfrac{p_z}{|E|+m} \\ \dfrac{p_x+ip_y}{|E|+m} \end{pmatrix} N^2 = \frac{2\,|E|}{|E|+m}\cdot N^2.$$

† The dimensions of equation (3.150) are L^3; in order to obtain a true probability, the Dirac wave function must also include a normalisation term with dimensions $L^{-3/2} \equiv V^{-1/2}$. Its presence is not necessary for our discussion.

Thus we find

$$N = \left(\frac{|E| + m}{2m}\right)^{\frac{1}{2}} \tag{3.151}$$

and u_1 of Table 3.3 becomes

$$u_1 = \begin{pmatrix} 1 \\ 0 \\ \dfrac{p_z}{|E| + m} \\ \dfrac{p_x + ip_y}{|E| + m} \end{pmatrix} \left(\frac{|E| + m}{2m}\right)^{\frac{1}{2}}. \tag{3.152}$$

Until now we have considered the positive energy states only

$$\psi = u\, \mathrm{e}^{ipx}.$$

The equations equivalent to (3.144) and (3.147) for the negative energy states

$$\psi = v\, \mathrm{e}^{-ipx}$$

are

$$\bar{v}\gamma_\lambda v = -\frac{p_\lambda}{im}\, \bar{v}v \tag{3.153}$$

$$v^\dagger v = -\frac{|E|}{m}\, \bar{v}v.$$

The normalisation of the negative energy states may then the carried out as before. We choose the definition

$$\bar{v}_r v_s = -\,\delta_{rs} \tag{3.154}$$

and therefore find

$$v_r^\dagger v_s = \frac{|E|}{m}\, \delta_{rs} \tag{3.155}$$

and

$$N = \left(\frac{|E| + m}{2m}\right)^{\frac{1}{2}}$$

as before.

It may be verified easily that the normalisation conditions (3.148) and (3.154), and the definitions of the spinors given in Table 3.3 and (3.151), lead to the following completeness relations:

$$\sum_{r=1}^{2} (\bar{u}_r u_r - \bar{v}_r v_r) = 4 \tag{3.156}$$

$$\sum_{r=1}^{2} (\bar{u}_{r\alpha} u_{r\beta} - \bar{v}_{r\alpha} v_{r\beta}) = \sum_{r=1}^{2} [u_{r\alpha}^\dagger (\gamma_4)_{\alpha\beta} u_{r\beta} - v_{r\alpha}^\dagger (\gamma_4)_{\alpha\beta} v_{r\beta}] = \delta_{\alpha\beta}.$$

3.3(j). *Orthogonality conditions for the Dirac spinors*

In writing equations (3.148) and (3.154)

$$\bar{u}_r u_s = \delta_{rs}, \quad \bar{v}_r v_s = -\delta_{rs}$$

we assumed the existence of orthogonality relationships between the spin states of the Dirac spinors. The correctness of this assumption may be shown by working out, for example $u_1^\dagger u_2$, with the aid of Table 3.3. We find that

$$u_1^\dagger u_2 = 0.$$

The orthogonality relationships between positive and negative energy states are also important. Consider the following Dirac equations and their solutions:

$$\gamma_\lambda \frac{\partial \psi}{\partial x_\lambda} + m\psi = 0, \quad \psi = u\, \mathrm{e}^{ipx}$$

$$\frac{\partial \overline{\psi}}{\partial x_\lambda} \gamma_\lambda - m\overline{\psi} = 0, \quad \overline{\psi} = \overline{v}\, \mathrm{e}^{ipx}$$

therefore

$$(i\gamma_\lambda p_\lambda + m)u = 0 \tag{3.157}$$

$$\overline{v}(i\gamma_\lambda p_\lambda - m) = 0.$$

If the first equation is multiplied on the left by $\overline{v}$ and the second on the right by u, and then the two equations are subtracted, we find

$$2m\overline{v}u = 0.$$

Thus if $m \neq 0$ it is apparent that

$$\overline{v}u = \bar{u}v = 0. \tag{3.158}$$

In writing this equation we have inserted an obvious extension to our proof.

The orthogonality and the normalisation conditions lead to the following operator equations:

$$\sum_{r=1}^{2} (u_{r\alpha}\bar{u}_{r\beta} - v_{r\alpha}\overline{v}_{r\beta}) = \delta_{\alpha\beta} \tag{3.159}$$

$$\sum_{r=1}^{2} (u_r\bar{u}_r - v_r\overline{v}_r) = \hat{1}$$

where the subscripts α and β refer to the individual terms composing the spinors and the term $\hat{1}$ is the unit operator.

The proof of the first of equations (3.159) is most easily accomplished by a direct comparison with Table 3.3. A rapid proof of the second may be made by introducing a function w which we define as

$$\sum_{r=1}^{2} (u_r c_r + v_r c_r) = w.$$

Equations (3.148), (3.154) and (3.158) then show that

$$c_s = \bar{u}_s w, \quad -c_s = \bar{v}_s w$$

and therefore we find that

$$\sum_{s=1}^{2} (u_s \bar{u}_s - v_s \bar{v}_s) w = w$$

which is thes econd of equations (3.159).

3.3(k). *Projection operators for the Dirac spinors*

The spinor forms of the Dirac equations (3.129)

$$(i\gamma_\lambda p_\lambda + m) u = 0$$

$$(-i\gamma_\lambda p_\lambda + m) v = 0$$

may be used to define a *projection operator*†

$$\Lambda^{\pm} = \frac{\mp i\gamma_\lambda p_\lambda + m}{2m} \equiv \frac{\mp i\gamma p + m}{2m}. \tag{3.160}$$

An inspection of equations (3.129) and (3.160) shows that Λ possesses the following properties

$$\Lambda^+ + \Lambda^- = \hat{1}, \quad \Lambda^{\pm}\Lambda^{\mp} = 0, \quad (\Lambda^{\pm})^2 = \Lambda^{\pm} \tag{3.161}$$

$$\Lambda^+ u = u, \quad \Lambda^+ v = 0 \tag{3.162}$$

$$\Lambda^- u = 0, \quad \Lambda^- v = v.$$

It can be seen that Λ^+ retains the positive frequency part of the Dirac wave function, and that Λ^- retains the negative part. Thus the energy projection operators Λ^+ may be used to reduce the four-component Dirac wave function to a two-component one. This enables many problems involving Dirac particles, for example β-decay, to be examined without the use of specific forms for the terms u and v. This point will be considered in greater detail in § 7.4(d).

In a similar manner to the method described above we may define spin projection operators

$$\Sigma^{\pm} = \frac{\pm \hat{\sigma} \cdot \mathbf{p} + |\mathbf{p}|}{2\,|\mathbf{p}|} = \frac{\pm \sigma_p + 1}{2} \tag{3.163}$$

† A projection operator is an operator whose components sum to give the unit operator. These components project out (that is eliminate some, retain others) sections of state functions (compare A.7 (Appendixes, p. 704)).

where

$$\sigma_p = \frac{\hat{\boldsymbol{\sigma}} \cdot \mathbf{p}}{|\mathbf{p}|}.$$

These operators possess similar properties to the $\Lambda^{\pm}$ operators; thus we find that

$$\Sigma^{+} + \Sigma^{-} = \hat{1}$$

$$\Sigma^{\pm}\Sigma^{\mp} = 0, \quad (\Sigma^{\pm})^2 = \Sigma^{\pm}. \tag{3.164}$$

The function of the operators $\Sigma^{\pm}$ may be understood by recalling equations (3.139)

$$\hat{\boldsymbol{\sigma}} \cdot \mathbf{p}\, u_1 = |\mathbf{p}|\, u_1, \quad \hat{\boldsymbol{\sigma}} \cdot \mathbf{p}\, u_2 = -\, |\mathbf{p}|\, u_2$$

where $\mathbf{p} = (0, 0, p)$; we therefore find that for this specialised coordinate system

$$\Sigma^{+}u_1 = u_1, \quad \Sigma^{+}u_2 = 0 \tag{3.165}$$

$$\Sigma^{-}u_1 = 0, \quad \Sigma^{-}u_2 = u_2$$

and

$$\Sigma^{+}v_1 = v_1, \quad \Sigma^{+}v_2 = 0 \tag{3.166}$$

$$\Sigma^{-}v_1 = 0, \quad \Sigma^{-}v_2 = v_2.$$

Thus the projection operator Σ^{+} retains states with the spin vector parallel to the momentum $\mathbf{p}$, whilst Σ^{-} does the opposite. In general we may define spinors $u_{\pm}$ and $v_{\pm}$ which are eigenfunctions of $\Sigma^{\pm}$, for example

$$\Sigma^{+}u_{+} = u_{+}, \quad \Sigma^{-}u_{-} = u_{-} \tag{3.167}$$

$$\Sigma^{-}u_{+} = 0, \quad \Sigma^{+}u_{-} = 0.$$

Let us now return to equation (3.160); the projection operators for the positive and negative energy states can be represented as a sum over spinors

$$\Lambda^{+} = \sum_{r=1}^{2} u_r\bar{u}_r, \quad \Lambda^{-} = -\sum_{r=1}^{2} v_r\bar{v}_r. \tag{3.168}$$

The proof of these equations is readily obtained with the aid of (3.159)

$$\sum_{r=1}^{2} (u_r\bar{u}_r - v_r\bar{v}_r) = \hat{1}.$$

Thus we find that

$$\sum_r u_r\bar{u}_r = \Lambda^{+} \sum_r u_r\bar{u}_r = \Lambda^{+} \sum_r (u_r\bar{u}_r - v_r\bar{v}_r) = \Lambda^{+}.$$

The relations (3.168) are dependent upon the choice of the normalisation conditions

$$\bar{u}_r u_s = \delta_{rs}, \quad \bar{v}_r v_s = -\,\delta_{rs}.$$

The choice

$$u_r^\dagger u_s = \delta_{rs}, \quad v_r^\dagger v_s = -\delta_{rs} \tag{3.169}$$

is made by some authors; equations (3.168) then become

$$\sum_r u_r \bar{u}_r = \frac{m}{|E|} \Lambda^+ = \frac{-i\gamma p + m}{2|E|} \tag{3.170}$$

$$\sum_r v_r \bar{v}_r = -\frac{m}{|E|} \Lambda^- = \frac{-i\gamma p - m}{2|E|}.$$

The second choice possesses certain advantages when dealing with neutrinos, which are massless. In this situation sums over spin states may be carried out without the embarrassment of infinite terms appearing. When this form of normalisation is adopted the term N appearing in (3.151) becomes

$$N = \left(\frac{m + |E|}{2|E|}\right)^{\frac{1}{2}}$$

so that

$$u_1 = \begin{pmatrix} 1 \\ 0 \\ \dfrac{p_z}{|E| + m} \\ \dfrac{p_x + ip_y}{|E| + m} \end{pmatrix} \left(\frac{m + |E|}{2|E|}\right)^{\frac{1}{2}}. \tag{3.171}$$

3.3(1). *The Dirac equation for particles of zero mass*

The equations (3.129)

$$(i\gamma_\lambda p_\lambda + m)\, u_r = 0$$

$$(-i\gamma_\lambda p_\lambda + m)\, v_r = 0$$

for spinors of the positive and negative energy states of the Dirac wave function, reduce to the same form when $m = 0$,

$$i\gamma_\lambda p_\lambda u = 0, \quad i\gamma_\lambda p_\lambda v = 0. \tag{3.172}$$

Thus fewer components are required for the Dirac wave function. This important point was first realised by Weyl (1929), but as the resulting wave functions failed to conserve parity, no attention was paid to them until Lee and Yang (1956a) suggested that parity was not conserved in weak interactions.

We will introduce a new spinor in (3.172)

$$i\gamma_\lambda p_\lambda u_\nu = 0$$

where the subscript ν refers to the neutrino, as this is the only known Dirac particle with zero mass. The equation can be re-written as

$$i\gamma_k p_k u_\nu = -i\gamma_4 p_4 u_\nu, \qquad k = 1, 2, 3$$

and upon multiplying both sides by $\gamma_4\gamma_5$ we obtain

$$i\gamma_4\gamma_5\gamma_k p_k u_\nu = -i\gamma_4\gamma_5\gamma_4 p_4 u_\nu = i\gamma_5 p_4 u_\nu$$

and therefore find, with the aid of (3.112), that

$$\hat{\sigma}\cdot\mathbf{p}u_\nu = -\gamma_5 E u_\nu \tag{3.173}$$

where E can take positive or negative values

$$E = \pm|\mathbf{p}|.$$

Now we have shown previously (3.117) that

$$\hat{\sigma}\cdot\mathbf{p} = \pm|\mathbf{p}|$$

where the positive sign implies a spin pointing along the direction of $\mathbf{p}$ and vice versa. This conclusion is unaltered if the particle has zero mass. Thus, since u_ν is an eigenfunction of $\hat{\sigma}\cdot\mathbf{p}$, equation (3.173) may be satisfied by the following solutions:

$$\gamma_5 u_\nu = \pm u_\nu. \tag{3.174}$$

We shall first consider the solution

$$\gamma_5 u_\nu = u_\nu$$

or

$$(1-\gamma_5)\, u_\nu = 0.$$

Equation (3.173) then becomes

$$-\gamma_5 E u_\nu = -\gamma_5 E\gamma_5 u_\nu = -E u_\nu = \hat{\sigma}\cdot\mathbf{p}\, u_\nu. \tag{3.175}$$

If the spinor has positive energy

$$E = +|\mathbf{p}|$$

equation (3.175) can only be satisfied if the spin points in the opposite direction to $\mathbf{p}$. On the other hand, for a negative energy state

$$E = -|\mathbf{p}|$$

equation (3.175) is satisfied only if the spin is parallel to the direction of motion. We will refer to the negative energy state as the *antineutrino*, and denote it by the symbol $\bar{\nu}$.

The reverse situation holds if we choose the alternative solution in (3.174)

$$\gamma_5 u_\nu = -u_\nu$$

or

$$(1+\gamma_5)\, u_\nu = 0.$$

We then obtain

$$E u_\nu = \hat{\boldsymbol{\sigma}} \cdot \mathbf{p}\, u_\nu.$$

This equation can be satisfied if the positive energy spinor points along the direction of motion and if the negative energy spinor is antiparallel.

A spin which points along the direction of motion is said to be *right-handed*, or to have *positive helicity*. When the spin and motion are antiparallel the system is said to be *left-handed*, or to have *negative helicity*.

The results obtained above may be summarised by Table 3.4:

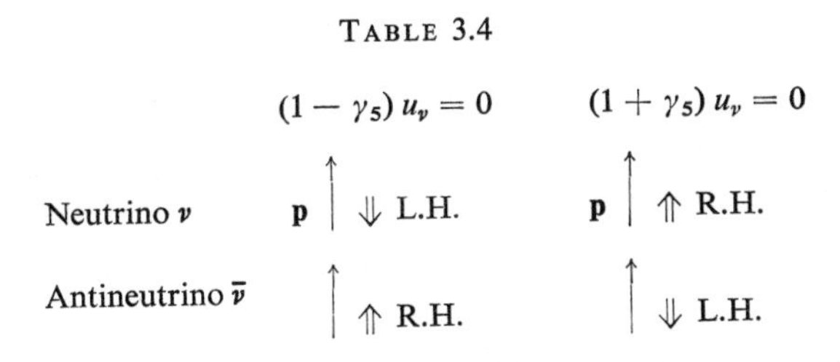

TABLE 3.4

	$(1-\gamma_5)\, u_\nu = 0$	$(1+\gamma_5)\, u_\nu = 0$
Neutrino ν	$\mathbf{p}$ ↑ ⇓ L.H.	$\mathbf{p}$ ↑ ⇑ R.H.
Antineutrino $\bar{\nu}$	↑ ⇑ R.H.	↑ ⇓ L.H.

The choice of spinor function

$$u_\nu = \pm \gamma_5\, u_\nu$$

is in turn reflected in the total wave function

$$\psi_\nu = \pm \gamma_5 \psi_\nu.$$

It is convenient, therefore, to represent the two possible states of the neutrino by the wave functions ψ_R and ψ_L with the following properties:

$$\psi_\nu = \psi_R + \psi_L$$

$$\gamma_5 \psi_R = -\psi_R, \quad \gamma_5 \psi_L = \psi_L \tag{3.176}$$

$$\psi_R = \tfrac{1}{2}(1-\gamma_5)\, \psi_\nu, \quad \psi_L = \tfrac{1}{2}(1+\gamma_5)\, \psi_\nu \tag{3.177}$$

and an inspection of Table 3.4 shows that ψ_R and ψ_L represent states of right- and left-handed helicity for the neutrino respectively. The terms $\frac{1}{2}(1 \pm \gamma_5)$ appearing in (3.177) will be denoted by the symbols

$$\Lambda_R = \tfrac{1}{2}(1-\gamma_5), \qquad \Lambda_L = \tfrac{1}{2}(1+\gamma_5). \tag{3.178}$$

They may be regarded as projection operators, and have the property that

$$\Lambda_R \psi_R = \psi_R, \qquad \Lambda_R \psi_L = 0 \qquad (3.179)$$
$$\Lambda_L \psi_L = \psi_L, \qquad \Lambda_L \psi_R = 0.$$

3.3(m). *Lorentz transformations and the Dirac equation*

In order to assist us in later problems it is worth considering some of the elementary aspects of the Lorentz transformation for the Dirac equation at this point. During a transformation we would expect the wave function ψ to change, but we would also expect the physical content of the theory to remain unaltered if the Dirac equation is truly covariant.

In a reference frame K the Dirac equation can be written as

$$\gamma_\lambda \frac{\partial \psi}{\partial x_\lambda} + m\psi = 0.$$

Now consider a second frame K'; if the Dirac equation is to remain Lorentz invariant it must assume the same form in the new coordinate system. If we assume that the transformation for ψ between the two reference frames is a linear one

$$\psi'(x') = S\psi(x) \qquad (3.180)$$

and that the γ operators remain unaltered, we find that in the frame K' the Dirac equation becomes

$$\gamma_\alpha \frac{\partial \psi'(x')}{\partial x'_\alpha} + m\psi'(x') = 0. \qquad (3.181)$$

Now in § 3.2(i) we obtained the relation (3.33)

$$x'_\lambda = a_{\lambda\alpha} x_\alpha$$

so that

$$\frac{\partial}{\partial x_\lambda} = a_{\alpha\lambda} \frac{\partial}{\partial x'_\alpha}.$$

But equation (3.180) can be re-formulated (compare § 3.2 (i))

$$\psi(x) = S^{-1}\psi'(x')$$
$$SS^{-1} = S^{-1}S = \hat{1} \qquad (3.182)$$

and so the Dirac equation in the reference frame K can be written as

$$\gamma_\lambda S^{-1} \frac{\partial \psi'(x')}{\partial x_\lambda} + mS^{-1}\psi'(x') = 0.$$

herefore

$$\gamma_\lambda S^{-1} a_{\alpha\lambda} \frac{\partial \psi'(x')}{\partial x'_\alpha} + m S^{-1} \psi'(x') = 0$$

nd

$$a_{\alpha\lambda} S \gamma_\lambda S^{-1} \frac{\partial \psi'(x')}{\partial x'_\alpha} + m \psi'(x') = 0 .$$

This equation is the same as (3.181) if we choose S so as to satisfy the condition

$$\gamma_\alpha = a_{\alpha\lambda} S \gamma_\lambda S^{-1} . \tag{3.183}$$

3.3 (n). *Behaviour of the linear transformation operator S*

We shall now examine the behaviour of S for an infinitesimal Lorentz trans-ormation of the form

$$x'_\lambda = a_{\lambda\alpha} x_\alpha = (\delta_{\lambda\alpha} + \varepsilon_{\lambda\alpha}) x_\alpha . \tag{3.184}$$

his equation corresponds to an infinitesimal rotation in four-dimensional space, nd $\varepsilon_{\lambda\alpha}$ represents an infinitesimal four-dimensional rotation tensor. During ro-ation the length of the four-vector must remain unaltered, that is

$$x'^2_\lambda = x^2_\alpha . \tag{3.185}$$

If equation (3.184) is substituted in (3.185), and if terms quadratic in ε are ropped as infinitesimals of higher order, we find that the condition

$$\varepsilon_{\lambda\alpha} = - \varepsilon_{\alpha\lambda} \tag{3.186}$$

nust be fulfilled if equation (3.185) is to be satisfied.

Now we may write the transformation operator S and its inverse as

$$S = \hat{1} + T, \quad S^{-1} = \hat{1} - T \tag{3.187}$$

vhere T is an infinitesimal operator. The product of these equations satisfies he relation (3.182)

$$S S^{-1} = S^{-1} S = \hat{1}$$

o first order in T. Thus if we re-write equation (3.183) as

$$S^{-1} \gamma_\alpha S = a_{\alpha\lambda} \gamma_\lambda \tag{3.188}$$

nd substitute into it the expressions given in (3.184) and (3.187), we find that

$$\left(\hat{1} - T\right) \gamma_\alpha \left(\hat{1} + T\right) = a_{\alpha\lambda} \gamma_\lambda = (\delta_{\alpha\lambda} + \varepsilon_{\alpha\lambda}) \gamma_\lambda$$

and so

$$\gamma_\alpha T - T \gamma_\alpha = \varepsilon_{\alpha\lambda} \gamma_\lambda \tag{3.189}$$

neglecting terms in T^2. A solution to this equation is given by

$$T = \tfrac{1}{4}\varepsilon_{\varrho\lambda}\gamma_\varrho\gamma_\lambda \qquad (\varrho \neq \lambda) \tag{3.190}$$

where terms with $\varrho = \lambda$ are omitted from T since $\varepsilon_{\varrho\varrho} = \varepsilon_{\lambda\lambda} = 0$ by virtue of (3.186). The proof that (3.190) is a solution of equation (3.189) is best made by direct substitution. The terms in T with $\varrho \neq \alpha$ and $\lambda \neq \alpha$ commute with γ_α in (3.189), and so vanish. That part of T which does not vanish in (3.189) is

$$T' = \tfrac{1}{4}\sum_{\varrho=1}^{4} \varepsilon_{\varrho\alpha}\gamma_\varrho\gamma_\alpha + \tfrac{1}{4}\sum_{\lambda=1}^{4} \varepsilon_{\alpha\lambda}\gamma_\alpha\gamma_\lambda$$

where summation signs have been inserted to make it clear that no summation is made over α.

Thus we may write

$$\gamma_\alpha T' = -\tfrac{1}{4}\sum_{\varrho=1}^{4} \varepsilon_{\varrho\alpha}\gamma_\varrho + \tfrac{1}{4}\sum_{\lambda=1}^{4} \varepsilon_{\alpha\lambda}\gamma_\lambda = \tfrac{1}{2}\sum_{\lambda=1}^{4} \varepsilon_{\alpha\lambda}\gamma_\lambda$$

$$T'\gamma_\alpha = -\tfrac{1}{2}\sum_{\lambda=1}^{4} \varepsilon_{\alpha\lambda}\gamma_\lambda$$

and therefore

$$\gamma_\alpha T - T\gamma_\alpha = \gamma_\alpha T' - T'\gamma_\alpha = \sum_{\lambda=1}^{4} \varepsilon_{\alpha\lambda}\gamma_\lambda = \varepsilon_{\alpha\lambda}\gamma_\lambda .$$

This result completes our proof of equation (3.190).

The solution (3.190) for T can now be inserted into (3.187) to give the following expression for S

$$S = \hat{1} + \tfrac{1}{4}\varepsilon_{\varrho\lambda}\gamma_\varrho\gamma_\lambda . \tag{3.191}$$

The Hermitian conjugate of this expression is

$$S^\dagger = \hat{1} + \tfrac{1}{4}\varepsilon^*_{\varrho\lambda}\gamma^\dagger_\lambda\gamma^\dagger_\varrho = \hat{1} + \tfrac{1}{4}\varepsilon^*_{\varrho\lambda}\gamma_\lambda\gamma_\varrho \tag{3.192}$$

where we have used the definition given in (A. 3.8) (p. 698) and also the Hermitian property of the γ-matrices A.5 (Appendixes, p. 701).

The properties of S will be examined further; now the complex conjugate form of equation (3.184) is

$$x'^*_\lambda = a^*_{\lambda\alpha}x^*_\alpha = (\delta_{\lambda\alpha} + \varepsilon^*_{\lambda\alpha})\, x^*_\alpha \tag{3.193}$$

and since we are using Minkowski coordinates

$$x^*_k = x_k, \quad x^*_4 = -x_4$$

it follows that

$$\varepsilon^*_{ik} = \varepsilon_{ik}, \quad \varepsilon^*_{i4} = -\varepsilon_{i4}, \quad \varepsilon^*_{4k} = -\varepsilon_{4k}, \quad \varepsilon^*_{44} = \varepsilon_{44}. \tag{3.194}$$

Now consider the following expansion:

$$\gamma_4 S^\dagger \gamma_4 = \gamma_4\left(\hat{1} + \tfrac{1}{4}\varepsilon^*_{\varrho\lambda}\gamma_\lambda\gamma_\varrho\right)\gamma_4$$
$$= \hat{1} + \tfrac{1}{4}(\varepsilon^*_{ik}\gamma_4\gamma_k\gamma_i\gamma_4 + \varepsilon^*_{4k}\gamma_4\gamma_k\gamma_4\gamma_4 + \varepsilon^*_{i4}\gamma_4\gamma_4\gamma_i\gamma_4 + \varepsilon^*_{44}\gamma_4\gamma_4\gamma_4\gamma_4)$$
$$= \hat{1} + \tfrac{1}{4}(\varepsilon_{ik}\gamma_k\gamma_i + \varepsilon_{4k}\gamma_k\gamma_4 + \varepsilon_{i4}\gamma_4\gamma_i + \varepsilon_{44}\gamma_4\gamma_4)$$
$$= \hat{1} + \tfrac{1}{4}\varepsilon_{\varrho\lambda}\gamma_\lambda\gamma_\varrho = \hat{1} - \tfrac{1}{4}\varepsilon_{\varrho\lambda}\gamma_\varrho\gamma_\lambda$$
$$= S^{-1}. \tag{3.195}$$

In deriving the above relation we have used equations (3.194) and the anti-commutation properties of the γ-matrices. Equation (3.195) can be inverted to give

$$S^\dagger = \gamma_4 S^{-1}\gamma_4. \tag{3.196}$$

This expression holds for any finite proper Lorentz transformation. It will be used in § 3.3(p), when we consider the Lorentz invariance of the Dirac bilinear covariants. The importance of the term $S^\dagger$ lies in the fact that it appears in the Lorentz transformations of the Hermitian conjugate form of the Dirac wave function

$$\overline{\psi}(x) \to \overline{\psi}'(x') = \psi^{\dagger\prime}(x')\,\gamma_4 = [S\psi(x)]^\dagger\,\gamma_4 = \psi^\dagger(x)\,S^\dagger\gamma_4. \tag{3.197}$$

3.3(o). *Parity and the Dirac wave function*

In the previous section the behaviour of S for an infinitesimal transformation was examined. We now consider the situation arising when a transformation is made from a reference frame K, with space and time coordinates $\mathbf{x}$ and t, to a frame K' in which the space coordinates are reversed

$$\mathbf{x} \to \mathbf{x}' = -\mathbf{x}, \quad t \to t' = t.$$

In this situation the elements of the matrix $a_{\alpha\lambda}$ of equation (3.183)

$$\gamma_\alpha = a_{\alpha\lambda} S\gamma_\lambda S^{-1}$$

can be written as

$$a_{\alpha\lambda} = \begin{pmatrix} -1 & 0 & 0 & 0 \\ 0 & -1 & 0 & 0 \\ 0 & 0 & -1 & 0 \\ 0 & 0 & 0 & 1 \end{pmatrix}$$

according to equation (3.50). Thus

$$\gamma_1 = -S\gamma_1 S^{-1}, \quad \gamma_3 = -S\gamma_3 S^{-1}$$
$$\gamma_2 = -S\gamma_2 S^{-1}, \quad \gamma_4 = S\gamma_4 S^{-1}.$$

Apart from an undetermined phase factor of modulus unity, a solution to these equations is provided by

$$S = \gamma_4 .$$

Thus equation (3.180)

$$\psi'(x') = S\psi(x)$$

can be written as

$$\psi'(x') = \psi'(-\mathbf{x}, t) = \xi_P \gamma_4 \psi(\mathbf{x}, t)$$

where ξ_P represents the phase factor. It may be verified easily that this solution satisfies the Dirac equation for the reference frame K' (3.181)

$$\gamma_\alpha \frac{\partial \psi'(x')}{\partial x'_\alpha} + m\psi'(x') = 0 .$$

Thus we may write the following linear relations between $\psi'(-\mathbf{x}, t)$ and $\psi(\mathbf{x}, t)$:

$$\psi'(x') = \psi'(-\mathbf{x}, t) = \xi_P \gamma_4 \psi(\mathbf{x}, t) \tag{3.198}$$

$$\psi(\mathbf{x}, t) = \xi_P \gamma_4 \psi'(-\mathbf{x}, t).$$

The second equation can be written down immediately, since the principle of relativity requires that no preference be given to any special reference frame. Thus we could have considered the Dirac equation for ψ' firstly in the reference frame K', and then reflected it to the frame K.

Returning to equations (3.198), if a second reflection is performed we may write

$$\psi''(x'') = \xi_P \gamma_4 \psi'(x') = \xi_P^2 \gamma_4 \gamma_4 \psi(\mathbf{x}, t) = \xi_P^2 \psi(\mathbf{x}, t).$$

But a second reflection returns the frame K' to the original frame K, and so

$$\psi''(x'') = \psi(\mathbf{x}, t) = \xi_P^2 \psi(\mathbf{x}, t). \tag{3.199}$$

An obvious solution to this equation is

$$\xi_P^2 = +1, \quad \xi_P = \pm 1. \tag{3.200}$$

It will be shown in § 5.4(c), however, that the absolute sign of a Dirac wave function has no physical significance, since a rotation through 360° yields $\psi = -\psi$. The solution $\xi_P^2 = -1$ is therefore also acceptable, but it appears to have little physical significance. For the rest of this chapter we shall therefore adopt the convention $\xi_P^2 = +1$. Thus we may write equation (3.180) as

$$\psi'(x') = S\psi(x) = \pm \gamma_4 \psi(x) \tag{3.201}$$

for a spatial reflection.

It should be noted that parity changes affect the structure of the Dirac spinor functions. In equations (3.135) and (3.137) it was shown that the following reations existed between the large and small components of the Dirac wave function:

$$u_s = \frac{\boldsymbol{\sigma} \cdot \mathbf{p}}{|E| + m} u_L, \quad v_s = \frac{\boldsymbol{\sigma} \cdot \mathbf{p}}{|E| + m} v_L.$$

These expressions change sign on the reflection of the coordinates; thus the large and small terms have opposite parity.

3.3.(p). *The Dirac bilinear covariants*

In § 1.3(a) it was shown that particles with half integral spin must enter any interaction in pairs. The Dirac particles fulfil this condition, and so terms for their wave functions are grouped in pairs in the description of any physical process. The wave functions transform as in (3.180) and (3.197)

$$\psi \to \psi' = S\psi, \quad \bar{\psi} \to \bar{\psi}' = \psi^\dagger S^\dagger \gamma_4.$$

From these quantities we can construct bilinear terms which have covariant transformation properties. One example which has already been encountered is that for the probability current density, which takes the form (3.92)

$$j_\lambda = i\bar{\psi}\gamma_\lambda\psi.$$

Altogether sixteen bilinear covariant quantities may be constructed. They are obtained by inserting the Γ_i terms of (3.106) between $\bar{\psi}$ and ψ. They are

$$\begin{array}{lll} \bar{\psi}\psi & (1) & S \\ \bar{\psi}\gamma_\lambda\psi & (4) & V \\ i\bar{\psi}\gamma_\lambda\gamma_\mu\psi \quad \lambda < \mu & (6) & T \\ i\bar{\psi}\gamma_\lambda\gamma_5\psi & (4) & A \\ \bar{\psi}\gamma_5\psi & (1) & P \end{array} \tag{3.202}$$

where the numbers in parentheses represent the number of components. The i terms appear because of our choice for the form of γ-matrices (see § 3.3(d) and (3.106)). The significance of the letters *SVTAP* will become clear when we examine the invariance properties of the bilinear covariants under Lorentz transformations.

We will consider two classes of transformations – proper finite Lorentz transformations and space inversions. In practical calculations $\bar{\psi}$ and ψ could refer to the same or different particles and be sub-labelled appropriately. This is irrelevant to our present considerations.

(1) $S = \bar{\psi}\psi$

For a Lorentz transformation we find that

$$\begin{aligned}\bar{\psi}(x)\,\psi(x) \to \bar{\psi}'(x')\,\psi'(x') &= \psi^{\dagger\prime}(x')\,\gamma_4\psi'(x')\\ &= \psi^{\dagger}(x)\,S^{\dagger}\gamma_4 S\psi(x)\\ &= \psi^{\dagger}(x)\,\gamma_4 S^{-1}\gamma_4\gamma_4 S\psi(x)\\ &= \psi^{\dagger}(x)\,\gamma_4\psi(x)\\ &= \bar{\psi}(x)\,\psi(x)\end{aligned}$$

where we have used equations (3.180), (3.196) and (3.197).

Thus $\bar{\psi}\psi$ behaves like a *scalar* under both proper Lorentz transformations and space inversions (compare § 3.2(j)).

(2) $P = \bar{\psi}\gamma_5\psi$.

It is now convenient to consider the final term in (3.202). We find that

$$\begin{aligned}\bar{\psi}(x)\,\gamma_5\psi(x) \to \bar{\psi}'(x')\,\gamma_5\psi'(x') &= \psi^{\dagger}(x)\,S^{\dagger}\gamma_4\gamma_5 S\psi(x)\\ &= \psi^{\dagger}(x)\,\gamma_4 S^{-1}\gamma_4\gamma_4\gamma_5 S\psi(x)\\ &= \bar{\psi}(x)\,S^{-1}\gamma_5 S\psi(x).\end{aligned}$$

For a finite proper Lorentz transformation we may use equations (3.187) and (3.190) to show that

$$\begin{aligned}S^{-1}\gamma_5 S &= \left(\hat{1} - \tfrac{1}{4}\,\varepsilon_{\varrho\lambda}\gamma_\varrho\gamma_\lambda\right)\gamma_5\left(\hat{1} + \tfrac{1}{4}\,\varepsilon_{\varrho\lambda}\gamma_\varrho\gamma_\lambda\right)\\ &= \gamma_5\left(\hat{1} - \tfrac{1}{4}\,\varepsilon_{\varrho\lambda}\gamma_\varrho\gamma_\lambda\right)\left(\hat{1} + \tfrac{1}{4}\,\varepsilon_{\varrho\lambda}\gamma_\varrho\gamma_\lambda\right)\\ &= \gamma_5 S^{-1}S = \gamma_5\end{aligned}$$

but under space inversion $S = \pm\gamma_4$, and so

$$S^{-1}\gamma_5 S = \gamma_4\gamma_5\gamma_4 = -\gamma_5.$$

Thus we may sum up

$$\bar{\psi}'(x')\,\gamma_5\psi'(x')\begin{cases} = \bar{\psi}(x)\,\gamma_5\psi(x) & \text{for proper finite Lorentz transformations}\\ = \bar{\psi}-(x)\,\gamma_5\psi(x) & \text{for space inversions.}\end{cases} \tag{3.203}$$

Thus the quantity $P = \bar{\psi}\gamma_5\psi$ behaves like a *pseudoscalar* (compare (3.54)).

(3) $V = \bar{\psi}\gamma_\lambda\psi$.

In §§ 3.3(b) and (c) a term of this type was shown to behave like a four-vector. Nevertheless, it is instructive to examine it under a formal Lorentz transformation

$$\bar{\psi}(x)\,\gamma_\lambda\psi(x) \to \bar{\psi}'(x')\,\gamma_\lambda\psi'(x') = \psi^\dagger(x)\,S^\dagger\gamma_4\gamma_\lambda S\psi(x)$$
$$= \psi^\dagger(x)\,\gamma_4 S^{-1}\gamma_4\gamma_4\gamma_\lambda S\psi(x)$$
$$= \bar{\psi}(x)\,S^{-1}\gamma_\lambda S\psi(x)$$
$$= a_{\lambda\alpha}\bar{\psi}(x)\,\gamma_\alpha\psi(x)$$

where we have used (3.188).

Thus $\bar{\psi}\gamma_\lambda\psi$ transforms like a four-vector (compare equation (3.33)). It is called a *polar vector* term.

(4) $A = i\bar{\psi}\gamma_\lambda\gamma_5\psi$.

Using equation (3.188) we may write

$$i\bar{\psi}(x)\,\gamma_\lambda\gamma_5\psi(x) \to i\bar{\psi}'(x')\,\gamma_\lambda\gamma_5\psi'(x') = i\psi^\dagger(x)S^\dagger\gamma_4\gamma_\lambda\gamma_5 S\psi(x)$$
$$= i\bar{\psi}(x)\,S^{-1}\gamma_\lambda\gamma_5 S\psi(x)$$
$$= i\bar{\psi}(x)\,S^{-1}\gamma_\lambda SS^{-1}\gamma_5 S\psi(x)$$
$$= ia_{\lambda\alpha}\bar{\psi}(x)\,\gamma_\alpha S^{-1}\gamma_5 S\psi(x).$$

We have already shown (compare (3.203)) that

$$S^{-1}\gamma_5 S = \gamma_5 \quad \text{for a proper Lorentz transformation}$$
$$S^{-1}\gamma_5 S = -\gamma_5 \quad \text{for a parity transformation.}$$

When the spatial coordinates are reflected, however, the coefficients $a_{ii} (i = 1, 2, 3)$ change sign (see equation (3.50)). Thus the term $A = i\bar{\psi}\gamma_\lambda\gamma_5\psi$ behaves like a polar vector, apart from failing to change sign upon reflection. It is called an *axial vector* (compare (3.56)).

(5) $T = i\bar{\psi}\gamma_\lambda\gamma_\mu\psi \qquad (\lambda < \mu)$

$$i\bar{\psi}(x)\,\gamma_\lambda\gamma_\mu\psi(x) \to i\bar{\psi}'(x')\,\gamma_\lambda\gamma_\mu\psi'(x') = i\psi^\dagger(x)\,S^\dagger\gamma_4\gamma_\lambda\gamma_\mu S\psi(x)$$
$$= i\bar{\psi}(x)\,S^{-1}\gamma_\lambda\gamma_\mu S\psi(x)$$
$$= i\bar{\psi}(x)\,S^{-1}\gamma_\lambda SS^{-1}\gamma_\mu S\psi(x)$$
$$= ia_{\lambda\alpha}a_{\mu\beta}\bar{\psi}(x)\,\gamma_\alpha\gamma_\beta\psi(x)$$

where we have again used equation (3.188). A comparison of the transformation for T with that given in equation (3.43) shows that T transforms like a tensor of the second rank. It is referred to, accordingly, as the *tensor* form of the Dirac bilinear covariants.

3.4. THE DEVELOPEMENT OF A CLASSICAL FIELD EQUATION

It is the practice in quantum mechanics to construct equations of motion which are based upon those of classical mechanics (in the limit $\hbar \to 0$). These equations, experience has taught us, are adequate for describing the behaviour of macroscopic matter.

A similar approach can be adopted in constructing a quantised field theory. A classical field equation is constructed and then quantised by converting its classical amplitudes to quantum mechanical operators. This procedure is called *second quantisation.*

A quantised field theory may also be constructed by using an axiomatic approach (see, for example, Haag, 1959). This method consists in taking certain basic axioms and constructing a field theory from them. The technique is elegant, and will be examined in a later chapter. For the present, however we shall use a mainly historical approach.

The development of a field theory may be carried through in the early stages by adopting a formalism which is familiar from classical mechanics – the Lagrangian formalism. This procedure possesses certain advantages in that the theory may be checked in its classical nonrelativistic limit by comparing it with the expressions from classical mechanics. A further advantage accrues from the Lagrangian formalism; the Lagrangian density is Lorentz invariant, and this property of the system can be used to generate a set of conserved quantities. For example invariance of the system under four-dimensional translation implies that energy and momentum are conserved in that system; conservation of angular momentum and charge may also be readily demonstrated. These features will be discussed in detail in § 3.4(f).

3.4(a). *The principle of least action and the Lagrangian equation of motion in classical mechanics*

3.4(a.1). *Justification of the Lagrangian method.* We start by considering the Lagrangian form of the equations of motion in classical mechanics, and the canonical equations of Hamilton. Both are formulations of the physical content of Newton's second law of motion. They possess certain advantages over the more elementary statements of the principles of Newtonian mechanics in terms of forces and their components. The approach to a physical problem by the latter method involves choosing a suitable coordinate system for the problem, writing down an equation of motion in this system and then solving the equation. In practice the labour involved in this approach may be considerable as the method often involves complicated geometrical reasoning and the resolution of forces. Much of this difficulty may be avoided by the use of the Lagrangian method, since it can be readily applied to any system of coordinates. It possesses the urther advantage that its formal structure may be used for the proof of general heorems.

The method involves writing down a *Lagrangian function* L, which is equal to he kinetic energy T of a system minus its potential energy V

$$L = L(q_i, \dot{q}_i) = T(\dot{q}_i, q_i) - V(q_i) \tag{3.204}$$

where we have written dq_i/dt as $\dot{q}_i$ in order to avoid a clumsy notation later in this chapter. The symbols V and T are expressed as functions of coordinates q_i

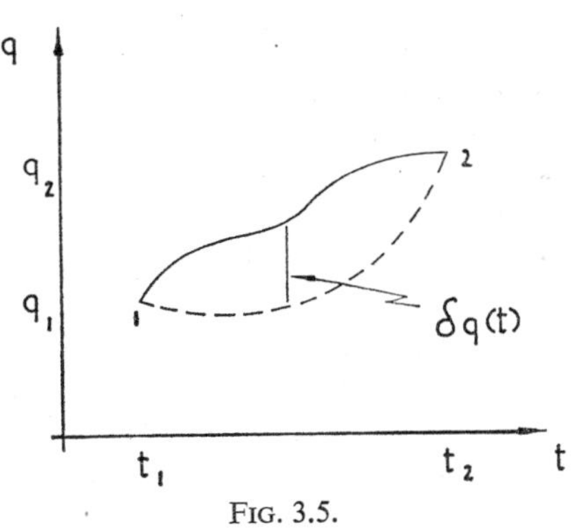

FIG. 3.5.

and velocity $\dot{q}_i$ in a form which is suitable for any system of coordinates. If the system under investigation possesses r particles and n degrees of freedom ($n = 3r$), then the configuration of the system may be completely specified by using n generalised coordinates $q_1 \ldots q_n$ to describe the system†.

3.4(a.2). *Derivation of the Euler–Lagrange equation of motion.* This equation will be obtained by using *Hamilton's principle of least action.* This principle states that for a conservative system, the action integral, defined by

$$A = \int_{t_1}^{t_2} dt\, L \tag{3.205}$$

is a constant.

We commence by integrating the system over the region from point 1 to point 2 in Fig. 3.5. For simplicity we have assumed that all except one of the generalised coordinates q_n is zero. The integration is carried out along the closed and dotted lines. Let the variations along the two lines be

$$q(t) \rightarrow q(t) + \delta q(t) \tag{3.206}$$

$$\dot{q}(t) \rightarrow \dot{q}(t) + \delta\dot{q}(t) = \dot{q}(t) + \frac{d}{dt}\,\delta q(t). \tag{3.207}$$

† We shall only consider systems without constraints, that is the q's may vary independently of each other without any limitation.

We will define

$$\delta q(t_1) = \delta q(t_2) = 0 \tag{3.208}$$

that is, both paths of integration must begin and end at the same point. Now the variation of equation (3.205) yields

$$\delta A = \int_{t_1}^{t_2} dt \left(\frac{\partial L}{\partial q} \, \delta q + \frac{\partial L}{\partial \dot{q}} \, \delta \dot{q} \right) \tag{3.209}$$

Consider a partial integration of the second term

$$\int_{t_1}^{t_2} dt \frac{\partial L}{\partial \dot{q}} \, \delta \dot{q} = \left[\frac{\partial L}{\partial \dot{q}} \, \delta q \right]_{t_1}^{t_2} - \int_{t_1}^{t_2} dt \, \delta q \frac{d}{dt} \frac{\partial L}{\partial \dot{q}_i} .$$

The first term on the right-hand side of this equation is zero by virtue of (3.208) and so equation (3.209) becomes

$$\delta A = \int_{t_1}^{t_2} dt \left[\frac{\partial L}{\partial q} - \frac{d}{dt} \frac{\partial L}{\partial \dot{q}} \right] \delta q(t) .$$

If we had assumed nonzero values for all q terms, then a summation sign would be inserted in this equation, yielding

$$\delta A = \sum_{i=1}^{n} \int_{t_1}^{t_2} dt \left[\frac{\partial L}{\partial q_i} - \frac{d}{dt} \frac{\partial L}{\partial \dot{q}_i} \right] \delta q_i(t) . \tag{3.210}$$

Now if Hamilton's principle is to be satisfied, δA must be zero, and for every component q_i we must have

$$\frac{\partial L}{\partial q_i} - \frac{d}{dt} \frac{\partial L}{\partial \dot{q}_i} = 0 . \tag{3.211}$$

This is the *Euler–Lagrange equation of motion:* for convenience it will be referred to as the Lagrange equation.

The ready application of this equation in classical mechanics may be seen, for example, in the simple harmonic oscillator, where

$$L = T - V = \tfrac{1}{2} m \dot{q}^2 - \tfrac{1}{2} k q^2$$

then

$$\frac{\partial L}{\partial q} = -kq , \qquad \frac{d}{dt} \frac{\partial L}{\partial \dot{q}} = m \ddot{q} .$$

Upon substitution in (3.211), we find the familiar equation for simple harmonic motion

$$m\ddot{q} + kq = 0\,.$$

3.4(b). *Generalised momenta and Hamilton's equation of motion*

The generalised momentum for particles in a system described by equation (3.211) is defined as

$$p_i = \frac{\partial L}{\partial \dot{q}_i}\,. \tag{3.212}$$

It can be readily verified that for the harmonic oscillator

$$p_i = \frac{\partial L}{\partial \dot{q}_i} = m\dot{q}_i$$

but in general p_i need not have the dimensions of momentum. This only occurs if $\dot{q}_i$ represents the time derivative of a length. In spatial polar coordinates, for example, p_i can possess the dimensions of angular momentum.

A further definition will be introduced, that for the Hamiltonian of a system

$$H = \sum_{i=1}^{n} p_i\dot{q}_i - L(q, \dot{q}) \equiv p_i\dot{q}_i - L. \tag{3.213}$$

This definition again will be examined for the case of the simple oscillator

$$\begin{aligned} H &= m\dot{q}_i^2 - \tfrac{1}{2}(m\dot{q}_i^2 - kq_i^2) \\ &= \tfrac{1}{2}m\dot{q}_i^2 + \tfrac{1}{2}kq_i^2 \end{aligned}$$

that is, H represents the total energy of this system.

If equation (3.213) is differentiated, we obtain

$$\begin{aligned} dH &= \dot{q}_i\,dp_i + p_i\,d\dot{q}_i - \frac{\partial L}{\partial \dot{q}_i}\,d\dot{q}_i - \frac{\partial L}{\partial q_i}\,dq_i \\ &= \dot{q}_i\,dp_i - \frac{\partial L}{\partial q_i}\,dq_i = \dot{q}_i\,dp_i - \dot{p}_i\,dq_i \end{aligned} \tag{3.214}$$

by virtue of (3.212). Thus H depends only on p and q, and so we may write

$$dH = \frac{\partial H}{\partial p_i}\,dp_i + \frac{\partial H}{\partial q_i}\,dq_i \tag{3.215}$$

or

$$H = H(p, q)\,.$$

Each equation is independent, and so, upon equating corresponding coefficients in (3.214) and (3.215) we find

$$\dot{q}_i = \frac{dq_i}{dt} = \frac{\partial H}{\partial p_i} \tag{3.216}$$

$$\dot{p}_i = \frac{dp_i}{dt} = -\frac{\partial H}{\partial q_i}.$$

These equations are called the *canonical form of the equations of motion* or *Hamilton's equations.*

We mentioned in § 3.4(a.1) that the formal presentation of H and L could be used readily for the proof of general theorems. As an example consider the derivative of H with respect to time

$$\frac{dH}{dt} = \frac{\partial H}{\partial q_i}\dot{q}_i + \frac{\partial H}{\partial p_i}\dot{p}_i = 0. \tag{3.217}$$

The constancy of H as a function of time follows from Hamilton's equations (3.216). This result implies that energy is conserved, since

$$H = p_i\dot{q}_i - L = p_i\dot{q}_i - T + V$$

and, if T is any quadratic function of velocity (true for any classical mechanical system), then by equations (3.212) and (3.204) we may write

$$p_i\dot{q}_i = \frac{\partial L}{\partial \dot{q}_i}\dot{q}_i = \frac{\partial T}{\partial \dot{q}_i}\dot{q}_i = 2T.$$

Thus we find that H may be identified as the total energy

$$H = T + V \tag{3.218}$$

for mechanical systems which satisfy the condition stated above, and the expression

$$\frac{dH}{dt} = 0$$

implies that energy is conserved in these systems.

3.4(c). *The equation for a classical relativistic field*

3.4(c.1). *The transition to a continuous system.* In the previous section we have indicated that in classical mechanical systems the Lagrangian L may be used to obtain an equation of motion for the system through the principle of least action

$$\delta A = \int_{t_1}^{t_2} dtL = 0. \tag{3.219}$$

Iere L depends upon the generalised coordinates q of the system and their erivative with respect to time.

The equations of motion for a field – *the field equations* – may be obtained in similar manner. The term field is taken to mean a physical system with an innite number of degrees of freedom. The generalised coordinates q_i for mehanical systems with a finite number of degrees of freedom, are replaced by *ield functions* $\eta(x)$, where the variable x is a continuous index.

Our problem is therefore to show how a mechanical system with a finite umber of degrees of freedom can be transformed to a continuous one. We will gain examine a classical nonrelativistic system, namely a one-dimensional sysem of N point particles each of mass m coupled elastically by a series of springs of length a, each with a force constant k. Let η_i represent the displacement of the particle i from its equilibrium position. The equation of motion for the system may be obtained by setting the variation of the action equal to zero

$$\delta A = \delta \int_{t_1}^{t_2} dt\, L(\eta_i, \dot{\eta}_i) = 0 \tag{3.220}$$

where, if end effects are neglected, we may write L as

$$L = \tfrac{1}{2} \sum_{i=1}^{N} [m\dot{\eta}_i^2 - k(\eta_{i+1} - \eta_i)^2].$$

This expression may be re-written as

$$L = \sum_{i=1}^{N} a\, \mathscr{L}_i$$

where

$$\mathscr{L}_i = \tfrac{1}{2}\left[\frac{m}{a}\dot{\eta}_i^2 - ka\left(\frac{\eta_{i+1} - \eta_i}{a}\right)^2\right]. \tag{3.221}$$

In order to make the transition to a continuous system, we will let the number of degrees of freedom tend to infinity. Then we may write

$$\sum_{i=1}^{N} a \to \int_{x_1}^{x_2} dx, \quad \frac{m}{a} \to \varrho, \quad ka \to Y, \quad \frac{\eta_{i+1} - \eta_i}{a} \to \frac{\partial \eta}{\partial x}$$

where ϱ is the linear mass density, and the term Y may be identified as Young's modulus. Thus the action equation becomes

$$\delta A = \delta \int_{t_1}^{t_2} dt\, L = \delta \int_{t_1}^{t_2}\int_{x_1}^{x_2} dt\, dx \left[\varrho\dot{\eta}^2 - Y\left(\frac{\partial \eta}{\partial x}\right)^2\right]$$

$$= \delta \int_{t_1}^{t_2}\int_{x_1}^{x_2} dt\, dx\, \mathscr{L} = 0. \tag{3.222}$$

The term $\mathscr{L}$ is called the *Lagrangian density*. It can be seen that, in additio to a dependence on η and $\dot{\eta}$, $\mathscr{L}$ is dependent upon the spatial derivative of th field function. The Lagrange equation may be derived by exploiting this de pendence, together with the principle of least action as stated in (3.222), in th manner indicated in § 3.4(a.2). The equation found is

$$\frac{\partial \mathscr{L}}{\partial \eta} - \frac{\partial}{\partial x}\left[\frac{\partial \mathscr{L}}{\partial(\partial \eta/\partial x)}\right] - \frac{\partial}{\partial t}\left[\frac{\partial \mathscr{L}}{\partial(\partial \eta/\partial t)}\right] = 0. \qquad (3.223$$

In deriving this equation, the invariance condition (compare (3.208))

$$\delta\eta(x_1) = \delta\eta(x_2) = 0$$

is employed; we have also ignored any explicit time dependence for the functior $\dot{\eta}$. The field equation, obtained by inserting the expression for $\mathscr{L}$ from (3.222 in equation (3.223), is

$$\varrho\frac{\partial^2 \eta}{\partial t^2} - Y\frac{\partial^2 \eta}{\partial x^2} = 0$$

which is the familiar expression for compressional waves travelling with a velocity $(Y/\varrho)^{1/2}$ in a linear system.

3.4(c.2). *The transition to a relativistic equation.* In transferring our attention from a nonrelativistic, one-dimensional formulation to a relativistic field theory extending in space–time, it is obvious we must use a covariant formulation, since a distinction between space and time would be artificial. If such a distinction was attempted at some coordinate point $\mathbf{x}$, t the application of a Lorentz transformation would immediately mix the space and time coordinates.

We will therefore represent the action integral in four-dimensional notation as

$$A = \int \mathrm{d}\mathbf{x}\, \mathrm{d}t\, \mathscr{L} = \int_{\Omega} \mathrm{d}^4x\, \mathscr{L} \qquad (3.224)$$

where the symbol Ω refers to a four-dimensional volume in space–time. The term $\mathscr{L}$ is again called the Lagrangian density.

Since we wish this definition to apply in any coordinate system, the integral should not depend upon the choice of reference frame. The integral A must therefore be a scalar quantity. Since the term d^4x is invariant under proper Lorentz transformations, the scalar property of A then implies that $\mathscr{L}$ should also be Lorentz invariant for proper transformations.

It should be noted as well that $\mathscr{L}$ should not contain differentials of order higher than one, since second order differentials and higher would leave the integral dependent upon its coordinates.

Thus Hamilton's principle of least action becomes

$$\delta A = \delta \int_{\Omega} \mathrm{d}^4x\, \mathscr{L}\left(\eta, \frac{\partial \eta}{\partial x_\lambda}\right) = 0. \qquad (3.225)$$

The statement $\delta A = 0$ implies that the variation of $\eta \to \eta + \delta\eta$ causes no alteration in A, where $\delta\eta$ is considered as an arbitrary variation with a boundary condition that it is equal to zero everywhere on the surface of Ω. The latter statement is the equivalent of that given in equation (3.208), namely

$$\delta q(t_1) = \delta q(t_2) = 0$$

at the boundary points t_1 and t_2 of a line integral.

In practice the field function $\eta(x)$ may possess more than one component so that we will write

$$\eta \to \eta_\sigma \tag{3.226}$$

where σ is a discontinuous index.

By using the same techniques as in § 3.4(a.2) we may write the variation of the action integral as

$$\begin{aligned}\delta A = 0 &= \int_\Omega d^4x \left[\frac{\partial \mathscr{L}}{\partial \eta_\sigma}\,\delta\eta_\sigma + \frac{\partial \mathscr{L}}{\partial(\partial\eta_\sigma/\partial x_\lambda)}\,\delta\,\frac{\partial \eta_\sigma}{\partial x_\lambda}\right] \\ &= \int_\Omega d^4x \left[\frac{\partial \mathscr{L}}{\partial \eta_\sigma} - \frac{\partial}{\partial x_\lambda}\,\frac{\partial \mathscr{L}}{\partial(\partial\eta_\sigma/\partial x_\lambda)}\right]\delta\eta_\sigma + \oint_\Sigma d\sigma_\lambda\,\frac{\partial \mathscr{L}}{\partial(\partial\eta_\sigma/\partial x_\lambda)}\,\delta\eta_\sigma \end{aligned} \tag{3.227}$$

where a partial integration has been performed (compare (3.65)), and Σ represents the boundary of Ω. We have also used the fact that since the variation $\delta(\partial\eta_\sigma/\partial x_\lambda)$ is obtained from the variation of η_σ, we may write

$$\delta\left(\frac{\partial \eta_\sigma}{\partial x_\lambda}\right) = \frac{\partial}{\partial x_\lambda}(\delta\eta_\sigma).$$

The last term in (3.227) is zero, since $\delta\eta_\sigma$ is zero everywhere on the boundary of Σ by definition. Thus we have obtained again the Euler–Lagrange equation, which is now referred to as the *field equation*

$$\sum_\sigma \left[\frac{\partial \mathscr{L}}{\partial \eta_\sigma} - \frac{\partial}{\partial x_\lambda}\,\frac{\partial \mathscr{L}}{\partial(\partial\eta_\sigma/\partial x_\lambda)}\right] = 0. \tag{3.228}$$

(The summation sign has been introduced for σ as it is not a repeated suffix in (3.228)).

The variation which we have performed is not the most general, since not only η but also the boundary of Ω could have been subjected to a variation. Thus the total variation could have been

$$\delta\eta_\sigma = \bar{\delta}\eta_\sigma + \frac{\partial \eta_\sigma}{\partial x_\lambda}\,\delta x_\lambda \tag{3.229}$$

where $\bar{\delta}\eta$ represents the intrinsic variation of η at a fixed point in space-time, and

$(\partial\eta/\partial x_\lambda/\delta x_\lambda)$ the variation due to the boundary. The latter variation is considered as well in the more sophisticated formulations of field theory (see, for example, Aharoni, 1959). The net result is the same as in the equation (3.228).

3.4(d). *The energy and momentum of a field*

In equation (3.213) the Hamiltonian of a mechanical system was defined as

$$H = p_i \dot{q}_i - L$$

where

$$p_i = \frac{\partial L}{\partial \dot{q}_i}$$

represented a generalised momentum canonically conjugate to q_i.

When the mechanical system is replaced by a field, then the field must become the carrier of energy, momentum and other observable dynamical quantities. These quantities spread out through the domain of the field, and one therefore considers densities, for example the energy density.

A quantity canonically conjugate to $\eta_\sigma(x)$, with the property of a generalised momentum, may be defined by the expression

$$\pi_\sigma = \frac{\partial \mathscr{L}}{\partial \dot{\eta}_\sigma} = \frac{\partial \mathscr{L}}{\partial(\partial\eta_\sigma/\partial t)}. \tag{3.230}$$

We then may define a Hamiltonian density $\mathscr{H}$ as

$$H = \int d\mathbf{x}\, \mathscr{H} = \int d\mathbf{x}(\pi_\sigma \dot{\eta}_\sigma - \mathscr{L}). \tag{3.231}$$

The term π_σ cannot be regarded as a momentum density; an expression which fulfils this condition dimensionally is

$$-\pi_\sigma \frac{\partial \eta_\sigma}{\partial x_k} = -\frac{\partial \mathscr{L}}{\partial \dot{\eta}_\sigma}\frac{\partial \eta_\sigma}{\partial x_k} \equiv -\frac{\partial \mathscr{L}}{\partial \dot{x}_k} \tag{3.232}$$

where k runs from 1 to 3, that is spatial coordinates only, and the negative sign has been introduced for later convenience.

Many of the concepts in quantum mechanics may be developed more easily with the aid of Hamiltonians and the Hamiltonian formalism rather than with Lagrangians. A basic difficulty of the Hamiltonian method, however, is that it treats time separately from the spatial coordinates, as we can see in equation (3.231) for example. The power of the Lagrangian method in relativistic problems lies in the fact that time appears symmetrically in the Lagrangian.

3.4(e). *Invariance principles and conservation laws – Noether's theorem*

The invariance of the action integral under continuous Lorentz transformations

$$\delta A = \delta \int_{\Omega} d^4x\, \mathscr{L} = \delta \int_{\Omega'} d^4x'\, \mathscr{L}' = 0 \tag{3.233}$$

eads to the appearance of certain conservation laws for the fields associated with he Lagrangian. This statement is known as *Noether's theorem.*

We commence by writing the Lorentz transformation of the coordinate system as

$$x_\lambda \to x'_\lambda = x_\lambda + \delta x_\lambda \tag{3.234}$$

where δx_λ represents an infinitesimal transformation. Similarly the transformation for the field function may be written as

$$\eta_\sigma(x) \to \eta'_\sigma(x') = \eta_\sigma(x) + \delta\eta_\sigma(x). \tag{3.235}$$

In this equation $\delta\eta_\sigma(x)$ represents the change in the field function arising both from the change in the form of the function and the change in its argument x. We shall require the variation in the form of the function for fixed argument, this term is written as (compare (3.229))

$$\bar{\delta}\eta_\sigma(x) = \eta'_\sigma(x) - \eta_\sigma(x) = \delta\eta_\sigma(x) - \frac{\partial \eta_\sigma}{\partial x_\lambda}\,\delta x_\lambda. \tag{3.236}$$

We may write the variation of the action integral as

$$\delta A = \delta \int d^4x\, \mathscr{L} = \int \delta(d^4x)\, \mathscr{L} + \int d^4x\, \delta\mathscr{L} = 0 \tag{3.237}$$

and the terms for insertion in this equation must now be constructed. We first transform the four-dimensional volume element d^4x to d^4x'. The Jacobian for this transformation leads to the following expression:

$$d^4x' = \left(1 + \frac{\partial}{\partial x_\lambda}\,\delta x_\lambda\right) d^4x. \tag{3.238}$$

Thus we find that

$$\delta(d^4x) = d^4x\, \frac{\partial}{\partial x_\lambda}\,\delta x_\lambda$$

and equation (3.237) becomes

$$\delta A = \int_{\Omega} d^4x \left(\delta\mathscr{L} + \mathscr{L}\, \frac{\partial}{\partial x_\lambda}\,\delta x_\lambda\right) = 0. \tag{3.239}$$

The total variation of the Lagrangian gives

$$\delta\mathscr{L} = \mathscr{L}'(x') - \mathscr{L}(x) = \bar{\delta}\mathscr{L} + \frac{\partial\mathscr{L}}{\partial x_\lambda}\,\delta x_\lambda$$

where $\bar{\delta}\mathscr{L}$ represents the intrinsic variation in the form of the Lagrangian. Thi equation may be inserted in (3.239), yielding

$$\begin{aligned}\delta A &= \int_\Omega d^4x \left(\bar{\delta}\mathscr{L} + \frac{\partial\mathscr{L}}{\partial x_\lambda}\,\delta x_\lambda + \mathscr{L}\,\frac{\partial}{\partial x_\lambda}\,\delta x_\lambda\right) \\ &= \int_\Omega d^4x \left[\bar{\delta}\mathscr{L} + \frac{\partial}{\partial x_\lambda}\,(\mathscr{L}\delta x_\lambda)\right] = 0 \end{aligned} \tag{3.240}$$

where the variation in the intrinsic form of the Lagrangian gives the expressior

$$\bar{\delta}\mathscr{L} = \frac{\partial\mathscr{L}}{\partial\eta_\sigma}\,\bar{\delta}\eta_\sigma + \frac{\partial\mathscr{L}}{\partial(\partial\eta_\sigma/\partial x_\mu)}\,\bar{\delta}\,\frac{\partial\eta_\sigma}{\partial x_\mu}\,.$$

Using the Euler–Lagrange equation (3.228) and the fact that the operators δ and $\partial/\partial x$ commute, this equation can be written as

$$\bar{\delta}\mathscr{L} = \frac{\partial}{\partial x_\mu}\,\frac{\partial\mathscr{L}}{\partial(\partial\eta_\sigma/\partial x_\mu)}\,\bar{\delta}\eta_\sigma + \frac{\partial\mathscr{L}}{\partial(\partial\eta_\sigma/\partial x_\mu)}\,\frac{\partial}{\partial x_\mu}\,\bar{\delta}\eta_\sigma = \frac{\partial}{\partial x_\mu}\left[\frac{\partial\mathscr{L}}{\partial(\partial\eta_\sigma/\partial x_\mu)}\,\bar{\delta}\eta_\sigma\right].$$

This expression can now be inserted in (3.240), yielding

$$\begin{aligned}\delta A &= \int_\Omega d^4x \left\{\frac{\partial}{\partial x_\mu}\left[\frac{\partial\mathscr{L}}{\partial(\partial\eta_\sigma/\partial x_\mu)}\,\bar{\delta}\eta_\sigma\right] + \frac{\partial}{\partial x_\lambda}\,(\mathscr{L}\delta x_\lambda)\right\} \\ &= \int_\Omega d^4x \left\{\frac{\partial}{\partial x_\mu}\left[\frac{\partial\mathscr{L}}{\partial(\partial\eta_\sigma/\partial x_\mu)}\left(\delta\eta_\sigma - \frac{\delta\eta_\sigma}{\delta x_\lambda}\,\delta x_\lambda\right)\right] + \frac{\partial}{\partial x_\mu}\,(\mathscr{L}\delta_{\mu\lambda}\delta x_\lambda)\right\} \\ &= \int_\Omega d^4x\,\frac{\partial}{\partial x_\mu}\left\{\left[\mathscr{L}\delta_{\mu\lambda} - \frac{\partial\mathscr{L}}{\partial(\partial\eta_\sigma/\partial x_\mu)}\,\frac{\partial\eta_\sigma}{\partial x_\lambda}\right]\delta x_\lambda + \frac{\partial\mathscr{L}}{\partial(\partial\eta_\sigma/\partial x_\mu)}\,\delta\eta_\sigma\right\} = 0\,. \end{aligned} \tag{3.241}$$

The term $\delta_{\mu\lambda}$ in (3.241) is the Kronecker symbol. Since the domain of the four-dimensional volume Ω is arbitrary and nonzero, the integrand in (3.241) may be written in the form

$$\frac{\partial f_\mu}{\partial x_\mu} = 0 \tag{3.242}$$

where

$$f_\mu = \left[\mathscr{L}\delta_{\mu\lambda} - \frac{\partial\mathscr{L}}{\partial(\partial\eta_\sigma/\partial x_\mu)}\,\frac{\partial\eta_\sigma}{\partial x_\lambda}\right]\delta x_\lambda + \frac{\partial\mathscr{L}}{\partial(\partial\eta_\sigma/\partial x_\mu)}\,\delta\eta_\sigma\,.$$

A comparison with equation (3.74) shows that (3.242) represents an equation of continuity for the four-vector f_μ, that is the differential form of the conservation law. A more convenient formulation may be obtained by introducing a term C_λ, defined as the integral of f_4 over ordinary three-dimensional space

$$C_\lambda = \frac{1}{i} \int d\mathbf{x}\, f_4 . \tag{3.243}$$

Then with the aid of equation (3.242) we can write

$$\frac{d}{dt} C_\lambda = \frac{1}{i} \frac{d}{dt} \int d\mathbf{x}\, f_4 = \int d\mathbf{x} \frac{\partial f_4}{dx_4} = - \int d\mathbf{x} \frac{\partial f_k}{\partial x_k} \equiv - \sum_{k=1}^{3} \int d\mathbf{x} \frac{\partial f_k}{\partial x_k} = 0 . \tag{3.244}$$

The final result follows from the fact that since the integrand on the right is a three-dimensional divergence, the integral can be replaced by a surface integral with the aid of Gauss' theorem. Since the field functions and their derivatives vanish, by definition, at spatial infinity the integral equals zero.

Thus the quantity

$$C_\lambda = \frac{1}{i} \int d\mathbf{x} \left\{ \left[\mathscr{L} \delta_{4\lambda} - \frac{\partial \mathscr{L}}{\partial(\partial \eta_\sigma / \partial x_4)} \frac{\partial \eta_\sigma}{\partial x_\lambda} \right] \delta x_\lambda + \frac{\partial \mathscr{L}}{\partial(\partial \eta_\sigma / \partial x_4)} \delta \eta_\sigma \right\} \tag{3.245}$$

is conserved as a function of time, that is it is a *constant of the motion.*

3.4(f). *Applications of Noether's theorem*

3.4(f.1). *Conservation of energy and momentum.* Our first application of Noether's theorem will be to the conservation of energy and momentum. We show this property by considering an infinitesimally small translation of the coordinate system in space and time

$$x_\lambda \to x'_\lambda = x_\lambda + a_\lambda, \qquad a_\lambda \to 0 . \tag{3.246}$$

The quantity $\delta \eta_\sigma$ in (3.245) is zero by definition under displacements of this nature. Now consider the term C_4 and recall equation (3.231)

$$\begin{aligned} C_4 &= \frac{1}{i} \int d\mathbf{x} \left[\mathscr{L} \delta_{44} - \frac{\partial \mathscr{L}}{\partial(\partial \eta_\sigma / \partial x_4)} \frac{\partial \eta_\sigma}{\partial x_4} \right] a_4 \\ &= \frac{1}{i} \int d\mathbf{x} \left(\mathscr{L} - \frac{\partial \mathscr{L}}{\partial \dot{\eta}_\sigma} \dot{\eta}_\sigma \right) a_4 = \frac{1}{i} \int d\mathbf{x}\, (\mathscr{L} - \pi_\sigma \dot{\eta}_\sigma)\, a_4 \\ &= i \int d\mathbf{x}\, \mathscr{H} a_4 = iHa_4 \equiv p_4 a_4 . \end{aligned}$$

Thus apart from the arbitrary displacement term a_4, C_4 is equivalent to the fourth component of the momentum four-vector. Using the expression (3.232) we find that the components C_k, where k runs from 1 to 3, are related to the components of linear momentum

$$C_k = -\frac{1}{i}\int d\mathbf{x} \frac{\partial \mathscr{L}}{\partial(\partial\eta_\sigma/\partial x_4)} \frac{\delta\eta_\sigma}{\partial x_k} a_k = -\int d\mathbf{x} \frac{\partial \mathscr{L}}{\partial\dot{\eta}_\sigma} \frac{\partial\eta_\sigma}{\partial x_k} a_k = p_k a_k .$$

Thus we may conclude that the energy and momentum associated with a field are conserved quantities, and that the conservation property follows from the invariance under displacements in time and space respectively.

The momentum four-vector for the classical relativistic field follows immediately from our expressions for total energy and momentum. It is

$$p_\lambda = i\int d\mathbf{x}\left[\frac{\partial \mathscr{L}}{\partial(\partial\eta_\sigma/\partial x_4)} \frac{\partial\eta_\sigma}{\partial x_\lambda} - \mathscr{L}\delta_{4\lambda}\right]. \tag{3.247}$$

3.4(f.2). *Invariance under gauge transformations.* In the previous section a group of transformations was considered which changed the space–time conditions for the field, but left its intrinsic structure unaltered. In the present section we will consider the reverse situation, namely

$$\begin{aligned} x_\lambda &\to x'_\lambda = x_\lambda \\ \eta_\sigma(x) &\to \eta'_\sigma(x') = \eta'_\sigma(x). \end{aligned} \tag{3.248}$$

The Lagrangian formalism is used to describe the real (that is measurable) dynamic properties of fields. Thus the Lagrangian must be a real function. On the other hand, since a field function is not necessarily a measurable quantity (this point will be discussed in greater detail in Chapter 4), no restriction exists requiring it to be a real function. We can conclude, therefore, that when the fields are complex functions of the form η and η^* (the complex conjugate of η), the Lagrangian must be formed from quadratic combinations of the type $\eta\eta^*$. Similar remarks apply to the derivatives of the functions. Thus the field functions may be multiplied by an arbitrary phase factor of the type $e^{i\alpha}$ (α real), without affecting the form of the quadratic functions appearing in the Lagrangian. We therefore introduce the simultaneous transformations

$$\begin{aligned} \eta_\sigma \to \eta'_\sigma &= \eta_\sigma\, e^{i\alpha} = \eta_\sigma + i\alpha\eta_\sigma \equiv \eta_\sigma + \delta\eta_\sigma \\ \eta^*_\sigma \to \eta^{*\prime}_\sigma &= \eta^*_\sigma\, e^{-i\alpha} = \eta^*_\sigma - i\alpha\eta^*_\sigma \equiv \eta^*_\sigma + \delta\eta^*_\sigma \\ \alpha &\to 0. \end{aligned} \tag{3.249}$$

The transformations are called *gauge transformation of the first kind.* The 'conserved current' f_μ, appearing in the continuity equation (3.242), is

$$j_\mu = f_\mu = i\alpha\left[\frac{\partial \mathscr{L}}{\partial(\partial\eta_\sigma/\partial x_\mu)}\eta_\sigma - \frac{\partial \mathscr{L}}{\partial(\partial\eta^*_\sigma/\partial x_\mu)}\eta^*_\sigma\right] \tag{3.250}$$

or a gauge transformation (we have introduced the symbol j_μ for later convenence, since j_μ is the conventional symbol for current density). The conserved quantity C_λ (3.245) in a gauge transformation is then given by

$$
\begin{aligned}
C_\lambda = C &= \frac{1}{i}\int d\mathbf{x}\, i\alpha\left[\frac{\partial\mathscr{L}}{\partial(\partial\eta_\sigma/\partial x_4)}\eta_\sigma - \frac{\partial\mathscr{L}}{\partial(\partial\eta_\sigma^*/\partial x_4)}\eta_\sigma^*\right] \\
&= \frac{\alpha}{i}\int d\mathbf{x}\left(\frac{\partial\mathscr{L}}{\partial\dot{\eta}_\sigma^*}\eta_\sigma^* - \frac{\partial\mathscr{L}}{\partial\dot{\eta}_\sigma}\eta_\sigma\right) \\
&= \frac{\alpha}{i}\int d\mathbf{x}(\pi_\sigma^*\eta_\sigma^* - \pi_\sigma\eta_\sigma)
\end{aligned}
\tag{3.251}
$$

where we have dropped the subscript λ since it does not appear in the transformation. The conserved quantity C depends upon the nature of the term α. It will be shown in §§ 4.3(i.1) and 4.5(d) that amongst other important quantities C may be made proportional to the total electrical charge of the field.

Thus the invariance of the Lagrangian under gauge transformations of the first kind can lead to charge conservation.

CHAPTER 4

QUANTUM THEORY OF NON-INTERACTING FIELDS

4.1. SOME FEATURES FROM QUANTUM MECHANICS

4.1 (a). *Introduction*

In the previous chapter a classical relativistic field theory was developed. In order to construct a relativistic quantised field theory, certain axioms and principles from quantum mechanics must be included. We start, therefore, by summarising certain aspects of quantum mechanics.

4.1 (b). *State vectors and Hilbert space*

In Chapter 3 (§ 3.4 (a)) a nonrelativistic classical system, with n degrees of freedom, was considered. It was stated that the properties of such a system could be described with the aid of n generalised coordinates $q_1, q_2, \ldots, q_n$.

In quantum mechanics a basically similar approach is made. Physical systems are represented by *state* or *wave functions* Ψ and it is assumed that a *complete knowledge of the wave function represents the maximum attainable information about the system.* A convenient method of representing the wave function is by means of a complex vector space, with either a finite or infinite number of components (dimensions). We shall assume that the vectors in this space have finite length and form finite scalar products; the space is then called a *Hilbert space.* We will assume that the space has n dimensions, and represent the directions by n linearly independent vectors. These vectors are called the *base vectors* and they will be given the symbol α. The set of n base vectors necessary for the description of an n-dimensional space is sometimes called a *complete set.* The state (wave) function describing a physical system is then represented by a linear vector in this space and is called a *state vector.*

$$\Psi = c_1\alpha_1 + c_2\alpha_1 + \cdots = \sum_{i=1}^{n} c_i\alpha_i \tag{4.1}$$

where α_i represents a base vector and c_i a component of the state vector along the 'direction' α_i. The terms $c_i\alpha_i$ thus represent the components of the individual tate functions which add together to give a complete description of a given physical state. The analogy of (4.1) with vectors in ordinary three-dimensional pace is a close one, for example we can write a vector **A** as

$$\mathbf{A} = A_x\mathbf{i} + A_y\mathbf{j} + A_z\mathbf{k}$$

where **i**, **j** and **k** are unit vectors. In Hilbert space, however, α_i and c_i can be complex.

We will use the *Dirac notation* to represent the state vectors (Dirac, 1947). In his notation a state is represented by a *ket vector* $|\Psi\rangle$. The notation lays emphasis on the relevant dynamical variables of a quantum mechanical system. If, for example, we were dealing with a problem in which the states of angular momentum assumed importance, then the state vector could be written as $|lm\rangle$. For simplicity we will restrict ourselves to a single parameter for our present discussion

$$|\Psi(\omega)\rangle = |\omega\rangle \equiv \Psi(\omega).$$

Thus we can write equation (4.1) as

$$|\Psi(\omega)\rangle = |\omega\rangle = \sum_{i=1}^{n} c_i(\omega)\,|\alpha_i\rangle. \tag{4.2}$$

The linearity of Hilbert space implies that the state vectors can be multiplied by complex numbers and added together. Thus we may write

$$\lambda_1\,|\omega\rangle + \lambda_2\,|\omega\rangle = (\lambda_1 + \lambda_2)\,|\omega\rangle = \lambda_3\,|\omega\rangle$$

where λ_1, λ_2 and λ_3 are complex numbers. It should be noted that no change in direction in Hilbert space has been introduced by this equation. It is obvious that the additive principle of the equation can be extended to an infinite sequence, and hence to the process of integration.

The linearity of the space also implies the *principle of superposition*, namely that dynamical states may be added to give new ones

$$\begin{aligned}|\omega_1\rangle + |\omega_2\rangle &= \sum_{i=1}^{n} c_i\,|\alpha_i\rangle + \sum_{i=1}^{n} d_i\,|\alpha_i\rangle \\ &= \sum_{i=1}^{n} (c_i + d_i)\,|\alpha_i\rangle = |\omega_3\rangle.\end{aligned} \tag{4.3}$$

This property is of great importance in quantum mechanics.

In relativistic systems scalar quantities are obviously important both in ordinary and in Hilbert space. The *scalar* or *inner product* of two vectors in ordinary space is given as

$$\mathbf{A}\cdot\mathbf{B} = A_xB_x + A_yB_y + A_zB_z \tag{4.4}$$

and in the Dirac notation it is written as

$$\langle \Psi_1(\omega) | \Psi_2(\omega) \rangle \equiv \langle \omega_1 | \omega_2 \rangle \tag{4.5}$$

where $\langle \;|$ is called a *bra vector*. It should be noted that some authors write the scalar product as (ω_1, ω_2).

If the scalar product equals zero

$$\langle \omega_1 \mid \omega_2 \rangle = 0$$

then the two state vectors are said to be *orthogonal* to each other. The scalar product of state vector with itself $\langle \omega \mid \omega \rangle$ is called the *norm*, and if the norm is equal to $+1$ the state vector is said to be *normalised*

$$\langle \omega \mid \omega \rangle = |\omega|^2 = +1.$$

If $\langle \omega \mid \omega \rangle \geqq 0$ the scalar product is said to have positive definiteness. This condition is also described as a Hilbert space with positive definite metric. Recently attempts have been made to construct field theories with indefinite metrics. No significant success has been achieved with this modification.

Let us return to the normalisation condition: if we represent $|\omega\rangle$ in the manner of (4.2)

$$|\omega\rangle = \sum_{i=1}^{n} c_i \, |\alpha_i\rangle$$

and assume that the coefficients c_i may be complex and that the base vectors $|\alpha_i\rangle$ form a complete set, then we can write

$$\langle \omega \mid \omega \rangle = \sum_i |c_i|^2 = +1$$

if the terms $|\alpha_i\rangle$ satisfy the condition of *orthonormality*

$$\langle \alpha_i \mid \alpha_j \rangle = \delta_{ij} \tag{4.6}$$

and if

$$\langle \omega | = \sum_{i=1}^{n} c_i^* \, \langle \alpha_i | .$$

Thus in general we might expect the scalar product to be of the form

$$\langle \Psi_1(\omega) | \Psi_2(\omega) \rangle = \sum_{i,j} c_i^* \, d_j \, \langle \alpha_i \mid \alpha_j \rangle$$

$$= \sum_i c_i^* \, d_i \tag{4.7}$$

which is the equivalent of the scalar product of two vectors in ordinary space (4.4); it must be emphasised that as in ordinary space a *scalar product is a pure number without direction.*

It is apparent from the above discussion that the component c_i of $|\omega\rangle$ in the 'direction' $|\alpha_i\rangle$ is given by $\langle\alpha_i \mid \omega\rangle$ by virtue of (4.2) and (4.6), furthermore

$$\langle\alpha_i \mid \omega\rangle = (\langle\omega \mid \alpha_i\rangle)^*.$$

It is also apparent that the *state vectors are indeterminate within a phase factor* since the transformation

$$|\omega\rangle \rightarrow |\omega'\rangle = e^{i\beta} |\omega\rangle$$

leaves the normalisation condition unaltered

$$\langle\omega' \mid \omega'\rangle = \langle\omega \mid \omega\rangle = 1.$$

Equations involving state vectors can be put in an easily calculable form if we pattern them in such a way as to conform to the notation of matrix algebra. In this notation a ket and a bra vector can be represented by a column and row vector respectively

$$|\omega\rangle = \begin{matrix} c_1 \\ c_2 \\ c_3 \\ \vdots \\ c_n \end{matrix} \tag{4.8}$$

$$\langle\omega| = c_1^* c_2^* c_3^* \ldots c_n^*.$$

A notation of this type satisfies equation (4.7); it will also be found to satisfy the requirements of operator notation in the next section. It is obvious that in matrix notation a ket vector is the Hermitian conjugate of a bra vector

$$\langle\Psi(\omega)| = \langle\omega| = |\omega\rangle^\dagger \tag{4.9}$$

where the symbol † represents a transposed and complex conjugated quantity. Using the notation of matrix algebra the following properties for the scalar product may be verified easily:

$$\langle\omega_2 \mid \omega_1\rangle = (\langle\omega_1 \mid \omega_2\rangle)^* \tag{4.10}$$

$$\langle\omega_2 \mid \lambda\omega_1\rangle = \lambda \langle\omega_2 \mid \omega_1\rangle$$

$$\langle\omega \mid (|\omega_1\rangle + |\omega_2\rangle) = \langle\omega \mid \omega_1\rangle + \langle\omega \mid \omega_2\rangle.$$

In these equations the symbol * refers to the complex conjugate of the entire scalar product, and λ is any complex number.

4.1(c). *Operators*

The state vectors of a physical system are not directly measurable quantities The measurable quantities, for example energy, momentum, charge, are calle *observables*, and are defined with the aid of operators as indicated below.

An *operator*, as the name suggests, operates on a state vector. A linear operator A operates on a ket vector from the left, or a bra from the right and trans forms then into another ket or bra

$$A\,|\omega\rangle \to |\omega'\rangle, \quad \langle\omega|\,A \to \langle\omega'|. \tag{4.11}$$

If the operation is of the form

$$A\,|\omega\rangle = a\,|\omega\rangle \tag{4.12}$$

where a is a number (which may be complex), then $|\omega\rangle$ is said to be an *eigen vector* or *eigenstate* of A, and the number a is called the *eigenvalue* of A.

If the vector space is a linear one, it is apparent that a linear operator shoul also possess the following properties.

$$A(|\omega_1\rangle + |\omega_2\rangle) = A\,|\omega_1\rangle + A\,|\omega_2\rangle$$

$$A\,|\lambda\omega\rangle = \lambda A\,|\omega\rangle$$

where λ is a complex number.

In the previous section it was mentioned that the bra and ket vectors may be regarded as matrices. An operator also can be given a convenient specification by constructing a set of matrix elements for it. These elements are specified with respect to a given set of base vectors in Hilbert space. Let

$$|\omega\rangle = \sum_{j=1}^{n} c_j\,|\alpha_j\rangle, \qquad |\omega'\rangle = \sum_{j=1}^{n} d_j\,|\alpha_j\rangle.$$

Upon substitution in equation (4.11), we find

$$|\omega'\rangle = \sum_{j=1}^{n} d_j\,|\alpha_j\rangle = A\sum_{j=1}^{n} c_j\,|\alpha_j\rangle = \sum_{j=1}^{n} A\,|\alpha_j\rangle\, c_j.$$

If the scalar product is now formed by multiplying from the left with $\langle\alpha_i|$ we obtain

$$\sum_{j=1}^{n} d_j\,\langle\alpha_i\,|\alpha_j\rangle = d_i = \sum_{j=1}^{n} \langle\alpha_i|\,A\,|\alpha_j\rangle\, c_j = \sum_{j=1}^{n} A_{ij}c_j$$

where we have used (4.6) and have introduced the notation

$$A_{ij} = \langle\alpha_i|\,A\,|\alpha_j\rangle. \tag{4.13}$$

This term is often called the *matrix element of the operator*; it is obvious that

$|\omega\rangle$ and $|\omega'\rangle$ each have n components then A must be a square matrix with n^2 omponents. It should be noted that equation (4.12) bears a strong formal reemblance to equation (3.33); this is not surprising since Hilbert space has been nodelled on ordinary space.

An operator A can be constructed from a ket and a bra vector written in the ollowing combination:

$$A = |\alpha\rangle \langle\beta| . \tag{4.14}$$

t is obvious that if $|\alpha\rangle$ and $\langle\beta|$ are regarded as column and row matrices respectively then A is also a matrix. A more satisfactory justification of (4.14) can e found by using it to operate on bras and kets with appropriate dimensions

$$A\,|\omega\rangle = |\alpha\rangle \langle\beta \mid \omega\rangle = \lambda_1\,|\alpha\rangle$$

$$\langle\omega| A = \langle\omega|\alpha\rangle \langle\beta| = \lambda_2\,\langle\beta|$$

where λ_1 and λ_2 represent numbers since a scalar product (4.5) is a pure number. t can be seen that these equations each satisfy the definition of an operator given n (4.11).

In § 4.1 (b) we stated that the set of n linearly independent base vectors, necesary for the description of an n-dimensional space, is sometimes called a complete et. The completeness condition may be given mathematical form with the aid of (4.14); thus we can write

$$\sum_{i=1}^{n} |\alpha_i\rangle \langle\alpha_i \mid \alpha_j\rangle = \sum_{i=1}^{n} |\alpha_i\rangle\, \delta_{ij} = |\alpha_j\rangle$$

where we have used the orthonormality condition (4.6). The *completeness* condiion is therefore given by

$$\sum_{i=1}^{n} |\alpha_i\rangle \langle\alpha_i| = \hat{1} \tag{4.15}$$

where the symbol $\hat{1}$ represents the unit operator; it is defined as

$$\hat{1}\,|\omega\rangle = |\omega\rangle$$

and possesses matrix elements δ_{ij}.†

Two important operators in quantum mechanics are the Hermitian and uniary operators. The *Hermitian operator* is one which is equal to its own Hermiian conjugate

$$A = A^{\dagger} \tag{4.16}$$

where the conjugate of an operator is given by

$$\langle\alpha_i|\, A^{\dagger}\, |\alpha_j\rangle = (\langle\alpha_j|\, A\, |\alpha_i\rangle)^* \tag{4.17}$$

† The definition of comp leteness can also be given in terms of continuous functions, e.g. $\int d\alpha\, |\alpha\rangle \langle\alpha| = \hat{1}$.

that is the symbol † implies that the matrix elements are transposed and comple conjugated.

Equations (4.16) and (4.17) imply that Hermitian operators must have rea eigenvalues, since if we have the relation $A|\alpha\rangle = a|\alpha\rangle$ (compare (4.12)), the

$$\langle\alpha| A |\alpha\rangle = a \langle\alpha \mid \alpha\rangle$$
$$= \langle\alpha| A^\dagger |\alpha\rangle = (\langle\alpha| A |\alpha\rangle)^* = a^* \langle\alpha \mid \alpha\rangle$$

and so $a^* = a$. The quantity $\langle\alpha| A |\alpha\rangle$ is often called the *expectation value* of A and is denoted by the symbol $\langle A\rangle$. It is apparent from the above argument tha the expectation value must be a real number if the operator A is Hermitian

$$\langle A\rangle^* = (\langle\alpha| A |\alpha\rangle)^* = \langle\alpha| A^\dagger |\alpha\rangle = \langle A\rangle. \quad (4.18)$$

Since the observable quantities of a physical system are real, they can be re presented by Hermitian operators.

Let us consider the expectation value for a system $\langle\omega| A |\omega\rangle$. If we assume tha $|\omega\rangle$ can be expanded into a complete set of eigenfunctions of A, and if we use th two conditions of orthonormality (4.6) and completeness (4.15), we find that

$$\langle\omega| A |\omega\rangle = \sum_i \langle\omega| A |\alpha_i\rangle \langle\alpha_i|\omega\rangle = \sum_i c_i a_i \langle\omega|\alpha_i\rangle = \sum_i |c_i|^2 a_i.$$

Now in § 4.1(b) we stated that the component of $|\omega\rangle$ in the 'direction' $|\alpha_i\rangle$ was given by $\langle\alpha_i \mid \omega\rangle$. If the state vector is normalised ($\langle\omega \mid \omega\rangle = 1$), we may therefore postulate that the probability of finding a system, described by a stat vector $|\omega\rangle$, with a value a_i for the Hermitian operator (observable) A is $|\langle\alpha_i|\omega\rangle|$

$$P_{a_i} = |\langle\alpha_i \mid \omega\rangle|^2. \quad (4.19)$$

In other words $\langle\alpha_i \mid \omega\rangle$ is the probability amplitude for observing the value a_i

A *unitary operator* has the property that its Hermitian conjugate is equal to its inverse

$$A^\dagger = A^{-1} \quad \text{or} \quad A^\dagger A = A^{-1}A = \hat{1} \quad (4.20)$$

where $\hat{1}$ is the unit operator. The unitary operator is frequently denoted by the symbol U

$$U^\dagger U = U^{-1}U = \hat{1}.$$

It should be noted that U is not necessarily Hermitian.

The simultaneous action of the unitary operator on a pair of kets

$$U|\omega\rangle = |\omega'\rangle, \qquad U|\alpha\rangle = |\alpha'\rangle$$

leaves the scalar product unaltered

$$\langle\alpha' \mid \omega'\rangle = \langle\alpha| U^\dagger U |\omega\rangle = \langle\alpha| \hat{1} |\omega\rangle = \langle\alpha \mid \omega\rangle.$$

Since the probability of finding a physical system in a particular state is proportional to $|\langle\alpha|\omega\rangle|^2$ according to (4.19), the combination $U^\dagger U$ therefore leaves the probability unaltered. We shall show later that an important function of the unitary operator is in the description of the conservation laws of physical systems.

4.1 (d). *The multiplication of operators*

The product of two operators A and B forms quantities which do not necessarily obey the commutation rule

$$AB - BA = [A, B] = 0.$$

A familiar example of this property in quantum mechanics is the canonical commutation relation

$$[x, p] = i \equiv i\hbar$$

where x and p represent position and momentum operators respectively.

The bracketed term $[A, B]$ will be referred to as the *commutator bracket*. An *anticommutator bracket* will be defined as

$$\{A, B\} = AB + BA.$$

Whilst the commutator bracket possesses certain analogies to the Poisson brackets of classical mechanics, no property is encountered in classical physics which requires the use of anticommutator brackets.

The statement that the product AB may not equal BA may be verified by writing the operators A and B in terms of their matrix elements (4.13). The matrix properties of operators may also be used to verify the following equations (compare A.3 (Appendixes, p. 696)):

$$\begin{aligned}(AB)^* &= A^*B^* \\ (AB)^\dagger &= B^\dagger A^\dagger \\ (AB)^T &= B^T A^T \\ (AB)^{-1} &= B^{-1}A^{-1} \\ \text{tr}\,(AB) &= \text{tr}\,(BA).\end{aligned} \tag{4.21}$$

These rules may be extended easily to an arbitrary number of operators, for example

$$\begin{aligned}(ABC)^\dagger &= C^\dagger B^\dagger A^\dagger \\ (ABC)^{-1} &= C^{-1}B^{-1}A^{-1}\end{aligned}$$

but it should be noted that in general $\text{tr}\,(ABC) \neq \text{tr}\,(CBA)$.

The rules given in (4.21) may be combined with (4.16) and (4.17) to yield the following important relation for the complex conjugate of matrix elements involving the commutator brackets of two Hermitian operators

$$(\langle\alpha_1| [A, B] |\alpha_2\rangle)^* = -\langle\alpha_2| [A, B] |\alpha_1\rangle. \tag{4.22}$$

4.1 (e). *Some properties of the unitary operator*

If an operator A has an eigenvector $|\alpha\rangle$ with eigenvalue a

$$A |\alpha\rangle = a |\alpha\rangle$$

then a new operator and eigenvector with the same eigenvalue may be defined with the aid of the unitary operator

$$A \to A' = UAU^{-1} \tag{4.23}$$

$$|\alpha\rangle \to |\alpha'\rangle = U |\alpha\rangle. \tag{4.24}$$

Then, using (4.20), we find that

$$A' |\alpha'\rangle = UAU^{-1}U |\alpha\rangle = UA |\alpha\rangle = a |\alpha'\rangle. \tag{4.25}$$

The terms A' and $|\alpha'\rangle$ are called the *unitary transforms* of A and $|\alpha\rangle$. If A is a Hermitian operator, so also is A'

$$A'^\dagger = (UAU^{-1})^\dagger = (U^{-1})^\dagger A^\dagger U^\dagger = (U^\dagger)^\dagger AU^{-1} = UAU^{-1} = A' \tag{4.26}$$

where we have used equations (4.20) and (4.21). Since we stated in § 4.1 (c) that physical observables may be represented by Hermitian operators, it follows that two systems are physically equivalent if they are related by the unitary transformations (4.23) and (4.24). As a further example of invariance under a unitary transformation we may quote the familiar canonical commutation relation of quantum mechanics

$$[x_i, p_j] = i\delta_{ij} \equiv i\hbar\delta_{ij}$$

which transforms as

$$\begin{aligned} [x_i, p_j] \to [x_i', p_j'] &= Ux_iU^{-1}Up_jU^{-1} - Up_jU^{-1}Ux_iU^{-1} \\ &= U[x_i, p_j]U^{-1} \\ &= i\delta_{ij}. \end{aligned} \tag{4.27}$$

The converse of the above statements also holds true, that is if two sets of operators describe the same physical system with the same eigenvalues and commutation relations, they are related by a unitary transformation.

We may summarise the above arguments by stating that if any two of the following conditions hold:

(1) operators transform as $A' = UAU^{-1}$,
(2) kets and bras transform as $|\alpha'\rangle = U|\alpha\rangle$

$$\langle\alpha'| = \langle\alpha| U^{-1},$$

(3) physical equations remain invariant in form

then the third condition is also satisfied. The choice of the two conditions is mainly a matter of taste, for example if we use conditions (2) and (3)

$$|\alpha'\rangle = U|\alpha\rangle, \qquad |\omega'\rangle = U|\omega\rangle$$

$$\langle\alpha'| A' |\omega'\rangle = \langle\alpha| A |\omega\rangle$$

then $A' = UAU^{-1}$, and if A is invariant under the transformation

$$A' = A = UAU^{-1} \quad \text{or} \quad [A, U] = 0. \tag{4.28}$$

In Chapter 3 the conservation of laws for observable quantities, such as energy and momentum, were introduced by considering the invariance of the Lagrangian density under certain transformations. We shall now show that the unitary transformation can also be used to define certain quantities which may be conserved. It should be noted that U itself is not necessarily an observable, since it does not need to be an Hermitian operator (see § 4.1 (c)).

If an infinitesimal unitary transformation is written in the form

$$U = \hat{1} - iF \tag{4.29}$$

where $F \to 0$, then we may also write

$$U^\dagger = \hat{1} + iF^\dagger. \tag{4.30}$$

The unitarity condition then gives us the following relation

$$U^\dagger U = (\hat{1} + iF^\dagger)(\hat{1} - iF) = \hat{1} + i(F^\dagger - F) = \hat{1}$$

where we have dropped the term of order F^2. Thus F is Hermitian and an observable

$$F^\dagger = F. \tag{4.31}$$

F is called the *generator* of the infinitesimal transformation. The effect of F on an arbitrary operator A may be demonstrated in the following manner:

$$\begin{aligned} A' = UAU^{-1} &= (\hat{1} - iF)A(\hat{1} + iF) \\ &= A - i[F, A] \end{aligned} \tag{4.32}$$

where we have used the Hermitian property of F. Thus if F commutes with A, then A remains invariant under the unitary transformation U and

$$A' = A.$$

It is frequently convenient to write F as the product of two terms

$$F = G\delta\lambda$$

where $\delta\lambda$ is assumed to be an infinitesimal real number and G is a Hermitia operator. G is also called the generator of the transformation. In this notatio equation (4.32) becomes

$$A' = A - i[G, A]\delta\lambda \tag{4.33}$$

and it can be seen that $A' = A$ if G commutes with A.

It is sometimes necessary to examine a finite rather than an infinitesimal trans formation. This may be achieved by considering a succession of infinitesima transformations.

We first note that if two unitary transformations are applied in succession, th result is a third unitary transformation. Let the first two transformations be

$$A'' = VA'V^{-1}, \qquad A' = UAU^{-1}$$

then

$$A'' = VUAU^{-1}V^{-1} = (VU)\,A(VU)^{-1}$$

by (4.21). Now VU is a unitary operator, since by (4.20) and (4.21)

$$(VU)^\dagger VU = U^\dagger V^\dagger VU = U^\dagger U = \hat{1}.$$

Thus our statement is proved; the result may be extended to an arbitrary numbe of unitary transformations.

Now consider an infinitesimal transformation of the form

$$U' = (\hat{1} - iG\delta\lambda) = \left(\hat{1} - iG\,\frac{\lambda}{n}\right)$$

where $\lambda = n\delta\lambda$ and is finite. Then by making successive transformations of the type given above we may construct a unitary operator with the form

$$U = (U')^n = \left(\hat{1} - iG\,\frac{\lambda}{n}\right)^n = e^{-iG\lambda}, \qquad n \to \infty. \tag{4.34}$$

This relation permits us to write the following operator equation:

$$\frac{\partial U}{\partial \lambda} = -iG\,e^{-iG\lambda} = -iGU$$

and, in general, we may write a finite unitary transformation for an operator as

$$A' = A(\lambda + \chi) = e^{-iG\lambda}A(\chi)\,e^{iG\lambda} \tag{4.35}$$

where the operator $A(\lambda + \chi)$ is dependent on the function $(\lambda + \chi)$. As an example of the use of this equation, it is sometimes necessary to displace operators in space–time (see, for example, § 8.9 (b)). If we write $\lambda = x$, then G can be identified as the four-momentum operator P since we require the exponential to be dimensionless. If we then set $\chi = 0$, for example, the operator A transforms as

$$A' = A(x) = e^{-iPx} A(0)\, e^{iPx}$$

under a displacement in space–time.

It should be noted that the signs $\pm$ appearing in both the infinitesimal and finite transformations are purely a matter of definition and can always be written to suit our convenience providing that the convention is applied systematically.

The form we have adopted for U and the properties we have found for it in this section will be used later to obtain a set of conservation laws. It is apparent from our discussion in this section, however, that the Lorentz transformation and the unitary transformation must have many points in common. This similarity will be examined again in Chapter 5.

4.1 (f). *The Schrödinger and Heisenberg formulations of quantum mechanics*

The equations of motion for quantum mechanical systems may be given in two forms, which are called the Schrödinger and Heisenberg representations. The Schrödinger equation is the more useful for many practical calculations about atomic or nuclear systems, as it yields the simpler equations of the two representations. The Heisenberg representation possesses certain advantages for examining the more formal aspects of quantum mechanics since it possesses close analogues with classical mechanics. It is therefore an extremely useful method for examining the conservation laws of quantum mechanical systems. We shall show in later chapters that it is also useful for describing systems of interacting particles.

4.1 (f.1). *The Schrödinger representation.* Let a quantum mechanical system be represented by a state vector $|\Psi_S(t)\rangle$ at a time t. The Schrödinger representation then assumes that any temporal changes which occur in the system will cause changes only in the state vectors and not in the operators. Thus at a time t' we find a new state vector $|\Psi_S(t')\rangle$ related to $|\Psi_S(t)\rangle$ by the expression

$$|\Psi_S(t)\rangle = T(t, t')\, |\Psi_S(t')\rangle. \tag{4.36}$$

The operator $T(t, t')$ obviously possesses the boundary condition

$$T(t, t) = \hat{1} \tag{4.37}$$

and if probability is to be conserved in the system it must be unitary since the condition

$$\langle \Psi_S(t) \mid \Psi_S(t)\rangle = \langle \Psi_S(t') \mid T^\dagger(t, t')\, T(t, t')\, |\Psi_S(t')\rangle = \langle \Psi_S(t') \mid \Psi_S(t')\rangle \tag{4.38}$$

requires a relationship of the form

$$T^\dagger(t, t')\, T(t, t') = \hat{1}. \tag{4.39}$$

As in the previous section we can express a unitary operator in terms of a Hermitian generator. A suitable solution for T, satisfying both (4.37) and(4.39), is given by the expression

$$T = e^{-iH(t-t')} \equiv e^{-i\frac{H}{\hbar}(t-t')} \tag{4.40}$$

where H is a Hermitian operator with the dimensions of energy.

If equation (4.36) is differentiated with respect to time we obtain

$$i \frac{\partial}{\partial t} |\Psi_S(t)\rangle = HT(t, t')\, |\Psi_S(t')\rangle = H\, |\Psi_S(t)\rangle. \tag{4.41}$$

This is Schrödinger's equation, and we may identify H as the total Hamiltonian for the system.

The equation (4.41) satisfies the following operator equation:

$$i \frac{\partial T}{\partial t} = HT. \tag{4.42}$$

This equation, together with its conjugate forms

$$-i \frac{\partial T^\dagger}{\partial t} = (HT)^\dagger = T^\dagger H \tag{4.43}$$

$$-i \frac{\partial T^{-1}}{\partial t} = T^{-1} H$$

will be used in later sections.

4.1 (f.2). *The Heisenberg representation.* In contrast to the Schrödinger representation, the Heisenberg scheme describes the temporal development of a physical system in terms of operators which are dependent upon time, and state vectors which are independent of time. In what follows we shall write the Heisenberg state vector as $|\Psi_H\rangle$; since it is independent of time, the functional term t is unnecessary. The operators will be labelled appropriately.

We shall assume that the Heisenberg and Schrödinger representations coincide at some arbitrary time t'. Then at any subsequent time t we may write

$$|\Psi_S(t)\rangle = T(t, t')\, |\Psi_S(t')\rangle = T(t, t')\, |\Psi_H\rangle. \tag{4.44}$$

Since the expectation value of an observable must be the same in any representation, we may write

$$\begin{aligned} \langle A\rangle &= \langle \Psi_S(t)|\, A_S\, |\Psi_S(t)\rangle \\ &= \langle \Psi_H|\, T^{-1}(t, t')\, A_S T(t, t')\, |\Psi_H\rangle \\ &= \langle \Psi_H|\, A_H(t)\, |\Psi_H\rangle \end{aligned} \tag{4.45}$$

here use has been made of the unitary property of T (equation (4.39)). Thus the relationship between Schrödinger and Heisenberg operators is

$$A_H(t) = T^{-1}(t, t')\, A_S T(t, t') = T^{-1}(t, t')\, A_H(t')\, T(t, t'). \tag{4.46}$$

It should be noted that the Hamiltonian operator is the same in both Heisenberg and Schrödinger representations. If the operators coincide at some arbitrary time t', as we have already assumed, then by (4.46)

$$H_H(t' + \delta t) = T^{-1}(t' + \delta t, t')H_S T(t' + \delta t, t') = T^{-1}H_H(t')T,$$

ut by (4.40)

$$T(t' + \delta t, t') = \hat{1} - iH_S\delta t$$

f

$$t - t' = \delta t \to 0.$$

n the above equations we have put a subscript S on the Hamiltonian of 4.40) to indicate that it arose in the Schrödinger representation. The equation or $H_H(t' + \delta t)$ now becomes

$$\begin{aligned} H_H(t' + \delta t) &= (\hat{1} + iH_S\delta t)\, H_S(\hat{1} - iH_S\delta t) \\ &= H_S + i(H_S - H_S)\, H_S\delta t \\ &= H_S = H_H(t') \end{aligned} \tag{4.47}$$

when terms of order $(\delta t)^2$ have been neglected. In future, therefore, we will designate the Hamiltonian in both Schrödinger and Heisenberg representations by he symbol H

$$H = H_H(t' + \delta t) = H_H(t') = H_S$$

and so

$$T^{-1}HT = H. \tag{4.48}$$

We now return to equation (4.46) to obtain the equation of motion for a quantum mechanical system in the Heisenberg representation. Consider the variation of the Heisenberg operator of equation (4.46) with time

$$\begin{aligned} i\hbar \frac{d}{dt} A_H(t) &\equiv i \frac{d}{dt} A_H(t) \\ &= i \frac{dT^{-1}}{dt} A_S T + iT^{-1} A_S \frac{dT}{dt} \\ &= -T^{-1}HA_S T + T^{-1}A_S HT \\ &= -T^{-1}HTT^{-1}A_S T + T^{-1}A_S TT^{-1}HT \\ &= -HA_H(t) + A_H(t)H \\ &= [A_H(t), H] \end{aligned} \tag{4.49}$$

where use has been made of equations (4.42), (4.43) and (4.48).

Equation (4.49) shows that providing A does not depend explicitly on time then $dA_H/dt = 0$ if A commutes with H. Thus if the commutation condition is fulfilled, A represents a quantity which is conserved in a given physical system and it is called a *constant of the motion*. The eigenvalues of A are called *good quantum numbers*.

4.1 (g). *Conservation laws in quantum mechanical systems*

Equation (4.49) is of great importance in determining the quantities which may be conserved in quantum mechanical systems. It represents the quantum mechanical equivalent of Noether's theorem (§ 3.4(e)).

The invariance property may also be linked with the unitary transformation operator U. A unitary transformation does not change the dynamical properties of a system, and hence the operator controlling the temporal development will not change under this transformation

$$UT(t, t')\, U^{-1} = T(t, t')$$

so that

$$UT = TU. \tag{4.50}$$

But with the aid of equation (4.46) we can also write

$$U(t) = T^{-1}(t, t')\, U(t')\, T(t, t')$$

this equation can only be compatible with (4.50) if

$$U(t') = U(t) \tag{4.51}$$

that is

$$i\frac{dU}{dt} = [U, H] = 0. \tag{4.52}$$

It was stated in § 4.1 (c) that U is not necessarily a Hermitian operator and therefore an observable. However, it was shown in § 4.1 (e) that U can be written as $U = \hat{1} - iF$, where F is a Hermitian generator; it is clear that if U commutes with H so also will F

$$[F, H] = 0. \tag{4.53}$$

In Chapter 3 it was shown that the principle of least action led to conservation laws. The commutation of the generator F with H also yields a conservation law, since dF/dt then equals zero. In Chapter 5 we will show that F and the term C_λ of equation (3.245) are equivalent.

4.2. QUANTISED FIELD THEORIES

4.2(a). *Introduction*

We are now in a position to merge the classical field theory developed in the previous chapter with the quantum mechanical principles to yield a quantised field theory. For a classical field we regarded the field function $\eta_\sigma(x)$ as a series of generalised coordinates expressing the propagation of a disturbance; we also showed that certain conserved quantities were associated with this disturbance, e.g. energy and momentum.

A quantised field theory may be constructed by making certain assumptions about $\eta_\sigma(x)$; the assumptions evolve, in part, from the principles which we find necessary for the satisfactory working of a quantum mechanical theory. The assumptions are the following.

(1) In a quantised field theory the field functions $\eta_\sigma(x)$ are no longer classical numbers (*c-numbers*), but instead are linear operators (*q-numbers*). They will be called, therefore, *field operators.*

(2) The field operators may act on state vectors representing an assembly of particles and cause the quantum states of that assembly to change.

(3) Certain commutation relations exist between the operators. The establishment of the commutation relations between field operators is referred to as the quantisation of the field.†

(4) The eigenvalues of Hermitian operators constructed from the field functions, for example energy and momentum, represent observable quantities which may be measured by experiment.

The establishment of a quantum field theory according to the above principles is sometimes called *second quantisation.*

4.2(b). *Types of quantised field theories*

The initial process of setting up a quantised field equation is similar to that described for a classical field in Chapter 3. A suitable Lagrangian is chosen, and then by a application of the principle of least action an Euler–Lagrange equation is obtained. The terms appearing however, in this equation are no longer c-numbers; instead they represent operators.

The form of the Lagrangian is determined by the properties which we wish to incorporate in the field equation. Certain restrictions are imposed, however, on the form of $\mathscr{L}$; they are as follows.

(1) The Lagrangian density $\mathscr{L}$ must be invariant under proper Lorentz transformations; that is, it behaves as a scalar under these transformations (§ 3.4(c.2)).

(2) The requirement of Lorentz invariance means that the field equations must be covariant. The Lagrangian density must assume a form which is appropriate to satisfy this condition.

† In fact assumption (3) follows from assumption (2) or vice-versa.

(3) The Lagrangian density $\mathscr{L}$ should not contain differentials of higher order than one (§ 3.4(c.2)).

(4) The terms appearing in $\mathscr{L}$ should give it the dimensions of an energy density; this statement follows from equation (3.231). This equation also contains the implication that $\mathscr{L}$ should be real since the energy associated with a field is an observable quantity.

In addition to these conditions, the 'hopeful' conditions are also used. These are based on the hope that the field equations will be simple.

(5) The field equations will be linear in the field operators $\eta_\sigma(x)$. Since these equations are obtained by means of the Euler–Lagrange equation, this statement implies that $\mathscr{L}$ should be constructed from at most bilinear combinations of the terms $\eta_\sigma(x)$ and their first derivatives. The combinations should be, of course scalar quantities (condition (1)).

(6) The fields are *local*, that is the state of the field and its interactions at a given space time point x are completely determined by the field operators $\eta_\sigma(x)$ and their derivatives evaluated at that point.

In addition to simplicity the linear field equation obeys the principle of superposition, that is the sum of the solutions of the field equation is also a solution – a principle which is used for the fields describing macroscopic and microscopic systems, and which appears to be borne out by experiment. Such a simple situation may not exist for sub-microscopic systems, and many workers have produced nonlinear field theories. These theories appear to offer no significant advantage over the linear theories in more common use.

Three fields are used normally to describe the properties of elementary particles. They are the vector field (which is used for photons and mesons with spin one), the Klein–Gordon field (for bosons of zero spin) and the Dirac field (for leptons, nucleons and hyperons).

Major problems arise in the description of particles and their interactions by fields; although a large measure of success has been achieved in describing electromagnetic and weak interactions, only partial success has been obtained for strong interactions. Although the difficulties may lie with the mathematical techniques at present available, the possibility exists that the field equations in use at present are not suitable. In what follows, however, it will be assumed that conventional field theory is adequate.

4.3. THE KLEIN–GORDON (SCALAR) FIELD

4.3(a). *The single component Hermitian field*

The field equation developed in this section is suitable for a spinless particle of charge zero which is, in every way, indistinguishable from its antiparticle. These properties will be demonstrated in Chapter 5. A field equation of this type is useful for describing π^0 mesons.

We start by considering a scalar or Klein–Gordon field $\varphi(x)$, which by denition of the word scalar implies that the field is invariant under proper Lorentz transformations

$$\varphi(x) \rightarrow \varphi'(x') = \varphi(x). \tag{4.54}$$

The Langrangian $\mathscr{L}$ must be constructed from scalar terms; the only ones which can be formed which satisfy the requirements of § 4.2 are $a(\partial\varphi/\partial x_\lambda \cdot \partial\varphi/\partial x_\lambda)$ and $b\varphi^2$, where a and b are arbitrary constants. Thus we write

$$\mathscr{L} = a\frac{\partial\varphi}{\partial x_\lambda}\frac{\partial\varphi}{\partial x_\lambda} + b\varphi^2. \tag{4.55}$$

Upon application of the Euler–Lagrange equation (3.228)

$$\frac{\partial\mathscr{L}}{\partial\varphi} - \frac{\partial}{\partial x_\lambda}\cdot\frac{\partial\mathscr{L}}{\partial\left(\frac{\partial\varphi}{\partial x_\lambda}\right)} = 0$$

we obtain

$$2b\varphi - 2a\frac{\partial^2\varphi}{\partial x_\lambda^2} = 0.$$

Thus if we set

$$b/a = m^2$$

we obtain the equivalent of the Klein–Gordon equation (3.73)

$$(\Box^2 - m^2)\varphi = 0 \tag{4.56}$$

It must be remembered that the term φ now represents an operator instead of a wave function.

4.3(b). *The plane wave representation of* φ

The simplest solution for φ which will satisfy equation (4.56) is

$$\varphi = Ce^{ikx} \tag{4.57}$$

where C represents a term independent of x, and k is a four-momentum ($\hbar = c = 1$ units). However, a more complete decomposition of φ into plane waves may be made. Since (4.56) is a linear equation, a linear combination of plane waves is also a solution. Consider the Fourier transform of $\varphi(x)$ in momentum (k-) space

$$\varphi(x) = \frac{1}{(2\pi)^2}\int_{-\infty}^{+\infty} d^4k C(k)\,\delta(k^2 + m^2)\,e^{ikx} \tag{4.58}$$

where

$$d^4k = \frac{1}{i}\, d\mathbf{k}\, dk_4 = d\mathbf{k}\, d\omega$$

and the δ-function is four-dimensional.

Since $\varphi(x)$ is a function of four-dimensional Minkowski space, equation (4.58) also has been made four-dimensional. It is a simple matter to verify that equation (4.58) satisfies (4.56)

$$(\Box^2 - m^2)\,\varphi = \frac{-1}{(2\pi)^2}\int_{-\infty}^{+\infty} d^4k C(k)\,(k^2 + m^2)\,\delta(k^2 + m^2)\, e^{ikx} = 0. \qquad (4.59)$$

This result follows from the fact that the integrand of (4.59) contains a term of the form $x\,\delta(x)$. This quantity is of magnitude zero (compare A.6 (Appendixes, p. 703)).

The presence of the δ-function in (4.58) may be removed by an integration over ω. We introduce the notation

$$k^2 = \mathbf{k}^2 + k_4^2 = \mathbf{k}^2 - \omega^2, \qquad kx = \mathbf{k}\cdot\mathbf{x} - \omega t \qquad (4.60)$$

then

$$\delta(k^2 + m^2) = \delta(\mathbf{k}^2 + k_4^2 + m^2) = \delta(\mathbf{k}^2 + m^2 - \omega^2)$$

$$= \delta(\omega_k^2 - \omega^2) = \frac{1}{2\omega_k}\,[\delta(\omega + \omega_k) + \delta(\omega - \omega_k)] \qquad (4.61)$$

where we have used $\omega_k = +\sqrt{(\mathbf{k}^2 + m^2)}$; the positive root is selected for ω_k by virtue of the property of the δ-function (compare A.6). The solution (4.58) for $\varphi(x)$ thus becomes

$$\varphi(x) = \frac{1}{(2\pi)^2}\int_{-\infty}^{+\infty} d\mathbf{k}\, d\omega\, \frac{1}{2\omega_k}\,[\delta(\omega + \omega_k) + \delta(\omega - \omega_k)]\, C(\mathbf{k}, \omega)\, e^{i(\mathbf{k}\cdot\mathbf{x} - \omega t)}$$

$$= \frac{1}{(2\pi)^2}\int_{-\infty}^{+\infty} \frac{d\mathbf{k}}{2\omega_k}\,[C(\mathbf{k}, -\omega_k)\, e^{i(\mathbf{k}\cdot\mathbf{x} + \omega_k t)} + C(\mathbf{k}, \omega_k)\, e^{i(\mathbf{k}\cdot\mathbf{x} - \omega_k t)}]. \qquad (4.62)$$

We note that since the integral over $d\mathbf{k}$ goes from $-\infty$ to $+\infty$ the change

$$C(\mathbf{k}, -\omega_k)\, e^{i\mathbf{k}\cdot\mathbf{x}} \to C(-\mathbf{k}, -\omega_k) e^{-i\mathbf{k}\cdot\mathbf{x}} \qquad (4.63)$$

may be made. In addition by virtue of equation (4.61) ω_k is a function of $\mathbf{k}$, and so we may write

$$C(\mathbf{k}, \omega_k) = C(\mathbf{k}) \qquad (4.64)$$

$$C(-\mathbf{k}, -\omega_k) = C(-\mathbf{k});$$

hus the term in the square brackets in (4.62) becomes

$$[C(-\mathbf{k})\, e^{-i(\mathbf{k}\cdot\mathbf{x}-\omega_k t)} + C(\mathbf{k})\, e^{i(\mathbf{k}\cdot\mathbf{x}-\omega_k t)}] \equiv [C(-\mathbf{k})\, e^{-ikx} + C(\mathbf{k})\, e^{ikx}]$$

vhere we have reverted to imaginary fourth components in the exponential. Thus equation (4.62) becomes

$$\varphi(x) = \frac{1}{(2\pi)^2} \int_{-\infty}^{+\infty} \frac{d\mathbf{k}}{2\omega_k} [C(-\mathbf{k})\, e^{-ikx} + C(\mathbf{k})\, e^{ikx}]. \tag{4.65}$$

Finally, let us consider the Hermitian character of $\varphi(x)$

$$\varphi^\dagger(x) = \frac{1}{(2\pi)^2} \int_{-\infty}^{+\infty} \frac{d\mathbf{k}}{2\omega_k} [C^\dagger(-\mathbf{k})\, e^{ikx} + C^\dagger(\mathbf{k})\, e^{-ikx}] = \varphi(x).$$

If this equation is compared with (4.65) we find the requirement

$$C^\dagger(\mathbf{k}) = C(-\mathbf{k}), \qquad C^\dagger(-\mathbf{k}) = C(\mathbf{k}) \tag{4.66}$$

if $\varphi = \varphi^\dagger$. For purposes of normalisation it is customary to write

$$C(\mathbf{k}) = \sqrt{(4\pi\omega_k)}\, a(\mathbf{k}), \qquad C^\dagger(\mathbf{k}) = \sqrt{(4\pi\omega_k)}\, a^\dagger(\mathbf{k}) \tag{4.67}$$

hence, upon assembling terms from (4.65), (4.66) and (4.67), we find

$$\varphi(x) = \frac{1}{(2\pi)^{3/2}} \int_{-\infty}^{+\infty} \frac{d\mathbf{k}}{\sqrt{(2\omega_k)}} [a(\mathbf{k})\, e^{ikx} + a^\dagger(\mathbf{k})\, e^{-ikx}]. \tag{4.68}$$

This equation represents a Lorentz invariant decomposition of the scalar field operator: it can be written in a number of ways; for example it is sometimes separated into its positive and negative frequency components

$$\varphi(x) = \varphi^{(-)}(x) + \varphi^{(+)}(x) \tag{4.69}$$

where

$$\varphi^{(-)}(x) = \frac{1}{(2\pi)^{3/2}} \int_{-\infty}^{+\infty} \frac{d\mathbf{k}}{\sqrt{(2\omega_k)}}\, a(\mathbf{k})\, e^{ikx} \tag{4.70}$$

$$\varphi^{(+)}(x) = \frac{1}{(2\pi)^{3/2}} \int_{-\infty}^{+\infty} \frac{d\mathbf{k}}{\sqrt{(2\omega_k)}}\, a^\dagger(\mathbf{k})\, e^{-ikx}.$$

It is easily verified that

$$\varphi^{(-)\dagger} = \varphi^{(+)}$$

as the Hermitian character of $\varphi(x)$ requires. The properties of the operators $\varphi^{(-)}(x)$ and $\varphi^{(+)}(x)$ will be specified in equations (4.88) to (4.90).

It should be noted that the definitions of $\varphi^{(+)}$ and $\varphi^{(-)}$ are not unique; some authors write $\varphi^{(+)}$ for the term containing e^{ikx}.

For convenience of computation the notation of equations (4.68) and (4.70) is changed frequently into that for a system in which the field is enclosed in a large box with volume V and sides of length L. If periodic boundary conditions are then imposed, the momentum $\mathbf{k}$ becomes discrete and obeys the relations

$$k_i = \frac{2\pi}{L} n_i \qquad (i = 1, 2, 3) \tag{4.71}$$

$$n_i = 0, 1, 2, 3, \ldots$$

$$\frac{1}{(2\pi)^{3/2}} \int d\mathbf{k} \to \frac{1}{\sqrt{V}} \sum_{\mathbf{k}}, \qquad \delta(\mathbf{k} - \mathbf{k}') \to \delta_{\mathbf{k}\mathbf{k}'}.$$

Thus equation (4.68) becomes

$$\varphi(x) = \frac{1}{\sqrt{V}} \sum_{\mathbf{k}} \frac{1}{\sqrt{(2\omega_k)}} [a(\mathbf{k})\, e^{ikx} + a^\dagger(\mathbf{k})\, e^{-ikx}]. \tag{4.72}$$

4.3(c). *Quantisation of the scalar field*

So far we have decomposed the scalar field into plane wave momentum states in a covariant manner. The quantisation of the field is achieved by using the second proposition in § 4.2(a), namely that the field operators can act upon state vectors representing an assembly of particles and cause the quantum states of that assembly to change.

It should be noted that we are carrying out the quantisation in reverse order to the method adopted by many authors. The quantisation is frequently carried out by considering the commutation relations and then determining the properties of the field operators in an occupation number space. The reverse procedure is adopted here in order to show at an early stage that the isolated fields behave like an assembly of non-interacting particles. We shall examine the commutation relations in more detail in § 4.6.

We start by defining a normalised state vector

$$|n_{\mathbf{k}_1} n_{\mathbf{k}_2} n_{\mathbf{k}_3} \ldots\rangle$$

in an *occupation number space*. In this ket the symbol $n_{\mathbf{k}_i}$ represents the (integral) number of particles with momentum $\mathbf{k}_i$. The normalisation condition is chosen so that

$$\langle n'_{\mathbf{k}_1} n'_{\mathbf{k}_2} \ldots n'_{\mathbf{k}_i} \ldots | n_{\mathbf{k}_1} n_{\mathbf{k}_2} \ldots n_{\mathbf{k}_i} \ldots\rangle = \delta_{n'_{\mathbf{k}_1} n_{\mathbf{k}_1}} \ldots \delta_{n'_{\mathbf{k}_i} n_{\mathbf{k}_i}} \ldots \tag{4.73}$$

For convenience we will consider only one momentum state, and write the ket as

$\ldots n_k \ldots\rangle$ The second quantisation of the field is now carried out by defining the :rms $a(\mathbf{k})$ and $a^\dagger(\mathbf{k})$ in $\varphi(x)$ (4.68) as operators, which have the unique property of ıducing the following transformations of the ket

$$a(\mathbf{k})\,|\ldots n_\mathbf{k}\ldots\rangle = \sqrt{n_\mathbf{k}}\,|\ldots (n-1)_\mathbf{k}\ldots\rangle \tag{4.74}$$

$$a^\dagger(\mathbf{k})\,|\ldots n_\mathbf{k}\ldots\rangle = \sqrt{(n+1)_\mathbf{k}}\,|\ldots (n+1)_\mathbf{k}\ldots\rangle. \tag{4.75}$$

'hus the operator $a(\mathbf{k})$ destroys one particle in the occupation number state ith momentum $\mathbf{k}$, whilst $a^\dagger(\mathbf{k})$ creates a particle in the same state. The terms $(\mathbf{k})$ and $a^\dagger(\mathbf{k})$ are called, respectively, *destruction* and *creation operators*.

Thus $a(\mathbf{k})$ and $a^\dagger(\mathbf{k})$ have the following matrix elements:

$$\begin{aligned}\langle\ldots n'_{\mathbf{k}_i}\ldots(n'_\mathbf{k}-1)\ldots|a(\mathbf{k})|\ldots n_{\mathbf{k}_i}\ldots n_\mathbf{k}\ldots\rangle \\ = \sqrt{n_\mathbf{k}}\ldots\delta_{n'_{\mathbf{k}_i}n_{\mathbf{k}_i}}\ldots\delta_{n'_\mathbf{k}n_\mathbf{k}}\ldots\end{aligned} \tag{4.76}$$

$$\begin{aligned}\langle\ldots n'_{\mathbf{k}_i}\ldots n'_\mathbf{k}\ldots|a^\dagger(\mathbf{k})|\ldots n_{\mathbf{k}_i}\ldots(n_\mathbf{k}-1)\ldots\rangle \\ = \sqrt{n_\mathbf{k}}\ldots\delta_{n'_{\mathbf{k}_i}n_\mathbf{k}}\ldots\delta_{n'_\mathbf{k}n_\mathbf{k}}\ldots.\end{aligned}$$

The double action of the operators induces the following changes:

$$\begin{aligned}a(\mathbf{k})a^\dagger(\mathbf{k})\,|\ldots n_\mathbf{k}\ldots\rangle &= \sqrt{(n+1)_\mathbf{k}}\,a(\mathbf{k})\,|\ldots(n+1)_\mathbf{k}\ldots\rangle \\ &= (n+1)_\mathbf{k}\,|\ldots n_\mathbf{k}\ldots\rangle \\ a^\dagger(\mathbf{k})\,a(\mathbf{k})\,|\ldots n_\mathbf{k}\ldots\rangle &= \sqrt{n_\mathbf{k}}\,a^\dagger(\mathbf{k})\,|\ldots(n-1)_\mathbf{k}\ldots\rangle \\ &= n_\mathbf{k}\,|\ldots n_\mathbf{k}\ldots\rangle.\end{aligned} \tag{4.77}$$

The combination of the two equations thus yields the relation

$$[a(\mathbf{k})\,a^\dagger(\mathbf{k}) - a^\dagger(\mathbf{k})\,a(\mathbf{k})]\,|\ldots n_\mathbf{k}\ldots\rangle = |\ldots n_\mathbf{k}\ldots\rangle.$$

On the other hand, the operations

$$a(\mathbf{k})\,a^\dagger(\mathbf{k}')\,|\ldots n_\mathbf{k}\ldots n_{\mathbf{k}'}\ldots\rangle = \sqrt{n_\mathbf{k}}\,\sqrt{(n+1)_{\mathbf{k}'}}\,|\ldots(n-1)_\mathbf{k}\ldots(n+1)_{\mathbf{k}'}\ldots\rangle$$

$$a^\dagger(\mathbf{k}')\,a(\mathbf{k})\,|\ldots n_\mathbf{k}\ldots n_{\mathbf{k}'}\ldots\rangle = \sqrt{(n+1)_{\mathbf{k}'}}\,\sqrt{n_\mathbf{k}}\,|\ldots(n-1)_\mathbf{k}\ldots(n+1)_{\mathbf{k}'}\ldots\rangle$$

show that $a(\mathbf{k})$ and $a^\dagger(\mathbf{k}')$ commute, since $n_\mathbf{k}$ and $n_{\mathbf{k}'}$ are entirely unrelated

$$[a(\mathbf{k})\,a^\dagger(\mathbf{k}') - a^\dagger(\mathbf{k}')\,a(\mathbf{k})] = 0.$$

Thus we obtain the relation

$$[a(\mathbf{k})\,a^\dagger(\mathbf{k}') - a^\dagger(\mathbf{k}')\,a(\mathbf{k})] = [a(\mathbf{k}), a^\dagger(\mathbf{k})'] = \delta_{\mathbf{k}\mathbf{k}'} \quad\text{or}\quad \delta(\mathbf{k}-\mathbf{k}') \tag{4.78}$$

depending on whether the system is a discrete or continuous one. In a simila manner the following relations can be obtained

$$[a(\mathbf{k}), a(\mathbf{k}')] = [a^\dagger(\mathbf{k}), a^\dagger(\mathbf{k}')] = 0. \qquad (4.79$$

4.3(d). *The occupation number operator*

The product of $a^\dagger(\mathbf{k})$ and $a(\mathbf{k})$ is called the *occupation number operator*

$$N(\mathbf{k}) = a^\dagger(\mathbf{k})a(\mathbf{k}). \qquad (4.80$$

The commutation relations (4.78) and (4.79) allow us to associate the follow ing properties with $N(\mathbf{k})$

$$[N(\mathbf{k}), N(\mathbf{k}')] = 0 \qquad (4.81$$

$$[a(\mathbf{k}), N(\mathbf{k})] = a(\mathbf{k})$$

$$[a^\dagger(\mathbf{k}), N(\mathbf{k})] = -a^\dagger(\mathbf{k}).$$

From equations (4.77) we can write

$$N(\mathbf{k})\,|\ldots n_\mathbf{k} \ldots\rangle = n_\mathbf{k}\,|\ldots n_\mathbf{k} \ldots\rangle$$

or in more symbolic language

$$N(\mathbf{k})\,|\Psi\rangle = n_\mathbf{k}\,|\Psi\rangle.$$

Now since $N(\mathbf{k})$ represents the product of an operator times its Hermitiar conjugate, it is positive definite. Thus the expectation value of $N(\mathbf{k})$ must als be positive definite if it operates on state vectors in a Hilbert space with positiv definite metric (§ 4.1(b))

$$\langle\Psi|\,N(\mathbf{k})\,|\Psi\rangle = n_\mathbf{k}\,\langle\Psi\,|\,\Psi\rangle \geqq 0$$

for

$$\langle\Psi\,|\,\Psi\rangle \geqq 0.$$

The positive definiteness of $N(\mathbf{k})$ allows us to define a vacuum state. The equations (4.81) can be made to yield the following relations:

$$N(\mathbf{k})\,(a^\dagger(\mathbf{k})\,|\Psi\rangle) = (n+1)_\mathbf{k}\,(a^\dagger(\mathbf{k})\,|\Psi\rangle) \qquad (4.82)$$

$$N(\mathbf{k})\,(a(\mathbf{k})\,|\Psi\rangle) = (n-1)_\mathbf{k}\,(a(\mathbf{k})\,|\Psi\rangle)$$

thus $a^\dagger(\mathbf{k})\,|\Psi\rangle$ and $a(\mathbf{k})\,|\Psi\rangle$ are eigenvectors of $N(\mathbf{k})$ with eigenvalues $(n+1)_\mathbf{k}$ and $(n-1)_\mathbf{k}$ respectively. But we have shown already that $N(\mathbf{k})$ must have positive eigenvalues. Therefore the eigenvalue $(n-1)$ must always remain positive: in order for this condition to be obeyed a state $|\Psi_0\rangle$ must exist with the property

$$a(\mathbf{k})\,|\Psi_0\rangle = 0 \qquad (4.83)$$

nd therefore

$$N(\mathbf{k})\,|\Psi_0\rangle = 0.$$

The state $|\Psi_0\rangle$ is called the *vacuum state*. From this state we can build up tates containing 1, 2, 3, ... particles. Thus the state containing 1 particle is given y

$$|\mathbf{k}\rangle \equiv |\Psi(1_\mathbf{k})\rangle = a^\dagger(\mathbf{k})\,|\Psi_0\rangle. \tag{4.84}$$

The simpler notation of the first term will normally be used when it does not ntroduce ambiguities. The state will have eigenvalue 1 for the operator $N(\mathbf{k})$

$$N(\mathbf{k})\,|\Psi(1_\mathbf{k})\rangle = 1_\mathbf{k}\,|\Psi(1_\mathbf{k})\rangle. \tag{4.85}$$

Similarly the two particle state

$$|\Psi(2_\mathbf{k})\rangle = a^\dagger(\mathbf{k})a^\dagger(\mathbf{k})\,|\Psi_0\rangle$$

satisfies the eigenvalue equation

$$N(\mathbf{k})\,|\Psi(2_\mathbf{k})\rangle = 2_\mathbf{k}\,|\Psi(2_\mathbf{k})\rangle.$$

For simplicity we will write the n-particle state as $|n_\mathbf{k}\rangle$

$$N(\mathbf{k})\,|n_\mathbf{k}\rangle = n_\mathbf{k}\,|n_\mathbf{k}\rangle. \tag{4.86}$$

This state is normalised, as we indicated in equation (4.73),

$$\langle n_\mathbf{k}|n_\mathbf{k}\rangle = 1$$

and the normalised eigenvectors are easily built up from the vacuum state with the aid of equations (4.75) and (4.84)

$$|n_\mathbf{k}\rangle = \frac{1}{\sqrt{(n_\mathbf{k}!)}}\,[a^\dagger(\mathbf{k})]^{n_\mathbf{k}}\,|\Psi_0\rangle. \tag{4.87}$$

In calculations involving real physical processes the operators $a(\mathbf{k})$ and $a^\dagger(\mathbf{k})$ normally appear associated with $\varphi^{(-)}(x)$ and $\varphi^{(+)}(x)$ respectively (4.70). It is apparent from equation (4.83) that

$$\varphi^{(-)}(x)\,|\Psi_0\rangle = 0 \tag{4.88}$$

for all values of x; the zero value also holds for the Hermitian conjugate form

$$\langle\Psi_0|\,\varphi^{(+)}(x) = 0. \tag{4.89}$$

We note, finally, that the operator $a^\dagger(\mathbf{k})$ may be subjected to a Fourier transformation, and hence it may be shown that the action of $\varphi^{(+)}(x)$ on the vacuum state

$$\varphi^{(+)}(x)\,|\Psi_0\rangle = \frac{1}{(2\pi)^{3/2}}\int d\mathbf{k}\,e^{-ikx}\,|\mathbf{k}\rangle \equiv \frac{1}{\sqrt{V}}\sum_\mathbf{k} e^{-ikx}\,|\mathbf{k}\rangle \tag{4.90}$$

creates a particle within a distance $\sim 1/m$ (the Compton wavelength) from th space point $\mathbf{x}$.

4.3(e). *The decomposition of the scalar field into spherical waves*

For many problems in the physics of elementary particles, for example pion-nucleon scattering, an expansion of the scalar field into spherical waves is more useful than one into plane waves. Consider the Klein–Gordon equation

$$(\Box^2 - m^2)\,\varphi = 0.$$

We may split this equation into space and time parts

$$(\Box^2 - m^2)\,\varphi = \left(\nabla^2 - \frac{\partial^2}{\partial t^2}\right)\varphi_s\varphi_t = (E^2 - \mathbf{k}^2)\,\varphi_s\varphi_t = 0 \tag{4.91}$$

where φ_s and φ_t represent the space and time components of φ respectively. In order to avoid a clumsy notation, we shall use the following symbols in the remainder of this section:

$$k \equiv |\mathbf{k}| \qquad kr \equiv |\mathbf{k}|\,|\mathbf{x}|. \tag{4.92}$$

The space part of φ satisfies the equation

$$(\nabla^2 + k^2)\,\varphi_s = 0. \tag{4.93}$$

If the operator ∇^2 is expressed in polar coordinates, a suitable solution for φ_s is given by an expression which behaves like

$$\varphi_s \sim g_{lk}(r)\,Y_l^m(\theta, \varphi) \tag{4.94}$$

where $g_{lk}(r)$ is a normalised spherical Bessel function

$$g_{lk}(r) = k\sqrt{(2/R)}\,j_l(kr) \tag{4.95}$$

$$\int_0^R dr r^2 [g_{lk}(r)]^2 = 1 \tag{4.96}$$

the normalisation being carried out in a large sphere of radius R. The term $Y_l^m(\theta, \varphi)$ is a spherical harmonic; its mathematical properties are examined in A.7 (Appendixes, p. 704). The angles θ and φ represent the polar direction of $\mathbf{r}$.

A suitable solution for the Klein–Gordon equation may now be developed. The solution to the time part will be of the form

$$\varphi_t \sim e^{ik_4x_4} \equiv e^{-i\omega_k t}.$$

This expression may be combined with equation (4.94); in addition an operator term must be introduced since $\varphi(x)$ is no longer a wave function, and finally we recall that $\varphi(x)$ should be Hermitian. A suitable form for $\varphi(x)$ is therefore

$$\varphi(x) = \sum_{klm} [C(klm)\, g_{lk}(r) Y_l^m(\theta, \varphi)\, e^{-i\omega_k t} + C^\dagger(klm)\, g_{lk}(r) Y_l^{m*}(\theta, \varphi)\, e^{i\omega_k t}]$$
$$\equiv \varphi^{(-)} + \varphi^{(+)} \tag{4.97}$$

where the summation sign $\sum_{klm}$ implies

$$\sum_{klm} = \sum_k \sum_{l=0}^{\infty} \sum_{m=-l}^{m=+l}.$$

Since

$$Y_l^{m*}(\theta, \varphi) = (-1)^m Y_l^{-m}(\theta, \varphi) \tag{4.98}$$

the Hermitian character of φ can be preserved only if

$$C^\dagger(klm) = (-1)^m C(kl - m). \tag{4.99}$$

It is convenient to introduce a denominator $\sqrt{(2\omega_k)}$ as before, so that we will write

$$C(klm) = \frac{1}{\sqrt{(2\omega_k)}}\, a(klm), \quad C^\dagger = \frac{1}{\sqrt{(2\omega_k)}}\, a^\dagger(klm).$$

Thus the expression for $\varphi(x)$ in spherical waves becomes

$$\varphi(x) = \sum_{klm} \frac{1}{\sqrt{(2\omega_k)}}\, g_{lk}(r)\, [a(klm) Y_l^m(\theta, \varphi)\, e^{-i\omega_k t} + a^\dagger(klm) Y_l^{m*}(\theta, \varphi)\, e^{i\omega_k t}] \tag{4.100}$$

where $a^\dagger$ and a are creation and destruction operators with equivalent commutation relations to (4.78) and (4.79). A reminder that the above expression is not covariant is scarcely necessary.

The behaviour of the spherical field operator has been examined in some detail by Hamilton (1956, 1959). Hamilton shows that a fairly simple relation may be established between $a(\mathbf{k})$ in the plane wave representation and $a(klm)$, if the space part of the exponential in $\varphi^{(-)}(x)$ is expanded into spherical harmonics

$$e^{i\mathbf{k}\cdot\mathbf{x}} = \sqrt{(8R)}\, \frac{\pi}{k} \sum_{l=0}^{\infty} \sum_{m=-l}^{m=+l} i^l g_{lk}(r) Y_l^m(\theta, \varphi) Y_l^{m*}(\alpha, \beta) \tag{4.101}$$

where α and β refer to the polar direction of $\mathbf{k}$. An integration over the volume V then leads to the following relation between $a(\mathbf{k})$ and $a(klm)$:

$$a(\mathbf{k}) = \frac{\pi}{k} \sqrt{\left(\frac{8R}{V}\right)} \sum_{l=0}^{\infty} \sum_{m=-l}^{m=+l} (-i)^l\, Y_l^m(\alpha, \beta)\, a(klm) \tag{4.102}$$

$$a^\dagger(\mathbf{k}) = \frac{\pi}{k} \sqrt{\left(\frac{8R}{V}\right)} \sum_{l=0}^{\infty} \sum_{m=-l}^{m=+l} i^l Y_l^{m*}(\alpha, \beta) a^\dagger(klm).$$

From the second equation it is apparent that single particle states in the plane and spherical wave representations are related by the expression

$$|\mathbf{k}\rangle = \frac{\pi}{k}\sqrt{\left(\frac{8R}{V}\right)}\sum_{l=0}^{\infty}\sum_{m=-l}^{m=+l} i^l Y_l^{m*}(\mathbf{k})\,|klm\rangle \tag{4.103}$$

where we have introduced the notation

$$Y_l^{m*}(\mathbf{k}) = Y_l^{m*}(\alpha, \beta) \tag{4.104}$$

since the angles α and β refer to the direction of the momentum vector.

One state which is of particular interest has been evaluated by Hamilton (1956). It is the state representing two particles in their centre of momentum frame (§ 3.2(h))

$$|\mathbf{k}, -\mathbf{k}\rangle = a^\dagger(\mathbf{k})\,a^\dagger(-\mathbf{k})\,|\Psi_0\rangle.$$

In spherical coordinates the equivalent state with angular momentum quantum numbers L and M is given by Hamilton as

$$|kLM\rangle = \int d\Omega_{\mathbf{k}} Y_L^M(\mathbf{k})\,a^\dagger(\mathbf{k})\,a^\dagger(-\mathbf{k})\,|\Psi_0\rangle \tag{4.105}$$

where $d\Omega_{\mathbf{k}}$ is an element of solid angle around $\mathbf{k}$.

The expansion of the field operators need not be restricted to operators for plane or spherical waves; from our construction it is apparent that $\varphi(x)$ can be expanded in any complete set of functions

$$\varphi(x) = \sum_\alpha [a_\alpha f_\alpha(x) + a_\alpha^\dagger f_\alpha^*(x)] \tag{4.106}$$

where $f_\alpha(x)$ is a normalised wave function

$$i\int d\mathbf{x}\left(f_\alpha^*\frac{\partial f_\beta}{\partial t} - \frac{\partial f_\alpha^*}{\partial t}f_\beta\right) = \delta_{\alpha\beta} \tag{4.107}$$

and $a_\alpha^\dagger$ and a_α create and destroy respectively particles in states with functions f_α^* and f_α.

4.3(f). *Total energy and momentum operators for the scalar field*

In § 3.4(f.1) the four-momentum for a classical relativistic field was shown to be an expression of the form

$$p_\lambda = i\int d\mathbf{x}\left[\frac{\partial\mathscr{L}}{\partial\left(\dfrac{\partial\eta_\sigma}{\partial x_4}\right)}\frac{\partial\eta_\sigma}{\partial x_\lambda} - \mathscr{L}\delta_{4\lambda}\right]. \tag{4.108}$$

The properties of this expression will now be examined for a quantised Klein–Gordon field by substituting the field operator φ for the c-number terms η_σ

$$\eta_\sigma \to \varphi.$$

In equation (4.55) the Lagrangian for the scalar field was written as

$$\mathscr{L} = a \frac{\partial\varphi}{\partial x_\lambda} \frac{\partial\varphi}{\partial x_\lambda} + b\varphi^2$$

nd it was found that

$$b/a = m^2.$$

hus a was left undetermined; for later convenience we choose $a = -\frac{1}{2}$; in ddition we shall replace the suffix λ by μ in order to avoid ambiguity, hence $\mathscr{L}$ beomes

$$\mathscr{L} = -\tfrac{1}{2}\left[\frac{\partial\varphi}{\partial x_\mu} \frac{\partial\varphi}{\partial x_\mu} + m^2\varphi^2\right].$$

This expression gives

$$\frac{\partial\mathscr{L}}{\partial\left(\dfrac{\partial\varphi}{\partial x_4}\right)} = -\frac{\partial\varphi}{\partial x_4}$$

o that

$$P_\lambda = i\int d\mathbf{x}\left[-\frac{\partial\varphi}{\partial x_4} \cdot \frac{\partial\varphi}{\partial x_4} - \mathscr{L}\,\delta_{4\lambda}\right] \tag{4.109}$$

where we use a capital letter for P_λ to indicate that P is an operator in q-number theory.

We will first consider the fourth component of P_λ

$$\begin{aligned} P_4 &= i\int d\mathbf{x}\left[-\left(\frac{\partial\varphi}{\partial x_4}\right)^2 + \tfrac{1}{2}\left(\frac{\partial\varphi}{\partial x_\mu}\right)^2 + \tfrac{1}{2}\, m^2\varphi\right] \\ &= \frac{i}{2}\int d\mathbf{x}\left[(\nabla\varphi)^2 - \left(\frac{\partial\varphi}{\partial x_4}\right)^2 + m^2\varphi^2\right] \\ &= \frac{i}{4}\int \frac{d\mathbf{x}}{V} \sum_{\mathbf{k},\mathbf{k}'} \frac{1}{\sqrt{(\omega_k\omega_k')}} \{(-\mathbf{k}\cdot\mathbf{k}' + k_4k_4')\,[a(\mathbf{k})\,e^{ikx} - a^\dagger(\mathbf{k})\,e^{-ikx}] \times \\ &\quad \times [a(\mathbf{k}')\,e^{ik'x} - a^\dagger(\mathbf{k}')\,e^{-ik'x}] \\ &\quad + m^2[a(\mathbf{k})\,e^{ikx} + a^\dagger(\mathbf{k})\,e^{-ikx}]\,[a(\mathbf{k}')\,e^{ik'x} + a^\dagger(\mathbf{k}')\,e^{-ik'x}]\}. \end{aligned}$$

Now P_4 is the equivalent of i times the Hamiltonian operator H, and we have shown already in § 4.1 (f.2) that H is independent of time. Thus we may fix the time in our expression for P_4 at any convenient value; we choose

$$t = 0$$

so that

$$e^{ikx} \rightarrow e^{i\mathbf{k}\cdot\mathbf{x}}.$$

(If the ωt terms had been left in they would have cancelled out in any case.) Upon writing $k_4 = i\omega_k$ and rearranging terms, P_4 becomes

$$P_4 = \frac{i}{4}\int \frac{d\mathbf{x}}{V} \sum_{\mathbf{k},\mathbf{k}'} \frac{1}{\sqrt{(\omega_k \omega'_k)}} \cdot [(m^2 - \mathbf{k}\cdot\mathbf{k}' - \omega_k\omega'_k)\, a(\mathbf{k})\, a(\mathbf{k}')\, e^{i(\mathbf{k}+\mathbf{k}')\cdot\mathbf{x}}$$

$$+ (m^2 + \mathbf{k}\cdot\mathbf{k}' + \omega_k\omega'_k)\, a(\mathbf{k})\, a^\dagger(\mathbf{k}')\, e^{i(\mathbf{k}-\mathbf{k}')\cdot\mathbf{x}}$$

$$+ (m^2 + \mathbf{k}\cdot\mathbf{k}' + \omega_k\omega'_k)\, a^\dagger(\mathbf{k})\, a(\mathbf{k}')\, e^{-i(\mathbf{k}-\mathbf{k}')\cdot\mathbf{x}}$$

$$+ (m^2 - \mathbf{k}\cdot\mathbf{k}' - \omega_k\omega'_k)\, a^\dagger(\mathbf{k})\, a^\dagger(\mathbf{k}')\, e^{-i(\mathbf{k}+\mathbf{k}')\cdot\mathbf{x}}].$$

The integration over volume may now be carried out; this process yields δ-functions of the type

$$\int \frac{d\mathbf{x}}{V}\, e^{i(\mathbf{k}-\mathbf{k}')\cdot\mathbf{x}} = \delta_{\mathbf{k}\mathbf{k}'}$$

so that

$$P_4 = \frac{i}{4}\sum_{\mathbf{k}} \frac{1}{\omega_k} [(m^2 + \mathbf{k}^2 - \omega_k^2)\, a(\mathbf{k})\, a(-\mathbf{k})$$

$$+ (m^2 + \mathbf{k}^2 + \omega_k^2)(N(\mathbf{k}) + \hat{1}) + (m^2 + \mathbf{k}^2 + \omega_k^2)\, N(\mathbf{k})$$

$$+ (m^2 + \mathbf{k}^2 - \omega_k^2)\, a^\dagger(\mathbf{k})\, a^\dagger(-\mathbf{k})]$$

$$= \frac{i}{4}\sum_{\mathbf{k}} \frac{1}{\omega_k} \cdot 2\omega_k^2 [2N(\mathbf{k}) + \hat{1}] = iH.$$

In obtaining the above expression use has been made of equations (4.78) and (4.80). Thus the Hamiltonian operator for a scalar field may be written as

$$\frac{1}{i} P_4 = H = \sum_{\mathbf{k}} [N(\mathbf{k}) + \tfrac{1}{2}]\, \omega_k \tag{4.110}$$

where for future convenience we have represented $\frac{1}{2}\hat{1}$ as $\frac{1}{2}$.

If this operator is set to work on a normalised state vector

$$|\Psi\rangle = |n_{\mathbf{k}_1} n_{\mathbf{k}_2} \ldots\rangle$$

describing an assembly of particles, where $n_{\mathbf{k}_1}$ denotes the number present with momentum $\mathbf{k}_1$, $n_{\mathbf{k}_2}$ the number with momentum $\mathbf{k}_2$ and so on, then the expectation value of H is

$$\langle H\rangle = E = \langle\Psi| \sum_{\mathbf{k}} [N(\mathbf{k}) + \tfrac{1}{2}]\, \omega_k |\Psi\rangle = \sum_{\mathbf{k}} (n_{\mathbf{k}} + \tfrac{1}{2})\, \omega_k. \tag{4.111}$$

A similar analysis can be carried out for the spatial components of P_λ. The analysis yields the following expression for the total momentum operator $\mathbf{P}$ of the field

$$\mathbf{P} = \sum_{\mathbf{k}} [N(\mathbf{k}) + \tfrac{1}{2}]\, \mathbf{k} = \sum_{\mathbf{k}} N(\mathbf{k})\, \mathbf{k}. \tag{4.112}$$

The second term inside the bracket may be dropped since $\mathbf{k}$ is summed over all directions, and so a value $\mathbf{k}$ in one direction is summed with a value $-\mathbf{k}$ in the opposite direction. Thus the expectation value of $\mathbf{P}$ is

$$\langle \mathbf{P} \rangle = \sum_{\mathbf{k}} n_{\mathbf{k}} \mathbf{k}.$$

These equations show clearly that $n_{\mathbf{k}}$ should be regarded as the number of particles with energy $\omega_{\mathbf{k}}$ and momentum $\mathbf{k}$. They provide some justification for the operator techniques which we have developed to describe the properties of the scalar field. When $n_{\mathbf{k}}$ is zero, that is there are no particles in the field, the total energy of the field becomes

$$\tfrac{1}{2} \sum_{\mathbf{k}} \omega_k.$$

This expression is known as the *zero point energy*, and upon summing over all energy states yields an infinite energy. Its presence represents one of the puzzles of field theory. In practice it can be safely ignored in most calculations on the behaviour of elementary particles, since it is a constant term and can be conveniently cancelled out.

The choice of the term

$$a = -\tfrac{1}{2}$$

in the Lagrangian density for the free scalar field (equation (4.55)), can be seen to be justified by the physically sensible form for equations (4.111) and (4.112). The Langrangian must take this form in order to yield a positive definite energy.

The choice of $a = -\frac{1}{2}$ also settles the question of dimensions for $\varphi(x)$. The Lagrangian density is an energy density, and so in $\hbar = c = 1$ units we have

$$[\mathscr{L}] = L^{-1} \cdot L^{-3} = \frac{1}{L^2} [\varphi^2]$$

by equation (4.55), and therefore $\varphi(x)$ has dimensions

$$[\varphi(x)] = L^{-1}.$$

In c.g.s. units we may write

$$[\mathscr{L}] = ML^2T^{-2} \cdot L^{-3} = \frac{1}{L^2} [\varphi]^2$$

and so

$$[\varphi(x)] = M^{1/2}L^{1/2}T^{-1}.$$

Therefore we may now find suitable powers for $\hbar$ and c with the aid of (4.72)

$$M^{1/2}L^{1/2}T^{-1} = \frac{1}{\sqrt{V}}\frac{1}{\sqrt{\omega}}\hbar^{\alpha}c^{\beta}.$$

We choose ω to be a frequency, and thus to have dimensions T^{-1}, and the dimensional equation then yields

$$\alpha = \tfrac{1}{2}, \qquad \beta = 1$$

so that

$$\varphi(x) = \frac{1}{\sqrt{V}}\sum_{\mathbf{k}}\sqrt{\left(\frac{\hbar c^2}{2\omega_k}\right)}\,[a(\mathbf{k})\,e^{ikx} + a^{\dagger}(\mathbf{k})\,e^{-ikx}] \tag{4.113}$$

in c.g.s. units.

4.3(g). *Displacement operators and the scalar field*

Until now we have considered the properties of the scalar field in the Heisenberg representation. This statement may be verified by recalling that the state vectors $|n_{\mathbf{k}}\rangle$ contain no time-dependent terms, and by showing that the scalar field operator satisfies equation (4.49)

$$i\frac{d\varphi(x)}{dt} = [\varphi(x), H]. \tag{4.114}$$

This equation by easily verified with the aid of equation (4.110)

$$H = \sum_{\mathbf{k}}[N(\mathbf{k}) + \tfrac{1}{2}]\,\omega_k$$

and the commutation relations (4.81). Thus we find

$$[\varphi(x), H] = \frac{1}{\sqrt{V}}\sum_{\mathbf{k}}\frac{\omega_k}{\sqrt{(2\omega_k)}}\{[a(\mathbf{k}), N(\mathbf{k})]\,e^{ikx} + [a^{\dagger}(\mathbf{k}), N(\mathbf{k})]\,e^{-ikx}\}$$

$$= \frac{1}{\sqrt{V}}\sum_{\mathbf{k}}\sqrt{\left(\frac{\omega_k}{2}\right)}\,[a(\mathbf{k})\,e^{ikx} - a^{\dagger}(\mathbf{k})\,e^{-ikx}] = i\frac{d\varphi(x)}{dt}$$

where we have used $\varphi(x)$ in the form given in equation (4.72). Equation (4.114) can be extended in a similar manner to all the field displacement operators

$$-\frac{\partial\varphi(x)}{dx_{\lambda}} = i[P_{\lambda}, \varphi(x)]. \tag{4.115}$$

It should be noted that the canonically conjugate momentum density (3.230), now in operator form, also equals $(d\varphi(x)/dt)$

$$\pi(x) = \frac{\partial \mathscr{L}}{\partial \dot{\varphi}(x)} = \dot{\varphi}(x) \equiv \frac{d\varphi(x)}{dt} \tag{4.116}$$

$$= \frac{-i}{\sqrt{V}} \sum_{\mathbf{k}} \sqrt{\left(\frac{\omega_k}{2}\right)} [a(\mathbf{k})\, e^{ikx} - a^{\dagger}(\mathbf{k})\, e^{-ikx}]. \tag{4.117}$$

4.3(h). *The non-Hermitian scalar field*

The Hermitian scalar field gives a suitable description of the behaviour of uncharged pions. In order to describe the properties of the charged spinless mesons, a non-Hermitian field must be introduced. Its construction follows closely along the lines we have already discussed for the Hermitian field.

As we indicated in § 3.4(f.2), a field function $\eta_\sigma(x)$ must be complex in order to describe a charged field. Accordingly the non-Hermitian field will have real and imaginary components

$$\varphi = \frac{1}{\sqrt{2}} (\varphi_1 - i\varphi_2) \tag{4.118}$$

$$\varphi^{\dagger} = \frac{1}{\sqrt{2}} (\varphi_1 + i\varphi_2)$$

and therefore

$$\varphi \neq \varphi^{\dagger}.$$

Both φ_1 and φ_2 are real and therefore Hermitian.

The field equations may be constructed from the Lagrangian density as in § 4.3(a). However, we invoke the further condition that the Lagrangian should be real. A suitable Lagrangian density for the non-Hermitian field is given by the expression

$$\mathscr{L} = -\frac{\partial \varphi^{\dagger}}{\partial x_\lambda} \frac{\partial \varphi}{\partial x_\lambda} - m^2 \varphi^{\dagger} \varphi. \tag{4.119}$$

Strictly speaking this Lagrangian is unsatisfactory as it is not invariant under charge conjugation (§ 5.5(b)). The non-invariance is associated with the commutation properties of quantised fields. For the present, however, this form of the Lagrangian is quite adequate for our discussion.

The Lagrangian yields the following field equations

$$\frac{\partial \mathscr{L}}{\partial \varphi} - \frac{\partial}{\partial x_\lambda} \frac{\partial \mathscr{L}}{\partial(\partial \varphi/\partial x_\lambda)} = (\Box^2 - m^2)\, \varphi^{\dagger} = 0 \tag{4.120}$$

$$\frac{\partial \mathscr{L}}{\partial \varphi^{\dagger}} - \frac{\partial}{\partial x_\lambda} \frac{\partial \mathscr{L}}{\partial(\partial \varphi^{\dagger}/\partial x_\lambda)} = (\Box^2 - m^2)\, \varphi = 0.$$

The composition of the field into plane waves will again be considered. As in equation (4.58), we may introduce the following solutions to the field equations:

$$\varphi(x) = \frac{1}{(2\pi)^2} \int d\mathbf{k}\, d\omega\, C(k)\, \delta(k^2 + m^2)\, e^{ikx}$$
$$\varphi^\dagger(x) = \frac{1}{(2\pi)^2} \int d\mathbf{k}\, d\omega\, C^\dagger(k)\, \delta(k^2 + m^2)\, e^{-ikx}. \tag{4.121}$$

Since the fields are non-Hermitian, equation (4.66)

$$C^\dagger(\mathbf{k}) = C(-\mathbf{k})$$

no longer holds. The developement of the field operators may still be carried through, however, by the methods of § 4.3(a). A simpler approach is to use the Hermitian properties of φ_1 and φ_2. Denoting both fields by $\varphi_r(x)$ ($r = 1, 2$), and expanding as for (4.72), we find that

$$\varphi_r(x) = \frac{1}{\sqrt{V}} \sum_{\mathbf{k}} \frac{1}{\sqrt{(2\omega_k)}} [a_r(\mathbf{k})\, e^{ikx} + a_r^\dagger(\mathbf{k})\, e^{-ikx}] \tag{4.122}$$

where the operators $a_r(\mathbf{k})$ and $a_r^\dagger(\mathbf{k})$ satisfy relations similar to (4.78) and (4.79)

$$[a_r(\mathbf{k}), a_s^\dagger(\mathbf{k}')] = \delta_{rs}\delta_{\mathbf{k}\mathbf{k}'} \tag{4.123}$$
$$[a_r(\mathbf{k}), a_s(\mathbf{k})] = [a_r(\mathbf{k}), a_r(\mathbf{k})] = 0$$
$$[a_r^\dagger(\mathbf{k}), a_s^\dagger(\mathbf{k})] = [a_r^\dagger(\mathbf{k}), a_r^\dagger(\mathbf{k})] = 0.$$

We now introduce the linear combinations.

$$a(\mathbf{k}) = \frac{1}{\sqrt{2}} [a_1(\mathbf{k}) - ia_2(\mathbf{k})], \qquad b^\dagger(\mathbf{k}) = \frac{1}{\sqrt{2}} [a_1^\dagger(\mathbf{k}) - ia_2^\dagger(\mathbf{k})]. \tag{4.124}$$

Then, with the aid of equations (4.118) and (4.122), we find that

$$\varphi(x) = \frac{1}{\sqrt{V}} \sum_{\mathbf{k}} \frac{1}{\sqrt{(2\omega_k)}} [a(\mathbf{k})\, e^{ikx} + b^\dagger(\mathbf{k})\, e^{-ikx}] \tag{4.125}$$
$$\varphi^\dagger(x) = \frac{1}{\sqrt{V}} \sum_{\mathbf{k}} \frac{1}{\sqrt{(2\omega_k)}} [a^\dagger(\mathbf{k})\, e^{-ikx} + b(\mathbf{k})\, e^{ikx}].$$

The commutation relations for $a(\mathbf{k})$, $b(\mathbf{k})$ and their adjoints are easily derived with the aid of equations (4.123) and (4.124). They are

$$[a(\mathbf{k}), a^\dagger(\mathbf{k}')] = [b(\mathbf{k}), b^\dagger(\mathbf{k}')] = \delta_{\mathbf{k}\mathbf{k}'} \tag{4.126}$$
$$[a(\mathbf{k}), b(\mathbf{k}')] = [a^\dagger(\mathbf{k}), b^\dagger(\mathbf{k}')] = [a(\mathbf{k}), b^\dagger(\mathbf{k}')] = [a^\dagger(\mathbf{k}), b(\mathbf{k}')] = 0.$$

f the continuous representation

$$\frac{1}{\sqrt{V}}\sum_{\mathbf{k}} \rightarrow \frac{1}{(2\pi)^{3/2}}\int d\mathbf{k}$$

ad been used, we would have replaced $\delta_{\mathbf{kk}'}$ by $\delta(\mathbf{k}-\mathbf{k}')$.

The expressions for $\varphi(x)$ and $\varphi^\dagger(x)$ can also be given in terms of spherical r cylindrical waves. In the former case the field operators are written as

$$\varphi(x) = \sum_{k}\sum_{l=0}^{\infty}\sum_{m=-l}^{+l} \frac{1}{\sqrt{(2\omega_k)}} g_{lk}(r)\,[a(klm)\,Y_l^m(\theta,\varphi)\,e^{-i\omega_k t} + b^\dagger(klm)\,Y_l^{m*}(\theta,\varphi)\,e^{i\omega_k t}] \quad (4.127)$$

vith a corresponding equation for $\varphi^\dagger(x)$.

4.3 (i). *The physical interpretation of the non-Hermitian scalar field*

We start by introducing new state vectors satisfying the conditions

$$a^\dagger(\mathbf{k})\,|n_{\mathbf{k}}^{+}\rangle = \sqrt{(n+1)_{\mathbf{k}}^{+}}\,|(n+1)_{\mathbf{k}}^{+}\rangle \qquad (4.128)$$

$$a(\mathbf{k})\,|n_{\mathbf{k}}^{+}\rangle = \sqrt{n_{\mathbf{k}}^{+}}\,|(n-1)_{\mathbf{k}}^{+}\rangle$$

$$b^\dagger(\mathbf{k})\,|n_{\mathbf{k}}^{-}\rangle = \sqrt{(n+1)_{\mathbf{k}}^{-}}\,|(n+1)_{\mathbf{k}}^{-}\rangle \qquad (4.129)$$

$$b(\mathbf{k})\,|n_{\mathbf{k}}^{-}\rangle = \sqrt{n_{\mathbf{k}}^{-}}\,|(n-1)_{\mathbf{k}}^{-}\rangle .$$

Equations of this form enable the commutation relations (4.126) to be satisfied. The physical significance of the signs + and − will become clear later.

New occupation number operators may also be defined

$$N^{+}(\mathbf{k}) = a^\dagger(\mathbf{k})\,a(\mathbf{k}), \qquad N^{-}(\mathbf{k}) = b^\dagger(\mathbf{k})\,b(\mathbf{k}). \qquad (4.130)$$

These operators possess the following properties

$$[a(\mathbf{k}), N^{+}(\mathbf{k})] = a(\mathbf{k}), \qquad [a^\dagger(\mathbf{k}), N^{+}(\mathbf{k})] = -a^\dagger(\mathbf{k}) \qquad (4.131)$$

$$[b(\mathbf{k}), N^{-}(\mathbf{k})] = b^\dagger(\mathbf{k}), \qquad [b^\dagger(\mathbf{k}), N^{-}(\mathbf{k})] = -b^\dagger(\mathbf{k}).$$

These equations are easily verified with the aid of (4.126). As in § 4.3 (d) one may easily show that vacuum states exist

$$a(\mathbf{k})\,|\Psi_0\rangle = 0, \qquad N^{+}(\mathbf{k})\,|\Psi_0\rangle = 0$$

and that the number operators have the following eigenvalues

$$N^{+}(\mathbf{k})\,|n_{\mathbf{k}}^{+}\rangle = n_{\mathbf{k}}^{+}\,|n_{\mathbf{k}}^{+}\rangle \qquad (4.132)$$

$$N^{-}(\mathbf{k})\,|n_{\mathbf{k}}^{-}\rangle = n_{\mathbf{k}}^{-}\,|n_{\mathbf{k}}^{-}\rangle .$$

4.3(i.1). *The charge operator for the field.* We are now in a position to offe physical interpretations of the field operators. We start by considering th electrical charge of the field†. In § 3.4(f.2) it was shown that the invariance c the Lagrangian density under gauge transformations of the first kind implie that a conserved quantity C existed. This term was found to be of the forr

$$C = \frac{1}{i} \int d\mathbf{x} \, i\alpha \left[\frac{\partial \mathscr{L}}{\partial(\partial \eta_\sigma / \partial x_4)} \eta_\sigma - \frac{\partial \mathscr{L}}{\partial(\partial \eta_\sigma^* / \partial x_4)} \eta_\sigma^* \right].$$

We transfer our attention from the classical to the quantised scalar field b making the substitutions

$$\eta_\sigma \to \varphi, \qquad \eta_\sigma^* \to \varphi^\dagger.$$

It is obvious from equation (4.119) that $\mathscr{L}$ will be invariant under the gauge trans formations

$$\varphi \to \varphi' = \varphi \, e^{i\alpha} = \varphi + i\alpha\varphi, \qquad \alpha \to 0$$

$$\varphi^\dagger \to \varphi^{\dagger\prime} = \varphi^\dagger \, e^{-i\alpha} = \varphi^\dagger - i\alpha\varphi^\dagger, \qquad \alpha \to 0.$$

We now examine the integrand for C; since $\mathscr{L}$ is of the form (4.119)

$$\mathscr{L} = -\frac{\partial \varphi^\dagger}{\partial x_\lambda} \frac{\partial \varphi}{\partial x_\lambda} - m^2 \varphi^\dagger \varphi$$

we find that

$$\frac{\partial \mathscr{L}}{\partial(\partial \varphi / \partial x_4)} = -\frac{\partial \varphi^\dagger}{\partial x_4}, \qquad \frac{\partial \mathscr{L}}{\partial(\partial \varphi^\dagger / \partial x_4)} = -\frac{\partial \varphi}{\partial x_4}$$

so that C becomes

$$C = -\alpha \int d\mathbf{x} \left(\frac{\partial \varphi^\dagger}{\partial x_4} \varphi - \frac{\partial \varphi}{\partial x_4} \varphi^\dagger \right). \tag{4.133}$$

It should be noted that this expression is frequently written in the form given ir equation (3.251)

$$C = i\alpha \int d\mathbf{x} [\pi\varphi - \pi^\dagger \varphi^\dagger]$$

where

$$\pi = \frac{\partial \mathscr{L}}{\partial(\partial \varphi / \partial t)} = \frac{\partial \mathscr{L}}{\partial \dot{\varphi}} = \dot{\varphi}^\dagger$$

$$\pi^\dagger = \frac{\partial \mathscr{L}}{\partial(\partial \dot{\varphi}^\dagger / \partial t)} = \frac{\partial \mathscr{L}}{\partial \dot{\varphi}^\dagger} = \dot{\varphi}.$$

† Although we will refer to electrical charge here, the principle being discussed is far more general. Other forms of charge (leptonic, baryonic) may be associated with the field. They will be examined in § 9.3.

The terms φ and $\varphi^\dagger$, in (4.133) can be expanded into the Fourier transforms of (4.125). The integrations over space may be carried out in the manner indicated in § 4.3(f), and one finally obtains the expression

$$C = -\alpha \sum_{\mathbf{k}} [a^\dagger(\mathbf{k})\, a(\mathbf{k}) - b^\dagger(\mathbf{k})\, b(\mathbf{k})]$$
$$= -\alpha \sum_{\mathbf{k}} [N^+(\mathbf{k}) - N^-(\mathbf{k})]. \tag{4.134}$$

We now assert that if we make the substitutions

$$\alpha = -\xi e, \qquad e = |e|$$
$$C = Q\xi$$

where $|e|$ is the value of the electronic charge (without sign) and ξ is a small arbitrary constant (compare the arbitrary displacement terms in a_4 and a_k in § 3.4(f.1)), then Q may be identified as the electric charge operator for the non-Hermitian scalar field

$$Q = e \sum_{\mathbf{k}} [N^+(\mathbf{k}) - N^-(\mathbf{k})]. \tag{4.135}$$

It is now obvious that the sign appearing in the number operators refers to the sign of the electric charge. If we consider the action of Q on a state vector $|\Psi\rangle$ we find

$$Q\, |\Psi\rangle = e \sum_{\mathbf{k}} [N^+(\mathbf{k}) - N^-(\mathbf{k})]\, |\ldots n_{\mathbf{k}}^+ \ldots n_{\mathbf{k}}^- \ldots\rangle = e \sum_{\mathbf{k}} (n_{\mathbf{k}}^+ - n_{\mathbf{k}}^-)\, |\Psi\rangle. \tag{4.136}$$

Thus $n_{\mathbf{k}}^+$ denotes the number of particles with charge e^+, and $n_{\mathbf{k}}^-$ is the number of particles with negative charge. Now consider the operation

$$Qa^\dagger(\mathbf{k})\, |\Psi\rangle = e \sum_{\mathbf{k}} [N^+(\mathbf{k}) - N^-(\mathbf{k})]\, a^\dagger(\mathbf{k})\, |\Psi\rangle$$
$$= e \sum_{\mathbf{k}} [(n+1)_{\mathbf{k}}^+ - n_{\mathbf{k}}^-]\, a^\dagger(\mathbf{k})\, |\Psi\rangle.$$

In deriving this expression, equations (4.131) have been used. The above equation, taken in conjunction with (4.136), obviously implies that the state $a^\dagger(\mathbf{k})\, |\Psi\rangle$ contains one more unit of positive charge than the state $|\Psi\rangle$. A similar deduction can be made about $b^\dagger(\mathbf{k})\, |\Psi\rangle$, although in this case the conclusion refers to negative charge. It should be noted that the net charge of the state can be altered by one unit by destroying particles as well as creating them.

In equation (4.70) the Hermitian scalar field was split into its positive and negative frequency parts. A similar division may be made for the non-Hermitian field as indicated below; once again, warning should be given that the convention adopted is not unique.

TABLE 4.1

$\varphi^{(-)}(x) = \frac{1}{\sqrt{V}} \sum_{\mathbf{k}} a(\mathbf{k})\, e^{ikx}$	destroys particles of charge e^+
$\varphi^{(+)}(x) = \frac{1}{\sqrt{V}} \sum_{\mathbf{k}} b^\dagger(\mathbf{k})\, e^{-ikx}$	creates particles of charge e^-
$\varphi^{\dagger(+)}(x) = \frac{1}{\sqrt{V}} \sum_{\mathbf{k}} a^\dagger(\mathbf{k})\, e^{-ikx}$	creates particles of charge e^+
$\varphi^{\dagger(-)}(x) = \frac{1}{\sqrt{V}} \sum_{\mathbf{k}} b(\mathbf{k})\, e^{ikx}$	destroys particles of charge e^-
$\varphi = \frac{\varphi_1 - i\varphi_2}{\sqrt{2}}$	creates particles of charge e^-, destroys e^+
$\varphi^\dagger = \frac{\varphi_1 + i\varphi_2}{\sqrt{2}}$	creates particles of charge e^+, destroys e^-.

Thus the operator $\varphi^{(+)}(x)$ can be used to describe the creation of π^- mesons for example, whilst $\varphi^{\dagger(+)}(x)$ creates π^+ mesons. The single particle states for particles of momentum $\mathbf{k}$ can be written as

$$
\begin{aligned}
|\mathbf{k}^-\rangle &= b^\dagger(\mathbf{k})\, |\Psi_0\rangle \quad \text{charge } e^- \\
|\mathbf{k}^+\rangle &= a^\dagger(\mathbf{k})|\, \Psi_0\rangle \quad \text{charge } e^+.
\end{aligned}
\tag{4.137}
$$

Finally, we note that equation (4.133)

$$C = -\alpha \int d\mathbf{x} \left(\frac{\partial \varphi^\dagger}{\partial x_4} \varphi - \frac{\partial \varphi}{\partial x_4} \varphi^\dagger \right)$$

can be written so that its integrand represents the fourth component of a current density

$$C = \frac{\xi}{i} \int d\mathbf{x}\, ie \left(\frac{\partial \varphi^\dagger}{\partial x_4} \varphi - \frac{\partial \varphi}{\partial x_4} \varphi^\dagger \right)$$

since

$$\frac{C}{\xi} = Q = \int d\mathbf{x}\, \varrho = \frac{1}{i} \int d\mathbf{x}\, j_4$$

where ϱ represents the charge density, therefore

$$j_4 = ie \left(\frac{\partial \varphi^\dagger}{\partial x_4} \varphi - \frac{\partial \varphi}{\partial x_4} \varphi^\dagger \right)$$

and in general

$$j_\mu(x) = ie \left(\frac{\partial \varphi^\dagger}{\partial x_\mu} \varphi - \frac{\partial \varphi}{\partial x_\mu} \varphi^\dagger \right) = e \left(\frac{\partial \varphi_1}{\partial x_\mu} \varphi_2 - \frac{\partial \varphi_2}{\partial x_\mu} \varphi_1 \right). \tag{4.138}$$

The integration of $j_\mu(x)$ over the spatial coordinates yields the total current operator for the non-Hermitian scalar field. Its components are given by

$$J_\mu = e \sum_{\mathbf{k}} \frac{k_\mu}{\omega} [N^+(\mathbf{k}) - N^-(\mathbf{k})] \tag{4.139}$$

$$\mathbf{J} = e \sum_{\mathbf{k}} \frac{\mathbf{k}}{\omega} [N^+(\mathbf{k}) - N^-(\mathbf{k})]$$

$$Q = e \sum_{\mathbf{k}} [N^+(\mathbf{k}) - N^-(\mathbf{k})].$$

4.3(i.2). *The total energy and momentum operators.* When the Lagrangian density (4.119)

$$\mathscr{L} = -\frac{\partial \varphi^\dagger}{\partial x_\lambda}\frac{\partial \varphi}{\partial x_\lambda} - m^2 \varphi^\dagger \varphi$$

is inserted into the expression for the four-momentum of a field (4.108)

$$p_\lambda = i \int d\mathbf{x} \left[\frac{\partial \mathscr{L}}{\partial(\partial \eta_\sigma / \partial x_4)} \frac{\partial \eta_\sigma}{\partial x_4} - \mathscr{L} \delta_{4\lambda} \right]$$

the following equation is obtained (the symbol P_λ denotes an operator)

$$P_\lambda = i \int d\mathbf{x} \left[-\frac{\partial \varphi^\dagger}{\partial x_4}\frac{\partial \varphi}{\partial x_\lambda} - \frac{\partial \varphi}{\partial x_4}\frac{\partial \varphi^\dagger}{\partial x_\lambda} - \mathscr{L}\delta_{4\lambda} \right].$$

Upon repeating the operations outlined in § 4.3(f) the following equations for the total Hamiltonian and momentum operators are obtained.

$$\frac{1}{i} P_4 = H \sum_{\mathbf{k}} [N^+(\mathbf{k}) + N^-(\mathbf{k}) + \hat{1}]\, \omega_k \tag{4.140}$$

$$\mathbf{P} = \sum_{\mathbf{k}} [N^+(\mathbf{k}) + N^-(\mathbf{k})]\, \mathbf{k}. \tag{4.141}$$

The expectation values for these operators assume appropriate values, namely

$$\langle H \rangle = \sum_{\mathbf{k}} (n_{\mathbf{k}}^+ + n_{\mathbf{k}}^- + \hat{1})\, \omega_k \tag{4.142}$$

$$\langle \mathbf{P} \rangle = \sum_{\mathbf{k}} (n_{\mathbf{k}}^+ + n_{\mathbf{k}}^-)\, \mathbf{k}. \tag{4.143}$$

It can be seen that a term for the zero point energy again appears in H.

The expressions (4.135), (4.140) and (4.141) for Q, H and $\mathbf{P}$ respectively, and their associated expectation values, amply justify our original assumptions regarding the properties of the operators $a^\dagger(\mathbf{k})$ and $a(\mathbf{k})$. The expectation values show that a quantised field theory describes an assembly of elementary particles in a physically sensible manner. The only difficulty about the equations is the existence of a zero point energy, and, as mentioned earlier (§ 4.3(f)), this term does affect calculations on the behaviour of the particles.

In § 3.3(a) it was pointed out that the current density arising from the Klein–Gordon equation

$$j_\lambda = \frac{i}{2m}\left(\varphi \frac{\partial \varphi^*}{\partial x_\lambda} - \varphi^* \frac{\partial \varphi}{\partial x_\lambda}\right)$$

lead to a physically unsatisfactory equation for the probability density ϱ. This equation bears a certain resemblance to the charge current density operator for the non-Hermitian field (4.138)

$$j_\mu = ie\left(\frac{\partial \varphi^\dagger}{\partial x_\mu}\varphi - \frac{\partial \varphi}{\partial x_\mu}\varphi^\dagger\right).$$

The similarity was first recognised by Pauli and Weisskopf (1934). They pointed out that with slight alterations the wave functions φ and φ^* in (3.75) could be re-interpreted as the operators φ and $\varphi^\dagger$ describing the behaviour of an assembly of charged particles. Thus the existence of negative values for ϱ merely implies that in certain assemblies more negatively than positively charged particles are present. This approach overcomes the problems set by ordinary wave equations, which attempt to describe the behaviour of a single particle.

4.4. THE ELECTROMAGNETIC (VECTOR) FIELD

4.4(a). *The classical electromagnetic field*

The classical theory of the electromagnetic field represents the first successful field theory. Many of the ideas of quantised field theories were first developed by considering analogies with the electromagnetic field in the classical limit $\hbar \to 0$.

We shall use *Heaviside–Lorentz* units, with

$$\hbar = c = 1, \quad \alpha = \frac{e^2}{4\pi\hbar c} \equiv \frac{e^2}{4\pi} \sim \frac{1}{137}$$

where α is the characteristic development parameter of electromagnetic interactions.

The classical electromagnetic field can be described by vectors **E** for the electric component of the field and **H** for the magnetic component. The vectors satisfy Maxwell's equations

$$\nabla \times \mathbf{H} = \frac{\partial \mathbf{E}}{\partial t} \tag{4.144}$$

$$\nabla \cdot \mathbf{H} = 0$$

$$\nabla \times \mathbf{E} = -\frac{\partial \mathbf{H}}{\partial t}$$

$$\nabla \cdot \mathbf{E} = 0$$

the absence of electrical charge. When charge is present they become

$$\nabla \times \mathbf{H} = \frac{\partial \mathbf{E}}{\partial t} + \mathbf{j} \tag{4.145}$$

$$\nabla \cdot \mathbf{H} = 0$$

$$\nabla \times \mathbf{E} = -\frac{\partial \mathbf{H}}{\partial t}$$

$$\nabla \cdot \mathbf{E} = \varrho$$

where ϱ and $\mathbf{j}$ refer to charge and current density respectively.

Since div·curl equals zero, we may re-write the first of the equations (4.145) and combine it with the last to obtain an equation of continuity

$$\nabla \cdot \nabla \times \mathbf{H} = 0 = \frac{\partial}{\partial t} \nabla \cdot \mathbf{E} + \nabla \cdot \mathbf{j} = \frac{\partial \varrho}{\partial t} + \nabla \cdot \mathbf{j}.$$

Thus charge is conserved in Maxwell's equations.

The quantities **E** and **H** can be combined in the four-vector A for the electromagnetic field. The first three components of A are called the *vector potential* **A**, whilst the fourth component is given by

$$A_4 = i\frac{\varphi}{c} \equiv i\varphi.$$

The real quantity φ is called the *scalar potential.* The electric and magnetic fields are related to A by the expressions

$$\mathbf{E} = -\frac{\partial \mathbf{A}}{\partial t} - \nabla\varphi \tag{4.146}$$

$$\mathbf{H} = \nabla \times \mathbf{A}.$$

The definitions given above satisfy the second and third equations of (4.145). The first and fourth equations may be written as

$$\frac{\partial^2 \mathbf{A}}{\partial t^2} - \nabla^2 \mathbf{A} + \nabla\left(\nabla \cdot \mathbf{A} + \frac{\partial \varphi}{\partial t}\right) = \mathbf{j} \tag{4.147}$$

$$\nabla^2 \varphi + \frac{\partial}{\partial t} \nabla \cdot \mathbf{A} = -\varrho.$$

It should be noted that the vector A is not a directly measurable quantity; the quantities which can be determined by experiment are the vectors **E** and **H**. This effect arises because the vector A is not completely determined by equations

(4.146). An inspection of the equations reveals that **E** and **H** are invariant unde the transformation

$$\mathbf{A}(x) \to \mathbf{A}'(x) = \mathbf{A}(x) + \nabla\chi(x) \tag{4.148}$$

$$\varphi(x) \to \varphi'(x) = \varphi(x) - \frac{\partial\chi(x)}{\partial t}$$

or in four-vector notation

$$A_\mu(x) \to A'_\mu(x) = A_\mu(x) + \frac{\partial\chi(x)}{\partial x_\mu}$$

where χ is an arbitrary scalar quantity. Insertion of the transformed quantitie (4.148) into (4.146) yields

$$\mathbf{E}' = -\frac{\partial\mathbf{A}}{\partial t} - \frac{\partial}{\partial t}\nabla\chi - \nabla\varphi + \frac{\partial}{\partial t}\nabla\chi = -\frac{\partial\mathbf{A}}{\partial t} - \nabla\varphi$$

$$\mathbf{H}' = \nabla \times \mathbf{A} + \nabla \times \nabla\chi = \nabla \times \mathbf{A}$$

and thus proves our statement.

The invariance of **E** and **H** for arbitrary values of **A** and φ is called *gauge in variance of the second kind.* Two gauges are normally encountered; they are th Lorentz and Coulomb gauges. Both lead to simplifications of (4.148).

The *Lorentz gauge* or *condition* imposes the following requirement:

$$\nabla \cdot \mathbf{A} + \frac{\partial\varphi}{\partial t} \equiv \frac{\partial A_\mu}{\partial x_\mu} = 0. \tag{4.149}$$

This condition is also known as the *subsidiary condition.* The Maxwell equation (4.147) then reduce to

$$\nabla^2\mathbf{A} - \frac{\partial^2\mathbf{A}}{\partial t^2} = -\mathbf{j}$$

$$\nabla^2\varphi - \frac{\partial^2\varphi}{\partial t^2} = -\varrho$$

or

$$\Box^2 A_\mu = -j_\mu \tag{4.150}$$

in four-vector notation. In the absence of electric charge this equation become

$$\Box^2 A_\mu = 0. \tag{4.151}$$

The restriction given in (4.149) then leads to the following requirement for χ:

$$\Box^2\chi = 0.$$

This result may be found by inserting (4.148) in (4.149).

The *Coulomb gauge* is defined by the condition

$$\nabla \cdot \mathbf{A} = 0. \tag{4.152}$$

quations (4.147) then become

$$-\Box^2 \mathbf{A} + \nabla \frac{\partial \varphi}{\partial t} = \mathbf{j}$$

$$\nabla^2 \varphi = -\varrho.$$

t is obvious from these equations that both **A** and φ are real (Hermitian for a uantised field). The second equation is Poisson's equation, and hence in the oulomb gauge the scalar potential φ may be obtained from the static charge ensity distribution. If $\mathbf{j} = \varrho = 0$ then φ can be set equal to zero, and the above quations reduce to

$$\Box^2 \mathbf{A} = 0. \tag{4.153}$$

his equation can be regarded as the wave equation for a classical free photon eld in the Coulomb gauge. A possible solution for **A**, which satisfies (4.153), is

$$\mathbf{A} = \mathbf{e} A_0 e^{i(\mathbf{k} \cdot \mathbf{x} - \omega_k t)} \tag{4.154}$$

here A_0 is an amplitude and **e** a unit vector pointing in the direction of **A**; **e** is alled the polarisation vector for the field. The condition (4.152)

$$\nabla \cdot \mathbf{A} = 0$$

s only satisfied if

$$\mathbf{e} \cdot \mathbf{k} = 0.$$

hus **e** must lie in the plane at right angles to **k**; the electric component of the eld also lies in this plane since by (4.146)

$$\mathbf{E} = -\frac{\partial \mathbf{A}}{\partial t} = i\omega \mathbf{A}.$$

n contrast, the direction of the magnetic field is given by

$$\mathbf{H} = \nabla \times \mathbf{A} = i\mathbf{k} \times \mathbf{A}.$$

hus, the vectors **E**, **H** and **k** are mutually orthogonal.

Let us finally consider the polarisation. Classically, polarisation is defined in erms of the vibration of the transverse electric field. Let a wave propagate along he z-axis, then $\mathbf{k} \cdot \mathbf{x} \to kz$ in (4.154), and equation (4.153) reduces to two inde- endent relations.

$$\Box^2 \mathbf{A} \to \begin{cases} \dfrac{\partial^2 A_x}{\partial z^2} - \dfrac{\partial^2 A_x}{\partial t^2} = 0 \\[2ex] \dfrac{\partial^2 A_y}{\partial z^2} - \dfrac{\partial^2 A_y}{\partial t^2} = 0. \end{cases} \tag{4.155}$$

Thus we can associate two linearly independent polarisation vectors with A_x an A_y. The nature of the polarisation is determined by the relative phases an amplitudes of A_x and A_y; for example, a phase shift zero yields plane polarisatio whilst a phase shift of $\pi/2$ and equal amplitudes represent circular polarisatio

For later convenience we will represent circular polarisation by the comple vectors

$$e_R = -\frac{1}{\sqrt{2}}(\mathbf{e}_x + i\mathbf{e}_y), \quad e_L = \frac{1}{\sqrt{2}}(\mathbf{e}_x - i\mathbf{e}_y) \tag{4.156}$$

where $\mathbf{e}_x$ and $\mathbf{e}_y$ are unit vectors and the subscripts R and L refer to right-hande and left-handed circular polarisation respectively (right circular polarisation im plies that the vector $\mathbf{E}$, and therefore $\mathbf{A}$, rotates in a clockwise direction abou the z-axis).

4.4(b). *The electromagnetic field tensor*

The properties of the electromagnetic field can be described more elegantly b the introduction of a tensor of rank 2.

Consider, for example, the x components of the electric and magnetic field (4.146)

$$iE_x = -i\frac{\partial A_x}{\partial t} - i\frac{\partial \varphi}{\partial x} \equiv \frac{\partial A_1}{\partial x_4} - \frac{\partial A_4}{\partial x_1}$$

$$H_x = \frac{\partial A_z}{\partial y} - \frac{\partial A_y}{\partial z} \equiv \frac{\partial A_3}{\partial x_2} - \frac{\partial A_2}{\partial x_3}.$$

Now the derivatives of a four-vector yield a four-tensor of rank 2 (§ 3.2(i)). W therefore introduce the antisymmetric tensor

$$f_{\mu\nu} = \frac{\partial A_\nu}{\partial x_\mu} - \frac{\partial A_\mu}{\partial x_\nu} = -f_{\nu\mu}.$$

A comparison of this equation with (4.146) shows that the components o $f_{\mu\nu}$ are

$$f_{\mu\nu} = \begin{pmatrix} 0 & H_z & -H_y & -iE_x \\ -H_z & 0 & H_x & -iE_y \\ H_y & -H_x & 0 & -iE_z \\ iE_x & iE_y & iE_z & 0 \end{pmatrix}$$

In this notation it may be easily shown that the first and last of Maxwell's equations (4.145)

$$\nabla \times \mathbf{H} = \frac{\partial \mathbf{E}}{\partial t} + \mathbf{j}, \quad \nabla \cdot \mathbf{E} = \varrho$$

duce to

$$\frac{\partial f_{\mu\nu}}{\partial x_\nu} = j_\mu$$

nd the second and third equations

$$\nabla \cdot \mathbf{H} = 0, \quad \nabla \times \mathbf{E} = -\frac{\partial \mathbf{H}}{\partial t}$$

ecome

$$\frac{\partial f_{\mu\nu}}{\partial x_\varrho} + \frac{\partial f_{\nu\varrho}}{\partial x_\mu} + \frac{\partial f_{\varrho\mu}}{\partial x_\nu} = 0.$$

4.4(c). *Lagrangians for the vector field*

The simplest Lagrangian density which can be constructed for a vector field ith the aid of the rules given in § 4.2(b) is

$$\mathscr{L} = a \frac{\partial A_\mu}{\partial x_\nu} \frac{\partial A_\mu}{\partial x_\nu} + bA_\mu A_\mu. \tag{4.158}$$

A slightly more complicated version of this equation will also fit the rules

$$\mathscr{L} = \frac{a}{2} \left(\frac{\partial A_\mu}{\partial x_\nu} - \frac{\partial A_\nu}{\partial x_\mu} \right)^2 + bA_\mu A_\mu. \tag{4.159}$$

Upon application of the Euler–Lagrange equation (3.228), which assumes the orm

$$\frac{\partial \mathscr{L}}{\partial A_\mu} - \frac{\partial}{\partial x_\nu} \frac{\partial \mathscr{L}}{\partial (\partial A_\mu / \partial x_\nu)} = 0$$

ve obtain the following field equations from (4.158) and (4.159) respectively

$$-a \frac{\partial^2 A_\mu}{\partial x_\nu^2} + bA_\mu \equiv \left(\Box^2 - \frac{b}{a} \right) A_\mu = 0 \tag{4.160}$$

$$-a \frac{\partial}{\partial x_\nu} \left(\frac{\partial A_\mu}{\partial x_\nu} - \frac{\partial A_\nu}{\partial x_\mu} \right) + bA_\mu \equiv \frac{\partial}{\partial x_\nu} \left(\frac{\partial A_\mu}{\partial x_\nu} - \frac{\partial A_\nu}{\partial x_\mu} \right) - \frac{b}{a} A_\mu = 0.$$

If we make the identification

$$b/a = m^2$$

t can be seen that the first of equations (4.160) is the vector equivalent of the Klein–Gordon equation. The second equation in (4.160) is called the *Proca equation*. The Proca field A_μ can be shown to describe the behaviour of particles of mass m and spin 1. We shall concentrate on the situation for $m = 0$. Bo equations then reduce to the Maxwell equation (4.151)

$$\Box^2 A_\mu = 0.$$

(We are, of course, assuming that no electric charges are present in the fielc It should be noted that the second equation in (4.160) yields the Maxwell equ tion by virtue of the Lorentz condition (4.149)

$$\frac{\partial}{\partial x_\nu}\frac{\partial A_\nu}{\partial x_\mu} = \frac{\partial}{\partial x_\mu}\frac{\partial A_\nu}{\partial x_\nu} = 0.$$

We will choose to work with the first of the Lagrangian densities (4.158), ar set $a = -\frac{1}{2}$ as in § 4.3(f). This value for a again leads to physically sensib values for the total energy and momentum operators for the field. Thus we wri

$$\mathscr{L} = -\tfrac{1}{2}\frac{\partial A_\mu}{\partial x_\nu}\frac{\partial A_\mu}{\partial x_\nu}. \tag{4.16}$$

This expression is also frequently written as

$$\mathscr{L} = -\tfrac{1}{4}f_{\mu\nu}f_{\mu\nu} - \tfrac{1}{2}\frac{\partial A_\mu}{\partial x_\mu}\frac{\partial A_\nu}{\partial x_\nu}.$$

The second Lagrangian (4.159)

$$\mathscr{L} = -\tfrac{1}{4}\left(\frac{\partial A_\mu}{\partial x_\nu} - \frac{\partial A_\nu}{\partial x_\mu}\right)^2 \tag{4.16}$$

has the unfortunate property that it yields a zero value for the fourth componer of the expression for the canonical conjugate momentum density

$$\pi_\mu = \frac{\partial \mathscr{L}}{\partial(\partial A_\mu/\partial t)} = -i\frac{\partial \mathscr{L}}{\partial(\partial A_\mu/\partial x_4)} = \tfrac{1}{2}i\left(\frac{\partial A_\mu}{\partial x_4} - \frac{\partial A_4}{\partial x_\mu}\right)$$

$$\pi_4 = 0.$$

We anticipate § 4.6 by stating that this result yields senseless commutatio relations, and is therefore not suitable for a quantised field.

4.4(d). *Plane wave representation for the electromagnetic field*

The free electromagnetic field is uncharged, and therefore must be represente by Hermitian operators (§ 4.3(h)). The Fourier expansion for the electromagneti field can be carried out, therefore, as in § 4.3(b) for the Hermitian scalar field

In the present situation, however, a little care must be exercised when re presenting the fourth component of the field, since we wish to use the notatio

$$A_4 = iA_0 \equiv i\varphi$$

and φ is real so that A_4 is pure imaginary. Following equation (4.58) we may write the Fourier expansion of A as

$$A_\mu(x) = \frac{1}{(2\pi)^2} \int_{-\infty}^{+\infty} d^4k C_\mu(k)\, \delta(k^2)\, e^{ikx}$$

$$= \frac{1}{(2\pi)^2} \int_{-\infty}^{+\infty} d\mathbf{k}\, d\omega C_\mu(k)\, \delta(k^2)\, e^{ikx} \qquad (\mu = 1, 2, 3, 0)$$

where the Hermitian condition

$$A_\mu^\dagger(x) = A_\mu(x)$$

implies that

$$C_\mu^\dagger(\mathbf{k}) = C_\mu(-\mathbf{k})$$

(compare (4.66)).

An integration over ω may now be performed by using the techniques given in equations (4.59) to (4.68), yielding

$$A_\mu(x) = \frac{1}{(2\pi)^2} \int_{-\infty}^{+\infty} d\mathbf{k} \frac{1}{2\omega_k} [C_\mu(\mathbf{k})\, e^{ikx} + C_\mu^\dagger(\mathbf{k})\, e^{-ikx}].$$

We will specify the terms $C_\mu(\mathbf{k})$ in the following form (compare (4.67))

$$C_\mu(\mathbf{k}) = \sqrt{(4\pi\omega_k)}\, a_\lambda(\mathbf{k})\, e_\mu^{(\lambda)}(\mathbf{k}) \tag{4.163}$$

$$C_\mu^\dagger(\mathbf{k}) = \sqrt{(4\pi\omega_k)}\, a_\lambda^\dagger(\mathbf{k})\, e_\mu^{(\lambda)}(\mathbf{k})$$

where the term $a_\lambda^\dagger(\mathbf{k})$ and $a_\lambda(\mathbf{k})$ will be shown to be creation and destruction operators respectively, and $e_\mu^{(\lambda)}(\mathbf{k})$ is a polarisation vector. The repeated suffix in these equations implies a summation over λ, and so the expression for the electromagnetic field becomes

$$A_\mu(x) = \frac{1}{(2\pi)^{3/2}} \int_{-\infty}^{+\infty} d\mathbf{k} \frac{1}{\sqrt{(2\omega_k)}} \sum_\lambda [a_\lambda(\mathbf{k})\, e^{ikx} + a_\lambda^\dagger(\mathbf{k})\, e^{-ikx}]\, e_\mu^{(\lambda)}(\mathbf{k})$$

$$\equiv \frac{1}{\sqrt{V}} \sum_{\mathbf{k},\lambda} \frac{1}{\sqrt{(2\omega_k)}} [a_\lambda(\mathbf{k})\, e^{ikx} + a_\lambda^\dagger(\mathbf{k})\, e^{-ikx}]\, e_\mu^{(\lambda)}(\mathbf{k}) \tag{4.164}$$

$$\mu = 1, 2, 3, 0; \quad \lambda = 1, 2, 3, 0.$$

For later convenience we will revert to $\mu = 1, 2, 3, 4$ and $\lambda = 1, 2, 3, 4$ with the condition

$$a_4(\mathbf{k}) = ia_0(\mathbf{k}), \qquad a_4^\dagger(\mathbf{k}) = ia_0^\dagger(\mathbf{k}),$$

that is the Hermitian conjugate of $a_4(\mathbf{k})$ is $-a_4^\dagger(\mathbf{k})$.

The term $e_\mu^{(\lambda)}(\mathbf{k})$ in (4.164) represents a four-dimensional set of unit polarisation vectors. The summation over λ is carried out for each $(\mathbf{k})$ term. The purpose of the unit vectors may be understood by considering a wave travelling along the z-axis so that $k_4 = i\omega_k$ and $A_4 = iA_0$ both point along the time axis

$$k_\mu = (0, 0, k, i\omega_k)$$

$$\omega_k = |\mathbf{k}| = k.$$

The four polarisation vectors can then be written as

$$\begin{aligned} e_\mu^{(1)} &= (1, 0, 0, 0) \\ e_\mu^{(2)} &= (0, 1, 0, 0) \\ e_\mu^{(3)} &= (0, 0, 1, 0) \\ e_\mu^{(4)} &= (0, 0, 0, 1) \end{aligned} \tag{4.165}$$

so that the scalar products of $e_\mu^{(\lambda)}$ with k_μ can be written as

$$\begin{aligned} k_\mu e_\mu^{(\lambda)} &= 0 \qquad &(\lambda = 1, 2) \\ &= k = \omega_k \qquad &(\lambda = 3) \\ &= i\omega_k \qquad &(\lambda = 4). \end{aligned} \tag{4.166}$$

The first two equations represent the condition $\mathbf{e} \cdot \mathbf{k} = 0$ for a classical field (§ 4.4(a)).

It is obvious that the z-axis is not unique and that $\mathbf{k}$ can point in any direction. In this case $e_\mu^{(1)}(\mathbf{k})$ and $e_\mu^{(2)}(\mathbf{k})$ are regarded as mutually orthogonal unit vectors pointing in directions at right angles to the $\mathbf{k}$-axis (*transverse polarisation vectors*), $e_\mu^{(3)}(\mathbf{k})$ points along the $\mathbf{k}$-axis (*longitudinal polarisation vector*), and $e_\mu^{(4)}(\mathbf{k})$ points in the time direction. Thus each component of the vector field $A_\mu(x)$ has four states of polarisation associated with it; we anticipate matters by saying that the fourth vector $e_\mu^{(4)}$ refers to *time-like* or *scalar photons*. It is apparent from (4.165) that the following orthonormality relations exist:

$$\begin{aligned} e_\mu^{(\lambda)} e_\mu^{(\lambda')} &= \delta_{\lambda\lambda'} \\ e_\mu^{(\lambda)} e_\nu^{(\lambda)} &= \delta_{\mu\nu}. \end{aligned} \tag{4.167}$$

4.4(e). *Quantisation of the electromagnetic field*

As in the case of the scalar field, $a_\lambda(\mathbf{k})$ and $a^\dagger_\lambda(\mathbf{k})$ represent destruction and reation operators respectively. The occupation number space on which they operate can be defined by the following expressions (compare (4.74))†:

$$a_\lambda(\mathbf{k})\,|\ldots n_{\lambda\mathbf{k}}\ldots\rangle = \sqrt{n_{\lambda\mathbf{k}}}\,|\ldots (n-1)_{\lambda\mathbf{k}}\ldots\rangle \tag{4.168}$$

$$a^\dagger_\lambda(\mathbf{k})\,|\ldots n_{\lambda\mathbf{k}}\ldots\rangle = \sqrt{(n+1)_{\lambda\mathbf{k}}}\,|\ldots (n+1)_{\lambda\mathbf{k}}\ldots\rangle.$$

These equations lead to the following commutation relations

$$[a_\lambda(\mathbf{k}), a^\dagger_{\lambda'}(\mathbf{k}')] = \delta_{\lambda\lambda'}\delta_{\mathbf{k}\mathbf{k}'} \tag{4.169}$$

since the polarisation states λ and λ' are unrelated. In addition one finds that the equivalent relations to (4.79) are

$$[a_\lambda(\mathbf{k}), a_\lambda(\mathbf{k}')] = [a^\dagger_\lambda(\mathbf{k}), a^\dagger_\lambda(\mathbf{k}')] = 0 \tag{4.170}$$

$$[a_\lambda(\mathbf{k}), a_{\lambda'}(\mathbf{k})] = [a^\dagger_\lambda(\mathbf{k}), a^\dagger_{\lambda'}(\mathbf{k})] = 0.$$

The occupation number operator

$$N_\lambda(\mathbf{k}) = a^\dagger_\lambda(\mathbf{k})\,a_\lambda(\mathbf{k}) \tag{4.171}$$

may also be formed with analogous properties to that given in § 4.3(d).

Finally, we show the action of the operators on the vacuum state, which was defined in equation (4.83) as

$$a(\mathbf{k})\,|\Psi_0\rangle = 0.$$

If the electromagnetic field is decomposed into positive and negative frequency parts

$$A^{(-)}_\mu(x) = \frac{1}{\sqrt{V}}\sum_{\mathbf{k},\lambda}\frac{1}{\sqrt{(2\omega_k)}}\,a_\lambda(\mathbf{k})\,e^{ikx}e^{(\lambda)}_\mu(\mathbf{k}) \tag{4.172}$$

$$A^{(+)}_\mu(x) = \frac{1}{\sqrt{V}}\sum_{\mathbf{k},\lambda}\frac{1}{\sqrt{(2\omega_k)}}\,a^\dagger_\lambda(\mathbf{k})\,e^{-ikx}e^{(\lambda)}_\mu(\mathbf{k})$$

$$(\mu = 1, 2, 3, 4)$$

their operation on the vacuum state produces the following equations:

$$A^{(-)}_\mu(x)\,|\Psi_0\rangle = 0 \tag{4.173}$$

$$\langle\Psi_0|\,A^{(+)}_\mu(x) = 0$$

$$A^{(+)}_\mu(x)\,|\Psi_0\rangle = \frac{1}{\sqrt{V}}\sum_{\mathbf{k},\lambda}\frac{e^{-ikx}}{\sqrt{(2\omega_k)}}\,e^{(\lambda)}_\mu(\mathbf{k})\,|\mathbf{k}_\lambda\rangle. \tag{4.174}$$

† There are certain formal problems associated with this approach; we shall not discuss them here as they are examined in detail in most advanced texts on field theory.

4.4(f). *Physical interpretation of the electromagnetic field operators*

The terms appearing in the Fourier transform (4.164) may be given physica meaning by again considering the total energy and momentum operators for th field. The Hamiltonian operator can be constructed as indicated in § 4.3(f) Using the Lagrangian (4.161) and equation (4.108), we obtain

$$H = \tfrac{1}{2} \sum_{i=1}^{3} \int d\mathbf{x} \left[\frac{\partial A_\mu}{\partial x_i} \frac{\partial A_\mu}{\partial x_i} - \frac{\partial A_\mu}{\partial x_4} \frac{\partial A_\mu}{\partial x_4} \right]$$

$$= \frac{-1}{4V} \sum_{\mathbf{k},\mathbf{k}'} \sum_{\lambda,\lambda'} \int \frac{d\mathbf{x}}{\sqrt{(\omega_k \omega_k')}} \left[a_\lambda(\mathbf{k})\, e^{ikx} - a_{\mathbf{k}}^\dagger(\mathbf{k})\, e^{-ikx} \right] \times$$

$$\times \left[a_{\lambda'}(\mathbf{k}')\, e^{ik'x} - a_{\lambda'}^\dagger(\mathbf{k}')\, e^{-ik'x} \right] \delta_{\lambda\lambda'} (\omega_k \omega_k' + \mathbf{k} \cdot \mathbf{k}'$$

$$= \tfrac{1}{2} \sum_{\mathbf{k},\lambda} \left[a_\lambda(\mathbf{k})\, a_\lambda^\dagger(\mathbf{k}) + a_\lambda^\dagger(\mathbf{k})\, a_\lambda(\mathbf{k}) \right] \omega_k . \qquad (4.175$$

A cursory inspection of this equation suggests that the Hamiltonian operator and in turn the total energy for the field, recieves contributions from the scala and longitudinal polarisation terms. This contradicts our experience in classica physics, where only the transverse polarisations contribute to the total energy of the field. In fact the first impression is wrong; we may show this by examining the effect of the subsidiary condition (4.149)

$$\frac{\partial A_\mu}{\partial x_\mu} = 0.$$

Some care must be taken in examining the full implications of this equation if the formal structure of the theory is being considered. The implications have been discussed in detail by Gupta (1950, 1951) and Bleuler (1950), and their work appears in most textbooks on quantum electrodynamics. Since the difficulties they discuss are of a formal nature, we will not repeat them. The equation (4.176) given below is based on their work. We also mention that there are many other ways of dealing with this problem.

We now return to the problem under discussion. In quantum mechanical language equation (4.149) becomes an expectation value

$$\left\langle \frac{\partial A_\mu}{\partial x_\mu} \right\rangle = 0.$$

We now construct this term. We start by postulating that only states which satisfy the condition

$$\frac{\partial A_\mu^{(-)}(x)}{\partial x_\mu} |\Psi\rangle = 0 \qquad (4.176)$$

e associated with the electromagnetic field. $A_\mu^{(-)}(x)$, is given by equation 172). The condition (4.176) leads to the relation

$$\frac{i}{\sqrt{V}} \sum_{\mathbf{k}} [a_3(\mathbf{k}) + ia_4(\mathbf{k})] \sqrt{\left(\frac{\omega_k}{2}\right)} e^{ikx} |\Psi\rangle = 0 \tag{4.177}$$

iere we have used equations (4.166).

It is apparent from (4.177) that $|\Psi\rangle$ must satisfy the relation

$$(a_3 + ia_4) |\Psi\rangle = 0 \tag{4.178}$$

r all values of $\mathbf{k}$. The expression conjugate to this is (see discussion following .164))

$$\langle\Psi| (a_3^\dagger + ia_4^\dagger) = 0 \tag{4.179}$$

$$\langle\Psi| \frac{\partial A_\mu^{(+)}}{\partial x_\mu} = 0.$$

Thus we find

$$\langle\Psi| \frac{\partial A_\mu^{(+)}}{\partial x_\mu} + \frac{\partial A_\mu^{(-)}}{\partial x_\mu} |\Psi\rangle = \langle\Psi| \frac{\partial A_\mu}{\partial x_\mu} |\Psi\rangle = \left\langle \frac{\partial A_\mu}{\partial x_\mu} \right\rangle = 0. \tag{4.180}$$

We may now use equations (4.178) and (4.179) to write

$$\langle\Psi| (a_3^\dagger - ia_4^\dagger)(a_3 + ia_4) |\Psi\rangle + \langle\Psi| (a_3^\dagger + ia_4^\dagger)(a_3 - ia_4) |\Psi\rangle = 0$$

$$\langle\Psi| a_3^\dagger a_3 + a_4^\dagger a_4 |\Psi\rangle = 0. \tag{4.181}$$

It should be noted that this result implies that

$$\langle a_3^\dagger a_3\rangle + \langle a_4^\dagger a_4\rangle = \langle N_3\rangle + \langle N_4\rangle = 0$$

ut not

$$\langle N_3\rangle = \langle N_4\rangle = 0.$$

We may now determine the eigenvalue of the Hamiltonian operator H (4.175) r the electromagnetic field. With the aid of the relation (4.169),

$$a_\lambda(\mathbf{k})\, a_\lambda^\dagger(\mathbf{k}) - a_\lambda^\dagger(\mathbf{k})\, a_\lambda(\mathbf{k}) = 1$$

nd (4.181) we obtain the following expression for the total energy of the ectromagnetic field:

$$E = \langle H\rangle = \tfrac{1}{2} \sum_{\mathbf{k},\lambda=1,2} \langle\Psi| [a_\lambda(\mathbf{k})\, a_\lambda^\dagger(\mathbf{k}) + a_\lambda^\dagger(\mathbf{k})\, a_\lambda(\mathbf{k})]\, \omega_k |\Psi\rangle$$

$$= \sum_{\mathbf{k},\lambda=1,2} \langle\Psi| [N_\lambda(\mathbf{k}) + \tfrac{1}{2}]\, \omega_k |\Psi\rangle = \sum_{\mathbf{k},\lambda=1,2} (n_{\lambda\mathbf{k}} + \tfrac{1}{2})\, \omega_k. \tag{4.182}$$

Here the term $n_{\lambda\mathbf{k}}$ represents the eigenvalue of the operator $N_\lambda(\mathbf{k})$. Thus it c be seen that if the zero point energy term is ignored, the total energy of the fie arises from terms which are associated with transverse polarisation. This res is in accordance with our knowledge of the classical field. It should be not that although the state with no transverse photons present is called the vacuu state of the electromagnetic field, the summation over **k** in (4.182) leads to infinite vacuum energy.

A similar procedure may be used to obtain the linear momentum associat with the electromagnetic field. The linear momentum operator may be shown be

$$\mathbf{P} = \tfrac{1}{2} \sum_{\mathbf{k},\lambda} [a_\lambda(\mathbf{k})\, a_\lambda^\dagger(\mathbf{k}) + a_\lambda^\dagger(\mathbf{k})\, a_\lambda(\mathbf{k})]\, \mathbf{k} \qquad (4.18$$

which possesses the expectation value

$$\langle \mathbf{P} \rangle = \sum_{\mathbf{k},\lambda=1,2} n_{\lambda\mathbf{k}} \mathbf{k} \qquad (4.18$$

on summing over all directions in space (compare (4.112)).

Thus we have shown that the form we have chosen for the electromagnet field produces physically sensible results for the expectation values for ener and momentum for the field. In Chapter 5 we shall show that the intrinsic sp associated with the electromagnetic field is one. Thus the field we have develope is suitable for describing photons.

4.5. THE DIRAC (SPINOR) FIELD

4.5(a). *The plane wave representation*

A suitable Lagrangian for the Dirac field can be constructed from the comb nation of a bilinear scalar $\bar{\psi}\psi$ (compare § 3.3(p)) and a term containing a deriv tive $\bar{\psi}\gamma_\mu(\partial\psi/\partial x_\mu)$. These two terms satisfy the conditions of § 4.2 and yield Lagrangian density

$$\mathscr{L} = a\bar{\psi}\gamma_\lambda \frac{\partial\psi}{\partial x_\lambda} + b\bar{\psi}\psi$$

$$= a\bar{\psi}\left(\gamma_\lambda \frac{\partial}{\partial x_\lambda} + \frac{b}{a}\right)\psi. \qquad (4.18$$

This Lagrangian is unsymmetrical; we will postpone discussion of this remar until § 4.5(d).

A comparison with equation (3.89) shows that we can make the identificatio

$$b/a = m.$$

A choice of $a = -1$ will be shown later to give physically sensible results for the energy and momentum of this field. Therefore we will write the Lagrangian density as

$$\mathscr{L} = -\bar{\psi}\left(\gamma_\lambda \frac{\partial}{\partial x_\lambda} + m\right)\psi. \tag{4.186}$$

The Lagrangian density has dimensions of energy per unit volume; in $\hbar = c = 1$ units the dimensions of $\mathscr{L}$ are therefore L^{-4} and those of ψ are given by the equations

$$[\mathscr{L}] = L^{-4} = \frac{1}{L}[\psi^2]$$

$$[\psi] = L^{-3/2}.$$

For historical reasons the dimensions of ψ are also normally given as $L^{-3/2}$ in c.g.s. units. This result may be achieved by writing

$$\mathscr{L} = -\hbar c\bar{\psi}\left(\gamma_\lambda \frac{\partial}{\partial x_\lambda} + \frac{mc}{\hbar}\right)\psi \equiv -\hbar c\bar{\psi}_\alpha\left((\gamma_\lambda)_{\alpha\beta}\frac{\partial}{\partial x_\lambda} + \frac{mc}{\hbar}\right)\psi_\beta$$

where we have included a reminder that ψ is a four-component field and that γ_λ is a 4×4 matrix.

The introduction of the Euler–Lagrange equations (3.228)

$$\frac{\partial \mathscr{L}}{\partial \eta_\sigma} - \frac{\partial}{\partial x_\lambda}\frac{\partial \mathscr{L}}{\partial\left(\dfrac{\partial \eta_\sigma}{\partial x_\lambda}\right)} = 0$$

into equation (4.186) leads to the covariant field equations for Dirac particles

$$\gamma_\lambda \frac{\partial \psi}{\partial x_\lambda} + m\psi = 0 \tag{4.187}$$

$$\frac{\partial \bar{\psi}}{\partial x_\lambda}\gamma_\lambda - m\bar{\psi} = 0.$$

The solutions of equations (4.187) again may be expressed as Fourier integrals (§ 4.3(b)). Thus taking the first equation, a suitable solution can be developed from the expression

$$\psi(x) = \frac{1}{(2\pi)^2}\int_{-\infty}^{+\infty} d\mathbf{p}\, dE\, \xi(p)\, e^{ipx} \tag{4.188}$$

where $\xi(p)$ is defined later. An equivalent expression to (4.188) can be written for the adjoint field, but we shall postpone discussion of the latter until after equation (4.188) has been developed.

The insertion of equation (4.188) into (4.187) yields the expression

$$\frac{1}{(2\pi)^2} \int_{-\infty}^{+\infty} d\mathbf{p}\, dE(i\gamma p + m)\, \xi(p)\, e^{ipx} = 0. \tag{4.189}$$

It is convenient to introduce the projection operator Λ^+ at this stage; the operator was defined in (3.160) as

$$\Lambda^{\pm} = \frac{\mp i\gamma_\lambda p_\lambda + m}{2m} \equiv \frac{\mp i\gamma p + m}{2m}.$$

The equation (4.189) then suggests the following form for $\xi(p)$ (compare (4.59))

$$\xi(p) = \delta(p^2 + m^2)\, \Lambda^+ \chi(p) \tag{4.190}$$

since

$$\frac{1}{(2\pi)^2} \int_{-\infty}^{+\infty} d\mathbf{p}\, dE_0(i\gamma p + m) \frac{(m - i\gamma p)}{2m}\, \delta(p^2 + m^2)\, \chi(p)\, e^{ipx} = 0$$

by virtue of the property of the δ-function

$$x\, \delta(x) = 0.$$

(compare A.6 (Appendixes, p. 703)).

The insertion of equation (4.190) into (4.188) gives the expression

$$\psi(x) = \frac{1}{(2\pi)^2} \int_{-\infty}^{+\infty} d\mathbf{p}\, dE\, \delta(p^2 + m^2)\, \Lambda^+ \chi(p)\, e^{ipx}.$$

This equation bears a strong similarity to (4.58), and an integration over E may now be performed by using the procedure outlined in equations (4.58) to (4.65). The integration yields the equation

$$\psi(x) = \frac{1}{(2\pi)^2} \int_{-\infty}^{+\infty} \frac{d\mathbf{p}}{2E_p} [\Lambda^+ \chi(p)\, e^{ipx} + \Lambda^- \chi(-p)\, e^{-ipx}] \tag{4.191}$$

where E_p represents the positive square root

$$E_p = + \sqrt{(\mathbf{p}^2 + m^2)}.$$

We recall from § 3.3(k) that Λ^+ and Λ^- represent projection operators for the positive and negative frequency states

$$\Lambda^+ u = u, \qquad \Lambda^+ v = 0$$

$$\Lambda^- u = 0, \qquad \Lambda^- v = v$$

ıd it is obvious that $\chi(p)$ must contain the spinors u and v. Therefore the spinor ɛld may be reduced to a form similar to that for the scalar and vector fields by ıtroducing the linear combinations

$$\Lambda^+\chi(p) = \sqrt{(8\pi m E_p)}\sum_r a_r(\mathbf{p})\,u_r(\mathbf{p}) \tag{4.192}$$

$$\Lambda^-\chi(-p) = \sqrt{(8\pi m E_p)}\sum_r b_r^\dagger(\mathbf{p})\,v_r(\mathbf{p}).$$

hus we find

$$\psi(x) = \frac{1}{(2\pi)^{3/2}}\sum_{r=1,2}\int_{-\infty}^{+\infty} d\mathbf{p}\sqrt{\left(\frac{m}{E_p}\right)}\,[a_r(\mathbf{p})\,u_r(\mathbf{p})\,e^{ipx} + b_r^\dagger(\mathbf{p})\,v_r(\mathbf{p})\,e^{-ipx}]$$

$$= \psi^{(-)}(x) + \psi^{(+)}(x) \tag{4.193}$$

$$\equiv \frac{1}{\sqrt{V}}\sum_{\mathbf{p},r=1,2}\sqrt{\left(\frac{m}{E_p}\right)}\,[a_r(\mathbf{p})\,u_r(\mathbf{p})\,e^{ipx} + b_r^\dagger(\mathbf{p})\,v_r(\mathbf{p})\,e^{-ipx}] \tag{4.194}$$

ʷhere

$$\psi^{(-)}(x) = \frac{1}{(2\pi)^{3/2}}\sum_r\int_{-\infty}^{+\infty} d\mathbf{p}\sqrt{\left(\frac{m}{E_p}\right)}\,a_r(\mathbf{p})u_r(\mathbf{p})e^{ipx} \tag{4.195}$$

$$\psi^{(+)}(x) = \frac{1}{(2\pi)^{3/2}}\sum_r\int_{-\infty}^{+\infty} d\mathbf{p}\sqrt{\left(\frac{m}{E_p}\right)}\,b_r^\dagger(\mathbf{p})v_r(\mathbf{p})e^{-ipx}$$

ınd their obvious equivalents in the discrete representation.

A similar type of equation may be developed for the adjoint field

$$\bar{\psi}(x) = \frac{1}{(2\pi)^{3/2}}\sum_{r=1,2}\int_{-\infty}^{+\infty} d\mathbf{p}\sqrt{\left(\frac{m}{E_p}\right)}\,[a_r^\dagger(\mathbf{p})\bar{u}_r(\mathbf{p})e^{-ipx} + b_r(\mathbf{p})\bar{v}_r(\mathbf{p})e^{ipx}]$$

$$= \bar{\psi}^{(+)}(x) + \bar{\psi}^{(-)}(x). \tag{4.196}$$

A comparison of the individual terms in the expansions (4.195) and (4.196) ;hows that

$$\bar{\psi}^{(+)} = \overline{\psi^{(-)}}, \qquad \bar{\psi}^{(-)} = \overline{\psi^{(+)}}. \tag{4.197}$$

It should be remembered that both $\psi(x)$ and $\bar{\psi}(x)$ are four-component fields, and that u and v also possess four components. Thus to be completely explicit we should write equation (4.194), for example, as

$$\psi_\alpha(x) = \frac{1}{\sqrt{V}}\sum_{\mathbf{p},r}\sqrt{\left(\frac{m}{E_p}\right)}\,[a_r(\mathbf{p})\,u_{r\alpha}(\mathbf{p})\,e^{ipx} + b_r^\dagger(\mathbf{p})\,v_{r\alpha}(\mathbf{p})\,e^{-ipx}]. \tag{4.198}$$

The spinors appearing in (4.194) and (4.196) are normalised as indicated in equations (3.148) and (3.154)

$$\bar{u}_r u_s = \delta_{rs}, \qquad \bar{v}_r v_s = -\delta_{rs}$$

that is

$$\sum_{\alpha=1}^{4} \bar{u}_{r\alpha}(\mathbf{p})\, u_{s\alpha}(\mathbf{p}) = \delta_{rs}.$$

If the normalisation condition (3.169)

$$u_r^\dagger u_s = \delta_{rs}, \qquad v_r^\dagger v_s = -\delta_{rs}$$

is used, a more suitable form for the field operators is

$$\psi(x) = \frac{1}{\sqrt{V}} \sum_{\mathbf{p},r} [a_r(\mathbf{p})\, u_r(\mathbf{p})\, e^{ipx} + b_r^\dagger(\mathbf{p})\, v_r(\mathbf{p})\, e^{-ipx}] \tag{4.199}$$

$$\bar{\psi}(x) = \frac{1}{\sqrt{V}} \sum_{\mathbf{p},r} [a_r^\dagger(\mathbf{p})\, \bar{u}_r(\mathbf{p})\, e^{-ipx} + b_r(\mathbf{p})\, \bar{v}_r(\mathbf{p})\, e^{ipx}].$$

4.5(b). *The Hamiltonian operator for the Dirac field*

We will consider the Hamiltonian for the Dirac field at this stage rather than the quantisation rules. The reason for this reversal of procedure compared with the scalar and electromagnetic fields will be shown to lie in the commutation laws for the Dirac field.

The Hamiltonian may be constructed according to the procedure already discussed in § 4.3(f). The Hamiltonian operator is (compare (4.108))

$$H = \int d\mathbf{x} \left[\frac{\partial \mathscr{L}}{\partial(\partial\psi/\partial x_4)} \frac{\partial \psi}{\partial x_4} - \mathscr{L} \right]$$

$$= \int d\mathbf{x} \left[-\bar{\psi}\gamma_4 \frac{\partial \psi}{\partial x_4} + \bar{\psi} \left(\gamma_\lambda \frac{\partial}{\partial x_\lambda} + m \right) \psi \right]. \tag{4.200}$$

Now by virtue of the field equation (4.187) the second term (the Lagrangian density) vanishes. This odd behaviour of the Lagrangian does not cause serious trouble in practice. We therefore write

$$H = -\int d\mathbf{x} \bar{\psi}\gamma_4 \frac{\partial \psi}{\partial x_4}.$$

In what follows it will be convenient to write

$$a_r^\dagger(\mathbf{p}) \equiv a^\dagger, \qquad a_{r'}(\mathbf{p}') \equiv a'$$

$$\bar{u}_r(\mathbf{p}) \equiv \bar{u}, \qquad u_{r'}(\mathbf{p}') \equiv u'.$$

Then, with the aid of (4.194) and (4.196), H becomes

$$H = -\frac{1}{V}\int d\mathbf{x} \sum_{\mathbf{p},\mathbf{p}'} \sum_{r,r'} \frac{m}{\sqrt{(E_p E_p')}} [a^\dagger \bar{u} e^{-ipx} + b\bar{v}e^{ipx}] \times$$
$$\times \gamma_4 i p_4' [a' u' e^{ip'x} - b^{\dagger\prime} v' e^{-ip'x}]$$
$$= \frac{1}{V}\int d\mathbf{x} \sum_{\mathbf{p},\mathbf{p}'} \sum_{r,r'} \frac{m E_p'}{\sqrt{(E_p E_p')}} [a^\dagger a' u^\dagger u' e^{-i(p-p')x} - a^\dagger b^{\dagger\prime} u^\dagger v' e^{-i(p+p')x}$$
$$+ b a' v^\dagger u' e^{i(p+p')x} - b b^{\dagger\prime} v^\dagger v' e^{i(p-p')x}]$$

where we have written $\bar{u} = u^\dagger \gamma_4$. We may now integrate over space and use the property of the δ-function

$$\frac{1}{V}\int d\mathbf{x}\, e^{i(\mathbf{p}-\mathbf{p}')\mathbf{x}} = \delta_{\mathbf{p}\mathbf{p}'}$$

giving

$$H = \sum_{\mathbf{p}} \sum_{r,r'} m[a_r^\dagger(\mathbf{p})\, a_{r'}(\mathbf{p})\, u_r^\dagger(\mathbf{p})\, u_{r'}(\mathbf{p}) - a_r^\dagger(\mathbf{p})\, b_{r'}^\dagger(-\mathbf{p})\, u_r^\dagger(\mathbf{p})\, v_{r'}(-\mathbf{p})$$
$$+ b_r(\mathbf{p})\, a_{r'}(-\mathbf{p})\, v_r^\dagger(\mathbf{p})\, u_{r'}(-\mathbf{p}) - b_r(\mathbf{p})\, b_{r'}^\dagger(\mathbf{p})\, v_r^\dagger(\mathbf{p})\, v_{r'}(\mathbf{p})]$$
$$= \sum_{\mathbf{p}} \sum_{r} m \cdot \frac{E_p}{m} [a_r^\dagger(\mathbf{p})\, a_r(\mathbf{p}) - b_r(\mathbf{p})\, b_r^\dagger(\mathbf{p})]. \tag{4.201}$$

In order to obtain this result we have made use of the relations given in equations (3.149), (3.155) and Table 3.3.

So far we have not considered the significance of the terms $a_r(\mathbf{p})$; they could be merely functions in a classical field theory. Since the terms $a_r(\mathbf{p})$ and $a_r^\dagger(\mathbf{p})$ are associated with the positive energy states $u_r(\mathbf{p})\, e^{ipx}$, our intention is to make them destruction and creation operators respectively. Similarly, $b_r(\mathbf{p})$ and $b_r^\dagger(\mathbf{p})$ have been constructed in such a way that they may be interpreted as destruction and creation operators for the antiparticle states.

In the case of electrons and positrons, the electrons are regarded as the positive energy states. We will therefore define the following occupation number operators (compare (4.130))

$$N_r^-(\mathbf{p}) = a_r^\dagger(\mathbf{p})\, a_r(\mathbf{p}), \qquad N_r^+(\mathbf{p}) = b_r^\dagger(\mathbf{p})\, b_r(\mathbf{p}) \tag{4.202}$$

where the positive and negative signs refer to the sign of the electrical charge. As in the case of the scalar field, the operators will have eigenvalues which give the number of particles present (compare (4.132)), for example

$$N_r^-(\mathbf{p})\, |n_{r\mathbf{p}}^-\rangle = n_{r\mathbf{p}}^-\, |\, n_{r\mathbf{p}}^-\rangle. \tag{4.203}$$

We now return to equation (4.201). If we assume that the operators a and b obey similar commutation relations to those for the scalar and electromagnetic fields, we could write

$$[b_r(\mathbf{p}), b_s^\dagger(\mathbf{p}')] = \delta_{rs}\delta_{\mathbf{pp}'} \quad (4.20$$

and equation (4.201) would then become

$$H = \sum_{\mathbf{p},r} E_p[N_r^-(\mathbf{p}) - N_r^+(\mathbf{p}) - 2]. \quad (4.20$$

The expectation values of this operator could yield all values between $+\infty$ and $-\infty$ for the energy of the free particles in a Dirac field. Negative energy eigenvalues clearly are unacceptable, and so our assumption is incorrect. The problem may be solved by adopting an anticommutation relation for the creation and destruction operators of the Dirac field

$$\{a_r(\mathbf{p}), a_s^\dagger(\mathbf{p}')\} \equiv [a_r(\mathbf{p}), a_s^\dagger(\mathbf{p}')]_+ = a_r(\mathbf{p})\, a_s^\dagger(\mathbf{p}') + a_s^\dagger(\mathbf{p}')\, a_r(\mathbf{p}) = \delta_{rs}\delta_{\mathbf{pp}'} \quad (4.206$$

$$\{b_r(\mathbf{p}), b_s^\dagger(\mathbf{p}')\} = \delta_{rs}\delta_{\mathbf{pp}'}.$$

Thus equation (4.201) becomes

$$\begin{aligned} H &= \sum_{\mathbf{p},r} E_p[a_r^\dagger(\mathbf{p})\, a_r(\mathbf{p}) - b_r(\mathbf{p})\, b_r^\dagger(\mathbf{p})] \\ &= \sum_{\mathbf{p},r} E_p[a_r^\dagger(\mathbf{p})\, a_r(\mathbf{p}) + b_r^\dagger(\mathbf{p})\, b_r(\mathbf{p}) - 2] \\ &= \sum_{\mathbf{p},r} E_p[N_r^-(\mathbf{p}) + N_r^+(\mathbf{p}) - 2]. \end{aligned} \quad (4.207$$

It can be seen that this expression makes physical sense, since the eigenvalue associated with it will always have positive definite energy (if we ignore the embarrassing term representing the zero point energy).

The total momentum operator may be developed according to the principle laid out above; by using the anticommutation relation we arrive at the following expression:

$$\begin{aligned} \mathbf{P} &= \sum_{\mathbf{p},r} \mathbf{p}[a_r^\dagger(\mathbf{p})\, a_r(\mathbf{p}) - b_r(\mathbf{p})\, b_r^\dagger(\mathbf{p})] \\ &= \sum_{\mathbf{p},r} \mathbf{p}[N_r^-(\mathbf{p}) + N_r^+(\mathbf{p}) - 2]. \end{aligned} \quad (4.208$$

On summing $\mathbf{p}$ over all directions the term $-2\sum_{\mathbf{p}} \mathbf{p}$ vanishes, and so the expectation value of $\mathbf{P}$ will be the sum of the momenta of all the particles in the field.

4.5(c). *The quantisation of the Dirac field*

In equation (4.206) an anticommutation relation was postulated between creation and destruction operators

$$\{a_r(\mathbf{p}), a_s^\dagger(\mathbf{p}')\} = \delta_{rs}\delta_{\mathbf{pp}'}.$$

Additional relations are necessary for a full description of the Dirac field

$$\begin{aligned}
\{a_r, a_r\} &= 0, & \{a_r^\dagger, a_r^\dagger\} &= 0 \\
\{a_r, a_s\} &= 0, & \{a_r^\dagger, a_s^\dagger\} &= 0 \\
\{a_r, a_r'\} &= 0, & \{a_r^\dagger, a_r^{\dagger\prime}\} &= 0 \\
\{a_r, b_s\} &= 0, & \{a_r, b_s^\dagger\} &= 0 \\
\{a_r, b_r'\} &= 0, & \{a_r, b_r^{\dagger\prime}\} &= 0 \\
\{a_r^\dagger, b_s^\dagger\} &= 0, & \{a_r^\dagger, b_r^{\dagger\prime}\} &= 0
\end{aligned} \tag{4.209}$$

where for simplicity we have written

$$a_r(\mathbf{p}) = a_r, \qquad a_r(\mathbf{p}') = a_r'.$$

The anticommutation rules (4.206) and (4.209) may be used to show that the occupation number operator has the following properties:

$$\begin{aligned}
[N_r^-(\mathbf{p})]^2 &= [a_r^\dagger a_r]^2 = a_r^\dagger a_r a_r^\dagger a_r \\
&= a_r^\dagger [1 - a_r^\dagger a_r]\, a_r = a_r^\dagger a_r = N_r^-(\mathbf{p}).
\end{aligned} \tag{4.210}$$

A similar relation holds for the operator $N_r^+(\mathbf{p})$. Thus $N_r^\pm(\mathbf{p})$ has eigenvalues 0 and 1 only; this result implies that each particle state has either 0 or 1 particle in it. This statement is, of course, the Pauli exclusion principle. Thus the requirement of a positive definite energy for the Hamiltonian of the Dirac field leads to the anticommutation relations and Fermi–Dirac statistics. We will return to this point in § 9.2(g) when we discuss the connection between spin and statistics.

The vacuum state of the Dirac field is again defined as the state with no particles present, that is a state with no momentum but infinite zero point energy. Mathematically it is defined by the relations

$$a_r(\mathbf{p})\,|\Psi_0\rangle = b_r(\mathbf{p})\,|\Psi_0\rangle = \langle\Psi_0|\, a_r^\dagger(\mathbf{p}) = \langle\Psi_0|\, b_r^\dagger(\mathbf{p}) = 0 \tag{4.211}$$

or

$$\begin{aligned}
\psi^{(-)}(x)\,|\Psi_0\rangle &= \langle\Psi_0|\, \overline{\psi}^{(+)}(x) = 0 \\
\overline{\psi}^{(-)}(x)\,|\Psi_0\rangle &= \langle\Psi_0|\, \psi^{(+)}(x) = 0.
\end{aligned}$$

The one particle states are given by

$$|\mathbf{p}_r^-\rangle = a_r^\dagger(\mathbf{p})\,|\Psi_0\rangle \tag{4.212}$$

(compare (4.84)) where the negative sign in the state vector refers to the sign of the electrical charge. If we wish to sum over momentum and spin we can write (compare (4.90))

$$\psi^{(+)}(x)\,|\Psi_0\rangle = \frac{1}{\sqrt{V}}\sum_{\mathbf{p},r}\sqrt{\left(\frac{m}{E_p}\right)}\,v_r(\mathbf{p})\,e^{-ipx}\,|\mathbf{p}_r\rangle. \qquad (4.213)$$

The relations (4.211), (4.212) and (4.213) allow us to draw up the followin useful table.

TABLE 4.2

$\psi^{(-)}(x) = \frac{1}{\sqrt{V}}\sum_{\mathbf{p},r}\sqrt{\left(\frac{m}{E_p}\right)}\,a_r(\mathbf{p})\,u_r(\mathbf{p})\,e^{ipx}$	absorbs electrons (particles)
$\psi^{(+)}(x) = \frac{1}{\sqrt{V}}\sum_{\mathbf{p},r}\sqrt{\left(\frac{m}{E_p}\right)}\,b_r^{\dagger}(\mathbf{p})\,v_r(\mathbf{k})\,e^{-ipx}$	creates positrons (antiparticles)
$\bar{\psi}^{(-)}(x) = \frac{1}{\sqrt{V}}\sum_{\mathbf{p},r}\sqrt{\left(\frac{m}{E_p}\right)}\,b_r(\mathbf{p})\,\bar{v}_r(\mathbf{p})\,e^{ipx}$	absorbs positrons
$\bar{\psi}^{(+)}(x) = \frac{1}{\sqrt{V}}\sum_{\mathbf{p},r}\sqrt{\left(\frac{m}{E_p}\right)}\,a_r^{\dagger}(\mathbf{p})\,\bar{u}_r(\mathbf{p})\,e^{-ipx}$	creates electrons

It must be remembered that the definitions are arbitrary; some authors re verse the roles of the positive and negative signs.

The Pauli exclusion principle and equations (4.209) lead to the followin equations, which we shall use later

$$a_r^{\dagger}(\mathbf{p})\,a_r^{\dagger}(\mathbf{p})\,|\Psi_0\rangle = 0 \qquad (4.214)$$

$$a_r^{\dagger}(\mathbf{p})\,a_s^{\dagger}(\mathbf{p}')\,|\Psi_0\rangle = -\,a_s^{\dagger}(\mathbf{p}')\,a_r^{\dagger}(\mathbf{p})\,|\Psi_0\rangle. \qquad (4.215)$$

4.5(d). *Charge and current operators for the Dirac field – symmetrisation*

In §§ 3.4(f.2) and 4.3(i.1) it was shown that the invariance of a field unde gauge transformations of the first kind led to the concept of the conservation o charge. We shall follow a similar developement in this section. The term

$$j_\mu = i\alpha\left[\frac{\partial\mathcal{L}}{\partial(\partial\eta_\sigma/\partial x_\mu)}\,\eta_\sigma - \frac{\partial\mathcal{L}}{\partial(\partial\eta_\sigma^*/\partial x_\mu)}\,\eta_\sigma^*\right]$$

of equation (3.250), assumes the following form:

$$j_\mu = -\,i\alpha\bar{\psi}\gamma_\mu\psi \qquad (4.216)$$

when the Lagrangian density (4.186) is inserted. This equation obviously bears a stong affinity to the probability current density of equation (3.92)

$$j_\lambda = i\bar{\psi}\gamma_\lambda\psi.$$

Thus if we make the tentative assignment

$$\alpha = -e$$

equation (4.216) would appear to be a satisfactory form for the electric current density for electrons. Similarly, we may show that for a Dirac field the conserved quantity

$$C = \alpha \int d\mathbf{x} \left[\frac{\partial \mathscr{L}}{\partial(\partial \eta_\sigma / \partial x_4)} \eta_\sigma - \frac{\partial \mathscr{L}}{\partial(\partial \eta_\sigma^* / \partial x_4)} \eta_\sigma^* \right]$$

of equation (3.251) becomes

$$C = -\alpha \int d\mathbf{x} \bar{\psi} \gamma_4 \psi = -\alpha \int d\mathbf{x} \psi^\dagger \psi. \tag{4.217}$$

The integrand of this equation is similar to that for the probability density of equation (3.77)

$$\varrho = \psi^\dagger \psi$$

and so we will write

$$C = Q = e \int d\mathbf{x} \psi^\dagger \psi \tag{4.218}$$

where Q represents the total electric charge operator for the Dirac field. The Fourier transforms (4.194) and (4.196) may be inserted in equation (4.218), and the resulting expression is of the form

$$\begin{aligned} Q &= e \sum_{\mathbf{p},r} [a_r^\dagger(\mathbf{p})\, a_r(\mathbf{p}) + b_r(\mathbf{p})\, b_r^\dagger(\mathbf{p})] \\ &= e \sum_{\mathbf{p},r} [N_r^-(\mathbf{p}) - N_r^+(\mathbf{p}) + 2] \end{aligned} \tag{4.219}$$

after the necessary spatial integrals have been performed (see § 4.5(b)).

The charge operator Q has similar properties to that developed in § 4.3(i.1) for the non-Hermitian scalar field. One difference, however, is that the individual charge states can only take the values 0 or $\pm e$, since the occupation number N only has eigenvalues 0, 1 (4.210). A further difference concerns the form of the operator; a comparison of equations (4.219) and (4.135) shows that the latter does not contain a vacuum charge term. The last term in (4.219) gives the physically senseless result that the vacuum possesses infinite charge.

The embarrassment caused by the appearance of this term can be circumvented to some extent by the process of *symmetrisation*. In a classical field this may be done by replacing all bilinear terms of the type $\eta_\sigma^* \eta_\sigma$ in the following manner:

$$\eta_\sigma^* \eta_\sigma \to \tfrac{1}{2}(\eta_\sigma^* \eta_\sigma + \eta_\sigma \eta_\sigma^*). \tag{4.220}$$

A substitution of this type does not alter a classical theory. The substitution adds a number, however, to the terms appearing in the Lagrangian in a quantised theory. The number arises by virtue of the commutation laws. It can be made to perform the useful function of cancelling out the infinite vacuum terms appearing in the expressions for the physical observables.

The expression symmetrisation is reserved in quantum field theory for those fields (particles) obeying commutation relations (or equivalently Bose–Einstein statistics). Heisenberg (1934) suggested the rule that *all Lagrangians and physical observables can be symmetrised in their boson fields and antisymmetrised in the fermion fields.* Thus in the case of the Dirac field all the bilinear covariants should be antisymmetrised. The prescription given by Heisenberg is

$$\bar{\psi}_\alpha(\Gamma_i)_{\alpha\beta}\psi_\beta \rightarrow \tfrac{1}{2}[\bar{\psi}_\alpha(\Gamma_i)_{\alpha\beta}\psi_\beta - \psi_\beta(\Gamma_i)_{\alpha\beta}\bar{\psi}_\alpha] \tag{4.221}$$

where the terms Γ_i are a product of γ-matrices as defined in (3.106). In addition to removing embarrassing (but not very important) infinite terms, the process of symmetrisation or antisymmetrisation fulfils a more important function. It provides field equations which remain physically sensible under the symmetry transformations discussed in Chapters 5 and 9.

As a result of the process of antisymmetrisation the current density operator (4.216)

$$j_\mu = ie\bar{\psi}\gamma_\mu\psi$$

becomes

$$j_\mu = \frac{ie}{2}[\bar{\psi}_\alpha(\gamma_\mu)_{\alpha\beta}\psi_\beta - \psi_\beta(\gamma_\mu)_{\alpha\beta}\bar{\psi}_\alpha] \tag{4.222}$$

whilst the charge operator (4.218) is written as

$$Q = \frac{e}{2}\int d\mathbf{x}[\psi_\alpha^\dagger\psi_\alpha - \psi_\alpha\psi_\alpha^\dagger]. \tag{4.223}$$

Upon using the techniques described previously (§§ 4.3(f) and 4.5(b)), the operator Q reduces to

$$Q = e\sum_{\mathbf{p},r}[N_r^-(\mathbf{p}) - N_r^+(\mathbf{p})] \tag{4.224}$$

and the charge current operator in four-vector notation is

$$J_\mu = e\sum_{\mathbf{p},r}\frac{p_\mu}{E_p}[N_r^-(\mathbf{p}) - N_r^+(\mathbf{p})]. \tag{4.225}$$

It should be noted that these definitions are not unique for the Dirac field. In later sections we wish to adopt the convention that the operator $a^\dagger(\mathbf{p})$ creates particles and $b^\dagger(\mathbf{p})$ creates antiparticles. Since both electrons and protons are regarded by convention as particles, we will then switch signs on the occupation number operators and write

$$Q = e\sum_{\mathbf{p},r}[\underset{\text{Protons}}{N_r^+(\mathbf{p})} - \underset{\text{Antiprotons}}{N_r^-(\mathbf{p})}]$$

for the proton field.

A similar treatment could have been used for the charged scalar field. The Lagrangian (4.119) can be replaced by one of the form

$$\mathscr{L} = -\tfrac{1}{2}\left[\left(\frac{\partial \varphi^\dagger}{\partial x_\lambda}\frac{\partial \varphi}{\partial x_\lambda} + \frac{\partial \varphi}{\partial x_\lambda}\frac{\partial \varphi^\dagger}{\partial x_\lambda}\right) + m^2(\varphi^\dagger\varphi + \varphi\varphi^\dagger)\right]. \quad (4.226)$$

The charge and current density operators obtainable from this expression are identical, however, with those given in § 4.3(i.1). No infinite terms appear.

We note finally that the equations for the current density operator are often given alternative forms; they are

$$j_\mu = \frac{ie}{2}[\bar{\psi}_\alpha(\gamma_\mu)_{\alpha\beta}\psi_\beta - \psi_\beta(\gamma_\mu)_{\alpha\beta}\bar{\psi}_\alpha]$$

$$= \frac{ie}{2}[\bar{\psi}_\alpha(\gamma_\mu)_{\alpha\beta}\psi_\beta - \psi_\beta(\gamma_\mu)^\dagger_{\beta\alpha}\bar{\psi}_\alpha] \quad (4.227)$$

$$= \frac{ie}{2}(\gamma_\mu)_{\alpha\beta}(\bar{\psi}_\alpha\psi_\beta - \psi_\beta\bar{\psi}_\alpha) \quad (4.228)$$

$$= \frac{ie}{2}[\bar{\psi}, \gamma_\mu\psi]. \quad (4.229)$$

4.6. THE COVARIANT COMMUTATION RELATIONS

4.6(a). *Introduction*

The commutation rules for quantised fields cannot be deduced in a rigorous manner. As a result there is no universally accepted procedure for constructing the commutation relations.

In this book we have chosen to set up commutation rules for annihilation and creation operators, and we will now use these rules to obtain commutation relations for the field operators. We could just as easily have proceeded in the reverse direction.

The latter approach can be made in a number of ways. In the older textbooks and articles on field theory the basic commutation law of quantum mechanics

$$[x_\alpha, p_\beta] = i\delta_{\alpha\beta} \quad (4.230)$$

was used as a starting-point. Field operators η and cells of action of volume $\delta\tau$ were then introduced, and a form of correspondence principle was then used to obtain a commutation relation. The correspondence is

$$q_\alpha(t) \equiv \eta_\alpha(\mathbf{x}, t) \quad (4.231)$$

$$p_\beta(t) \equiv \delta\tau(\mathbf{x})\,\pi_\beta(\mathbf{x}, t)$$

so that

$$[\eta_\alpha(\mathbf{x}, t), \pi_\beta(\mathbf{x}, t)]\, \delta\tau = i\delta_{\alpha\beta}.$$

The transition to the continuum is then made by summing over the cells

$$\sum_{\mathbf{x}} \delta\tau(\mathbf{x}')\, [\eta_\alpha(\mathbf{x}, t), \pi_\beta(\mathbf{x}', t)] = i\delta_{\alpha\beta}\delta_{\mathbf{x}\mathbf{x}'} = \lim_{\delta\tau\to 0} \int d\mathbf{x}' [\eta_\alpha(\mathbf{x}, t), \pi_\beta(\mathbf{x}', t)]$$

so that

$$[\eta_\alpha(\mathbf{x}, t), \pi_\beta(\mathbf{x}', t)] = i\delta_{\alpha\beta}\delta(\mathbf{x} - \mathbf{x}'). \tag{4.232}$$

The commutation rules then have to be extended to cover nonidentical times. The method has been criticised because essentially noncovariant techniques are used.

An alternative method starts by using the relativistic principle that two Hermitian operators (observables), which we will represent by the symbols $A(x)$ and $B(x')$, cannot interfere with each other when their separation is space-like (3.9)

$$[A(x), B(x')] = 0 \quad \text{for} \quad (x - x')^2 > 0. \tag{4.233}$$

This is known as *the principle of microcausality*. The method works in a logical and consistent manner for Boson fields, but the approach to Fermi–Dirac fields is less satisfactory.

The main justification for the method used in this book is that the mathematics are comparatively simple, and that the physical concepts can be clearly stated. Comparatively few assumptions have to be made in order to yield a consistent set of commutation rules for quantised fields.

4.6(b). *Scalar fields*

We recall equations (4.72), (4.78) and (4.79):

$$\varphi(x) = \frac{1}{\sqrt{V}} \sum_{\mathbf{k}} \frac{1}{\sqrt{(2\omega_k)}} [a(\mathbf{k})\, e^{ikx} + a^\dagger(\mathbf{k})\, e^{-ikx}] \tag{4.72}$$

$$[a(\mathbf{k}), a^\dagger(\mathbf{k}')] = \delta_{\mathbf{k}\mathbf{k}'} \tag{4.78}$$

$$[a(\mathbf{k}), a(\mathbf{k}')] = [a^\dagger(\mathbf{k}), a^\dagger(\mathbf{k}')] = 0. \tag{4.79}$$

A commutator $[\varphi(x), \varphi(x')]$ may be developed with the aid of these relations

$$[\varphi(x), \varphi(x')]$$

$$= \frac{1}{2V} \sum_{\mathbf{k},\mathbf{k}'} \frac{1}{\sqrt{(\omega_k \omega_{k'})}} \{[a(\mathbf{k}), a^\dagger(\mathbf{k}')]\, e^{i(kx - k'x')} + [a^\dagger(\mathbf{k}), a(\mathbf{k}')]\, e^{-i(kx - k'x')}\}$$

$$= \frac{1}{2V} \sum_{\mathbf{k}} \frac{1}{\omega_k} [e^{ik(x - x')} - e^{-ik(x - x')}]. \tag{4.234}$$

At this point it is convenient to turn to a continuous representation (4.71)

$$\frac{1}{\sqrt{V}} \sum_{\mathbf{k}} \rightarrow \frac{1}{(2\pi)^{3/2}} \int d\mathbf{k}$$

so that

$$\begin{aligned}
[\varphi(x), \varphi(x')] &= \frac{1}{(2\pi)^3} \int \frac{d\mathbf{k}}{2\omega_k} [e^{ik(x-x')} - e^{-ik(x-x')}] \\
&= \frac{i}{(2\pi)^2} \int \frac{d\mathbf{k}}{\omega_k} \sin k(x - x') \\
&= i\Delta(x - x') \equiv i\Delta(y) \qquad (4.235)
\end{aligned}$$

where

$$x - x' = y$$

$$\Delta(y) = \frac{-i}{(2\pi)^3} \int \frac{d\mathbf{k}}{2\omega_k} [e^{iky} - e^{-iky}] = \frac{1}{(2\pi)^3} \int \frac{d\mathbf{k}}{\omega_k} \sin ky. \qquad (4.236)$$

4.6(c). *Some properties of the function $\Delta(x - x')$*

The function $\Delta(x - x')$ and related quantities are of importance in the theory of interacting fields, and it is therefore necessary to examine some of its properties. For convenience of writing we shall sometimes use the form given in equation (4.236).

The properties we shall need later are as follows:

(1)
$$\Delta(y) = -\Delta(-y). \qquad (4.237)$$

This property is obvious from the form of equation (4.236).

(2) Δ is invariant under proper Lorentz transformations. This is obvious since Δ is formed from a product of the scalar fields $\varphi(x)$ and $\varphi(x')$. It is worth proving the statement explicitly, however, as an introduction to certain other theorems.

The function Δ may be written as

$$\Delta(y) = \frac{-i}{(2\pi)^3} \int d\mathbf{k}\, d\omega\, e^{iky}\, \varepsilon(k)\, \delta(k^2 + m^2) \qquad (4.238)$$

where

$$\begin{aligned}
\varepsilon(k) = \frac{\omega}{|\omega|} &= +1 \quad (\omega > 0) \\
&= -1 \quad (\omega < 0)
\end{aligned}$$

and from (4.61)

$$\begin{aligned}
\delta(k^2 + m^2) &= \delta(\omega_k^2 - \omega^2) \\
&= \frac{1}{2\omega_k} [\delta(\omega - \omega_k) + \delta(\omega + \omega_k)].
\end{aligned}$$

It is obvious that when (4.238) is integrated over all values of ω the definition of Δ given in (4.236) is obtained. Now $\varepsilon(k)$ is invariant since a proper Lorentz transformation cannot interchange past and future events. Thus equation (4.238) is obviously invariant under proper Lorentz transformations since each term in it is an invariant quantity.

(3) Δ satisfies the Klein–Gordon equation

$$(\Box_y^2 - m^2)\,\Delta(y) = 0. \tag{4.239}$$

This equation follows immediately from (4.238)

$$(\Box_y^2 - m^2)\,\Delta(y) = \frac{i}{(2\pi)^3}\int d\mathbf{k}\, d\omega\, e^{iky}\varepsilon(k)\,(k^2 + m^2)\,\delta(k^2 + m^2);$$

now terms of the form $x\delta(x)$ equal zero and so equation (4.239) is proved.

(4)
$$\Delta(y) = 0 \quad \text{for} \quad y^2 > 0. \tag{4.240}$$

We start the proof of this equation by introducing the *equal times commutator* $\Delta(x - x')_t$ which is defined as

$$i\Delta(x - x')_t = [\varphi(\mathbf{x}, t), \varphi(\mathbf{x}', t)]. \tag{4.241}$$

Now consider the situation when $t = t'$ in equation (4.235). In this circumstance we may write

$$e^{ik_4(x_4 - x'_4)} = e^{-i\omega(t-t')} = 1$$

and

$$[\varphi(x), \varphi(x')] \to i\Delta(x - x')_t = \frac{i}{(2\pi)^3}\int \frac{d\mathbf{k}}{\omega_k}\sin \mathbf{k}\cdot(\mathbf{x} - \mathbf{x}') = 0 \tag{4.242}$$

since $\sin \mathbf{k}\cdot(\mathbf{x} - \mathbf{x}')$ is an odd function of $\mathbf{k}$, and $\mathbf{k}$ is integrated over both positive and negative values.

Now at $t = t'$ the relation between events (or fields) located at x and x' is space-like (see equation (3.9))

$$(x - x')^2 > 0$$

and by a suitable Lorentz transformation any space-like four-vector $(x - x')$ can be transformed so that its fourth component vanishes in the new reference frame (Fig. 4.1).

Thus equation (4.240) is proved. Since $\varphi(x)$ and $\varphi(x')$ are both Hermitian operators, equation (4.240) is simply a restatement of (4.233)

$$[A(x), B(x')] = 0 \quad \text{for} \quad (x - x')^2 > 0.$$

As stated earlier, one can start from this equation, which is simply a statement of the second axiom of the principle of relativity (§ 3.2(a)), and obtain commutation

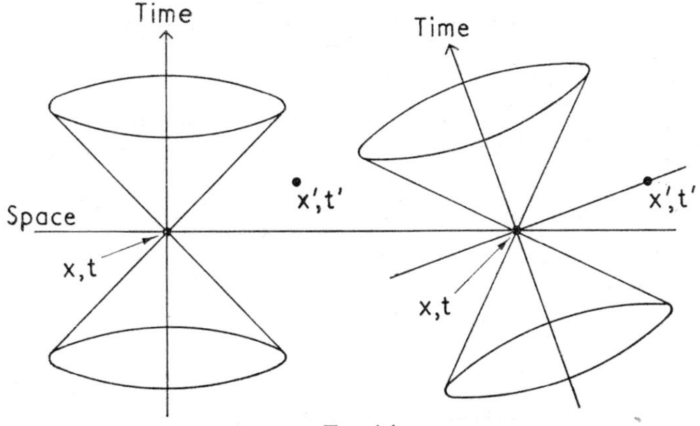

FIG. 4.1.

relations. The derivation of (4.233) implies that our starting-point is also a satisfactory one.

(5)
$$\frac{\partial}{\partial t} \Delta(x - x')_t = -\delta(\mathbf{x} - \mathbf{x}'). \tag{4.243}$$

If equation (4.235) is differentiated with respect to t we find

$$\frac{\partial}{\partial t} [\varphi(x), \varphi(x')] = i \frac{\partial}{\partial t} \Delta(x - x')$$

$$= \frac{-i}{(2\pi)^3} \int \frac{d\mathbf{k}}{2\omega_k} \omega_k [e^{i\mathbf{k}\cdot(\mathbf{x}-\mathbf{x}') - i\omega_k(t-t')} + e^{-i\mathbf{k}\cdot(\mathbf{x}-\mathbf{x}') + i\omega_k(t-t')}]$$

thus at $t = t'$ we obtain

$$\frac{\partial}{\partial t} \Delta(x - x') = \frac{-1}{(2\pi^3)} \int d\mathbf{k}\, e^{i\mathbf{k}\cdot(\mathbf{x}-\mathbf{x}')} = -\delta(\mathbf{x} - \mathbf{x}')$$

which is equation (4.243).

Equation (4.232) also follows from this relationship. In equation (4.116) it was shown that

$$\pi(x) = \frac{\partial \varphi(x)}{\partial t}$$

and so

$$[\pi(\mathbf{x}, t), \varphi(\mathbf{x}', t)] = \left[\frac{\partial \varphi(x)}{\partial t}, \varphi(x')\right]_{t=t'} = i \frac{\partial}{\partial t} \Delta(x - x')_t = -i\delta(\mathbf{x} - \mathbf{x}').$$

This equation is the same as (4.232) with terms reversed and only one component considered. The extension to scalar equations with many components is obvious.

4.6(d). *Properties of the associated Δ-functions*

An alternative form for equation (4.236) is given by considering $\Delta(y)$ as a complex function (see § 10.2(a)) and using contour integration. The method involves the use of ω as a complex variable; $\Delta(y)$ can then be expressed as

$$\Delta(y) = \frac{1}{(2\pi)^4} \int_C d\mathbf{k}\, d\omega \frac{e^{iky}}{k^2 + m^2} = \frac{1}{(2\pi)^4} \int_C d^4k \frac{e^{iky}}{k^2 + m^2} \tag{4.244}$$

where the denominator

$$k^2 + m^2 = \omega_k^2 - \omega^2$$

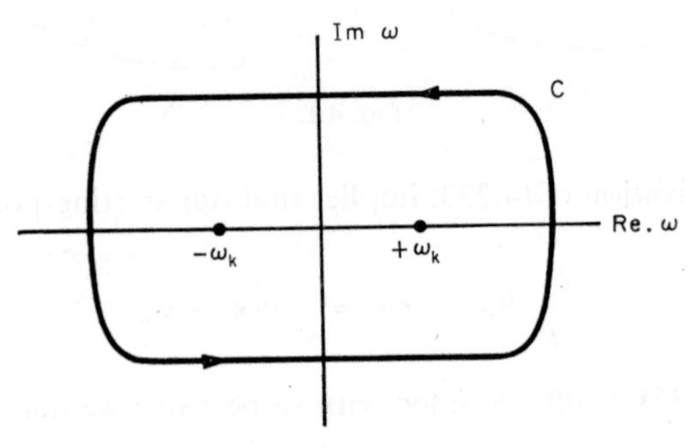

FIG. 4.2.

has poles at $\omega = \pm\omega_k$ (compare (4.61)). The integration of equation (4.244) along the contour C in the complex ω plane (Fig. 4.2) then yields (4.236)†

$$\begin{aligned}\Delta(y) &= \frac{-1}{(2\pi)^4} \int d\mathbf{k}\, e^{i\mathbf{k}\cdot\mathbf{y}} \int_C d\omega \frac{e^{-i\omega t}}{\omega^2 - \omega_k^2} \\ &= \frac{-1}{(2\pi)^4} \int d\mathbf{k}\, e^{i\mathbf{k}\cdot\mathbf{y}}\, 2\pi i \left(\frac{e^{-i\omega_k t}}{2\omega_k} + \frac{e^{i\omega_k t}}{-2\omega_k} \right) \\ &= \frac{-i}{(2\pi)^3} \int \frac{d\mathbf{k}}{2\omega_k} (e^{iky} - e^{-iky})\end{aligned}$$

where we have made use of the relations

$$\int_{-a}^{+a} dx\, e^{ibx} = \int_{-a}^{+a} dx\, e^{-ibx} \tag{4.245}$$

and

$$\frac{1}{\omega^2 - \omega_k^2} = \frac{1}{2\omega} \left[\frac{1}{\omega - \omega_k} + \frac{1}{\omega + \omega_k} \right].$$

† The technique of contour integration is outlined in § 10.2(a).

The delta functions obtainable from (4.244) are dependent upon the contour C. Now consider the contours given in Fig. 4.3.

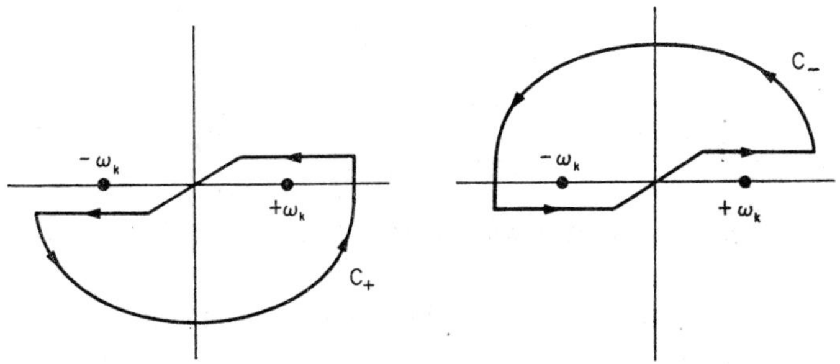

FIG. 4.3.

They yield the associated Δ-functions

$$\Delta^{\pm}(y) = \frac{1}{(2\pi)^4} \int_{C\pm} d\mathbf{k}\, d\omega \frac{e^{iky}}{k^2 + m^2}. \tag{4.246}$$

A straightforward integration over ω shows that

$$\Delta^{+}(y) = \frac{-i}{(2\pi)^3} \int \frac{d\mathbf{k}}{2\omega_k} e^{iky} \tag{4.247}$$

$$\Delta^{-}(y) = \frac{i}{(2\pi)^3} \int \frac{d\mathbf{k}}{2\omega_k} e^{-iky}$$

and so

$$\Delta(y) = \Delta^{+}(y) + \Delta^{-}(y).$$

Equations (4.247) can be cast in an alternative form which will be useful later

$$\Delta^{\pm}(y) = \frac{\mp i}{(2\pi)^3} \int d^4k e^{\pm iky} \theta(\omega)\, \delta(k^2 + m^2) \tag{4.248}$$

where

$$\theta(\omega) = 1 \quad (\omega > 0)$$
$$0 \quad (\omega < 0).$$

Considering only $\Delta^{+}(y)$ and writing $d^4k = d\mathbf{k}\, d\omega$ we find (compare (4.61))

$$\Delta^{+}(y) = \frac{-i}{(2\pi)^3} \int d\mathbf{k}\, d\omega\, e^{iky} \theta(\omega)\, \delta(\omega_k^2 - \omega^2)$$

$$= \frac{-i}{(2\pi)^3} \int \frac{d\mathbf{k}}{2\omega_k} d\omega\, e^{iky} \theta(\omega) [\delta(\omega - \omega_k) + \delta(\omega + \omega_k)]$$

$$= \frac{-i}{(2\pi)^2} \int \frac{d\mathbf{k}}{2\omega_k} e^{iky} \qquad (\omega_k > 0)$$

which is the first of equations (4.247).

The terms $\Delta^{\pm}(y)$ are related to the commutators of the expressions

$$\varphi^{(+)}(x) = \frac{1}{(2\pi)^{3/2}} \int \frac{d\mathbf{k}}{\sqrt{2\omega_k}}\, a^{\dagger}(\mathbf{k})\, e^{-ikx}$$

$$\varphi^{(-)}(x) = \frac{1}{(2\pi)^{3/2}} \int \frac{d\mathbf{k}}{\sqrt{2\omega_k}}\, a(\mathbf{k})\, e^{ikx}$$

of equation (4.70). A straightforward calculation shows that

$$\begin{aligned} [\varphi^{(+)}(x), \varphi^{(+)}(x')] &= [\varphi^{(-)}(x), \varphi^{(-)}(x')] = 0 \\ [\varphi^{(-)}(x), \varphi^{(+)}(x')] &= i\Delta^{+}(x - x') \\ [\varphi^{(+)}(x), \varphi^{(-)}(x')] &= i\Delta^{-}(x - x'). \end{aligned} \tag{4.249}$$

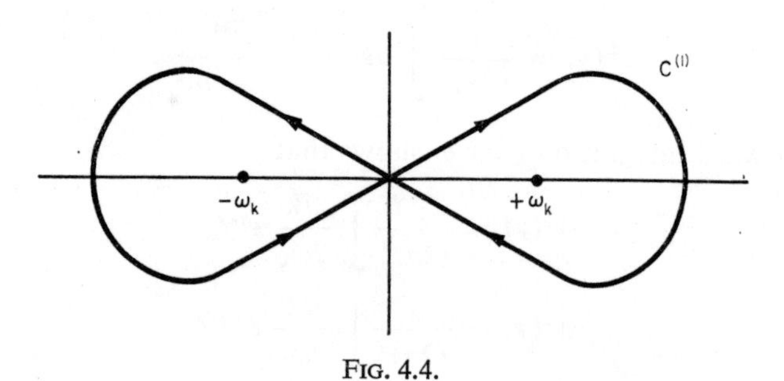

FIG. 4.4.

One further useful Δ-function may be defined. It is obtained by taking the differences of Δ^{+} and Δ^{-}

$$-i\Delta^{(1)} = \Delta^{+} - \Delta^{-}. \tag{4.250}$$

The expression $-i\Delta^{(1)}(y)$ is a mathematical function of the form

$$-i\Delta^{(1)}(y) = \frac{1}{(2\pi)^4} \int_{C(1)} d\mathbf{k}\, d\omega\, \frac{e^{iky}}{k^2 + m^2} = \frac{1}{(2\pi)^4} \int_{C(1)} d^4k\, \frac{e^{iky}}{k^2 + m^2} \tag{4.251}$$

and may be obtained by taking the contour given in Fig. 4.4.

4.6(e). *Vacuum expectation values and the Δ-functions*

The covariant commutation relations (4.235) and (4.249) are of great importance for calculations involving the interaction of fields. They provide simple and quick techniques for the analysis of quite complicated mathematical expressions. Frequent examples will be given in later chapters; as a simple introduction to the techniques we consider the *vacuum expectation value* of two scalar field functions

$$\langle \Psi_0 | \varphi(x)\,\varphi(x')\,|\Psi_0\rangle = \langle \Psi_0 | \varphi^{(+)}(x)\,\varphi^{(+)}(x') + \varphi^{(+)}(x)\,\varphi^{(-)}(x') + \varphi^{(-)}(x)\,\varphi^{(+)}(x') + \varphi^{(-)}(x)\,\varphi^{(-)}(x')\,|\Psi_0\rangle. \quad (4.252)$$

Now, from equations (4.88) and (4.89)

$$\varphi^{(-)}\,|\Psi_0\rangle = 0, \qquad \langle \Psi_0 |\,\varphi^{(+)} = 0$$

we know that only the third term in equation (4.252) gives a nonzero contribution, and so we can write

$$\begin{aligned}\langle\Psi_0| \varphi(x)\,\varphi(x')\,|\Psi_0\rangle &= \langle \Psi_0 | \varphi^{(-)}(x)\,\varphi^{(+)}(x')\,|\Psi_0\rangle \\ &= \frac{1}{(2\pi)^3}\int \frac{d\mathbf{k}\,d\mathbf{k}'}{2\sqrt{(\omega_k\omega_k')}}\,e^{i(kx-k'x')}\,\langle\Psi_0|\,a(\mathbf{k})\,a^\dagger(\mathbf{k}')\,|\Psi_0\rangle \\ &= \frac{1}{(2\pi)^3}\int \frac{d\mathbf{k}}{2\omega_k}\,e^{ik(x-x')}\,\langle\Psi_0|\,\hat{1} + a^\dagger(\mathbf{k})\,a(\mathbf{k})\,|\Psi_0\rangle \\ &= \frac{1}{(2\pi)^3}\int \frac{d\mathbf{k}}{2\omega_k}\,e^{ik(x-x')} = i\Delta^+(x-x'). \end{aligned} \quad (4.253)$$

In deriving this equation we have used equations (4.70), (4.73), the commutation rule (4.78) and the property (4.83) that

$$a(\mathbf{k})\,|\Psi_0\rangle = 0.$$

In a similar manner one may prove that

$$\langle\Psi_0|\,\varphi(x')\,\varphi(x)\,|\Psi_0\rangle = -i\Delta^-(x-x').$$

This equation leads to the relation

$$\langle\Psi_0|\,\{\varphi(x), \varphi(x')\}\,|\Psi_0\rangle = \Delta^{(1)}(x-x')$$

where

$$\{\varphi(x), \varphi(x')\} = \varphi(x)\,\varphi(x') + \varphi(x')\,\varphi(x).$$

In making practical calculations it is frequently necessary to sort out the temporal sequence of the processes involved in a given interaction. This evaluation may be performed with the aid of the Dyson *chronological* or *time-ordered product* of operators. It is defined in the following manner

$$\begin{aligned} P[\varphi(x)\,\varphi(x')] &= \varphi(x)\,\varphi(x') \quad (t > t') \\ &= \varphi(x')\,\varphi(x) \quad (t < t'). \end{aligned} \quad (4.254)$$

In the product time flows in such a manner that earliest operators are on the right, so that they are the first to operate on the state functions. The product not restricted to two operators, for example

$$P[\varphi(x)\,\varphi(x')\,\varphi(x'')\,\varphi(x''')] = \varphi(x)\,\varphi(x')\,\varphi(x'')\,\varphi(x''')$$

if

$$t > t' > t'' > t'''.$$

The chronological ordering of two operators can be written conveniently wit the aid of the symbol $\varepsilon(x - x')$, which will be defined by its possession of th following properties

$$\begin{aligned} \varepsilon(x - x') = {} & +1 \quad (t > t') \\ & -1 \quad (t < t'). \end{aligned} \tag{4.25}$$

Thus we find that equation (4.254) can be written as

$$P[\varphi(x)\,\varphi(x')] = \frac{1 + \varepsilon(x - x')}{2}\,\varphi(x)\,\varphi(x') + \frac{1 - \varepsilon(x - x')}{2}\,\varphi(x')\,\varphi(x) \tag{4.256}$$

and the vacuum expectation value of the time-ordered product of operators be comes

$$\begin{aligned} &\langle \Psi_0|\, P[\varphi(x)\,\varphi(x')]\,|\Psi_0\rangle \\ &= \frac{1 + \varepsilon(x + x')}{2}\, i\Delta^+(x - x') - \frac{1 - \varepsilon(x - x')}{2}\, i\Delta^-(x - x') \\ &= \tfrac{1}{2}\Delta_F(x - x'). \end{aligned} \tag{4.257}$$

In deriving the second line of this equation we have made use of equatio (4.253). The symbol

$$\Delta_F(y) = [1 + \varepsilon(y)]\, i\Delta^+(y) - [1 - \varepsilon(y)]\, i\Delta^-(y) \tag{4.258}$$

is called *Feynman's Δ-function*. It is apparent from equation (4.258) that Δ_F ha the following properties:

$$\begin{aligned} \Delta_F(y) &= 2i\Delta^+(y) \qquad (t_y > 0 \equiv t > t') \\ \Delta_F(y) &= -2i\Delta^-(y) \qquad (t_y < 0 \equiv t < t'). \end{aligned} \tag{4.259}$$

The Δ_F function has the following integral representation

$$\Delta_F(y) = -\frac{2i}{(2\pi)^4} \int_{C_F} d^4k \, \frac{e^{iky}}{k^2 + m^2} \tag{4.260}$$

where the contour C_F goes below the pole at $\omega = -\omega_k$ and above it at $\omega = \omega_k$ (Fig. 4.5).

The proof that (4.260) is a satisfactory form for Δ_F may be readily demonstrated. The fourth component of the term e^{iky} is $e^{-i\omega t_y}$ which may be broken down into its components for the real and imaginary parts of ω as

$$e^{-i\omega t_y} = e^{-i(\mathrm{Re}\,\omega + i\,\mathrm{Im}\,\omega)\,t_y} = e^{-i\,\mathrm{Re}\,\omega t_y} e^{\mathrm{Im}\,\omega t_y}. \tag{4.261}$$

Now for $t_y > 0$ we can close the contour of Fig. 4.5 by means of a circle of infinite radius in the lower half plane, and for $|\omega| \to \infty$ and $\mathrm{Im}\,\omega < 0$ it is apparent from (4.261) that there is no contribution to the integral from the semi-

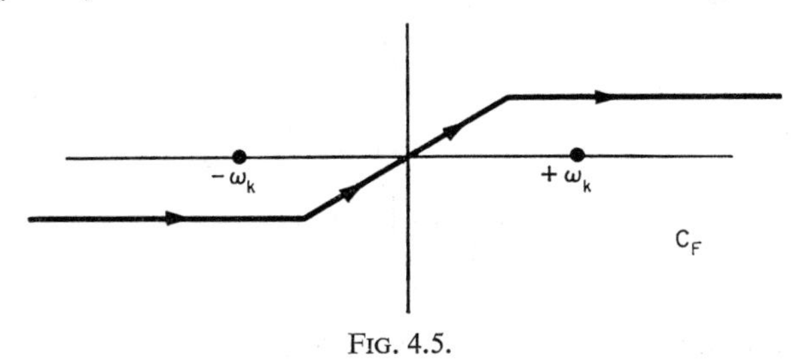

FIG. 4.5.

circle. The contour C_F in Fig. 4.5 encircles the pole $\omega = \omega_k$ only in a clockwise direction. Thus from Fig. 4.3 and equation (4.246) we may infer that

$$\Delta_F(y) = -2i(-\Delta^+(y)) = 2i\Delta^+(y) \quad \text{for} \quad t_y > 0$$

which is equation (4.259). The second part of equation (4.259) follows by closing the contour C_F by an infinite semicircle in the upper half plane, $\mathrm{Im}\,\omega > 0$, when $t_y < 0$, that is $t' > t$.

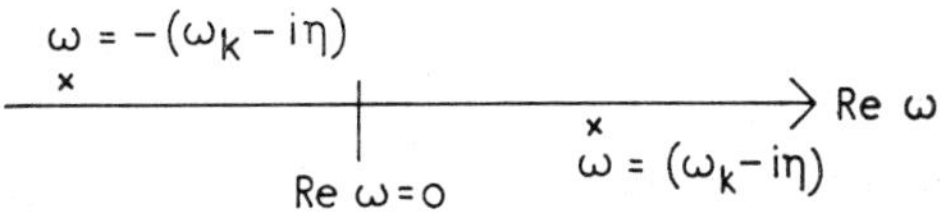

$\omega = -(\omega_k - i\eta)$

Re ω

Re $\omega = 0$

$\omega = (\omega_k - i\eta)$

FIG. 4.6.

An alternative technique is worth mentioning, as it is frequently used in quantum electrodynamics and in dispersion relations. The integration of (4.260) was performed by deforming the contours to pass around the poles. Alternatively, the same result may be obtained by integrating along the real ω-axis to form $-\infty$ to $+\infty$ and by displacing the poles an infinitesimal distance from the real axis (Fig. 4.6).

$$\begin{aligned}\Delta_F(y) &= -\frac{2i}{(2\pi)^4}\int_{-\infty}^{+\infty} d^4k\,\frac{e^{iky}}{-[\omega^2 - (\omega_k - i\eta)^2]} \\ &= -\frac{2i}{(2\pi)^4}\int_{-\infty}^{+\infty} d^4k\,\frac{e^{iky}}{k^2 + m^2 - i\varepsilon}\end{aligned} \tag{4.262}$$

where $\varepsilon = 2\omega_k\eta$ is a small positive number, which is allowed to tend to zero after integration. The contours of integration may again be closed by taking semicircles of infinite radius in the upper and lower half planes.

4.6(f). *Advanced and retarded Δ-functions*

Two further Δ-functions, which appear in the theory of elementary particles, are the advanced and retarded Δ-function, Δ_A and Δ_R. They are defined by the expressions

$$\Delta_A(y) = \frac{1}{(2\pi)^4}\int_{C_A} d^4k \frac{e^{iky}}{k^2+m^2} \tag{4.263}$$

$$\Delta_R(y) = \frac{1}{(2\pi)^4}\int_{C_R} d^4k \frac{e^{iky}}{k^2+m^2}.$$

The contours C_A and C_R are shown in Fig. 4.7.

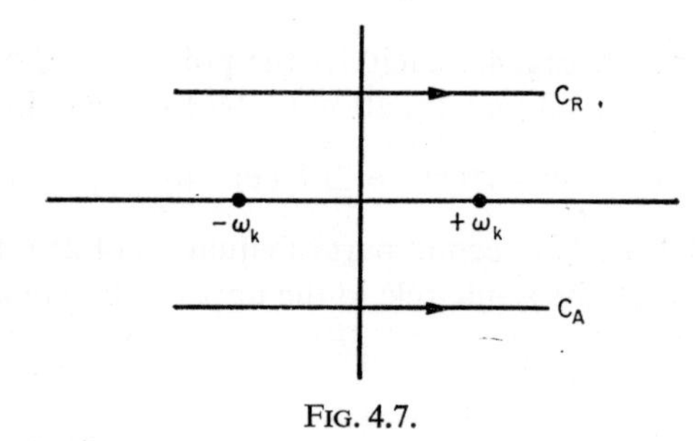

FIG. 4.7.

If $t_y > 0$ the contours can be closed by a large semicircle in the lower half plane (see the argument following equation (4.261)). We then obtain

$$\Delta_A(y) = 0 \qquad \text{for} \quad t_y > 0 \equiv t > t' \tag{4.264}$$

$$\Delta_R(y) = -\Delta(y)$$

where the definition (4.244) has been used. The negative sign appears because C_R is taken in the clockwise direction. When $t_y < 0$, the contours can be closed in the upper half plane. We then find

$$\Delta_A(y) = \Delta(y) \quad \text{for} \quad t_y < 0 \equiv t' > t \tag{4.265}$$

$$\Delta_R(y) = 0.$$

The definitions given above then lead to the relations

$$\Delta(y) = \Delta_A(y) - \Delta_R(y) \tag{4.266}$$

$$\Delta_A(y) = \Delta_R(-y).$$

We finally mention one further property of the Δ-functions. In equation (4.239) it was shown that

$$(\Box_y^2 - m^2)\,\Delta(y) = 0.$$

We record without proof the following relations:

$$(\Box_y^2 - m^2)\,\Delta^{\pm}(y) = 0 \tag{4.267}$$

$$(\Box_y^2 - m^2)\,\Delta^{(1)}(y) = 0$$

and

$$(\Box_y^2 - m^2)\,\Delta_F(y) = 2i\delta(y) \tag{4.268}$$

$$(\Box_y^2 - m^2)\,\Delta_A(y) = -\,\delta(y)$$

$$(\Box_y^2 - m^2)\,\Delta_R(y) = -\,\delta(y)$$

where

$$\delta(y) = \delta(x - x')\,\delta(y - y')\,\delta(z - z')\,\delta(t - t'). \tag{4.269}$$

4.6(g). *Commutation relations for the electromagnetic field*

The commutation relations for the electromagnetic field are similar to those for the scalar field. The starting-point is the commutator $[A_\mu(x), A_\nu(x')]$. This expression may be treated by the technique described in § 4.6(b). In this case the relations (4.167), (4.169) and (4.170) are used

$$e_\mu^{(\lambda)} e_\nu^{(\lambda)} = \delta_{\mu\nu} \tag{4.167}$$

$$[a_\lambda(\mathbf{k}),\, a_{\lambda'}^\dagger(\mathbf{k}')] = \delta_{\lambda\lambda'}\delta_{\mathbf{kk}'} \tag{4.169}$$

$$[a_\lambda(\mathbf{k}),\, a_\lambda(\mathbf{k}')] = [a_\lambda^\dagger(\mathbf{k}),\, a_\lambda^\dagger(\mathbf{k}')] = 0. \tag{4.170}$$

We then find that

$$[A_\mu(x),\, A_\nu(x')] = \frac{1}{(2\pi)^3}\int \frac{d\mathbf{k}}{2\omega_k}\,[e^{ik(x-x')} - e^{-ik(x-x')}]\,\delta_{\mu\nu}. \tag{4.270}$$

This equation is obviously equivalent to (4.235). We will introduce the symbol

$$D(y) \equiv D(x - x') = \frac{-i}{(2\pi)^3}\int \frac{d\mathbf{k}}{2\omega_k}\,[e^{ik(x-x')} - e^{-ik(x-x')}]$$

$$= \frac{-1}{(2\pi)^3}\int \frac{d\mathbf{k}}{\omega_k}\,e^{i\mathbf{k}\cdot\mathbf{y}}\sin\omega t_y. \tag{4.271}$$

Thus equation (4.270) becomes

$$[A_\mu(x),\, A_\nu(x')] = iD(x - x')\,\delta_{\mu\nu} \tag{4.272}$$

which bears a close resemblance to (4.235).

The function $D(x - x')$ vanishes everywhere except on the light cone. This result may be shown by integrating over all spatial directions in (4.271); writing

$$|\mathbf{x} - \mathbf{x}'| = r, \qquad t = (t - t')$$

equation (4.271) then becomes

$$D(x - x') = \frac{1}{8\pi^2 r} \int_0^\infty d\omega_k [e^{i\omega_k(r+t)} + e^{-i\omega_k(r+t)} - e^{i\omega_k(r-t)} - e^{-i\omega_k(r-t)}]$$

$$= \frac{1}{8\pi^2 r} \int_{-\infty}^{+\infty} d\omega_k \, [e^{i\omega_k(r+t)} - e^{i\omega_k(r-t)}]$$

$$= \frac{1}{4\pi r} [\delta(r + t) - \delta(r - t)].$$

It should be noted that this equation is correct only if the mass of the particles associated with the field is zero.

The function D closely resembles the Δ-function of § 4.6(a); in fact we may define $D(x - x')$ as

$$D(x - x') = \lim_{m \to 0} \Delta(x - x', m) = \frac{1}{(2\pi)^4} \int_C d^4k \, \frac{e^{ik(x-x')}}{k^2} \tag{4.273}$$

(compare (4.244)). The associated forms may be obtained by integrating over the appropriate contours. The most important of these terms is again the function D_F which is equivalent to Δ_F of equation (4.262)

$$D_F(x - x') = \lim_{m \to 0} \Delta_F(x - x', m) = -\frac{2i}{(2\pi)^4} \int_{-\infty}^{+\infty} d^4k \, \frac{e^{ik(x-x')}}{k^2 - i\varepsilon}. \tag{4.274}$$

D_F appears when the vacuum expectation value for the chronological product of operators is considered (compare (4.257))

$$\langle \Psi_0 | \, P[A_\mu(x) \, A_\nu(x')] \, | \Psi_0 \rangle = \tfrac{1}{2} D_F(x - x') \, \delta_{\mu\nu}. \tag{4.275}$$

The proof of this equation may be developed in exactly the same manner as for the scalar fields.

4.6(*h*). *Commutation relations for the Dirac field*

The rules for constructing commutation relations for the Dirac field are similar to those for the scalar field, except that commutators are replaced by anticommutators. We first consider an expression of the form

$$\{\psi_\alpha(x), \bar{\psi}_\beta(x')\} = \psi_\alpha(x)\,\bar{\psi}_\beta(x') + \bar{\psi}_\beta(x')\,\psi_\alpha(x).$$

This expression may be broken down into the more manageable components, $\psi^{(-)}(x)$ ...

$$\begin{aligned}\{\psi_\alpha(x), \bar{\psi}_\beta(x')\} &= \{[\psi_\alpha^{(-)}(x) + \psi_\alpha^{(+)}(x)], [\bar{\psi}_\beta^{(+)}(x') + \bar{\psi}_\beta^{(-)}(x')]\}\\ &= \{\psi_\alpha^{(-)}(x), \bar{\psi}_\beta^{(+)}(x')\} + \{\psi_\alpha^{(-)}(x), \bar{\psi}_\beta^{(-)}(x')\} + \{\psi_\alpha^{(+)}(x), \bar{\psi}_\beta^{(+)}(x')\}\\ &\quad + \{\psi_\alpha^{(+)}(x), \bar{\psi}_\beta^{(-)}(x')\}\end{aligned} \tag{4.276}$$

where the separate terms are defined in equations (4.195) and (4.196). Let us examine the first term

$$\begin{aligned}&\{\psi_\alpha^{(-)}(x), \bar{\psi}_\beta^{(+)}(x')\}\\ &= \frac{1}{(2\pi)^3}\int d\mathbf{p}\, d\mathbf{p}'\, \frac{m}{\sqrt{(E_p E_p')}} \sum_{r,r'} \{a_r(\mathbf{p}), a_{r'}^\dagger(\mathbf{p}')\}\, u_{r\alpha}(\mathbf{p})\, \bar{u}_{r'\beta}(\mathbf{p}')\, e^{i(px-p'x')}\\ &= \frac{1}{(2\pi)^3}\int d\mathbf{p}\, \frac{m}{E_p} \sum_r u_{r\alpha}(\mathbf{p})\, \bar{u}_{r\beta}(\mathbf{p})\, e^{ip(x-x')}\\ &= \frac{1}{(2\pi)^3}\int d\mathbf{p}\, \frac{m}{E_p}\, \Lambda_{\alpha\beta}^{+}\, e^{ip(x-x')}\\ &= \frac{1}{(2\pi)^3}\int \frac{d\mathbf{p}}{2E_p}\, (-i\gamma_\lambda p_\lambda + m)_{\alpha\beta}\, e^{ip(x-x')}\\ &= -\left(\gamma_\lambda \frac{\partial}{\partial x_\lambda} - m\right)_{\alpha\beta} \frac{1}{(2\pi)^3}\int \frac{d\mathbf{p}}{2E_p}\, e^{ip(x-x')}\\ &= -i\left(\gamma_\lambda \frac{\partial}{\partial x_\lambda} - m\right)_{\alpha\beta} \Delta^{+}(x-x').\end{aligned} \tag{4.277}$$

In deriving this expression we have used equations (4.206), (3.168), (3.160) and (4.247). It is customary to represent the expression following $-i$ in (4.277) by the symbol $S_{\alpha\beta}^{+}(x-x')$

$$S_{\alpha\beta}^{+}(x-x') = \left(\gamma_\lambda \frac{\partial}{\partial x_\lambda} - m\right)_{\alpha\beta} \Delta^{+}(x-x'). \tag{4.278}$$

When the remaining terms in (4.276) are examined it is found that only the final term gives a nonzero result. The second and third vanish because they contain anticommutators of the type $\{a^\dagger, b^\dagger\}$ (see equations (4.209)). The last term in

(4.276) may be treated in the same manner as the first. We then find that

$$\{\psi_\alpha^{(+)}(x), \overline{\psi}_\beta^{(-)}(x')\} = -i\left(\gamma_\lambda \frac{\partial}{\partial x_\lambda} - m\right)_{\alpha\beta} \Delta^-(x - x') = -iS^-_{\alpha\beta}(x - x'). \quad (4.279)$$

The final expression for the anticommutation relation (4.276) therefore become[s]

$$\begin{aligned}\{\psi_\alpha(x), \overline{\psi}_\beta(x')\} &= -iS^+_{\alpha\beta}(x - x') - iS^-_{\alpha\beta}(x - x') \\ &= -iS_{\alpha\beta}(x - x'). \end{aligned} \quad (4.280)$$

In a similar manner one can show that

$$\{\psi_\alpha(x), \psi_\beta(x')\} = \{\overline{\psi}_\alpha(x), \overline{\psi}_\beta(x')\} = 0. \quad (4.281)$$

This result arises because the expressions written above only contain anti-commutators of the type $\{a^\dagger, a^\dagger\}$, which vanish by virtue of (4.209). Equations (4.277) and (4.279) are in fact the only nonvanishing commutators for the Dirac field.

The terms $S^+_{\alpha\beta}(x - x')$ and $S^-_{\alpha\beta}(x - x')$ appear in the expressions for the vacuum expectation values for the products of Dirac operators. Consider, for example, the expression $\langle\Psi_0|\, \psi_\alpha(x)\, \overline{\psi}_\beta(x')\, |\Psi_0\rangle$; following the treatment given in equations (4.252), (4.253) and (4.277) we may write

$$\begin{aligned}\langle\Psi_0|\, \psi_\alpha(x)\, \overline{\psi}_\beta(x')\, |\Psi_0\rangle &= \langle\Psi_0|\, \psi_\alpha^{(-)}\, \overline{\psi}_\beta^{(+)}(x')\, |\Psi_0\rangle \\ &= \frac{1}{(2\pi)^3}\int d\mathbf{p}\, d\mathbf{p}' \frac{m}{\sqrt{(E_p E_p')}} \sum_{r,r'} \langle\Psi_0|\, a_r(\mathbf{p})\, a^\dagger_{r'}(\mathbf{p}')\, |\Psi_0\rangle\, u_{r\alpha}(\mathbf{p})\, \bar{u}_{r'\beta}(\mathbf{p}')\, e^{i(px - p'x')} \\ &= \frac{1}{(2\pi)^3}\int d\mathbf{p}\, \frac{m}{E_p} \sum_r \langle\Psi_0 \mid \Psi_0\rangle\, u_{r\alpha}(\mathbf{p})\, \bar{u}_{r\beta}(\mathbf{p})\, e^{ip(x-x')} \\ &= \frac{1}{(2\pi)^3}\int d\mathbf{p}\, \frac{(-i\gamma_\lambda p_\lambda + m)_{\alpha\beta}}{2E_p}\, e^{ip(x-x')} = -iS^+_{\alpha\beta}(x - x'). \end{aligned} \quad (4.282)$$

The properties of the term $S(x - x')$ and its associated functions may be readily obtained from equation (4.280). We may write

$$\begin{aligned} S(x - x') = S^+(x - x') + S^-(x - x') &= \left(\gamma_\lambda \frac{\partial}{\partial x_\lambda} - m\right)[\Delta^+(x - x') + \Delta^-(x - x')] \\ &= \left(\gamma_\lambda \frac{\partial}{\partial x_\lambda} - m\right)\Delta(x - x') \end{aligned} \quad (4.283)$$

(compare § 4.6(d)). The properties which we have established for the functions for scalar fields in § 4.6(b), (c) and (d) now allow us to write down the mathematical properties of S. If S is written as a contour integral we find, by virtue o[f]

.244), that

$$S(x - x') = \left(\gamma_\lambda \frac{\partial}{\partial x_\lambda} - m\right) \frac{1}{(2\pi)^4} \int_C d^4p \frac{e^{ip(x-x')}}{p^2 + m^2}$$

$$= \frac{i}{(2\pi)^4} \int_C d^4p \frac{\gamma_\lambda p_\lambda + im}{p^2 + m^2} e^{ip(x-x')} \tag{4.284}$$

here the contour is that of Fig. 4.2.

Similarly the application of the differential operator $(\gamma_\lambda \partial/\partial x_\lambda - m)$ to the ssociated Δ-functions given in § 4.6(c), (d) and (e) yields the appropriate S-unctions. The most important of these functions is the S_F term

$$S_F(x - x') = \left(\gamma_\lambda \frac{\partial}{\partial x_\lambda} - m\right) \Delta_F(x - x')$$

$$= \frac{2}{(2\pi)^4} \int_{C_F} d^4p \frac{\gamma_\lambda p_\lambda + im}{p^2 + m^2} e^{ip(x-x')} \tag{4.285}$$

$$= \frac{2}{(2\pi)^4} \int_{-\infty}^{+\infty} d^4p \frac{\gamma_\lambda p_\lambda + im}{p^2 + m^2 - i\varepsilon} e^{ip(x-x')}$$

$$= \frac{-2i}{(2\pi)^4} \int_{-\infty}^{+\infty} d^4p \frac{i\gamma_\lambda p_\lambda - m}{p^2 + m^2 - i\varepsilon} e^{ip(x-x')} \tag{4.286}$$

here we have used equations (4.260), (4.261) and (4.262); the contour in (4.285) that of Fig. 4.5. Equation (4.286) represents the situation when the poles have een displaced from the real E-axis, as in Fig. 4.6.

It is sometimes convenient to use the equation

$$\frac{1}{\gamma_\lambda p_\lambda - im} = \frac{\gamma_\lambda p_\lambda + im}{p^2 + m^2} \tag{4.287}$$

o re-write (4.285) in the following form:

$$S_F(x - x') = \frac{2}{(2\pi)^4} \int_{C_F} d^4p \frac{e^{ip(x-x')}}{\gamma_\lambda p_\lambda - im}. \tag{4.288}$$

The S_F term appears in vacuum expectation values for the time-ordered products of Dirac fields. A second chronological product will now be introduced; it was first used by Wick (1950) and will be denoted by the symbol T. It differs from the Dyson chronological product by changing signs for Dirac operators; is identical with the P-product for boson fields, however.

$$T[\varphi(x)\,\varphi(x')] = P[\varphi(x)\,\varphi(x')] \qquad (4.28$$

$$T[A_\mu(x)\,A_\nu(x')] = P[A_\mu(x)\,A_\nu(x')]$$

$$T[C(x)\,D(x')] = \varepsilon(x-x')\,P[C(x)\,D(x')] = C(x)\,D(x') \quad (t > t') \qquad (4.29$$

$$= -D(x')C(x) \quad (t < t')$$

where $\varepsilon(x - x')$ is defined in (4.255) and $C(x)$ represents an operator $\psi_\alpha(x)$ $\overline{\psi}_\alpha(x)$, whilst $D(x')$ is equivalent to $\psi_\beta(x')$ or $\overline{\psi}_\beta(x')$.

The T-product is more useful than the P-product for Dirac fields because the anticommutation properties of fermion operators.

The vacuum expectation value for a T-product may be found without difficulty. The method used is the same as in § 4.6(e), and equation (4.282) may be employed. The result obtained is given by the following equations:

$$\langle\Psi_0|\,T[\psi_\alpha(x)\,\overline{\psi}_\beta(x')]\,|\Psi_0\rangle = -\tfrac{1}{2}\,S_{F\alpha\beta}(x-x') \qquad (4.29$$

$$\langle\Psi_0|\,T[\overline{\psi}_\beta(x')\,\psi_\alpha(x)]\,|\Psi_0\rangle = \tfrac{1}{2}\,S_{F\alpha\beta}(x-x').$$

4.6(i). *Vacuum expectation values for the Dirac current operator*

The companion expression to equation (4.282) may be written as

$$\langle\Psi_0|\,\overline{\psi}_\beta(x')\,\psi_\alpha(x)\,|\Psi_0\rangle = -iS^-_{\alpha\beta}(x-x'). \qquad (4.29$$

This equation can be used to obtain an expression for the vacuum expectation value of the current operator. If the current density is defined by (3.92)

$$j_\lambda = i\overline{\psi}\gamma_\lambda\psi$$

we find, with the aid of (4.279) and (4.277),

$$\begin{aligned}\langle\Psi_0|\,j_\lambda\,|\Psi_0\rangle &= i\lim_{x\to x'}\sum_{\alpha\beta}\langle\Psi_0|\,\overline{\psi}_\beta(x')\,(\gamma_\lambda)_{\beta\alpha}\,\psi_\alpha(x)\,|\Psi_0\rangle\\ &= i\lim_{x\to x'}\sum_{\alpha\beta}(\gamma_\lambda)_{\beta\alpha}\,\langle\Psi_0|\,\overline{\psi}_\beta(x')\,\psi_\alpha(x)\,|\Psi_0\rangle\\ &= \lim_{x\to x'}\sum_{\alpha\beta}(\gamma_\lambda)_{\beta\alpha}\,S^-_{\alpha\beta}(x-x')\\ &= \lim_{x\to x'}\operatorname{tr}\left[\gamma_\lambda\left(\gamma_\mu\frac{\partial}{\partial x_\mu} - m\right)\Delta^-(x-x')\right]\\ &= \lim_{x\to x'}\frac{4}{(2\pi)^3}\int\frac{d\mathbf{p}}{2E_p}\,p_\lambda\,e^{-ip(x-x')}\\ &= \frac{4}{(2\pi)^3}\int\frac{d\mathbf{p}}{2E_p}\,p_\lambda. \end{aligned} \qquad (4.293$$

This integral obviously is divergent and physically meaningless. Physically one would expect that

$$\langle \Psi_0 | j_\lambda | \Psi_0 \rangle = 0 \tag{4.294}$$

since the expectation value for a current (for example a charge current) in a vacuum in the absence of external fields would be expected to be zero.

The condition stated in equation (4.294) may easily be fulfilled by using the antisymmetrised form of the current operator (4.229)

$$j_\lambda = \frac{ie}{2} [\overline{\psi}, \gamma_\lambda \psi].$$

One may then show that

$$\langle \Psi_0 | j_\lambda(x) | \Psi_0 \rangle = \frac{ie}{2} \sum_{\alpha\beta} \langle \Psi_0 | [\overline{\psi}_\beta(x), (\gamma_\lambda)_{\beta\alpha} \psi_\alpha(x)] | \Psi_0 \rangle = 0. \tag{4.295}$$

Although this equation has been written in a form which is suitable for the charge current density, similar remarks apply to the probability current density.

One further property of the Dirac current operator is worthy of note. Equation (4.229) may be expanded in the following manner:

$$\begin{aligned} j_\lambda(x) &= \frac{ie}{2} [\overline{\psi}(x), \gamma_\lambda \psi(x)] \\ &= \frac{ie}{2} (\gamma_\lambda)_{\beta\alpha} [\overline{\psi}_\beta(x), \psi_\alpha(x)] \\ &= \frac{ie}{2} (\gamma_\lambda)_{\beta\alpha} \{ [\overline{\psi}_\beta^{(+)}(x), \psi_\alpha^{(-)}(x)] + [\overline{\psi}_\beta^{(+)}(x), \psi_\alpha^{(+)}(x)] \\ &\qquad + [\overline{\psi}_\beta^{(-)}(x), \psi_\alpha^{(-)}(x)] + [\overline{\psi}_\beta^{(-)}(x), \psi_\alpha^{(+)}(x)] \} \\ &= \frac{ie}{2} (\gamma_\lambda)_{\beta\alpha} \{ [\overline{\psi}_\beta^{(+)}(x), \psi_\alpha^{(-)}(x)] - [\psi_\alpha^{(+)}(x), \overline{\psi}_\beta^{(-)}(x)] \\ &\qquad + [\overline{\psi}_\beta^{(+)}(x), \psi_\alpha^{(+)}(x)] + [\overline{\psi}_\beta^{(-)}(x), \psi_\alpha^{(-)}(x)] \}. \end{aligned} \tag{4.296}$$

Consider the term

$$[\overline{\psi}_\beta^{(+)}(x), \psi_\alpha^{(+)}(x)]. \tag{4.297}$$

Since equations (4.277) and (4.279) are the only nonzero anticommutators for the components of the Dirac field we can write

$$\{\overline{\psi}_\beta^{(+)}(x), \psi_\alpha^{(+)}(x)\} = 0 \tag{4.298}$$

and so (4.297) becomes

$$[\overline{\psi}_\beta^{(+)}(x), \psi_\alpha^{(+)}(x)] = 2\overline{\psi}_\beta^{(+)}(x)\, \psi_\alpha^{(+)}(x). \tag{4.299}$$

Similarly we may write

$$[\overline{\psi}_\beta^{(-)}(x), \psi_\alpha^{(-)}(x)] = 2\overline{\psi}_\beta^{(-)}(x)\, \psi_\alpha^{(-)}(x). \tag{4.300}$$

Now we turn to the first two brackets in the expansion of $j_\lambda(x)$. If

$$x - x' = 0$$

then equations (4.277) and (4.279) are equal and so

$$\{\bar{\psi}_\beta^{(+)}(x), \psi_\alpha^{(-)}(x)\} = \{\psi_\alpha^{(+)}(x), \bar{\psi}_\beta^{(-)}(x)\} \tag{4.301}$$

hence

$$\bar{\psi}_\beta^{(+)}(x)\,\psi_\alpha^{(-)}(x) + \psi_\alpha^{(-)}(x)\,\bar{\psi}_\beta^{(+)}(x) = \psi_\alpha^{(+)}(x)\,\bar{\psi}_\beta^{(-)}(x) + \bar{\psi}_\beta^{(-)}(x)\,\psi_\alpha^{(+)}(x)$$

so that

$$\bar{\psi}_\beta^{(+)}(x)\,\psi_\alpha^{(-)}(x) - \psi_\alpha^{(+)}(x)\,\bar{\psi}_\beta^{(-)}(x) = \bar{\psi}_\beta^{(-)}(x)\,\psi_\alpha^{(+)}(x) - \psi_\alpha^{(-)}(x)\,\bar{\psi}_\beta^{(+)}(x).$$

With the aid of this result and (4.299) and (4.300), equation (4.296) becomes

$$j_\lambda(x) = ie(\gamma_\lambda)_{\beta\alpha}[\bar{\psi}_\beta^{(+)}(x)\,\psi_\alpha^{(-)}(x) - \psi_\alpha^{(+)}(x)\,\psi_\beta^{(-)}(x) + \bar{\psi}_\beta^{(+)}(x)\,\psi_\alpha^{(+)}\psi(x)$$
$$+ \bar{\psi}_\beta^{(-)}(x)\,\psi_\alpha^{(-)}(x)]. \tag{4.302}$$

It can be seen that the creation operators are always on the left of the destruction operators in this equation. In § 8.4(a) a normal ordering operator will be introduced. The function of this operator is to place creation operators to the left of destruction operators. If the operators are fermion operators, this interchange involves a change of sign

$$N[\bar{\psi}_\beta^{(-)}(x)\,\psi_\alpha^{(+)}(x')] = -\psi_\alpha^{(+)}(x')\,\bar{\psi}_\beta^{(-)}(x) \tag{4.303}$$

but no change of sign occurs for the operators associated with boson fields

$$N[A_\mu^{(-)}(x)\,A_\nu^{(+)}(x')] = A_\nu^{(+)}(x')\,A_\mu^{(-)}(x). \tag{4.304}$$

It is therefore apparent that (4.302) may be written as

$$j_\lambda(x) = ieN[\bar{\psi}\gamma_\lambda\psi] = ie(\gamma_\lambda)_{\beta\alpha}\,N[\bar{\psi}_\beta(x)\,\psi_\alpha(x)] \tag{4.305}$$

where the expression in square brackets in (4.302) defines the normal product of two Dirac operators.

CHAPTER 5

THE SYMMETRY PROPERTIES OF FREE FIELDS

5.1. THE LORENTZ TRANSFORMATION AND THE INVARIANCE PROPERTIES OF PHYSICAL SYSTEMS

5.1 (a). *Introduction*

The statement that a physical system is invariant under a Lorentz transformation implies, in the language of field theory, that a system described by a field $\eta(x)$ yields precisely the same amount of information as one described by $\eta'(x')$ – the transformed field. The field function η, its argument x and their transformed counterparts are related in the following manner:

$$x \to x', \qquad x'_\lambda = a_{\lambda\alpha} x_\alpha \tag{5.1}$$

$$\begin{aligned} \eta_\varrho(x) \to \eta'_\sigma(x'), \quad \eta'_\sigma(x') &= S_{\sigma\varrho}\eta_\varrho(x) \\ &\equiv S_{\sigma\varrho}\eta_\varrho(a^{-1}x') \end{aligned} \tag{5.2}$$

where $a_{\lambda\alpha}$ and $S_{\sigma\varrho}$ are matrices. Equation (5.1) is merely a restatement of (3.33); an equivalent expression to (5.2) was used in discussing the Lorentz transformation of the Dirac equation in § 3.3 (m).

In quantum field theory two systems may be linked by a unitary transformation. In § 4.1 it was shown that if the operators and state vectors of a quantum mechanical system transformed as

$$\begin{aligned} A' &= UAU^{-1} \\ |\alpha'\rangle &= U\,|\alpha\rangle \end{aligned} \tag{5.3}$$

then the eigenvalues a and a' for the equations

$$A\,|\alpha\rangle = a\,|\alpha\rangle, \quad A'\,|\alpha'\rangle = a'\,|\alpha'\rangle \tag{5.4}$$

were the same. The unitary transformation also leaves the commutation (or anticommutation) rules unaltered; consider the relation

$$[A, B] = a$$

where a is a c-number, then

$$[A, B] \rightarrow [A', B'] = U[A, B]\, U^{-1} = UU^{-1}a = a. \quad (5.5)$$

The unitary transformation for the quantised field operators and state vector will be expressed in the following manner:

$$\eta_\sigma(x) \rightarrow \eta'_\sigma(x) = U\eta_\sigma(x)\, U^{-1} \quad (5.6)$$

$$|\omega\rangle \rightarrow |\omega'\rangle = U\,|\omega\rangle \quad (5.7)$$

where it will be noticed that the transformation of the operator has been chosen in such a manner that the argument x is unchanged. Thus the effects of change in x must be absorbed in the state vector $|\omega\rangle$. This procedure is not a unique one.

A relationship must obviously exist between equations (5.6) and (5.2). We shall explore it later in this chapter.

If an operator is invariant under a unitary transformation, the condition

$$A' = UAU^{-1} = A$$

holds. An important extension of this relationship involves the Lagrangian density which is a function of the field operators. The transformation may be written as

$$\mathscr{L}' = U\mathscr{L}\,[\eta(x)]\, U^{-1} = \mathscr{L}\,[U\eta(x)\, U^{-1}] = \mathscr{L}\,[\eta'(x)]. \quad (5.8)$$

Now the Lagrangian density determines the form of the field equations, and the physical properties associated with the field. Thus if a physical system is invariant under a transformation, the operation must leave the Lagrangian density unaltered, and so the condition must be

$$U\mathscr{L}\,[\eta(x)]\, U^{-1} = \mathscr{L}\,[\eta(x)] \quad (5.9)$$

$$[U, \mathscr{L}] = 0.$$

5.1(b). *Continuous transformations*

The transformations of fields divide into two main classes – continuous and discrete. The former may be made by repeated applications of an infinitesimal transformation operator. This operator is unitary, and its form was given in equation (4.29)

$$U = \hat{1} - iF$$

where $F \rightarrow 0$ and the unitarity condition implies that

$$F = F^\dagger$$

(compare (4.31)). Thus the operator F (the generator) is Hermitian and observable. In § 4.1(g) it was shown that if F satisfied the relation

$$[F, H] = 0 \quad (4.53)$$

where H is the total Hamiltonian for the system, then the expectation value for F is a conserved quantity. F is then called a constant of the motion. It is sometimes convenient to split F into two components

$$F = G\,\delta\lambda \tag{5.10}$$

where $\delta\lambda \to 0$ and is real, so that

$$G = G^{\dagger} \tag{5.11}$$

and equation (4.53) becomes

$$[F, H] = [G, H] = 0 \tag{5.12}$$

so that G is also a constant of the motion. We shall occasionally refer to F and G (and their equivalents) as conserved generators.

We return, finally, to equation (5.6)

$$U\eta_\sigma(x)\,U^{-1} = \eta'_\sigma(x).$$

Under a continuous transformation this becomes

$$\eta'_\sigma(x) = (\hat{1} - iF)\,\eta_\sigma(x)\,(\hat{1} + iF)$$

and so the local variation may be written as

$$\bar{\delta}\eta_\sigma(x) = \eta'_\sigma(x) - \eta_\sigma(x) = -i\,[F, \eta_\sigma(x)] = -i[G, \eta_\sigma(x)]\,\delta\lambda. \tag{5.13}$$

The conserved quantity C_λ (3.245), associated with Noether's theorem, satisfies this expression if the substitution

$$C_\lambda = F = G\,\delta\lambda$$

is made. Noether's theorem and the unitary transformation are alternative methods of approaching the same problem.

5.1 (c). *Discrete transformations*

These transformations possess the property that the double application of the transformation operator leads to the original system

$$UU\eta U^{-1}U^{-1} = \eta.$$

This result implies that

$$U^2 = \hat{1} \tag{5.14}$$

and so the eigenvalues of U are ± 1. Therefore U is both unitary and Hermitian. This result contrasts with that obtained for continuous transformations where U was unitary but not Hermitian; the Hermitian property was associated with the generator in a continuous transformation. Thus U itself is a dynamical variable for discrete transformations.

If the system is an invariant one, then equations (5.9) and (4.52) may be applie as well

$$[U, \mathscr{L}] = [U, H] = 0. \tag{5.15}$$

Thus U is both an observable and a constant of the motion. Its eigenvalue is 'good' quantum number (§ 4.1 (f)).

In the discussion of the classes of Lorentz transformation in § 3.2(j), th following discrete transformations were introduced:

(1) Space reflection $\mathbf{x} \rightarrow \mathbf{x}' = -\mathbf{x}$
(2) Time reflection $t \rightarrow t' = -t$.

In this chapter the implications of these transformations will be examined fo quantised fields. In addition we shall examine one which does not depend on th space-time properties of the field, namely charge conjugation. In this operation particle is changed to an antiparticle.

5.2. GAUGE TRANSFORMATIONS

Examples of continuous transformations were quoted in § 3.4(f) for a classica relativistic field theory; they were:

(1) invariance under four-vector displacement leading to conservation of four-momentum;
(2) invariance under gauge transformation giving charge conservation.

These examples are representative of the two categories of continuous transformations. We will re-examine them and consider gauge invariance first, as the mathematics involved are simpler. The principles developed can then be used for space–time transformations.

In § 3.4(f.2) it was stated that under a gauge transformation the space–time conditions of the field are left unaltered and only its internal structure changes

$$x_\lambda \rightarrow x'_\lambda = x_\lambda \tag{3.248}$$

$$\eta_\sigma(x) \rightarrow \eta'_\sigma(x') = \eta'_\sigma(x).$$

Thus equations (5.6) and (5.2) can be written as

$$\eta'_\sigma(x) = U\eta_\sigma(x)\, U^{-1} = S_{\sigma\varrho}\eta_\varrho(x). \tag{5.16}$$

It is convenient to split S into a unit and an infinitesimal operator

$$S = \hat{1} + T \tag{5.17}$$

$$S_{\sigma\varrho} = \delta_{\sigma\varrho} + T_{\sigma\varrho} = \delta_{\sigma\varrho} + \Lambda_{\sigma\varrho}\,\delta\lambda \qquad (\delta\lambda \rightarrow 0)$$

here $\delta\lambda$ is an infinitesimal arbitrary number, and $\Lambda_{\sigma\varrho}$ is an operator; it is ometimes called the transforming operator. Equation (5.16) then becomes

$$\eta'_\sigma(x) = (\delta_{\sigma\varrho} + \Lambda_{\sigma\varrho}\,\delta\lambda)\,\eta_\varrho(x) = \eta_\sigma(x) + \Lambda_{\sigma\varrho}\eta_\varrho(x)\,\delta\lambda. \quad (5.18)$$

Ve now write the unitary operator as

$$U = \hat{1} - iG\,\delta\lambda \quad (5.19)$$

compare § 5.1 (b)) so that equation (5.16) becomes

$$\eta'_\sigma(x) = (\hat{1} - iG\,\delta\lambda)\,\eta_\sigma(x)\,(\hat{1} + iG\,\delta\lambda)$$

$$= \eta_\sigma(x) - i[G, \eta_\sigma(x)]\,\delta\lambda.$$

ince the local variation of the field is given by

$$\bar{\delta}\eta_\sigma(x) = \eta'_\sigma(x) - \eta_\sigma(x)$$

ve find that

$$\bar{\delta}\eta_\sigma(x) = \Lambda_{\sigma\varrho}\eta_\varrho(x)\,\delta\lambda = -i[G, \eta_\varrho(x)]\,\delta\lambda. \quad (5.20)$$

his equation possesses the following solution:

$$G = -\int d\mathbf{x}'\,\pi_\mu(x')\,\Lambda_{\mu\varrho}\eta_\varrho(x')$$

which may be demonstrated by inserting it into (5.20) and using (4.232)

$$-i[G, \eta_\sigma(x)] = i\int d\mathbf{x}'[\pi_\mu(x'), \eta_\sigma(x)]\,\Lambda_{\mu\varrho}\eta_\varrho(x')$$

$$= \int d\mathbf{x}'\,\delta_{\mu\sigma}\,\delta(\mathbf{x} - \mathbf{x}')\,\Lambda_{\mu\varrho}\eta_\varrho(x')$$

$$= \Lambda_{\sigma\varrho}\eta_\varrho(x).$$

The symbols x' and μ in the equation for G have no special significance, and so we re-write G as

$$G = -\int d\mathbf{x}\,\pi_\sigma(x)\,\Lambda_{\sigma\varrho}\eta_\varrho(x). \quad (5.21)$$

Now let us consider the effect of an infinitesimal gauge transformation § 3.4(f.2)) on a non-Hermitian field

$$\eta_\sigma(x) \to \eta'_\sigma(x) = \eta_\sigma(x)\,e^{i\alpha} \qquad (\alpha \to 0). \quad (5.22)$$

It is convenient to split the term α into

$$\alpha = -\varepsilon\,\delta\lambda \qquad (\delta\lambda \to 0) \quad (5.23)$$

so that

$$\eta'_\sigma(x) = \eta_\sigma(x)\,e^{-i\varepsilon\delta\lambda} = \eta_\sigma(x) - i\varepsilon\eta_\sigma(x)\,\delta\lambda. \quad (5.24)$$

A comparison with (5.18) then shows that

$$\Lambda_{\sigma\varrho} = -i\varepsilon\,\delta_{\sigma\varrho}. \quad (5.25)$$

The conjugate field operator $\eta_\sigma^\dagger(x)$ therefore transforms as

$$\eta_\sigma^\dagger(x) \to \eta_\sigma^{\dagger\prime}(x) = \eta_\sigma^\dagger(x) + i\varepsilon\eta_\sigma^\dagger(x)\,\delta\lambda \tag{5.26}$$

$$\Lambda_{\sigma\varrho}^\dagger = i\varepsilon\delta_{\sigma\varrho} \tag{5.27}$$

and so the generator G may be written as

$$G = i\varepsilon\,\delta\lambda \int d\mathbf{x}[\pi_\sigma(x)\,\eta_\sigma(x) - \pi_\sigma^\dagger(x)\,\eta_\sigma^\dagger(x)]. \tag{5.28}$$

This term is obviously equivalent to the term C obtained when a gauge transformation was performed on a classical field (3.251). If we dispense with the infinitesimal arbitrary number $\delta\lambda$, then G may be recognised as the total electric charge operator for a field (§ 4.3(i.1))

$$\frac{G}{\delta\lambda} = Q = i\varepsilon \int d\mathbf{x}[\pi_\sigma(x)\,\eta_\sigma(x) - \pi_\sigma^\dagger(x)\,\eta_\sigma^\dagger(x)]. \tag{5.29}$$

Gauge transformations can lead to other conservation laws than that for electric charge; these will be discussed in § 9.3.

The pattern for performing continuous unitary transformations with quantised fields can now be understood. The application of the transformation operator yields equations of the type given in (5.13) and (5.20)

$$\bar{\delta}\eta_\sigma(x) = -i[G, \eta_\sigma(x)]\,\delta\lambda$$

$$\Lambda_{\sigma\varrho}\eta_\varrho(x) = -i[G, \eta_\sigma(x)].$$

(The sign $\pm i$ is unimportant since it depends on the particular form chosen for the generator.) The solution of (5.20) yields an expression for the generator of the continuous transformation; this solution was given in (5.21)

$$G = -\int d\mathbf{x}\,\pi_\sigma\Lambda_{\sigma\varrho}\eta_\varrho(x).$$

The integration may be carried out with the aid of the field operators given in Chapter 4, and the quantity associated with the generator may be identified. Furthermore if the generator commutes with the Hamiltonian the quantity is conserved in the transformation.

5.3. CONTINUOUS ROTATIONS AND DISPLACEMENTS

5.3(a). *Introduction*

It was stated in Chapter 3 that the most general form of a proper Lorentz transformation can be resolved into a displacement plus a rotation of the coordinate frame. It was also shown in that chapter that the invariance of the action integral under displacements led to the conservation of the energy–momentum four-vector for the classical field. In this section we shall show that a rotation leads to the generator for angular momentum.

We now consider an infinitesimal transformation of the form

$$x_\alpha \to x'_\lambda = a_\lambda + a_{\lambda\alpha} x_\alpha$$
$$= a_\lambda + (\delta_{\lambda\alpha} + \varepsilon_{\lambda\alpha})\, x_\alpha \tag{5.30}$$

where the term $\varepsilon_{\lambda\alpha}$ represents an infinitesimal four-dimensional rotation tensor with the following property (compare (3.186));

$$\varepsilon_{\lambda\alpha} = -\varepsilon_{\alpha\lambda}.$$

The transformation (5.30) causes the field operators to change as in (5.2)

$$\eta_\varrho(x) \to \eta'_\sigma(x') = S_{\sigma\varrho}\eta_\varrho(x) = (\delta_{\sigma\varrho} + T_{\sigma\varrho})\, \eta_\varrho(x) \tag{5.31}$$

and for later convenience we will split the operator $T_{\sigma\varrho}$ into two factors

$$T_{\sigma\varrho} = \tfrac{1}{2}\Sigma^{\lambda\alpha}_{\sigma\varrho}\varepsilon_{\lambda\alpha}$$

hence

$$\eta'_\sigma(x') = (\delta_{\sigma\varrho} + \tfrac{1}{2}\Sigma^{\lambda\alpha}_{\sigma\varrho}\varepsilon_{\lambda\alpha})\, \eta_\varrho(x). \tag{5.32}$$

Before proceeding we shall seek a solution for $S_{\sigma\varrho}$ for finite transformations. For simplicity we examine a single component field and perform a rotation through a small angle $\delta\theta_{\lambda\alpha}$ in the $\lambda\alpha$ plane. Thus

$$\delta\theta_{\lambda\alpha} = \varepsilon_{\lambda\alpha} = -\varepsilon_{\alpha\lambda}$$

and so

$$S(\delta\theta_{\lambda\alpha}) = \hat{1} + \tfrac{1}{2}\Sigma^{\lambda\alpha}\varepsilon_{\lambda\alpha} = \hat{1} + \tfrac{1}{2}(\Sigma^{\lambda\alpha} - \Sigma^{\alpha\lambda})\, \delta\theta_{\lambda\alpha} \tag{5.33}$$

where we recall the rule that a repeated suffix implies a summation (§ 3.2(b)). An explicit form for S for finite rotations can now be found by employing the method used in deriving (4.34). The following relation is then obtained

$$S = e^{\frac{1}{2}(\Sigma^{\lambda\alpha} - \Sigma^{\alpha\lambda})\theta_{\lambda\alpha}}. \tag{5.34}$$

We now return to equation (5.31); the variation of the coordinates in that equation can be linked to the change in the field operator through a Taylor expansion and the use of (5.30)

$$\eta'_\sigma(x') = \eta'_\sigma(x) + \frac{\partial \eta'_\sigma}{\partial x_\lambda}(a_\lambda + \varepsilon_{\lambda\alpha}x_\alpha). \tag{5.35}$$

If this equation is inserted in (5.32) we obtain

$$\eta'_\sigma(x) + \frac{\partial \eta'_\sigma}{\partial x_\lambda}(a_\lambda + \varepsilon_{\lambda\alpha}x_\alpha) = (\delta_{\sigma\varrho} + \tfrac{1}{2}\Sigma^{\lambda\alpha}_{\sigma\varrho}\varepsilon_{\lambda\alpha})\, \eta_\varrho(x).$$

Thus the local variation of the field becomes

$$\bar{\delta}\eta_\sigma(x) = \eta'_\sigma(x) - \eta_\sigma(x) = -\frac{\partial \eta'_\sigma}{\partial x_\lambda}(a_\lambda + \varepsilon_{\lambda\alpha}x_\alpha) + \tfrac{1}{2}\Sigma^{\lambda\alpha}_{\sigma\varrho}\varepsilon_{\lambda\alpha}\eta_\varrho(x)$$

$$= -\frac{\partial \eta_\sigma}{\partial x_\lambda}(a_\lambda + \varepsilon_{\lambda\alpha}x_\alpha) + \tfrac{1}{2}\Sigma^{\lambda\alpha}_{\sigma\varrho}\varepsilon_{\lambda\alpha}\eta_\varrho(x) \tag{5.36}$$

to first order in the variation of x_λ. Using the property

$$\varepsilon_{\lambda\alpha} = -\varepsilon_{\alpha\lambda}$$

of the tensor $\varepsilon_{\lambda\alpha}$, we may re-write (5.36) as

$$\bar{\delta}\eta_\sigma(x) = -a_\lambda\frac{\partial \eta_\sigma}{\partial x_\lambda} + \tfrac{1}{2}\varepsilon_{\lambda\alpha}\left[\left(x_\lambda\frac{\partial}{\partial x_\alpha} - x_\alpha\frac{\partial}{\partial x_\lambda}\right)\eta_\sigma(x) + \Sigma^{\lambda\alpha}_{\sigma\varrho}\eta_\varrho(x)\right] \tag{5.37}$$

An inspection of this equation shows that the first term is a displacement of the field operator, and is therefore probably associated with the four-momentum. The expression

$$\left(x_\lambda\frac{\partial}{\partial x_\alpha} - x_\alpha\frac{\partial}{\partial x_\lambda}\right)\eta_\sigma(x)$$

bears a formal resemblance to the function for orbital angular momentum encountered in nonrelativistic quantum mechanics. The final term can be associated with the properties of the field itself. We will show later that it is linked with the intrinsic spin of the particles comprising the field.

We now introduce the infinitesimal unitary transformation

$$U = \hat{1} - iF$$

and write F as

$$F = -a_\lambda P_\lambda + \tfrac{1}{2}\varepsilon_{\lambda\alpha}M_{\lambda\alpha}$$

so that

$$U = \hat{1} + ia_\lambda P_\lambda - \tfrac{1}{2}i\varepsilon_{\lambda\alpha}M_{\lambda\alpha}. \tag{5.38}$$

By (5.13) the local variation of the field can then be written as

$$\bar{\delta}\eta_\sigma(x) = -i[F, \eta_\sigma(x)]$$

$$= ia_\lambda[P_\lambda, \eta_\sigma(x)] - \tfrac{1}{2}i\varepsilon_{\lambda\alpha}[M_{\lambda\alpha}, \eta_\sigma(x)]. \tag{5.39}$$

If the coefficients in this equation are equated with those of (5.37), we obtain

$$-\frac{\partial \eta_\sigma}{\partial x_\lambda} = i[P_\lambda, \eta_\sigma(x)] \tag{5.40}$$

$$\left[\left(x_\lambda\frac{\partial}{\partial x_\alpha} - x_\alpha\frac{\partial}{\partial x_\lambda}\right)\eta_\sigma(x) + \Sigma^{\lambda\alpha}_{\sigma\varrho}\eta_\varrho(x)\right] = -i[M_{\lambda\alpha}, \eta_\sigma(x)]. \tag{5.41}$$

The conserved generating operators are therefore P_λ and $M_{\lambda\alpha}$. The former term epresents the four-momentum operator; substitution of the components P_λ iven in Chapter 4 show that equation (5.40) is satisfied for any field (see in articular equations (4.114) and (4.115)). The angular momentum term will be reated in more detail in the following sections.

5.3 (b). *The angular momentum operator*

Equation (5.41) is concerned with rotations in both space and time. This is too general since we are primarily interested in ordinary (that is spatial) angular nomentum. We therefore consider pure rotations in space and replace the generator $M_{\lambda\alpha}$ by J_{ij}

$$M_{\lambda\alpha} \to J_{ij} = J_k \qquad (i, j, k \text{ cyclic}).$$

Thus for an infinitesimal rotation about the k-axis (that is rotation in the ij plane), the field operator transforms in the manner of (5.32)

$$\eta'_\sigma(x) = (\delta_{\sigma\varrho} + \tfrac{1}{2}\Sigma^{ij}_{\sigma\varrho}\,\varepsilon_{ij})\,\eta_\varrho(x)$$

and the generator for angular momentum is given by comparing (5.20) with (5.41) and using (5.21)†

$$J_k = J_{ij} = -\int \mathrm{d}\mathbf{x}\,\pi_\sigma(x)\left[\left(x_i\frac{\partial}{\partial x_j} - x_j\frac{\partial}{\partial x_i}\right)\eta_\sigma(x) + \Sigma^{ij}_{\sigma\varrho}\eta_\varrho(x)\right]. \tag{5.42}$$

The first term inside the square brackets is again recognisable as a component of an orbital angular momentum operator, and in subsequent sections we shall show that the second term represents the spin operator. We will therefore split J_k into two terms

$$J_k = L_k + S_k$$

where

$$L_k = L_{ij} = -\int \mathrm{d}\mathbf{x}\,\pi_\sigma(x)\left(x_i\frac{\partial}{\partial x_j} - x_j\frac{\partial}{\partial x_i}\right)\eta_\sigma(x) \tag{5.43}$$

$$S_k = S_{ij} = -\int \mathrm{d}\mathbf{x}\,\pi_\sigma(x)\,\Sigma^{ij}_{\sigma\varrho}\eta_\varrho(x).$$

The components J_k can be shown to obey the commutation rule for angular momentum

$$[J_i, J_j] = iJ_k \tag{5.44}$$

where i, j and k are taken cyclically. The general proof of the above expression in terms of field operators is too complex for this book; a specific example for the photon field will be given in equation (5.54).

The relation $[J_i, J_j] = iJ_k$ implies that it is impossible to construct state vectors which are simultaneously eigenvectors of J_x, J_y and J_z. Supposing such

† An identical expression could have been obtained by considering the invariance of the action integral under rotations and applying Noether's theorem (§ 3.4(e))

terms existed with eigenvalues m_i, then we could write

$$J_x |\omega\rangle = m_x |\omega\rangle, \quad J_y |\omega\rangle = m_y |\omega\rangle, \quad J_z |\omega\rangle = m_z |\omega\rangle$$

but these equations are inconsistent with (5.44), for example

$$[J_x, J_y] |\omega\rangle = (m_x m_y - m_y m_x) |\omega\rangle = 0$$

but by (5.44)

$$[J_x, J_y] |\omega\rangle = iJ_z |\omega\rangle = im_z |\omega\rangle.$$

Thus m_z is zero, and cyclic combinations give zero values for m_x and m_y as well this is nonsensical if the system possesses non-zero angular momentum. It is therefore customary to construct state vectors which are eigenstates of J_z only and to leave J_x and J_y unspecified.

The total angular momentum operator for the field is given by

$$\mathbf{J}^2 = J_x^2 + J_y^2 + J_z^2.$$

This term commutes with all three components of $\mathbf{J}$, for example

$$\begin{aligned}[J_z, \mathbf{J}^2] &= J_z J_x^2 - J_x^2 J_z + J_z J_y^2 - J_y^2 J_z \\ &= i(J_x J_y + J_y J_x) - i(J_y J_x + J_x J_y) = 0\end{aligned}$$

where we have used (5.44). Thus it is possible to construct functions which are simultaneously eigenstates of J_z and $\mathbf{J}^2$, and since both operators are conserved generators they may be regarded as constants of the motion.

The properties of the operators and state vectors for angular momentum are given in most textbooks on quantum mechanics, and so will not be repeated here We shall, however, introduce certain features which will be of use later.

Let the simultaneous eigenvector of J_z and $\mathbf{J}^2$ be $|jm\rangle$

$$J_z |jm\rangle = m |jm\rangle, \quad \mathbf{J}^2 |jm\rangle = k |jm\rangle$$

where k is some function of j. We will also introduce operators J_+ and J_- defined as

$$\begin{aligned} J_+ &= J_x + iJ_y \quad \text{then} \quad [J_z, J_+] = J_+ \\ J_- &= J_x - iJ_y \qquad\qquad\quad [J_z, J_-] = -J_- \end{aligned}$$

and so

$$J_z J_- |jm\rangle = (J_- J_z - J_-) |jm\rangle = (m - 1) J_- |jm\rangle.$$

Thus $J_- |jm\rangle$ is an eigenvector of J_z with eigenvalue $(m - 1)$, but

$$J_z |j, m - 1\rangle = (m - 1) |j, m - 1\rangle$$

and so

$$J_- |jm\rangle = c |j, m - 1\rangle$$

where c is a complex number.

We may normalise our vectors in the following manner:

$$\sum_{j=1}^{n} \langle jm \mid jm \rangle = 1 \quad \text{for all } m$$

and so

$$\frac{1}{c^*c} \sum_j \langle jm | \, J_-^\dagger J_- \, | jm \rangle = \frac{1}{c^*c} \sum_j \langle jm | \, J_+ J_- \, | jm \rangle = 1$$

but

$$J_+ J_- = J_x^2 + J_y^2 + J_z = \mathbf{J}^2 - J_z^2 + J_z$$

and so

$$J_+ J_- \, | jm \rangle = (k - m^2 + m) \, | jm \rangle = k - m(m - 1) \, | jm \rangle$$

and since the phase term in c is arbitrary, a suitable solution for c is given by

$$c = \sqrt{[k - m(m - 1)]}.$$

Now the maximum and minimum eigenvalues for J_z are reached when $\mathbf{J}^2 = J_z^2$; let the eigenvalue of the lowest state be given by $m = -j$; we fail to get another state if

$$J_- \, | j, m = -j \rangle = c \, | j, -j - 1 \rangle = 0.$$

This condition is satisfied if $c = 0$, and as $m = -j$ then

$$k = -j(-j - 1) = j(j + 1)$$

and so

$$J_- \, | jm \rangle = \sqrt{[j(j + 1) - m(m - 1)]} \, | j, m - 1 \rangle.$$

Similarly, using J_+, it is possible to show that if the largest value of m is j' then $k = j'(j' + 1)$, hence $j = j'$ and there are $2j + 1$ values of m. We can therefore summarise the action of the operators J_z, $\mathbf{J}^2$, J_+ and J_- on the state $|jm\rangle$ as follows:

$$J_z \, | jm \rangle = m \, | jm \rangle, \quad \mathbf{J}^2 \, | jm \rangle = j(j + 1) \, | jm \rangle \tag{5.45}$$

$$J_- \, | jm \rangle = \sqrt{[j(j + 1) - m(m - 1)]} \, | j, m - 1 \rangle \tag{5.46}$$

$$J_+ \, | jm \rangle = \sqrt{[j(j + 1) - m(m + 1)]} \, | j, m + 1 \rangle.$$

The operators J_+ and J_- are sometimes called respectively the *raising* and *lowering operators* for the angular momentum states.

5.3(c). *The angular momentum operators for the Klein–Gordon field*

For simplicity we consider a Hermitian scalar field. By definition of the word scalar, a field of this type transforms into itself under a rotation

$$x_\alpha \to x'_\lambda = a_{\lambda\alpha} x_\alpha = (\delta_{\lambda\alpha} + \varepsilon_{\lambda\alpha}) \, x_\alpha$$

$$\varphi(x) \to \varphi'(x') = \varphi(x).$$

Thus the transforming operator in (5.32) must equal zero, and so in (5.43)

$$\Sigma_{\sigma\varrho}^{ij} = 0$$

and the intrinsic spin associated with a scalar (or a pseudoscalar) field is zero

We next consider the orbital angular momentum operator (5.43)

$$L_k = -\int d\mathbf{x}\, \pi(x) \left(x_i \frac{\partial}{\partial x_j} - x_j \frac{\partial}{\partial x_i} \right) \varphi(x). \tag{5.47}$$

The operators L_z and $\mathbf{L}^2$ can be evaluated by the same methods used for obtaining the Hamiltonian operator in § 4.3(f). In the present case the problem is simpler if spherical harmonics are used (see A.7 (Appendixes, p. 704)). The field operator is then expanded as in (4.100)

$$\varphi(x) = \sum_{klm} \frac{1}{\sqrt{(2\omega_k)}} g_{lk}(r) \left[a(klm)\, Y_l^m(\theta, \varphi)\, e^{-i\omega_k t} + a^\dagger(klm)\, Y_l^{m*}(\theta, \varphi)\, e^{i\omega_k t} \right].$$

The evaluation of (5.47) then leads to the relation

$$L_z = \sum_{klm} [N(klm) + \tfrac{1}{2}]\, m_l$$

where $N(klm)$ represents the occupation number operator and the $\frac{1}{2}$ is a zero-point term. We may dispense with the latter term (compare § 4.3(f)); using a commutation relation of the type given in (4.81), we then find that the eigenvalue of L_z for the single particle state $|klm\rangle$ is given by

$$\begin{aligned} L_z\, |klm\rangle &= \sum_{k'l'm'} [N(k'l'm')\, m_{l'}]\, a^\dagger(klm)\, |\Psi_0\rangle \\ &= [a^\dagger(klm) + a^\dagger(klm)\, N(klm)]\, m_l\, |\Psi_0\rangle \\ &= m_l\, |klm\rangle. \end{aligned} \tag{5.48}$$

It can be seen, therefore, that the operator L_z fulfils the same function in both field theory and quantum mechanics. The same remark can be made about the orbital angular momentum operator $\mathbf{L}^2$.

5.3(d). *Angular momentum operators for the electromagnetic field*

By definition of the word vector, a vector field transforms in the same way as its coordinates:

$$\begin{aligned} x_\varrho \to x'_\sigma &= (\delta_{\sigma\varrho} + \varepsilon_{\sigma\varrho})\, x_\varrho \\ A_\varrho(x) \to A'_\sigma(x') &= (\delta_{\sigma\varrho} + \varepsilon_{\sigma\varrho})\, A_\varrho(x) \\ &= (\delta_{\sigma\varrho} + \tfrac{1}{2} \Sigma_{\sigma\varrho}^{\lambda\alpha}\, \varepsilon_{\lambda\alpha})\, A_\varrho(x). \end{aligned} \tag{5.49}$$

Thus we find

$$\varepsilon_{\sigma\varrho} = \tfrac{1}{2} \Sigma_{\sigma\varrho}^{\lambda\alpha}\, \varepsilon_{\lambda\alpha}. \tag{5.50}$$

This equation can be satisfied if

$$\Sigma^{\lambda\alpha}_{\sigma\varrho} = \delta_{\sigma\lambda}\delta_{\varrho\alpha} - \delta_{\sigma\alpha}\delta_{\varrho\lambda} \tag{5.51}$$

with spatial components

$$\Sigma^{ij}_{\sigma\varrho} = \delta_{\sigma i}\delta_{\varrho j} - \delta_{\sigma j}\delta_{\varrho i}$$

hence, from equation (5.43), the spin operator for the electromagnetic field has a component $S_k = S_{ij}$ given by

$$\begin{aligned} S_k &= -\int d\mathbf{x}\, \pi_\sigma(x)\, (\delta_{\sigma i}\delta_{\varrho j} - \delta_{\sigma j}\delta_{\varrho i})\, A_\varrho(x) \\ &= -\int d\mathbf{x}\, (\pi_i A_j - \pi_j A_i). \end{aligned} \tag{5.52}$$

This equation obviously represents the component of a vector product, and it is apparent from the above expression that $\mathbf{S}$ can be written as

$$\mathbf{S} = -\int d\mathbf{x}\, \boldsymbol{\pi} \times \mathbf{A}. \tag{5.53}$$

The components of S satisfy the commutation relations for angular momentum. If we write $A(x)$ as A, $A(x')$ and A' and so on, then by (4.232) we find

$$\begin{aligned} [S_1, S_2] &= [\int d\mathbf{x}(\pi_2 A_3 - \pi_3 A_2), \int d\mathbf{x}'(\pi_3' A_1' - \pi_1' A_3')] \\ &= \iint d\mathbf{x}\, d\mathbf{x}'\, \{[A_3, \pi_3']\, \pi_2 A_1' + [\pi_3, A_3']\, \pi_1' A_2\} \\ &= -i \int d\mathbf{x}(\pi_1 A_2 - \pi_2 A_1) = iS_3. \end{aligned} \tag{5.54}$$

The above equation can be generalised and written as

$$[S_i, S_j] = iS_k. \tag{5.55}$$

The component S_k will now be calculated; according to equations (3.230) and (4.161) we can write

$$\pi_\sigma(x) = \frac{\partial \mathscr{L}}{\partial \dot{A}_\sigma} = \frac{\partial A_\sigma}{\partial t} \tag{5.56}$$

and so by (4.164)

$$\pi_i(x) = \frac{i}{\sqrt{V}} \sum_{\mathbf{k},\lambda} \sqrt{\left(\frac{\omega_k}{2}\right)}\, [-a_\lambda(\mathbf{k})\, e^{ikx} + a_\lambda^\dagger(\mathbf{k})\, e^{-ikx}]\, e_i^{(\lambda)}(\mathbf{k}). \tag{5.57}$$

If we then use the techniques of § 4.3(f) and equation (5.52) we may show that S_k becomes

$$\begin{aligned} S_k &= -\int d\mathbf{x}\, (\pi_i A_j - \pi_j A_i) \\ &= i \sum_{\mathbf{k},\lambda,\lambda'} e_i^{(\lambda)} e_j^{(\lambda')} (a_\lambda a_{\lambda'}^\dagger - a_{\lambda'} a_\lambda^\dagger) \end{aligned} \tag{5.58}$$

where we have written a_λ for $a_\lambda(\mathbf{k})$ for simplicity of notation.

The elements of this operator can be readily evaluated with respect to the direction of propagation of the photon. Consider a photon beam moving along the z-axis

$$\mathbf{k} = (0, 0, k).$$

If we now use the values of $e_\mu^{(\lambda)}$ given in (4.165) we find that the components of $\mathbf{S}$ are given by

$$S_k = S_{ij} = i \sum_{\mathbf{k},\lambda,\lambda'} e_i^{(\lambda)} e_j^{(\lambda')} (a_\lambda a_{\lambda'}^\dagger - a_{\lambda'} a_\lambda^\dagger)$$
$$= i \sum_{\mathbf{k}} [a_i(\mathbf{k})\, a_j^\dagger(\mathbf{k}) - a_j(\mathbf{k})\, a_i^\dagger(\mathbf{k})]. \tag{5.59}$$

Using the commutation relations (4.169) the matrix elements of S_k for a photon of momentum $\mathbf{k}$ are then given by

$$(S_k)_{\alpha\beta} = i \langle \Psi_0 |\, a_\alpha [a_i a_j^\dagger - a_j a_i^\dagger]\, a_\beta^\dagger \,| \Psi_0 \rangle$$
$$= i \langle \Psi_0 |\, a_\alpha a_i a_j^\dagger a_\beta^\dagger \,| \Psi_0 \rangle - i \langle \Psi_0 |\, a_\alpha a_j a_i^\dagger a_\beta^\dagger \,| \Psi_0 \rangle$$
$$= i(\delta_{\alpha j}\delta_{i\beta} - \delta_{\alpha i}\delta_{j\beta}). \tag{5.60}$$

In this representation, the matrix elements of S_k are

$$S_1 \equiv S_x = \begin{pmatrix} 0 & 0 & 0 \\ 0 & 0 & -i \\ 0 & i & 0 \end{pmatrix} \quad S_2 \equiv S_y = \begin{pmatrix} 0 & 0 & i \\ 0 & 0 & 0 \\ -i & 0 & 0 \end{pmatrix} \tag{5.61}$$

$$S_3 \equiv S_z = \begin{pmatrix} 0 & -i & 0 \\ i & 0 & 0 \\ 0 & 0 & 0 \end{pmatrix}$$

$$\mathbf{S}^2 = S_x^2 + S_y^2 + S_z^2 = \begin{pmatrix} 2 & 0 & 0 \\ 0 & 2 & 0 \\ 0 & 0 & 2 \end{pmatrix}. \tag{5.62}$$

If we represent the photon spin function by χ_s, then from our discussion of the operators for angular momentum (5.45)

$$\mathbf{S}^2 \chi_s = s(s+1)\, \chi_s = 2\chi_s, \quad s = 1. \tag{5.63}$$

Thus, the spin operator for the vector field is suitable for describing a system of particles with spin 1. Suitable spin functions for the photon may be introduced by writing

$$\chi_s^{m_s} = \chi_1^{\pm 1} = \mp \frac{1}{\sqrt{2}} \begin{pmatrix} 1 \\ \pm i \\ 0 \end{pmatrix} \tag{5.64}$$

compare (4.156). Our choice of sign has been made to achieve conformity with the spherical harmonic $Y_1^{\pm 1}$ (equation (A.7.10, p. 706)). Because of the Lorentz condition (4.149) a spin component $m_s = 0$ does not exist for the photon.

The results obtained suggest an alternative form for the destruction and creation operators for the electromagnetic field. If we introduce operators

$$a_R = \frac{a_1 - ia_2}{\sqrt{2}}, \qquad a_L = \frac{a_1 + ia_2}{\sqrt{2}} \tag{5.65}$$

$$a_R^\dagger = \frac{a_1^\dagger + ia_2^\dagger}{\sqrt{2}}, \qquad a_L^\dagger = \frac{a_1^\dagger - ia_2^\dagger}{\sqrt{2}}$$

where R refers to right circular polarisation (compare (4.156)) or a spin pointing parallel to $\mathbf{k}$, and L refers to the reverse conditions, then we may re-write (5.59) as

$$S_3 = i(a_1 a_2^\dagger - a_2 a_1^\dagger) = i(a_2^\dagger a_1 - a_1^\dagger a_2)$$
$$= a_R^\dagger a_R - a_L^\dagger a_L. \tag{5.66}$$

Thus $a_R^\dagger a_R$ represents an occupation number operator (§ 4.3 (d)) for photons with spin parallel to $\mathbf{k}$, and $a_L^\dagger a_L$ the operator for antiparallel photons. Thus we may represent the single photon states of right and left circular polarisation by the terms

$$|\mathbf{k}R\rangle = a_R^\dagger(\mathbf{k})\,|\Psi_0\rangle, \qquad |\mathbf{k}L\rangle = a_L^\dagger(\mathbf{k})\,|\Psi_0\rangle \tag{5.67}$$

which satisfy the relations (compare (5.48))

$$S_3\,|\mathbf{k}R\rangle = |\mathbf{k}R\rangle, \qquad S_3\,|\mathbf{k}L\rangle = -\,|\mathbf{k}L\rangle. \tag{5.68}$$

5.3 (e). *Vectorial addition of angular momentum*

In the previous section a spin angular momentum function $\chi_1^{\pm 1}$ was introduced for the photon. Now the term $e^{\pm ikx}$ appearing in the electromagnetic field operator can be expanded in terms of orbital angular momentum functions $Y_l^{m_l}(\theta, \varphi)$ as in § 4.3 (e), and we are therefore presented with the problem of how the angular momentum functions combine. Now the total angular momentum and its component along the axis of quantisation are simultaneously conserved quantities, and so if we combine, say, χ_1^{+1} with $Y_l^{m_l}$, we may expect states with $m = m_l + 1$ and $j = l \pm 1$, l depending on the relative orientations of $\mathbf{S}$ and $\mathbf{L}$. Now a state, say, with $m = m_l + 1$ and $j = l + 1$ could also be formed by the combination of an orbital function $Y_{l+1}^{m_l+2}$ with a spin function χ_1^{-1}, and so we must ask ourselves about the relative amplitudes for the two functions.

The present problem is obviously a special aspect of the more general one of combining angular momentum functions. Consider two functions $Y_{j_1}^{m_1}$ and $Y_{j_2}^{m_2}$

with the following properties (5.45):

$$\mathbf{J}_1^2 Y_j^m = j_1(j_1 + 1)\, Y_{j_1}^{m_1}$$
$$J_{1z} Y_{j_1}^{m_1} = m_1 Y_{j_1}^{m_1}$$
$$\mathbf{J}_2^2 Y_{j_2}^{m_2} = j_2(j_2 + 1)\, Y_{j_2}^{m_2}$$
$$J_{2z} Y_{j_2}^{m_2} = m_2 Y_{j_2}^{m_2}.$$

The angular momentum operators for the combined system are given by **J** and J_z, where

$$\mathbf{J}^2 = (\mathbf{J}_1 + \mathbf{J}_2)^2 = \mathbf{J}_1^2 + \mathbf{J}_2^2 + 2\mathbf{J}_1 \cdot \mathbf{J}_2.$$

$$J_z = J_{1z} + J_{2z}.$$

The simultaneous eigenfunction of these operators will be written as $\mathcal{Y}_J^M$ where

$$\mathbf{J}^2 \mathcal{Y}_J^M = J(J + 1)\, \mathcal{Y}_J^M, \qquad J_z \mathcal{Y}_J^M = M \mathcal{Y}_J^M = (m_1 + m_2)\, \mathcal{Y}_J^M.$$

Now the product of the functions of $Y_{j_1}^{m_1}$ and $Y_{j_2}^{m_2}$ is an eigenfunction of J_z

$$\begin{aligned} J_z Y_{j_1}^{m_1} Y_{j_2}^{m_2} &= (J_{1z} + J_{2z})\, Y_{j_1}^{m_1} Y_{j_2}^{m_2} \\ &= (m_1 + m_2)\, Y_{j_1}^{m_1} Y_{j_2}^{m_2} \\ &= M Y_{j_1}^{m_1} Y_{j_2}^{m_2} \end{aligned} \tag{5.69}$$

but it is not, in general, an eigenfunction of $\mathbf{J}^2$ since the scalar product $\mathbf{J}_1 \cdot \mathbf{J}_2$ contains the unspecified operators J_x and J_y.

The relationship between $\mathcal{Y}_J^M$ and $Y_{j_2}^{m_1} Y_{j_2}^{m_2}$ is therefore not a simple one. Let us consider a specific example, let†

$$Y_{j_1}^{m_1} = Y_1^{0, \pm 1}, \qquad Y_{j_2}^{m_2} = Y_{\frac{1}{2}}^{\pm\frac{1}{2}}$$

that is a combination of states with angular momentum 1 and $\frac{1}{2}$. This problem is encountered in meson–nucleon scattering, the possible states for J are $\frac{3}{2}$ and $\frac{1}{2}$ and those for M are given in Table 5.1.

TABLE 5.1

m_1	m_2	M
1	$\frac{1}{2}$	$\frac{3}{2}$
1	$-\frac{1}{2}$	$\frac{1}{2}$
0	$\frac{1}{2}$	$\frac{1}{2}$
0	$-\frac{1}{2}$	$-\frac{1}{2}$
-1	$\frac{1}{2}$	$-\frac{1}{2}$
-1	$-\frac{1}{2}$	$-\frac{3}{2}$

† For convenience of notation we shall use Y for both spin and orbital angular momentum functions.

Thus we have the states

$$J = \tfrac{3}{2}, \qquad M = \tfrac{3}{2}, \tfrac{1}{2}, -\tfrac{1}{2}, -\tfrac{3}{2}$$

$$J = \tfrac{1}{2}, \qquad M = \tfrac{1}{2}, -\tfrac{1}{2}.$$

Obviously the states with $M = \pm\frac{3}{2}$ can only be formed in one way

$$\mathscr{Y}_J^M = \mathscr{Y}_{3/2}^{3/2} = Y_1^1 Y_{1/2}^{1/2}, \qquad \mathscr{Y}_{3/2}^{-3/2} = Y_1^{-1} Y_{1/2}^{-1/2}$$

but the relationship between the others is less obvious. A relationship between states can be constructed with the aid of the raising and lowering operators of equation (5.46)

$$J_- \,|jm\rangle \equiv J_- Y_j^m = \sqrt{[j(j+1) - m(m-1)]}\; Y_j^{m-1}$$

$$J_+ \,|jm\rangle \equiv J_+ Y_j^m = \sqrt{[j(j+1) - m(m+1)]}\; Y_j^{m+1}.$$

Then

$$J_- Y_{1/2}^{1/2} = Y_{1/2}^{-1/2}, \qquad J_- Y_{1/2}^{-1/2} = 0,$$

$$J_- Y_1^1 = \sqrt{2}\, Y_1^0, \qquad J_- Y_1^0 = \sqrt{2}\, Y_1^{-1}, \qquad J_- Y_1^{-1} = 0.$$

Now let us consider the state $\mathscr{Y}_{3/2}^{3/2} = Y_1^1 Y_{1/2}^{1/2}$ and operate on both sides with J_-

$$J_- \mathscr{Y}_{3/2}^{3/2} = \sqrt{3}\, \mathscr{Y}_{3/2}^{1/2} = J_- Y_1^1 Y_{1/2}^{1/2} = \sqrt{2}\, Y_1^0 Y_{1/2}^{1/2} + Y_1^1 Y_{1/2}^{-1/2}$$

so that

$$\mathscr{Y}_{3/2}^{1/2} = \sqrt{\frac{2}{3}} \cdot Y_1^0 Y_{1/2}^{1/2} + \frac{1}{\sqrt{3}} \cdot Y_1^1 Y_{1/2}^{-1/2}.$$

The whole operation may then be repeated to yield the coefficients for $\mathscr{Y}_{3/2}^{-1/2}$ and so on. Now consider the state $\mathscr{Y}_{1/2}^{1/2}$; this can be constructed from the products $Y_1^1 Y_{1/2}^{-1/2}$ and $Y_1^0 Y_{1/2}^{1/2}$

$$\mathscr{Y}_{1/2}^{1/2} = a\, Y_1^1 Y_{1/2}^{-1/2} + b Y_1^0 Y_{1/2}^{1/2}.$$

where a and b are numerical coefficients which satisfy the condition

$$|a|^2 + |b|^2 = 1$$

if $\mathscr{Y}_{1/2}^{1/2}$ is properly normalised. If we apply J_+ to the above relation we find

$$J_+ \mathscr{Y}_{1/2}^{1/2} = 0 = a Y_1^1 Y_{1/2}^{1/2} + b\sqrt{2}\, Y_1^1 Y_{1/2}^{1/2}$$

$$= (a + b\sqrt{2})\, Y_1^1 Y_{1/2}^{1/2}.$$

The simultaneous conditions

$$|a|^2 + |b|^2 = 1, \qquad a + \sqrt{2}\, b = 0$$

have ambiguity of sign in their solution; we shall use the convention

$$a = \sqrt{\frac{2}{3}}, \qquad b = -\frac{1}{\sqrt{3}}$$

so that

$$\mathscr{Y}_{1/2}^{1/2} = \sqrt{\frac{2}{3}}\, Y_1^1 Y_{1/2}^{-1/2} - \frac{1}{\sqrt{3}}\, Y_1^0 Y_{1/2}^{1/2}.$$

Similarly a solution for $Y_{1/2}^{-1/2}$ can be found by using J_-. The complete set of re lations for the combinations of $Y_1^{0,\pm 1}$ and $Y_{1/2}^{\pm 1/2}$ are given in Table 5.2.

TABLE 5.2

	$\mathscr{Y}_{3/2}^{3/2}$	$\mathscr{Y}_{3/2}^{1/2}$	$\mathscr{Y}_{3/2}^{-1/2}$	$\mathscr{Y}_{3/2}^{-3/2}$	$\mathscr{Y}_{1/2}^{1/2}$	$\mathscr{Y}_{1/2}^{-1/2}$
$Y_1^1\ Y_{1/2}^{1/2}$	1					
$Y_1^0\ Y_{1/2}^{1/2}$		$\sqrt{\frac{2}{3}}$			$-\sqrt{\frac{1}{3}}$	
$Y_1^{-1} Y_{1/2}^{1/2}$			$\sqrt{\frac{1}{3}}$			$-\sqrt{\frac{2}{3}}$
$Y_1^1\ Y_{1/2}^{-1/2}$		$\sqrt{\frac{1}{3}}$			$\sqrt{\frac{2}{3}}$	
$Y_1^0\ Y_{1/2}^{-1/2}$			$\sqrt{\frac{2}{3}}$			$\sqrt{\frac{1}{3}}$
$Y_1^{-1} Y_{1/2}^{-1/2}$				1		

It is apparent from Table 5.2 that the product $Y_{j_1}^{m_1} Y_{j_2}^{m_2}$ can be expressed as a linear combination of $\mathscr{Y}_J^M$ functions in addition to the reverse procedure.

The techniques described above can be applied to other combinations of angular momentum functions, and the general expression may be given as

$$\mathscr{Y}_J^M = \sum_{m_1=-j_1}^{m_1=+j_1} \sum_{m_2=M-m_1} C_{J\ j_1\ j_2}^{M\ m_1\ m_2} Y_{j_1}^{m_1} Y_{j_2}^{m_2} \tag{5.70}$$

$$Y_{j_1}^{m_1} Y_{j_2}^{m_2} = \sum_{J=|j_1-j_2|}^{J=j_1+j_2} C_{J\ j_1\ j_2}^{M\ m_1\ m_2} \mathscr{Y}_J^M .$$

The coefficients C represent amplitudes for the combination $Y_{j_1}^{m_1}$ and $Y_{j_2}^{m_2}$ in such a way as to yield Y_J^M. They are variously called *Clebsch–Gordan*, *Wigner* or *vector addition coefficients*. We shall use the first name. The above equations are frequently given in Dirac notation, for example

$$\langle \gamma_1 \gamma_2 \mid JM \rangle = \sum_{m_1 m_2} \langle JM | j_1 m_1 j_2 m_2 \rangle \langle \gamma_1 \mid j_1 m_1 \rangle \langle \gamma_2 \mid j_2 m_2 \rangle \tag{5.71}$$

where $\langle JM \mid j_1 m_1 j_2 m_2 \rangle$ represents the Clebsch–Gordan coefficient, and γ_1 and γ_2 additional degrees of freedom.† There are many variants on the forms for

† A simple example of the notation $\langle \gamma | jm \rangle$ is $\langle \theta\varphi | lm \rangle = Y_l^m(\theta, \varphi)$

$C_{Jj_1j_2}^{Mm_1m_2}$ and $\langle JM \mid j_1m_1j_2m_2\rangle$ given in the literature. Our use of either of the above orms will be determined by convenience. Tables for the Clebsch–Gordan coefficients, together with certain additional properties, are given in A.7 (Appenixes, p. 704).

5.3(f). *Angular momentum functions for the electromagnetic field*

The technique described in the previous section may be used to evaluate the ngular momentum functions for the electromagnetic field. Consider a beam ropagating in the z-direction; the exponential $e^{ikx} = e^{i(\mathbf{k}\cdot\mathbf{x}-\omega t)}$ can then be vritten as in equation (A.7.1)

$$e^{ikx} = e^{i(kz-\omega t)} = \sum_{l=0}^{\infty} i^l \sqrt{[4\pi(2l+1)]}\, j_l(kr)\, Y_l^0(\theta)\, e^{-i\omega t} = \sum_{l=0}^{\infty} \alpha_l Y_l^0. \quad (5.72)$$

The superscript 0 appears on the polynomial Y, as the z-component of orbital ngular momentum is zero for a wave propagating along the z-axis. This can be een by writing

$$L_z = xp_y - yp_x = 0$$

ince both p_y and p_x are zero. Thus the eigenvalues of L_z are $m_l = 0$.

It can be seen from (5.63) that if the spin of the photon is included we must onsider terms of the following form

$$\chi_s^{m_s} e^{ikx} = \sum_{l=0}^{\infty} \alpha_l \chi_s^{m_s} Y_l^0 \quad (5.73)$$

vhere α_l is given by (5.72). This product can be expressed in terms of spherical armonics relating to the total angular momentum $\mathbf{J}$ by using the relation

$$\mathbf{J} = \mathbf{L} + \mathbf{S}$$

nd the Clebsch–Gordan coefficients given in (5.70). It is apparent from our revious discussion of the properties of $\chi_s^{m_s}$ that we can write

$$\chi_s^{m_s} \equiv Y_1^{\pm 1}$$

nd so from (5.70)

$$\chi_s^{m_s} Y_l^0 = \sum_{j=|l-s|}^{j=l+s} C_{j\ l\ s}^{m_\gamma\, 0\, m_s}\, \mathscr{Y}_j^{m_\gamma}$$

vhere

$$m_\gamma = m_l + m_s = 0 + m_s = \pm 1$$

ince the photon has components of spin ± 1 along its axis of propagation (5.68). The total angular momentum therefore takes on values

$$j = l+1,\, l,\, l-1$$

and conversely

$$l = j - 1, j, j + 1.$$

Thus equation (5.73) becomes

$$\chi_s^{m_s} e^{ikx} = \sum_{l=0}^{\infty} \alpha_l \sum_{j=l-1}^{j=l+1} C_j^{m_\gamma}{}_l^0{}_s^{m_\gamma} \mathscr{Y}_j^{m_\gamma}.$$

It is obvious from our discussion that the only $\mathscr{Y}$ terms of interest are

$$\mathscr{Y}_j^{m_\gamma} = \mathscr{Y}_{l-1}^{\pm 1}, \mathscr{Y}_l^{\pm 1}, \mathscr{Y}_{l+1}^{\pm 1} \quad (5.74)$$

and since $m_\gamma = \pm 1$

$$j \neq 0. \quad (5.75)$$

Destruction and creation operators with appropriate coefficients can be used in the expansion of the electromagnetic field in spherical harmonics, as in § 4.3(e) for the scalar field.

5.3(g). *Angular momentum operators for the spinor field*

A transformation of the type

$$x_\alpha \to x'_\lambda = (\delta_{\lambda\alpha} + \varepsilon_{\lambda\alpha})\, x_\alpha, \quad \psi'(x') = S\psi(x)$$

has already been discussed in §§ 3.3(m) and (n) for the Dirac wave equation. The following solution was obtained for S in (3.191):

$$S = (\hat{1} + \tfrac{1}{4}\, \varepsilon_{\varrho\lambda}\gamma_\varrho\gamma_\lambda) = (\hat{1} + \tfrac{1}{4}\, \varepsilon_{\lambda\alpha}\gamma_\lambda\gamma_\alpha)$$

where we have changed the suffices to suit the present expressions.

It is obvious that a similar solution will apply to the operators for the Dirac field and so the transforming operator $\Sigma_{\sigma\varrho}^{\lambda\alpha}$ of equations (5.32)

$$\psi_\varrho(x) \to \psi'_\sigma(x') = (\delta_{\sigma\varrho} + \tfrac{1}{2}\, \Sigma_{\sigma\varrho}^{\lambda\alpha}\varepsilon_{\lambda\alpha})\, \psi_\varrho(x)$$

can be written as

$$\Sigma_{\sigma\varrho}^{\lambda\alpha} = \tfrac{1}{2}\,(\gamma_\lambda\gamma_\alpha)_{\sigma\varrho} = \tfrac{1}{4}\,(\gamma_\lambda\gamma_\alpha - \gamma_\alpha\gamma_\lambda)_{\sigma\varrho}. \quad (5.76)$$

We again restrict ourselves to the spatial components. In § 3.3(e) it was shown that the spin operator for the Dirac equation could be written as

$$\hat{\sigma} = -\frac{i}{2}\, \gamma \times \gamma$$

or

$$\hat{\sigma}_k = \hat{\sigma}_{ij} = -\frac{i}{2}\,(\gamma_i\gamma_j - \gamma_i\gamma_j) \quad (3.110)$$

and, if we compare this equation with (5.76), we find

$$\Sigma^{ij} = \tfrac{1}{2} i\hat{\sigma}_k. \tag{5.77}$$

Thus the spin operator (5.43) becomes

$$S_k = S_{ij} = -\tfrac{1}{2} i \int d\mathbf{x}\, \pi(x)\, \hat{\sigma}_k \psi(x). \tag{5.78}$$

Now the Lagrangian for the Dirac field was given in (4.186):

$$\mathscr{L} = -\bar{\psi}\left(\gamma_\lambda \frac{\partial}{\partial x_\lambda} + m\right)\psi$$

and so

$$\pi(x) = \frac{\partial \mathscr{L}}{\partial \dot{\psi}} = i\bar{\psi}\gamma_4 = i\psi^\dagger. \tag{5.79}$$

Upon inserting this result in (5.78) we obtain

$$S_k = \tfrac{1}{2} \int d\mathbf{x} \psi^\dagger(x)\, \hat{\sigma}_k \psi(x) \tag{5.80}$$

or

$$\mathbf{S} = \tfrac{1}{2} \int d\mathbf{x} \psi^\dagger(x)\, \hat{\boldsymbol{\sigma}} \psi(x). \tag{5.81}$$

The properties of this operator may be inferred, without further development, from our treatment of the Dirac equation in §§ 3.3(e) and (g), namely that the spin of the Dirac particle is $\frac{1}{2}$, and that $\mathbf{S} \cdot \mathbf{p}$ remains a constant of the motion

$$[\mathbf{S} \cdot \mathbf{p}, H] = 0. \tag{5.82}$$

If we choose the coordinate frame

$$\mathbf{p} = 0, 0, p \tag{5.83}$$

then it is not difficult to show that

$$S_p = \frac{\mathbf{S} \cdot \mathbf{p}}{|\mathbf{p}|} = \tfrac{1}{2}(a_1^\dagger a_1 - a_2^\dagger a_2 - b_1^\dagger b_1 + b_2^\dagger b_2). \tag{5.84}$$

The Dirac field operators can be expanded in spherical harmonics, but except in the nonrelativistic limit the expansion is more difficult than for the electromagnetic field. In this limit only the large components of the spinors remain. These satisfy the relations (3.142)

$$\sigma_z \omega_1 = \omega_1, \qquad \sigma_2 \omega_2 = -\omega_2$$

where σ_z is the Pauli spin operator, and if we denote the spin function in spherical coordinates as $\chi_s^{m_s}$ we may write

$$\omega_1 \equiv \chi_{1/2}^{1/2}, \qquad \omega_2 \equiv \chi_{1/2}^{-1/2}. \tag{5.85}$$

These functions can be combined with the orbital angular momentum functions as in (5.70) and yield the relation

$$Y_l^{m_l}\chi_{1/2}^{m_s} = \sum_{j=l-1/2}^{j=l+1/2} C_{j}^{m_l+m_s}{}_{l}^{m_l}{}_{1/2}^{m_s}\,\mathscr{Y}_j^{m_l+m_s}. \tag{5.86}$$

Radial functions for the small components of the spinors have been given by Hamilton (1956, 1959).

5.4. SPACE REFLECTION AND PARITY

5.4(a). *Introduction*

We now turn to the study of discrete transformations, and first examine the effects of spatial reflections

$$\mathbf{x} \to \mathbf{x}' = -\mathbf{x}, \qquad x_4 \equiv t \to t' = t \tag{5.87}$$

on state vectors and field operators.

If a vector changes sign under this transformation it is said to be a *polar vector*, if it fails to change sign it is said to be an *axial vector*. Frequently encountered examples of polar and axial vectors are, respectively, linear and angular momentum. It is obvious that scalars also should not change sign under the transformation given in (5.87). Thus we would expect the quantities listed below to behave in the manner indicated for a reflection of spatial coordinates in a classical system:

space	$\mathbf{x}$	$-\mathbf{x}$	(5.88)
time	t	t	
momentum	$\mathbf{p}$	$-\mathbf{p}$	
energy	E	E	
angular momentum	$\mathbf{L}$	$\mathbf{L}$	
current density	$\mathbf{j}(\mathbf{x}, t)$	$-\mathbf{j}(-\mathbf{x}\ t)$	
charge density	$\varrho(\mathbf{x}, t)$	$\varrho(-\mathbf{x}, t)$.	

The spatial reflection of a quantised field

$$\eta_\sigma(\mathbf{x}, t) \to \eta_\sigma(-\mathbf{x}, t)$$

may be carried out with the aid of a unitary operator with eigenvalues ± 1 (see (5.14)). We shall designate this operator as the *parity operator P*. We shall show that its action on a field operator $\eta_\sigma(x)$ is of the following form:

$$P\eta_\sigma(x)P^{-1} = P\eta_\sigma(\mathbf{x}, t)P^{-1} = \xi_P S_P \eta_\sigma(-\mathbf{x}, t) \tag{5.89}$$

where the terms $\xi_P S_P$ have the following property

$$|\xi_P S_P|^2 = 1.$$

5.4(b). *Scalar and pseudoscalar fields*

The action of the parity operator on a field may be defined in two ways. Either we can define the action of the operator on the field and go on to demonstrate its action on creation and destruction operators and state vectors, or we can employ the reverse procedure. The latter method will be used here since it brings out our physical objectives more clearly.

Classical experience tells us that the inversion (5.87)

$$\mathbf{x} \to -\mathbf{x}, \qquad t \to t$$

causes a momentum vector to change sign. We therefore define the parity operation by the following equation

$$P\,|\mathbf{k}\rangle = \xi_P\,|-\mathbf{k}\rangle$$
$$\equiv Pa^\dagger(\mathbf{k})\,|\Psi_0\rangle = \xi_P a^\dagger(-\mathbf{k})\,|\Psi_0\rangle. \qquad (5.90)$$

This equation implies the conversion of a single particle state with momentum $\mathbf{k}$ (equation (4.84)) to a state with momentum $-\mathbf{k}$. The factor ξ_P has been included since it was noted in § 4.1(b) that a state vector can only be defined up to a complex phase factor of modulus unity. If the space is inverted a second time, we find

$$P^2 a^\dagger(\mathbf{k})\,|\Psi_0\rangle = \xi_P P a^\dagger(-\mathbf{k})\,|\Psi_0\rangle = \xi_P^2 a^\dagger(\mathbf{k})\,|\Psi_0\rangle.$$

Now the double application of P returns the system to its original coordinate frame, since $P^2 = \hat{1}$ (5.14). Thus ξ_P possesses the following solutions

$$\xi_P^2 = 1, \qquad \xi_p = \pm 1. \qquad (5.91)$$

The term ξ_P is called the *intrinsic parity of the particle* (*field*). Particles (fields) which transform with the value $\xi_P = +1$ are said to be *scalar;* those with $\xi_P = -1$ are called *pseudoscalar.*

Equation (5.90) can be used to define the action of the parity operator on the creation and destruction operators. Now the parity operator is unitary (§ 5.1(c)) and so $P^{-1}P = \hat{1}$, therefore equation (5.90) can be re-written as

$$Pa^\dagger(\mathbf{k})\,|\Psi_0\rangle = Pa^\dagger(\mathbf{k})P^{-1}P\,|\Psi_0\rangle = \xi_P a^\dagger(-\mathbf{k})\,|\Psi_0\rangle.$$

It is customary to adopt the convention that the vacuum state is an eigenstate of P with positive intrinsic parity

$$P\,|\Psi_0\rangle = |\Psi_0\rangle \qquad (5.92)$$

and so we find that

$$Pa^\dagger(\mathbf{k})P^{-1} = \xi_P a^\dagger(-\mathbf{k}) \tag{5.93}$$

similarly,

$$Pa(\mathbf{k})P^{-1} = \xi_P a(-\mathbf{k}). \tag{5.93}$$

The suitability of equations (5.93) can be demonstrated by considering their action on the operators for the momentum (4.112) and energy (4.110) of the scalar field

$$\mathbf{P} = \sum_{\mathbf{k}} N(\mathbf{k})\,\mathbf{k} = \sum_{\mathbf{k}} a^\dagger(\mathbf{k})a(\mathbf{k})\,\mathbf{k}$$

$$H = \sum_{\mathbf{k}} [N(\mathbf{k}) + \tfrac{1}{2}]\,\omega_k = \sum_{\mathbf{k}} [a^\dagger(\mathbf{k})a(\mathbf{k}) + \tfrac{1}{2}]\,\omega_k.$$

Using equations (5.93) we find that the parity operation causes the following transformations:

$$P\mathbf{P}P^{-1} = \xi_P^2 \sum_{\mathbf{k}} a^\dagger(-\mathbf{k})\,a(-\mathbf{k})\mathbf{k} = -\sum_{\mathbf{k}} a^\dagger(\mathbf{k})\,a(\mathbf{k})\mathbf{k} = -\mathbf{P} \tag{5.94}$$

$$PHP^{-1} = \xi_P^2 \sum_{\mathbf{k}} [a^\dagger(-\mathbf{k})\,a(-\mathbf{k}) + \tfrac{1}{2}]\,\omega_k = \sum_{\mathbf{k}} [a^\dagger(\mathbf{k})\,a(\mathbf{k}) + \tfrac{1}{2}]\,\omega_k = H$$

where we have taken advantage of the sum over all momentum states from $-\infty$ to $+\infty$ to change the signs of terms as required. Equations (5.94) are in satisfactory accord with classical experience. A similar argument can be applied to the current operator for the field (4.139), and it gives

$$P\mathbf{J}P^{-1} = -\mathbf{J}, \qquad PQP^{-1} = Q \tag{5.95}$$

$$P\mathbf{j}P^{-1} = -\mathbf{j}, \qquad P\varrho P^{-1} = \varrho.$$

We now examine the implications of equations (5.93) for the Klein–Gordon field operator (4.72)

$$\varphi(x) = \varphi(\mathbf{x}, t) = \frac{1}{\sqrt{V}} \sum_{\mathbf{k}} \frac{1}{\sqrt{(2\omega_k)}} [a(\mathbf{k})\, e^{ikx} + a^\dagger(\mathbf{k})\, e^{-ikx}]$$

then, using the summation over momentum states, we find that

$$P\varphi(\mathbf{x}, t)P^{-1} = \frac{1}{\sqrt{V}} \sum_{\mathbf{k}} \frac{1}{\sqrt{(2\omega_k)}} [Pa(\mathbf{k})P^{-1}\, e^{ikx} + Pa^\dagger(\mathbf{k})P^{-1}\, e^{-ikx}]$$

$$= \xi_P \frac{1}{\sqrt{V}} \sum_{\mathbf{k}} \frac{1}{\sqrt{(2\omega_k)}} [a(-\mathbf{k})\, e^{i(\mathbf{k}\cdot\mathbf{x}-\omega t)} + a^\dagger(-\mathbf{k})\, e^{-i(\mathbf{k}\cdot\mathbf{x}-\omega t)}]$$

$$= \xi_P \frac{1}{\sqrt{V}} \sum_{\mathbf{k}} \frac{1}{\sqrt{(2\omega_k)}} [a(\mathbf{k})\, e^{i(-\mathbf{k}\cdot\mathbf{x}-\omega t)} + a^\dagger(\mathbf{k})\, e^{-i(-\mathbf{k}\cdot\mathbf{x}-\omega t)}]$$

$$= \xi_P \varphi(-\mathbf{x}, t) = \varphi_P(x). \tag{5.96}$$

This equation gives us the transformation for the Hermitian Klein–Gordon field; those for the non-Hermitian fields (4.125) can be obtained in a similar manner

$$P\varphi(\mathbf{x}, t)P^{-1} = \xi_P \varphi(-\mathbf{x}, t) = \varphi_P(x) \tag{5.97}$$
$$P\varphi^\dagger(\mathbf{x}, t)P^{-1} = \xi_P \varphi^\dagger(-\mathbf{x}, t) = \varphi_P^\dagger(x).$$

In pion and kaon physics it is often necessary to examine the angular momentum states associated with specific reactions. The parity of these states can be important. We will therefore examine the effect of inverting the coordinates of the Klein–Gordon field when it has been expanded in spherical waves. The expansion was given in equation (4.100)

$$\varphi(x) = \sum_{klm} \frac{1}{\sqrt{(2\omega_k)}} g_{lk}(r)\,[a(klm)Y_l^m(\theta, \varphi)\,e^{-i\omega_k t} + a^\dagger(klm)Y_l^{m*}(\theta, \varphi)\,e^{i\omega_k t}].$$

An inversion of the spatial coordinates causes the following changes in a system expressed in spherical coordinates

$$r \to r' = r, \qquad \theta \to \theta' = \pi - \theta, \qquad \varphi \to \varphi' = \varphi + \pi$$
$$Y_l^m(\theta, \varphi) \to Y_l^m(\theta', \varphi') = (-1)^l\, Y_l^m(\theta, \varphi). \tag{5.98}$$

The derivation of this theorem concerning Y_l^m is straightforward, and can be made by inspecting equations (A.7.7) and (A.7.8). Equation (5.98) thus shows that the inversion operation can be written as

$$P\varphi(\mathbf{x}, t)P^{-1} = \xi_P \varphi(-\mathbf{x}, t) = \xi_P(-1)^l\, \varphi(x)$$

for the component of φ with orbital angular momentum l and so

$$\sum_{klm} \frac{1}{\sqrt{2\omega_k}} g_{lk}(r)\,[Pa(klm)P^{-1}Y_l^m(\theta, \varphi)\,e^{-i\omega_k t} + Pa^\dagger(klm)P^{-1}Y_l^{m*}(\theta, \varphi)\,e^{i\omega_k t}]$$
$$= \xi_P \sum_{klm} (-1)^l \frac{1}{\sqrt{2\omega_k}} g_{lk}(r)\,[a(klm)Y_l^m(\theta, \varphi)\,e^{-i\omega_k t} + a^\dagger(klm)Y_l^{m*}(\theta, \varphi)\,e^{i\omega_k t}].$$

Thus the relation between the destruction and creation operators under space inversion becomes

$$Pa(klm)P^{-1} = \xi_P(-1)^l\, a(klm), \qquad Pa^\dagger(klm)P^{-1} = \xi_P(-1)^l\, a^\dagger(klm). \tag{5.99}$$

The eigenstates of P can be found by applying the arguments contained in equations (5.90–5.93) in reverse; the result obtained for a single particle state is

$$P\,|klm\rangle = \xi_P(-1)^l\,|klm\rangle. \tag{5.100}$$

Therefore one particle states with orbital angular momentum l are eigenstates of P with parity $(-1)^l$. The parity of a scalar or pseudoscalar particle therefore obeys the following rules (compare remark after equation (5.91)):

TABLE 5.3

	Scalar particle	Pseudoscalar particle
Parity	$(-1)^l$	$(-1)^{l+1}$
Even parity states	$s, d, \ldots$	$p, f, \ldots$
Odd parity states	$p, f, \ldots$	$s, d, \ldots$

Equation (5.100) may be extended to two spinless particles with intrinsi parities ξ_{P1} and ξ_{P2}, forming a state of relative angular momentum L in thei centre of momentum system. An expression for this state was given in (4.105

$$|kLM\rangle = \int d\Omega_{\mathbf{k}} Y_L^M(\mathbf{k})\, a_1^\dagger(\mathbf{k})\, a_2^\dagger(-\mathbf{k})\, |\Psi_0\rangle$$

hence since $\int d\Omega_{\mathbf{k}}$ covers all directions of $\mathbf{k}$

$$\begin{aligned} P\,|kLM\rangle &= \xi_{P_1}\xi_{P_2} \int d\Omega_{\mathbf{k}} Y_L^M(\mathbf{k})\, a_1^\dagger(-\mathbf{k})\, a_2^\dagger(\mathbf{k})\, |\Psi_0\rangle \\ &= \xi_{P_1}\xi_{P_2} \int d\Omega_{\mathbf{k}} Y_L^M(-\mathbf{k})\, a_1^\dagger(\mathbf{k})\, a_2^\dagger(-\mathbf{k})\, |\Psi_0\rangle \\ &= \xi_{P_1}\xi_{P_2}(-1)^L\, |kLM\rangle. \end{aligned} \quad (5.101$$

5.4(c). *The electromagnetic field*

Before examining the action of the parity operation on an electromagneti field, we shall consider the general problem of the effects of reflections on par- ticles with spin. Consider a particle moving in any direction in space; a spatia

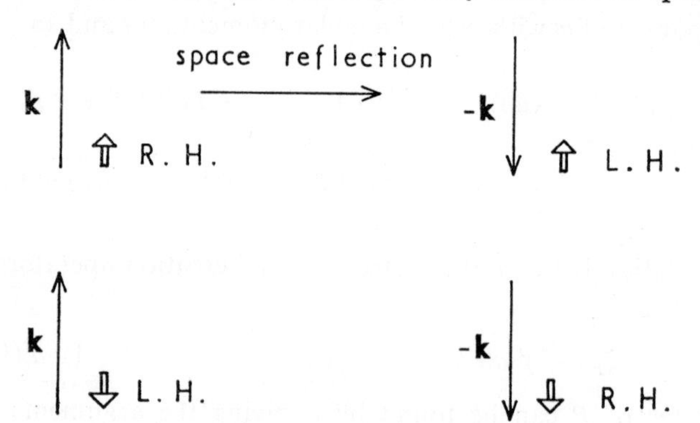

FIG. 5.1. Effect of spatial reflections on linear momentum and spin. The long and short arrows refer to momentum and spin respectively.

reflection causes the momentum, which is a vector quantity, to change sign, but since angular momentum is an axial vector quantity the reflection leaves it un- changed. The effect of space reflection on particles with, say, right-handed spin, therefore to change them into particles with left-handed spin moving in a eversed direction. This effect is illustrated in Fig. 5.1.

We now consider photons. A photon in a state of right circular polarisation an be represented by the expression (5.67)

$$|\mathbf{k}R\rangle = a_R^\dagger(\mathbf{k})\,|\Psi_0\rangle.$$

n a similar manner to equation (5.90) we define the parity operation by the xpression

$$P\,|\mathbf{k}R\rangle = Pa_R^\dagger(\mathbf{k})\,|\Psi_0\rangle = \xi_P\,|-\mathbf{k}L\rangle \tag{5.102}$$

where we have used Fig. 5.1. The term ξ_P obviously fulfils the condition

$$\xi_P^2 = +1 \tag{5.103}$$

ince a double reflection returns the system to its original state.

If the convention (5.92), that space has positive intrinsic parity, is again adopted hen we can show, as in equations (5.90–5.93), that the equation (5.102) is satisied if the operator $a_R^\dagger(\mathbf{k})$ obeys the following condition

$$Pa_R^\dagger(\mathbf{k})P^{-1} = \xi_P a_L^\dagger(-\mathbf{k}). \tag{5.104}$$

The extension of equation (5.104) to the other creation and destruction operators s obvious

$$Pa_R^\dagger(\mathbf{k})P^{-1} = \xi_P a_L^\dagger(-\mathbf{k}), \qquad Pa_L^\dagger(\mathbf{k})P^{-1} = \xi_P a_R^\dagger(-\mathbf{k}) \tag{5.105}$$

$$Pa_R(\mathbf{k})P^{-1} = \xi_P a_L(-\mathbf{k}), \qquad Pa_L(\mathbf{k})P^{-1} = \xi_P a_R(-\mathbf{k}).$$

The techniques used in the previous section can then be used to obtain the ransformation properties of the field; they yield the following equation

$$PA_\mu(\mathbf{x}, t)P^{-1} = \xi_{P\mu} A_\mu(-\mathbf{x}, t). \tag{5.106}$$

The term $\xi_{P\mu}$ for an electromagnetic field cannot be obtained directly by considering the field in isolation; its sign can be found, however, by examining the interaction of an electromagnetic field with a charged current (4.150)

$$\Box^2 A_\mu = -j_\mu.$$

According to equation (5.95) the charge current density changes from $\mathbf{j}$ to $-\mathbf{j}$ under an inversion of coordinates, whilst the fourth component remains unchanged. Thus we can write

$$\xi_{P\mu} = -1 \quad (\mu = 1, 2, 3) \tag{5.107}$$

$$\xi_{P\mu} = +1 \quad (\mu = 4).$$

For free photons the Coulumb gauge (§ 4.4(a)) with $A_4 = 0$ can always be chosen to specify the field, and so in this gauge $A_\mu(x)$ transforms like a real polar vector

$$P\mathbf{A}(\mathbf{x}, t)P^{-1} = -\mathbf{A}(-\mathbf{x}, t).$$

Thus the intrinsic parity of a photon is -1.

A system of particular interest, which we shall have occasion to use later (§§ 9.5(b) and 11.2(d)), is that for two photons travelling in opposite directions with equal momentum. An inversion of axes causes the following transformation

$$P\,|\mathbf{k}R; -\mathbf{k}R\rangle = |-\mathbf{k}L; \mathbf{k}L\rangle, \qquad P\,|\mathbf{k}L; -\mathbf{k}L\rangle = |-\mathbf{k}R; \mathbf{k}R\rangle \tag{5.108}$$

$$P\,|\mathbf{k}R; -\mathbf{k}L\rangle = |-\mathbf{k}L; \mathbf{k}R\rangle, \qquad P\,|\mathbf{k}L; -\mathbf{k}R\rangle = |-\mathbf{k}R; \mathbf{k}L\rangle$$

$$\equiv |\mathbf{k}R; -\mathbf{k}L\rangle, \qquad \equiv |\mathbf{k}L; -\mathbf{k}R\rangle.$$

Thus $|\mathbf{k}R; -\mathbf{k}L\rangle$ and $|\mathbf{k}L; -\mathbf{k}R\rangle$ are both eigenstates of P with eigenvalues $+1$; their components of angular momentum along the $\mathbf{k}$-axis are $+2$ and -2 respectively. States with zero component of angular momentum along the $\mathbf{k}$-axis may be formed by taking the linear combinations

$$|\mathbf{k}R; -\mathbf{k}R\rangle \pm |\mathbf{k}L; -\mathbf{k}L\rangle. \tag{5.109}$$

An examination of the first two equations in (5.108) shows that the upper and lower combinations in (5.109) are eigenstates of P with eigenvalues $+1$ and -1 respectively.

We finally consider the expansion of the electromagnetic field in spherical waves. In § 5.3(f) it was shown that the Fourier components in the electromagnetic field could be written as $\chi_s^{m_s} e^{ikz}$, where $\chi_s^{m_s}$ represented the spin function of the photon. This expression can be expanded into a sum of terms $Y_j^{\pm 1}$, where $j = l, l \pm 1$. Since the field has negative intrinsic parity the components with $j = l$ have parity

$$-(-1)^l = -(-1)^j = (-1)^{j+1} \quad (M \text{ field}) \tag{5.110}$$

whilst those with $j = l \pm 1$ have parity

$$-(-1)^l = (-1)^{l+1} = (-1)^j \qquad (E \text{ field}). \tag{5.111}$$

If the field $A_\mu(x)$ is expanded into its electric and magnetic multipole components, the first condition can be identified with the magnetic 2^j pole field and the second with the electric 2^j pole field.† The expansion performed in this manner may be found in Blatt and Weisskopf (1952).

We can therefore draw up Table 5.4 for the angular momentum and parity states for the components of the electromagnetic field; the extension beyond $j = 2$ is obvious.

† For example the component of the field with $j = 1$ is called a dipole field, with $j = 2$ a quadrupole field and so on; $E\,2$ represents an electric quadrupole.

TABLE 5.4

j	m_γ	Type	Parity	Class
0		Never		
1	± 1	Dipole $\nearrow l = j \pm 1$ $\searrow l = j$	Odd Even	$E1$ $M1$
2	± 1	Quadrupole $\nearrow l = j \pm 1$ $\searrow l = j$	Even Odd	$E2$ $M2$

5.4(d). *The inversion of the Dirac field*

The inversion of the Dirac field is more complicated than that for the scalar and vector fields. This because the Dirac wave function (and field operator) is uncertain up to a change in sign. In § 3.3(o) the effect of reversing the coordinates in the Dirac equation was examined, and it was shown that a double inversion of the Dirac wave function led to the relation

$$\psi = \xi_P^2 \psi. \tag{5.112}$$

An obvious solution to this equation is $\xi_P^2 = +1$; an alternative solution, however, can be shown to exist. If a rotation is performed the Dirac field transforms as

$$\begin{aligned} \psi'(x') &= S\psi(x) \\ &= e^{\frac{1}{2}(\Sigma^{\alpha\lambda} - \Sigma^{\lambda\alpha})\theta_{\lambda\alpha}}\psi(x) \end{aligned} \tag{5.113}$$

where we have used (5.31) and (5.34). The operators $\Sigma^{\lambda\alpha} - \Sigma^{\alpha\lambda}$ can be obtained from (5.76)

$$\Sigma^{\lambda\alpha} - \Sigma^{\alpha\lambda} = \tfrac{1}{2}(\gamma_\lambda \gamma_\alpha - \gamma_\alpha \gamma_\lambda).$$

If the rotation is about the k-axis, then by equation (5.77)

$$\Sigma^{ij} - \Sigma^{ji} = i\hat{\sigma}_k$$

and so

$$\begin{aligned} S(\theta) &= e^{i\frac{1}{2}\hat{\sigma}_k\theta} \\ &= 1 + i\hat{\sigma}_k \frac{\theta}{2} - \frac{1}{2!}\left(\frac{\theta}{2}\right)^2 - i\frac{1}{3!}\hat{\sigma}_k\left(\frac{\theta}{2}\right)^3 + \cdots \\ &= \cos\frac{\theta}{2} + i\hat{\sigma}_k \sin\frac{\theta}{2} \end{aligned} \tag{5.114}$$

where θ represents a rotation in the ij plane, and we have made use of the fact that $\hat{\sigma}_k^2 = 1$ for $k = 1, 2, 3$ (see equations (3.110) and (3.104)).

Now for a rotation through $\theta = 360°$, equations (5.113) and (5.114) imply that

$$\psi'(x') = S\psi(x) = -\psi(x). \tag{5.115}$$

But a rotation through 360° returns the system to its original position, and so we can also write

$$\psi'(x') = \psi(x). \tag{5.116}$$

A double inversion also returns the system to its original position and is equivalent to a rotation through 360°. We are therefore forced to conclude that the equation (5.112)

$$\psi = \xi_P^2 \psi$$

can have solutions

$$\xi_P^2 = \pm 1 \tag{5.117}$$

and that the parity of a single spinor field has no significance.

The implications of this result have been discussed in a number of papers (see, for example, Yang and Tiomno, 1950; Morpurgo and Touschek, 1955). No great physical significance appears to be associated with the result, and it is customary to choose $\xi_P^2 = +1$ so that $\xi_P = \pm 1$.

The spatial reversal of the Dirac wave equation was considered in § 3.3(o) and the following equation derived

$$\psi(\mathbf{x}, t) = \xi_P \gamma_4 \psi(-\mathbf{x}, t).$$

If this equation is compared with the general parity operation for a field (5.89)

$$P\eta_\sigma(\mathbf{x}, t)P^{-1} = \xi_P S_P \eta_\sigma(-\mathbf{x}, t)$$

we may conclude that $S_P = \gamma_4$. We therefore define the parity operation for the Dirac field as

$$P\psi(\mathbf{x}, t)P^{-1} = \xi_P \gamma_4 \psi(-\mathbf{x}, t) = \psi_P(x) \tag{5.118}$$

$$P\bar{\psi}(\mathbf{x}, t)P^{-1} = \xi_P \bar{\psi}(-\mathbf{x}, t)\, \gamma_4 = \bar{\psi}_P(x).$$

note we do this

The effect of the parity operation on the Dirac field will now be examined. Expanding the Dirac field as in (4.194)

$$\psi(x) = \frac{1}{\sqrt{V}} \sum_{\mathbf{p},r} \sqrt{\frac{m}{E_p}}\, [a_r(\mathbf{p}) u_r(\mathbf{p})\, e^{ipx} + b_r^\dagger(\mathbf{p})\, v_r(\mathbf{p})\, e^{-ipx}]$$

the inversion operation yields

$$P\psi(x)P^{-1} = \frac{1}{\sqrt{V}} \sum_{\mathbf{p},r} \sqrt{\frac{m}{E_p}}\, [Pa_r(\mathbf{p})P^{-1} u_r(\mathbf{p})\, e^{ipx} + Pb_r^\dagger(\mathbf{p})P^{-1} v_r(\mathbf{p})\, e^{-ipx}] \tag{5.119}$$

$$= \xi_P \gamma_4 \psi(-\mathbf{x}, t)$$

$$= \xi_P \frac{1}{\sqrt{V}} \sum_{\mathbf{p},r} \sqrt{\frac{m}{E_p}} [\gamma_4 a_r(\mathbf{p}) u_r(\mathbf{p}) e^{i(-\mathbf{p}\cdot\mathbf{x}-Et)} + \gamma_4 b_r^\dagger(\mathbf{p}) v_r(\mathbf{p}) e^{-i(-\mathbf{p}\cdot\mathbf{x}-Et)}]$$

$$= \xi_P \frac{1}{\sqrt{V}} \sum_{\mathbf{p},r} \sqrt{\frac{m}{E_p}} [\gamma_4 a_r(-\mathbf{p})\, u_r(-\mathbf{p})\, e^{i(\mathbf{p}\cdot\mathbf{x}-Et)}$$

$$+ \gamma_4 b_r^\dagger(-\mathbf{p})\, v_r(-\mathbf{p})\, e^{-i(\mathbf{p}\cdot\mathbf{x}-Et)}]. \qquad (5.120)$$

We now consider the terms† $\gamma_4 u_r(-\mathbf{p})$ and $\gamma_4 v_r(-\mathbf{p})$. If the γ_4 matrix is applied to the spinors u and v, we find (compare equations (3.136) and (3.137))

$$\gamma_4 u_r(-\mathbf{p}) = u_r(\mathbf{p}), \qquad \gamma_4 v_r(-\mathbf{p}) = -v_r(\mathbf{p}). \qquad (5.121)$$

The application of the chirality operator σ_p (3.140) shows that the direction of the spins do not change during this transformation – a result which is in accord with Fig. (5.1). Thus a state with spin parallel to momentum becomes a state with spin antiparallel after spatial inversion and vice versa.

We now return to equation (5.120) and insert these results, yielding

$$P\psi(x)P^{-1} = \xi_P \frac{1}{\sqrt{V}} \sum_{\mathbf{p},r} \sqrt{\frac{m}{E_p}} [a_r(-\mathbf{p})\, u_r(\mathbf{p})\, e^{ipx} - b_r^\dagger(-\mathbf{p})\, v_r(\mathbf{p})\, e^{-ipx}]. \qquad (5.122)$$

A comparison of this equation with (5.119) shows that

$$Pa_r(\mathbf{p})P^{-1} = \xi_P a_r(-\mathbf{p}), \qquad Pb_r^\dagger(\mathbf{p})P^{-1} = -\xi_P b_r^\dagger(-\mathbf{p}) \qquad (5.123)$$

and similarly one may show for the $\bar{\psi}$-field

$$Pa_r^\dagger(\mathbf{p})P^{-1} = \xi_P a_r^\dagger(-\mathbf{p}), \qquad Pb_r(\mathbf{p})P^{-1} = -\xi_P b_r(-\mathbf{p}). \qquad (5.124)$$

The properties of the single particle states under spatial inversion follow immediately from (5.123) and (5.124)

$$Pa_r^\dagger(\mathbf{p})\,|\Psi_0\rangle = Pa_r^\dagger(\mathbf{p})P^{-1}P\,|\Psi_0\rangle = \xi_P a_r^\dagger(-\mathbf{p})\,|\Psi_0\rangle \qquad (5.125)$$

$$Pb_r^\dagger(\mathbf{p})_r^\dagger\,|\Psi_0\rangle = Pb_r^\dagger(\mathbf{p})P^{-1}P\,|\Psi_0\rangle = -\xi_P b_r^\dagger(-\mathbf{p})\,|\Psi_0\rangle.$$

In the next section (§ 5.5(d)) we shall describe the state $b_r^\dagger(\mathbf{p})\,|\Psi_0\rangle$ as the charge conjugate or antiparticle state relative to $a_r^\dagger(\mathbf{p})\,|\Psi_0\rangle$. Thus we may conclude from (5.125) that a particle, obeying the Dirac equation, and its antiparticle have opposite intrinsic parity.

† Some care must normally be taken in examining the transformation properties of Dirac fields. It should always be remembered that the inversion operators P, C, T act in Hilbert space, and the γ-matrices in spinor space. The behaviour of the wave functions and field operators coincides for spatial reflections.

Although it is meaningless to ascribe an intrinsic parity to a specific fermion field, it is meaningful to discuss the relative parity of two fermions. As we have mentioned previously, fermions always appear in pairs in any process because of their half integral spin and the necessity to conserve angular momentum. It is therefore customary to assign a parity by definition to a specified particle, and then to discuss the parity of the second particle relative to the first. For example, nucleons are ascribed even parity.

The pairing of the Dirac particles occurs in the bilinear covariants as described in § 3.3(p). In that section the problem of Lorentz transformations was examined for two identical particles. We shall now examine the situation for two fields which are not necessarily the same; we can then re-write equation (3.202) in the following bilinear forms.

$$\begin{aligned} \overline{\psi}_a \Gamma_i \psi_b \equiv \psi_a \psi_b & \qquad S \\ \psi_a \gamma_\lambda \psi_b & \qquad V \\ i\psi_a \gamma_\lambda \gamma_\mu \psi_b \quad \lambda < \mu & \qquad T \\ i\overline{\psi}_a \gamma_\lambda \gamma_5 \psi_b & \qquad A \\ \overline{\psi}_a \gamma_5 \psi_b & \qquad P \end{aligned} \tag{5.126}$$

Under a space inversion these expressions transform as follows

$$P\overline{\psi}_a \Gamma_i \psi_b P^{-1} = \xi_{Pa} \xi_{Pb} \varepsilon_{Pi} \overline{\psi}_a \Gamma_i \psi_b . \tag{5.127}$$

The coefficients ε_{Pi} can be worked out in the manner given in § 3.3(p), and yield the results given in Table 5.5

TABLE 5.5

	S	V	T	A	P
ε_{Pi}	$+1$	$-1(\lambda = 1, 2, 3)$ $+1(\lambda = 4)$	$+1(\lambda, \mu = 1, 2, 3)$ $-1(\lambda$ or $\mu = 4)$	$+1(\lambda = 1, 2, 3)$ $-1(\lambda = 4)$	-1

It is obvious from this table that the spatial components of currents in the Dirac field (3.72)

$$j_k = i\overline{\psi}\gamma_k\psi \quad (k = 1, 2, 3)$$

transform as

$$Pj_k P^{-1} = -j_k \tag{5.128}$$

and so

$$P\mathbf{j}P^{-1} = -\mathbf{j}. \tag{5.129}$$

5.5 CHARGE CONJUGATION

5.5(a). *Introduction*

The charge conjugation operation is one which transforms particles to antiparticles, by changing the sign of all the intrinsic properties, for example charge and strangeness, whilst leaving the mechanical properties unaltered. The topic was first expressed in modern form by Kramers (1937), but the roots of the subject lie in the work of Dirac on the relativistic wave equation, and in Pauli–Weisskopf equations for charged scalar fields.

True neutral particles are those which have no distinction between particle and antiparticle; these are called *self-conjugate particles*, and examples are provided by the π^0 meson and the photon.

5.5(b). *Scalar and pseudoscalar fields*

In equations (4.137) the single particle states of the Klein–Gordon field for particles of opposite charge were written as

$$|\mathbf{k}^+\rangle = a^\dagger(\mathbf{k})\,|\Psi_0\rangle \quad \text{charge } e^+$$

$$|\mathbf{k}^-\rangle = b^\dagger(\mathbf{k})\,|\Psi_0\rangle \quad \text{charge } e^-.$$

For example, these states could represent single pions of opposite charge but with the same momentum. They may be regarded, therefore, as particle and antiparticle with respect to each other.

We now assume the existence of a unitary operator C with the property of transforming particle to antiparticle

$$Cb^\dagger(\mathbf{k})\,|\Psi_0\rangle = \xi_C^* a^\dagger(\mathbf{k})\,|\Psi_0\rangle \tag{5.130}$$

$$Ca^\dagger(\mathbf{k})\,|\Psi_0\rangle = \xi_C b^\dagger(\mathbf{k})\,|\Psi_0\rangle.$$

The term ξ_C is called *the intrinsic charge parity of the field.* Since a double application of the operator C leads to the original state, ξ_C must fulfil the condition

$$|\xi_C|^2 = 1.$$

If it is assumed that the vacuum state is an eigenstate of C with eigenvalue one

$$C\,|\Psi_0\rangle = |\Psi_0\rangle$$

then the unitary property of C can be used to re-write (5.130) as

$$Cb^\dagger(\mathbf{k})C^{-1}C\,|\Psi_0\rangle = \xi_C^* a^\dagger(\mathbf{k})\,|\Psi_0\rangle$$

or

$$Cb^\dagger(\mathbf{k})C^{-1} = \xi_C^* a^\dagger(\mathbf{k}).$$

In a similar manner the following relations may be derived

$$Ca(\mathbf{k})C^{-1} = \xi_C^* b(\mathbf{k}), \qquad Cb^\dagger(\mathbf{k})C^{-1} = \xi_C^* a^\dagger(\mathbf{k}) \tag{5.131}$$

$$Cb(\mathbf{k})C^{-1} = \xi_C a(\mathbf{k}), \qquad Ca^\dagger(\mathbf{k})C^{-1} = \xi_C b^\dagger(\mathbf{k}).$$

These rules may be used to examine the effect of charge conjugation on the total charge operator for the field (4.135).

$$Q = e \sum_{\mathbf{k}} [N^+(\mathbf{k}) - N^-(\mathbf{k})] = e \sum_{\mathbf{k}} [a^\dagger(\mathbf{k})\, a(\mathbf{k}) - b^\dagger(\mathbf{k})\, b(\mathbf{k})].$$

The application of the charge conjugation operator then causes Q to change to an operator of opposite sign

$$CQC^{-1} = -Q, \tag{5.132}$$

a physically sensible result.

We next consider the charge conjugation relations for the scalar fields. In equations (4.125) the plane wave expansions for non-Hermitian scalar fields were given as

$$\varphi(x) = \frac{1}{\sqrt{V}} \sum_{\mathbf{k}} \frac{1}{\sqrt{(2\omega_k)}} [a(\mathbf{k})\, e^{ikx} + b^\dagger(\mathbf{k})\, e^{-ikx}]$$

$$\varphi^\dagger(x) = \frac{1}{\sqrt{V}} \sum_{\mathbf{k}} \frac{1}{\sqrt{(2\omega_k)}} [a^\dagger(\mathbf{k})\, e^{-ikx} + b(\mathbf{k})\, e^{ikx}].$$

The effect of the charge conjugation operator on $\varphi(x)$ may be evaluated easily, with the aid of (5.131)

$$C\varphi(x)C^{-1} = \frac{1}{\sqrt{V}} \sum_{\mathbf{k}} \frac{1}{\sqrt{(2\omega_k)}} [Ca(\mathbf{k})C^{-1}e^{ikx} + Cb^\dagger(\mathbf{k})C^{-1}e^{-ikx}]$$

$$= \xi_C^* \frac{1}{\sqrt{V}} \sum_{\mathbf{k}} \frac{1}{\sqrt{(2\omega_k)}} [b(\mathbf{k})\, e^{ikx} + a^\dagger(\mathbf{k})\, e^{-ikx}]$$

$$= \xi_C^* \varphi^\dagger(x).$$

Thus we can write

$$C\varphi(x)C^{-1} = \xi_C^* \varphi^\dagger(x) = \varphi_C(x) \tag{5.133}$$

and similarly

$$C\varphi^\dagger(x)C^{-1} = \xi_C \varphi(x) = \varphi_C^\dagger(x). \tag{5.134}$$

It is apparent from these equations that bilinear combinations of the type $\varphi^+\varphi$ transform as

$$\varphi^\dagger\varphi \rightarrow \varphi\varphi^\dagger.$$

Thus the Lagrangian given in (4.119)

$$\mathscr{L} = -\frac{\partial\varphi^\dagger}{\partial x_\lambda}\frac{\partial\varphi}{\partial x_\lambda} - m^2\varphi^\dagger\varphi$$

transforms as

$$C\mathscr{L}C^{-1} = -\frac{\partial\varphi}{\partial x_\lambda}\frac{\partial\varphi^\dagger}{\partial x_\lambda} - m^2\varphi\varphi^\dagger .$$

These equations would be equal for a classical field, but not for a quantised field, since the operators for the latter do not commute (see § 4.6(a), in particular (4.235)). The problem may be overcome by *symmetrising* all bilinear terms. This is done by making the substitutions

$$\varphi^\dagger\varphi \to \tfrac{1}{2}(\varphi^\dagger\varphi + \varphi\varphi^\dagger) \tag{5.135}$$

$$\varphi^\dagger\frac{\partial\varphi}{\partial x} \to \tfrac{1}{2}\left(\varphi^\dagger\frac{\partial\varphi}{\partial x} + \frac{\partial\varphi}{\partial x}\varphi^\dagger\right)$$

$$\frac{\partial\varphi}{\partial x_\lambda}\frac{\partial\varphi^\dagger}{\partial x_\lambda} \to \tfrac{1}{2}\left(\frac{\partial\varphi}{\partial x_\lambda}\frac{\partial\varphi^\dagger}{\partial x_\lambda} + \frac{\partial\varphi^\dagger}{\partial x_\lambda}\frac{\partial\varphi}{\partial x_\lambda}\right).$$

This process, or rather that of antisymmetrisation, has already been mentioned for the Dirac field in § 4.5(d). A Lagrangian for the scalar field was given in that section; it was

$$\mathscr{L} = -\tfrac{1}{2}\left[\left(\frac{\partial\varphi^\dagger}{\partial x_\lambda}\frac{\partial\varphi}{\partial x_\lambda} + \frac{\partial\varphi}{\partial x_\lambda}\frac{\partial\varphi^\dagger}{\partial x_\lambda}\right) + m^2(\varphi^\dagger\varphi + \varphi\varphi^\dagger)\right]. \tag{4.226}$$

It is evident that for this equation

$$C\mathscr{L}C^{-1} = \mathscr{L}.$$

The operators associated the symmetrised Lagrangian have the added advantage that they have no zero-point terms. However, they are twice as long as the unsymmetrised form and should be avoided whenever symmetrisation is not essential.

We finally consider two-particle systems. Only states with equal numbers of particles and antiparticles can be eigenstates of C. The simplest two-particle state fulfilling this condition is

$$a^\dagger(\mathbf{k})\, b^\dagger(\mathbf{k})\, |\Psi_0\rangle$$

since

$$\begin{aligned} Ca^\dagger(\mathbf{k})\, b^\dagger(\mathbf{k})\, |\Psi_0\rangle &= \xi_{Ca}\xi^*_{Cb} b^\dagger(\mathbf{k})\, a^\dagger(\mathbf{k})\, |\Psi_0\rangle \\ &= \xi_{Ca}\xi^*_{Cb} a^\dagger(\mathbf{k})\, b^\dagger(\mathbf{k})\, |\Psi_0\rangle \end{aligned}$$

where we have made use of the fact that $a^\dagger(\mathbf{k})$ and $b^\dagger(\mathbf{k})$ commute.

A two-particle state which is of some importance is that for two spinless particles in their mutual c-system with orbital angular momentum L. Consider two pions with opposite charge. Using equation (4.105) we may write this state as

$$|kLm\rangle = \int d\Omega_k Y_L^M(\mathbf{k})\, b^\dagger(\mathbf{k})\, a^\dagger(-\mathbf{k})\, |\Psi_0\rangle \tag{5.136}$$

and under charge conjugation it transforms as

$$C\,|kLM\rangle = |\xi_C|^2 \int d\Omega_{\mathbf{k}} Y_L^M(\mathbf{k})\, a^\dagger(\mathbf{k})\, b^\dagger(-\mathbf{k})\, |\Psi_0\rangle .$$

If the variable of integration is changed from the direction $\mathbf{k}$ to $-\mathbf{k}$, the integra will be unchanged since we are integrating over all directions. Thus sinc $|\xi_C|^2 = 1$ the integral can be written as

$$\begin{aligned} C\,|kLM\rangle &= \int d\Omega_{\mathbf{k}} Y_L^M(-\mathbf{k})\, a^\dagger(-\mathbf{k})\, b^\dagger(\mathbf{k})\, |\Psi_0\rangle \\ &= (-1)^L \int d\Omega_{\mathbf{k}} Y_L^M(\mathbf{k})\, b^\dagger(\mathbf{k})\, a^\dagger(-\mathbf{k})\, |\Psi_0\rangle \\ &= (-1)^L\, |kLM\rangle \end{aligned} \qquad (5.137$$

where we have used the fact that $a^\dagger$ and $b^\dagger$ commute. This result have could beer obtained by the following simple physical argument. Charge conjugation inter- changes π^+ and π^- mesons in a two-pion system; this is equivalent to space inver- sion and so the P and C operators have the same consequences for the systen (compare (5.101)).

5.5(c). *The charge conjugation of self-conjugate fields*

Since the π^0 meson and the photon are self-conjugate particles, their field transform as

$$\begin{aligned} C\varphi(x)C^{-1} &= \xi_C\varphi(x) \qquad (5.138) \\ Ca^\dagger(\mathbf{k})\,|\Psi_0\rangle &= \xi_C a^\dagger(\mathbf{k})\,|\Psi_0\rangle \\ CA_\mu(x)C^{-1} &= \xi_C A_\mu(x) \\ Ca_\lambda^\dagger(\mathbf{k})\,|\Psi_0\rangle &= \xi_C a_\lambda^\dagger(\mathbf{k})\,|\Psi_0\rangle \\ \xi_C^{\,2} &= +1 . \end{aligned}$$

Nothing can be said about the sign of ξ_C from the behaviour of the isolated fields. The signs may be established by examining the interaction of the fields (invariance of interactions under charge conjugation is discussed more fully in § 9.2). The invariance of interactions under charge conjugation lays the following requirements on ξ_C

$$\begin{aligned} \xi_C &= +1 \quad \text{Hermitian scalar field} \\ \xi_C &= -1 \quad \text{electromagnetic field.} \end{aligned} \qquad (5.139)$$

The last requirement can be seen easily if we consider the basic equation for electromagnetic interaction (4.150)

$$\Box^2 A_\mu = -j_\mu .$$

It is obvious that

$$Cj_\mu C^{-1} = -j_\mu$$

(the result will be proved in the next section), and so A_μ must change sign under charge conjugation.

5.5(d). *The Dirac field*

We commence the charge conjugation operation for the Dirac field by defining the process for the destruction and creation operators. If we require the single particle states to transform from particle to antiparticle (or vice versa)†

$$Ca_r^\dagger(\mathbf{k})\,|\Psi_0\rangle = \xi_C b_{r'}^\dagger(\mathbf{k})\,|\Psi_0\rangle \tag{5.140}$$

then the creation and destruction operators must transform as

$$Ca_r(\mathbf{p})C^{-1} = \xi_C^* b_{r'}(\mathbf{p}), \qquad Cb_r^\dagger(\mathbf{p})C^{-1} = \xi_C^* a_{r'}^\dagger(\mathbf{p}) \tag{5.141}$$

$$Ca_r^\dagger(\mathbf{p})C^{-1} = \xi_C b_{r'}^\dagger(\mathbf{p}), \qquad Cb_r(\mathbf{p})C^{-1} = \xi_C a_{r'}(\mathbf{p})$$

with

$$|\xi_C|^2 = 1. \tag{5.142}$$

The proof of equations (5.141) and (5.142) may be carried out in the manner indicated in § 5.5(b). These definitions imply that the total charge operator for the Dirac field (4.224) changes sign under charge conjugation

$$CQC^{-1} = -Q. \tag{5.143}$$

This result may be easily obtained by writing out the occupation number operator in (4.224) in terms of creation and destruction operators.

Before constructing the expressions for the charge conjugation of the Dirac field operators, we will first show that the spinors of the Dirac equation are related in the following manner:

$$u = S_C\bar{v}^T, \qquad v = S_C\bar{u}^T \tag{5.144}$$

where the symbols $\bar{u}^T$ and $\bar{v}^T$ indicate that $\bar{u}$ and $\bar{v}$ are written as column vectors and the term S_C in defined by the equation

$$-\gamma_\lambda^T = S_C^{-1}\gamma_\lambda S_C. \tag{5.145}$$

In the representation given in § 3.3(d) this equation can be satisfied if

$$S_C = \gamma_2\gamma_4. \tag{5.146}$$

It should be noted that this solution leaves any phase factors associated with S_C undetermined.

We now prove equation (5.144); consider the following equations for the Dirac spinors (3.157)

$$(i\gamma_\lambda p_\lambda + m)\,u = 0$$

$$\bar{v}(i\gamma_\lambda p_\lambda - m) = 0.$$

† In this expression $r \neq r'$, because the spin of a hole and a particle point in opposite directions; consequently v_2 and v_1 are the corresponding antiparticle states to u_1 and u_2 in Table 3.3.

If the second equation is transposed

$$-i\gamma_{\lambda}^{T}\bar{v}^{T}p_{\lambda}+m\bar{v}^{T}=0$$

and multiplied by a function S_C from the left

$$-iS_C\gamma_{\lambda}^{T}\bar{v}^{T}p_{\lambda}+mS_C\bar{v}^{T}=0$$

then this equation may be re-written as

$$i\gamma_{\lambda}S_C\bar{v}^{T}p_{\lambda}+mS_C\bar{v}^{T}=(i\gamma_{\lambda}p_{\lambda}+m)\,S_C\bar{v}^{T}=0 \tag{5.147}$$

if

$$-S_C\gamma_{\lambda}^{T}=\gamma_{\lambda}S_C.$$

The latter equation is the same as (5.145). A comparison of (5.147) and (3.129) shows that $S_C\bar{v}^T$ satisfies the relation

$$u=S_C\bar{v}^{T}$$

which is the first of equations (5.144). The second equation may be proved in a similar manner.

We may use (5.144) to obtain the charge conjugation relation for the Dirac field

$$\psi(x)=\frac{1}{\sqrt{V}}\sum_{\mathbf{p},r}\sqrt{\frac{m}{E_p}}[a_r(\mathbf{p})\,u_r(\mathbf{p})\,e^{ipx}+b_r^{\dagger}(\mathbf{p})\,v_r(\mathbf{p})\,e^{-ipx}]$$

and so

$$C\psi(x)C^{-1}=\xi_C^{*}\frac{1}{\sqrt{V}}\sum_{\mathbf{p},r}\sqrt{\frac{m}{E_p}}[b_{r'}(\mathbf{p})\,u_r(\mathbf{p})\,e^{ipx}+a_{r'}^{\dagger}(\mathbf{p})\,v_r(\mathbf{p})\,e^{-ipx}]$$

$$=\xi_C^{*}S_C\frac{1}{\sqrt{V}}\sum_{\mathbf{p},r}\sqrt{\frac{m}{E_p}}\,[b_r(\mathbf{p})\,\bar{v}_r^{T}(\mathbf{p})\,e^{ipx}+a_r^{\dagger}(\mathbf{p})\,\bar{u}_r^{T}(\mathbf{p})\,e^{-ipx}]$$

$$=\xi_C^{*}S_C\bar{\psi}^{T}(x)=\psi_C(x). \tag{5.148}$$

Similarly†

$$C\bar{\psi}(x)C^{-1}=-\xi_C\psi^{T}(x)S_C^{-1}=\bar{\psi}_C(x). \tag{5.149}$$

Alternatively this equation can be derived by writing

$$\bar{\psi}_C(x)=\psi_C^{\dagger}(x)\,\gamma_4=(\xi_C^{*}S_C\bar{\psi}^{T})^{\dagger}\gamma_4=\xi_C\bar{\psi}^{T\dagger}S_C^{\dagger}\gamma_4$$

$$=\xi_C(\psi^{\dagger}\gamma_4)^{T\dagger}S_C^{-1}\gamma_4=\xi_C(\gamma_4\psi)^{T}S_C^{-1}\gamma_4$$

$$=\xi_C\psi^{T}\gamma_4^{T}S_C^{-1}\gamma_4$$

$$=-\xi_C\psi^{T}(x)S_C^{-1}$$

† It should be noted that the transpose sign in equations (5.148) and (5.149) applies to the spinors only. A similar situation arises in equations (5.186) and (5.187).

where we have used the relation $S_C^{-1} = S_C^{\dagger}$, and an adaptation of (5.145)

$$-\gamma_\lambda^T S_C^{-1} = S_C^{-1}\gamma_\lambda.$$

The implications of this algebraic manipulation may be understood more clearly by means of the following example. In § 8.3(a) we shall show that the equations for the interaction of electrons with an electromagnetic field can be written as

$$\gamma_\mu\left(\frac{\partial}{\partial x_\mu} - ieA_\mu\right)\psi + m\psi = 0 \tag{8.44}$$

$$\left(\frac{\partial}{\partial x_\mu} + ieA_\mu\right)\overline{\psi}\gamma_\mu - m\overline{\psi} = 0.$$

If the second equation is transposed

$$\gamma_\mu^T\left(\frac{\partial}{\partial x_\mu} + ieA_\mu\right)\overline{\psi}^T - m\overline{\psi}^T = 0$$

and if the equation is multiplied on the left by a function S_C, then by (5.145)

$$S_C\gamma_\mu^T\left(\frac{\partial}{\partial x_\mu} + ieA_\mu\right)\overline{\psi}^T - mS_C\overline{\psi}^T$$

$$= -\gamma_\mu\left(\frac{\partial}{\partial x_\mu} + ieA_\mu\right)S_C\overline{\psi}^T - mS_C\overline{\psi}^T = 0$$

or

$$\gamma_\mu\left(\frac{\partial}{\partial x_\mu} + ieA_\mu\right)S_C\overline{\psi}^T + mS_C\overline{\psi}^T = 0. \tag{5.150}$$

This equation is the same as the first of equations (8.44) if the substitutions

$$\psi \rightarrow \psi_C = S_C\overline{\psi}^T, \quad -e \rightarrow +e$$

are made. Thus the charge conjugate field describes electrons of charge $+e$, that is positrons.

We next consider the behaviour of the bilinear covariants (§ 3.3(p)) under charge conjugation. It must be remembered that a properly quantised field theory, that is one which obeys the commutation or anticommutation laws, must be properly symmetrised or antisymmetrised (see the remarks in § 5.5(b) on this topic). Thus all the bilinear terms of the Dirac field must be properly antisymmetrised, even if they are not explicitly written out in full. The prescription for antisymmetrisation was given in (4.221)

$$\overline{\psi}_\alpha(\Gamma_i)_{\alpha\beta}\psi_\beta = \tfrac{1}{2}[\overline{\psi}_\alpha(\Gamma_i)_{\alpha\beta}\psi_\beta - \psi_\beta(\Gamma_i)_{\alpha\beta}\overline{\psi}_\alpha].$$

Now if equations (5.148) and (5.149) are expressed in terms of field components, they become

$$C\bar{\psi}_\alpha C^{-1} = -\xi_C \psi_\sigma (S_C^{-1})_{\sigma\alpha}, \qquad C\psi_\beta C^{-1} = \xi_C^* (S_C)_{\beta\varrho} \bar{\psi}_\varrho \tag{5.151}$$

where we have dropped the transposed signs since they relate to the complete field operator, not to its components

$$\psi_\alpha^* = \psi_\alpha^\dagger, \qquad \psi_\beta = \psi_\beta^T.$$

Thus under charge conjugation the term representing the scalar bilinear covariant

$$S = \tfrac{1}{2}[\bar{\psi}_\alpha \delta_{\alpha\beta} \psi_\beta - \psi_\beta \delta_{\alpha\beta} \bar{\psi}_\alpha] = \tfrac{1}{2}[\bar{\psi}_\alpha \psi_\alpha - \psi_\alpha \bar{\psi}_\alpha]$$

transforms as

$$\begin{aligned} CSC^{-1} &= -\tfrac{1}{2}\xi_C \xi_C^* [\psi_\sigma (S_C^{-1})_{\sigma\alpha} (S_C)_{\alpha\varrho} \bar{\psi}_\varrho - (S_C)_{\alpha\varrho} \bar{\psi}_\varrho \psi_\sigma (S_C^{-1})_{\sigma\alpha}] \\ &= -\tfrac{1}{2}[\psi_\sigma \bar{\psi}_\varrho \delta_{\sigma\varrho} - \delta_{\sigma\varrho} \bar{\psi}_\varrho \psi_\sigma] = \tfrac{1}{2}[-\psi_\alpha \bar{\psi}_\alpha + \bar{\psi}_\alpha \psi_\alpha] = S \end{aligned} \tag{5.152}$$

where we have altered the subscripts on the field terms because they have no special significance, and we have used relation

$$(S_C^{-1})_{\sigma\alpha} (S_C)_{\alpha\varrho} = \delta_{\sigma\varrho}. \tag{5.153}$$

The vector term

$$V = \tfrac{1}{2}[\bar{\psi}_\alpha (\gamma_\lambda)_{\alpha\beta} \psi_\beta - \psi_\beta (\gamma_\lambda)_{\alpha\beta} \bar{\psi}_\alpha] \tag{5.154}$$

can be treated in the same manner as the scalar term

$$\begin{aligned} CVC^{-1} &= -\tfrac{1}{2}[\psi_\sigma (S_C^{-1})_{\sigma\alpha} (\gamma_\lambda)_{\alpha\beta} (S_C)_{\beta\varrho} \bar{\psi}_\varrho - (S_C)_{\beta\varrho} \bar{\psi}_\varrho (\gamma_\lambda)_{\alpha\beta} \psi_\sigma (S_C^{-1})_{\sigma\alpha}] \\ &= +\tfrac{1}{2}[\psi_\sigma (\gamma_\lambda^T)_{\sigma\varrho} \bar{\psi}_\varrho - \bar{\psi}_\varrho (\gamma_\lambda^T)_{\sigma\varrho} \psi_\sigma] \\ &= \tfrac{1}{2}[\psi_\sigma (\gamma_\lambda)_{\varrho\sigma} \bar{\psi}_\varrho - \bar{\psi}_\varrho (\gamma_\lambda)_{\varrho\sigma} \psi_\sigma] = -V. \end{aligned} \tag{5.155}$$

CVC⁻¹ = −V

In general we can write the charge conjugate transformation of the bilinear covariants as

$$C\bar{\psi}_a \Gamma_i \psi_b C^{-1} \equiv \tfrac{1}{2} C[\bar{\psi}\Gamma_i \psi - \psi \Gamma_i \bar{\psi}] C^{-1} = \xi_{Ca} \xi_{Cb}^* \varepsilon_{Ci} \bar{\psi}_a \Gamma_i \psi_b \tag{5.156}$$

where

$$\begin{aligned} \varepsilon_{Ci} &= 1, & i &= S, A, P \\ \varepsilon_{Ci} &= -1, & i &= V, T \end{aligned} \tag{5.157}$$

and the subscripts a and b indicate whether or not the fields (particles) are of the same type; they have no connection with the subscripts α and β used in equations (5.151).

The vector form of the bilinear covariant (5.155) appears in the expression for the current density of the Dirac field (4.227). It is therefore obvious that the current density operator for this field transforms as

$$C j_\lambda C^{-1} = -j_\lambda. \tag{5.158}$$

This result is in accordance with that obtained for the total charge operator (5.143). Thus the charge conjugation operator converts, for example, an electron current into a positron current.

5.6. TIME REVERSAL

5.6(a). *Introduction*

In classical physics any operation which reverses the order of time in a physical system, induces the following changes:

$$\begin{array}{lccc} \text{time} & t & \xrightarrow{T} & -t \\ \text{space} & \mathbf{x} & & \mathbf{x} \\ \text{momentum} & \mathbf{p} & & -\mathbf{p} \\ \text{energy} & E & & E \\ \text{current density} & \mathbf{j} & & -\mathbf{j} \\ \text{angular momentum} & \mathbf{J} & & -\mathbf{J} \end{array} \tag{5.159}$$

In quantum theory the operator which generates a reversal of time must possess more complicated properties than the space reflection and charge conjugation operators.

The difficulty which arises in time reversal can be seen in the following simple example. It was shown in § 4.1(e) that the canonical commutation relation (4.27) remains invariant under a unitary transformation

$$[x_i, p_j] = [x_i', p_j'] = i\delta_{ij} \equiv i\hbar\delta_{ij}.$$

Now consider the effect of time reversal on this relation according to (5.159)

$$[x_i, p_j] \xrightarrow{T} [x_i, -p_j] = -i\delta_{ij}. \tag{5.160}$$

It is obvious from this example that the commutation relation is not invariant under the direct transformation of $t \to -t$. This difficulty can be avoided if the time reversal operation also reverses the order of the operators. The procedure for establishing a satisfactory time reversal operation was first introduced by Wigner (1932); a different approach leading to the same result was developed by Schwinger (1951). We shall employ an adaptation of the latter method. The method employs an operator T which causes transformations of the vector $|\omega\rangle$ in Hilbert space to its dual $\langle\omega_T|$. T acts as a unitary transformation on any single operator, but reverses the order of any product of operators. This statement may be given algebraic expression by the definitions†

$$TA(\mathbf{x}, t)\, T^{-1} = S_T A(\mathbf{x}, -t) = A_T(x) \tag{5.161}$$

$$TABT^{-1} = (AB)_T = TBT^{-1}TAT^{-1} = B_T A_T$$

$$T\,|\omega\rangle = \langle\omega_T|, \quad \langle\omega|\, T^{-1} = |\omega_T\rangle$$

where the term S_T is determined by the algebraic structure of $A(\mathbf{x}, t)$; for example we expect

$$T\mathbf{x}T^{-1} = \mathbf{x} = \mathbf{x}_T \qquad (S_T = 1)$$

$$T\mathbf{p}T^{-1} = -\mathbf{p} = \mathbf{p}_T \quad (S_T = -1) \tag{5.162}$$

if $\mathbf{x}$ and $\mathbf{p}$ refer to position and momentum operators respectively.

It is obvious that the rules given in (5.161) satisfy the canonical commutation relation (4.27)

$$[x_i, p_j] = i\delta_{ij}$$

$$T[x_i, p_j]\, T^{-1} = [p_{iT}, x_{iT}] = [-p_j, x_i] = [x_i, p_j] = i\delta_{ij}.$$

The rules (5.161) also illustrate the behaviour of the angular momentum operator $\mathbf{J}$ under time reversal. In nonrelativistic theory the components of the orbital angular momentum operator are given by

$$J_i = [x_j p_k - p_j x_k] \quad (i, j, k \text{ cyclic}).$$

This transforms according to (5.161) and (5.162) as

$$TJ_iT^{-1} = T(x_j p_k - p_j x_k)\, T^{-1} = -p_k x_j + x_k p_j = -J_i$$

and

$$T\mathbf{J}^2T^{-1} = T\left(\sum_i J_i^2\right) T^{-1} = \mathbf{J}^2.$$

The above equations therefore can be written in the following manner:

$$\{J_z, T\} = 0, \quad [\mathbf{J}^2, T] = 0. \tag{5.163}$$

The same rules can be shown to apply to the angular momentum operators in field theory. Now let us consider the action of T on an angular momentum state $|jm\rangle$, where (5.45)

$$J_z\,|jm\rangle = m\,|jm\rangle, \qquad \mathbf{J}^2\,|jm\rangle = j(j+1)\,|jm\rangle.$$

Then by (5.163) we find

$$J_zT\,|jm\rangle = -TJ_z|jm\rangle = -mT|jm\rangle$$

† The transformation of bras to kets and vice-versa, which is equivalent to complex conjugation of the wave function, is necessary for our subsequent examination of time reversed matrix elements for interactions.

o that $T|jm\rangle$ is an eigenstate of J_z with eigenvalue $-m$, but

$$J_z|j, -m\rangle = -m|j, -m\rangle$$

and since $\mathbf{J}^2$ is invariant under T we may conclude, with the aid of (5.161), that

$$T|jm\rangle = C_{jm}\langle j, -m| \tag{5.164}$$

where

$$|C_{jm}|^2 = 1.$$

We finally note the action of the time reversal operation on a scalar product

$$\langle\omega_1|\omega_2\rangle = \langle\omega_1|\,\hat{1}\,|\omega_2\rangle = \langle\omega_1|\,T^{-1}T\,|\omega_2\rangle = \langle\omega_{2T}|\omega_{1T}\rangle. \tag{5.165}$$

The time reversal transformation is sometimes called an *antiunitary* one, since the order of the product has been reversed. The expression is not appropriate, since T itself is unitary. Some authors prefer the expression *antilinear transformation.*

5.6(b). *Time reversal and scalar fields*

Consider the third of equations (5.161)

$$T|\omega\rangle = \langle\omega_T|.$$

We postulate that this rule holds for all states, *including the vacuum state*

$$T|\Psi_0\rangle = \langle\Psi_0| \tag{5.166}$$

$$\langle\Psi_0|\,T^\dagger = \langle\Psi_0|\,T^{-1} = |\Psi_0\rangle.$$

Thus the single particle state $a^\dagger(\mathbf{k})\,|\Psi_0\rangle$ must transform as

$$Ta^\dagger(\mathbf{k})\,|\Psi_0\rangle = \langle\Psi_0|\,a_T^\dagger(\mathbf{k}). \tag{5.167}$$

Now the reversal of the direction of time flow implies that a momentum $\mathbf{k}$ should change to $-\mathbf{k}$, and so we define $a_T^\dagger(\mathbf{k})$ as

$$Ta^\dagger(\mathbf{k})\,T^{-1} = a_T^\dagger(\mathbf{k}) = \xi_T^* a(-\mathbf{k}). \tag{5.168}$$

Thus a single particle state with momentum $\mathbf{k}$ transforms to one with momentum $-\mathbf{k}$ upon time reversal

$$Ta^\dagger(\mathbf{k})\,|\Psi_0\rangle = Ta^\dagger(\mathbf{k})T^{-1}T\,|\Psi_0\rangle = \xi_T^*\langle\Psi_0|\,a(-\mathbf{k}). \tag{5.169}$$

Equation (5.168) can be extended in an obvious manner to the other creation and destruction operations

$$\left.\begin{aligned} Ta(\mathbf{k})\,T^{-1} &= \xi_T a^\dagger(-\mathbf{k}), \qquad & Ta^\dagger(\mathbf{k})\,T^{-1} &= \xi_T^* a(-\mathbf{k}) \\ Tb(\mathbf{k})\,T^{-1} &= \xi_T^* b^\dagger(-\mathbf{k}), \qquad & Tb^\dagger(\mathbf{k})\,T^{-1} &= \xi_T b(-\mathbf{k}) \end{aligned}\right\} \tag{5.170}$$

where

$$|\xi_T|^2 = 1.$$

Thus the field operator $\varphi(x)$

$$\varphi(x) = \frac{1}{\sqrt{V}} \sum_{\mathbf{k}} \frac{1}{\sqrt{(2\omega_k)}} [a(\mathbf{k})\, e^{ikx} + b^\dagger(\mathbf{k})\, e^{-ikx}] \tag{4.125}$$

transforms as

$$\begin{aligned} T\varphi(\mathbf{x}, t)\, T^{-1} &= \frac{1}{\sqrt{V}} \sum_{\mathbf{k}} \frac{1}{\sqrt{(2\omega_k)}} [Ta(\mathbf{k})\, T^{-1}\, e^{ikx} + Tb^\dagger(\mathbf{k})\, T^{-1}\, e^{-ikx}] \\ &= \xi_T \frac{1}{\sqrt{V}} \sum_{\mathbf{k}} \frac{1}{\sqrt{(2\omega_k)}} [a^\dagger(-\mathbf{k})\, e^{i(\mathbf{k}\cdot\mathbf{x} - \omega t)} + b(-\mathbf{k})\, e^{-i(\mathbf{k}\cdot\mathbf{x} - \omega t)}] \\ &= \xi_T \frac{1}{\sqrt{V}} \sum_{\mathbf{k}} \frac{1}{\sqrt{(2\omega_k)}} [a^\dagger(\mathbf{k})\, e^{-i(\mathbf{k}\cdot\mathbf{x} + \omega t)} + b(\mathbf{k})\, e^{i(\mathbf{k}\cdot\mathbf{x} + \omega t)}] \\ &= \xi_T\, \varphi^\dagger(\mathbf{x}, -t) = \varphi_T(x) \end{aligned} \tag{5.171}$$

In deriving (5.171) we have made use of the fact that the summation over **k** extends over all states between $-\infty$ and $+\infty$. Similarly, it is easy to show that

$$T\varphi^\dagger(\mathbf{x}, t)\, T^{-1} = \xi_T^* \varphi(\mathbf{x}, -t) = \varphi_T^\dagger(x).$$

Equations (5.171) lead to physically sensible results for the behaviour of the four-vectors for current density and momentum. Consider first the operator for current density (4.138)

$$j_\mu = ie \left[\frac{\partial \varphi^\dagger}{\partial x_\mu} \varphi - \frac{\partial \varphi}{\partial x_\mu} \varphi^\dagger \right]$$

where $j_\mu = j_1, j_2, j_3, i\varrho$; this expression transforms in the following manner under time reversal by use of equation (5.171) and its Hermitian conjugate

$$T j_k(\mathbf{x}, t)\, T^{-1} = -j_k(\mathbf{x}, -t) \qquad (k = 1, 2, 3) \tag{5.172}$$

$$T\varrho(\mathbf{x}, t)\, T^{-1} = \varrho(\mathbf{x}, -t).$$

Similar arguments can be applied to the operators for four-momentum (4.109); alternatively, the techniques applied in obtaining equation (5.94) can be used. The results obtained by either method are, of course, the same; they are

$$T\mathbf{P}T^{-1} = -\mathbf{P}, \qquad THT^{-1} = H. \tag{5.173}$$

Thus the quantised field operators transform in accord with classical experience (5.159). It should be noted that j_λ and P_λ transform with the opposite sign to that for the coordinates x_λ. P_λ and j_λ are therefore said to be pseudovectors under time reversal.

5.6(c). *The electromagnetic field*

The behaviour of the electromagnetic field under time reversal can be most easily inferred from the requirement that the electromagnetic interaction should be invariant under this operation. In fact the whole problem of time reversal invariance can be understood most easily by the examination of the interaction of fields rather than by examining isolated fields. An examination of the

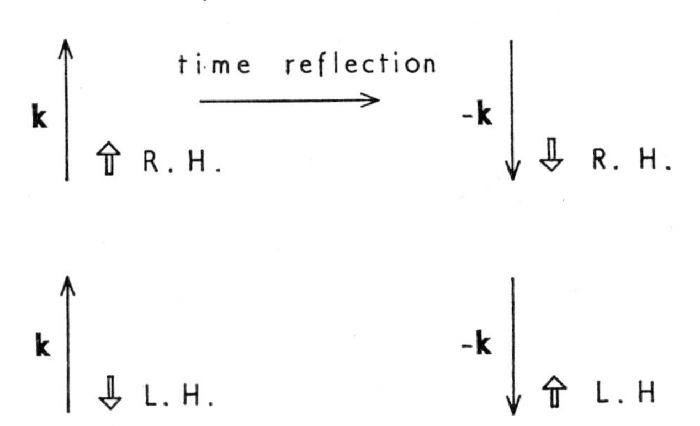

FIG. 5.2. Effect of reflection of time on linear momentum and spin. The long and short arrows refer to momentum and spin respectively.

formal structure of time reversal in isolated fields, however, will prove to be of some use later in this chapter.

We return to the electromagnetic field. Consider the interaction of an electromagnetic field with a current of charged mesons. According to (4.150) we may write

$$\Box^2 A_\mu = -j_\mu$$

and since (5.172)

$$T\mathbf{j}T^{-1} = -\mathbf{j}, \qquad T\varrho T^{-1} = \varrho$$

the electromagnetic field must transform as

$$T\mathbf{A}(\mathbf{x}, t)\, T^{-1} = -\mathbf{A}(\mathbf{x}, -t), \quad T\varphi(\mathbf{x}, t)\, T^{-1} = \varphi(\mathbf{x}, -t) \tag{5.174}$$

in order to preserve time reversal invariance. This equation can be shown to lead to the following rules for the creation and destruction operators (5.65)

$$Ta_R(\mathbf{k})\, T^{-1} = -a_R^\dagger(-\mathbf{k}), \qquad Ta_R^\dagger(\mathbf{k})\, T^{-1} = -a_R(-\mathbf{k}) \tag{5.175}$$

$$Ta_L(\mathbf{k})\, T^{-1} = -a_L^\dagger(-\mathbf{k}), \qquad Ta_L^\dagger(\mathbf{k})\, T^{-1} = -a_L(-\mathbf{k}). \tag{5.176}$$

Thus time reversal changes both the momentum and spin of the photon. This result is indicated schematically in Fig. 5.2.

5.6(d). *The Dirac field*

The procedure of § 3.3(m) can be used to produce a time reversed form of the Dirac *wave function*. This form does not necessarily coincide with that obtained for a Dirac field acting in Hilbert space. In the case of space reversal the two forms do remain the same. As we emphasised earlier, two separate spaces must be considered for the Dirac field operator – spinor space and Hilbert space.

The time reversal operation for the Dirac field can be carried out in the same manner as for the scalar field. The single particle state transforms as (compare equation (5.169) and Fig. 5.2)

$$Ta_r^\dagger(\mathbf{p})\,|\Psi_0\rangle = \varepsilon_r \xi_T^* \,\langle\Psi_0|\, a_{r'}(-\mathbf{p}) \tag{5.177}$$

where r' and r refer to spin functions which point in opposite directions. The operator $a_r^\dagger(\mathbf{p})$ therefore transforms as

$$Ta_r^\dagger(\mathbf{p})\,T^{-1} = a_{Tr}^\dagger(\mathbf{p}) = \varepsilon_r \xi_T^* a_{r'}(-\mathbf{p})$$

and the other operators behave as follows

$$Ta_r(\mathbf{p})\,T^{-1} = \varepsilon_r \xi_T a_{r'}^\dagger(-\mathbf{p}), \qquad Ta_r^\dagger(\mathbf{p}) = \varepsilon_r \xi_T^* a_{r'}(-\mathbf{p}) \tag{5.178}$$

$$Tb_r(\mathbf{p})\,T^{-1} = \varepsilon_r \xi_T^* b_{r'}(-\mathbf{p}), \qquad Tb_r^\dagger(\mathbf{p})\,T^{-1} = \varepsilon_r \xi_T b_{r'}(-\mathbf{p})$$

where

$$|\xi_T|^2 = 1.$$

Thus the field operator $\psi(x)$ (4.194)

$$\psi(x) = \frac{1}{\sqrt{V}} \sum_{\mathbf{p},r} \sqrt{\frac{m}{E_p}}\,[a_r(\mathbf{p})\,u_r(\mathbf{p})e^{ipx} + b_r^\dagger(\mathbf{p})\,v_r(\mathbf{p})e^{-ipx}]$$

transforms as

$$T\psi(x)\,T^{-1} = \xi_T \frac{1}{\sqrt{V}} \sum_{\mathbf{p},r} \varepsilon_r \sqrt{\frac{m}{E_p}}\,[a_{r'}^\dagger(-\mathbf{p})\,u_r(\mathbf{p})\,e^{i(\mathbf{p}\cdot\mathbf{x}-Et)} + b_{r'}(-\mathbf{p})\,v_r(\mathbf{p})\,e^{-i(\mathbf{p}\cdot\mathbf{x}-Et)}]$$

$$= \xi_T \frac{1}{\sqrt{V}} \sum_{\mathbf{p},r} \varepsilon_r \sqrt{\frac{m}{E_p}}\,[a_{r'}^\dagger(\mathbf{p})\,u_r(-\mathbf{p})\,e^{-i(\mathbf{p}\cdot\mathbf{x}+Et)} + b_{r'}(\mathbf{p})\,v_r(-\mathbf{p})\,e^{i(\mathbf{p}\cdot\mathbf{x}+Et)}]. \tag{5.179}$$

We now postulate the existence of an operator S_T which has the following property†

$$u(-\mathbf{p}) = S_T \bar{u}^T(\mathbf{p}), \qquad v(-\mathbf{p}) = S_T \bar{v}^T(\mathbf{p}) \tag{5.180}$$

† Using Table 3.3 and $S_T = \gamma_1\gamma_3\gamma_4$ the explicit forms for the transformation are

$$S_T\bar{u}_1^T(\mathbf{p}) = u_2(-\mathbf{p}) \qquad S_T\bar{u}_2^T(\mathbf{p}) = -u_1(-\mathbf{p})$$
$$S_T\bar{v}_1^T(\mathbf{p}) = v_2(-\mathbf{p}) \qquad S_T\bar{v}_2^T(\mathbf{p}) = -v_1(-\mathbf{p}).$$

These signs imply that $\varepsilon_r = \pm 1$ in (5.178) in order that the required transformation property for the field (5.281) is obtained.

where the superscript T refers to the transpose of the spinor. Thus upon dispensing with the dummy index r', equation (5.179) becomes

$$T\psi(x)\,T^{-1} = \xi_T S_T \frac{1}{\sqrt{V}} \sum_{\mathbf{p},r} \sqrt{\frac{m}{E_p}}\,[a_r^\dagger(\mathbf{p})\,\bar{u}_r^T(\mathbf{p})\,e^{-i(\mathbf{p}\cdot\mathbf{x}+Et)} + b_r(\mathbf{p})\,\bar{v}_r^T(\mathbf{p})\,e^{i(\mathbf{p}\cdot\mathbf{x}+Et)}]$$

$$= \xi_T S_T \bar{\psi}^T(\mathbf{x}, -t) = \psi_T(x). \qquad (5.181)$$

We now identify S_T; consider equation (3.146)

$$\bar{u}(i\gamma_\lambda p_\lambda + m) = 0, \qquad \bar{u} = \bar{u}(\mathbf{p}).$$

This equation may be transposed, yielding

$$(i\gamma_\lambda^T p_\lambda + m)\,\bar{u}^T = (i\gamma_1^T p_1 + i\gamma_2^T p_2 + i\gamma_3^T p_3 + i\gamma_4^T p_4 + m)\,\bar{u}^T = 0. \qquad (5.182)$$

We now introduce a matrix S_T, such that

$$S_T\gamma_i^T = -\gamma_i S_T \quad (i = 1, 2, 3) \qquad (5.183)$$

$$S_T\gamma_4^T = \gamma_4 S_T$$

or

$$\gamma_\lambda^T = \varepsilon_\lambda S_T^{-1}\gamma_\lambda S_T \qquad (\varepsilon_\lambda = -1;\ \lambda = 1, 2, 3)$$

$$(\varepsilon_\lambda = +1;\ \lambda = 4)$$

and if (5.182) is multiplied on the left by S_T we obtain

$$(iS_T\gamma_1^T p_1 + iS_T\gamma_2^T p_2 + iS_T\gamma_3^T p_3 + iS_T\gamma_4^T p_4 + mS_T)\,\bar{u}^T$$

$$= (-i\boldsymbol{\gamma}\cdot\mathbf{p} + i\gamma_4 p_4 + m)\,S_T\bar{u}^T = 0. \qquad (5.184)$$

Now if equation (3.129)

$$(i\gamma_\lambda p_\lambda + m)\,u = (i\boldsymbol{\gamma}\cdot\mathbf{p} + i\gamma_4 p_4 + m)\,u = 0$$

has its momentum term reversed, the equation becomes

$$(-i\boldsymbol{\gamma}\cdot\mathbf{p} + i\gamma_4 p_4 + m)\,u(-\mathbf{p}) = 0 \qquad (5.185)$$

and upon comparing this equation with (5.184) we find

$$u(-\mathbf{p}) = S_T\bar{u}^T(\mathbf{p})$$

which is equation (5.180). The condition (5.183)

$$\gamma_i^T = -S_T^{-1}\gamma_i S_T \quad (i = 1, 2, 3)$$

$$\gamma_4^T = S_T^{-1}\gamma_4 S_T$$

may be satisfied if $S_T = \gamma_1\gamma_3\gamma_4$ to within a phase factor, when the representation for the γ-matrices given in (3.103) is used. Thus equation (5.181) becomes

$$T\psi(\mathbf{x}, t)T^{-1} = \xi_T S_T \bar{\psi}^T(\mathbf{x}, -t) = \xi_T \gamma_1\gamma_3\gamma_4 \bar{\psi}^T(\mathbf{x}, -t) = \psi_T(x). \qquad (5.186)$$

By way of contrast, the time reversed form of the Dirac *wave function* is given b $\psi(\mathbf{x}, -t) = \gamma_1\gamma_2\gamma_3\psi(\mathbf{x}, t)$ (compare § 3.3(o)).

Similarly we find that

$$\begin{aligned} T\overline{\psi}(\mathbf{x}, t)\, T^{-1} &= \overline{\psi}_T(x) = \psi_T^\dagger\gamma_4 = [\xi_T S_T \overline{\psi}^T(x)]^\dagger\gamma_4 \\ &= [\xi_T S_T(\psi^\dagger\gamma_4)^T]^\dagger\gamma_4 = \xi_T^*\psi^T\gamma_4^* S_T^{-1}\gamma_4 \\ &= \xi_T^*\psi^T\gamma_4^T S_T^{-1}\gamma_4 = \xi_T^*\psi^T S_T^{-1}\gamma_4\gamma_4 \\ &= \xi_T^*\psi^T S_T^{-1} \end{aligned}$$

where we have used the fact that in the representation (3.103)

$$\gamma_4 = \gamma_4^* = \gamma_4^T.$$

Thus the adjoint field transforms as

$$T\overline{\psi}(x)\, T^{-1} = \xi_T^*\psi^T(\mathbf{x}, -t)\, S_T^{-1} = \overline{\psi}_T(x). \qquad (5.187$$

We finally consider the transformation properties of the bilinear covariants As in the case of charge conjugation the transformation is demonstrated mor conveniently with the components of the Dirac field. Equations (5.186) an (5.187) can then be written as

$$\begin{aligned} T\psi_\beta T^{-1} &= \psi_{\beta T} = \xi_T(S_T)_{\beta\sigma}\overline{\psi}_\sigma \\ T\overline{\psi}_\alpha T^{-1} &= \overline{\psi}_{\alpha T} = \xi_T^*\psi_\varrho(S_T^{-1})_{\varrho\alpha} \end{aligned} \qquad (5.188$$

where we have dispensed with the transpose sign since we are dealing with singl components of a row or column vector.

As an illustration of the technique we consider the vector term for the bilinea covariants. We recall that the covariant form should be antisymmetrised (5.154

$$V = \overline{\psi}\gamma_\lambda\psi \equiv \tfrac{1}{2}\,[\overline{\psi}_\alpha(\gamma_\lambda)_{\alpha\beta}\psi_\beta - \psi_\beta(\gamma_\lambda)_{\alpha\beta}\overline{\psi}_\alpha].$$

Under time reversal this expression becomes (with the aid of the rule for th products of operators (5.161))

$$\begin{aligned} TVT^{-1} &= \tfrac{1}{2}\, T[\overline{\psi}_\alpha(\gamma_\lambda)_{\alpha\beta}\psi_\beta - \psi_\beta(\gamma_\lambda)_{\alpha\beta}\overline{\psi}_\alpha]\, T^{-1} \\ &= \tfrac{1}{2}\,[\psi_{\beta T}(\gamma_\lambda)_{\alpha\beta}\overline{\psi}_{\alpha T} - \overline{\psi}_{\alpha T}(\gamma_\lambda)_{\alpha\beta}\psi_{\beta T}] \\ &= \tfrac{1}{2}\,|\xi_T|^2\, [(S_T)_{\beta\sigma}\overline{\psi}_\sigma(\gamma_\lambda)_{\alpha\beta}\psi_\varrho(S_T^{-1})_{\varrho\alpha} \\ &\qquad\qquad - \psi_\varrho(S_T^{-1})_{\varrho\alpha}(\gamma_\lambda)_{\alpha\beta}(S_T)_{\beta\sigma}\overline{\psi}_\sigma] \\ &= \tfrac{1}{2}\,[\overline{\psi}_\sigma(S_T)_{\sigma\beta}(\gamma_\lambda^T)_{\beta\alpha}(S_T^{-1})_{\alpha\varrho}\psi_\varrho - \varepsilon_\lambda\psi_\varrho(\gamma_\lambda)_{\sigma\varrho}\overline{\psi}_\sigma] \\ &= \tfrac{1}{2}\,\varepsilon_\lambda\,[\overline{\psi}_\sigma(\gamma_\lambda)_{\sigma\varrho}\psi_\varrho - \psi_\varrho(\gamma_\lambda)_{\sigma\rho}\overline{\psi}_\sigma] \\ &= \varepsilon_\lambda V \end{aligned} \qquad (5.189)$$

where from (5.183)

$$\varepsilon_\lambda = -1 \quad (\lambda = 1, 2, 3)$$
$$= +1 \quad (\lambda = 4).$$

The method given above can be used to show that the bilinear covariants listed in (5.126) transform as

$$T\overline{\psi}_a \Gamma_i \psi_b T^{-1} = \xi_{Ta}^* \xi_{Tb} \varepsilon_i \overline{\psi}_a \Gamma_i \psi_b. \tag{5.190}$$

It should be understood that strictly speaking the covariant must be written in antisymmetrised form – see § 5.5(d)). The coefficients ε_i can be evaluated as indicated for the vector term (5.189) and yield the results given in Table 5.6.

TABLE 5.6

i	S	V	T	A	P
ε_i	$+1$	$-1(\lambda = 1, 2, 3)$ $+1(\lambda = 4)$	$-1(\lambda, \mu = 1, 2, 3)$ $+1(\lambda$ or $\mu = 4)$	$-1(\lambda = 1, 2, 3)$ $+1(\lambda = 4)$	-1

It can be seen from this table that the spatial components of currents change sign

$$T\mathbf{j}T^{-1} = -\mathbf{j} \tag{5.191}$$
$$\mathbf{j} = i\overline{\psi}\boldsymbol{\gamma}\psi$$

under time reversal. This is the result we would expect from classical experience.

5.7. COMBINED REFLECTIONS

The operations P, C and T can be combined into a single reflection – this is known as a *strong reflection* R_S

$$R_S = CTP. \tag{5.192}$$

Consider the action of R_S on a scalar field

$$R_S \varphi(x) R_S^{-1} = CTP\varphi(x)P^{-1}T^{-1}C^{-1}; \tag{5.193}$$

with the aid of equations (5.96), (5.171) and (5.134) we obtain

$$\varphi(\mathbf{x}, t) \xrightarrow{P} \xi_p \varphi(-\mathbf{x}, t) \xrightarrow{T} \xi_p \xi_T \varphi^\dagger(-\mathbf{x}, -t) \xrightarrow{C} \xi_P \xi_T \xi_C \varphi(-\mathbf{x}, -t) \tag{5.194}$$

or

$$R_S \varphi(x) R_S^{-1} = \xi_S \varphi(-x) \tag{5.195}$$

where

$$\xi_S = \xi_P \xi_T \xi_C .$$

Similarly

$$R_S \varphi^\dagger(x)\, R_S^{-1} = \xi_S^* \varphi^\dagger(-x) \tag{5.196}$$

$$\xi_S^* = \xi_P \xi_T^* \xi_C^* .$$

A similar operation can be performed for the Dirac field

$$\psi(\mathbf{x}, t) \xrightarrow{P} \xi_P \gamma_4 \psi(-\mathbf{x}, t) \xrightarrow{T} \xi_P \xi_T \gamma_4 \gamma_1 \gamma_3 \gamma_4 \bar{\psi}^T(-\mathbf{x}, -t). \tag{5.197}$$

Now

$$\bar{\psi}^T = \frac{1}{\sqrt{V}} \sum_{\mathbf{p},r} \sqrt{\frac{m}{E_p}}\, [a_r^\dagger(\mathbf{p})\, \bar{u}_r^T(\mathbf{p})\, e^{-ipx} + b_r(\mathbf{p})\, \bar{v}_r^T(\mathbf{p})\, e^{ipx}]$$

and so from equations (5.149) and (5.146)

$$\bar{\psi}_C^T(-x) = -\xi_C(\psi^T S_C^{-1})^T = -\xi_C S_C^{-1T} \psi = -\xi_C(\gamma_4\gamma_2)^T \psi$$
$$= -\xi_C \gamma_2^T \gamma_4^T \psi = -\xi_C \gamma_2 \gamma_4 \psi(-x).$$

Therefore under charge conjugation the expression (5.197) becomes

$$\xrightarrow{C} -\xi_P \xi_T \xi_C \gamma_4 \gamma_1 \gamma_3 \gamma_4 \gamma_2 \gamma_4 \psi(-\mathbf{x}, -t)$$
$$= \xi_P \xi_T \xi_C \gamma_5 \psi(-\mathbf{x}, -t) = \xi_S \gamma_5 \psi(-\mathbf{x}, -t).$$

Thus we can write

$$R_S \psi(x)\, R_S^{-1} = \xi_S \gamma_5 \psi(-x) \tag{5.198}$$

and similarly

$$R_S \bar{\psi}(x)\, R_S^{-1} = -\xi_S^* \bar{\psi}(-x)\, \gamma_5 .$$

The action of strong reflection on (spinless) single particle states switches kets to bras and converts particles to antiparticles. Consider, for example, the state $a^\dagger(\mathbf{k})\, |\Psi_0\rangle$, then (compare (5.196))

$$R_S a^\dagger(\mathbf{k})\, |\Psi_0\rangle = CTPa^\dagger(\mathbf{k})\, |\Psi_0\rangle$$
$$= \xi_P CTa^\dagger(-\mathbf{k})\, |\Psi_0\rangle = \xi_P \xi_T^* \langle\Psi_0|\, a(\mathbf{k}) C$$
$$= \xi_P \xi_T^* \xi_C^* \langle\Psi_0|\, b(\mathbf{k}) = \xi_S^* \langle\Psi_0|\, b(\mathbf{k}). \tag{5.199}$$

It can be seen that the momentum of the particle is unchanged during the operation. This result is to be expected since we have reflected both time and space. On the other hand, spin states change sign during strong reflections. We illustrate this point by the following symbolic transformation: (where **p**, **s** and e represent respectively the momentum, spin orientation and 'charge' states, and ε_s depends n the specific properties of the field)

$$CTP\,|\mathbf{p}, \mathbf{s}, e\rangle = \xi_P CT\,|-\mathbf{p}, \mathbf{s}, e\rangle$$
$$= \varepsilon_s \xi_P \xi_T^* \,\langle e, -\mathbf{s}, \mathbf{p}|\, C$$
$$= \varepsilon_s \xi_S^* \,\langle -e, -\mathbf{s}, \mathbf{p}|$$

The bilinear covariants transform in a simple manner under strong reflections. he technique has already been described in previous sections (see, in particular, quations (5.152), (5.154) and (5.155)). Symbolically, we may write

$$\begin{aligned} R_S\overline{\psi}\Gamma_i\psi R_S^{-1} &\equiv \tfrac{1}{2} R_S[\overline{\psi}\Gamma_i\psi - \psi\Gamma_i\overline{\psi}]R_S^{-1} \\ &= -\tfrac{1}{2}[\psi\gamma_5\Gamma_i\gamma_5\overline{\psi} - \overline{\psi}\gamma_5\Gamma_i\gamma_5\psi] \\ &= \tfrac{1}{2}[\overline{\psi}\gamma_5\Gamma_i\gamma_5\psi - \psi\gamma_5\Gamma_i\gamma_5\overline{\psi}] \\ &= \tfrac{1}{2}\varepsilon_i[\overline{\psi}\Gamma_i\psi - \psi\Gamma_i\overline{\psi}] \equiv \varepsilon_i\overline{\psi}\Gamma_i\psi \end{aligned} \tag{5.200}$$

where we have used the rule (5.161) that time reversal inverts the order of operators. The anticommutation relations for the γ-matrices (3.97) then show hat

$$\begin{aligned} \varepsilon_i &= +1 \quad \text{for} \quad S, T, P \\ \varepsilon_i &= -1 \quad \text{for} \quad V, A. \end{aligned} \tag{5.201}$$

The combined reflection operation PTC leads to the important CPT theorem which we shall discuss in Chapter 9.

CHAPTER 6

THE INTERACTION OF FIELDS I WAVE FUNCTIONS, PHASE SHIFTS AND POTENTIALS

THIS chapter will be concerned mainly with the description of the elastic sca tering of particles by using the techniques of wave and matrix mechanics. I forms a useful introduction to subsequent chapters which deal with the inte action of fields, since it will become apparent in the examples we discuss late that once the appropriate application of creation and destruction operators ha been made in the theory of interacting fields, mathematical functions are le which are basically similar to the wave functions discussed in Chapter 3.

The problem of elastic scattering can be approached in two ways by wav mechanics. The first method involves the phenomenological analysis of th phase shifts induced in the partial waves by a scattering process. The second metho assumes the existence of a potential between the beams of scattered particles, an uses the methods of perturbation theory to derive scattering amplitudes an cross-sections. In this chapter the two methods will be examined and linked to gether.

6.1. THE METHOD OF PARTIAL WAVES

6.1 (a). *The scattering of spinless particles*

We start by making a phenomenological analysis of a relatively simple proces – that of the scattering of a beam of particles of spin zero. We assume that th scattering centre is also spinless. The principle used is to construct a solutio to the wave equation well outside the region of interaction so that the particle are free and obey the wave equation

$$\nabla^2\psi + k^2\psi = 0 \tag{6.1}$$

where ψ represents the wave function and k the momentum (in the c-system) o the scattered particle. If we define the z-axis to be parallel to the incoming beam the wave function ψ has the asymptotic form

$$\psi(\mathbf{r}) \underset{r\to\infty}{=} \frac{1}{\sqrt{v}}\left[e^{ikz} + f(\theta)\,\frac{e^{ikr}}{r}\right] = \frac{1}{\sqrt{v}}\,e^{ikz} + \psi_{sc} \tag{6.2}$$

here v is the velocity of the particles in the c-system, and

$$\psi_{sc} = \frac{1}{\sqrt{v}} f(\theta) \frac{e^{ikr}}{r}$$

$$kr = |\mathbf{k}|\,|\mathbf{r}|.$$

The expression (6.2) represents an incoming plane wave (e^{ikz}) and an out-oing scattered wave (ψ_{sc}). The factor $1/\sqrt{v}$ is for purposes of normalisation – its unction will become obvious later. The inclusion of the term $1/r$ in ψ_{sc} means at the physical requirement of an inverse square law for the scattered particles ccurs in an obvious manner. The term $f(\theta)$ is called the *scattering amplitude.*

Now consider an undisturbed plane wave, normalised in the manner given bove; it may be written in the following form with the aid of equation (A. 7.1)). 704)†

$$\frac{e^{ikz}}{\sqrt{v}} = \frac{1}{\sqrt{v}} \sum_{l=0}^{\infty} \sqrt{[4\pi(2l+1)]}\, i^l j_l(kr)\, Y_l^0(\theta)$$

$$= \frac{1}{kr}\frac{1}{\sqrt{v}} \sum_{l=0}^{\infty} \sqrt{[4\pi(2l+1)]}\, i^l \sin\left(kr - \frac{l\pi}{2}\right) Y_l^0(\theta) \qquad (r \to \infty)$$

$$= \frac{1}{kr}\sqrt{\left(\frac{4\pi}{v}\right)} \sum_{l=0}^{\infty} \sqrt{(2l+1)}\, \frac{i^{l+1}}{2} \left[e^{-i\left(kr-\frac{l\pi}{2}\right)} - e^{i\left(kr-\frac{l\pi}{2}\right)}\right] Y_l^0(\theta) \qquad (6.3)$$

where l represents the orbital angular momentum quantum number and we have sed the asymptotic form of $j\,(kr)$ (A. 7.6) p. 705)

$$j_l(kr) = \frac{1}{kr} \sin\left(kr - \frac{l\pi}{2}\right) \qquad (r \to \infty).$$

The superscript 0 appears in the function $Y_l^m(\theta)$ since both k_x and k_y are zero and ence the z-component of orbital angular momentum is zero

$$m_z = xk_y - yk_x = 0.$$

The first exponential inside the square brackets of (6.3) represents an ingoing wave to the coordinate position $z = 0$, and the second term represents an outgoing wave; in combination they represent an undisturbed system.

We will now locate the scattering act at $z = 0$. In the presence of the scatterer, only the outgoing part of the wave is disturbed. We will therefore insert a complex) *scattering coefficient* η_l before the term for the outgoing wave in order

† In order to simplify notation we have written $Y(\theta, \varphi) = Y(\theta)$.

to describe its modification by the scatterer. Thus the asymptotic form of ψ ca now be written as

$$\psi = \frac{1}{kr}\sqrt{\left(\frac{4\pi}{v}\right)}\sum_{l=0}^{\infty}\sqrt{(2l+1)}\,\frac{i^{l+1}}{2}\left[e^{-i\left(kr-\frac{l\pi}{2}\right)} - \eta_l e^{i\left(kr-\frac{l\pi}{2}\right)}\right] Y_l^0(\theta)$$

$$= \psi_{\text{in}} + \psi_{\text{out}}. \qquad (6.$$

If this equation is compared with (6.2) and (6.3), the asymptotic form of th scattered wave is found to be

$$\psi_{\text{sc}} = \psi - \frac{1}{\sqrt{v}}\,e^{ikz}$$

$$= \frac{1}{kr}\sqrt{\left(\frac{4\pi}{v}\right)}\sum_{l=0}^{\infty}\sqrt{(2l+1)}\,\frac{i^{l+1}}{2}\left[e^{-i\left(kr-\frac{l\pi}{2}\right)} - \eta_l e^{i\left(kr-\frac{l\pi}{2}\right)} - e^{-i\left(kr-\frac{l\pi}{2}\right)} + e^{i\left(kr-\frac{l\pi}{2}\right)}\right] Y_l^0(\theta)$$

$$= \frac{1}{kr}\sqrt{\left(\frac{4\pi}{v}\right)}\sum_{l=0}^{\infty}\sqrt{(2l+1)}\,\frac{(\eta_l - 1)}{2i}\,e^{ikr}\,Y_l^0(\theta)$$

$$= \frac{1}{\sqrt{v}}\,f(\theta)\,\frac{e^{ikr}}{r} \qquad (6.$$

since

$$i^l e^{-i\frac{l\pi}{2}} = 1.$$

Thus the scattering amplitude is given by

$$f(\theta) = \sqrt{(4\pi)}\sum_{l=0}^{\infty}\sqrt{(2l+1)}\,\frac{(\eta_l - 1)}{2ik}\,Y_l^0(\theta) \qquad (6.$$

$$= \sum_{l=0}^{\infty}(2l+1)\,\frac{(\eta - 1)}{2ik}\,P_l^0(\theta) \qquad (6.$$

where the second version uses the relation given in (A.7.7) $Y_l^0 = \sqrt{[2l+1)/4\pi]}\, P_l^0$ The term $(\eta_l - 1)/2ik$ appearing in both equations is sometimes called the *sca tering function*; we shall denote it by f_l.

Now the flux j of particles scattered in the direction (θ, φ) into an element o area $dA = r^2\, d\Omega$ is given by the expression

$$j\,dA = \frac{1}{2im}\left(\psi_{\text{sc}}^* \frac{\partial\psi_{\text{sc}}}{\partial r} - \psi_{\text{sc}}\frac{\partial\psi_{\text{sc}}^*}{\partial r}\right) r^2\,d\Omega = v\,|\psi_{\text{sc}}|^2\, r^2\, d\Omega = |f(\theta)|^2\, d\Omega$$

where we have used equation (3.75) for the flux density; we have also used (6.5) Since the incoming beam has been normalised to unity by inclusion of the facto

$\sqrt{v}$, the cross-section for scattering into an element of solid angle $d\Omega$ can be itten as

$$\sigma_{sc}(\theta) = \frac{d\sigma_{sc}}{d\Omega} = |f(\theta)|^2 = \frac{\pi}{k^2} |\sum_{l=0}^{\infty} \sqrt{(2l+1)}\, (\eta_l - 1)\, Y_l^0(\theta)|^2. \tag{6.8}$$

nce the functions Y_l^0 are orthogonal and normalised

$$\int d\Omega\, Y_l^{m*} Y_{l'}^{m'} = \delta_{mm'} \delta_{ll'} \tag{A.7.11}$$

e total scattering cross-section becomes

$$\sigma_{sc} = \int d\Omega\, |f(\theta)|^2 = \frac{\pi}{k^2} \sum_{l=0}^{\infty} (2l+1)\, |\eta_l - 1|^2. \tag{6.9}$$

This cross-section refers to particles *elastically* scattered in the interaction. addition particles may be removed from the incident beam by other processes. 'e will define, therefore, a total *reaction cross-section* to represent the particles hich vanish from the incident beam. This can be done by taking the difference intensities of the particles entering and leaving the scattering system according equation (6.4)

$$\sigma_r = \int d\Omega\, vr^2(|\psi_{in}|^2 - |\psi_{out}|^2)$$

$$= \frac{\pi}{k^2} \sum_{l=0}^{\infty} (2l+1)\, (1 - |\eta_l|^2). \tag{6.10}$$

Equations (6.9) and (6.10) bear an interesting relationship to each other. The aximum value of the scattering cross-section for a given l value is realised if $_l = -1$; σ_r then vanishes for the same l value. The maximum of σ_r occurs for $_l = 0$; we then find for a given value of l

$$\sigma_{r,l} = \sigma_{sc,l} = \frac{\pi}{k^2} (2l+1).$$

'he permitted range of values for σ_{sc} and σ_r for a given l value are shown in 'ig. 6.1; they are required to lie inside the shaded area.

The sum of the elastic and reaction cross-sections is called the total cross-ection

$$\sigma_T = \sigma_{sc} + \sigma_r = \frac{\pi}{k^2} \sum_{l=0}^{\infty} (2l+1)\, (|\eta_l - 1|^2 + 1 - |\eta_l|^2)$$

$$= \frac{2\pi}{k^2} \sum_{l=0}^{\infty} (2l+1)\, (1 - \mathrm{Re}\,\eta_l). \tag{6.11}$$

The symbol Re means "real part of ".) The total cross-section is related to the

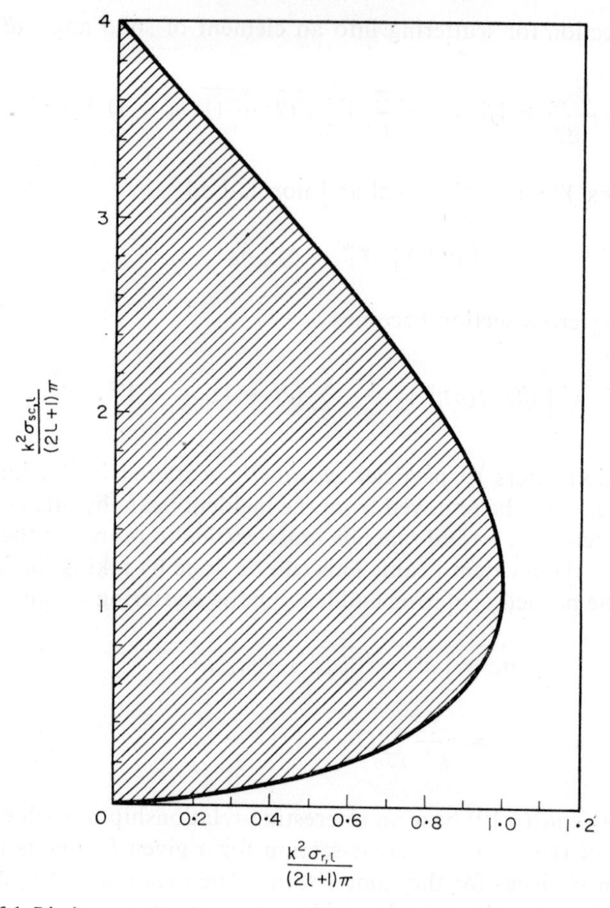

FIG. 6.1. Limits on scattering and reaction cross-sections for given value of l. Allowed cross-sections must lie in the shaded area (Blatt and Weisskopf, 1952).

imaginary part of the forward-scattering amplitude which we shall write a Im $f(0)$. The relationship may be established with the aid of equation (6.7)

$$f(\theta) = \sum_{l=0}^{\infty} (2l+1) \frac{(\eta_l - 1)}{2ik} P_l^0(\theta)$$

and so

$$\operatorname{Im} f(0) = \frac{1}{2k} \sum_{l=0}^{\infty} (2l+1)(1 - \operatorname{Re} \eta_l)$$

since $P_l^0(0) = 1$. Thus we find that

$$\frac{4\pi}{k} \operatorname{Im} f(0) = \frac{2\pi}{k^2} \sum_{l=0}^{\infty} (2l+1)(1 - \operatorname{Re} \eta_l) = \sigma_T. \quad (6.12)$$

his important equation is known as the *optical theorem.* It also holds when the articles have spin.

6.1 (b). *Phase shifts and scattering functions*

The scattering process changes both the amplitudes and the phases of the utgoing waves. The change is commonly expressed in one of two ways, either y using a complex phase shift δ_l

$$\eta_l = e^{2i\delta_l}, \qquad \delta_l = \alpha_l + i\beta_l \tag{6.13}$$

r by using a real phase shift δ_l and an inelasticity parameter ϱ

$$\eta_l = \varrho_l e^{2i\delta_l}, \quad \varrho_l, \delta_l \text{ real} \quad (0 < \varrho_l < 1). \tag{6.14}$$

he two definitions are obviously equivalent and we shall normally use the latter ne. It leads to the following relation for the reaction cross-section (6.10):

$$\sigma_r = \frac{\pi}{k^2} \sum_{l=0}^{\infty} (2l+1)(1-\varrho_l^2). \tag{6.15}$$

It is obvious from this expression that pure elastic scattering occurs for $\varrho_l = \pm 1$; e shall use $\varrho_l = 1$ and η_l then becomes

$$\begin{aligned} \eta_l = e^{2i\delta_l} &= \cos 2\delta_l + i \sin 2\delta_l \\ &= 1 - 2\sin^2\delta_l + 2i \sin\delta_l \cos\delta_l \\ &= 1 + i\,2 \sin\delta_l e^{i\delta_l}. \end{aligned} \tag{6.16}$$

The scattering function appearing in (6.6) and (6.7)

$$f_l = \frac{\eta_l - 1}{2ik} = \frac{\varrho_l e^{2i\delta_l} - 1}{2ik} \tag{6.17}$$

herefore becomes

$$f_l = \frac{\sin\delta_l\, e^{i\delta_l}}{k} \tag{6.18}$$

or elastic scattering,† and leads to the following expressions for the scattering mplitude:

$$f(\theta) = \sqrt{(4\pi)} \sum_{l=0}^{\infty} \sqrt{(2l+1)}\, \frac{\sin\delta_l e^{i\delta_l}}{k}\, Y_l^0(\theta) \tag{6.19}$$

$$= \sum_{l=0}^{\infty} (2l+1) \frac{\sin\delta_l e^{i\delta_l}}{k} P_l^0(\theta). \tag{6.20}$$

† If complex phase shifts (6.13) are used, f_l assumes the form given in equation (6.18) even f $\varrho_l \neq 1$.

Using the notation given above it is a simple matter to show that the total cros section for elastic scattering (6.9) assumes the following form:

$$\sigma_{sc} = \int d\Omega\, |f(\theta)|^2 = \frac{4\pi}{k^2} \sum_{l=0}^{\infty} (2l+1) \sin^2 \delta_l.$$

Thus the scattering cross-section associated with any l value becomes a maximu if

$$\delta_l = n\frac{\pi}{2} \quad (n = 1, 3, 5, 7, \ldots). \tag{6.2}$$

This condition is called a *resonance.*

It should be noted that the expression $(e^{2i\delta_l} - 1)/2i$ is sometimes called th scattering amplitude of the wave; we prefer to keep this definition for $f(\theta)$.

6.1 (c). *Scattering from a 'black sphere'*

At the end of the previous section pure elastic scattering ($\varrho_l = 1$) was con sidered. We will now examine the opposite extreme where the scattering system highly absorptive. This may be done by introducing the condition of maximum inelasticity $\eta_l = \varrho_l = 0$ (compare (6.10)). We assume that there is some radiu of interaction R for the system, and that the wavelength of the incoming particle are much smaller than R, so that $kR \gg 1$†. Thus there will be many l value participating in the interaction. We assume that the region inside R is complete absorptive or 'black'. If we regard the system semi-classically we may then argu that waves with l values up to a maximum set by

$$l_{max} = kR$$

will be completely absorbed, and that waves with $l > l_{max}$ are completely un affected. We then have the conditions

$$\begin{aligned} \varrho_l = \eta_l &= 0 \quad (l < kR) \\ \eta_l &= 1 \quad (l > kR) \end{aligned} \tag{6.22}$$

and so we find

$$\sigma_{sc} = \sigma_r = \frac{\pi}{k^2} \sum_{l=0}^{l_{max}} (2l+1)$$

$$\sigma_T = \frac{2\pi}{k^2} \sum_{l=0}^{l_{max}} (2l+1)$$

† This assumption is not strictly necessary in order to describe absorptive scattering. An ade quate assumption would be that $\eta_l = 0$ for certain l values. The introduction of the radius R is fo later convenience.

om equations (6.9) and (6.10). The standard formula for an arithmetic pro-
ession can now be used to show that

$$\sum_{l=0}^{l_{max}} (2l+1) = l_{max}^2 = k^2R^2$$

ıd so

$$\sigma_{sc} = \sigma_r = \pi R^2, \quad \sigma_T = 2\pi R^2. \tag{6.23}$$

ow let us consider elastic scattering distributions in this approximation; it is
ɔparent from equations (6.7) and (6.22) that $f(\theta)$ is purely imaginary

$$f(\theta) = \frac{i}{k}\sum_{l=0}^{l_{max}} (l+\tfrac{1}{2})\, P_l^0(\theta)$$

ıd so

$$\frac{d\sigma_{sc}}{d\Omega} = |f(\theta)|^2 = \frac{1}{k^2}\Big[\sum_l (l+\tfrac{1}{2})\, P_l^0(\theta)\Big]^2.$$

We now assume that a large number of l values are required for the scattering,
ıd so we may replace the summation over l by an integration. At the same time
ıe scattering becomes confined to small angles and it is convenient to introduce
ıe following approximations:†

$$l+\tfrac{1}{2} = kb, \qquad P_l^0(\theta) = J_0(kb\sin\theta)$$

here J_0 is a Bessel function and b is an impact parameter – that is the distance
f closest approach to the scattering centre. Thus the differential cross-section
ecomes

$$\frac{d\sigma_{sc}}{d\Omega} = \frac{1}{k^2}\left[\int_0^R db\, bk^2 J_0(kb\sin\theta)\right]^2 = R^2\left[\frac{J_1(kR\sin\theta)}{\sin\theta}\right]^2. \tag{6.24}$$

his is the classical formula for diffraction scattering from a totally absorbing
black') sphere of radius R; it has a minimum at

$$\sin\theta = \frac{3{\cdot}8}{kR}.$$

he diffraction formula is a useful one for scattering at high energies, since then
ıe physical conditions approximate to the assumptions made at the beginning
f this section. It is useful even if the particles possess spin, since if l is much
eater than the spin values, the effect of the latter is small in differential and
ɔtal cross-sections.

† A more detailed discussion of this procedure may be found in an article by Glauber (1958).
etailed properties of the mathematical functions $P_l^0(\theta)$ and $J_0(kb\sin\theta)$ may be found in
hnke and Emde (1945).

In practice the interactions at high energies are often not completely absorp-tive. An opacity factor a may then be introduced by writing

$$a \equiv |1 - \eta_l| \quad (l < kR)$$
$$\eta_l = 1 \qquad (l > kR).$$

If the further assumption is made that the (real) phase shifts are small, then it is simple matter to show that

$$\sigma_{sc} = a^2 \pi R^2, \qquad \sigma_T = 2a\pi R^2 \tag{6.25}$$
$$\frac{d\sigma_{sc}}{d\Omega} = a^2 R^2 \left[\frac{J_1(kR \sin \theta)}{\sin \theta} \right]^2.$$

Further refinements may be made to this treatment by removing the sharp boundary conditions at R (see, for example, Levy, 1962).

6.1(d). *The scattering of particles with spin*

We now examine the more complex situation which arises when both the scattered particle and the scattering centre possesss spin.

We will use the following notation:

$$\begin{aligned} l &= \text{orbital angular momentum of incident particle} \\ s_a &= \text{spin of incident particle} \\ s_b &= \text{spin of scattering centre} \\ j &= \text{total angular momentum.} \end{aligned} \tag{6.26}$$

All the angular momenta are given in units of $\hbar$. The numbers l, s_a, s_b and are called *channel numbers* (Blatt and Weisskopf, 1952). We now introduce *channel spin operator* **S**, defined as

$$\mathbf{S} = \mathbf{s}_a + \mathbf{s}_b.$$

Each value of the channel spin has $2s + 1$ orientations, given by

$$m_s = s, s - 1, \ldots, -s$$

and therefore the total number of possible orientations of the channel spin is given by the relation

$$\begin{aligned} &[2(s_a + s_b) + 1] + [2(s_a + s_b - 1) + 1] + \cdots + [2|s_a - s_b| + 1] \\ &= (2s_a + 1)(2s_b + 1) + 2[s_a + (s_a - 1) + \cdots - (s_a - 1) - s_a] \\ &= (2s_a + 1)(2s_b + 1). \end{aligned}$$

An unpolarised beam of particles arriving at a target can be considered as an incoherent mixture of waves in all $(2s_a + 1)(2s_b + 1)$ channels. Each possibl ɛt of spin orientations of incident and target particles have the same statistical ɾobability, that is

$$\frac{1}{(2s_a + 1)(2s_b + 1)}. \tag{6.27}$$

The contributions of the individual states add incoherently, since the phase ɛlations between them are random. Thus the relative probability of occurrence f a channel spin s is

$$p_s = \frac{2s + 1}{(2s_a + 1)(2s_b + 1)} \tag{6.28}$$

nce there are $(2s + 1)$ states associated with each value of s.

The channel spin s must now be combined with the orbital angular momentum to obtain the total angular momentum of the system. The angular momentum perators combine vectorially

$$\mathbf{J} = \mathbf{L} + \mathbf{S}$$

ɩus the quantum number j can take on the following values:

$$|l - s| \leqq j \leqq (l + s).$$

hus each partial wave of orbital angular momentum l can be considered as a near combination of waves with different values of j. For example consider a ystem with $s_a = s_b = \frac{1}{2}$ and $l = 1$

$$s_a = s_b = \tfrac{1}{2} \quad \text{gives} \quad s = 0, 1,$$

$$s = 0, \quad l = 1 \qquad j = 1$$

$$s = 1, \quad l = 1 \qquad j = 0, 1, 2.$$

hus j can take the values 0, 1, 2. Systems of this type are encountered in nucleon–ucleon scattering problems. The spin system $s = 0$ is called the *singlet* state, nd $s = 1$ the *triplet* state. The reasons for this appellation is obvious if we onsider the statistical weights of the two states according to equation (6.28)

$$p_s = \tfrac{1}{4} \text{ for } s = 0, \quad p_s = \tfrac{3}{4} \text{ for } s = 1.$$

It is apparent from the above discussion that the scattering of particles with ɔin is far more complex than in the spinless case. A further complicating ɩctor is caused by the fact that the orbital angular momentum quantum number can change. The conservation laws require that the total angular momentum j nd its component m_j are invariant but place no restriction on l. If the interction is strong or electromagnetic parity is also preserved, and since the parity ɛrm associated with l is given by $(-1)^l$, a further restriction can be imposed. 'onsider, for example, the elastic scattering of two protons in the state $j = 2$,

$s = 1, l = 1$; a transition to the state $j = 2, s = 1, l = 3$ is permissible but not one to $j = 2, s = 0, l = 2$.

The construction of scattering amplitudes for systems with arbitrary channel spin is complicated (see, for example, Blatt and Weisskopf, 1952). In practice, in elementary particle physics two important systems occur – the scattering of spin 0, spin $\frac{1}{2}$ particles; and spin $\frac{1}{2}$, spin $\frac{1}{2}$ scattering; they can be treated individually.

We start by considering the first example, which is encountered in pion–nucleon scattering. Since the channel spin $s = \frac{1}{2}$, changes in l values cannot occur without violating parity conservation. Thus l is conserved as well as j in the scattering. We will assume the protons are unpolarised and consider the case when the proton spin points along the z-axis so that the wave function for the incident system is

$$\frac{1}{\sqrt{v}}\, e^{ikz}\, \chi_{1/2}^{1/2}$$

where $\chi_{1/2}^{1/2}$ represents the spin function for the proton in the $+z$-direction. The term e^{ikz} can be expanded as in (6.3) and leads to the orbital angular momentum function $Y_l^0(\theta)$. This function and that for the spin must be combined in such a manner that the angular momentum operators combine vectorially

$$\mathbf{J} = \mathbf{L} + \mathbf{S}.$$

The technique for performing this operation was developed in § 5.3(e), and we may adapt equation (5.70) to suit our present problem

$$\begin{aligned} Y_l^0 \chi_s^{m_s} &= \sum_{j=|l-s|}^{j=l+s} C_{j\ \ l\ \ s}^{m_j\ 0\ m_s}\, \mathscr{Y}_j^{m_j} \\ &= \sqrt{\left(\frac{l+1}{2l+1}\right)}\, \mathscr{Y}_{l+1/2}^{1/2} - \sqrt{\left(\frac{l}{2l+1}\right)}\, \mathscr{Y}_{l-1/2}^{1/2} \end{aligned} \tag{6.29}$$

where

$$m_j = m_l + m_s = m_s = \tfrac{1}{2}$$

since $m_l = 0$ (see discussion following (6.3)). The Clebsch–Gordan coefficients have been evaluated with the aid of Table A.7.1 (p. 708). The term $\mathscr{Y}_j^{m_j}$ is an eigenfunction of $\mathbf{J}^2$ and J_z with eigenvalues $j(j+1)$ and m_j respectively. The construction of the wave function for the scattered wave now proceeds as in equations (6.4) and (6.5), with the replacement of $Y_l^0(\theta)$ by the functions $\mathscr{Y}_j^{1/2}$ together with their weighting values. We therefore obtain

$$\psi_{\text{sc}} = \frac{1}{kr} \sqrt{\left(\frac{4\pi}{v}\right)} \sum_{l=0}^{\infty} \left[\sqrt{(l+1)}\, \frac{\eta_{l+} - 1}{2i}\, \mathscr{Y}_{l+}^{1/2} + \sqrt{l}\, \frac{\eta_{l-} - 1}{2i}\, \mathscr{Y}_{l-}^{1/2}\right] e^{ikr}$$

where the subscripts $l+$ and $l-$ imply $j = l + \frac{1}{2}$ and $j = l - \frac{1}{2}$ respectively. I e write the scattering function f_{lj} as

$$f_{lj} = f_{l\pm} = \frac{\eta_{l\pm} - 1}{2ik} \tag{6.30}$$

he scattered wave assumes the form

$$\psi_{sc} = \frac{e^{ikr}}{r} \sqrt{\left(\frac{4\pi}{v}\right)} \sum_{l=0}^{\infty} \left[\sqrt{(l+1)} f_{l+} \mathscr{Y}_{l+}^{1/2} - \sqrt{l} f_{l-} \mathscr{Y}_{l-}^{1/2}\right]. \tag{6.31}$$

he functions $\mathscr{Y}_{l\pm}^{1/2}$ can be evaluated in terms of Y and χ with the aid of Table .7.1 (p. 708), and yield

$$\mathscr{Y}_{l+}^{1/2} = \sqrt{\left(\frac{l+1}{2l+1}\right)}\, Y_l^0 \chi_{1/2}^{1/2} + \sqrt{\left(\frac{l}{2l+1}\right)}\, Y_l^1 \chi_{1/2}^{-1/2}$$

$$\mathscr{Y}_{l-}^{1/2} = -\sqrt{\left(\frac{l}{2l+1}\right)}\, Y_l^0 \chi_{1/2}^{1/2} + \sqrt{\left(\frac{l+1}{2l+1}\right)}\, Y_l^1 \chi_{1/2}^{-1/2}.$$

Vith the aid of equation (A.7.7) (p. 705), the scattered wave (6.31) therefore becomes

$$\begin{aligned}\psi_{sc} &= \frac{e^{ikr}}{r} \sqrt{\left(\frac{4\pi}{v}\right)} \sum_{l=0}^{\infty} \frac{1}{\sqrt{(2l+1)}} \{[(l+1) f_{l+} + l f_{l-}]\, Y_l^0 \chi_{1/2}^{1/2} \\ &\qquad + \sqrt{[l(l+1)]}\,(f_{l+} - f_{l-}) Y_l^1 \chi_{1/2}^{-1/2}\} \\ &= \frac{1}{\sqrt{v}} \frac{e^{ikr}}{r} \sum_{l=0}^{\infty} \{[(l+1) f_{l+} + l f_{l-}]\, P_l^0(\theta)\, \chi_{1/2}^{1/2} + (f_{l+} - f_{l-})\, P_l^1(\theta)\, e^{i\varphi} \chi_{1/2}^{-1/2}\}.\end{aligned} \tag{6.32}$$

Now for a scattering system with spin we can write the asymptotic wave unction in a form equivalent to (6.2)

$$\psi(\mathbf{r}) \underset{r\to\infty}{=} \frac{1}{\sqrt{v}} \left[e^{ikz} \chi_s^{m_s} + \frac{e^{ikr}}{r} \sum_{m_s'} M^{m_s m_s'}(\theta, \varphi)\, \chi_s^{m_s'}\right] = \frac{1}{\sqrt{v}}\, e^{ikz} \chi_s^{m_s} + \psi_{sc} \tag{6.33}$$

here $M^{m_s m_s'}$ is a scattering matrix in spin space connecting the initial and final pin states χ. If this equation is compared with (6.32), and it is recalled that we ssumed an initial spin state $\chi_{1/2}^{1/2}$, then it is apparent that we have two terms—ne corresponding to no spin flip and the other to spin flip; we shall designate hem by the symbols $g(\theta)$ and $h(\theta)$ respectively

$$g(\theta) = \sum_{l=0}^{\infty} [(l+1) f_{l+} + l f_{l-}]\, P_l^0(\theta) \tag{6.34}$$

$$h(\theta) = -\sum_{l=0}^{\infty} (f_{l+} - f_{l-})\, P_l^1(\theta)$$

here the minus sign has been inserted for later convenience.

Since $\chi^{1/2}_{1/2}$ and $\chi^{-1/2}_{1/2}$ are orthogonal functions, the scattering cross-sectio (6.8) therefore becomes†

$$\frac{d\sigma_{sc}}{d\Omega} = |g(\theta)|^2 + |h(\theta)|^2. \quad (6.3$$

The cross-section is therefore complicated unless one state dominates the sca tering amplitude, then considerable simplifications can be made. If we write f_{lj} a in (6.17)

$$f_{lj} = \frac{\eta_{lj} - 1}{2ik} = \frac{\varrho_{lj}\, e^{2i\delta_{lj}} - 1}{2ik} \quad (6.3$$

where δ_{lj} is the phase shift for the channel with quantum numbers l and j, the one may show (compare (6.8) to (6.10)) that the total cross-sections for elasti and inelastic scattering in a single channel are given by

$$\sigma_{sc} = \frac{\pi}{k^2}(j + \tfrac{1}{2})\,|\eta_{lj} - 1|^2 \quad (6.3$$

$$\sigma_r = \frac{\pi}{k^2}(j + \tfrac{1}{2})(1 - |\eta_{lj}|^2) = \frac{\pi}{k^2}(j + \tfrac{1}{2})(1 - \varrho^2_{lj}).$$

The scattering of spin $\frac{1}{2}$, spin $\frac{1}{2}$ systems, for example nucleon–nucleon sca tering, is considerably more complicated than the spin $\frac{1}{2}$, spin 0 case. The spi systems can form singlet ($s = 0$) and triplet ($s = 1$) states, and in the latte states, transitions between different values can occur if $l = j \pm 1$. Suitabl matrices for dealing with this problem have been constructed by a number c workers (Stapp, Ypsilantis and Metropolis, 1957; Macgregor, Moravcsik an Stapp, 1960; Nigam, 1963). The following parameters are required (pure elasti scattering is considered):

$$\begin{aligned}
\alpha_l &= e^{2i\delta_l} - 1 && s = 0 \quad (6.38\\
\alpha_{lj} &= e^{2i\delta_{lj}} - 1 && s = 1;\ l = j\\
\alpha_{j\pm1,j} &= \cos^2\varepsilon_j\, e^{2i\delta_{j\pm1,j}} + \sin^2\varepsilon_j\, e^{2i\delta_{j\mp1,j}} - 1 && s = 1;\ l = j \pm 1\\
\alpha^j &= \tfrac{1}{2}\sin 2\varepsilon_j (e^{2i\delta_{j\mp1,j}} - e^{2i\delta_{j\pm1,j}}) && s = 1;\ l = j \pm 1\\
& && l \to (l \pm 2)
\end{aligned}$$

where ε_j represents the parameter which couples states of angular momentun $l = j + 1$ and $l = j - 1$ for a given j value.

† As we are considering an unpolarised system, strictly speaking we should also have con sidered the functions $|g(\theta)|^2$ and $|h(\theta)|^2$ for a system with the spin initially pointing opposite t the z-axis, and averaged over the two systems by dividing by $2s + 1 = 2$. Since the terms $|g(\theta)|$ and $|h(\theta)|^2$ are the same in both cases, equation (6.35) is adequate.

6.2. THE SCATTERING OF POLARISED PARTICLES

6.2(a). *Definitions*

Consider a beam of particles of, say, spin $\frac{1}{2}$. The spin part of their wave function may be written as a linear combination of two orthogonal components, each giving an amplitude and phase of a certain spin orientation

$$\chi = a_1\chi_{1/2}^{1/2} + a_2\chi_{1/2}^{-1/2} \tag{6.39}$$

where the superscripts indicate whether the spins lie parallel or antiparallel to a certain direction (which is chosen to suit the conditions of an experiment).

We will refer to the two alternative spin states as the *states of polarisation of the beam*. The intensity of the beam can be represented as the sum of the probabilities of the two states of polarisation

$$I = |a_1|^2 + |a_2|^2$$

and the *polarisation* of the beam is defined as the net probability that the spins point parallel to our chosen direction

$$P = \frac{|a_1|^2 - |a_2|^2}{|a_1|^2 + |a_2|^2} = \frac{1}{I}\left(|a_1|^2 - |a_2|^2\right). \tag{6.40}$$

6.2(b). *The quantum mechanics of polarised beams*

The description of polarisation is most easily accomplished with the aid of density matrix techniques.

Let us consider a pure spin state and represent it by the state vector $|\chi_s\rangle$; we may expand this term into a complete set of orthonormal basis states $|m\rangle$ (compare equation (4.15))

$$|\chi_s\rangle = \sum_m |m\rangle\langle m|\chi_s\rangle = \sum_m a_{sm}|m\rangle. \tag{6.41}$$

The expectation value of an operator A with respect to χ_s is then given by

$$\langle A_s\rangle = \langle\chi_s|A|\chi_s\rangle = \sum_{l,m} a_{sl}^* a_{sm}\langle l|A|m\rangle.$$

Now let us consider an incoherent mixture of pure states, say the singlet and triplet states in a nucleon–nucleon system, each occurring with a statistical weight p_s (6.28). The average expectation value of the operator A for this system is given by

$$\langle\bar{A}\rangle = \frac{\sum_s p_s\langle A_s\rangle}{\sum_s p_s\langle\chi_s|\chi_s\rangle} = \frac{\sum_{l,m}\sum_s p_s a_{sl}^* a_{sm}\langle l|A|m\rangle}{\sum_{l,m}\sum_s p_s a_{sl}^* a_{sm}\langle l|m\rangle}$$

$$= \frac{\sum_{l,m}\sum_s \langle l|A|m\rangle p_s a_{sl}^* a_{sm}}{\sum_m\sum_s p_s a_{sm}^* a_{sm}} \tag{6.42}$$

Now the terms $\sum_s p_s a_{sl}^* a_{sm}$ form the elements of a matrix $\sum_s |\chi_s\rangle p_s \langle\chi_s|$. Th

matrix is called *the density matrix*† and is symbolised by ϱ; its matrix element are given by

$$\varrho_{ml} = \sum_s p_s a_{sl}^* a_{sm} = \sum_s p_s a_{sm} a_{sl}^* = \sum_s \langle m|\chi_s\rangle p_s \langle\chi_s|l\rangle. \quad (6.4$$

From this expression it is apparent that the denominator in (6.42) can be writte as tr ϱ, and the average expectation value becomes

$$\langle\overline{A}\rangle = \frac{\sum_{l,m} \langle l|A|m\rangle \varrho_{ml}}{\text{tr}\,\varrho} = \frac{\text{tr}\,A\varrho}{\text{tr}\,\varrho} = \frac{\text{tr}\,\varrho A}{\text{tr}\,\varrho} \quad (6.4$$

where the last term arises because the algebraic rules for traces allow us to writ tr ϱA = tr $A\varrho$.

Some properties of the density matrix are worth noting. First, since $\langle\overline{A}\rangle$ is real quantity A is a Hermitian operator, and so ϱ must be Hermitian

$$\varrho_{ml} = \varrho_{lm}^*.$$

Secondly, it is apparent from (6.43) that tr ϱ represents the intensity of the bea

$$\text{tr}\,\varrho = \sum_m \varrho_{mm} = I \quad (6.4$$

(this is written often as tr $\varrho = 1$ since the appropriate normalisation can alway be carried out).

In deriving the above equations no specific properties of the spins were use and so it is apparent that any system whose wave functions can be expande into an orthogonal set can be treated by the techniques described above.

As an example of the use of the density matrix we return to our system wit spin $\frac{1}{2}$ (§ 6.2(a)). Here we have a pure spin state and so $p_s = 1$; if we drop o the subscript s in order to conform to the notation of (6.39), the matrix ele ments for ϱ are then given by (6.43)

$$\varrho = \begin{pmatrix} a_1 a_1^* & a_1 a_2^* \\ a_2 a_1^* & a_2 a_2^* \end{pmatrix}. \quad (6.4$$

Thus

$$\text{tr}\,\varrho = I = |a_1|^2 + |a_2|^2$$

as in (6.45) and in § 6.2(a). Now if we choose the z-axis as the axis of quantisatio then the average value of the Pauli spin operator in the z-direction can be writte as

$$\langle\sigma_z\rangle = \frac{\text{tr}\,\sigma_z\varrho}{\text{tr}\,\varrho} = \frac{|a_1|^2 - |a_2|^2}{|a_1|^2 + |a_2|^2} = P_z.$$

† An excellent introduction to density matrices may be found in Dirac's *Quantum Mechani* (1947). A detailed account of their application to polarisation problems has been given by McMa ter (1961) and by Hamilton (1959).

This is the same result as that given in (6.40). Similar expressions can be written for the x and y components so that the expectation value for the polarisation in the directions x, y and z is given by

$$\langle \bar{\sigma}_i \rangle = \frac{\operatorname{tr} \sigma_i \varrho}{\operatorname{tr} \varrho} = P_i \qquad (i = x, y, z). \tag{6.47}$$

It is apparent from (6.46) that the elements of the density matrix can be constructed directly by writing χ_s in the form of a column matrix which has as elements the coefficients a_{sm} of (6.41) and (6.43) with m running between s and $-s$

$$\chi_s = \begin{matrix} a_{ss} \\ a_{ss-1} \\ \vdots \\ a_{s-s} \end{matrix}.$$

Thus for pure spin states

$$\varrho = \chi_s \chi_s^\dagger \tag{6.48}$$

(compare (6.46)) and if the state is a mixed one

$$\varrho = \sum_s p_s \chi_s \chi_s^\dagger .$$

It is often convenient to split the density matrix for a spin $\frac{1}{2}$ system into two parts since a partially polarised beam can be regarded as an incoherent superposition of unpolarised and totally polarised states. This can be done by writing the density matrix as

$$\varrho = \tfrac{1}{2}I(\hat{1} + \mathbf{P} \cdot \boldsymbol{\sigma}) = \tfrac{1}{2}I(\hat{1} + P_x \sigma_x + P_y \sigma_y + P_z \sigma_z) \tag{6.49}$$

where $\hat{1}$ is a 2×2 unit matrix and $\mathbf{P}$ is the polarisation vector. If we use the Pauli spin matrices (3.104), ϱ assumes the form

$$\varrho = \tfrac{1}{2}I \begin{bmatrix} 1 + P_z & P_x + iP_y \\ P_x - iP_y & 1 - P_z \end{bmatrix} \tag{6.50}$$

and as in equations (6.45) and (6.47)

$$\operatorname{tr} \varrho = I$$

$$\langle \bar{\sigma}_z \rangle = \frac{\operatorname{tr} \sigma_z \varrho}{\operatorname{tr} \varrho} = P_z .$$

6.2(c). *Composite spin space*

Before discussing the scattering matrix it is convenient to consider the problems associated with combining the spin spaces of two particles. Consider the elastic scattering process

$$a + b \rightarrow a + b$$

if the spin of the particle a is s_a and that of b is s_b then, according to equation (6.27) and (6.28), the composite spin system is a combination with $(2s_b + 1)$ $(2s_a + 1)$ components. For example, in the scattering of two nucleons, a single and a triplet state can occur for each l value.

The equivalent of the state vector (6.41) therefore becomes

$$|\chi\rangle = \sum_{m,n} c_{mn} |m, n\rangle \tag{6.51}$$

where c is a numerical coefficient. We shall work in a space where the state vectors $|\chi\rangle$ possess $(2s_a + 1)(2s_b + 1)$ components and the operators have $(2s_a + 1)(2s_b + 1)$ rows and colums (in matrix representation), that is $[(2s_a + 1)(2s_b + 1)]^2$ elements. We will represent the spin operators by the symbol S. If we again consider two nucleons, the matrix elements of S assume the form

$$\langle m', n'| S |m, n\rangle = \langle m', n'| \sigma_{\mu a}\sigma_{\nu b} |m, n\rangle = (\sigma_\mu)_{m'm}(\sigma_\nu)_{n'n} \tag{6.52}$$

where σ_μ and σ_ν represent the 2×2 Pauli spin matrices and the unit matrix (that is, we allow for partially polarised beams as in (6.49)). Now it is a property of matrix algebra that a matrix with N^2 elements can be expressed as a set of N^2 linearly independent matrices each with N^2 elements. For example with two nucleons we can construct the following sets of spin matrices:

$$\hat{1}_a\hat{1}_b(1), \quad \hat{1}_a\boldsymbol{\sigma}_b(3), \quad \boldsymbol{\sigma}_a\hat{1}_b(3), \quad \boldsymbol{\sigma}_a\boldsymbol{\sigma}_b(9)$$

thus giving a total of sixteen matrices (the number in brackets indicates the number of matrices of each type); a simpler example can be seen in equation (6.49). A typical matrix in composite spin space would have the following form

$$\hat{1}_a\sigma_{xb} = \begin{pmatrix} 0 & 1 & 0 & 0 \\ 1 & 0 & 0 & 0 \\ 0 & 0 & 0 & 1 \\ 0 & 0 & 1 & 0 \end{pmatrix}.$$

We will denote the linearly independent matrices by S_μ; it is evident that each matrix S_μ is Hermitian, and by considering specific examples it is easy to show that

$$\operatorname{tr} S_\mu S_\nu = (2s_a + 1)(2s_b + 1)\,\delta_{\mu\nu}. \tag{6.53}$$

Now we have shown in (6.49) and (6.50) that it is possible to expand the density matrix ϱ for a spin $\frac{1}{2}$ particle into four terms involving the unit matrix and the Pauli spin operators. We will make the same expansion in the present problem

$$\varrho = b_\mu S_\mu \equiv \sum_\mu b_\mu S_\mu \tag{6.54}$$

ut now the expansion contains $[(2s_a + 1)(2s_b + 1)]^2$ terms. If we operate on oth sides of this equation with S_ν and take the trace we find

$$\langle \bar{S_\nu} \rangle \operatorname{tr} \varrho = (2s_a + 1)(2s_b + 1)\, b_\nu$$

here we have used (6.44) and (6.53). We may solve this equation for b_ν and sert the result in (6.54) yielding

$$\varrho = \frac{\operatorname{tr} \varrho}{(2s_a + 1)(2s_b + 1)} \sum_\mu \langle \bar{S_\mu} \rangle S_\mu. \tag{6.55}$$

is evident that (6.49) is a specific example of this equation

$$\varrho = \frac{\operatorname{tr} \varrho}{(2s_a + 1)} \left[\frac{(\operatorname{tr} \hat{1}\varrho)\,\hat{1}}{\operatorname{tr} \varrho} + \sum_i \frac{(\operatorname{tr} \sigma_i \varrho)\,\sigma_i}{\operatorname{tr} \varrho} \right] = \tfrac{1}{2} I(\hat{1} + \mathbf{P} \cdot \boldsymbol{\sigma}).$$

6.2(d). *The scattering matrix*

The elastic scattering of two particles with spin will now be considered. The nalysis of polarisation experiments may be made in a convenient manner by ntroducing a scattering matrix M (Wolfenstein and Ashkin, 1952). This matrix cts in the composite spin space of the two particles; it is defined by writing he asymptotic form of the amplitude and spin condition of the outgoing beam s

$$\chi_{sc} = M\chi_i \equiv f(\theta)$$

where χ_i represent the spin functions for the ingoing beam. Now the density matrix is given by $\varrho = \chi\chi^\dagger$ (6.48), and so that for the scattered beam can be written as

$$\varrho_{sc} = \chi_{sc}\chi_{sc}^\dagger = (M\chi_i)(M\chi_i)^\dagger = M\chi_i\chi_i^\dagger M^\dagger = M\varrho_i M^\dagger \tag{6.56}$$

where ϱ_i represents the density matrix for the incident beam.

If we denote the intensities of the scattered and incident beams by I_{sc} and I_i espectively, then the scattering cross-section is given by

$$\frac{d\sigma_{sc}}{d\Omega} = \frac{I_{sc}}{I_i} = \frac{\operatorname{tr} \varrho_{sc}}{\operatorname{tr} \varrho_i} = \frac{\operatorname{tr} M\varrho_i M^\dagger}{\operatorname{tr} \varrho_i}. \tag{6.57}$$

n general $\operatorname{tr} \varrho_i$ can be normalised to unity without difficulty, and so the scattering cross-section is often set equal to the numerator of the above equation in the iterature.

Let us finally relate equation (6.57) to the expectation values of the polari- ation operators. Using (6.44) we can write

$$\langle \bar{S_\mu} \rangle_{sc} \operatorname{tr} \varrho_{sc} = \operatorname{tr} \varrho_{sc} S_\mu$$

hence

$$\frac{d\sigma_{sc}}{d\Omega}\langle\overline{S_\mu}\rangle_{sc} = \frac{1}{(2s_a+1)(2s_b+1)}\sum_\nu \langle\overline{S_\nu}\rangle_i \operatorname{tr} MS_\nu M^\dagger S_\mu \quad (6.5$$

where we have used (6.55) for ϱ_i and $\langle\overline{S_\nu}\rangle_i$ refers to the average expectation valu for the incident beam.

6.2(e). *The polarisation induced in the scattering of spin $\frac{1}{2}$ particles by a spin 0 target*

The elastic scattering of a beam of particles of spin $\frac{1}{2}$ by a spin zero target wi be examined. The method is applicable to the scattering of pions or kaons b nucleons, and its principle is capable of extension to more complex system We will assume that the incident beam of particles is unpolarised ($P_i = 0$). I this case all the S_ν terms in equation (6.58) are zero except for the unit matri and it therefore reduces to

$$\frac{d\sigma_{sc}}{d\Omega}\langle\overline{\sigma}\rangle_{sc} = \tfrac{1}{2}\operatorname{tr}(MM^\dagger\boldsymbol{\sigma}). \quad (6.5$$

The term $d\sigma_{sc}/d\Omega$ will be considered first; using equations (6.57) and (6.4 with $\mathbf{P}_i = 0$ it can be written as

$$\frac{d\sigma_{sc}}{d\Omega} = \frac{\operatorname{tr} M\varrho_i M}{\operatorname{tr}\varrho_i} = \frac{1}{2}\frac{I\operatorname{tr} MM^\dagger}{I} = \tfrac{1}{2}\operatorname{tr} MM^\dagger. \quad (6.6$$

General invariance arguments may be used to restrict the form which M ca assume in this equation. The most general form for M should contain a ter which is independent of relative spin orientations and one which is depender on spin. Now the spin is an axial vector, but the matrix M should be a scala since the scattering process should be independent of the coordinate systen Thus we must combine $\boldsymbol{\sigma}$ with a vector. The only available vectors are $\mathbf{k}$ an $\mathbf{k}'$, the initial and final momenta respectively, but terms of the type $\boldsymbol{\sigma}\cdot\mathbf{k}$ chang sign upon reflections of space coordinates and so the only possible form for th scattering matrix is

$$M = f_1(k,\theta)\,\hat{1} + \frac{(\boldsymbol{\sigma}\cdot\mathbf{k})(\boldsymbol{\sigma}\cdot\mathbf{k}')}{k^2}f_2(k,\theta)$$

where we assume that we are working in the c-system so that $|\mathbf{k}| = |\mathbf{k}'| = k$. If w now use the general relation

$$(\boldsymbol{\sigma}\cdot\mathbf{A})(\boldsymbol{\sigma}\cdot\mathbf{B}) = \mathbf{A}\cdot\mathbf{B} + i\boldsymbol{\sigma}\cdot\mathbf{A}\times\mathbf{B}$$

we can write

$$\frac{(\boldsymbol{\sigma}\cdot\mathbf{k})(\boldsymbol{\sigma}\cdot\mathbf{k}')}{k^2} = \cos\theta + i\boldsymbol{\sigma}\cdot\mathbf{n}\sin\theta \quad (6.61$$

$$\mathbf{n} = \frac{\mathbf{k}\times\mathbf{k}'}{|\mathbf{k}\times\mathbf{k}'|}.$$

Thus we find†

$$M = (f_1 + f_2 \cos\theta)\,\hat{1} + i\boldsymbol{\sigma}\cdot\mathbf{n} f_2 \sin\theta = g(\theta)\,\hat{1} + ih(\theta)\boldsymbol{\sigma}\cdot\mathbf{n}. \qquad (6.62)$$

This equation may be inserted in (6.60) and (6.59) to obtain expressions for ıe scattering cross-section and polarisation. Before doing so it is worth noting ıe following useful formulae (they may be checked by writing out their com- onents explicitly):

$$\begin{aligned} \operatorname{tr} \mathbf{A}\cdot\boldsymbol{\sigma} &= 0 \\ \operatorname{tr} (\mathbf{A}\cdot\boldsymbol{\sigma})\,\boldsymbol{\sigma} &= 2\mathbf{A} \\ \operatorname{tr} \boldsymbol{\sigma}(\mathbf{A}\cdot\boldsymbol{\sigma}) &= 2\mathbf{A} \\ \operatorname{tr} (\mathbf{A}\cdot\boldsymbol{\sigma})\,\boldsymbol{\sigma}(\mathbf{B}\cdot\boldsymbol{\sigma}) &= -2i\mathbf{A}\times\mathbf{B} \\ \operatorname{tr} (\mathbf{A}\cdot\boldsymbol{\sigma})\,(\mathbf{B}\cdot\boldsymbol{\sigma})\,\boldsymbol{\sigma} &= 2i\mathbf{A}\times\mathbf{B} \end{aligned} \qquad (6.63)$$

vhere we have assumed that **A** and **B** are vector operators and do not contain $\boldsymbol{\sigma}$. Thus the scattering cross-section (6.60), becomes

$$\begin{aligned} \frac{d\sigma_{sc}}{d\Omega} &= \tfrac{1}{2}\operatorname{tr} MM^{\dagger} = \tfrac{1}{2}\operatorname{tr}(g\,\hat{1} + ih\boldsymbol{\sigma}\cdot\mathbf{n})\,(g\,\hat{1} + ih\boldsymbol{\sigma}\cdot\mathbf{n})^{\dagger} \\ &= |g|^2 + |h|^2 \end{aligned} \qquad (6.64)$$

and the polarisation may be obtained from (6.59)

$$\begin{aligned} \frac{d\sigma_{sc}}{d\Omega}\langle\bar{\boldsymbol{\sigma}}\rangle_{sc} \equiv \frac{d\sigma_{sc}}{d\Omega}\mathbf{P} &= (|g|^2 + |h|^2)\,\mathbf{P} \\ &= \tfrac{1}{2}\operatorname{tr} MM^{\dagger}\boldsymbol{\sigma} = 2\,\mathrm{Im}\,(gh^*)\,\mathbf{n}. \end{aligned} \qquad (6.65)$$

Thus the polarisation vector of the scattered beam is given by

$$\mathbf{P} = \langle\bar{\boldsymbol{\sigma}}\rangle_{sc} = \frac{2\,\mathrm{Im}\,(gh^*)\,\mathbf{n}}{|g|^2 + |h|^2} \qquad (6.66)$$

and by (6.61) it lies perpendicularly to the scattering plane. Equation (6.66) can be re-written for later convenience as

$$\mathbf{P} = P\mathbf{n} \qquad (6.67)$$

where

$$P = \frac{2\,\mathrm{Im}\,(gh^*)}{|g|^2 + |h|^2}.$$

We finally note the physical significance of the terms $g(\theta)$ and $h(\theta)$ in (6.62). The vector **n** points perpendicularly to the plane defined by the directions of the

† This term is also often written as $M = g + \boldsymbol{\sigma}\cdot\mathbf{n}h$.

incident and scattered particles, and so if we assume the incident particle trave[ls] in the direction of the z-axis, and that the scattered particle has components

$$k'_x = k' \sin\theta \cos\varphi \qquad (k' = |\mathbf{k}'|)$$
$$k'_y = k' \sin\theta \sin\varphi$$
$$k'_z = k' \cos\theta$$

then the vector $\mathbf{n} = \mathbf{k} \times \mathbf{k}'/|\mathbf{k} \times \mathbf{k}'|$ has components

$$\mathbf{n} = (-\sin\varphi, \quad \cos\varphi, 0). \tag{6.68}$$

Thus if we quantise along the z-axis and let the spin functions be

$$\chi_{1/2}^{1/2} = \begin{pmatrix}1\\0\end{pmatrix}, \qquad \chi_{1/2}^{-1/2} = \begin{pmatrix}0\\1\end{pmatrix}$$

then from (6.62) and (3.104)

$$\begin{aligned} M\chi_{1/2}^{1/2} &= (g\hat{1} + ih\boldsymbol{\sigma}\cdot\mathbf{n})\begin{pmatrix}1\\0\end{pmatrix} \\ &= g\begin{pmatrix}1\\0\end{pmatrix} + h\begin{pmatrix}0 & \cos\varphi - i\sin\varphi \\ -\cos\varphi - i\sin\varphi & 0\end{pmatrix}\begin{pmatrix}1\\0\end{pmatrix} \\ &= g\begin{pmatrix}1\\0\end{pmatrix} - he^{i\varphi}\begin{pmatrix}0\\1\end{pmatrix} \\ &= g\chi_{1/2}^{1/2} - he^{i\varphi}\chi_{1/2}^{-1/2}. \end{aligned} \tag{6.69}$$

Therefore g and h may be recognised as the scattering amplitudes for no spinfli[p] and spinflip respectively. Their relationship to the phase shift analysis may b[e] understood by comparing the above equation with (6.32) and (6.34).

6.2(f). *The principle of the double scattering experiment*

We will examine the effects induced in the following experiment. An unpolar-ised beam of particles of spin $s_a = \frac{1}{2}$ strikes an unpolarised target with spin s_b the scattered beam then strikes a second target of the same type and the fina[l] angular distribution is examined.

The relevant equations may be constructed with the aid of (6.58). In the firs[t] experiment all the terms $\langle \overline{S}_\nu \rangle_i$ are zero except for the unit matrix, and so the polarisation of the scattered beam is given by

$$\frac{d\sigma_{sc}}{d\Omega}\langle \overline{\sigma} \rangle_{sc} = \frac{1}{2(2s_b + 1)} \operatorname{tr} MM^\dagger \boldsymbol{\sigma}. \tag{6.70}$$

This equation is the same as (6.59) when $s_b = 0$. In the second scattering the polarisation $\mathbf{P} \equiv \langle \sigma \rangle_{sc}$ of the first scattering will be written as $\langle \bar{\sigma} \rangle_1$ since it is now the polarisation of the incident beam, and the angular distribution is given by

$$\frac{d\sigma_{sc}}{d\Omega} \equiv I_2 = \frac{\text{tr } MM^\dagger}{2(2s_b + 1)} + \langle \bar{\sigma} \rangle_1 \cdot \frac{\text{tr } M\boldsymbol{\sigma} M^\dagger}{2(2s_b + 1)} \tag{6.71}$$

where the first term on the right-hand side comes from letting S_ν equal the unit matrix and the second from the terms $\sigma_a \hat{1}_b$ in S_ν. All other terms vanish since they give $\langle S_\nu \rangle = 0$ for an unpolarised target. The first term on the right in the above expression is simply the scattering cross-section for an unpolarised beam (compare (6.60)), and we will denote it by the symbol I_{02}, so that (6.71) becomes

$$I_2 = I_{02} + \langle \bar{\sigma} \rangle_1 \cdot \frac{\text{tr } M\boldsymbol{\sigma} M^\dagger}{2(2s_b + 1)}. \tag{6.72}$$

It was shown by Wolfenstein that the traces in equation (6.70) and (6.72) are equal

$$\text{tr } M\boldsymbol{\sigma} M^\dagger = \text{tr } MM^\dagger \boldsymbol{\sigma} \tag{6.73}$$

by arguments based on invariance under space and time reversal. They may be briefly stated as follows (Gammel and Thaler, 1960). The scattering matrix M can be written as

$$M = g + \mathbf{h} \cdot \boldsymbol{\sigma}$$

where g and $\mathbf{h}$ are no longer numbers as in equation (6.62) but are matrices in the spin space of the target. Evaluation of the traces with the aid of (6.63) yields

$$\text{tr } M\boldsymbol{\sigma} M^\dagger = 2\,\text{tr}'(\mathbf{h}g^\dagger + g\mathbf{h}^\dagger - i\mathbf{h} \times \mathbf{h}^\dagger)$$

$$\text{tr } MM^\dagger \boldsymbol{\sigma} = 2\,\text{tr}'(\mathbf{h}^\dagger g + g^\dagger \mathbf{h} - i\mathbf{h}^\dagger \times \mathbf{h})$$

where tr′ represents the trace in the target spin space. Now possible terms for $\mathbf{h} \times \mathbf{h}^\dagger$ are $\mathbf{k} \times \mathbf{k}'$ and $\mathbf{k} - \mathbf{k}'$, since $\text{tr}'\, \boldsymbol{\sigma}_b = 0$ and $\mathbf{k} + \mathbf{k}'$ is not allowed by Galilean invariance. However, both $\mathbf{k} \times \mathbf{k}'$ and $\mathbf{k} - \mathbf{k}'$ change sign under space and time inversion (compare Chapter 9), and so the last terms in the above equations must vanish. Finally, since the trace of a product of two matrices is independent of the order, equation (6.73) is proved.

We may therefore combine equations (6.70) and (6.72), yielding

$$I_2(\theta, \varphi) = I_{02} + \langle \bar{\sigma} \rangle_1 \cdot \langle \bar{\sigma} \rangle_2 I_{02} \tag{6.74}$$

$$= I_{02}(1 + P_1 P_2(\theta)\, \mathbf{n}_1 \cdot \mathbf{n}_2)$$

$$= I_{02}(1 + P_1 P_2 \cos \varphi_2) \tag{6.75}$$

where we have made appropriate alterations to the notation in equation (6.71) and have written

$$\langle\bar{\sigma}\rangle_1 = P_1\mathbf{n}_1, \qquad \langle\bar{\sigma}\rangle_2 = P_2(\theta)\,\mathbf{n}_2, \qquad \mathbf{n}_1 \cdot \mathbf{n}_2 = \cos\varphi_2.$$

Thus if we consider an experiment in which the two scattering planes are parallel and, say, scatter the particles to the left in the first scatterer and left in the second ($\varphi_2 = 0$) and then to the left in the first scatterer and right in the second scatterer, then, in effect, we will have reversed the second scattering plane ($\varphi_2 = \pi$) since $\mathbf{n}$ reverses direction as $\mathbf{k}'$ goes from left to right (compare (6.60)). If both left and right scatterings are performed at the same angle θ an asymmetry is found, given by

$$e_2(\theta) = \frac{I_2(\theta, 0) - I_2(\theta, \pi)}{I_2(\theta, 0) + I_2(\theta, \pi)} = P_1 P_2(\theta). \tag{6.76}$$

It is apparent from the form of the above equations that a scattering experiment of the type we have discussed above cannot yield information about the components of polarisation which lie along the direction of the momentum vector. This point will be raised again in § 14.2.

6.2(g). *The scattering of relativistic particles*

In the formalism developed in previous sections, the particles have been treated in a nonrelativistic manner. In the case of nucleons, for example, we should have used Dirac rather than Pauli spinors if we wished to develop scattering matrices in covariant forms. This neglect is not too serious in nucleon–nucleon scattering up to kinetic energies of ~ 500 MeV (in the laboratory system) since departures from nonrelativistic kinematics are still fairly small.

Detailed treatments of the polarisation arising in the scattering of relativistic particles have been made by a number of workers. One method (Stapp, 1956) involves use of the Lorentz transformations

$$\psi'(x') = S\psi(x) \tag{6.77}$$

(§ 5.4(d)) to relate spin state amplitudes in the rest frame of the particle to those in the frame where the particle has momentum $\mathbf{k}$. The scattering matrix then possesses the schematic form

$$M_{fi} = \bar{u}'\bar{u}Ou'u \tag{6.78}$$

where the subscripts f and i refer to final and initial states respectively, and O represents a Lorentz invariant combination of γ-matrices and kinematic factors, and $\bar{u}'$, $\bar{u}$, u' and u are Dirac spinors (compare § 9.4(d) for pion–nucleon scattering). Explicit formulae relating to polarisation parameters in nucleon–nucleon scattering are given in the paper mentioned above. An alternative method (Jacob and Wick, 1959) employs helicity amplitudes.

6.3. SCATTERING BY A POTENTIAL—THE BORN APPROXIMATION

6.3(a). *Introduction*

Until this point the scattering problem has been examined in terms of waves entering and leaving a scattering centre. Obviously the behaviour of the outgoing waves is determined by the nature of the interaction at the scattering centre. Thus the form of the interaction between the particles must be specified. One method of describing the interaction is to use a potential, which is assumed to act between scattered and scattering particles. The form of this potential normally can be constructed with the aid of physical arguments.

It should be noted, however, that the calculation of scattering cross-sections and phase shifts by the use of a potential can never be made covariant since potentials depend on the distance of separation of the particles, and distance by itself is not a relativistically covariant quantity.

6.3(b). *The scattering amplitude in the Born approximation*

The presence of a potential $V(\mathbf{r})$ causes the Schrödinger equation for the interaction of two particles with masses m_1 and m_2 to be written in the following manner:

$$\nabla^2\psi + \frac{2\mu}{\hbar^2}[E - V(\mathbf{r})]\psi = 0$$

or

$$\nabla^2\psi + [k^2 - 2\mu V(\mathbf{r})]\psi = 0 \tag{6.79}$$

in $\hbar = c = 1$ units, where

$$k^2 = \frac{2\mu E}{\hbar^2}, \qquad \mu = \frac{m_1 m_2}{m_1 + m_2}.$$

If the potential is assumed to be of fairly short range, then at large distances equation (6.79) reduces to

$$\nabla^2\psi + k^2\psi = 0, \qquad V(\mathbf{r}) \to 0$$

with the asymptotic solution (6.2)

$$\psi(\mathbf{r}) \underset{r\to\infty}{=} \frac{1}{\sqrt{v}}\left[e^{ikz} + f(\theta)\frac{e^{ikr}}{r}\right].$$

Exact solutions to equation (6.79) are not always feasible. Approximation methods are then used. One method is the Born approximation; it assumes that the spatial part of ψ can be written as

$$\psi(\mathbf{r}) = \frac{1}{\sqrt{v}} [e^{ikz} + \xi(\mathbf{r})] \tag{6.80}$$

where $\xi(\mathbf{r})$ represents the scattered wave and the assumption is made that $\xi(\mathbf{r})$ represents a very small addition to the unperturbed solution e^{ikz}

$$|e^{ikz}| \gg |\xi(\mathbf{r})|. \tag{6.81}$$

The substitution of equation (6.80) into (6.79) and use of (6.81) then yields

$$(\nabla^2 + k^2)\, \xi(\mathbf{r}) = 2\mu V(\mathbf{r})\, [e^{ikz} + \xi(\mathbf{r})] = 2\mu V(\mathbf{r})\, e^{ikz}. \tag{6.82}$$

The solution of this equation may be given with the aid of a Green's function (Mott and Massey, 1949)

$$\xi(\mathbf{r}) = 2\mu \int d\mathbf{r}\, V(\mathbf{r}')\, G(\mathbf{r}, \mathbf{r}')\, e^{ikz'} \tag{6.83}$$

where the Green's function is

$$G(\mathbf{r}, \mathbf{r}') = \frac{-1}{4\pi\, |\mathbf{r} - \mathbf{r}'|}\, e^{ik|\mathbf{r}-\mathbf{r}'|}.$$

Thus equation (6.80) can be written as

$$\psi(\mathbf{r}) = \frac{1}{\sqrt{v}} \left[e^{ikz} - \frac{2\mu}{4\pi} \int d\mathbf{r}'\, V(\mathbf{r}')\, \frac{e^{ik|\mathbf{r}-\mathbf{r}'|}}{|\mathbf{r} - \mathbf{r}'|}\, e^{ikz'} \right]. \tag{6.84}$$

The second term inside the brackets represents a variation of Kirchhoff's formula for the diffraction of light. Physically it states that the wave arriving

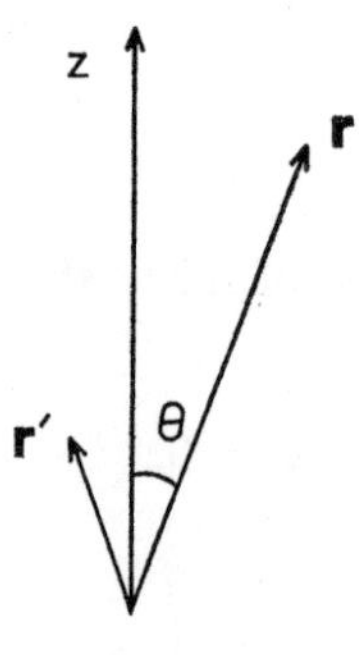

FIG. 6.2.

at some point $\mathbf{r}$ (Fig. 6.2) represents the sum of wavelets scattered at points $\mathbf{r}'$, and that their amplitudes at $\mathbf{r}$, are proportional to the scattering potential $V(\mathbf{r}')$. If $V(\mathbf{r}')$ is assumed to vanish sufficiently rapidly, then at distances $r = |\mathbf{r}|$ well beyond the range of the potential we can write

$$|\mathbf{r}-\mathbf{r}'| \xrightarrow[r\to\infty]{} r - \frac{\mathbf{r}\cdot\mathbf{r}'}{r}$$

$$\frac{1}{|\mathbf{r}-\mathbf{r}'|} \xrightarrow[r\to\infty]{} \frac{1}{r} + \frac{\mathbf{r}\cdot\mathbf{r}'}{r^3} \sim \frac{1}{r}.$$

Thus the equation (6.84) becomes

$$\underset{r\to\infty}{\psi(\mathbf{r})} = \frac{1}{\sqrt{v}}\left[e^{ikz} - \frac{2\mu}{4\pi}\frac{e^{ikr}}{r}\int d\mathbf{r}'\,V(\mathbf{r}')e^{ik\left(z'-\frac{\mathbf{r}\cdot\mathbf{r}'}{r}\right)}\right]. \tag{6.85}$$

If the incoming wave is assumed to move along the direction of the z-axis in Fig. (6.2) and if we represent the initial and final momenta as $\mathbf{k}_i$ and $\mathbf{k}_f$ respectively, then we can write

$$\mathbf{k}_i\cdot\mathbf{r}' = kz', \quad \mathbf{k}_f\cdot\mathbf{r}' = k\frac{\mathbf{r}\cdot\mathbf{r}'}{r}$$

since $\mathbf{r}$ represents the direction of the outgoing wave and elastic scattering is assumed to occur. Thus equation (6.85) becomes

$$\underset{r\to\infty}{\psi(\mathbf{r})} = \frac{1}{\sqrt{v}}\left[e^{ikz} - \frac{2\mu}{4\pi}\frac{e^{ikr}}{r}\int d\mathbf{r}'\,V(\mathbf{r}')\,e^{i(\mathbf{k}_i-\mathbf{k}_f)\cdot\mathbf{r}'}\right].$$

If this equation is compared with (6.2)

$$\underset{r\to\infty}{\psi(\mathbf{r})} = \frac{1}{\sqrt{v}}\left[e^{ikz} + f(\theta)\frac{e^{ikr}}{r}\right]$$

it can be seen that

$$f(\theta) = -\frac{2\mu}{4\pi}\int d\mathbf{r}\,V(\mathbf{r})\,e^{i(\mathbf{k}_i-\mathbf{k}_f)\cdot\mathbf{r}} \tag{6.86}$$

where the dashes have been dropped. This equation represents the amplitude of the scattered wave in the Born approximation and gives a scattering cross-section

$$\frac{d\sigma_{sc}}{d\Omega} = |f(\theta)|^2 = \left(\frac{\mu}{2\pi}\right)^2\left|\int d\mathbf{r}\,V(\mathbf{r})e^{i(\mathbf{k}_i-\mathbf{k}_f)\cdot\mathbf{r}}\right|^2. \tag{6.87}$$

The validity of (6.86) is dependent upon the assumption made in (6.81)

$$|e^{ikz}| \gg |\xi(\mathbf{r})|.$$

Using the solution for $\xi(\mathbf{r})$ given in (6.83), it can be shown (see, for example, Schiff, 1955) that if the potential extends over a range a and has a strength V_0, then the Born approximation is valid if

$$\mu V_0 a^2 \ll 1 \quad ka \ll 1 \quad \text{low energy limit} \tag{6.88}$$

$$\frac{V_0 a}{v} \ll 1 \quad ka \gg 1 \quad \text{high energy limit.}$$

The exact form of $f(\theta)$ will depend, of course, on the potential $V(\mathbf{r})$. If it is *central* (a function of r alone), then equation (6.86) reduces to

$$f(\theta) = -\frac{2\mu}{4\pi}\int_0^{2\pi} d\beta \int_0^{\pi} d\alpha \sin\alpha \int_0^{\infty} dr\, r^2 V(r)\, e^{iKr\cos\alpha}$$

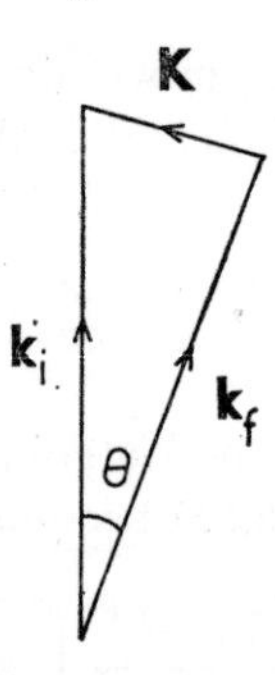

FIG. 6.3.

where we have taken a direction $\mathbf{K} = \mathbf{k}_i - \mathbf{k}_f$ as the polar axis (Fig. 6.3) and α and β represent polar angles about this direction. The magnitude of the vector $\mathbf{K}$ is

$$|\mathbf{K}| = |\mathbf{k}_i - \mathbf{k}_f| = 2k \sin \tfrac{1}{2}\theta.$$

The integrations over α and β then yield

$$f(\theta) = -2\mu \int_0^{\infty} dr\, r^2 V(r) \frac{\sin Kr}{Kr}. \tag{6.89}$$

6.3(c). *The relationship between phase shifts and potentials*

In equation (6.20) the elastic scattering amplitude by the method of partial waves was given as

$$f(\theta) = \frac{1}{k}\sum_{l=0}^{\infty} (2l+1) \sin\delta_l\, e^{i\delta_l} P_l^0(\theta) \tag{6.90}$$

where δ_l represented the phase shift for the partial wave associated with orbital angular momentum l.

Now consider equation (6.89) which represents the scattering by a central potential

$$f(\theta) = -2\mu \int_0^\infty dr\, r^2 V(r) \frac{\sin Kr}{Kr}$$

where

$$K = 2k \sin\tfrac{1}{2}\theta .$$

The function $\sin Kr/Kr$ can be expressed in the following form (Watson, 1944)

$$\frac{\sin Kr}{Kr} = \sum_{l=0}^{\infty} (2l+1)\, P_l^0(\theta)\, [j_l(kr)]^2$$

so that

$$f(\theta) = -2\mu \int_0^\infty dr\, r^2 V(r) \sum_{l=0}^{\infty} (2l+1) P_l^0(\theta)\, [j_l(kr)]^2. \tag{6.91}$$

Thus upon equating coefficients in (6.90) and (6.91) we find

$$\sin \delta_l e^{i\delta_l} = -2\mu k \int_0^\infty dr\, r^2 V(r)\, [j_l(kr)]^2. \tag{6.92}$$

If the phase shift is small this equation reduces to

$$\delta_l \sim -2\mu k \int_0^\infty dr\, r^2 V(r)\, [j_l(kr)]^2. \tag{6.93}$$

Since $[j(kr)]^2$ is always positive, equation (6.93) implies that small negative phase shifts are associated with positive (repulsive) potentials and positive phase shifts with negative (attractive) potentials, provided that the Born approximation is valid.

6.3(d). *The behaviour of the scattering amplitudes and phase shifts at low momenta*†

Equations (6.91) and (6.93) allow us to indicate the behaviour of the scattering amplitude and phase shifts at low momenta (if the potential is of short range). If $kr \ll l$ the term $j_l(kr)$ can be written as in (A.7.6) (p. 705)

$$j_l(kr) \sim \frac{(kr)^l}{(2l+1)!!}. \tag{6.94}$$

† Although we shall refer to equations in the Born approximation in this section, the results we finally quote can be proved without the application of this approximation and have a more general validity.

The behaviour of $j_l(kr)$ in this approximation allows us to write

$$f(\theta) \propto k^{2l} \tag{6.95}$$

$$\sigma_{sc} \propto k^{4l}$$

if $k \to 0$ from equation (6.91), and

$$\delta_l \propto k^{2l+1} \tag{6.96}$$

from (6.93).

Another important relation, which is often used in strong interactions at low energies, is obtained by using the inverse scattering function (compare (6.18))

$$f_l^{-1} = \frac{ke^{-i\delta_l}}{\sin \delta_l} = k \cot \delta_l - ik. \tag{6.97}$$

Thus all the dynamic features of the scattering are contained in the real part of the inverse scattering function. Now in the limit $k \to 0$ we may use (6.96) to write

$$\delta_l = Ak^{2l+1}$$

and since this approximation has been based upon the assumption that δ_l is small, we may write

$$\lim_{k\to 0} f_l^{-1} = \frac{k}{Ak^{2l+1}}(1 - iAk^{2l+1}) = \frac{1}{Ak^{2l}} - ik. \tag{6.98}$$

If we now compare this relation with (6.97) we find

$$k \cot \delta_l = \frac{1}{Ak^{2l}}. \tag{6.99}$$

Usually only *s*-waves are important in the low energy limit, and we may then write

$$k \cot \delta_0 = \frac{1}{A}. \tag{6.100}$$

Thus the scattering function for *s*-waves becomes

$$\lim_{k\to 0} f_0 = \frac{A}{1 - ikA} \tag{6.101}$$

in the low energy limit (compare (6.98)). It should be noted that A may be complex, for example in K^-p scattering both elastic scattering and absorptive channels occur, and so, in general, A may be represented as

$$A = a + ib \tag{6.102}$$

which corresponds to real and imaginary phase shifts (6.13).

The term A is called the *scattering length.* A more exact relation for *s*-waves at low energies is given by the *effective range formula* (Blatt and Jackson, 1949)

$$k \cot \delta_0 = \frac{1}{A} + \tfrac{1}{2} k^2 r_0 + 0(k^4) \tag{6.103}$$

where the last term implies 'of order k^4'. The term r_0 is called the *effective range.*

CHAPTER 7

THE INTERACTION OF FIELDS II THE S-MATRIX

7.1. INTRODUCTION

In Chapters 4 and 5 theorems were developed for the properties of isolated fields. These fields must now be coupled in order to describe the interaction of particles. An interaction implies that observable quantities, for example energy and momentum in an elastic scattering process, can be transferred from one system to another.

Considerable restrictions may be placed upon the coupling term which is used to describe an interaction. The condition of reality, for example, implies that the coupling expression must be Hermitian (§ 4.1 (c)). The invariance properties of free fields under the various forms of transformation, described in Chapters 4 and 5, should also apply to coupled fields. Thus we must consider factors like the conservation of energy and momentum. The conservation laws and their important exceptions will be discussed in § 7.3 and Chapter 9.

Even after the restrictions have been applied, however, almost limitless possibilities exist for the forms for the expressions describing the coupling of fields. When the possible forms of equations for the free fields were discussed, the hope was expressed that nature was kind and permitted their construction from the simplest possible mathematical terms (§ 4.2 (b)). Similar desires can be expressed for the coupled fields, but it must be borne in mind that the acid test for an interaction is not its simplicity but whether it leads to agreement with experiment.

This chapter will be concerned mainly with the formal description of interactions. The interaction of specific fields will be considered in later chapters.

7.2. THE S-MATRIX

The S-matrix was developed and applied to the interaction of elementary particles by Heisenberg (1943), but the principles behind it were first introduced by Wheeler (1937) in connection with problems in nuclear scattering and structure.

Heisenberg's basic tenet was that an 'exact' field theory could predict exactly certain basic observable quantities, namely:

(1) the energy and momentum of the free particles,
(2) the discrete energy levels of stationary systems,
(3) the asymptotic behaviour of wave functions in collision, emission and absorption processes, and thus the rate of interaction of these processes.

In order to carry out this programme Heisenberg introduced a certain unitary matrix, which he called the S-matrix.

$$SS^{\dagger} = S^{\dagger}S = S^{-1}S = \hat{1}. \tag{7.1}$$

This matrix was made the starting-point of his theory; his idea was that the rest of the theory of fields should develop from the basic description of the interaction of particles.

The S-matrix was defined by Heisenberg in the following manner: consider an asymptotic solution for the stationary state of a scattering process

$$\lim_{r\to\infty} \Psi = \Psi_f + \Psi_i \tag{7.2}$$

where Ψ_i represents the incoming waves and Ψ_f the outgoing (scattered) waves. The elements of the S-matrix were then defined as the terms relating the magnitudes and phases of the outgoing and incoming waves. Since the incoming and outgoing waves are associated respectively with the state of affairs before and after the scattering, it is intuitively evident (and can be explicitly demonstrated) that an equivalent definition of the S-matrix operator is

$$|\Psi_f\rangle = S\,|\Psi_i\rangle. \tag{7.3}$$

The states $|\Psi_i\rangle$ and $|\Psi_f\rangle$ in this equation refer to times in the remote past and future with respect to the interaction

$$\begin{aligned} |\Psi_i\rangle &\equiv t \to -\infty \\ |\Psi_f\rangle &\equiv t \to +\infty. \end{aligned} \tag{7.4}$$

Thus the particles represented by $|\Psi_i\rangle$ and $|\Psi_f\rangle$ behave as free particles.

The requirement of unitarity for the S-matrix operator automatically requires that probability should be conserved in the transition since

$$\langle\Psi_f \mid \Psi_f\rangle = \langle\Psi_i|\, S^{\dagger}S\,|\Psi_i\rangle = \langle\Psi_i \mid \Psi_i\rangle. \tag{7.5}$$

This result implies if 'something' goes into the interaction then 'something' must come out. The word *unitarity* is in fact frequently used in elementary particle physics to indicate that probability is conserved in a given process. Thus if $|\Psi_i\rangle$ is normalised $\langle\Psi_f \mid \Psi_f\rangle$ should equal unity. It should be noted that $|\Psi_f\rangle$ covers all possible final states.

7.3. THE TRANSITION AMPLITUDE AND THE S-MATRIX

7.3(a). *Construction of a transition amplitude*

We now formulate a matrix element for an interaction, which yields the transition amplitude for that interaction. Let the ket $|i\rangle$ represent a single initial state which exists before the interaction occurs. The operator S then takes $|i\rangle$ from time $t = -\infty$ to $t = +\infty$, that is through the interaction, which we will assume to be located near $t = 0$. The term $S|i\rangle$ then represents a superposition of all possible final states. Thus the amplitude for a transition from $|i\rangle$ to a particular final state $|f\rangle$ is given by the matrix element

$$S_{fi} = \langle f| S |i\rangle \tag{7.6}$$

where the S-matrix operator covers the infinite time interval $-\infty < t < +\infty$.

7.3(b). *The invariance properties of the transition amplitude*

In Chapters 3, 4 and 5 examples were given of the invariance properties of isolated fields, for example invariance under displacements in time implied conservation of energy. In a similar manner one may demonstrate that the invariance of the S-matrix leads to certain conservation laws.

Let us consider a unitary transformation which causes the states $|f\rangle$ and $|i\rangle$ to change as follows (compare § 4.1(e))

$$|f'\rangle = U|f\rangle, \qquad |i'\rangle = U|i\rangle \tag{7.7}$$

where U is a unitary operator. If the transformation is such that

$$\langle f'| S |i'\rangle = \langle f| S |i\rangle \tag{7.8}$$

(apart from the possible existence of an undetectable phase factor), then the physical operation corresponding to U leaves the transition probability unchanged. Upon inserting (7.7) into (7.8) we find that the invariance condition implies that

$$U^{-1}SU = S \quad \text{or} \quad [S, U] = 0. \tag{7.9}$$

This equation represents a compact statement of the physical conservation laws appearing in the interaction of fields. It implies that a theoretical study of the invariance properties of interactions must concentrate upon the unitary (and antiunitary) operators which commute with S.

As an example of the use of the argument given above we will consider the proper Lorentz transformation. Consider an infinitesimally small transformation of the type discussed in § 5.3(a) for free fields; this was given as

$$U = \hat{1} + ia_\lambda P_\lambda - \tfrac{1}{2} i\varepsilon_{\lambda\alpha} M_{\lambda\alpha} \tag{5.38}$$

where P and M represent linear and angular momentum operators respectively, and a and ε are infinitesimally small displacements and rotations.

The invariance property of equation (7.9) now becomes

$$[P, S] = 0, \qquad [M, S] = 0. \tag{7.10}$$

These equations express the conservation of energy–momentum and angular momentum in transitions involving elementary particles. Consider, for example, the operation of P on the state vectors $|f\rangle$ and $|i\rangle$

$$P\,|f\rangle = p_f\,|f\rangle, \qquad P\,|i\rangle = p_i\,|i\rangle \tag{7.11}$$

where the momentum four-vectors p_f and p_i represent eigenvalues of the operator P for the eigenstates $|f\rangle$ and $|i\rangle$ respectively. But since

$$[P, S] = 0$$

then

$$\langle f|\,[P, S]\,|i\rangle = 0 \tag{7.12}$$

and so

$$\langle f|\,[P, S]\,|i\rangle = \langle f|\,PS\,|i\rangle - \langle f|\,SP\,|i\rangle$$

$$= (p_f - p_i)\,\langle f|\,S\,|i\rangle = 0 \tag{7.14}$$

hence

$$p_f = p_i \tag{7.15}$$

and energy–momentum is conserved in interactions. The proof may be extended in an obvious manner to angular momentum with the aid of (7.10).

The method described above will be used in Chapter 9 for examining the invariance properties of interactions of various types.

7.4. THE TRANSITION PROBABILITY

7.4(a). *Basic equations*

In order to calculate the transition probability between an initial state $|i\rangle$ and a final state $|f\rangle$ it is convenient to introduce a series of reduced matrix operators which may be applied under suitable circumstances. They are the R, M and T matrices. The R-matrix operator (the *reactance operator*) is defined by equation

$$S = \hat{1} + iR \tag{7.16}$$

so that

$$S_{fi} = \langle f|\,S\,|i\rangle = \langle f|\,\hat{1}\,|i\rangle + i\,\langle f|\,R\,|i\rangle = \delta_{fi} + iR_{fi}. \tag{7.17}$$

The unitary condition for S then yields the relation

$$S^\dagger S = (\hat{1} - iR^\dagger)(\hat{1} + iR) = \hat{1}$$

$$(R - R^\dagger) = iR^\dagger R. \tag{7.18}$$

It was shown in the previous section that the invariance of the S-matrix under proper Lorentz transformations implies that energy and momentum are conserved in interactions

$$p_f = p_i.$$

It is sometimes convenient to take this conservation law out of the matrix element $\langle f| R |i\rangle$ by defining a new matrix element $\langle f| M |i\rangle$

$$\langle f| R |i\rangle = (2\pi)^4\, \delta(p_i - p_f)\, \langle f| M |i\rangle = (2\pi)^4\, \delta(p_i - p_f) M_{fi} \tag{7.19}$$

where the term $(2\pi)^4$ has been extracted for later convenience, and the δ-function is defined as

$$\delta(p_i - p_f) = \delta(\mathbf{p}_i - \mathbf{p}_f)\, \delta(E_i - E_f) \tag{7.20}$$

where E_i and $\mathbf{p}_i$ represent the total energy and momentum of the initial state, and E_f and $\mathbf{p}_f$ define similar terms for the final state.

Upon inserting equation (7.19) in (7.17) we obtain

$$S_{fi} = \delta_{fi} + i(2\pi)^4\, \delta(p_i - p_f) M_{fi} \tag{7.21}$$

where the term δ_{fi} in this equation and in (7.17) implies that 'nothing happens'. The probability for the transition $|i\rangle \to |f\rangle$ for the whole of space-time is then given by

$$|R_{fi}|^2 = (2\pi)^8\, [\delta(p_i - p_f)]^2\, |M_{fi}|^2. \tag{7.22}$$

The double δ-function can be written as†

$$\delta(p_i - p_f) \lim_{\substack{V\to\infty \\ T\to\infty}} \frac{1}{(2\pi)^4} \int_{VT} d^4x\, e^{i(p_i - p_f)x} = \delta(p_i - p_f)\, \frac{VT}{(2\pi)^4}. \tag{7.23}$$

where use has been made of the definition of the δ-function (see appendix A.6 p. 703) and the condition

$$p_i - p_f = 0$$

imposed by the remaining δ-function. Thus equation (7.22) becomes

$$|R_{fi}|^2 = (2\pi)^4\, \delta(p_i - p_f) VT\, |M_{fi}|^2. \tag{7.24}$$

Now $|R_{fi}|^2$ represents a transition probability over all space-time and is therefore not a useful quantity since, in practice, measurements are normally carried out for a finite time in a restricted volume of space, that is in an element of space-

† The derivation of equation (7.25) illustrates the main points in an argument but is not entirely rigorous. A more complete derivation may be found in a paper of Gasiorowicz (1960).

time. A more meaningful quantity is the transition probability per unit of space time, which is defined as

$$\omega = \frac{1}{VT} |R_{fi}|^2 = (2\pi)^4 \, \delta(p_i - p_f) \, |M_{fi}|^2. \tag{7.25}$$

In this equation the initial, and therefore the final, states belong to a continuous spectrum (or rather a set of discrete states, since we are using box normalisation). In practice the initial conditions in an experiment are defined, and the transitions therefore go to well-defined final states. One of two initial conditions is normally encountered – the interaction of two particles and the decay of a single particle. In the former case the quantity of interest is the cross-section which may be defined as

$$d\sigma_{fi} = \frac{\omega}{J_i} \, dN_f \tag{7.26}$$

where J_i is the flux density of the incoming particles and dN_f is the number of final states. The flux density is given by

$$J_i = \varrho_a \varrho_b v \tag{7.27}$$

where ϱ = density of the incoming particles a and $b = 1/V$, and v = relative velocity of the incoming particles.

The unitarity condition for the S-matrix implies that the total transition probability to all final states must equal unity – see remarks following (7.5). In practice we are interested in transitions to specific final states, for example states for a single particle with momenta between $\mathbf{p}$ and $\mathbf{p} + d\mathbf{p}$; this result may be achieved by treating dN_f as a weighting factor. Now the number of momentum states between $\mathbf{p}$ and $\mathbf{p} + d\mathbf{p}$ available to a single particle in a box of volume V is given by†

$$dn_f = \frac{V}{(2\pi)^3} \, d\mathbf{p} \equiv \frac{V}{(2\pi\hbar)^3} \, d\mathbf{p} \tag{7.28}$$

and $d\mathbf{p}$ has the following meaning in spherical coordinates:

$$d\mathbf{p} = p^2 \, dp \, d\Omega \equiv \mathbf{p}^2 \, dp \, d\Omega \tag{7.29}$$

† We make the usual assumption of quantisation in a cubic box with sides of length L so that the allowed momenta form a lattice of points in momentum space with coordinates

$$\frac{2\pi\hbar}{L} n_x, \qquad \frac{2\pi\hbar}{L} n_y, \qquad \frac{2\pi\hbar}{L} n_z$$

where n is an integer, hence the density of states is $(L/2\pi\hbar)^3$ and the number of points in an element of volume of momentum space is

$$\left(\frac{L}{2\pi\hbar}\right)^3 d\mathbf{p} \equiv \frac{V}{(2\pi)^3} \, d\mathbf{p}.$$

where $p = |\mathbf{p}|$ and $d\Omega$ indicates an element of solid angle. Thus if we have an ssembly of q particles in the final state dN_f becomes

$$dN_f = dn_{f1}dn_{f2} \ldots dn_{fq} = \frac{V^q}{(2\pi)^{3q}} \prod_{j=1}^{q} d\mathbf{p}_{fj} \tag{7.30}$$

where $\prod_{j=1}^{q}$ indicates a product of the terms $d\mathbf{p}_{fj}$.

In addition to energy and momentum, other degrees of freedom must be considered. If the reaction involves unpolarised particles with spin, the calculated cross-section must be averaged over the possible spin orientations in the initial tates and summed over the possible final states, and so terms of the type

$$\frac{1}{2s+1} \sum_{m_s} \tag{7.31}$$

are necessary in the expression for the cross-section, if an initial particle possesses spin. It is customary, therefore, to include in the formula for the cross-section a term

$$\overline{\sum_i} \sum_f \tag{7.32}$$

where $\overline{\sum_i}$ indicates a suitable average over the initial states, and $\sum_f$ a sum over all final states compatible with the initial conditions and with the required form for the cross-section.

Thus equation (7.26) becomes

$$d\sigma_{fi} = \frac{(2\pi)^4}{v} V^2 \overline{\sum_i} \sum_f |M_{fi}|^2 \frac{V^q}{(2\pi)^{3q}} \prod_{j=1}^{q} d\mathbf{p}_{fj}\, \delta(p_i - p_f) \tag{7.33}$$

where equations (7.27), (7.25), (7.32) and (7.30) have been inserted. The above equation is sometimes written in a somewhat different form. The operator M is Lorentz invariant, but its matrix elements $\langle f| M |i\rangle$ are not invariant because $|f\rangle$ and $|i\rangle$ are not invariantly normalised. We shall show in the next chapter that the normalisation terms appearing in $\langle f| M |i\rangle$ are $1/\sqrt{(2\omega V)}$ for bosons and $\sqrt{m/EV}$ for fermions; we therefore define a Lorentz invariant matrix element $T_{fi} = = \langle f| T |i\rangle$ by

$$S_{fi} = \delta_{fi} + i(2\pi)^4\, \delta(p_i - p_f) \frac{T_{fi}}{N} \tag{7.34}$$

where N is a suitable product of normalisation factors given by

$$N^2 = VN_a^2 VN_b^2 VN_1^2 \ldots VN_q^2 = V^{q+2} N_i^2 N_1^2 \ldots N_q^2 \tag{7.35}$$

where the term

$$N_i^2 = N_a^2 N_b^2$$

refers to the normalisation factors for the incoming particles, and $N_a^2 \ldots N_j^2 \ldots N$ assume the form

$$N_j^2 = 2\omega \quad \text{for bosons} \tag{7.36}$$

$$= E/m \quad \text{for fermions.}$$

The formula for the cross-section (7.33) now becomes

$$d\sigma_{fi} = \frac{(2\pi)^4}{v} \frac{1}{N_i^2} \overline{\sum_i} \sum_f |T_{fi}|^2 \frac{1}{(2\pi)^{3q}} \prod_{j=1}^{q} \frac{d\mathbf{p}_{fj}}{N_j^2} \delta(p_i - p_f) \tag{7.37}$$

and the factors

$$\frac{d\mathbf{p}_{fj}}{N_j^2} \tag{7.38}$$

in the final states are Lorentz invariants. The properties of the terms v and $\sum_f$ will be examined in greater detail in the next section.

The choice between using (7.33) or (7.37) in any problem is mainly a matter of taste. In general (7.33) is often used for perturbation theory and (7.37) for dispersion relations.

The equation for a decay process can be obtained immediately from the above expression. We will represent the transition probability per unit time to a specific final state by the symbol $d\lambda_{fi}$. Now the term ω of equation (7.25) represented the transition per unit volume of space-time; we are now interested in the transition probability per unit time. Thus $d\lambda$ may be written as

$$d\lambda_{fi} = V\omega \, dN_f$$

$$= V(2\pi)^4 \overline{\sum_i} \sum_f |M_{fi}|^2 \frac{V^q}{(2\pi)^{3q}} \prod_{j=1}^{q} d\mathbf{p}_{fj} \, \delta(p_i - p_f) \tag{7.39}$$

$$= \frac{1}{N_i^2} (2\pi)^4 \overline{\sum_i} \sum_f |T_{fi}|^2 \frac{1}{(2\pi)^{3q}} \prod_{j=1}^{q} \frac{d\mathbf{p}_{fj}}{N_j^2} \delta(p_i - p_f). \tag{7.40}$$

The mean lifetime τ is then given by the *inverse of the total transition probability per unit time*, and so it can be represented as

$$\frac{1}{\tau} = \lambda = \int_f d\lambda_{fi} \tag{7.41}$$

where the subscript f indicates an integration over all possible final states.

7.4(b). *The kinematic factors*

The terms appearing in (7.33), (7.37), (7.39) and (7.40) can become very complicated expressions if many particles are involved. In general we shall be considering fairly simple physical situations, and the principles used in evaluating the equations can be displayed in a straightforward manner. Rules for evaluating the kinematic factors in typical situations will be given in the following sections.

7.4(b.1). *The relative velocity.* In practice collisions observed in laboratory experiments involve only two particles; thus the term v, indicating relative velocity, in equation (7.27) may be written as

$$v = |v_a - v_b| = \left|\frac{\mathbf{p}_a}{E_a} - \frac{\mathbf{p}_b}{E_b}\right| \tag{7.42}$$

where v_a and v_b indicate the velocities of each particle. In the centre of momentum system this equation assumes a particularly simple form, since

$$\mathbf{p}_a = -\mathbf{p}_b$$

so that we may write

$$|\mathbf{p}_a| = |\mathbf{p}_b| = p_c$$

and (7.42) becomes

$$v = p_c\left(\frac{1}{E_a} + \frac{1}{E_b}\right) = p_c\frac{E_c}{E_a\,E_b} \tag{7.43}$$

where E_c is the total energy in the c-system.

Møller (1945) has shown that v may be written in a form which is valid in any reference frame

$$v = \frac{1}{E_a E_b}|F| \tag{7.44}$$

where

$$F = [(p_a p_b)^2 - m_a^2 m_b^2]^{\frac{1}{2}}$$

and p_a and p_b are both four-momenta. The scalar product of two four-momenta was given in equation (3.18). Equation (7.44) assumes particularly simple forms if one of the particles is stationary or is massless. They are as follows:

(1) Particle b stationary

$$v = \frac{1}{E_a m_b} m_b\,|\mathbf{p}_a| = |\mathbf{v}_a|. \tag{7.45}$$

(2) Particle b is massless $m_b = 0$

$$v = \frac{1}{E_a E_b}|p_a p_b|. \tag{7.46}$$

7.4(b.2). *The density of final states.* We now consider the expression†

$$\sum_f = \sum_f \prod_{j=1}^{q} d\mathbf{p}_{fj}\, \delta(p_i - p_f) \tag{7.47}$$

contained in equations (7.33) and (7.39). In this equation $\sum_f$ can assume various forms, depending on the nature of the final state. If, for example, we require the probability for a final state with particle 1 in the momentum interval $d\mathbf{p}_{f1}$, particle 2 in the momentum interval $d\mathbf{p}_{f2}, \ldots,$ then

$$\sum_f = \sum_\alpha d\mathbf{p}_{f1} d\mathbf{p}_{f2} \ldots d\mathbf{p}_{fq} \delta(p_i - p_f) \tag{7.48}$$

where α represents any additional degrees of freedom in the matrix element.

If, on the other hand, we are interested in one particle only, say, particle 1, then we can sum over all the momentum states $d\mathbf{p}_{f2} \ldots d\mathbf{p}_{fq}$ compatible with the δ-function

$$\begin{aligned} \sum_f &\equiv \sum_\alpha d\mathbf{p}_{f1} \int d\mathbf{p}_{f2} \int d\mathbf{p}_{f3} \ldots \int d\mathbf{p}_{fq}\, \delta(p_i - p_f) \\ &= \sum_\alpha p_1^2\, dp_1\, d\Omega_1 \int d\mathbf{p}_2 \ldots \int d\mathbf{p}_q\, \delta(p_i - p_f) \end{aligned} \tag{7.49}$$

where we have dropped the subscripts f for convenience of writing.

In practice the quantities measured in experiments are normally momentum spectra or angular distributions. In the former case a further integration over $d\Omega_1$ is necessary for equation (7.49); in the latter an integration over dp_1 is carried out. A problem of the former type is examined for Compton scattering in § 11.1(a).

If total cross-sections or lifetimes for decaying particles are examined, then all final states must be considered. Equation (7.47) then becomes

$$\sum_f = \sum_\alpha \int d\mathbf{p}_1 \int d\mathbf{p}_2 \ldots \int d\mathbf{p}_q\, \delta(p_i - p_f). \tag{7.50}$$

The integrals appearing in the above equations are often called *momentum* or *phase space integrals.*

The equations given above are sometimes written in a Lorentz invariant manner by associating them with the T-matrix (7.34). Factors of the type (7.36)

$$\begin{aligned} \frac{d\mathbf{p}_{fj}}{N_j^2} &= \frac{d\mathbf{p}_{fj}}{2\omega_j} \qquad \text{bosons} \\ &= m_j \frac{d\mathbf{p}_{fj}}{E_j} \qquad \text{fermions} \end{aligned}$$

† It must be remembered that the term $\sum_f$ in equations (7.33), (7.37), (7.39) and (7.40) covers the matrix elements as well as the momentum space terms. In the examples which we shall discuss, we shall examine the integrals over the δ-functions only, and so the presence of the matrix element is not essential.

then appear. We shall ignore the difference between bosons and fermions†, since simple compensatory factors $2m$ can always be inserted in the expression for the transition probability, and write the function given above as $d\mathbf{p}_j/2E_j$. Now with the aid of equation (4.61)

$$
\begin{aligned}
\int dE\, \delta(p^2 + m^2) &= \int dE\, \delta(\mathbf{p}^2 + m^2 - E^2) \\
&= \int dE\, \delta(E_p^2 - E^2) \\
&= \int \frac{dE}{2E_p} [\delta(E_p - E) + \delta(E_p + E)] \quad (E_p > 0) \\
&= \frac{1}{2E_p}.
\end{aligned} \tag{7.51}
$$

Thus the expression for $\sum_f$ appearing in (7.37) or (7.40) can be written in the form

$$
\begin{aligned}
\sum_f \equiv \sum_f \prod_{j=1}^{q} t_j \frac{d\mathbf{p}_{fj}}{2E_j} \delta(p_i - p_f) \\
= \sum_f \prod_{j=1}^{q} t_j\, d\mathbf{p}_{fj}\, dE_j\, \delta(p_j^2 + m_j^2)\, \delta(p_i - p_f)\, \theta(E_j) \\
= \sum_f \prod_{j=1}^{q} t_j\, d^4p_j\, \delta(p_j^2 + m_j^2)\, \delta(p_i - p_f)\, \theta(E_j)
\end{aligned} \tag{7.52}
$$

where

$$
\begin{aligned}
\theta(E_j) &= 1 \quad (E_j > 0) \\
&\ 0 \quad (E_j < 0) \\
t_j &= 1 \quad \text{bosons} \\
&\ 2m_j \quad \text{fermions.}
\end{aligned}
$$

Equivalent expressions to equations (7.48)–(7.50) can then be easily constructed.

The calculation of the terms $\sum_f$ can become extremely complicated if many particles are present in the final state. In order to illustrate the techniques involved in the calculation of the phase space integral, we shall consider situations involving two or three particles. Even these problems can become complicated in certain coordinate systems.

We start by considering a final state containing two particles. Equation (7.47) then becomes

$$
\sum_f \equiv \sum_f d\mathbf{p}_1\, d\mathbf{p}_2\, \delta(p_i - p_f). \tag{7.53}
$$

We will consider the term arising when particle 1 travels into an element of solid

† We shall also use the symbol E for the total energies of both bosons and fermions in this section.

angle $d\Omega_1$. We then must integrate over all the limits of $\mathbf{p}_1$ and $\mathbf{p}_2$ consisten with this condition (we shall often use the notation $|\mathbf{p}_j| \equiv p_j$ in this section)

$$\sum_f \equiv \sum_\alpha d\Omega_1 \int dp_1 p_1^2 \int d\mathbf{p}_2\, \delta(p_i - p_f)$$
$$= \sum_\alpha d\Omega_1 \int dp_1 p_1^2 \int d\mathbf{p}_2\, \delta(\mathbf{p}_i - \mathbf{p}_1 - \mathbf{p}_2)\, \delta(E_i - E_1 - E_2)$$
$$= \sum_\alpha d\Omega_1 \int dp_1 p_1^2\, \delta[E_i - E_1 - \sqrt{(m_2^2 + |\mathbf{p}_i - \mathbf{p}_1|^2)}]$$
$$= \sum_\alpha d\Omega_1 \int dE_1\, E_1 p_1\, \delta[E_i - E_1 - \sqrt{(m_2^2 + |\mathbf{p}_i - \mathbf{p}_1|^2)}] \tag{7.54}$$

where we have used the relation

$$E\, dE = p\, dp, \qquad p = |\mathbf{p}|$$

since

$$E^2 = \mathbf{p}^2 + m^2.$$

We now use the following relation (see appendix A.6, p. 703)

$$\int dx\, \delta[f(x)] = \left|\frac{\partial f}{\partial x}\right|^{-1}_{x = x_0}$$

where x_0 is a solution of the equation $f(x) = 0$. In the present situation

$$f = E_i - E_1 - \sqrt{(m_2^2 + |\mathbf{p}_i - \mathbf{p}_1|^2)}$$
$$= E_i - E_1 - E_2 \tag{7.55}$$

and

$$\left|\frac{\partial f}{\partial E_1}\right| = 1 + \frac{\partial E_2}{\partial E_1}$$

now

$$E_2^2 = m_2^2 + \mathbf{p}_i^2 - 2\mathbf{p}_i \cdot \mathbf{p}_1 + E_1^2 - m_1^2$$

and so

$$2E_2 \frac{\partial E_2}{\partial E_1} = -2\mathbf{p}_i \cdot \mathbf{p}_1 \frac{E_1}{\mathbf{p}_1^2} + 2E_1$$

$$\frac{\partial E_2}{\partial E_1} = \frac{\mathbf{p}_1^2 E_1 - \mathbf{p}_i \cdot \mathbf{p}_1 E_1}{\mathbf{p}_1^2 E_2} \tag{7.56}$$

$$\left|\frac{\partial f}{\partial E_1}\right| = 1 + \frac{\partial E_2}{\partial E_1}$$
$$= \frac{\mathbf{p}_1^2 E_i - \mathbf{p}_i \cdot \mathbf{p}_1 E_1}{\mathbf{p}_1^2 E_2}. \tag{7.57}$$

Thus equation (7.54) becomes

$$\sum_f \equiv \sum_\alpha d\Omega_1 \int dE_1 E_1 p_1\, \delta[f(E_1)]$$
$$= \sum_\alpha d\Omega_1 E_1 p_1 \left|\frac{\partial f}{\partial E_1}\right|^{-1}$$
$$= \sum_\alpha d\Omega_1 \frac{E_1 E_2\, |\mathbf{p}_1|^3}{\mathbf{p}_1^2 E_i - \mathbf{p}_i \cdot \mathbf{p}_1 E_1}. \tag{7.58}$$

It is usual to evaluate (7.58) in the c-system, where

$$|\mathbf{p}_1| = |\mathbf{p}_2| = p_c^f, \qquad \mathbf{p}_i = 0, \qquad E_i = E_c$$

so that (7.58) becomes

$$\sum_f \equiv \sum_\alpha d\Omega_1 E_1 E_2 \frac{p_c^f}{E_c}. \tag{7.59}$$

An expression for p_c^f is given in equation (A.8.2) (p. 713).

We next consider the evaluation of a phase space integral for a three-body process. Problems of this type arise, for example, in the β-decay of the neutron.

$$n \to p + e^- + \bar{\nu}.$$

In such problems the probability that certain particles emerge with specified energies or momenta is often required. The phase space integral can therefore be written as

$$\sum_f \equiv \sum_\alpha d\mathbf{p}_1 \int d\mathbf{p}_2 \int d\mathbf{p}_3\, \delta(\mathbf{p}_i - \mathbf{p}_1 - \mathbf{p}_2 - \mathbf{p}_3)\, \delta(E_i - E_1 - E_2 - E_3)$$
$$= \sum_\alpha d\mathbf{p}_1 \int d\mathbf{p}_2\, \delta[E_i - E_1 - E_2 - \sqrt{(m_3^2 + |\mathbf{p}_i - \mathbf{p}_1 - \mathbf{p}_2|^2)}] \tag{7.60}$$
$$= \sum_\alpha d\mathbf{p}_1 \int d\Omega_2 \int dE_2 E_2 p_2\, \delta(E_i - E_1 - E_2 - E_3). \tag{7.61}$$

The same procedure as in (7.55) is now used. Writing

$$f = E_i - E_1 - E_2 - E_3$$

we find that

$$\left|\frac{\partial f}{\partial E_2}\right| = 1 + \frac{\partial E_3}{\partial E_2} \tag{7.62}$$

and since in (7.60)

$$E_3^2 = m_3^2 + |\mathbf{p}_i - \mathbf{p}_1|^2 - 2(\mathbf{p}_i - \mathbf{p}_1)\cdot \mathbf{p}_2 + \mathbf{p}_2^2$$

$$E_3 \frac{\partial E_3}{\partial E_2} = -(\mathbf{p}_i - \mathbf{p}_1)\cdot \mathbf{p}_2 \frac{E_2}{\mathbf{p}_2^2} + E_2. \tag{7.63}$$

Thus (7.62) becomes

$$\left|\frac{\partial f}{\partial E_2}\right| = \frac{\mathbf{p}_2^2(E_2 + E_3) - (\mathbf{p}_i - \mathbf{p}_1)\cdot \mathbf{p}_2 E_2}{\mathbf{p}^2 E_3}$$

$$= \frac{\mathbf{p}_2^2(E_i - E_1) - (\mathbf{p}_i - \mathbf{p}_1)\cdot \mathbf{p}_2 E_2}{\mathbf{p}_2^2 E_3}. \tag{7.64}$$

This equation may be inserted in (7.61) yielding

$$\sum_f \equiv \sum_\alpha d\mathbf{p}_1 \int d\Omega_2 \int dE_2 E_2 p_2\, \delta[f(E_2)]$$

$$= \sum_\alpha d\mathbf{p}_1 \int d\Omega_2 E_2 p_2 \left|\frac{\partial f}{\partial E_2}\right|^{-1}$$

$$= \sum_\alpha d\mathbf{p}_1 \int \frac{d\Omega_2 E_2 E_3\, |\mathbf{p}_2|^3}{\mathbf{p}_2^2(E_i - E_1) - (\mathbf{p}_i - \mathbf{p}_1)\cdot \mathbf{p}_2 E_2}. \tag{7.65}$$

If the mass of one of the particles tends to infinity, considerable simplifications arise. Say $m_3 \to \infty$, then

$$\sum_f \equiv \sum_\alpha d\mathbf{p}_1 \int d\Omega_2 E_2 p_2 = 4\pi \sum_\alpha d\mathbf{p}_1 E_2 p_2 \tag{7.66}$$

where $p_2 = |\mathbf{p}_2|$.

If the integration in (7.61) had been performed over $d\Omega_2$ rather than dE_2, then the following equations would have been obtained in the c-system:

$$\sum_f \equiv 2\pi \sum_\alpha d\mathbf{p}_1 \int dp_2 p_2^2 \frac{E_3}{p_1 p_2}$$

$$= 8\pi^2 \sum_\alpha dE_1 \int dE_2 E_1 E_2 E_3 \tag{7.67}$$

where we have written $p\,dp = E\,dE$.

An inspection of equations (7.59) and (7.67) shows that they have a linear momentum dependence of p and p^4 respectively for a summation over all momentum states in the nonrelativistic limit. It is not difficult to show that if a system contains q particles in the final state then the momentum dependence of its phase space factors in the nonrelativistic limit is p^{3q-5}.

It is instructive to consider the Lorentz invariant equivalent of equation (7.67). If we consider three bosons in their c-system then we can write the first form of equations (7.52) as

$$\sum_f = \sum_f \frac{d\mathbf{p}_2}{2E_1}\frac{d\mathbf{p}_2}{2E_2}\frac{d\mathbf{p}_3}{2E_3}\,\delta(\mathbf{p}_1 + \mathbf{p}_2 + \mathbf{p}_3)\,\delta(E_c - E_1 - E_2 - E_3)$$

$$= \int \frac{d\Omega_1}{8E_1E_2E_3} p_1^2\, dp_1\, d\Omega_2\, p_2^2\, dp_2\, \delta(E_c - E_1 - E_2 - E_3)$$

$$= 8\pi^2 \int \frac{E_1 p_1\, dE_1}{8E_1E_2E_3}\, d(\cos\theta_{12})\, E_2 p_2\, dE_2\, \delta(E_c - E_1 - E_2 - E_3).$$

Now

$$\mathbf{p}_3^2 = \mathbf{p}_1^2 + \mathbf{p}_2^2 + 2\mathbf{p}_1 \cdot \mathbf{p}_2$$

hence

$$p_3\, dp_3 = E_3\, dE_3 = p_1 p_2\, d(\cos\theta_{12})$$

and upon substituting for $d(\cos\theta_{12})$ in $\sum_f$ and cancelling unwanted terms we find

$$\begin{aligned}\sum_f &= \pi^2 \int dE_1\, dE_2\, dE_3\, \delta(E_c - E_1 - E_2 - E_3) \\ &= \pi^2 \int dE_1\, dE_2 \qquad (7.68)\end{aligned}$$

This equation might have been inferred immediately from the form of (7.67). (We have presented equation (7.67) in the form suitable for an energy spectrum – the difference is of course trivial.)

The calculation of momentum space integrals for systems involving many particles is generally easier if the Lorentz invariant expressions (7.52) can be used. Consider the first function given in equations (7.52)

$$\varrho_q = \int \prod_{j=1}^{q} t_j \frac{d\mathbf{p}_j}{2E_j}\, \delta(p_i - p_f).$$

Srivastava and Sudarshan (1958) have pointed out that the property of Lorentz invariance can be exploited so that this equation can be written as a recurrence relation

$$\varrho_q = \int t_q \frac{d\mathbf{p}_q}{2E_q}\, \varrho_{q-1} \qquad (7.69)$$

where ϱ_{q-1} is the density integral for a system of $q-1$ particles with total energy ε_c in their c-system

$$\varrho_{q-1} = \int \prod_{k=1}^{q-1} t_k \frac{d\mathbf{p}_k}{2E_k}\, \delta\left(\varepsilon_c - \sum_{k=1}^{q-1} E_k\right) \delta\left(\sum_{k=1}^{q-1} \mathbf{p}_k\right)$$

$$\varepsilon_c = [(E_c - E_q)^2 - \mathbf{p}_q]^{\frac{1}{2}}.$$

In the equation for ϱ_{q-1} the terms p_k and E_k refer respectively to momentum and energy in the c-system of $q-1$ particles.

It is apparent that the reduction of equation (7.69) can be continued until the two body state is reached. Its evaluation is straightforward (compare equations (7.54) to (7.59)).

Momentum space integrals are important in ultra-high energy physics, where many particles may be produced in a single collision; an excellent review of the problems which then arise may be found in an article by Kretzschmar (1961).

7.4(c). *Cross-sections for processes of the type $a + b \to d + e$*

Consider a process in which there are two particles in both initial and final states

$$a + b \to d + e.$$

The cross-section can be written as

$$d\sigma_{fi} = \frac{(2\pi)^4}{v_{ab}} V^2 \overline{\sum_i} \sum_f |M_{fi}|^2 \frac{V^2}{(2\pi)^6} d\mathbf{p}_{fd} d\mathbf{p}_{fe}\, \delta(p_i - p_f) \tag{7.70}$$

from (7.33). We can consider either energy or angular intervals for the final states. The latter are generally more informative and so we rewrite (7.70) as a differential cross-section into angular intervals with the aid of (7.58)

$$\frac{d\sigma}{d\Omega} = \frac{V^4}{(2\pi)^2} \frac{1}{v_{ab}} \overline{\sum_i} \sum_\alpha |M_{fi}|^2 \frac{E_d E_e |\mathbf{p}_d|^3}{\mathbf{p}_d^2 E_i - \mathbf{p}_i \cdot \mathbf{p}_d E_d}. \tag{7.71}$$

The c-system is generally the most convenient one in which to work and, if necessary, cross-sections (or amplitudes) can always be transformed to any other system from it. In the c-system we find that

$$\mathbf{p}_i = 0 \qquad |\mathbf{p}_a| = |\mathbf{p}_b| = p_c^i \qquad E_a + E_b = E_c$$

$$\mathbf{p}_f = 0 \qquad |\mathbf{p}_d| = |\mathbf{p}_e| = p_c^f \qquad E_d + E_e = E_c$$

$$v_{ab} = \frac{p_c^i E_c}{E_a E_b}$$

where the last equation comes from (7.43). Thus equation (7.71) becomes

$$\frac{d\sigma}{d\Omega} = \frac{V^4}{(2\pi)^2} \frac{E_a E_b}{p_c^i E_c} \overline{\sum_i} \sum_\alpha |M_{fi}|^2 \frac{E_d E_e}{E_c} p_c^f$$

$$= \frac{V^4}{(2\pi)^2} \frac{E_a E_b E_d E_e}{E_c^2} \frac{p_c^f}{p_c^i} \overline{\sum_i} \sum_\alpha |M_{fi}|^2. \tag{7.72}$$

If the particles a and b possess spin s_a and s_b respectively, then we may write with the aid of equations (6.27) and (7.31)

$$\overline{\sum_i} = \frac{1}{(2s_a + 1)(2s_b + 1)} \sum_i. \tag{7.73}$$

An alternative formulation of (7.72) may be given if we use the matrix elements T_{fi} of (7.34) in place of M_{fi}. Using (7.36) and (7.37) it is not difficult to show that

$$\frac{d\sigma}{d\Omega} = \frac{1}{(2\pi E_c)^2} \frac{1}{(2s_a + 1)(2s_b + 1)} \frac{1}{n^2} \frac{p_c^f}{p_c^i} \sum_i \sum_\alpha |T_{fi}|^2 \tag{7.74}$$

where n^2 represents an appropriate combination of factors of two and masses prescribed by (7.36)

$$n^2 = 16 \qquad \text{four bosons} \tag{7.75}$$

$$= \frac{4}{m_1 m_2} \qquad \text{two bosons and two fermions}$$

$$= \frac{1}{m_a m_b m_d m_e} \qquad \text{four fermions.}$$

7.4(d). *The summation over states of polarisation*

The term $\sum_i \sum_\alpha$ in equations (7.72) and (7.74) implies that the polarisation states must be summed in calculating the transition probability. We start first by considering Dirac spinors; let us assume that the spinors in the matrix element give us a term of the form

$$T_{r'r} = \bar{u}_{r'} O u_r = \bar{u}_{r'}(\mathbf{p}') O u_r(\mathbf{p}) \tag{7.76}$$

where r' and r refer to the spin orientations of the final and initial particles respectively, $\mathbf{p}'$ and $\mathbf{p}$ represent their momenta and O is some 4×4 matrix in spinor space. The term $T_{r'r}$ will be taken to be equivalent to the spin parts of the matrix elements of both M_{fi} and T_{fi}

$$\left\{ \begin{matrix} M_{fi} \\ T_{fi} \end{matrix} \right\} \equiv T_{r'r}.$$

In calculating the transition probability we consider terms of the following form (compare (7.74))

$$\sum_i \sum_\alpha |T_{fi}|^2 \equiv \sum_{r,r'} |T_{r'r}|^2 = \sum_{r,r'} T_{r'r} T^\dagger_{r'r}. \tag{7.77}$$

Using equation (7.76) we may write the product $T_{r'r}\, T^\dagger_{r'r}$ as

$$\begin{aligned} T_{r'r} T^\dagger_{r'r} &= [\bar{u}_{r'}(\mathbf{p}') O u_r(\mathbf{p})]\, [\bar{u}_{r'}(\mathbf{p}') O u_r(\mathbf{p})]^\dagger \\ &= [\bar{u}_{r'} O u_r]\, [u^\dagger_r O^\dagger \bar{u}^\dagger_{r'}] \\ &= \bar{u}_{r'} O u_r u^\dagger_r \gamma_4 \gamma_4 O^\dagger (u^\dagger_{r'} \gamma_4)^\dagger \\ &= \bar{u}_{r'} O u_r \bar{u}_r \tilde{O} u_{r'} \end{aligned}$$

where

$$\tilde{O} = \gamma_4 O^\dagger \gamma_4 \tag{7.78}$$

and we have used (4.21) and the relations (3.93) and (3.105)

$$\gamma_4 \gamma_4 = \hat{1}, \qquad \gamma_4^\dagger = \gamma_4.$$

The sum of $|T|^2$ over all spin states then becomes

$$\sum_{r,r'} |T_{r'r}|^2 = \sum_{r'} \bar{u}_{r'}(\mathbf{p}) O \Lambda^+(\mathbf{p}) \tilde{O} u_{r'}(\mathbf{p}')$$

$$= \sum_{r'} [\bar{u}_{r'}(\mathbf{p}') O \Lambda^+(\mathbf{p}) \tilde{O} \Lambda^+(\mathbf{p}') u_{r'}(\mathbf{p}') - \bar{v}_{r'}(\mathbf{p}') O \Lambda^+(\mathbf{p}) \tilde{O} \Lambda^+(\mathbf{p}') v_{r'}(\mathbf{p}')] \quad (7.79)$$

where we have used the following relations:

$$\sum_{r} u_r(\mathbf{p})\, \bar{u}_r(\mathbf{p}) = \Lambda^+(\mathbf{p}) \quad (3.168)$$

$$\Lambda^+ u = u, \qquad \Lambda^+ v = 0. \quad (3.162)$$

Next consider a term of the form

$$\sum_{r'} [\bar{u}_{r'}(\mathbf{p}')\, Q u_{r'}(\mathbf{p}') - \bar{v}_{r'}(\mathbf{p}')\, Q v_{r'}(\mathbf{p}')]$$

where Q is any 4×4 matrix; this expression can be written as

$$\sum_{r'} \sum_{\alpha,\beta=1}^{4} [\bar{u}_{r'\alpha} Q_{\alpha\beta} u_{r'\beta} - \bar{v}_{r'\alpha} Q_{\alpha\beta} v_{r'\beta}] = \sum_{\alpha,\beta} Q_{\alpha\beta} \delta_{\alpha\beta} = \sum_{\alpha,\beta=1}^{4} Q_{\alpha\alpha} = \operatorname{tr} Q$$

where we have used (3.156). If this equation is then inserted into (7.79) the following general result is obtained with the aid of (3.160) and (7.78)

$$\sum_{r,r'} |T_{r'r}|^2 \equiv \sum_{i} \sum_{\alpha} |T_{fi}|^2 \equiv \sum_{i} \sum_{\alpha} |M_{fi}|^2$$

$$= \operatorname{tr} [O \Lambda^+(\mathbf{p})\, \tilde{O} \Lambda^+(\mathbf{p}')] \quad (7.80)$$

$$= \operatorname{tr} \left[O \frac{m - i\gamma p}{2m} \gamma_4 O^\dagger \gamma_4 \frac{m - i\gamma p'}{2m} \right]. \quad (7.81)$$

The extension of (7.81) to situations involving more than two spinors may be carried out in a straightforward manner. As an example we will consider two initial spinor states (unpolarised) and two final states. The appropriate expression is

$$\sum_{i} \sum_{\alpha} |T_{fi}|^2 = \sum_{r,s} \sum_{r',s'} |T_{r's'rs}|^2 \quad (7.82)$$

where r and s refer to the initial spin orientations. Let

$$T_{r's'rs} = \bar{u}_{r'}(\mathbf{p}_1')\, O u_r(\mathbf{p}_1)\, \bar{u}_{s'}(\mathbf{p}_2')\, P u_s(\mathbf{p}_2) \quad (7.83)$$

then by using the same techniques as those used for developing (7.80), one finds that

$$\sum_{r,s} \sum_{r',s'} |T_{r's'rs}|^2 = \operatorname{tr} [O \Lambda^+(\mathbf{p}_1)\, \tilde{O} \Lambda^+(\mathbf{p}_1')] \operatorname{tr} [P \Lambda^+(\mathbf{p}_2)\, \tilde{P} \Lambda^+(\mathbf{p}_2')]. \quad (7.84)$$

We next consider the sum over polarisation states for the photon. The photon possesses two polarisation states which are observable and two which are not § 4.4(f)). Only the former states appear in the ingoing or outgoing photons in a transition process. In order to evaluate the sum we consider some simple process involving a photon (Fig. 7.1). Here the lines p and p' represent an ingoing and outgoing electron respectively and k represents a photon.

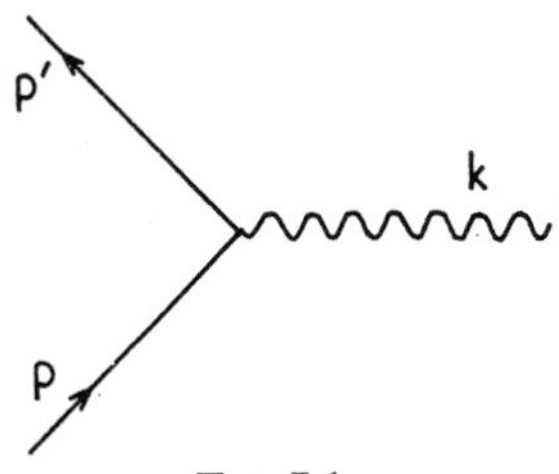

FIG. 7.1.

We shall show in Chapter 8 that the matrix element, M, for a process of this type is linear in the product $\gamma_\mu e_\mu^{(\lambda)}$ where γ is a Dirac γ-matrix and $e_\mu^{(\lambda)}$ is the photon polarisation vector. We can therefore factor the product $\gamma_\mu e_\mu^{(\lambda)}$ out of the matrix O in equation (7.76) and write

$$O = \gamma_\mu e_\mu^{(\lambda)} Q \tag{7.85}$$

where Q is a matrix which is independent of the photon polarisation. Thus the term $\tilde{O}$ of equation (7.78) can be written in the Coulomb gauge ($e_4^{(\lambda)} = 0$) as

$$\tilde{O} = -\tilde{Q}\gamma_\mu e_\mu^{(\lambda)}$$

since $e_\mu^{(\lambda)}$ is real and transverse (4.165); equation (7.80) can therefore be rewritten in the following form for a single state of photon polarisation

$$\sum_i \sum_\alpha |T_{fi}|^2 \equiv \sum_i \sum_\alpha |M_{fi}|^2 = -\text{tr}\,[\gamma_\mu e_\mu^{(\lambda)} Q \Lambda^+(\mathbf{p})\, \tilde{Q}\gamma_\mu e_\mu^{(\lambda)} \Lambda^+(\mathbf{p}')]. \tag{7.86}$$

If we are considering unpolarised photons this equation must be modified to include the necessary sums over the initial and final states of photon polarisation. We will use the same technique as in § 4.4(d), that is we arrange the coordinate system so that the photon travels along the z-axis

$$k_\mu = 0, 0, \omega, i\omega.$$

The two (observable) transverse polarisations can then be written as (4.165)

$$e_\mu^{(1)} = 1, 0, 0, 0$$

$$e_\mu^{(2)} = 0, 1, 0, 0$$

and equation (7.86) becomes

$$\sum_{i}\sum_{\alpha}|T_{fi}|^2 = \sum_{\lambda=1}^{2}\sum_{r,r'}|T_{r'r}|^2$$

$$= -\sum_{\lambda=1}^{2}\mathrm{tr}\,[\gamma_\mu e_\mu^{(\lambda)} Q\Lambda^+(\mathbf{p})\,\tilde{Q}\gamma_\mu e_\mu^{(\lambda)}\Lambda^+(\mathbf{p}')]$$

$$= -\sum_{\lambda=1}^{2}\mathrm{tr}\,[\gamma_\lambda Q\Lambda^+(\mathbf{p})\,\tilde{Q}\gamma_\lambda\Lambda^+(\mathbf{p}')]. \qquad (7.87)$$

The summation in this relation can be extended to cover all four states of polarisation without altering the result. We may demonstrate this by making use of the gauge invariance of the equations in quantum electrodynamics (§ 9.3 and § 4.4(f)). Feynman (1949) has shown that gauge invariance implies that if the four-dimensional polarisation vector e_μ is replaced by the four-momentum k_μ in a matrix element then that matrix element vanishes. Thus if we replace one of the vectors e by k in equation (7.86) and remember that the photon propagation vector is pointing along the z-axis, we obtain with the aid of (4.166)

$$-\omega\,\mathrm{tr}\,[(\gamma_3+i\gamma_4)\,Q\Lambda^+(\mathbf{p})\,\tilde{Q}\gamma_\mu e_\mu^{(\lambda)}\Lambda^+(\mathbf{p}')] = 0 \qquad (7.88)$$

$$-\omega\,\mathrm{tr}\,[\gamma_\mu e_\mu^{(\lambda)} Q\Lambda^+(\mathbf{p})\,\tilde{Q}(\gamma_3+i\gamma_4)\,\Lambda^+(\mathbf{p}')] = 0.$$

The remaining terms $\gamma_\mu e_\mu^{(\lambda)}$ can be chosen arbitrarily in each equation; they will be taken to be taken to be γ_3 and $-i\gamma_4$ in the first and second equations respectively. The irrelevant terms $-\omega$ will be dropped, and we then obtain

$$\mathrm{tr}\,[(\gamma_3+i\gamma_4)\,Q\Lambda^+(\mathbf{p})\,Q\gamma_3\Lambda^+(\mathbf{p}')] = 0 \qquad (7.89)$$

$$\mathrm{tr}\,[-i\gamma_4 Q\Lambda^+(\mathbf{p})\,Q(\gamma_3+i\gamma_4)\,\Lambda^+(\mathbf{p}')] = 0.$$

If these two equations are added the cross products of the γ-matrices vanish and we find that

$$\sum_{\lambda=3}^{4}\mathrm{tr}\,[\gamma_\lambda Q\Lambda^+(\mathbf{p})\,\tilde{Q}\gamma_\lambda\Lambda^+(\mathbf{p}')] = 0. \qquad (7.90)$$

Therefore we can rewrite equation (7.87) as

$$\sum_{i}\sum_{\alpha}|T_{fi}|^2 = -\sum_{\lambda=1}^{4}\mathrm{tr}\,[\gamma_\lambda Q\Lambda^+(\mathbf{p})\,\tilde{Q}\gamma_\lambda\Lambda^+(\mathbf{p}')]. \qquad (7.91)$$

This covariant description of the polarisation states of the photon leads to considerable simplifications in the evaluation of the matrix elements for transition probabilities. The extension of equation (7.91) to states involving more than one photon is straightforward. Let there be two photons with polarisations $e_\mu^{(\lambda)}$ and $e_\nu^{(\lambda)}$ then the equivalent expression to (7.85) may be written as

$$O = \gamma_\mu e_\mu^{(\lambda)} Q \gamma_\nu e_\nu^{(\lambda')} + \gamma_\nu e_\nu^{(\lambda')} R \gamma_\mu e_\mu^{(\lambda)} \tag{7.92}$$

$$\tilde{O} = \gamma_\nu e_\nu^{(\lambda')} \tilde{Q} \gamma_\mu e_\mu^{(\lambda)} + \gamma_\mu e_\mu^{(\lambda)} \tilde{R} \gamma_\nu e_\nu^{(\lambda')} \tag{7.93}$$

and the sum over polarisation states becomes

$$\sum_i \sum_\alpha |T_{fi}|^2 = \sum_{\lambda, \lambda' = 1}^{4} \mathrm{tr}\, [(\gamma_\lambda Q \gamma_{\lambda'} + \gamma_{\lambda'} R \gamma_\lambda)\, \Lambda^+(\mathbf{p})\, (\gamma_{\lambda'} \tilde{Q} \gamma_\lambda + \gamma_\lambda \tilde{R} \gamma_{\lambda'})\, \Lambda^+(\mathbf{p}')]. \tag{7.94}$$

Matrix elements similar to (7.92) are found in Compton scattering (§ 11.1).

7.5. SOME PROPERTIES OF THE TRANSITION AMPLITUDE

7.5(a). *The relationship between the scattering and transition amplitudes*

We wish to obtain a relationship between the elastic scattering amplitude of Chapter 6 and the transition amplitude of this chapter; this may be readily established with the aid of equations (7.74) and (6.8). If we consider spinless particles the former equation may be written as

$$\frac{d\sigma}{d\Omega} = \frac{1}{(2\pi E_c)^2} \frac{1}{n^2} \frac{p_c^f}{p_c^i} |T_{fi}|^2$$

where we have dropped the summation terms for spin states. In the case of elastic scattering $p_c^i = p_c^f$, and so the above equation becomes

$$\frac{d\sigma}{d\Omega} = \frac{1}{(2\pi E_c)^2} \frac{1}{n_i^4} |T_{fi}|^2$$

where n_i refers to the normalisation factors for the incident particles, and $n^2 = n_i^4$ since initial and final particles are the same in an elastic scattering process. If we compare the above equation with that obtained by partial wave analysis (6.8)

$$\frac{d\sigma_{sc}}{d\Omega} = |f(\theta)|^2$$

it is clear that we can write

$$f(\theta) = \frac{1}{2\pi E_c} \frac{1}{n_i^2} T_{fi} \tag{7.95}$$

(if we ignore an arbitrary phase factor). If necessary this expression can be extended into channels for angular momentum, isospin or any other appropriate variable; a specific example will be examined in § 14.1(c).

Additional relationships between the matrix operators and the functions appearing in the phase shift analysis may be established. An inspection of § 6.1 (a), and in particular equation (6.4), shows that for a single channel process

$$\eta_l = \varrho_l e^{2i\delta_l} \equiv S_l \tag{7.96}$$

where S_l represents the S-matrix element for the channel with orbital angular momentum l. Thus the scattering function f_l

$$f_l = \frac{\eta_l - 1}{2ik}$$

of equation (6.17) can be written as

$$f_l = \frac{S_l - 1}{2ik}. \tag{7.97}$$

In addition pure elastic scattering implies $\varrho_l = 1$ and so by (6.16)

$$\eta_l = 1 + 2\,i \sin \delta_l e^{i\delta_l}.$$

The elements of reactance matrix operator R of equation (7.16), $S = \hat{1} + iR$ then possess the simple interpretation

$$R_l \equiv 2 \sin \delta_l e^{i\delta_l}. \tag{7.98}$$

7.5(b). *Multichannel reactions*

The elements S_{ab} of the S-matrix, in a subsystem with total angular momentum j and a specified parity, normally obey the condition

$$S_{ab} = S_{ba} \tag{7.99}$$

if the processes are invariant under time reversal (compare § 9.2(a)). This condition implies that the S-matrix is symmetrical, and that the unitarity condition (7.1)

$$S^{\dagger} S = \hat{1}$$

reduces to

$$S^* S = \hat{1} \tag{7.100}$$

since by (7.99)

$$(S^{\dagger})_{ab} = S^*_{ba} = S^*_{ab}.$$

The relation (7.100) leads to a useful application in a situation in which there is a multichannel reaction and the elastic scattering channel dominates the others (Fermi, 1955). It is easy to arrange for the elastic scattering channel to be represented by an element which is on the diagonal of the S-matrix, and then to a good approximation we can represent the S-matrix as

$$S = S_0 + i\varepsilon \tag{7.101}$$

where S_0 is a diagonal matrix and ε a matrix whose elements are zero along the diagonal and small off the diagonal. Equation (7.100) then yields the conditions

$$\varepsilon S_0^* = S_0 \varepsilon^*, \; S_0^* S_0 = \hat{1}$$

and so for the matrix element S_{ab} we have

$$\varepsilon_{ab}(S_0^*)_{bb} = (S_0)_{aa}\varepsilon_{ab}^*$$

$$\varepsilon_{ab}e^{-2i\delta_b} = e^{2i\delta_a}\varepsilon_{ab}^*$$

where we have used the approximation $S_0 = e^{2i\delta}$ and δ represents the elastic scattering phase shift (7.96). If we multiply through by $e^{i(\delta_b-\delta_a)}$ we obtain

$$\varepsilon_{ab}e^{-i(\delta_a+\delta_b)} = e^{i(\delta_a+\delta_b)}\varepsilon_{ab}^*$$

hence

$$\varepsilon_{ab} = \varrho_{ab}e^{i(\delta_a+\delta_b)} \tag{7.102}$$

where ϱ_{ab} is real and $\varrho_{ab} = \varrho_{ba}$. As an example of the application of this principle let us consider the processes

$$\begin{array}{c}\gamma + p \rightarrow \gamma + p \\ \times \\ \pi + N \rightarrow \pi + N.\end{array}$$

Now the Compton scattering process is electromagnetic and has a negligible amplitude in comparison with that for meson photoproduction, hence we may represent the S-matrix as

$$\begin{array}{cc} & \begin{array}{cc}\gamma p & \pi N\end{array} \\ & \begin{pmatrix} 0 & \varrho e^{i\delta} \\ \varrho e^{i\delta} & e^{2i\delta}\end{pmatrix} \begin{array}{c}\gamma p \\ \pi N\end{array}\end{array} \tag{7.103}$$

so that the photoproduction amplitude is determined by ϱ and the phase shift involved in pion–nucleon scattering. It should be noted that a complete treatment should involve isospin channels (Fermi, 1955). The general problems involved in multichannel interactions have been extensively studied by many workers; (Dalitz (1962) has given an excellent review of the situation).

7.5(c). *The unitarity of the S-matrix and the optical theorem*

In § 7.2 we stated that the S-matrix operator was a unitary operator. We now explore the consequences of this unitarity property in terms of the R and T matrices. Since $S = \hat{1} + iR$

$$SS^\dagger = (\hat{1} + iR)(\hat{1} - iR^\dagger) = \hat{1}$$

and so

$$i(R - R^{\dagger}) = -RR^{\dagger} \tag{7.104}$$

$$i \langle f| R - R^{\dagger} |i\rangle = -\langle f| RR^{\dagger} |i\rangle.$$

Now by equation (4.17)

$$\langle f| R^{\dagger} |i\rangle = (\langle i| R |f\rangle)^* = R^*_{if}$$

and so if we use the completeness relation (4.15), we may modify (7.104) in the following manner:

$$i(R_{fi} - R^*_{if}) = -\sum_n \langle f| R |n\rangle \langle n| R^{\dagger} |i\rangle \tag{7.105}$$

where $\sum_n |n\rangle \langle n|$ represents a complete set of intermediate states. Now let us introduce the T-matrix (7.34); the above equation then becomes

$$i \frac{(2\pi)^4}{N_f N_i} \delta(p_i - p_f)(T_{fi} - T^*_{if}) = -\sum_n \frac{(2\pi)^8}{N_f N_i N_n^2} \delta(p_n - p_f)\, \delta(p_i - p_n)\, T_{fn} T^*_{in}$$

and since

$$\delta(p_n - p_f)\, \delta(p_i - p_n) = \delta(p_i - p_f)\, \delta(p_i - p_n)$$

we obtain the following unitarity condition, if invariance under time reversal holds (compare § 9.2(a))

$$\begin{aligned} i(T_{fi} - T^*_{if}) &= -2ImT_{fi} \\ &= -\sum_n \frac{(2\pi)^4}{N_n^2} \delta(p_i - p_n)\, T_{fn} T^*_{in} \end{aligned} \tag{7.106}$$

where the sum $\sum_n$ implies (compare (7.30))

$$\sum_n = \int V \frac{d\mathbf{p}_1}{(2\pi)^3} \dots V \frac{d\mathbf{p}_n}{(2\pi)^3} \equiv \frac{1}{(2\pi)^{3n}} \int d\mathbf{p}_1\, d\mathbf{p}_2 \dots d\mathbf{p}_n \tag{7.107}$$

where we have dropped the volume normalisation terms V since they inevitably cancel in any practical problem.

In the case of forward scattering we may write $i = f$ in the above equation and obtain

$$2 \text{ Im } T_{ii} = \sum_n \frac{(2\pi)^4}{N_n^2} \delta(p_i - p_n)\, |T_{in}|^2. \tag{7.108}$$

Now in equation (6.12) we obtained the optical theorem

$$\frac{4\pi}{p} \text{Im} f(0) = \sigma_T$$

where σ_T represented the total cross-section. In the case of forward scattering equation (7.95) becomes

$$f(0) = \frac{1}{2\pi E_c} \frac{1}{n_i^2} T_{ii}$$

so that

$$2 \operatorname{Im} T_{ii} = p_c E_c n_i^2 \sigma_T. \tag{7.109}$$

We may verify this relation by using it and (7.108) to evaluate σ_T

$$\sigma_T = \frac{1}{p_c E_c n_i^2} \sum_n \frac{(2\pi)^4}{N_n^2} \delta(p_i - p_n) |T_{in}|^2 = \frac{(2\pi)^4}{v N_i^2} \sum_n |T_{in}|^2 \frac{\delta(p_i - p_n)}{N_n^2} \tag{7.110}$$

which is equivalent to (7.37) if we recall that the symbol n represents states compatible with the initial conditions, and so it is equivalent to f in a summation. In the above equation we have written (compare equations (7.70) to (7.75))

$$n_i^2 = \frac{E_a E_b}{E_a E_b} n_i^2 = \frac{N_i^2}{E_a E_b}$$

and hence obtained the relative velocity v for the incoming particles.

A further important relation can be obtained by application of the unitarity principle. It concerns the inverse T-matrix – a term which is related to the scattering length of § 6.3(d). Let us return to equation (7.106) and consider a submatrix with definite angular momentum and parity (§ 9.2(a)); we may write it as

$$i \langle f| T - T^\dagger |i\rangle = -\sum_n \frac{(2\pi)^4}{N_n^2} \delta(p_i - p_n) \langle f| T |n\rangle \langle n| T^\dagger |i\rangle.$$

If we now multiply on the left by T^{-1} and on the right by $(T^\dagger)^{-1}$ we find

$$2 \operatorname{Im} T_{fi}^{-1} = -\sum_n \frac{(2\pi)^4}{N_n^2} \delta(p_i - p_n) \delta_{fn} \delta_{ni} = -\sum_f \frac{(2\pi)^4}{N_f^2} \delta(p_i - p_f) \delta_{fi}.$$

If we assume two-body final states only we may replace the summation $\sum_f$ by equations (7.30) and (7.59) and obtain

$$\operatorname{Im} T_{fi}^{-1} = -\frac{1}{2} \frac{4\pi}{(2\pi)^2} \frac{p_c^f}{E_c} \frac{E_{1f} E_{2f}}{N_f^2} \delta_{fi} = -\frac{1}{2\pi E_c} \frac{p_c^f}{n_f^2} \delta_{fi}. \tag{7.111}$$

Thus the imaginary component of the inverse T-matrix contains kinematic factors only, all the dynamic features are contained in the real part

$$T_{fi}^{-1} = \operatorname{Re} T_{fi}^{-1} - i p_c^f \delta_{fi} \frac{1}{2\pi E_c n_f^2}. \tag{7.112}$$

The above equation appears in many forms in the literature (see, for example,

Feldman, Matthews and Salam, 1960). Its form might have been conjectured almost directly from (7.95)

$$f(\theta) = \frac{1}{2\pi E_c} \frac{1}{n_i^2} T_{fi}.$$

Consider for example elastic s-wave scattering, then from equation (6.7)

$$f(\theta) = \frac{\sin \delta e^{i\delta}}{k}$$

so that

$$f^{-1}(\theta) = k \cot \delta - ik \equiv p \cot \delta - ip \tag{7.113}$$

and hence by equation (7.95)

$$T_{fi}^{-1} = \frac{1}{2\pi E_c} \frac{1}{n_f^2} (p \cot \delta - ip)$$

where $n_i^2 = n_f^2$ since we are considering elastic scattering.

CHAPTER 8

THE INTERACTION OF FIELDS III SPECIFIC FORMS FOR THE S-MATRIX

In the previous chapter the S-matrix operator was defined to be unitary

$$SS^\dagger = S^\dagger S = S^{-1}S = \hat{1}$$

and to possess the property of relating initial and final states

$$|\Psi_f\rangle = S\,|\Psi_i\rangle .$$

In developing the S-matrix elements for practical calculations, specific representations are used. The choice of representation used is partly a matter of taste and partly dictated by the nature of the problem being examined. It is obvious that the final results obtained in any physical problem should be independent of the representation used.

In this chapter we will develop the S-matrix elements in two forms. First, we will use perturbation theory and develop expressions which may be conveniently handled by Feynman graph techniques. Secondly, we will use the S-matrix operator in a Heisenberg representation which is useful for dispersion relations.

8.1. THE INTERACTION REPRESENTATION

8.1(a). *Introduction*

A covariant form of perturbation theory may be developed with the aid of the interaction representation. This representation is intermediate in form between the Schrödinger and Heisenberg representations (§ 4.1(f)). In it the operators have the time dependence of Heisenberg operators for a non-interacting system, and any changes occurring as a function of time in the state vectors arise solely from the interaction.

The scheme was first introduced by Stückleberg (1934), and has been used extensively in the development of quantum electrodynamics (Schwinger, 1948b).

Let us consider a quantum mechanical system which describes an interaction. We start with the Schrödinger formulation given in equation (4.41)

$$i\frac{\partial}{\partial t}\,|\Psi_S(t)\rangle = H\,|\Psi_S(t)\rangle$$

and assume that the Hamiltonian for the system can be split into two parts – one representing the Hamiltonian H_0 of the unperturbed fields and one representing the interaction H'

$$H = H_0 + H'. \tag{8.1}$$

The term H_0 is sometimes called the *bare field* (or *particle*) *Hamiltonian.*

A state vector $|\Psi_I(t)\rangle$ in the interaction representation will now be introduced. It will be defined as the term appearing in the following operation

$$|\Psi_S(t)\rangle = R(t, t')\,|\Psi_I(t)\rangle \tag{8.2}$$

where R represents the solution of the equation

$$i\frac{\partial}{\partial t} R\,(t, t') = H_0 R(t, t') \tag{8.3}$$

namely,

$$R(t, t') = e^{-iH_0(t-t')} \tag{8.4}$$

(compare equations (4.36–4.41)). Thus t' represents the time at which the two state vectors are equal

$$R(t', t') = \hat{1}.$$

Equations (8.1) and (8.2) may be inserted in (4.41) yielding

$$\begin{aligned} i\frac{\partial}{\partial t}\,[R\,|\Psi_I(t)\rangle] &= i\frac{\partial R}{\partial t}\,|\Psi_I(t)\rangle + iR\,\frac{\partial}{\partial t}\,|\Psi_I(t)\rangle \\ &= (H_0 + H')\,R\,|\Psi_I(t)\rangle. \end{aligned} \tag{8.5}$$

This equation reduces to

$$iR\,\frac{\partial}{\partial t}\,|\Psi_I(t)\rangle = H'R\,|\Psi_I(t)\rangle \tag{8.6}$$

by virtue of equation (8.3). If this equation is operated on from the left by R^{-1} we obtain

$$i\frac{\partial}{\partial t}\,|\Psi_I(t)\rangle = H_I(t)\,|\Psi_I(t)\rangle \tag{8.7}$$

where

$$H_I(t) = R^{-1}(t, t')\,H'R(t, t'). \tag{8.8}$$

Thus the variation of the state vector with time in the interaction representation is due solely to the interaction. The term $H_I(t)$ describes the effect of the interaction Hamiltonian in the new representation. In general, operators in the interaction representation are related to those in the Schrödinger representation by the expansion

$$A_I(t) = R^{-1}(t, t')\,A_S R(t, t') \tag{8.9}$$

ince

$$\langle \Psi_I(t)| A_I(t) |\Psi_I(t)\rangle = \langle \Psi_S(t)| A_S |\Psi_S(t)\rangle.$$

It can be seen that the operators are now explicitly dependent upon time, in contrast to the Schrödinger operators, and that their time dependence is the same as that for operators in the Heisenberg representation for a noninteracting system, since we have defined R in terms of H_0, the Hamiltonian for an unperturbed system. Therefore the operators used for the equations of motion and commutators in the interaction scheme are those for the free fields in the Heisenberg representation.

8.1 (b). *The relationship between the Heisenberg and the interaction representations*

This relationship may be readily established with the aid of equations (8.2) and (4.44). For example we may write

$$\begin{aligned} |\Psi_H\rangle &= T^{-1}(t, t') |\Psi_S(t)\rangle \\ &= T^{-1}(t, t') R(t, t') |\Psi_I(t)\rangle. \end{aligned} \tag{8.10}$$

We now introduce the symbol $U(t, t')$ defined as†

$$U(t, t') = R^{-1}(t, t') T(t, t') \tag{8.11}$$

so that (8.10) may be re-written as

$$|\Psi_I(t)\rangle = U(t, t') |\Psi_H\rangle. \tag{8.12}$$

It is apparent from this equation that

$$\frac{\partial}{\partial t} |\Psi_I(t)\rangle = \frac{\partial}{\partial t} U(t, t') |\Psi_H\rangle$$

and when this relation and (8.12) are inserted in equation (8.7) we obtain

$$i\frac{\partial}{\partial t} |\Psi_I(t)\rangle = i\frac{\partial}{\partial t} U(t, t') |\Psi_H\rangle = H_I(t) U(t, t') |\Psi_H\rangle.$$

Thus U satisfies the operator equation

$$i\frac{\partial}{\partial t} U(t, t') = H_I(t) U(t, t') \tag{8.13}$$

† A cursory inspection of (8.11) and comparison with (4.40) and (8.4) might suggest that $U(t, t') = \hat{1}$. This is not true, since the term H appearing in (4.40) covers both free and interacting fields in the case of an interaction.

and also by (8.11) U possesses the boundary value

$$U(t', t') = R^{-1}(t', t')\, T(t', t') = \hat{1} \tag{8.14}$$

since

$$T(t, t) = R^{-1}(t, t) = \hat{1}\,.$$

A solution for $U(t, t')$ satisfying these equations is given by the expression

$$e^{-iH_I(t)(t-t')} \tag{8.15}$$

provided that H_I does not change significantly in the time interval $(t - t')$. In general a simple solution for U is not feasible; iteration solutions for U will be discussed in § 8.2(a).

The operator U defines the temporal development of the state vectors in the interaction representation. The form of equation (8.12) shows that $|\Psi_H\rangle$ and $|\Psi_I(t)\rangle$ coincide at t'

$$|\Psi_I(t')\rangle = |\Psi_H\rangle$$

and so we may write

$$|\Psi_I(t)\rangle = U(t, t')\, |\Psi_I(t')\rangle\,. \tag{8.16}$$

The relation between operators in the Heisenberg and interaction representations may be written in a simple form with the aid of the U operator. From equations (8.9) and (4.46) we obtain the relation

$$A_I(t) = R^{-1} A_S R = R^{-1} T A_H(t)\, T^{-1} R = U(t, t')\, A_H(t)\, U^{-1}(t, t'). \tag{8.17}$$

It is evident from these equations that the field variables and their related expressions in the Heisenberg and interaction representations coincide at time t'. It is also obvious from the form of (8.15) that the two representations coincide at all times in the absence of an interaction.

8.1(c). *The S-matrix in the interaction representation*

In this section the S-matrix will be examined with the aid of the interaction representation. We will start by showing that the interaction representation leads to an operator which has the properties of Heisenberg's S-matrix.

In § 8.1(b) it was shown that the operator $U(t, t')$ linked states in the interaction representation in the following manner (8.16):

$$|\Psi_I(t)\rangle = U(t, t')\, |\Psi_I(t')\rangle\,.$$

We will now define an operator S with the following property:

$$S = \lim_{\substack{t \to +\infty \\ t' \to -\infty}} U(t, t') = U(\infty, -\infty). \tag{8.18}$$

The operator S must have properties similar to Heisenberg's S-matrix, since t links initial and final states

$$|\Psi_I(+\infty)\rangle = S\,|\Psi_I(-\infty)\rangle. \tag{8.19}$$

This equation obviously possesses a similar structure to (7.3).

In order to complete the proof that $U(+\infty, -\infty)$ is identical with the S-matrix operator we must show that U is unitary. This property can be demonstrated with the aid of equation (8.13) and its Hermitian conjugate

$$i\frac{\partial U}{\partial t} = H_I U, \qquad -i\frac{\partial U^\dagger}{\partial t} = U^\dagger H_I$$

where we have used the Hermitian property of H_I. These equations can be combined in the following relation

$$\begin{aligned} i\frac{\partial}{\partial t}[U^\dagger(t,t')\,U(t,t')] &= i\frac{\partial U^\dagger}{\partial t}U + iU^\dagger\frac{\partial U}{\partial t} \\ &= -U^\dagger H_I U + U^\dagger H_I U = 0 \end{aligned}$$

thus $U^\dagger U$ is constant, but the boundary condition (8.14) implies that $U(t,t) = U^\dagger(t,t) = \hat{1}$, and therefore the constant must be $\hat{1}$

$$U^\dagger(t,t')\,U(t,t') = \hat{1}. \tag{8.20}$$

In order to complete the proof that U is unitary we must also show that

$$U(t,t')\,U^\dagger(t,t') = \hat{1}.$$

Rather than use the rigorous proof, which may be found in the standard textbooks on field theory, we will use the following theorem from matrix algebra. If two square matrices A and B of the same finite dimension n satisfy the equation

$$AB = \hat{1}$$

then they also obey the relation

$$BA = \hat{1}$$

and in fact $B = A^{-1}$, where A^{-1} is uniquely determined by the property $AA^{-1} = \hat{1}$ which also implies $A^{-1}A = \hat{1}$.

This simple argument is adequate for most purposes, and if we apply it to equation (8.20) we may conclude that

$$U^\dagger(t,t')U(t,t') = U(t,t')U^\dagger(t,t') = \hat{1}. \tag{8.21}$$

Thus U is a unitary operator and the operator S defined by (8.18)

$$S = \lim_{\substack{t\to+\infty \\ t'\to-\infty}} U(t,t') = U(\infty,-\infty)$$

is equivalent to the S-matrix operator.

8.2. THE ITERATION SOLUTION OF THE S-MATRIX, COVARIANT PERTURBATION THEORY

8.2(a). *Introduction*

Let us return to equation (8.13)

$$i\frac{\partial}{\partial t}U(t, t') = H_I(t)\, U(t, t').$$

It was remarked after (8.15) that this equation does not yield a simple solution. One method of treating it is to expand U into a perturbation series by iterative methods and hope that succeeding terms converge rapidly to zero. Let us explore this method; for convenience of writing we set

$$U(t) = U(t, t')$$

so that equation (8.13) becomes

$$i\frac{\partial U}{\partial t} = H_I(t)\, U(t). \tag{8.22}$$

Let us now seek a solution to this equation by assuming that $H_I(t)$ is a small quantity of first order and that $U(t)$ may be expanded as

$$U(t) = \hat{1} + U_1(t) + U_2(t) + U_3(t) + \cdots = \sum_{n=0}^{\infty} U_n(t). \tag{8.23}$$

If we substitute this equation in (8.22) and equate terms of equal order, we find

$$i\frac{\partial U_1}{\partial t} = H_I(t) \tag{8.24}$$

$$i\frac{\partial U_2}{\partial t} = H_I(t)\, U_1$$

$$\cdots\cdots\cdots\cdots\cdots\cdots$$

These equations possess the solutions

$$U_1(t) = -i\int_{t'}^{t} dt_1 H_I(t_1)$$

$$U_2(t) = (-i)^2 \int_{t'}^{t} dt_1 \int_{t'}^{t_1} dt_2 H_I(t_1)\, H_I(t_2)$$

$$\cdots\cdots\cdots\cdots\cdots\cdots\cdots\cdots\cdots\cdots\cdots\cdots$$

and the general term in the expansion (8.23) is

$$U_n(t) = (-i)^n \int_{t'}^{t} dt_1 \int_{t'}^{t_1} dt_2 \dots \int_{t'}^{t_{n-1}} dt_n H_I(t_1)\, H_I(t_2) \dots H_I(t_n). \tag{8.25}$$

8.2(b). *Chronological products and the S-matrix*

It was pointed out by Dyson (1949) that equation (8.25) is essentially an integration over the time interval from t' to t, with the condition that any limit t_j in the integrals should be earlier than $t_{j-1} (j \leqq n)$. Dyson showed that this condition could be removed by introducing a *chronological ordering operator P*. The action of P for two fields has already been defined in equation (4.254); in general P may be defined by the following property

$$P[H_I(t_1) H_I(t_2) \dots H_I(t_n)] = H_I(t_i) H_I(t_j) \dots H_I(t_k) \tag{8.26}$$

where

$$t_i > t_j > \dots > t_k$$

that is P arranges any product of time dependent operators in chronological sequence so that the last one in time occurs first in the product and the earliest in time occurs last.

The effect of introducing this operator will now be discussed. Consider the $n = 2$ term of equation (8.25)

$$U_2(t) = (-i)^2 \int_{t'}^{t} dt_1 \int_{t'}^{t_1} dt_2 H_I(t_1)\, H_I(t_2). \tag{8.27}$$

The region of integration for this term is shown schematically in Fig. 8.1. This figure represents the $t_1 t_2$ plane. The second integral in (8.27) implies that the region of integration must lie below the line $t_1 = t_2$ which bisects the angle

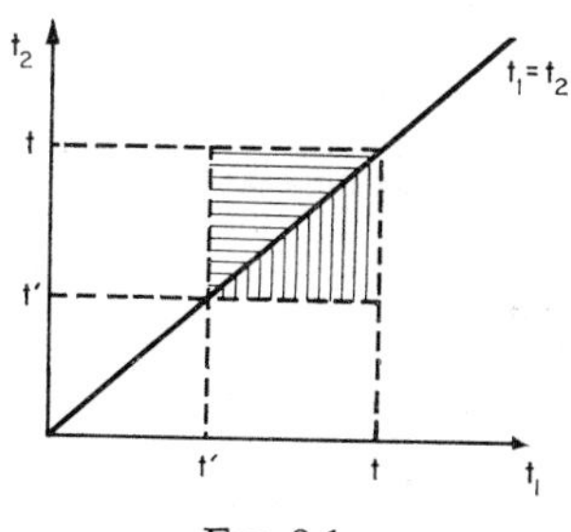

FIG. 8.1.

between the coordinates axes. The first integral takes us from t' to t along the abcissa. Therefore the region of integration must be represented by the vertically shaded triangle in Fig. 8.1. Now if the terms t_1 and t_2 are interchanged in equation (8.27) we find that

$$U_2(t) = (-i)^2 \int_{t'}^{t} dt_2 \int_{t'}^{t_2} dt_1 H_I(t_2)\, H_I(t_1)$$

and so the region of integration becomes the horizontally shaded area above the line in Fig. 8.1. The integrand in the above equation differs from that in (8.27) only in the inversion of the operators $H_I(t)$ and $H_I(t_2)$. If $H_I(t_1)$ and $H_I(t_2)$ commute, then both integrands are the same and $U_2(t)$ can be written as one half of the integral over the whole square. A similar result can be obtained with noncommuting operators if the chronological product P is introduced:

$$\begin{aligned} P[H_I(t_1)\, H_I(t_2)] &= H_I(t_1)\, H_I(t_2) \quad (t_1 > t_2) \\ &= H_I(t_2)\, H_I(t_1) \quad (t_2 > t_1). \end{aligned}$$

Thus we may write

$$U_2(t) = \frac{(-i)^2}{2} \int_{t'}^{t} dt_1 \int_{t'}^{t} dt_2\, P[H_I(t_1)\, H_I(t_2)]$$

and in general

$$U_n(t) = \frac{(-i)^n}{n!} \int_{t'}^{t} dt_1 \int_{t'}^{t} dt_2 \ldots \int_{t'}^{t} dt_n\, P[H_I(t_1)\, H_I(t_2) \ldots H_I(t_n)]. \tag{8.28}$$

A rigorous proof of this expression is complicated; its form can be arrived at by intuition if one considers the treatment given above for $n = 2$ and realises that the term $U_n(t)$ is completely symmetric with respect to $t_1, t_2, \ldots, t_n$ and so there will be $n!$ ways in which to make the ordering.

We now return to the S-matrix. If it is recalled that

$$U(t) = U(t, t')$$

and

$$S = \lim_{\substack{t \to +\infty \\ t' \to -\infty}} U(t, t')$$

then the iteration solution to the S-matrix becomes

$$\begin{aligned} S &= \lim_{\substack{t \to +\infty \\ t' \to -\infty}} \sum_{n=0}^{\infty} U_n(t) = \sum_{n=0}^{\infty} S_n \\ &= \hat{1} + \sum_{n=1}^{\infty} \frac{(-i)^n}{n!} \int_{-\infty}^{+\infty} dt_1 \int_{-\infty}^{+\infty} dt_2 \ldots \int_{-\infty}^{+\infty} dt_n\, P[H_I(t_1)\, H_I(t_2) \ldots H_I(t_n)]. \end{aligned} \tag{8.29}$$

Since the term $H_I(t)$ represents an integration over configuration space of a Hamiltonian density we may write

$$\int dt\, H_I(t) = \int dt\, d\mathbf{x}\, \mathscr{H}_I(\mathbf{x}, t) = \int d^4x\, \mathscr{H}_I(x)$$

and when the above relation is inserted in (8.29) we obtain

$$S = \sum_{n=0}^{\infty} S_n = \hat{1} + \sum_{n=1}^{\infty} \frac{(-i)^n}{n!} \int_{-\infty}^{+\infty} d^4x_1 \int_{-\infty}^{+\infty} d^4x_2 \ldots \int_{-\infty}^{+\infty} d^4x_n \times$$

$$\times P[\mathscr{H}_I(x_1)\,\mathscr{H}_I(x_2) \ldots \mathscr{H}_I(x_n)]. \tag{8.30}$$

This expression is covariant in form since the terms $\mathscr{H}_I(x)$ are scalars and the volume elements d^4x are Lorentz invariants. For some problems the lowest order in $\mathscr{H}_I$ is adequate, and then we may write

$$S = \hat{1} - i \int_{-\infty}^{+\infty} dt\, H_I(t) = \hat{1} - i \int_{-\infty}^{+\infty} d^4x\, \mathscr{H}_I(x). \tag{8.31}$$

8.2(c). *Bare and physical states — the adiabatic hypothesis*

The S-matrix operator developed in the previous section may be used to calculate the amplitude for a transition between two states i and f

$$S_{fi} = \langle f|\, S\, |i\rangle$$

(compare (7.6)). In ordinary quantum mechanics the unperturbed wave functions would be used to specify these states. If field theory is applied logically and consistently one may argue that this approach is inadequate, since account should be taken of the fact that the particles in the states $|f\rangle$ and $|i\rangle$, although well separated and no longer interacting with each other, are real particles interacting with their self-fields. For example we may regard the nucleon as consisting of a 'bare' core 'dressed' in a meson cloud.

This problem has been examined by a number of workers (see, for example, Dyson, 1951). It may be avoided by introducing the *adiabatic hypothesis*, in which the interaction is switched on and off at infinity. This result may be achieved technically by rewriting $H_I(t)$ as

$$H_I(t, \varepsilon) = e^{-\varepsilon|t|}\, H_I(t), \quad \varepsilon \to 0^+ \tag{8.32}$$

where the symbol $\varepsilon \to 0^+$ indicates that ε is an infinitesimally small but positive quantity; equation (8.22) then becomes

$$i \frac{\partial}{\partial t}\, U(t, \varepsilon) = H_I(t, \varepsilon)\, U(t, \varepsilon)$$

with the boundary condition

$$U(t', \varepsilon) = \hat{1}\,.$$

The S-matrix operator is then defined as follows:

$$\lim_{\substack{t\to+\infty \\ t'\to-\infty}} U(t, t', \varepsilon) = S_\varepsilon$$

$$S = \lim_{\varepsilon\to 0^+} S_\varepsilon .$$

For a rigorous proof of these conditions see Belinfante and Møller (1954), and Coester, Hammermesh and Tanaka (1954).

The presence of the term $e^{-\varepsilon|t|}$ also overcomes certain technical problems associated with the order of integration in terms such as (8.28) (in fact the integration can be performed in $n!$ ways). These points are discussed by Jauch and Rohrlich (1955). An example of the effect of a damping term is discussed in the next section in connection with equation (8.34).

The effect of introducing (8.32)

$$H_I(t, \varepsilon) = H_I(t)\, e^{-\varepsilon|t|}$$

is to make $H_I = 0$ at $t = \pm\infty$ so that the states $|i\rangle$ and $|f\rangle$ are simply eigenfunctions of H_0 and are thus bare particle states. In this condition ($H_I = 0$), both states are time independent. The development of an interaction can then be described as follows.

(1) At $t = -\infty$ the state $|i\rangle$ consists of a definite number of 'bare' particles, with spins and momenta defined to be equal to those of the real particles. (Although we are now discussing the particles as if they were localised, in practical calculations we shall use a more convenient approach involving plane or spherical waves.)

(2) At $t = -T$ (long before the interaction takes place) the interaction is adiabatically switched on and the state $|i\rangle$ is converted into a state

$$|\Psi_I(-T)\rangle = U(-T, -\infty)\,|i\rangle$$

corresponding to real physical particles with the same spins and momenta. The change in state is feasible since H_I allows for self-interactions in which the bare particle becomes dressed with these associated virtual quanta or mesons.

(3) During a time interval $-\tau < t < +\tau$, where $|T| \gg \tau$, the particles are in close proximity to each other and the interaction takes place.

(4) Real physical particles, satisfying the condition $p^2 = -m^2$, emerge from the interaction and remain in a dressed state until a time T has elapsed, so that at T the state is defined by the condition

$$|\Psi_I(T)\rangle = U(T, -T)|\Psi_I(-T)\rangle$$

(5) At $t = T$ the interaction is adiabatically switched off so that at $t = \infty$ only bare particle states exist

$$|f\rangle = U(\infty, T)\,|\Psi_I(T)\rangle .$$

The whole process is indicated schematically in Fig. 8.2.

The conditions discussed above are of formal rather than practical significance, ince many types of interaction can be calculated, with a fair degree of success, o lowest order only in perturbation theory. In the strong interactions, for example

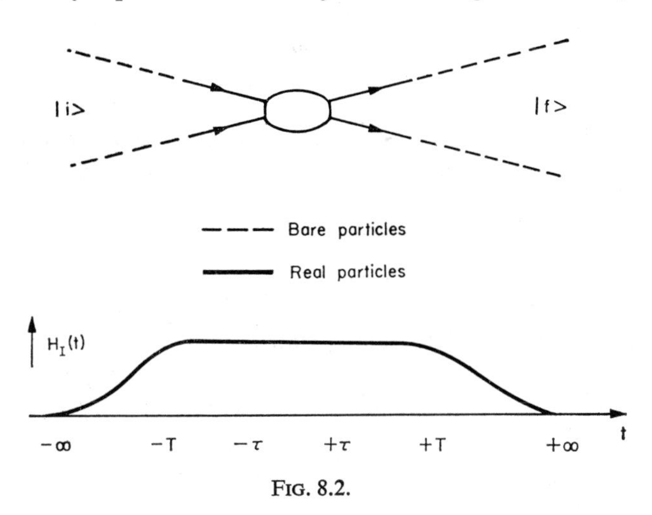

FIG. 8.2.

he pion–nucleon interaction, where first order perturbation theory fails, higher rder expansions also fail because H_I is not small and so the perturbation tech- ique is in any case not a useful one.

8.2(d). *A comparison with time-dependent perturbation theory*

The nth order matrix element $\langle f| S_n |i\rangle$ of covariant perturbation theory can e shown to be equivalent to the nth order element of time dependent perturba- ion theory. The latter expression appears in most textbooks on quantum me- hanics (see, for example, Schiff, 1955).

In equation (8.8) it was shown that the following relation existed between nteraction Hamiltonians in the interaction and Schrödinger representations

$$H_I(t) = R^{-1}(t, t')H'R(t, t')$$

vhere R has the solution given in equation (8.4)

$$R = e^{-iH_0(t-t')}$$

o that the representations coincide at some arbitrary time t'. We choose this ime to be at $t' = 0$ so that (8.8) becomes

$$H_I(t) = e^{iH_0t}H'e^{-iH_0t}.$$

Now according to equation (8.29)

$$S = \sum_{n=0}^{\infty} S_n$$

and so

$$\langle f| S |i\rangle = \sum_{n=0}^{\infty} \langle f| S_n |i\rangle .$$

We will consider the expansion of this expression for the first few orders an assume that $|f\rangle$ and $|i\rangle$ are eigenstates of the free field Hamiltonian, so that w may write

$$H_0 |i\rangle = E_i |i\rangle$$

$$H_0 |f\rangle = E_f |f\rangle$$

and similarly for virtual (intermediate) states

$$H_0 |m\rangle = E_m |m\rangle .$$

Thus if we take the first three orders of the expansion

$$S = \sum_{n=0}^{\infty} S_n$$

we find the following results.

(1) $n = 0$

$$\langle f| S_0 |i\rangle = \langle f|i\rangle = 0$$

when $f \neq i$

(2) $n = 1$

$$\begin{aligned} \langle f| S_1 |i\rangle &= -i \int_{-\infty}^{+\infty} dt \, \langle f| H_I(t) |i\rangle \\ &= -i \int_{-\infty}^{+\infty} dt \, \langle f| e^{iH_0 t} H' e^{-iH_0 t} |i\rangle \\ &= -i \langle f| H' |i\rangle \int_{-\infty}^{+\infty} dt \, e^{i(E_f - E_i)t} \\ &= -2\pi i \langle f| H' |i\rangle \, \delta(E_f - E_i) \end{aligned} \qquad (8.3$$

where use has been made of the expansion

$$H_I(t) = e^{iH_0 t} H' e^{-iH_0 t}$$

and the fact the operator $e^{iH_0 t}$ may be expressed as a series of the form $\hat{1} + iH_0 t + i^2/2!(H_0 t)^2 \ldots$ The action of these terms on the state $|f\rangle$ then yields the eigen value given in the third line of the above expansion; a similar remark applie to the operator $e^{-iH_0 t}$.

Equation (8.33) represents the matrix element of first order perturbation theory. This is an equation of some importance and we will return to it later.

(3) $n = 2$

The Dyson P product of equation (8.29) is an unnecessary complication for the $n = 2$ term, which possesses a simple structure; we will therefore write S_2 in the form of equation (8.27)

$$\langle f|\, S_2\, |i\rangle = (-i)^2 \int_{-\infty}^{+\infty} dt_1 \int_{-\infty}^{t_1} dt_2\, H_I(t_1)\, H_I(t_2).$$

This problem is best handled by considering intermediate virtual states, and by using the completeness relation given in equation (4.15)

$$\sum_m |m\rangle\langle m| = \hat{1}.$$

Thus we find that

$$\begin{aligned}\langle f|\, S_2\, |i\rangle &= (-i)^2 \sum_m \int_{-\infty}^{+\infty} dt_1 \int_{-\infty}^{t_1} dt_2\, \langle f|\, H_I(t_1)\, |m\rangle\langle m|\, H_I(t_2)\, |i\rangle \\ &= (-i)^2 \sum_m \langle f|\, H'\, |m\rangle\langle m|\, H'\, |i\rangle \int_{-\infty}^{+\infty} dt_1 e^{i(E_f - E_m)t_1} \int_{-\infty}^{t_1} dt_2 e^{i(E_m - E_i)t_2}.\end{aligned} \tag{8.34}$$

The integral over t_2 is an oscillatory function. In order to make it convergent at its lower limit an infinitesimal positive term ε can be inserted in the exponential (compare § 8.2(c))

$$\int_{-\infty}^{t_1} dt_2 e^{i(E_m - E_i)t_2} \rightarrow \int_{-\infty}^{t_1} dt_2 e^{i(E_m - E_i - i\varepsilon)t_2}$$

where ε is allowed to tend to zero after the integration

$$\varepsilon \rightarrow 0^+.$$

The integration over t_2 therefore yields the relation

$$\int_{-\infty}^{t_1} dt_2 e^{i(E_m - E_i - i\varepsilon)t_2} = i \frac{e^{i(E_m - E_i - i\varepsilon)t_1}}{E_i - E_m + i\varepsilon}$$

so that (8.34) becomes

$$\begin{aligned}\langle f|\, S_2\, |i\rangle &= -i \sum_m \frac{\langle f|\, H'\, |m\rangle\langle m|\, H'\, |i\rangle}{E_i - E_m + i\varepsilon} \int_{-\infty}^{+\infty} dt_1 e^{i(E_f - E_i)t_1} \\ &= -2\pi i \sum_m \frac{\langle f|\, H'\, |m\rangle\langle m|\, H'\, |i\rangle}{E_i - E_m + i\varepsilon}\, \delta(E_f - E_i).\end{aligned} \tag{8.35}$$

This is the relation which normally appears in the textbooks for the matrix element of the second order term in time dependent perturbation theory. The extension to the nth order is obvious, and so we may write

$$\langle f|\, S\, |i\rangle = -2\pi i\delta(E_f - E_i) \times$$
$$\times \left[\langle f|\, H'\, |i\rangle + \sum_m \frac{\langle f|\, H'\, |m\rangle\langle m|\, H'\, |i\rangle}{E_i - E_m + i\varepsilon} \right.$$
$$\left. + \sum_{m,m'} \frac{\langle f|\, H'\, |m\rangle\langle m|\, H'\, |m'\rangle\langle m'|\, H'\, |i\rangle}{(E_i - E_m + i\varepsilon)\,(E_i - E_{m'} + i\varepsilon)} + \cdots \right]. \quad (8.36)$$

8.2(e). *The transition probability in time-dependent perturbation theory—Fermi's 'Golden Rule'*

We return now to equation (8.33)

$$\langle f|\, S_1\, |i\rangle = -2\pi i\, \langle f|\, H'\, |i\rangle\; \delta(E_f - E_i).$$

The presence of the δ-function $\delta(E_f - E_i)$ guarantees that the matrix element i on the *energy shell.* The exact conservation of energy between initial and fina states also implies an infinite time interval between them according to the un certainty principle. This condition is only to be expected, however, as it wen into our original formulation of the S-matrix element.

The net result of this consideration is the same as in § 7.4(a); it is that th term $|\langle f|\, S_1\, |i\rangle|^2$ is not a meaningful quantity since it is proportional to a transition probability which extends over an infinite time. A more useful quan tity is the transition probability per unit time, which may be defined as

$$\lambda = \lim_{T\to\infty} \frac{|\langle f|\, S_1\, |i\rangle|^2}{T}$$
$$= \lim_{T\to\infty} (2\pi)^2\, [\delta(E_f - E_i)]^2\, \frac{|\langle f|\, H'\, |i\rangle|^2}{T}$$
$$= \lim_{T\to\infty} (2\pi)^2\, \delta(E_f - E_i) \left[\frac{1}{2\pi} \int_{-T/2}^{T/2} dt e^{i(E_f - E_i)t} \right] \frac{|\langle f|\, H'\, |i\rangle|^2}{T}$$
$$= \lim_{T\to\infty} 2\pi\, \delta(E_f - E_i) \left[\int_{-T/2}^{T/2} dt \right] \frac{|\langle f|\, H'\, |i\rangle|^2}{T}$$
$$= 2\pi\, |\langle f|\, H'\, |i\rangle|^2\, \delta(E_f - E_i). \quad (8.37)$$

In deriving this equation we have used a similar technique to that described in § 7.4(a). Equation (8.37) is formulated normally in a slightly different manner. The δ-function has a definite meaning only if an integration over the states E_i or E_f can be performed. The problems in the physics of elementary particles usually ave unique initial states and the interest lies in certain final states. The formula iven in (8.37) is then modified by multiplying by the term $(dn/dE_f)\, dE_f$ (where n/dE_f prepresents the number of states per unit energy interval), and then ntegrating over E_f to yield all the final states consistent with energy conservation

$$\lambda = 2\pi\, |\langle f|\, H'\, |i\rangle|^2 \left.\frac{dn}{dE_f}\right|_{E_f=E_i}. \tag{8.38}$$

This equation is known as 'Fermi's Golden Rule' (Fermi, 1951). For many roblems in nuclear and elementary particle physics, a calculation using the above ormula is sufficient to yield useful results. The formula may be modified for alculating cross-sections or decay rates in the manner indicated in Chapter 7.

8.3. A SPECIFIC EXAMPLE – THE PHOTON–ELECTRON INTERACTION

8.3(a). *Equations of motion for interacting systems*

In calculating the transition amplitude for an interacting system it is convenient o start with the Lagrangian for the interacting fields rather than with the Hamiltonian. This procedure is preferable because the Lagrangian leads naturally o Lorentz invariance.

Consider the interaction of two fields† η_1 and η_2 whose Lagrangian densities or the free fields are given by $\mathscr{L}_1(\eta_1, \partial\eta_1/\partial x)$ and $\mathscr{L}_2(\eta_2, \partial\eta_2/\partial x)$ respectively. The total Lagrangian density for the interacting system is expressed as

$$\mathscr{L} = \mathscr{L}_1 + \mathscr{L}_2 + \mathscr{L}_{\text{int}} \tag{8.39}$$

where $\mathscr{L}_{\text{int}}$ represents the interaction Lagrangian density; for the present no representation will be specified. It is obvious from this equation that if $\mathscr{L}_{\text{int}} = 0$, he variation of the action integral leads to two independent Euler–Lagrange quations (3.228), one for the field η_1, and the other for η_2. Thus the two fields ct as if they are free.

The specific form chosen for $\mathscr{L}_{\text{int}}$ will be governed by invariance principles nd hopes for simplicity. It must contain products of the two fields since the Lagrangian is formed from bilinear terms (§ 4.2), and frequently contains derivtives as well as ordinary field operators.

Consider, for example, the interaction of electromagnetic radiation with electrons

$$\mathscr{L} = \mathscr{L}_\gamma + \mathscr{L}_e + \mathscr{L}_{\text{int}}$$

† In practice more than two fields may be involved in a given interaction, for example in the photoproduction of pions from nucleons we must consider the action of the electromagnetic field s well as the fields associated with the pions and nucleons.

where $\mathscr{L}_\gamma$ = Lagrangian density for free photon field, and $\mathscr{L}_e$ = Lagrangian density for free electron field. Now the Lagrangian $\mathscr{L}_{\text{int}}$ must be a scalar quantity but the electromagnetic field operator is a four-vector. The simplest scalar which can be constructed is, therefore, the scalar product of the electron current vector with the electromagnetic field vector

$$\mathscr{L}_{\text{int}} = j_\mu A_\mu \tag{8.40}$$

so that the Lagrangian density becomes

$$\mathscr{L} = -\tfrac{1}{2}\left(\frac{\partial A_\mu}{\partial x_\nu}\frac{\partial A_\mu}{\partial x_\nu}\right) + \mathscr{L}_e + j_\mu A_\mu . \tag{8.41}$$

If a variation with respect to A_μ is carried out with the aid of the Euler-Lagrange equation (3.228) we obtain

$$\frac{\partial \mathscr{L}}{\partial A_\mu} - \frac{\partial}{\partial x_\nu}\frac{\partial \mathscr{L}}{\partial(\partial A_\mu/\partial x_\nu)} = j_\mu + \frac{\partial^2 A_\mu}{\partial x_\nu^2} = 0$$

or

$$\Box^2 A_\mu = -j_\mu .$$

This is the classical Maxwell equation for an electromagnetic field with source present (4.150).

Expressions for the charge density current for the electron–positron field were given in equations (4.229) and (4.305) in a fully antisymmetrised form

$$j_\mu = \frac{ie}{2}[\overline{\psi}, \gamma_\mu \psi] = ieN[\overline{\psi}\gamma_\mu\psi].$$

However, it is adequate for most purposes to write the charge current density operator as

$$j_\mu = ie\overline{\psi}\gamma_\mu\psi . \tag{8.42}$$

If this equation and the Lagrangian for the free electron–positron field (4.186) are inserted in equation (8.41) we obtain

$$\mathscr{L} = -\tfrac{1}{2}\left(\frac{\partial A_\mu}{\partial x_\nu}\frac{\partial A_\mu}{\partial x_\nu}\right) - \overline{\psi}\left(\gamma_\mu \frac{\partial}{\partial x_\mu} + m\right)\psi + ie\overline{\psi}\gamma_\mu\psi A_\mu .$$

If variations with respect to $\overline{\psi}$ and ψ are made with the aid of the Euler–Lagrange equation, we find that

$$\frac{\partial \mathscr{L}}{\partial \overline{\psi}} - \frac{\partial}{\partial x_\mu}\frac{\partial \mathscr{L}}{\partial(\partial \overline{\psi}/\partial x_\mu)} = -\gamma_\mu \frac{\partial \psi}{\partial x_\mu} - m\psi + ie\gamma_\mu\psi A_\mu = 0$$

$$\frac{\partial \mathscr{L}}{\partial \psi} - \frac{\partial}{\partial x_\mu}\frac{\partial \mathscr{L}}{\partial(\partial \psi/\partial x_\mu)} = -\overline{\psi}m + ie\overline{\psi}\gamma_\mu A_\mu + \frac{\partial \overline{\psi}}{\partial x_\mu}\gamma_\mu = 0$$

or in tidier form

$$\left(\gamma_\mu \frac{\partial}{\partial x_\mu} + m\right)\psi = ie\gamma_\mu A_\mu \psi \tag{8.43}$$

$$\frac{\partial \overline{\psi}}{\partial x_\mu}\gamma_\mu - m\overline{\psi} = -ieA_\mu \overline{\psi}\gamma_\mu .$$

An alternative arrangement, which we shall use occasionally, is given by

$$\gamma_\mu\left(\frac{\partial}{\partial x_\mu} - ieA_\mu\right)\psi + m\psi = 0 \tag{8.44}$$

$$\left(\frac{\partial}{\partial x_\mu} + ieA_\mu\right)\overline{\psi}\gamma_\mu - m\overline{\psi} = 0.$$

The first of equations (8.44) was obtained in § 3.3(f) (3.119), by a semi-classical argument. It is apparent from (8.43) that charge conservation occurs in the interaction

$$\begin{aligned}\frac{\partial j_\mu}{\partial x_\mu} &= ie\frac{\partial}{\partial x_\mu}(\overline{\psi}\gamma_\mu\psi)\\ &= ie(-ieA_\mu\overline{\psi}\gamma_\mu + m\overline{\psi})\psi + ie\overline{\psi}(ie\gamma_\mu A_\mu\psi - m\psi)\\ &= 0.\end{aligned} \tag{8.45}$$

8.3(b). *The relationship between the Hamiltonian and Lagrangian densities for interacting systems*

In the perturbation expansions for the *S*-matrix operator (§ 8.2), Hamiltonians were used rather than Lagrangians. These quantities are related through the expression (3.231)

$$\mathscr{H} = \pi_\sigma\dot{\eta}_\sigma - \mathscr{L}$$

so that we may write, in the manner of equation (8.39)

$$\begin{aligned}\mathscr{H}_0 + \mathscr{H}_{\text{int}} &= \pi_\sigma\dot{\eta}_\sigma - (\mathscr{L}_0 + \mathscr{L}_{\text{int}})\\ &= \frac{\partial\mathscr{L}}{\partial\dot{\eta}_\sigma}\dot{\eta}_\sigma - (\mathscr{L}_0 + \mathscr{L}_{\text{int}})\\ &= \frac{\partial}{\partial\dot{\eta}_\sigma}(\mathscr{L}_0 + \mathscr{L}_{\text{int}})\dot{\eta}_\sigma - (\mathscr{L}_0 + \mathscr{L}_{\text{int}})\end{aligned} \tag{8.46}$$

where

$$\mathscr{L}_0 = \mathscr{L}_1 + \mathscr{L}_2$$

$$\mathscr{H}_0 = \mathscr{H}_1 + \mathscr{H}_2 .$$

Now it is reasonable to assume that the interacting and non-interacting parts in this equation are satisfied separately, and so we may write

$$\mathscr{H}_{\text{int}} = \frac{\partial \mathscr{L}_{\text{int}}}{\partial \dot{\eta}_\sigma} \dot{\eta}_\sigma - \mathscr{L}_{\text{int}}. \tag{8.47}$$

In practice it is frequently found that $\mathscr{L}_{\text{int}}$ does not depend upon the time derivative of the field. This statement is true, for example, in the interaction of the electrons with the electromagnetic field quoted above. In this situation equation (8.47) reduces to

$$\mathscr{H}_{\text{int}} = -\mathscr{L}_{\text{int}}. \tag{8.48}$$

This relation holds true in all representations for most interactions, for example we will consider the Lagrangian density for the photon–electron interaction in the Schrödinger representation

$$\mathscr{L}' = j_{S\mu} A_{S\mu}$$

where the subscript S refers to the Schrödinger representation. It is obvious from the form of j_μ and A_μ that no time derivatives of fields occur in this expression, and so

$$\mathscr{H}' = -\mathscr{L}' = -j_{S\mu} A_{S\mu}.$$

Now from equations (8.8) and (8.9) we can obtain the Hamiltonian density in the interaction representation

$$\begin{aligned} \mathscr{H}_I &= R^{-1} \mathscr{H}' R \\ &= -R^{-1} j_{S\mu} R R^{-1} A_{S\mu} R = -j_{I\mu} A_{I\mu}. \end{aligned} \tag{8.49}$$

Thus the interaction term assumes the same form in Schrödinger and interaction representations.

The S-matrix operator assumes a particularly simple form if lowest order perturbation theory is adequate and equation (8.48) holds. Then equation (8.31) becomes

$$S = \hat{1} + i \int_{-\infty}^{+\infty} dt\, L_I(t) = \hat{1} + i \int_{-\infty}^{+\infty} d^4x\, \mathscr{L}_I(x). \tag{8.50}$$

8.3(c). *The S-matrix expansion for the photon–electron interaction*

The formulation of the S-matrix operator for a specific interaction can now be undertaken. In practice the techniques developed for the perturbation solution of the S-matrix operator are satisfactory only for weakly interacting fields—the weak and electromagnetic interactions. For the former class of interactions only the first integral in the expansion of the S-matrix is necessary for satisfactory lculations; higher order terms can be important for electromagnetic inter-tions. For this reason quantum electrodynamics will provide most of our amples of the treatment of the S-matrix by perturbation methods.

The operator $\mathscr{H}_I$ was shown to be

$$\mathscr{H}_I = -j_{I\mu}A_{I\mu} = -j_\mu A_\mu = -ie\overline{\psi}\gamma_\mu\psi A_\mu \tag{8.51}$$

equation (8.49); for convenience of writing the subscripts I will not be attached field terms. Thus since the operator S is a function of $\mathscr{H}_I$ it will consist of series of products of the operators $\overline{\psi}$, ψ and A, which in turn may be broken into creation and destruction operators. Consider, for example, the S_2 term in e expansion (8.30)

$$S_2 = \frac{(-i)^2}{2}\int_{-\infty}^{+\infty} d^4x_1 \int_{-\infty}^{+\infty} d^4x_2\, P[\mathscr{H}_I(x_1)\,\mathscr{H}_I(x_2)].$$

he part containing the field terms in this expression will be written as

$$O = \frac{1}{(ie)^2}\mathscr{H}_I(x_1)\,\mathscr{H}_I(x_2) = \overline{\psi}(x_1)\,\gamma_\mu\psi(x_1)\,A_\mu(x_1)\,\overline{\psi}(x_2)\,\gamma_\nu\psi(x_2)\,A_\nu(x_2).$$

xpressions if this type are frequently written as

$$O = \overline{\psi}(x_1)\,\gamma_\mu A_\mu(x_1)\,\psi(x_1)\,\overline{\psi}(x_2)\,\gamma_\nu A_\nu(x_2)\,\psi(x_2) \tag{8.52}$$

as

$$O = \overline{\psi}(x_1)\,A\!\!\!/_\mu(x_1)\,\overline{\psi}(x_2)\,A\!\!\!/_\nu(x_2)\,\psi(x_2), \qquad (A\!\!\!/_\mu \equiv \gamma_\mu A_\mu)$$

nce the operators $\overline{\psi}$ and ψ both commute with A, and so the ordering of A with spect to them is not important.

The operator O must now be so arranged that it provides a complete evaluation the amplitudes between the states $|i\rangle$ and $|f\rangle$. This is done by picking out om O the terms which contain the right combination of creation and destruc-on operators to remove particles from the state $|i\rangle$ and to create them in $|f\rangle$. or example, $|i\rangle$ could contain an electron with momentum $\mathbf{p}$ and polarisation r hich is being scattered by the interaction into the states $\mathbf{p}'$ and r'

$$|i\rangle \equiv |e_r^-(\mathbf{p})\rangle$$

$$|f\rangle \equiv |e_{r'}^-(\mathbf{p}')\rangle.$$

would then be necessary to pick out from O a term containing the operators $(\mathbf{p})$ and $a_{r'}^\dagger(\mathbf{p}')$. This analysis could be carried out in a systematic manner by panding O into a sum of terms

$$O = \sum_c O_c \tag{8.53}$$

where each term O_c represents a different product of emission and absorptic operators, arranged in such a way that the creation operators are all on the le and destruction operators are all on the right, so that they may act on the ap propriate states. This arrangement is referred to as the *normal product*, and th terms O_c as *normal constituents*. Thus the matrix element for the transition

$$|i\rangle \to |f\rangle \equiv |e_r^-(\mathbf{p})\rangle \to |e_{r'}^-(\mathbf{p}')\rangle$$

could be found by taking the coefficient of term $a_{r'}^\dagger(\mathbf{p}')\, a_r(\mathbf{p})$ in the expansion $\sum_c O_c$.

The full development of O into a suitable combination of creation and de struction operators can be done with the aid of commutation and anticommuta tion relations. For example the operator O could contain a term of the form

$$a_{r'}^\dagger(\mathbf{p}')\, a_{r''}(\mathbf{p}'')\, a_{r''}^\dagger(\mathbf{p}'')\, a_r(\mathbf{p}). \quad (8.5$$

An expression of this form could contribute to the matrix element since th operator $a_{r''}^\dagger(\mathbf{p}'')$, could create an electron in an intermediate state with para meters $\mathbf{p}''$ and r''; this electron could then be eliminated by $a_{r''}(\mathbf{p}'')$. Now b equation (4.206) we can rewrite (8.54) as

$$a_{r'}^\dagger(\mathbf{p}')\, a_r(\mathbf{p}) - a_{r'}^\dagger(\mathbf{p}')\, a_{r''}^\dagger(\mathbf{p}'')\, a_{r''}(\mathbf{p}'')\, a_r(\mathbf{p}).$$

The term on the right picks out states corresponding to the absorption of tw electrons with degrees of freedom r, $\mathbf{p}$ and r'', $\mathbf{p}''$, and creates two electrons i states r', $\mathbf{p}'$ and r'', $\mathbf{p}''$. This process is different from the one under consideratio and so only the left-hand term contributes to the matrix element for the proce

$$|e_r^-(\mathbf{p})\rangle \to |e_{r'}^-(\mathbf{p}')\rangle.$$

8.4. ORDERING THEOREMS

8.4(a). *The normal ordering operator and normal products*

In practice the separation of O into its normal constituents with the aid o commutation and anticommutation relations can become extremely complicate The procedure can be avoided, however, by introduced a *normal ordering opera tor N*. This technique allows the terms O_c to be written out directly.

Let a mixed group of creation and destruction operators be represented by th symbols $A, B, C, \ldots, Z$. The *normal product* of these operators is defined as

$$N[ABC \ldots Z] = (-1)^P [CDK \ldots W] \quad (8.5$$

where C, D, K are the same operators as in the original sequence, but ordered s that the creation operators are on the left of the destruction operators. The ter P represents the number of permutations of electron and positron operators r uired to go from the original to the normal ordering sequence; the factor -1 rises from the anticommutation relations for fermions. If the operator sequence ontains sums of products, each term in the sum must be treated in the same way

$$N[AB \ldots + PQ \ldots] = N[SB \ldots] + N[PQ \ldots].$$

In §§ 4.4(e) and 4.5(a) the operators $A_\mu(x)$, $\psi_\alpha(x)$ and $\overline{\psi}_\alpha(x)$ were written as

$$A_\mu(x) = A_\mu^{(+)}(x) + A_\mu^{(-)}(x) \tag{4.172}$$

$$\psi_\alpha(x) = \psi_\alpha^{(+)}(x) + \psi_\alpha^{(-)}(x) \tag{4.195}$$

$$\overline{\psi}_\alpha(x) = \overline{\psi}_\alpha^{(+)}(x) + \overline{\psi}_\alpha^{(-)}(x) \tag{4.196}$$

vhere the symbols $(+)$ and $(-)$ refer to the parts containing creation and de-truction operators respectively. The following expressions represent simple ex-mples of the normal ordering of creation and destruction operators

$$\begin{aligned} N[A_\mu(x_1)\, A_\nu(x_2)] &= A_\mu^{(+)}(x_1)\, A_\nu^{(+)}(x_2) + A_\mu^{(+)}(x_1)\, A_\nu^{(-)}(x_2) + A_\nu^{(+)}(x_2)\, A_\mu^{(-)}(x_1) \\ &\quad + A_\mu^{(-)}(x_1)\, A_\nu^{(-)}(x_2) \end{aligned} \tag{8.56}$$

$$\begin{aligned} N[\psi_\alpha(x_1)\, \overline{\psi}_\beta(x_2)] &= \psi_\alpha^{(+)}(x_1)\, \overline{\psi}_\beta^{(+)}(x_2) + \psi_\alpha^{(+)}(x_1)\, \overline{\psi}_\beta^{(-)}(x_2) - \overline{\psi}_\beta^{(+)}(x_2)\, \psi_\alpha^{(-)}(x_1) \\ &\quad + \psi_\alpha^{(-)}(x_1)\, \overline{\psi}_\beta^{(-)}(x_2) \end{aligned}$$

$$\begin{aligned} N[\overline{\psi}_\alpha(x_1)\, \psi_\beta(x_2)] &= \overline{\psi}_\alpha^{(+)}(x_1)\, \psi_\beta^{(+)}(x_2) + \overline{\psi}_\alpha^{(+)}(x_1)\, \psi_\beta^{(-)}(x_2) - \psi_\beta^{(+)}(x_2)\, \overline{\psi}_\alpha^{(-)}(x_1) \\ &\quad + \overline{\psi}_\alpha^{(-)}(x_1)\, \psi_\beta^{(-)}(x_2). \end{aligned}$$

Consider the first of equations (8.56); from this definition of the normal product of two electromagnetic field operators $N[A_\mu(x_1)\, A_\nu(x_2)]$ it follows that the ordi-ary product of two operators $A_\mu(x_1)\, A_\nu(x_2)$ may be written as

$$\begin{aligned} A_\mu(x_1)\, A_\nu(x_2) &= [A_\mu^{(-)}(x_1),\, A_\nu^{(+)}(x_2)] + N[A_\mu(x_1)\, A_\nu(x_2)] \\ &= \langle \Psi_0 |\, A_\mu(x_1)\, A_\nu(x_2)\, | \Psi_0 \rangle + N[A_\mu(x_1)\, A_\nu(x_2)] \end{aligned} \tag{8.57}$$

vhere use has been made of equations (4.249), (4.252) and (4.253), and equiva-ent properties of the commutators for the electromagnetic and scalar fields compare (4.273))

$$D = \lim_{m \to 0} \Delta\,.$$

Similar equations to (8.57) can be developed for any pair of field operators, for xample

$$\begin{aligned} \psi_\alpha(x_1)\, \overline{\psi}_\beta(x_2) &= \{\psi_\alpha^{(-)}(x_1),\, \overline{\psi}_\beta^{(+)}(x_2)\} + N[\psi_\alpha(x_1)\, \overline{\psi}_\beta(x_2)] \\ &= \langle \Psi_0 |\, \psi_\alpha(x_1)\, \overline{\psi}_\beta(x_2)\, | \Psi_0 \rangle + N[\psi_\alpha(x_1)\, \overline{\psi}_\beta(x_2)] \end{aligned}$$

which may be verified with the aid of § 4.6(h). In general, therefore, we may wri the following equation for the product of any two operators $A(x_1)$ and $B(x_2)$ a

$$A(x_1)\,B(x_2) = \langle \Psi_0 |\, A(x_1)\,B(x_2)\,| \Psi_0 \rangle + N[A(x_1)\,B(x_2)]. \qquad (8.5$$

When the two fields are different, for example an electromagnetic and a Dira field, the first term on the right-hand side vanishes by virtue of the first of equa tions (8.57), and so

$$A_\mu(x_1)\,\psi_\alpha(x_2) = N[A_\mu(x_1)\,\psi_\alpha(x_2)].$$

So far we have considered only the product of the two field operators. I practice the expressions are much more complicated, and so the procedure mu be extended to deal with this situation. The operators in the general produ $(ABCD \ldots Z)$ satisfy the following theorem (the proof will be given in § 8.4(b

$$\begin{aligned} O = \sum_c O_c &= ABCD \ldots Z \\ &= N[ABCD \ldots Z] \\ &\quad + \delta_p \langle AB \rangle_0 N[CD \ldots Z] + \delta_p \langle AC \rangle_0 N[BD \ldots Z] + \cdots \\ &\quad + \delta_p \langle AB \rangle_0 \langle CD \rangle_0 N[E \ldots Z] + \delta_p \langle AC \rangle_0 \langle BE \rangle_0 N[DF \ldots Z] + \cdots \\ &\quad + \delta_p \langle AB \rangle_0 \langle CD \rangle_0 \langle EF \rangle_0 N[G \ldots Z] + \cdots \\ &\quad + \delta_p \langle AB \rangle_0 \langle CD \rangle_0 \ldots N[PQ \ldots Z] + \cdots \end{aligned} \qquad (8.5$$

where $\delta_p = (-1)^P$ and P is the number of permutations required to bring th fermion operators to the requisite sequence; the terms $\langle AB \rangle_0$ are called *facto pairs*

$$\langle AB \rangle_0 = \langle \Psi_0 |\, AB \,| \Psi_0 \rangle.$$

The terms appearing in the factor pairs have the same relative ordering as in th original product.

Before offering proof of (8.59) it is instructive to write out a specific exampl For simplicity we will consider a product of four operators

$$\begin{aligned} O &= ABCD \\ &= N[ABCD] + \delta_p \langle AB \rangle_0 N[CD] + \delta_p \langle AC \rangle_0 N[BD] \\ &\quad + \delta_p \langle AD \rangle_0 N[BC] + \delta_p \langle BC \rangle_0 N[AD] + \delta_p \langle BD \rangle_0 N[AC] \\ &\quad + \delta_p \langle CD \rangle_0 N[AB] + \delta_p \langle AB \rangle_0 \langle CD \rangle_0 \\ &\quad + \delta_p \langle AC \rangle_0 \langle BD \rangle_0 + \delta_p \langle AD \rangle_0 \langle BC \rangle_0 \\ &= \sum_{c=1}^{10} O_c. \end{aligned} \qquad (8.6$$

In practice many of the terms listed in (8.59) and (8.60) would vanish. For example, using the techniques given in equations (4.252) and (4.253), and recalling (4.281), one may show that

$$\langle \Psi_0 | \, \psi_\alpha(x_1) \, \psi_\beta(x_2) \, | \Psi_0 \rangle = 0 \qquad (8.61)$$

$$\langle \Psi_0 | \, \overline{\psi}_\alpha(x_1) \, \overline{\psi}_\beta(x_2) \, | \Psi_0 \rangle = 0$$

$$\langle \Psi_0 | \, \overline{\psi}_\alpha(x_1) \, A_\mu(x_2) \, | \Psi_0 \rangle = 0 .$$

The immediate consequence of these relations is that the only nonzero values of O_c will be those which contain a ψ and $\overline{\psi}$ operator or two A operators.

8.4(b). *Proof of the theorem for normal ordering*

The proof of the theorem for normal ordering can be carried out by writing out the expression for O in full as in equation (8.52); it is simpler, however, to use an inductive method when the product contains a large number of operators (Dyson, 1951). Let the number of operators in the product be m; the theorem will be proved for m operators by assuming it is true for a product containing $m - 2$ terms. We have already shown it to be true when $m = 2$; it is also obviously true when m equals unity, and so an inductive proof would hold equally well for even or odd products of operators.

Let the m factors be originally in the sequence

$$O = ABCD \dots PQ \dots Z .$$

Any adjacent factors wrongly ordered in the sequence can be altered by writing

$$PQ = \pm QP + (PQ \mp QP) \qquad (8.62)$$

where the upper sign is used unless P and Q are both fermion operators. The term $(PQ \mp QP)$ is not an operator.

Repeated application of equation (8.62) to the product O yields an expression of the form

$$O = ABC \dots Z = N(O) + \varphi \qquad (8.63)$$

where φ represents a sum of terms of the form

$$(PQ \mp QP) \, O'$$

and O' represents a product of $(m - 2)$ operators. The equation (8.59) therefore holds if it also holds for φ.

Now consider a single term

$$\varphi_c = (PQ \mp QP) \, O'_c$$

and let us assume that equation (8.59) holds for O'_c. The normal constituents of PQO' and QPO' will be identical, if P and Q are not paired together. Thus the insertion of φ_c into equation (8.59) yields (compare (8.58))

$$\varphi_c = [\langle\Psi_0|\, PQ\, |\Psi_0\rangle \mp \langle\Psi_0|\, QP\, |\Psi_0\rangle]O'_c.$$

Thus equation (8.59) is true for φ_c and since φ in equation (8.63) is a sum of terms of the form φ_c, equation (8.59) is also true for φ – providing it holds for O'_c. We may show it holds for O'_c by considering a product smaller by two operators and then by arguing as above. We may carry the sequence back to $m = 2$ or 1, and since we know that the theorem holds for this number of operators, the general theorem (8.59) is proved.

8.5. GRAPHS

8.5(a). *Introduction*

The formulae representing the normal constituents of a product of operators may be visualised in an elegant manner by the use of graphs (Feynman, 1949; Dyson, 1949). A graph G consists of a number of vertices with lines joining them. Each vertex represents the points at which the various factors in a normal constituent O_c interact. Thus for a term of the type given in equation (8.52)

$$O = \overline{\psi}(x_1)\, \gamma_\mu A_\mu(x_1)\, \psi(x_1)\, \overline{\psi}(x_2)\, \gamma_\nu A_\nu(x_2)\, \psi(x_2)$$

each graph possesses two vertices at the space time points x_1 and x_2. The lines drawn to and between vertices are given a physical representation according to the following rules.

(1) For each factor pair (contraction) $\langle\Psi_0|\, \overline{\psi}(x_1)\, \psi(x_2)\, |\Psi_0\rangle$ a solid line is drawn from x_1 to x_2, the direction being marked by an arrow (Fig. 8.3).

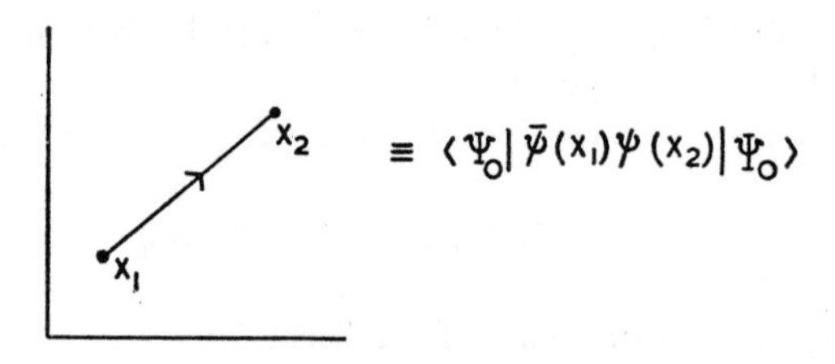

FIG. 8.3.

(2) A solid line running into or away from a vertex is drawn for the unpaired operators $\psi(x)$ and $\overline{\psi}(x)$ respectively. The other end of the line is not joined to a vertex (Fig. 8.4).

(3) For each factor pair $\langle\Psi_0|\, A_\mu(x_1)\, A\,(x_2)\, |\Psi_0\rangle$ a wavy line (without direction) joins x_1 and x_2 (Fig. 8.5).

The reason for the lack of an arrow for the photon lines will become apparent n the next section.

(4) A wavy line without direction, and with one end at the point x whilst the ther is free, is used to represent each unpaired operator $A_\mu(x)$ (Fig. 8.6).

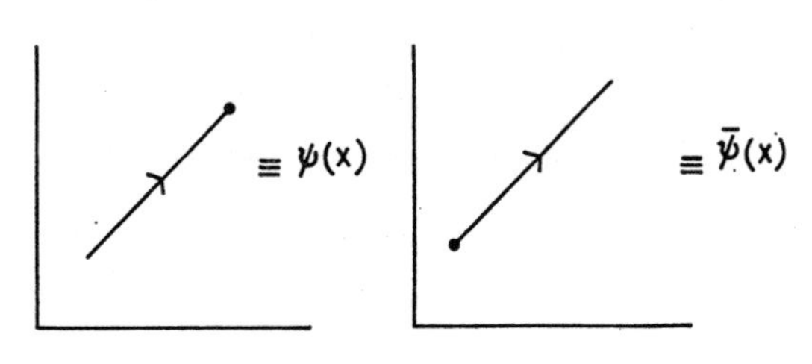

FIG. 8.4.

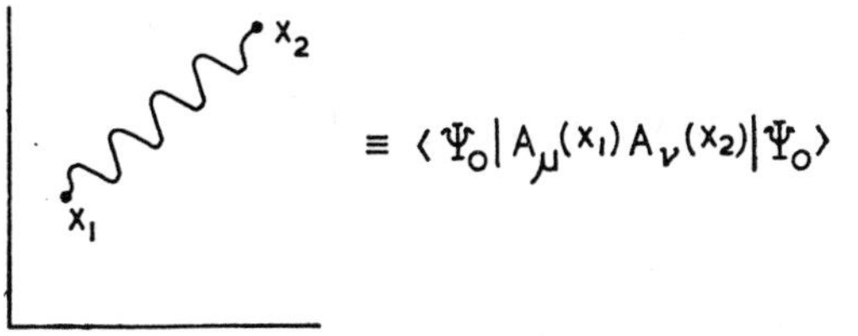

FIG. 8.5.

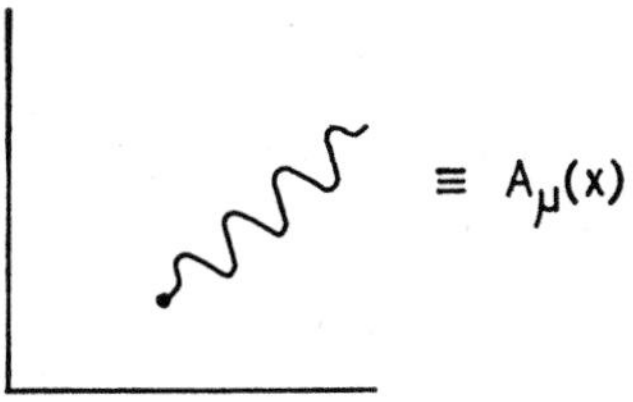

FIG. 8.6.

(5) Lines joining a point to itself are forbidden. This rule arises because operators of the type $\psi(x)\,\gamma_\mu\,\overline{\psi}(x)$ give rise to factor pairings with the following form:

$$\langle \Psi_0\, |\overline{\psi}(x)\;\gamma_\mu \psi(x)|\,\Psi_0\rangle .$$

In § 4.6(i) it was shown that this term equals zero. Thus normal constituents containing a factor pairing of the above type make zero contribution to the S-matrix element, and may be ignored.

The operator product (8.52)

$$O = \overline{\psi}(x_1)\,\gamma_\mu A_\mu(x_1)\,\psi(x_1)\,\overline{\psi}(x_2)\,\gamma_\nu A_\nu(x_2)\,\psi(x_2)$$

can now be broken down into its normal constituents with the aid of equations (8.59) and (8.60), and can be given both algebraic and graphical form. The physical content of the graphs will be indicated below. Some of the constituents listed in (8.59) give zero contributions by virtue of equations (8.61) and (4.294). The remainder take the following forms:

$$O = \sum_{c=1}^{8} O_c$$

$$O_1 = \delta_p N[\overline{\psi}(x_1)\,\gamma_\mu A_\mu(x_1)\,\psi(x_1)\,\overline{\psi}(x_2)\,\gamma_\nu A_\nu(x_2)\,\psi(x_2)] = G_1 .$$

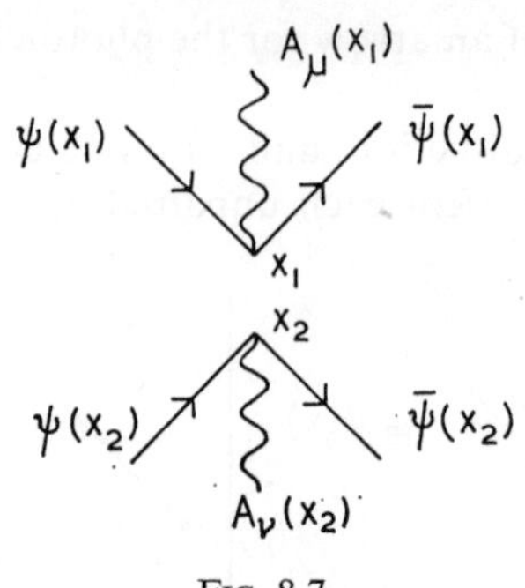

FIG. 8.7.

This graph (Fig. 8.7) represents a scattering process which fails to give a contribution to any observable physical process, because it fails to conserve energy and momentum when integrations over x_1 and x_2 are performed.

$$O_2 = \delta_p \langle \Psi_0 | \overline{\psi}(x_1)\, \psi(x_2)\, | \Psi_0 \rangle N[\gamma_\mu A_\mu(x_1)\, \psi(x_1)\, \overline{\psi}(x_2)\, \gamma_\nu A_\nu(x_2)] = G_2 . \quad (8.64)$$

Graph 2 (Fig. 8.8) corresponds to Compton scattering.

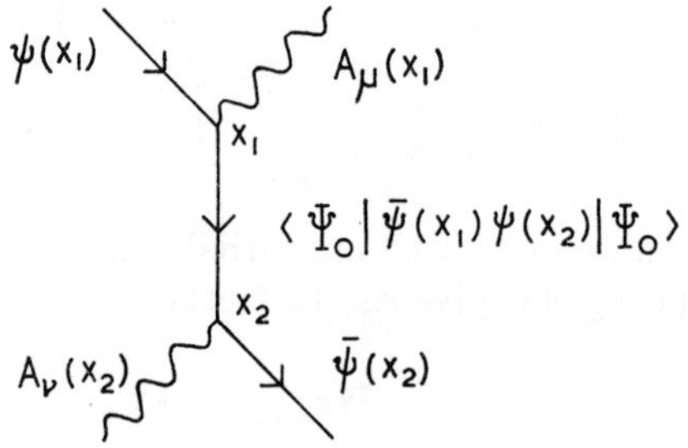

FIG. 8.8.

$$O_3 = \delta_p \langle \Psi_0 | \overline{\psi}(x_2) \psi(x_1)\, | \Psi_0 \rangle N[\overline{\psi}(x_1)\, \gamma_\mu A_\mu(x_1)\, \gamma_\nu A_\nu(x_2)\, \psi(x_2)] = G_3 . \quad (8.65)$$

Graph 3 (Fig. 8.9) is obviously equivalent to G_2 with x_1 and x_2 interchanged. Graphs G_2 and G_3 are said, therefore, to be *topologically indistinguishable.*

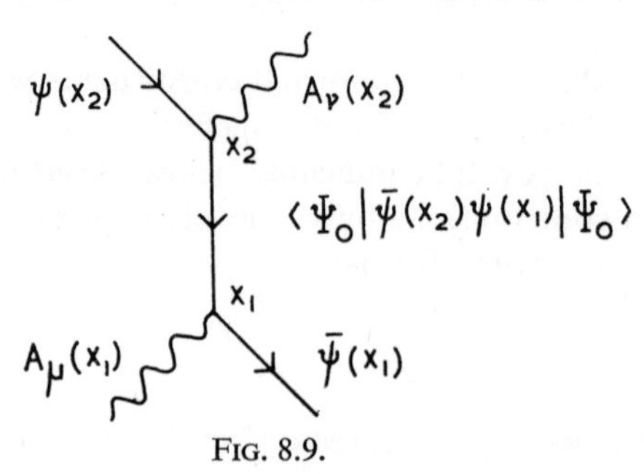

FIG. 8.9.

$$O_4 = \delta_p \langle \Psi_0 | \, \gamma_\mu A_\mu(x_1) \, \gamma_\nu A_\nu(x_2) \, | \Psi_0 \rangle N[\overline{\psi}(x_1) \, \psi(x_1) \, \overline{\psi}(x_2) \, \psi(x_2)] = G_4. \quad (8.66)$$

This contribution represents Møller scattering (Fig. 8.10).

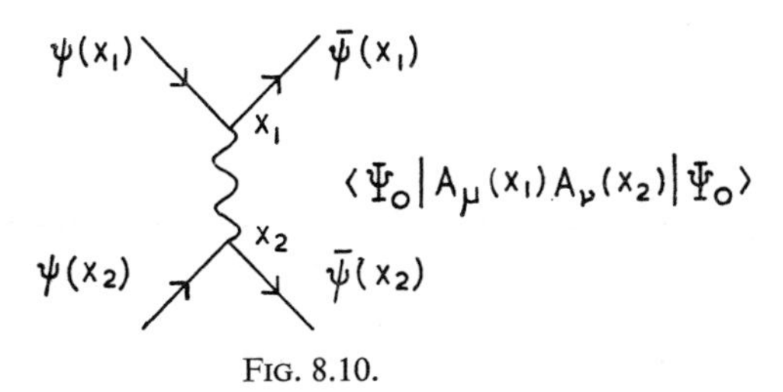

FIG. 8.10.

$$O_5 = \delta_p \langle \Psi_0 \, | \overline{\psi}(x_1) \, \psi(x_2) \, | \Psi_0 \rangle \langle \Psi_0 | \gamma_\mu A_\mu(x_1) \, \gamma_\nu A_\nu(x_2) \, | \Psi_0 \rangle N[\overline{\psi}(x_2) \, \psi(x_1)] = G_5. \quad (8.67)$$

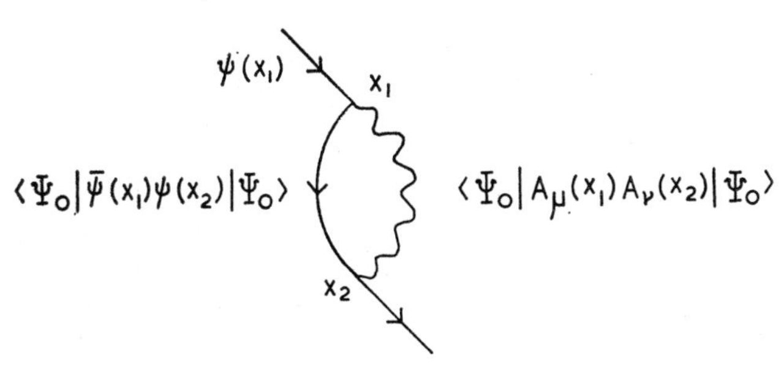

FIG. 8.11.

This term is equivalent to the *self-energy* of a fermion. It will be discussed in ome detail in § 11.3. The next contribution is topologically indistinguishable rom G_5:

$$O_6 = \delta_p \langle \Psi_0 | \, \overline{\psi}(x_2) \, \psi(x_1) \, | \Psi_0 \rangle \langle \Psi_0 | \, \gamma_\mu A_\mu(x_1) \, \gamma_\nu A_\nu(x_2) \, | \Psi_0 \rangle N[\overline{\psi}(x_1) \, \psi(x_2)] = G_6. \quad (8.68)$$

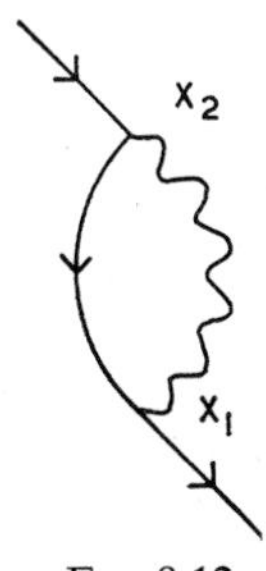

FIG. 8.12.

$$O_7 = \delta_p \langle \Psi_0 | \, \overline{\psi}(x_1) \, \psi(x_2) \, | \Psi_0 \rangle \langle \Psi_0 | \, \overline{\psi}(x_2) \, \psi(x_1) \, | \Psi_0 \rangle N \, [\gamma_\mu A_\mu(x_1) \, \gamma_\nu A_\nu(x_2)] = G_7. \quad (8.69)$$

Graph G_7 represents the self-energy of a photon:

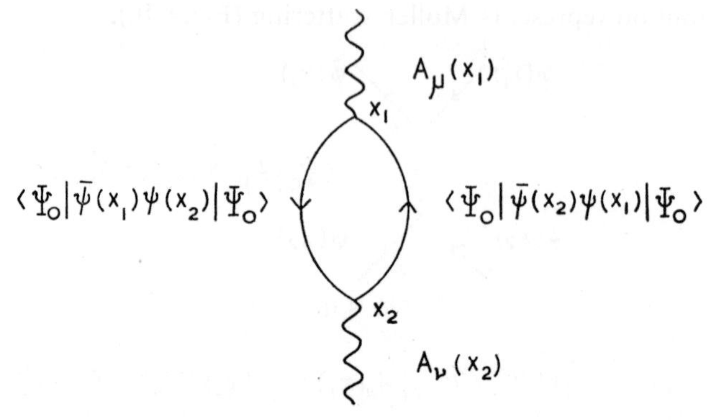

FIG. 8.13.

$$O_8 = \delta_p \langle \Psi_0 | \bar{\psi}(x_1)\,\psi(x_2)\,|\Psi_0\rangle\langle\Psi_0|\,\bar{\psi}(x_2)\,\psi(x_1)|\Psi_0\rangle\langle\Psi_0|\,\gamma_\mu A_\mu(x_1)\,\gamma_\nu A_\nu(x_2)\,|\Psi_0\rangle$$
$$= G_8 . \qquad (8.70)$$

This graph (Fig. 8.14) represents a spontaneous fluctuation which is not re lated to any particular initial or final state. This type of process is said to cor respond to a *vacuum fluctuation*.

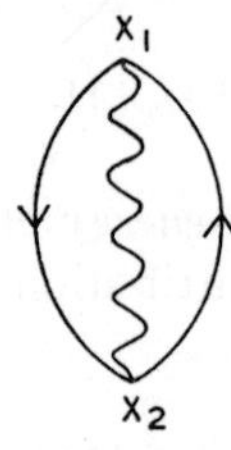

FIG. 8.14.

The above examples show that there is a one to one correspondence between th graphs and the normal constituents of the operator product. It is also apparen that *the factor pairs create and destroy virtual states* (*or particles*), *whilst th unpaired operators create and destroy the incoming and outgoing particles in an physical process.*

So far we have only considered the interaction of field operators at two fiel points or vertices. This occurred because the operator we started with (equa tion (8.52)) was equivalent to the second order S-matrix operator S_2 for th interaction of electron and photon fields. In general it may be shown that th nth order S-matrix S_n contains n vertices.

The importance of the graph technique is that it allows us to write dow quickly the appropriate matrix element for a process. Thus supposing we wishe o consider the effect on the Compton scattering cross-section if a virtual photon as exchanged between ingoing and outgoing electrons. This process could be epicted graphically as in Fig. 8.15.

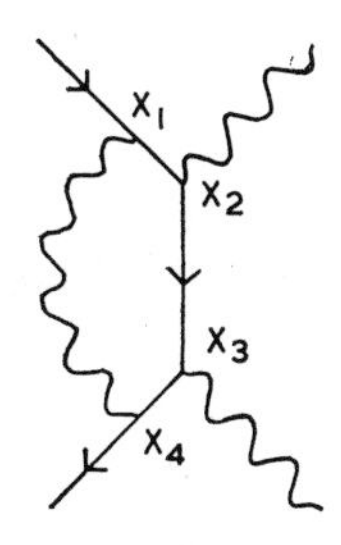

FIG. 8.15.

hen if the rules given at the beginning of this section are followed, the relevant mathematical expression is

$$\delta_p \langle \Psi_0 | \bar{\psi}(x_1)\,\psi(x_2)\,|\Psi_0\rangle \langle \Psi_0 | \bar{\psi}(x_2)\,\psi(x_3)\,|\Psi_0\rangle \langle \Psi_0 | \bar{\psi}(x_3)\,\psi(x_4)\,|\Psi_0\rangle \times$$
$$\times \langle \Psi_0 | \gamma_\mu A_\mu(x_1)\,\gamma_\nu A_\nu(x_4)\,|\Psi_0\rangle N[\psi(x_1)\,\bar{\psi}(x_4)\,\gamma_\varrho A_\varrho(x_2)\,\gamma_\sigma A_\sigma(x_3)].$$

8.5(b). *Feynman graphs*

The operator $\psi(x)$ (4.194) fulfils two functions – it destroys electrons and reates positrons. Similarly the operator $\bar{\psi}(x)$ creates electrons and annihilates ositrons. Thus Fig. 8.16 represents either an electron in an initial state or a positron in a final state.

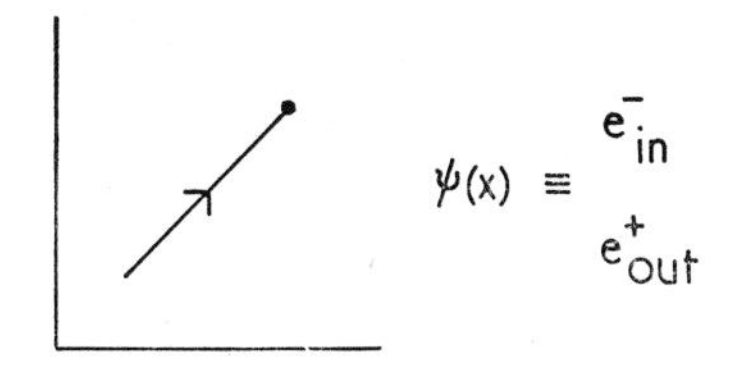

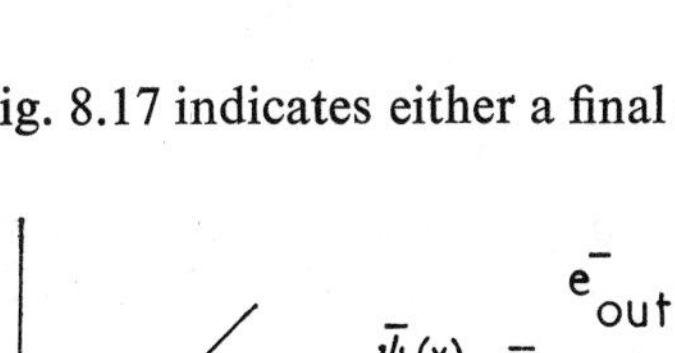

FIG. 8.16.

imilarly the symbol in Fig. 8.17 indicates either a final state electron or a positron in an initial state.

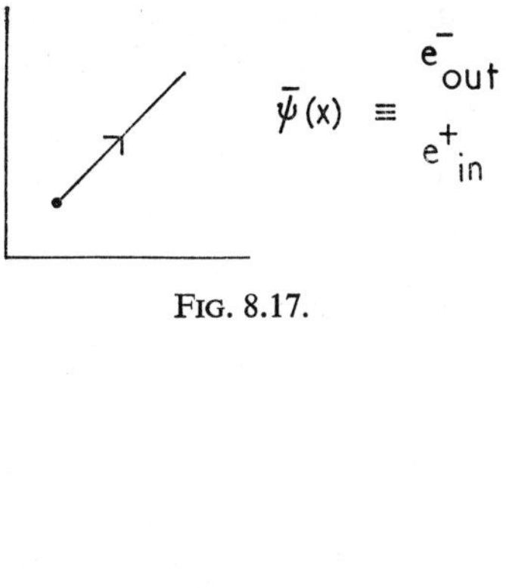

FIG. 8.17.

Additional information can be displayed in a graph by using an approach developed by Feynman (1949). In the diagrams drawn above (Figs. 8.16 and 8.17) the convention will be adopted that time increases up the page, and that lines directed up the page represent electrons and those directed down the page represent positrons. This approach (*the space-time approach*) leads to the conception of positrons as electrons moving backwards in time, and permits a graphical distinction between electrons and positrons.

Consider the graph G_4 (Fig. 8.10); this diagram may be broken down into four Feynman graphs (Fig. 8.18).

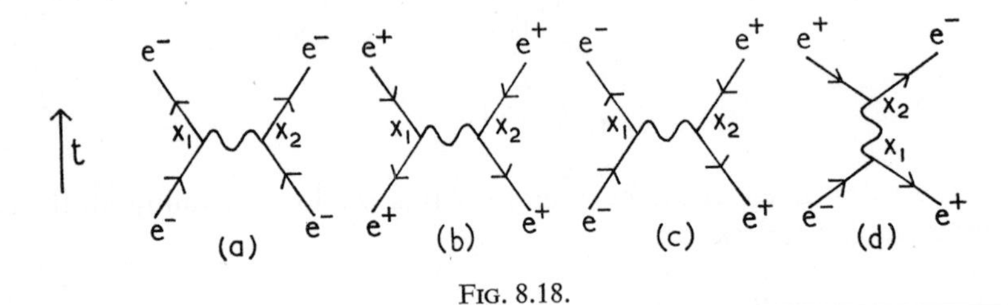

FIG. 8.18.

Examples (a) and (b) of Fig. 8.18 represent electron–electron and positron–positron scattering respectively, whilst (c) and (d) indicate electron–positron scattering; the last examples differ in their intermediate states; in (d) the electron–positron pair annihilate to produce a virtual photon at x_1 and are then formed again at x_2.

The physical content of graphs G_2 and G_3 (Figs. 8.8 and 8.9) may be extended as well, thus G_2 and G_3 become as in Fig. 8.19.

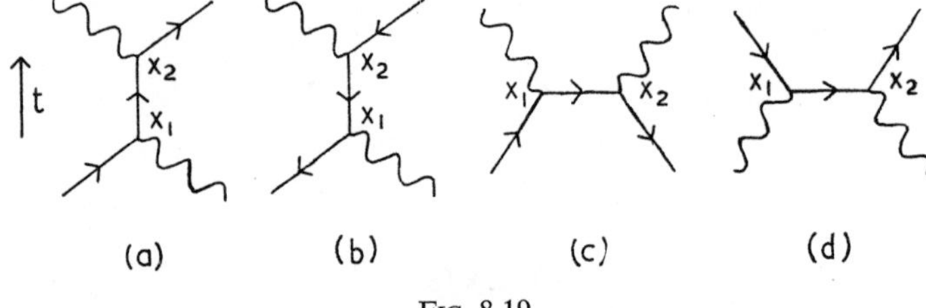

FIG. 8.19.

Diagrams (a) and (b) (Fig. 8.19) represent Compton scattering by an electron and positron respectively; diagram (c) indicates the annihilation of an electron–positron pair yielding two photons, whilst diagram (d) represents the inverse process of pair production by two photons.

8.5(c). *Chronological ordering of the operators*

In equation (8.30) the expression for the S-matrix element

$$S = \hat{1} + \sum_{n=1}^{\infty} \frac{(-i)^n}{n!} \int_{-\infty}^{+\infty} d^4x_1 \ldots \int_{-\infty}^{+\infty} d^4x_n \, P[\mathscr{H}_I(x_1) \ldots \mathscr{H}_I(x_n)]$$

contained the chronological ordering operator P (equation (4.254)). Now the chronological product of any two operators $A(x_1)$ and $B(x_2)$ can be written as

$$\begin{aligned} P[A(x_1)B(x_2)] &= \langle \Psi_0 | \, P[A(x_1)B(x_2)] \, | \Psi_0 \rangle + PN[A(x_1)B(x_2)] \\ &= \langle \Psi_0 | \, P[A(x_1)B(x_2)] \, | \Psi_0 \rangle + N[A(x_1)B(x_2)] \end{aligned} \tag{8.71}$$

where use has been made of equation (8.58). The P-term has been omitted from the normal product, since the normal ordering implies that the creation operator (which is latest in time) occurs on the left whilst the destruction operator removing the initial (earliest) state is on the right. Thus the time ordering of the operators is automatic in the normal product.

The factors in the general product of operators (8.59) can be treated in a similar manner. Thus we may write a typical term in the expansion as

$$O_c = \delta_p \langle AB \rangle_0 \langle CD \rangle_0 \ldots N(R) \tag{8.72}$$

where R represents the residue of operators appearing in the product after the factor pairings have been made. The chronological product can then be written as

$$\begin{aligned} PO_c &= \delta_p \langle P[AB] \rangle_0 \langle P[CD] \rangle_0 \ldots N(R) \\ &= \delta_p \langle T[AB] \rangle_0 \langle T[CD] \rangle_0 \ldots N(R) \end{aligned} \tag{8.73}$$

where the Dyson P-product has been replaced by a T-product (4.290) without causing alteration of the results. This change can be made because P and T products are indistinguishable for boson fields (equation (4.289)), and because all the fermion fields appear in pairs in the expansion of S_n, and so both terms must be changed simultaneously.

Values for the time-ordered products of the electromagnetic and Dirac fields were given in equations (4.275), (4.274), (4.291) and (4.286). They are

$$\begin{aligned} \langle \Psi_0 | \, T[A_\mu(x_1) \, A_\nu(x_2)] \, | \Psi_0 \rangle &\equiv \langle \Psi_0 | \, P[A_\mu(x_1) \, A_\nu(x_2)] \, | \Psi_0 \rangle \\ &= \tfrac{1}{2} D_F(x_1 - x_2) \delta_{\mu\nu} \\ &= \frac{-i}{(2\pi)^4} \delta_{\mu\nu} \int_{-\infty}^{+\infty} d^4k \, \frac{e^{ik(x_1 - x_2)}}{k^2 - i\varepsilon} . \end{aligned} \tag{8.74}$$

$$\langle \Psi_0 | T[\bar{\psi}_\beta(x_2)\,\psi_\alpha(x_1)]\,|\Psi_0\rangle = -\langle \Psi_0 | T\,[\psi_\alpha(x_1)\,\bar{\psi}_\beta(x_2)]\,|\Psi_0\rangle$$

$$= \tfrac{1}{2}\,S_{F\alpha\beta}(x_1 - x_2)$$

$$= \frac{-i}{(2\pi)^4} \int_{-\infty}^{+\infty} d^4p\, \frac{(i\gamma p - m)_{\alpha\beta}}{p^2 + m^2 - i\varepsilon}\, e^{ip(x_1 - x_2)}. \tag{8.75}$$

These expressions are called the *contractions* of the operators. The contraction of Dirac operators belonging to the same point x gives the vacuum expectation value for the current $\langle \Psi_0 | j_\mu(x)\,|\Psi_0\rangle$ when the correct combination of operators is used for $j_\mu(x)$. This was shown to be zero (equation (4.294)). Thus variables for which $x_1 \neq x_2$ are the only ones which need to be contracted. It is therefore adequate to write

$$j_\mu(x_1) \equiv ie\bar{\psi}(x_1)\,\gamma_\mu\psi(x_1)$$

since the difference between this term and the full definition of $j_\mu(x)$ given in equation (4.305)

$$j_\mu(x) = ieN[\bar{\psi}(x_1)\,\gamma_\mu\psi(x_1)]$$

has no effect on contractions with variables at space-time points x_2, when $x_1 \neq x_2$.

8.5(d). *Graphs in momentum space*

In practice, calculations of the transitions are normally carried out in momentum space rather than in configuration space, since the former offers a more useful physical description of a given process.

Diagrams in momentum space are constructed in the same manner as those in position space. There are certain differences however, which should be noted.

(1) Since a four-vector of momentum is associated with each line, that line should be given a direction. This rule applies to photon lines as well as to electron and positron lines. The direction of the arrow is arbitrary for internal photon lines – its function is to obtain the correct sign for k in two δ-functions associated with the vertices at each end of the line.

(2) In position space, positrons move in the opposite direction to their arrows. In momentum space the momentum of the positron is $-p$ in the direction of its arrow, $+p$ in the direction of the positron.

As an example of the method we will consider Compton scattering in lowest order perturbation theory. It is apparent from Figs. 8.8 and 8.9 that the relevant graphs (G_2 and G_3) are equivalent apart from the interchange of dummy indices x_1 and x_2. Thus we can write the appropriate S-matrix operator $S_2(\gamma e)$

for Compton scattering as

$$S_2(\gamma e^-) = \frac{(-i)^2}{2!} \int_{-\infty}^{+\infty} d^4x_1 \int_{-\infty}^{+\infty} d^4x_2 (ie)^2 \, P[O_2 + O_3]$$

$$= \frac{2e^2}{2!} \int_{-\infty}^{+\infty} d^4x_1 \int_{-\infty}^{+\infty} d^4x_2 \, P[O_2] \tag{8.76}$$

where use has been made of equations (8.30) and (8.51). Equation (8.76) can now be written more explicitly with the aid of (8.73) and (8.64)

$$S_2(\gamma e^-) = e^2 \int_{-\infty}^{+\infty} d^4x_1 \int_{-\infty}^{+\infty} d^4x_2 \delta_p \langle \Psi_0 | \, T[\overline{\psi}(x_1) \, \psi(x_2)] \, | \Psi_0 \rangle \times$$
$$\times \, N[\gamma_\mu A_\mu(x_1) \, \psi(x_1) \, \overline{\psi}(x_2) \, \gamma_\nu A_\nu(x_2)]. \tag{8.77}$$

Now for Compton scattering we are interested in an electron going into and out of the scattering process. Thus the appropriate combination of field operators will be of the form $\overline{\psi}^{(+)}(x) \, \psi^{(-)}(x')$, where $x \neq x'$. Equations (8.56) show that

$$N[\psi(x_1) \, \overline{\psi}(x_2)] = -\overline{\psi}^{(+)}(x_2) \, \psi^{(-)}(x_1) + \text{irrelevant terms}. \tag{8.78}$$

We now consider the photon operators. Equations (8.56) show that we have two products with final and initial photon operators in the appropriate sequence

$$N[A_\mu(x_1) \, A_\nu(x_2)] = A_\mu^{(+)}(x_1) \, A_\nu^{(-)}(x_2) + A_\nu^{(+)}(x_2) \, A_\mu^{(-)}(x_1) + \text{irrelevant terms}. \tag{8.79}$$

Thus equation (8.77) becomes

$$S_2(\gamma e^-) = -e^2 \int_{-\infty}^{+\infty} d^4x_1 \int_{-\infty}^{+\infty} d^4x_2 \, \langle \Psi_0 | \, T[\overline{\psi}(x_1) \, \psi(x_2)] \, | \Psi_0 \rangle \times$$
$$\times \, [\gamma_\mu A_\mu^{(+)}(x_1) \, \overline{\psi}^{(+)}(x_2) \, \psi^{(-)}(x_1) \, \gamma_\nu A_\nu^{(-)}(x_2)$$
$$+ \, \gamma_\nu A_\nu^{(+)}(x_2) \, \overline{\psi}^{(+)}(x_2) \, \psi^{(-)}(x_1) \, \gamma_\mu A_\mu^{(-)}(x_2)] = S_2(G_a) + S_2(G_b). \tag{8.80}$$

It is customary to arrange the terms in the S-matrix operators so that there is a convenient one to one correspondence between graph and algebraic expression. The vacuum expectation value in (8.80) is a number and so can be placed anywhere; the operators $A(x)$ and $\psi(x)$ can be permuted freely with each other, and so the components of equation (8.80) can be written in the following tidier form:

$$S_2(G_a) = -e^2 \int_{-\infty}^{+\infty} d^4x_1 \int_{-\infty}^{+\infty} d^4x_2 [\overline{\psi}^{(+)}(x_2) \, \gamma_\nu A_\nu^{(-)}(x_2) \langle \Psi_0 | \, T[\overline{\psi}(x_1) \, \psi(x_2)] \, | \Psi_0 \rangle \times$$
$$\times \, \gamma_\mu A_\mu^{(+)}(x_1) \, \psi^{(-)}(x_1)]$$

$$S_2(G_b) = -e^2 \int_{-\infty}^{+\infty} d^4x_1 \int_{-\infty}^{+\infty} d^4x_2 [\overline{\psi}^{(+)}(x_2)\, \gamma_\nu A_\nu^{(+)}(x_2)\, \langle \Psi_0 |\, T[\overline{\psi}(x_1)\psi(x_2)]\, |\Psi_0\rangle \times$$
$$\times\, \gamma_\mu A_\mu^{(-)}(x_1)\, \psi^{(-)}(x_1)]. \tag{8.81}$$

The appropriate graphs are shown in Fig. 8.20. We use p and k to represent the four-momenta of the incoming photon and electron respectively; the outgoing terms are dashed appropriately. The correspondence between graph and algebra can be clearly seen.

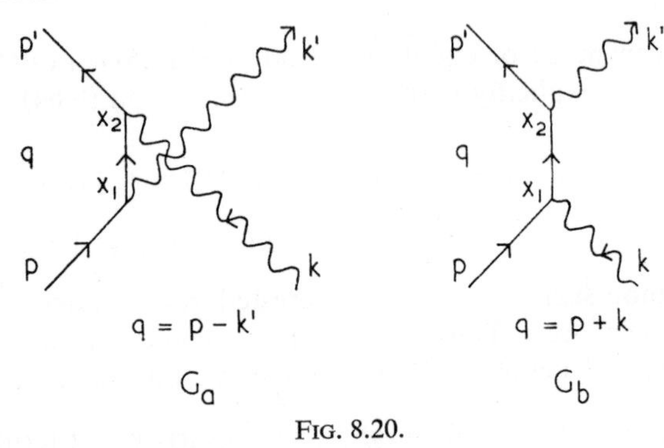

FIG. 8.20.

The intermediate electron line between the points x_1 and x_2 is called the *propagator*. The term is used both for the line on the graph and the appropriate algebraic function; the name propagator is used for any internal line.

8.6. EVALUATION OF THE S-MATRIX ELEMENTS

8.6(a). *A specific example – Compton scattering*

The rules for the evaluation of the S-matrix elements in covariant perturbation theory can be worked out most conveniently by first considering a specific example. Equations (7.6) and (8.80) show us that the appropriate matrix element for Compton scattering in lowest order perturbation theory is given by the relation

$$S_{fi} = \langle f |\, S_2(\gamma e^-)\, | i\rangle = \langle p'k' |\, S_2(G_a)\, | kp\rangle + \langle p'k' |\, S_2(G_b)\, | kp\rangle. \tag{8.82}$$

The initial and final states can be constructed by use of the Fourier integral expansion for field operators given in Chapter 4. We will assume that the incident electron is in a spin state $u_r(\mathbf{p})$, $r = 1, 2$, and that the initial photon is in a state of transverse polarisation given by $e_\mu^{(\lambda)}(\mathbf{k})$, $\lambda = 1, 2$. Then by equations (4.174) and (4.212) we can write

$$|i\rangle = |kp\rangle = a_\lambda^\dagger(\mathbf{k})\, a_r^\dagger(\mathbf{p})\, |\Psi_0\rangle \tag{8.83}$$
$$(\lambda = 1,2) \qquad (r = 1,2).$$

In a similar manner we can represent the final state as

$$|f\rangle = |k'p'\rangle = a^{\dagger}_{\lambda'}(\mathbf{k}')\, a^{\dagger}_{r'}(\mathbf{p}')\, |\Psi_0\rangle \qquad (\lambda' = 1,2) \qquad (r' = 1,2). \tag{8.84}$$

Now consider the graph G_b of Fig. (8.20). The ingoing operators corresponding to this graph contain terms $\psi^{(-)}(x_1)$ and $A^{(-)}_{\mu}(x_1)$ (compare (8.81)) which will absorb the initial state, for example

$$\begin{aligned} A^{(-)}_{\mu}(x_1)\,|k\rangle &= A^{(-)}_{\mu}(x_1)\, a^{\dagger}_{\lambda}(\mathbf{k})\,|\Psi_0\rangle \\ &= \frac{1}{\sqrt{V}} \sum_{\mathbf{k}'',\lambda''} \frac{e^{ik''x_1}}{\sqrt{(2\omega''_k)}}\, e^{(\lambda'')}_{\mu}(\mathbf{k}'')\, a_{\lambda''}(\mathbf{k}'')\, a^{\dagger}_{\lambda}(\mathbf{k})\,|\Psi_0\rangle \\ &= \frac{1}{\sqrt{V}} \frac{e^{ikx_1}}{\sqrt{(2\omega_k)}}\, e^{(\lambda)}_{\mu}(\mathbf{k})\,[1 - a^{\dagger}_{\lambda}(\mathbf{k})\, a_{\lambda}(\mathbf{k})]\,|\Psi_0\rangle \\ &= \frac{1}{\sqrt{V}} \frac{e^{ikx_1}}{\sqrt{(2\omega_k)}}\, e^{(\lambda)}_{\mu}(\mathbf{k})\,|\Psi_0\rangle \end{aligned} \tag{8.85}$$

where use has been made of equations (4.172), (4.169) and the following property (4.173):

$$A^{(-)}_{\mu}(x)\,|\Psi_0\rangle \equiv a_{\lambda}(\mathbf{k})\,|\Psi_0\rangle = 0.$$

When all the creation and destruction operators in $S_2(G_b)$ are taken into account, we find that

$$\langle p'k'|\, a^{\dagger}_{\lambda'}(\mathbf{k})\, a^{\dagger}_{r'}(\mathbf{p}')\, a_{\lambda}(\mathbf{k})\, a_r(\mathbf{p})\,|kp\rangle = \langle \Psi_0 \mid \Psi_0\rangle = 1 \tag{8.86}$$

where the proof follows that given above for (8.85).

Thus the matrix element corresponding to the graph G_b (equation (8.82)) becomes

$$\begin{aligned} &\langle p'k'|\, S_2(G_b)\,|kp\rangle \\ &= \langle p'k'| - e^2 \int_{-\infty}^{+\infty} d^4x_1 \int_{-\infty}^{+\infty} d^4x_2 \{\bar{\psi}^{(+)}(x_2)\,\gamma_{\nu} A^{(+)}_{\nu}(x_2)\,\langle \Psi_0|\, T[\bar{\psi}(x_1)\,\psi(x_2)]\,|\Psi_0\rangle \times \\ &\qquad \times \gamma_{\mu} A^{(-)}_{\mu}(x_1)\,\psi^{(-)}(x_1)\}\,|kp\rangle \\ &= -e^2 \int_{-\infty}^{+\infty} d^4x_1 \int_{-\infty}^{+\infty} d^4x_2 \left[\frac{1}{\sqrt{V}} \sqrt{\left(\frac{m}{E'_p}\right)}\, \bar{u}_{r'}(\mathbf{p}')\, e^{-ip'x_2}\right] \times \\ &\quad \times \left[\frac{1}{\sqrt{V}} \frac{e^{-ik'x_2}}{\sqrt{(2\omega'_k)}}\, \gamma_{\nu} e^{(\lambda')}_{\nu}(\mathbf{k}')\right] \left[\frac{-i}{(2\pi)^4} \int_{-\infty}^{+\infty} d^4q\, \frac{i\gamma q - m}{q^2 + m^2 - i\varepsilon}\, e^{iq(x_2 - x_1)}\right] \times \\ &\quad \times \left[\frac{1}{\sqrt{V}} \frac{e^{ikx_1}}{\sqrt{(2\omega_k)}}\, \gamma_{\mu} e^{(\lambda)}_{\mu}(\mathbf{k})\right] \left[\frac{1}{\sqrt{V}} \sqrt{\left(\frac{m}{E_p}\right)}\, u_r(\mathbf{p})\, e^{ipx_1}\right]. \end{aligned} \tag{8.87}$$

In writing this equation we have used (4.172), (4.195), (4.196), (8.81), (8.75) and (8.85). Integration of (8.87) over x_1 and x_2 yields δ-functions for the momentum four-vectors

$$\int_{-\infty}^{+\infty} d^4x_1 e^{i(p+k-q)x_1} \int_{-\infty}^{+\infty} d^4x_2 e^{i(q-p'-k')x_2}$$
$$= (2\pi)^4 \, \delta(p + k - q) \, (2\pi)^4 \, \delta(q - p' - k') \tag{8.88}$$

where the fourth component of the δ-function implies conservation of energy. The further integration of (8.88) over q yields the δ-function representing overall energy–momentum conservation

$$\int d^4q \, \delta(p + k - q) \, \delta(q - p' - k') \, F(q) = \delta(p + k - p' - k') \, F(p + k)$$

where $F(q)$ is some function of q. Thus the matrix element (8.87) becomes

$$i(2\pi)^4 \frac{e^2}{V^2} \delta(p + k - p' - k') \left(\frac{m^2}{4\omega_k \omega_k' E_p E_p'} \right)^{\frac{1}{2}} \times$$
$$\times \, \bar{u}_{r'}(\mathbf{p}') \left[\gamma_\nu e_\nu^{(\lambda')} \frac{i\gamma(p + k) - m}{(p + k)^2 + m^2 - i\varepsilon} \gamma_\mu e_\mu^{(\lambda)} \right] u_r(\mathbf{p}) \, . \tag{8.89}$$

The propagator in the Compton scattering process has the properties of a Dirac particle with four-momentum q according to the above equations. From equation (8.88) it is apparent that energy and momentum are conserved at each vertex. It should be noted, however, that the mass associated with the four-momentum q is not necessarily that of a real particle. According to (8.88) we can write

$$q = (p + k)$$

so that

$$q^2 = (p + k)^2 = p^2 + k^2 + 2pk$$
$$= -m^2 + 2(\mathbf{p} \cdot \mathbf{k} - E_p \omega_k)$$

where we have used (3.18). The term in brackets in the above expression is always negative and is variable. Thus q does not correspond to the mass of a physical particle

$$q^2 \neq -m^2$$

and the four-momentum q is therefore said to be *off the mass shell*, that is the surface in energy–momentum space available to a physical particle, and defined by the relation

$$m^2 = E_p^2 - \mathbf{p}^2 \, .$$

The term $i\varepsilon$ appearing in the denominator of (8.89) was originally introduced to avoid divergence difficulties. These do not arise here since the propagator is off the mass shell, so henceforward $i\varepsilon$ will be dropped.

8.6(b). *Summary of rules for evaluating graphs in momentum space*

The developement of the lowest order S-matrix element for Compton scattering in the previous section allows us to infer the general rules for constructing matrix elements corresponding to any given graph.

Let us consider the expression (8.89) in conjunction with its associated graph (Fig. 8.21)

$$i(2\pi)^4 \frac{e^2}{V^2} \delta(p + k - p' - k') \left(\frac{m^2}{4\omega_k \omega'_k E_p E'_p} \right)^{\frac{1}{2}} \times$$
$$\times \bar{u}_{r'}(\mathbf{p}') \left[\gamma_\nu e_\nu^{(\lambda')} \frac{i\gamma(p + k) - m}{(p + k)^2 + m^2} \gamma_\mu e_\mu^{(\lambda)} \right] u_r(\mathbf{p})$$

where we have dropped the term $i\varepsilon$ (see remark at end of the last section).

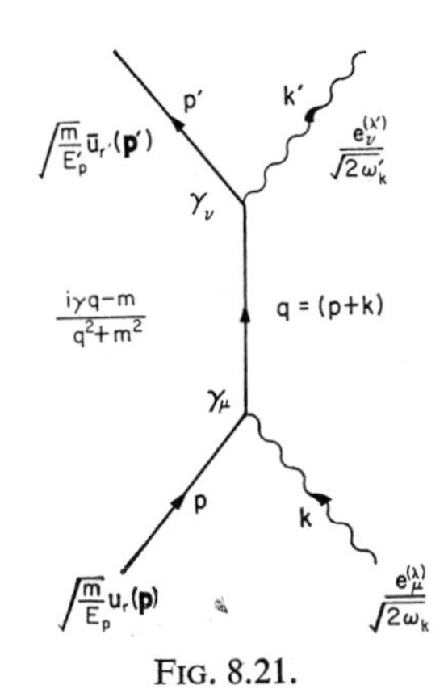

FIG. 8.21.

An inspection of the matrix element and graph shows a one to one correspondence in the physical variables. It is now a straightforward task to link components of diagrams with the algebraic terms in matrix elements. This is done in Table 8.1.

The techniques associated with this table can be used for other interactions than the electromagnetic one, for example a suitably invariant expression for the pion–nucleon interaction is given by

$$\mathscr{H}_I = -\mathscr{L}_I = -ig\bar{\psi}\gamma_5\psi\varphi \tag{8.90}$$

(compare (8.51)), where the term γ_5 appears because the pion field is pseudoscalar. We have indicated suitable graph terms for the pseudoscalar field in Table 8.1 by a broken line. The mathematical properties of the scalar propagator have already been given in §§ 4.6(d) and (e).

TABLE 8.1

Component of diagram	Factor in S-matrix element	
p	$\sqrt{\left(\frac{m}{E_p}\right)}\, u_r(\mathbf{p})$	annihilation of fermion (electron)
p	$\sqrt{\left(\frac{m}{E_p}\right)}\, \bar{u}_r(\mathbf{p})$	creation of fermion
p	$\sqrt{\left(\frac{m}{E_p}\right)}\, \bar{v}_r(\mathbf{p})$	annihilation of antifermion (positron)
p	$\sqrt{\left(\frac{m}{E_p}\right)}\, v_r(\mathbf{p})$	creation of antifermion
k, μ	$\frac{e_\mu^{(\lambda)}(\mathbf{k})}{\sqrt{(2\omega_k)}}$	annihilation of photon
k, μ	$\frac{e_\mu^{(\lambda)}(\mathbf{k})}{\sqrt{(2\omega_k)}}$	creation of photon
k	$\frac{1}{\sqrt{(2\omega_k)}}$	annihilation of meson
k	$\frac{1}{\sqrt{(2\omega_k)}}$	creation of meson
	$\frac{i\gamma p - m}{p^2 + m^2}$	fermion propagator
μ, ν	$\frac{1}{k^2}\,\delta_{\mu\nu}$	photon propagator
	$\frac{1}{k^2 + m^2}$	meson propagator
p′, p, k, γ_μ	$\gamma_\mu \delta(p - p' - k)$	vertices, photon–fermion interaction
p′, p, k, γ_5	$\gamma_5 \delta(p - p' - k)$	vertices pseudoscalar meson–nucleon interaction

We have dropped the terms $i\varepsilon$ from the propagators appearing in Table 8.1, since the four-momenta are always off the mass shell (see the end of the previous section).

It is obvious that Table 8.1 does not exhaust all the possible graphical rules. However, it does contain sufficient information for our immediate requirements†.

We now consider the numerical factor associated with the components of the S-matrix element. For convenience we will consider electromagnetic interactions only. Other examples of interactions will be considered later (§ 8.7). The numerical factor appearing in the matrix element S_n has the following form:

$$C_n = \frac{(-i)^n}{V^\beta}(-ie)^n(-i)^I\,\theta_p(-1)^l\,(2\pi)^\alpha\,\delta(p_i - p_f)$$

$$= e^n\frac{(-i)^I}{V^\beta}\,\theta_p(-1)^{l+n}\,(2\pi)^\alpha\,\delta(p_i - p_f) \qquad (8.91)$$

where n = number of vertices, I = number of internal lines, E = number of external lines, $\alpha = 4n - 4I$, $\beta = \frac{1}{2}E$, l = number of closed loops, and $\theta_P = (-1)^P$ sign of permutation for final electron states (compare (8.55) and (8.59)).

The derivation of these terms is obvious with the exception of the factor $(-1)^l$; it will be discussed below. The term $(-i)^n$ appears in the expansion for the S-matrix (8.30); a factor $1/n!$ also appears in (8.30) but this term vanishes as there are $n!$ topologically indistinguishable graphs associated with each process (compare (8.76)). The term $(-ie)^n$ arises because we are considering a general matrix element with n vertices ($S_n \equiv n$ vertices). The term θ_P is associated with the ordering of the operators ((8.55) and (8.59)). The expression $(-i)^I$ is associated with the mathematical structure of the propagators (8.74, 8.75). The factors $(2\pi)^\alpha$ and $1/V^\beta$ arise in the manner indicated in (8.87) and (8.89).

The final term in (8.91) arises from the integration over the four-momenta of the internal and external lines.

We now mention the factor $(-1)^l$; this term arises from the presence of closed loops in a graph, where l represents the number of closed electron loops. An example of a closed loop was given in Fig. 8.13. They sometimes appear in problems involving high order perturbation expansions, but will not be of any great concern to us. They are discussed in some detail in Jauch and Rohrlich (1955).

The above rules enable a rapid evaluation of the S-matrix element to be made. First, the graphs associated with any particular physical process may be drawn and the appropriate algebraic terms written down with the aid of Table 8.1. This expression must be multiplied by the coefficient C_n given by equation (8.91), and so, finally, the nth order S-matrix element can be written as

$$\langle f|\,S_n\,|i\rangle = C_n\,(\text{graph terms}). \qquad (8.92)$$

† Graph terms for a vector boson are given in § 12.11. Weinberg (1964) has given rules for evaluating graphs containing particles with arbitrary spin.

8.6(c). *Examples of the formation of matrix elements*

The procedure outlined in the previous section enables the transference from graph to algebra to be done in a rapid fashion. Two examples will be considered – Compton scattering by electrons, and electron–electron (Møller) scattering.

Compton scattering in lowest order provides a convenient starting-point since the matrix element was calculated for one graph in § 8.6(a). The appropriate graphs were given in Fig. 8.20. They yield the following data:

$$n = 2 \qquad \alpha = 4 \qquad p_f = p' + k'$$

$$E = 4 \qquad \beta = 2 \qquad p_i = p + k$$

$$I = 1 \qquad l = 0$$

Both graphs have $\theta_P = -1$. The matrix element for Compton scattering therefore can be written as

$$\langle p'k'|\, S_2(\gamma e^-)\, |kp\rangle = \langle p'k'|\, S_2(G_a)\, |kp\rangle + \langle p'k'|\, S_2(G_b)\, |kp\rangle$$

$$= i(2\pi)^4 \frac{e^2}{V^2}\left(\frac{m^2}{4\omega_k\omega_k' E_p E_p'}\right)^{\frac{1}{2}} \delta(p + k - p' - k') \times$$

$$\times\, \bar{u}_{r'}(\mathbf{p}) \left[\gamma_\mu e_\mu^{(\lambda)} \frac{i\gamma(p - k') - m}{(p - k')^2 + m^2} \gamma_\nu e_\nu^{(\lambda')} + \gamma_\nu e_\nu^{(\lambda')} \frac{i\gamma(p + k) - m}{(p + k)^2 + m^2} \gamma_\mu e_\mu^{(\lambda)}\right] u_r(\mathbf{p}). \tag{8.93}$$

The second term can be seen to correspond to equation (8.89).

We now consider electron–electron scattering in lowest ($n = 2$) order. The appropriate graphs are shown in Fig. 8.22.

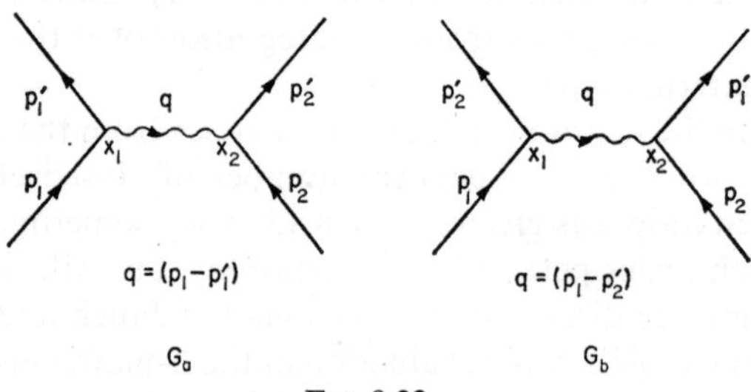

FIG. 8.22.

It is evident that in graph G_b spin states have been interchanged. The relevant parameters are

$$n = 2 \qquad \alpha = 4 \qquad p_f = p_1' + p_2'$$

$$E = 4 \qquad \beta = 2 \qquad p_i = p_1 + p_2$$

$$I = 1 \qquad l = 0$$

An examination of (8.66) shows that

$$\theta_P = +1 \quad \text{for} \quad G_a, \qquad \theta_P = -1 \quad \text{for} \quad G_b$$

so that the matrix element becomes

$$\langle p_2' p_1' |\, S_2(e^- e^-)\, | p_1 p_2 \rangle = -i(2\pi)^4 \frac{e^2}{V^2} \left(\frac{m^4}{E_1' E_2' E_1 E_2} \right)^{\frac{1}{2}} \delta(p_1 + p_2 - p_1' - p_2') \times$$

$$\times \left[\frac{\bar{u}_{r'}(\mathbf{p}_1')\, \gamma_\nu u_r(\mathbf{p}_1)\, \bar{u}_{s'}(\mathbf{p}_2')\, \gamma_\nu u_s(\mathbf{p}_2)}{(p_1 - p_1')^2} - \frac{\bar{u}_{s'}(\mathbf{p}_2')\, \gamma_\nu u_r(\mathbf{p}_1)\, \bar{u}_{r'}(\mathbf{p}_1')\, \gamma_\nu u_s(\mathbf{p}_2)}{(p_1 - p_2')^2} \right]. \tag{8.94}$$

8.6(d). *The substitution law*

This law is based upon the principle that the annihilation of particles and creation of antiparticles are equivalent processes. The concept is implicit in Dirac's theory of holes for relativistic particles, and was used by Fermi in his theory of β-decay (§ 12.2)

$$n \rightarrow p + e^- + \bar{\nu} \equiv \nu + n \rightarrow p + e^-$$

that is, the creation of an antineutrino is equivalent to the destruction of a neutrino.

The substitution law assumes a simple and powerful form when the graph technique is used. It allows us to use effectively the same diagram for calculations involving both particles and antiparticles by a judicious switching of the directions of arrows and, in places, signs of terms. For example, the same graph (Fig. 8.18(a) and (c)) can effectively be used to represent electron–electron and electron–positron scattering

$$e + e^- \rightarrow e^- + e^- \equiv e^- + e^- \equiv e^- + e^- \equiv e^- + e^+ \rightarrow e^+ + e^-.$$

Similarly, a reversal of arrows in Fig. 8.19(a) and (c) relate Compton scattering with the annihilation of an electron–positron pair to yield two photons

$$e^- + \gamma \rightarrow e^- + \gamma \equiv e^- + e^+ \rightarrow \gamma + \gamma.$$

It is also apparent that the matrix element can be used to obtain the time reversed process $\gamma + \gamma \rightarrow e^+ \rightarrow e^-$; in fact it is not difficult to see that any diagram with four external lines can be made to correspond to six different processes if we assume that there are two ingoing and two outgoing particles.

If matrix elements for several processes are written out with the aid of Table 8.1, in the manner discussed above, certain simple rules for substitution emerge. Let us assume we have two graphs G and G' and that an outgoing line in G' has become an incoming line in G. The substitution rules which then link the two graphs are summarised in Table 8.2.

TABLE 8.2

Graph G'	Graph G	Substitution rule
k' out	k in	$k' \leftrightarrow -k \quad e' \leftrightarrow e$
p' out	q in	$p' \leftrightarrow -q \quad \bar{u}(\mathbf{p}) \leftrightarrow \bar{v}(\mathbf{q})$
q' out	p in	$q' \leftrightarrow -p \quad v(\mathbf{q}) \leftrightarrow u(\mathbf{p})$
Right circular $\leftrightarrow$ left circular polarisation		

where k, p and q refer to bosons, fermions and antifermions respectively. The double arrow indicates that the relations are reversible.

The substitution law may be carried a stage further. In considering transition probabilities, squares of matrix elements occur, and these in turn involve summations over spin states as indicated in (7.80). The substitution laws then cause the projection operators to transform in the following manner:

$$p' \leftrightarrow q, \qquad \Lambda^+(p') \leftrightarrow -\Lambda^-(q) = -\Lambda^+(-q) \tag{8.95}$$

$$q' \leftrightarrow p, \qquad \Lambda^-(q') \leftrightarrow -\Lambda^+(p) = -\Lambda^-(-p)$$

(compare § 3.3(k)). Thus the net effect is a change of sign.

The result of the manipulations described above may be summarised by the following two rules:

(1) make the necessary substitutions of the four-momenta;

(2) if an odd number of fermion (or antifermion) lines have undergone substitutions, change the sign of the trace.

Applications of the substitution law to specific problems will be made in §§ 11.2(b) and (c). Their use in dispersion relations will be discussed in § 10.4(c).

8.7. PARTICLE INTERACTIONS AND COUPLING STRENGTHS

8.7(a). *Introduction*

In order to clarify certain techniques which will be used frequently in subsequent sections, rough calculations will now be made for two well-known interactions. These calculations will also help classify the three main types of interaction. As we have stated in Chapter 1, these are the electromagnetic, weak and strong interactions.

The matrix elements developed in the previous section represent typical expressions for the electromagnetic interaction. The associated transition probabilities will be calculated in detail in Chapter 11. These transitions are characterised by a coupling strength

$$\alpha = \frac{e^2}{4\pi} \equiv \frac{e^2}{4\pi\hbar c} \sim \frac{1}{137}.$$

For example the total cross-section for Compton scattering, which may be calculated from (8.93), gives

$$\sigma_T = \frac{8\pi}{3}\left(\frac{\alpha}{m}\right)^2 \tag{8.96}$$

in the nonrelativistic limit (compare (11.40)). In this expression m represents the mass of the electron; as a further example of the appearance of α we may quote the lifetime for the annihilation of a free electron–positron pair into two photons (11.109)

$$\tau_{2\gamma} = \frac{2}{\alpha^5 m}.$$

The parameter α appears, in fact, in all processes involving photons. As final examples we quote the following useful expressions:

$$\text{Radius of Bohr atom} = \frac{1}{\alpha m} \tag{8.97}$$

$$\text{Compton wavelength} = \frac{1}{m}$$

$$\text{Thompson radius} = \frac{\alpha}{m}$$

Velocity of electron in lowest Bohr orbit $= \alpha$ $(c = 1$ units$)$.

The weak and strong interactions possess coupling strengths which differ considerably from α. We shall derive rough magnitudes for the coupling strengths of these interactions by considering the following two examples:

(1) pion–nucleon scattering;
(2) β-decay of the neutron.

8.7(b). *Pion–nucleon scattering*

There is a formal similarity between the Lagrangians for the basic electromagnetic interaction and that for the pion–nucleon interaction as can be seen when we compare equations (8.51) and (8.90)

$$\mathscr{H}_I = -\mathscr{L}_I = -ie\bar{\psi}\gamma_\mu\psi A_\mu$$

$$\mathscr{H}_I = -\mathscr{L}_I = -ig\bar{\psi}\gamma_5\psi\varphi.$$

Since A and φ have the same dimensions (Chapter 4) g has the dimensions of electric charge. The simplest diagram for pion–nucleon scattering is given in Fig. 8.23, and is obviously equivalent to Fig. 8.21. Using Table 8.1 and equations (8.92) and (8.91) the S-matrix element becomes

$$\langle k'p'|\, S_2(\pi N)\, |kp\rangle$$

$$= \frac{(-i)^I}{V^\beta}\, \theta_P(-1)^{l+n}\, g^n(2\pi)^\alpha\, \delta(k + p - k' - p')\ \text{(graph terms)}$$

$$= \frac{i}{V^2}\, g^2(2\pi)^4\, \delta(k + p - k' - p')\ \text{(graph terms)}$$

$$= i(2\pi)^4\, \frac{g^2}{V^2}\, \delta(k + p - k' - p')\left(\frac{M^2}{4\omega\omega' EE'}\right)^{\frac{1}{2}} \bar{u}(\mathbf{p}')\, \gamma_5\, \frac{i\gamma q - M}{q^2 + M^2}\, \gamma_5 u(\mathbf{p}) \tag{8.98}$$

where we have substituted g for e, and M represents the nucleon mass. Strictly speaking, we should also consider a graph with the pion lines crossed and its

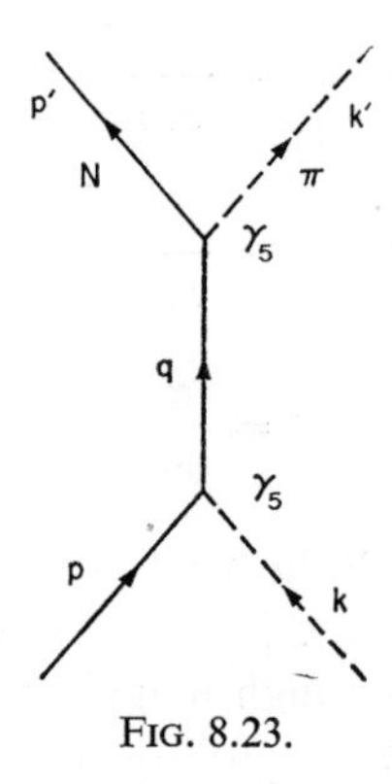

FIG. 8.23.

associated expression (compare Fig. 8.20 and equation (8.93)), but we intend to make an order of magnitude calculation, and so the above expression is adequate.

The propagator in the above expression may be considerably simplified. First, we write

$$\gamma_5(i\gamma q - M)\gamma_5 = -i\gamma q - M$$

and, secondly, we will work in the c-system so that in the notation of § 7.4(c)

$$\mathbf{k} = -\mathbf{p}, \qquad p_c^f = p_c^i = |\mathbf{k}|$$

hence

$$\omega = \omega', \qquad E = E'$$

$$q = (p + k) = i(E + \omega)$$

$$i\gamma q = i\gamma(p + k) = -\gamma_4(E + \omega).$$

Now let us make the further assumption that we are working in the nonrelativistic limit so that

$$E \sim M \gg \omega$$

and since only the large components in the spinor are important

$$\gamma_4 u \sim u. \tag{8.99}$$

The last relation may be easily verified with the aid of equations (3.131) and (3.103). Thus we finally obtain

$$\bar{u}\gamma_5 \frac{i\gamma q - M}{q^2 + M^2} \gamma_5 u = \bar{u} \frac{\gamma_4(E + \omega) - M}{-(E + \omega)^2 + M^2} u \sim -\frac{\bar{u}u}{2E}.$$

Thus the matrix element (8.98) becomes

$$\langle k'p'|\, S_2(\pi N)\, |kp\rangle \sim -i(2\pi)^4 \frac{g^2}{V^2} \delta(k + p - k' - p') \frac{1}{2\omega} \frac{\bar{u}u}{2E}.$$

If this expression is compared with (7.21)

$$\begin{aligned} S_{fi} &= \delta_{fi} + i(2\pi)^4 \, \delta(p_i - p_f) \, M_{fi} \\ &= i(2\pi)^4 \, \delta(p_i - p_f) \, M_{fi} \qquad (f \neq i) \end{aligned} \tag{8.100}$$

it is evident that

$$M_{fi} \sim -\frac{g^2}{V^2} \frac{1}{2\omega} \frac{\bar{u}u}{2E}$$

which may be inserted in equation (7.72)

$$\frac{d\sigma}{d\Omega} = \frac{V^4}{(2\pi)^2} \frac{\omega E \omega' E'}{E_c^2} \frac{p_c^f}{p_c^i} \overline{\sum_i} \sum_\alpha |M_{fi}|^2$$

to yield a cross-section. For our present purposes we may write

$$\overline{\sum_i} \sum_\alpha |\bar{u}u|^2 \sim 1$$

and so

$$\frac{d\sigma}{d\Omega} \sim \frac{V^4}{(2\pi)^2} \frac{\omega^2 E^2}{(\omega + E)^2} \frac{g^4}{V^4} \frac{1}{16\omega^2 E^2} \sim \frac{g^4}{(4\pi)^2} \frac{1}{4M^2}. \tag{8.101}$$

Thus the total cross-section for pion–nucleon scattering might be expected to be

$$\sigma_T \sim \frac{g^4}{(4\pi)^2} \frac{4\pi}{4M^2}. \tag{8.102}$$

We may compare this result with that for Compton scattering (8.96)

$$\sigma_T = \frac{8\pi}{3} \left(\frac{\alpha}{m}\right)^2.$$

Now the measured cross-sections for pion–nucleon scattering and for Compton scattering from an electron are about the same, hence

$$\frac{g^2}{4\pi} \sim \alpha \frac{M}{m} = \frac{e^2}{4\pi} \frac{M}{m} \sim 10. \tag{8.103}$$

More exact approaches show that $g^2/4\pi \sim 14$ (§ 10.3(g)). The coupling strength is often written in the form

$$\frac{f^2}{4\pi} \equiv \frac{f^2}{4\pi\hbar c} = \frac{g^2}{4\pi\hbar c} \left(\frac{m_\pi}{2M}\right)^2 \sim 0{\cdot}08. \tag{8.104}$$

It should be noted that the theory given above has led to a dominance of *s*-wave scattering, since the spin terms are only associated with the small components of the Dirac spinors (3.136), and these have been neglected; it is these spin terms which lead to *p*-wave effects. In contrast, experiment shows a dominant *p*-wave process just above threshold (Chapter 14), and so the theory given above predicts the wrong results. It is not improved by more careful calculation and higher order terms. Nevertheless, it provides a useful introduction to the order of magnitude of the coupling strength in the pion–nucleon problem.

8.7(c). *The β-decay of the neutron*

The β-decay process

$$n \to p + e^- + \bar{\nu}$$

involves four fermion fields, and so the Lagrangian requires at least four terms. It was shown originally by Fermi that the problem is more easily handled as

$$\nu + n \to p + e^-$$

that is, the substitution law is used. The simplest possible Lagrangian containing the requisite operators is of the form

$$\mathcal{L} = C\bar{\psi}_p \psi_n \bar{\psi}_e \psi_\nu. \tag{8.105}$$

More complex terms are possible and these will be discussed in § 9.2(b); the one given above is adequate for an order of magnitude approach. In contrast to the roblems of Compton scattering and pion–nucleon scattering, which required cond order terms in the S-matrix expansion in order to ensure overall conrvation of energy and momentum, β-decay can be treated in first order ap-

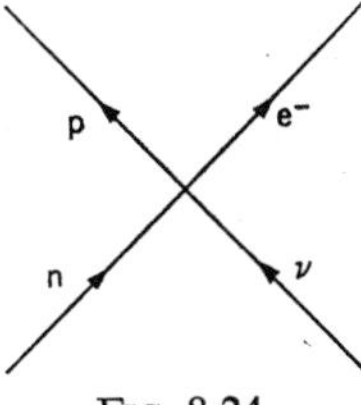

FIG. 8.24.

roximation and hence the relevant Feynman diagram (Fig. 8.24) has only a ngle vertex. Thus the parameters appearing in (8.91) have the following values:

n = number of vertices = 1,
E = number of external lines = 4,
I = number of internal lines = 0,
$\alpha = 4n - 4I = 4$,
$\beta = \frac{1}{2}E = 2$.

Ience by (8.91) and (8.92)†

$$\langle f|\, S_1\, |i\rangle = i(2\pi)^4 \frac{C}{V^2}\, \delta(p_i - p_f) \text{ (graph terms)}. \tag{8.106}$$

he graph terms are all of the type $\sqrt{(m/E)}\, u$, and we will assume they contribute actors of order unity. This may seem at first nonsensical in the case of the neuino where $m_\nu = 0$, but if we assume the polite fiction that m_ν has a small but nite value, then the method of normalisation employed for u causes the m term o vanish (compare (7.80) and (7.81)), and so no problem arises. This procedure vill be demonstrated in greater detail in § 12.2(b).

Thus, if the graph terms are of order unity, the matrix element M_{fi} of equaion (7.21) is

$$M_{fi} \sim \frac{C}{V^2} \tag{8.107}$$

ince $\delta_{fi} = 0$ in (7.21) for $f \neq i$. The lifetime of the neutron is then given by 7.39) and (7.41)

$$\frac{1}{\tau_n} = V(2\pi)^4 \sum_i \sum_f \frac{C^2}{V^4} \frac{V^3}{(2\pi)^9} \prod_{j=1}^{3} d\mathbf{p}_{fj}\, \delta(p_i - p_f)$$

$$= \frac{C^2}{(2\pi)^5} \int d\mathbf{p}_e\, d\mathbf{p}_\nu\, d\mathbf{p}_p\, \delta(p_i - p_f) \tag{8.108}$$

† No i was used in the Lagrangian (8.105) and consequently i appears in the S-matrix by virtue f (8.50).

where we have again assumed that the averaging and summing over spin states i of order unity. Since the mass of the proton is much greater than that of th electron or neutrino we may use (7.66) to evaluate the momentum space integra

$$\int d\mathbf{p}_e\, d\mathbf{p}_\nu\, d\mathbf{p}_p\, \delta(p_i - p_f) = 4\pi \int d\mathbf{p}_e\, E_\nu p_\nu. \tag{8.109}$$

Let us denote the difference in mass of the neutron and proton by Δ, then since the proton is virtually stationary and the neutrino is massless, the abov integral becomes

$$4\pi \int d\mathbf{p}_e(\Delta - E_e)^2 = (4\pi)^2 \int_m^\Delta dE_e E_e p_e(\Delta - E_e)^2$$

where m is the electron mass and we have used the relations $E\, dE = p\, dp$ an $p_e = |\mathbf{p}_e|$. If we set $E_e = \eta\Delta$ we may rewrite the integral as

$$(4\pi)^2 \Delta^5 \int_{m/\Delta}^{1} d\eta\eta \sqrt{\left(\eta^2 - \frac{m^2}{\Delta^2}\right)}(1 - \eta)^2. \tag{8.110}$$

In the limit $(m/\Delta) \to 0$ this expression becomes

$$(4\pi)^2 \Delta^5 \int_0^1 d\eta\, \eta^2(1 - \eta)^2 = (4\pi)^2 \frac{\Delta^5}{30}$$

and so

$$\frac{1}{\tau_n} = \frac{C^2}{(2\pi)^5} \frac{(4\pi)^2}{30} \Delta^5 = \frac{C^2}{60\pi^3} \Delta^5.$$

If the integration of equation (8.110) is performed exactly a 'blocking factor' o 0·47 must be included

$$\frac{1}{\tau_n} = \frac{0{\cdot}47}{60\pi^3} C^2 \Delta^5. \tag{8.111}$$

This relationship is in units $\hbar = c = 1$; it is apparent from the presence of th energy term Δ^5 that C^2 cannot be dimensionless. It is frequently convenient t express C^2 in a dimensionless combination. This may be done by introducing th 'natural life' of the nucleon as a convenient unit (M is the nucleon mass)

$$\tau_0 = \frac{1}{M} \equiv \frac{\hbar}{Mc^2} \sim 10^{-24} \text{sec}.$$

The lifetime for the β-decay of the neutron is $\sim 10^3$ sec, and so

$$\frac{\tau_0}{\tau_n} \sim 10^{-27} = \frac{C^2}{60\pi^3} \frac{\Delta^5}{M} = \frac{C^2 M^4}{60\pi^3} \left(\frac{\Delta}{M}\right)^5;$$

since $\Delta \sim 1{\cdot}3$ MeV and $M \sim 1000$ MeV we find

$$CM^2 \sim 10^{-5}. \tag{8.112}$$

If C is expressed in c.g.s. units it has dimensions erg cm^3, since $\mathscr{L}$ and ψ have dimensions of erg cm^{-3} and cm$^{-3/2}$ respectively in (8.105) (compare § 4.5(a)). It is then a simple matter to show that $C \sim 10^{-49}$ erg cm^3. This relationship can written as

$$C \sim 10^{-49} \text{ erg cm}^3 \sim 10^{-5}\, \hbar c \left(\frac{\hbar}{Mc}\right)^2 \sim 10^{-7}\, \hbar c \left(\frac{\hbar}{m_\pi c}\right)^2. \quad (8.113)$$

Since the probability of β-decay is proportional to C^2 a reasonable figure for the dimensionless coupling strength of a weak interaction is $\sim 10^{-12}$.

8.7(d). *The particles associated with the coupling terms*

Either of the values $C^2M^4 \sim 10^{-10}$ or $(g^2/4\pi) \sim 10$ could have been duplicated by examining other interactions in elementary particle physics. For example the lifetime of the process $\mu \to e + \nu + \bar{\nu}$ is $\sim 10^{-6}$ sec, that is $\sim 10^{-9}$ times shorter than that of the neutron, but when the kinematic terms are removed a value of $C^2M^4 \sim 10^{-10}$ is again found. In a similar manner an analysis of the reaction

$$\pi^- + p \to \Sigma^- + K^+$$

produces a value

$$\frac{g^2}{4\pi} \sim 10.$$

An examination of many reactions reveals the following rule relating particles and interactions. All interactions are strong unless

(1) photons are present (electromagnetic interactions),

(2) neutrinos are present and/or a change in strangeness occurs (weak interactions).

It is, of course, possible to have reactions which involve a combination of the interactions, for example $\gamma + p \to n + \pi^+$.

It should be noted that all strong interactions do not appear to have exactly the same coupling strength; a similar remark applies to weak interactions.

8.8. FIELD THEORY—A REFORMULATION

8.8(a). *Introduction*

It is apparent from the considerations of the previous section that a perturbation theory for the S-matrix might work satisfactorily for weak and electromagnetic interactions, but the presence of the developement parameter $(g^2/4\pi) \sim 10$ obviously causes serious trouble for strong interactions, since succeeding terms in the perturbation expansion involve increasing powers of $g^2/4\pi$.

An alternative approach is therefore required. The failures and successes of field theory have led to a critical re-assessment of its basic tenets in recent years.

Certain features have been retained and emphasized, for example symmetry properties; other features, for example equations of motion and Hamiltonians, are avoided whenever possible. The attempt to reformulate field theory, in a manner which is rigorously suitable in all problems, has arisen from the work of Haag (1955), Lehmann, Symanzik and Zimmerman (1955, 1957) and Wightman (1956). Their formulation is often called *the axiomatic approach to field theory*. This formulation has led to important results, notably rigorous proofs of the *CPT* theorem (Jost, 1957) and the spin–statistics relation. It is also used as a basis for the application of dispersion relations in relativistic field theory and for the discussion of the general properties and structure of the transition amplitude. In this formalism it is the interacting field operators which are examined, and not some approximation to them which is reached by the methods of perturbation theory.

However, it should be emphasised from the start that the formalism possesses no great advantages over the Hamiltonian–Lagrangian method if specific solutions of the scattering problem are being sought. In fact if one wants certain types of answers to problems then perturbation theory is still the only method available.

8.8(b). *The axiomatic formulation of field theory – the 'in' and 'out' operators*

It is customary to use the Heisenberg representation. We recall from § 4.1 (f.2) that in this representation the operators are time dependent whilst the state vectors describing physical systems are constant and have a positive definite metric.

Now let us consider a physical system in which an interaction occurs. We start with a set of isolated (and therefore non-interacting) particles which come together, interact, and then separate again into a set of noninteracting particles. The sets may not necessarily be the same. If we introduce a time scale and locate the interaction in the region of $t \sim 0$, then the initial and final sets are at $t = -\infty$ and $t = +\infty$ respectively. The system we have just described is obviously time dependent, and in the Heisenberg representation the time dependence lies in the operators. We must, therefore, formulate our description of the system in operator language. We therefore introduce for each type of particle two fields to describe its properties in the initial and final systems – an *incoming field* φ_{in} and an *outgoing field* φ_{out} (for simplicity we consider neutral spinless partices here). The fields are regarded as free, and obey the Klein–Gordon equation and the free field commutation relations (compare (4.235)).

$$(\square^2 - m^2)\,\varphi_{\substack{\text{in}\\\text{out}}} = 0 \tag{8.114}$$

$$\left[\varphi_{\substack{\text{in}\\\text{out}}}(x),\ \varphi_{\substack{\text{in}\\\text{out}}}(x')\right] = i\Delta(x - x').$$

The mathematical properties of the Δ-function were given in § 4.6(b). The fields are also subject to the usual symmetry laws (Chapters 5 and 9).

The incoming and outgoing particles associated with the fields can be given terms of annihilation and creation operators as before (compare (4.72))

$$\varphi_{\substack{\text{in}\\\text{out}}}(x) = \frac{1}{\sqrt{V}} \sum_{\mathbf{k}} \frac{1}{\sqrt{(2\omega_k)}} \left[a_{\substack{\text{in}\\\text{out}}}(\mathbf{k})\, e^{ikx} + a^{\dagger}_{\substack{\text{in}\\\text{out}}}(\mathbf{k})\, e^{-ikx}\right] \tag{8.115}$$

$$\varphi(x) = \frac{1}{\sqrt{V}} \sum_{\mathbf{k}} \frac{1}{\sqrt{(2\omega_k)}} \left[a(\mathbf{k}, t)\, e^{ikx} + a^{\dagger}(\mathbf{k}, t)\, e^{-ikx}\right].$$

hese equations are equivalent to the usual expansion of fields in plane waves; e field $\varphi(x)$ is known as the interpolating field, its function will become apparent ter. The creation and destruction operators obey the commutation relations

$$\left[a_{\substack{\text{in}\\\text{out}}}(\mathbf{k}),\ a^{\dagger}_{\substack{\text{in}\\\text{out}}}(\mathbf{k}')\right] = [a(\mathbf{k}, t), a^{\dagger}(\mathbf{k}', t)] = \delta_{\mathbf{kk}'}. \tag{8.116}$$

The equations (8.115) yield particle states of the form

$$|i_{\text{in}}\rangle = |\mathbf{k}_1\mathbf{k}_2 \ldots \mathbf{k}_{n\ \text{in}}\rangle = a^{\dagger}_{\text{in}}(\mathbf{k}_1) \ldots a^{\dagger}_{\text{in}}(\mathbf{k}_n)\ |0_{\text{in}}\rangle \tag{8.117}$$

$$|f_{\text{out}}\rangle = |\mathbf{k}'_1\mathbf{k}'_2 \ldots \mathbf{k}_{m\ \text{out}}\rangle = a^{\dagger}_{\text{out}}(\mathbf{k}'_1) \ldots a^{\dagger}_{\text{out}}(\mathbf{k}'_m)\ |0_{\text{out}}\rangle$$

here the conditions

$$a_{\text{in}}\ |0_{\text{in}}\rangle = 0, \qquad a_{\text{out}}\ |0_{\text{out}}\rangle = 0 \tag{8.118}$$

efine the vacuum states.

The expansion of the field in plane waves is not always convenient for specific oblems. We therefore recall the wave functions $f_\alpha(x)$ which were introduced in uation (4.106)

$$\varphi(x) = \sum_{\alpha} [a_\alpha f_\alpha(x) + a^{\dagger}_\alpha f^*_\alpha(x)]$$

here

$$[a_\alpha, a^{\dagger}_\beta] = \delta_{\alpha\beta} \tag{8.119}$$

th all other commutators vanishing, and where $f_\alpha(x)$ is normalised as in uation (4.107)

$$i \int d\mathbf{x} \left(f^*_\alpha \frac{\partial f_\beta}{\partial t} - \frac{\partial f_\alpha}{\partial t} f^*_\beta \right) = \delta_{\alpha\beta}.$$

he integrand of this equation is frequently written as

$$f^*_\alpha \overleftrightarrow{\frac{\partial}{\partial t}} f_\beta = f^*_\alpha \frac{\partial f_\beta}{\partial t} - \frac{\partial f^*_\alpha}{\partial t} f_\beta \tag{8.120}$$

that (4.107) becomes

$$i \int d\mathbf{x}\, f^*_\alpha \overleftrightarrow{\frac{\partial}{\partial t}} f_\beta = \delta_{\alpha\beta}. \tag{8.121}$$

The functions $f_\alpha(x)$ satisfy the Klein–Gordon equation

$$(\Box^2 - m^2) f_\alpha(x) = 0 \quad (8.12$$

and in the plane wave representation can be written as

$$f_\alpha(x) = f_k(x) = \frac{1}{\sqrt{V}} \frac{e^{ikx}}{\sqrt{(2\omega_k)}}, \qquad f_k^*(x) = \frac{1}{\sqrt{V}} \frac{e^{-ikx}}{\sqrt{(2\omega_k)}}. \quad (8.12$$

The functions $f_\alpha(x)$ are also sometimes given without the square root sign on 2 (Lehmann, 1959); equation (8.119) then becomes

$$[a_\alpha, a_\beta^\dagger] = 2\omega_k \delta_{\alpha\beta}. \quad (8.12$$

Finally, using the notation of equations (8.120) and (8.123), we may note that t creation and destruction operators can be written as

$$a_\alpha^\dagger = i \int d\mathbf{x}\, \varphi(x) \frac{\overleftrightarrow{\partial}}{\partial t} f_\alpha(x) = -i \int d\mathbf{x}\, f_\alpha(x) \frac{\overleftrightarrow{\partial}}{\partial t} \varphi(x) \quad (8.12$$

$$a_\alpha = -i \int d\mathbf{x}\, \varphi(x) \frac{\overleftrightarrow{\partial}}{\partial t} f_\alpha^*(x) = i \int d\mathbf{x}\, f_\alpha^*(x) \frac{\overleftrightarrow{\partial}}{\partial t} \varphi(x)$$

so that a single particle state can be written as

$$|\mathbf{k}_{\text{in}}\rangle = a_{\text{in}}^\dagger(\mathbf{k})\, |0_{\text{in}}\rangle = i \int d\mathbf{x}\, \varphi_{\text{in}}(x) \frac{\overleftrightarrow{\partial}}{\partial t} f_k(x)\, |0_{\text{in}}\rangle \quad (8.12$$

(compare (8.117) and (4.84)). These equations apply to both 'in' and 'out' fi operators.

8.8(c). *The axiomatic formulation – definitions*

As we have mentioned earlier, the general interaction problem involves initial configuration of states at time $t = -\infty$ and a second at time $t = +\infty$ These states are the ones detected by an experimental system, and can be describ by the 'in' and 'out' systems. The configurations can be related by the followi definitions.

(1) There exist states with positive definite metric in Hilbert space (§ 4.1(b which can be constructed by applying the free field operators φ_{in} and φ_{out} to t vacuum. Thus the state $|i_{\text{in}}\rangle$ describes a noninteracting set of particles at ti $t = -\infty$, and $|f_{\text{out}}\rangle$ a system at $t = +\infty$

$$|i_{\text{in}}\rangle = |\mathbf{k}_1 \mathbf{k}_2 \ldots \mathbf{k}_{n\,\text{in}}\rangle = a_{\text{in}}^\dagger(\mathbf{k}_1) \ldots a_{\text{in}}^\dagger(\mathbf{k}_n)\, |0_{\text{in}}\rangle$$

$$|f_{\text{out}}\rangle = |\mathbf{k}_1' \mathbf{k}_2' \ldots \mathbf{k}'_{m\,\text{out}}\rangle = a_{\text{out}}^\dagger(\mathbf{k}_1') \ldots a_{\text{out}}^\dagger(\mathbf{k}_m')\, |0_{\text{out}}\rangle.$$

These states will differ in general, $|i_{\text{in}}\rangle$ corresponds in ordinary scattering ıeory to states with plane plus outgoing spherical waves, whilst $|f_{\text{out}}\rangle$ corresponds to plane plus incoming spherical waves.

(2) The incoming and outgoing systems form complete sets of states

$$\sum_{\alpha} |\alpha_{\substack{\text{in}\\\text{out}}}\rangle \langle\alpha_{\substack{\text{in}\\\text{out}}}| = \hat{1} \quad (\alpha = i, f) \tag{8.127}$$

:ompare equation (4.15)).

(3) The 'in' and 'out' vacuum states are identical (if an appropriate choice of hase factors is made)

$$|0_{\text{in}}\rangle = |0_{\text{out}}\rangle = |0\rangle. \tag{8.128}$$

Arising from this condition is the result that the matrix elements of $\varphi(x)$ beveen the vacuum and the in and out states for single particles are identical

$$\langle 0|\, \varphi(x)\, |\alpha_{\text{in}}\rangle = \langle 0|\, \varphi(x)\, |\alpha_{\text{out}}\rangle = f_{\alpha}(x). \tag{8.129}$$

'his result follows from the fact that the matrix elements depend on x only ırough $f_{\alpha}(x)$; equation (8.129) can be easily proved if the plane wave representtion (8.115) and (8.116) is used. Thus equation (8.129) implies that no essential ifference exists for the in and out states for a single isolated particle

$$|\mathbf{k}_{\text{in}}\rangle = |\mathbf{k}_{\text{out}}\rangle = |\mathbf{k}\rangle. \tag{8.130}$$

'his result could have been reached by physical argument alone.

(4) The transition amplitude is described by means of the S-matrix operator. 'he operator may be defined as

$$S = \sum_{\alpha} |\alpha_{\text{in}}\rangle \langle\alpha_{\text{out}}| \quad (\alpha = i, f) \tag{8.131}$$

vhere the presence of the symbol α in both in and out states implies that the ame particles comprise the two states. The S-matrix operator defined above is ınitary, since

$$\begin{aligned} S^{\dagger}S &= \sum_{\alpha,\beta} |\alpha_{\text{out}}\rangle \langle\alpha_{\text{in}} \mid \beta_{\text{in}}\rangle \langle\beta_{\text{out}}| \\ &= \sum_{\alpha,\beta} |\alpha_{\text{out}}\rangle\, \delta_{\alpha\beta}\, \langle\beta_{\text{out}}| \\ &= \sum_{\alpha} |\alpha_{\text{out}}\rangle \langle\alpha_{\text{out}}| = \hat{1} \end{aligned} \tag{8.132}$$

›y (8.127), similarly,

$$SS^{\dagger} = \hat{1}.$$

The S-matrix element may be written as

$$\begin{aligned} S_{fi} &= \langle f_{\text{in}}|\, S\, |i_{\text{in}}\rangle \\ &= \sum_{n} \langle f_{\text{in}} \mid n_{\text{in}}\rangle \langle n_{\text{out}} \mid i_{\text{in}}\rangle \\ &= \sum_{n} \delta_{fn} \langle n_{\text{out}} \mid i_{\text{in}}\rangle \\ &= \langle f_{\text{out}} \mid i_{\text{in}}\rangle \end{aligned} \tag{8.133}$$

and

$$\langle f_{\text{out}}| = \sum_i \langle f_{\text{out}} \mid i_{\text{in}}\rangle \langle i_{\text{in}}| = \sum_i S_{fi} \langle i_{\text{in}}|. \tag{8.13}$$

It should be noted that although the S-matrix elements have been given th same formal structure in both interaction and Heisenberg representations

$$S_{fi} = \langle f| S |i\rangle$$

the vectors $|f\rangle$ and $|i\rangle$ change with time in the interaction picture.

The in and out fields may be related through the S-matrix. Using equation (8.131) and (8.127) we may write

$$\begin{aligned} S\varphi_{\text{out}} &= \sum_\alpha \sum_\beta |\alpha_{\text{in}}\rangle \langle \alpha_{\text{out}}| \varphi_{\text{out}} |\beta_{\text{out}}\rangle \langle \beta_{\text{out}}| \\ &= \sum_\alpha \sum_\beta |\alpha_{\text{in}}\rangle \langle \alpha_{\text{in}}| \varphi_{\text{in}} |\beta_{\text{in}}\rangle \langle \beta_{\text{out}}| \\ &= \varphi_{\text{in}} S. \end{aligned} \tag{8.135}$$

The unitarity of the S-operator then gives us the relation

$$\varphi_{\text{out}} = S^\dagger \varphi_{\text{in}} S. \tag{8.136}$$

This equation is frequently used as a definition of the S-matrix operator.

(5) An *interpolating operator* $\varphi(x)$ exists which equals $\varphi_{\text{in}}(x)$ and $\varphi_{\text{out}}(x)$ fo $t \to -\infty$ and $t \to +\infty$ respectively. Thus the operators a and $a^\dagger$ in (8.115) ca be written as

$$\lim_{t\to\pm\infty} a(\mathbf{k}, t) = a_{\substack{\text{out}\\ \text{in}}}(\mathbf{k}) \tag{8.137}$$

$$\lim_{t\to\pm\infty} a^\dagger(\mathbf{k}, t) = a^\dagger_{\substack{\text{out}\\ \text{in}}}(\mathbf{k})$$

and for two arbitrary states $|\beta\rangle$ and $|\gamma\rangle$

$$\lim_{t\to\pm\infty} \langle\gamma| a(\mathbf{k}, t) |\beta\rangle = \langle\gamma| a_{\substack{\text{out}\\ \text{in}}}(\mathbf{k}) |\beta\rangle. \tag{8.138}$$

This relation is called the *asymptotic condition*. The following useful relation ca be associated with it

$$\int_{-\infty}^{+\infty} dt \frac{\partial}{\partial t} \langle\alpha| a(\mathbf{k}, t) |\beta\rangle = \langle\alpha| a_{\text{out}}(\mathbf{k}) |\beta\rangle - \langle\alpha| a_{\text{in}}(\mathbf{k}) |\beta\rangle. \tag{8.139}$$

(6) Invariance under the inhomogenous Lorentz transformation (5.30)

$$x_\alpha \to x'_\lambda = a_\lambda + a_{\lambda\alpha} x_\alpha$$

s postulated. This postulate requires the existence of a unitary operator U for both the in and out fields.

$$\varphi_{\substack{\text{in}\\ \text{out}}}(x') = U\varphi_{\substack{\text{in}\\ \text{out}}}(x)\, U^{-1}. \tag{8.140}$$

Thus the equations (4.115) and (4.35) (in suitable form)

$$i\frac{\partial \varphi}{\partial x_\lambda} = [P_\lambda, \varphi(x)] \tag{8.141}$$

$$\varphi(x + a) = e^{-iPa}\, \varphi(x)\, e^{iPa}$$

apply equally well to φ_{in} and φ_{out}.

(7) A unique vacuum state $|0\rangle$ exists which is invariant under the Lorentz group (§ 5.3)

$$U\,|0\rangle = |0\rangle. \tag{8.142}$$

This relation implies that

$$P_\lambda\,|0\rangle = 0, \qquad M_{\lambda\alpha}\,|0\rangle = 0 \tag{8.143}$$

where $M_{\lambda\alpha}$ is the angular momentum operator (5.41).

(8) The expectation value of the energy operator H is positive for all states except the vacuum

$$\langle\alpha|\, H\,|\alpha\rangle \geqq 0.$$

This postulate implies that the expectation value of the energy–momentum operator is a time-like vector (§ 3.2(d)).

(9) The interpolating field operators $\varphi(x)$ commute at space-like distances (compare (4.233))

$$[\varphi(x), \varphi(x')] = 0 \text{ if } (x - x')^2 > 0. \tag{8.144}$$

This is known as *local commutivity* or *causality.*

It can be seen that many of the above postulates have appeared in various forms in earlier chapters. We may summarise the above postulates by stating that a field theory should satisfy the following basic requirements:

(1) positive definite metric in Hilbert space,

(2) unitarity,

(3) asymptotic condition,

(4) Lorentz invariance,

(5) positive energies,

(6) causality.

8.9 THE REDUCTION OF THE S-MATRIX ELEMENT

8.9(a). *Introduction*

We shall now consider an interaction in which a neutral scalar (or pseudoscalar) particle with four-momentum k collides with a particle of four-momentum p in the initial (i) state to lead to final (f) states with momenta k' and p' respectively (Fig. 8.25). With appropriate modifications this idealised interaction could be used to describe the scattering of mesons by nucleons.

Previously a similar interaction was described by a Lagrangian (8.90). This is now unnecessary; we shall, however, formally define an interaction current $j(x)$ by means of an inhomogenous Klein–Gordon equation

$$(\Box^2 - m^2)\,\varphi(x) = -j(x). \tag{8.145}$$

This equation has the same formal structure as that for the electromagnetic interaction (§ 8.3(a)). It leads to a simpler notation in later sections. It should be noted that the matrix elements of the currents defined in (8.145) vanish between vacuum and single particle states. We may show this by using equation (4.115)

$$\frac{\partial\varphi(x)}{\partial x_\lambda} = -i[P_\lambda, \varphi(x)]$$

and the relation

$$P_\lambda\,|k\rangle = k_\lambda\,|k\rangle$$

so that

$$\Box^2\varphi(x) = -[P_\lambda, [P_\lambda, \varphi(x)]]$$

$$\langle 0|\,\Box^2\varphi(x)\,|k\rangle = -\langle 0|\,[P_\lambda, [P_\lambda, \varphi(x)]\,|k\rangle$$

$$= -k^2\,\langle 0|\,\varphi(x)\,|k\rangle = m^2\,\langle 0|\,\varphi(x)\,|k\rangle.$$

Thus we find

$$\langle 0|\,(\Box^2 - m^2)\,\varphi(x)\,|k\rangle = -\langle 0|\,j(x)\,|k\rangle = 0. \tag{8.146}$$

8.9(b). *The S-matrix in terms of retarded commutators*

The S-matrix describing the process shown in Fig. 8.25

$$\underbrace{k + p}_{i} \to \underbrace{k' + p'}_{f}$$

can be written in the following manner by using (8.133):

$$S_{fi} = \langle f_{\text{out}}\,|\,i_{\text{in}}\rangle = \langle p'k'_{\text{out}}\,|\,pk_{\text{in}}\rangle. \tag{8.147}$$

Our task is to write this equation in a form suitable for dispersion relations. It was shown by Low (1955) and by Lehmann, Symanzik and Zimmermann (1955) that the state vectors in the S-matrix can be rewritten in terms of operators. This is known as the reduction technique; we follow the description of it given by Lehmann (1959) and by Jackson (1961).

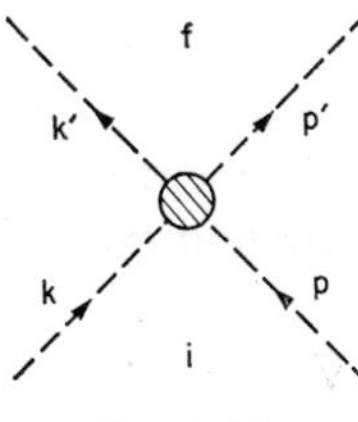

FIG. 8.25.

Using equations (8.117) and (8.139) we can write

$$S_{fi} = \langle p'k'_{\text{out}} \mid pk_{\text{in}}\rangle = \langle p'|\, a_{\text{out}}(\mathbf{k}')\,|pk_{\text{in}}\rangle$$

$$= \langle p'|\, a_{\text{in}}(\mathbf{k}')\,|pk_{\text{in}}\rangle + \int_{-\infty}^{+\infty} dt\,\frac{\partial}{\partial t}\,\langle p'|\, a(\mathbf{k}', t)\,|pk_{\text{in}}\rangle. \qquad (8.148)$$

Where we have written the single particle state $\langle p'_{\text{out}}|$ as $\langle p'|$ by virtue of (8.130). Now the term $\langle p'|\, a_{\text{in}}(\mathbf{k}')\,|pk_{\text{in}}\rangle$ can be written as

$$\langle p'k'_{\text{in}}|\, pk_{\text{in}}\rangle = \delta_{p'p}\delta_{k'k} = \delta_{fi} \qquad (8.149)$$

so that (8.148) becomes

$$S_{fi} - \delta_{fi} = -i\int d^4x\,\frac{\partial}{\partial t}\left[\langle p'|\,\varphi(x)\,|pk_{\text{in}}\rangle\,\frac{\overleftrightarrow{\partial}}{\partial t}\,f^*_{k'}(x)\right] \qquad (8.150)$$

where we have used (8.125) and written $d\mathbf{x}\,dt$ as d^4x.

We shall now consider the behaviour of the term on the right-hand side of the equation

$$\frac{\partial}{\partial t}\left[\varphi(x)\,\frac{\overleftrightarrow{\partial}}{\partial t}\,f^*_{k'}(x)\right] = \frac{\partial}{\partial t}\left[\varphi\,\frac{\partial f^*}{\partial t} - \frac{\partial \varphi}{\partial t}\,f^*\right]$$

$$= \varphi\,\frac{\partial^2 f^*}{\partial t^2} - \frac{\partial^2 \varphi}{\partial t^2}\,f^*. \qquad (8.151)$$

Now the term f^* satisfies the Klein–Gordon equation (8.122)

$$(\Box^2 - m^2)\,f^* = \left(\nabla^2 - \frac{\partial}{\partial t^2} - m^2\right)f^* = 0$$

so that (8.151) becomes

$$-\frac{\partial^2 \varphi}{\partial t^2} f^* + \varphi(\nabla^2 - m^2) f^*. \tag{8.152}$$

This relation must be integrated over all space-time in (8.150). Consider first the middle part of the integral from the above expression. If we make a double integration by parts of the spatial component and make the usual assumption that the fields vanish at infinity, then we find that

$$\int_{-\infty}^{+\infty} d^4x\, \varphi \nabla^2 f^* = \int_{-\infty}^{+\infty} d^4x\, f^* \nabla^2 \varphi. \tag{8.153}$$

Thus the right side of equation (8.150) becomes

$$-i \int d^4x \left[-\frac{\partial^2 \varphi}{\partial t^2} f^* + f^* \nabla^2 \varphi - m^2 \varphi f^* \right]$$

$$= -i \int d^4x\, [f^*(\Box^2 - m^2)\, \varphi] = i \int d^4x\, f^* j(x) \tag{8.154}$$

where we have ignored the state vectors for purposes of simplicity. Upon reinserting them the equation for the S-matrix element (8.150) becomes

$$S_{fi} = \delta_{fi} + i \int d^4x\, f^*_{k'}(x) \langle p' \,|\, j(x) \,|\, pk_{\text{in}} \rangle. \tag{8.155}$$

A comparison of this equation with (8.147)

$$S_{fi} = \langle p'k'_{\text{out}} |\, pk_{\text{in}} \rangle$$

shows that we have converted a state vector into a current operator. This process may be continued and we next take a meson from the 'in' state vector in (8.155)

$$\langle p' |\, j(x) \,|\, pk_{\text{in}} \rangle = \langle p' |\, j(x)\, a^\dagger_{\text{in}}(\mathbf{k}) \,|\, p \rangle. \tag{8.156}$$

Although we wish to proceed as before some extra care must now be taken over the time variables since x is integrated over all time in (8.155). This can be done by using either the chronological product (4.254) or a retarded commutator.† The latter expression is more suitable for the imposition of the causality requirement. It is defined as

$$R[A(x)\, B(y)] = -i\theta(t_x - t_y)\, [A(x), B(y)] \tag{8.157}$$

where $A(x)$ and $B(y)$ are two Heisenberg operators and

$$\theta(t_x - t_y) = \begin{cases} 1 & (t_x > t_y) \\ 0 & (t_x < t_y) \end{cases}. \tag{8.158}$$

† It can be shown that the chronological and retarded products both lead to the same matrix elements in the physical region of the scattering matrix.

The $-i$ term is present in (8.157) for convenience, since it causes $R[AB]$ to be Hermitian if A and B are Hermitian.

We can write (8.156) as

$$\begin{aligned}\langle p'|\, j(x)\,|pk_{\text{in}}\rangle &= \langle p'|\, j(x)\, a^{\dagger}_{\text{in}}(\mathbf{k})\,|p\rangle \\ &= \langle p'|\, [j(x), a^{\dagger}_{\text{in}}(\mathbf{k})]\,|p\rangle + \langle p'|\, a^{\dagger}_{\text{in}}(\mathbf{k})\, j(x)\,|p\rangle . \end{aligned} \tag{8.159}$$

The last term in this expression vanishes since

$$\langle p'|\, a^{\dagger}_{\text{in}}(\mathbf{k})\, j(x)\,|p\rangle = \delta_{p'k}\langle 0|\, j(x)\,|p\rangle = 0 \tag{8.160}$$

by virtue of (8.117) and (8.146). Thus (8.159) becomes

$$\begin{aligned}\langle p'|\, j(x)\,|pk_{\text{in}}\rangle &= \lim_{t_y\to -\infty} \langle p'|\, [j(x), a^{\dagger}(\mathbf{k}, t_y)]\,|p\rangle \\ &= \lim_{t_y\to -\infty} i\,\langle p'|\, R[j(x)\, a^{\dagger}(\mathbf{k}, t_y)]\,|p\rangle \\ &= \lim_{t_y\to +\infty} i\,\langle p'|\, R[j(x)\, a^{\dagger}(\mathbf{k}, t_y)]\,|p\rangle \\ &\quad - i\int_{-\infty}^{+\infty} dt_y \frac{\partial}{\partial t_y} \langle p'|\, R[j(x)\, a^{\dagger}(\mathbf{k}, t_y)]\,|p\rangle \end{aligned} \tag{8.161}$$

where we have applied the asymptotic condition (8.139) and have used the fact that we can insert $\theta(t_x - t_y)$ since $t_x > t_y$ in (8.159).

The boundary term at $t_y \to +\infty$ vanishes in (8.161) because of the presence of the retarded commutator, and so with the aid of (8.125) equation (8.161) becomes

$$\langle p'|\, j(x)\,|pk_{\text{in}}\rangle = \int_{-\infty}^{+\infty} d^4y \frac{\partial}{\partial t_y}\left[\langle p'|\, R[j(x)\, \varphi(y)]\,|p\rangle \frac{\overleftrightarrow{\partial}}{\partial t_y} f_{\mathbf{k}}(y)\right]. \tag{8.162}$$

With some work the matrix element can then be manipulated into the form

$$\begin{aligned} S_{fi} = \delta_{fi} - i\int_{-\infty}^{+\infty} d^4x\, d^4y\, f^{*}_{\mathbf{k}'}(x)\, f_{\mathbf{k}}(y)\, (\Box^2_x - m^2)\, (\Box^2_y - m^2) \times \\ \times \langle p'|\, R[\varphi(x)\, \varphi(y)]\,|p\rangle . \end{aligned} \tag{8.163}$$

This is essentially the form of the S-matrix element first given by Low and by Lehmann, Symanzik and Zimmerman; the reduction process can be carried even further if required until the S-matrix becomes a vacuum expectation value (§ 4.6(e)) of a suitable product of field operators (Lehmann, 1959). A more useful formulation of the matrix element for present purposes can be made if we express it in the terms of retarded current operators rather than field operators. We therefore return to (8.162) and examine the integrand (ignoring state vectors for the moment)

$$\frac{\partial}{\partial t_y}\left\{R[j(x)\,\varphi(y)]\,\frac{\overleftrightarrow{\partial}}{\partial t_y}\,f_k(y)\right\}$$

$$= -i\frac{\partial}{\partial t_y}\left\{\theta(t_x - t_y)\left[j(x), \left(\varphi(y)\,\frac{\partial}{\partial t_y}\,f_k(y) - \frac{\partial\varphi(y)}{\partial t_y}\,f_k(y)\right)\right]\right\}. \quad (8.164)$$

This expression contains two terms; one involves the differential of $\theta(t_x - t_y)$ whilst the second involves the differentiation of the commutator. Now from A.6 (Appendixes, p. 703) the differential of $\theta(t_x - t_y)$ can be written as an equal time commutator

$$\frac{\partial}{\partial t_y}\,\theta(t_x - t_y) = -\delta(t_x - t_y) \quad (8.165)$$

and so the first term involves $\delta(t_x - t_y)$ and the commutators. The second term can be treated by the same techniques as given in equations (8.150–8.155) and so the method will not be repeated. Thus equation (8.164) becomes

$$-R[j(x)\,j(y)]\,f_k(y) + i\delta(t_x - t_y)\,[j(x), \varphi(y)]\,\frac{\partial}{\partial t_y}\,f_k(y)$$

$$= -i\delta(t_x - t_y)\left[j(x), \frac{\partial\varphi(y)}{\partial t_y}\right]f_k(y). \quad (8.166)$$

The first of the equal time commutators vanishes in general (compare (4.242), for example), since the current $j(x)$ does not involve a first derivative of the field $\varphi(x)$. The second does not vanish; from local commutivity one would expect it to be proportional to $\delta(\mathbf{x} - \mathbf{y})$ and its derivatives (compare § 4.6(b)). It is sometimes called a *contact term*; it can be evaluated and be shown to be a real function. However, the S-matrix element is complex, and in later work we shall only require the imaginary part of the matrix element. We shall therefore ignore the equal time commutators in future. If, therefore, the first term in (8.166) is inserted into (8.162) and then into (8.155), the following expression is obtained for the S-matrix element:

$$S_{fi} = \delta_{fi} - i\int_{-\infty}^{+\infty} d^4x\,d^4y\,f_{k'}^*(x)\,\langle p'|\,R[j(x)\,j(y)]\,|p\rangle\,f_k(y). \quad (8.167)$$

In the plane wave representation (8.123) this becomes

$$S_{fi} = \delta_{fi} - \frac{i}{V\sqrt{(4\omega_k\omega_k')}}\int_{-\infty}^{+\infty} d^4x d^4y\,e^{i(ky - k'x)}\langle p'|\,R[j(x)\,j(y)]\,|p\rangle. \quad (8.168)$$

This expression can be given a tidier form by using the invariance properties f the S-matrix under displacements (equations (8.140) and (8.141)). Introducing ew coordinates

$$y = X - \frac{z}{2}, \qquad x = X + \frac{z}{2} \tag{8.169}$$

e find

$$j(x)\,j(y) = e^{-iPX} j\left(\frac{z}{2}\right) e^{iPX} e^{-iPX} j\left(-\frac{z}{2}\right) e^{iPX}$$

$$= e^{-iPX} j\left(\frac{z}{2}\right) j\left(-\frac{z}{2}\right) e^{iPX} \tag{8.170}$$

nd since

$$\langle p'|\, e^{-iPX} = \langle p'|\, e^{-ip'X} \tag{8.171}$$

he integrand of (8.168) becomes

$$e^{i(k+p-k'-p')X - i\frac{(k+k')}{2}z} \langle p'|\, R\left[j\left(\frac{z}{2}\right) j\left(-\frac{z}{2}\right)\right] |p\rangle. \tag{8.172}$$

hus upon integrating over X, equation (8.168) assumes the form

$$S_{fi} = \delta_{fi} - \frac{i(2\pi)^4\, \delta(k + p - k' - p')}{V\sqrt{(4\omega_k\omega_k')}} \int_{-\infty}^{+\infty} d^4z\, e^{-i\frac{(k+k')}{2}z} \times$$

$$\times \langle p'|\, R\left[j\left(\frac{z}{2}\right) j\left(-\frac{z}{2}\right)\right] |p\rangle. \tag{8.173}$$

This equation may be compared with (7.34)

$$S_{fi} = \delta_{fi} + i(2\pi)^4\, \delta(p_i - p_f)\, \frac{T_{fi}}{N}$$

$$= \delta_{fi} + i(2\pi)^4\, \frac{\delta(p_i - p_f)\, T_{fi}}{V^2 N_p N_{p'} \sqrt{(4\omega_k\omega_k')}}$$

where we have inserted appropriate normalisation factors from (7.35) and 7.36). A comparison of the above equation with (8.173) shows that we can write

$$T_{fi} = -N_p N_{p'} \int_{-\infty}^{+\infty} d^4z\, e^{-i\frac{(k+k')}{2}z} \langle p'|\, R\left[j\left(\frac{z}{2}\right) j\left(-\frac{z}{2}\right)\right] |p\rangle \tag{8.174}$$

where we have set $V = 1$ since it always cancels out eventually.

8.9(c). *Spinor fields*

Essentially the same procedure as that given above can be operated for the spinor fields. The form of the field equations (4.194) and (4.196)

$$\psi(x) = \frac{1}{\sqrt{V}} \sum_{\mathbf{p},r} \sqrt{\frac{m}{E_p}} \, [a_r(\mathbf{p})\, u_r(\mathbf{p})\, e^{ipx} + b_r^\dagger(\mathbf{p})\, v_r(\mathbf{p})\, e^{-ipx}]$$

$$\overline{\psi}(x) = \frac{1}{\sqrt{V}} \sum_{\mathbf{p},r} \sqrt{\frac{m}{E_p}} \, [a_r^\dagger(\mathbf{p})\, \overline{u}_r(\mathbf{p})\, e^{-ipx} + b_r(\mathbf{p})\, \overline{v}_r(\mathbf{p})\, e^{ipx}]$$

implies that the equivalent terms to $f_\alpha(x)$ and $f_\alpha^*(x)$ (8.123) are (in plane wave representation)

$$u_{pr}(x) = \frac{1}{\sqrt{V}} \sqrt{\left(\frac{m}{E_p}\right)} \, u_r(\mathbf{p})\, e^{ipx} \tag{8.175}$$

$$\overline{u}_{pr}(x) = \frac{1}{\sqrt{V}} \sqrt{\left(\frac{m}{E_p}\right)} \, \overline{u}_r(\mathbf{p})\, e^{-ipx}$$

with corresponding terms for $v_{pr}(x)$. Thus the destruction and creation operators become (compare (8.125))

$$a_\alpha = \int d\mathbf{x}\, \overline{u}_\alpha(x)\, \gamma_4 \psi(x) \tag{8.176}$$

$$a_\alpha^\dagger = \int d\mathbf{x}\, \overline{\psi}(x)\, \gamma_4 u_\alpha(x).$$

These formulae may be easily verified in the plane wave representation by using the normalisation conditions for spinors (equations (3.147) and (3.148))

$$u^\dagger u = \frac{E}{m}\, \overline{u}u, \qquad \overline{u}_r u_s = \delta_{rs}.$$

The interacting spinor field satisfies the same formal equations as for the electromagnetic interaction (compare (8.43))

$$\gamma_\mu \frac{\partial \psi}{\partial x_\mu} + m\psi = J(x) \tag{8.177}$$

$$\frac{\partial \overline{\psi}}{\partial x_\mu} \gamma_\mu - m\overline{\psi} = \overline{J}(x).$$

The above equations define the spinor current operator $J(x)$. The reduction of the S-matrix is carried out in the same way as for boson fields. For example consider the scattering amplitude (8.148)

$$S_{fi} = \langle f_{\text{out}} | i_{\text{in}} \rangle = \langle p'k'_{\text{out}} | pk_{\text{in}} \rangle$$

and assume p' and p are positive energy fermions. Then a straightforward calculation shows that the equation

$$S_{fi} = \langle p'k'_{\text{in}}|\, pk_{\text{in}}\rangle + i \int_{-\infty}^{+\infty} d^4x\, \langle k'|\, \bar{u}_{p'r'}(x)\, J(x)\, |pk_{\text{in}}\rangle$$

$$= \delta_{fi} + i \int_{-\infty}^{+\infty} d^4x\, \langle k'|\, \bar{u}_{p'r'}(x)\, J(x)\, |pk_{\text{in}}\rangle \tag{8.178}$$

is the spinor equivalent of (8.150). The p state can then be extracted from the in state to yield the equivalent of (8.163)

$$S_{fi} = \delta_{fi} - \int_{-\infty}^{+\infty} d^4x\, d^4y\, \bar{u}_{p'r'}(x)\, \overrightarrow{D}_x\, \langle k'|\, \theta(t_x - t_y)\, \{\psi(x), \bar{\psi}(y)\}\, |k\rangle\, \overleftarrow{D}_y u_{pr}(y) \tag{8.179}$$

where

$$\overrightarrow{D}_x = \gamma_\mu \frac{\partial}{\partial x_\mu} + m, \qquad \overleftarrow{D}_y = -\frac{\partial}{\partial y_\nu}\gamma_\nu + m \tag{8.180}$$

act in the directions indicated.

CHAPTER 9

THE INTERACTION OF FIELDS IV THE INVARIANCE PROPERTIES OF INTERACTING SYSTEMS

9.1. INTRODUCTION

In this chapter we shall examine the effects of limiting the forms of the transition matrix elements (Chapter 7), by exploiting the symmetry requirements for physical systems. We have already introduced this topic in § 7.3(b), where we showed that if the transition amplitude $\langle f| S |i\rangle$ (7.6) was invariant (apart from an undetectable phase factor) under the unitary transformation

$$|f'\rangle = U|f\rangle, \qquad |i'\rangle = U|i\rangle \tag{9.1}$$

then

$$[U, S] = 0, \tag{9.2}$$

$$U^{-1}SU = S, \qquad USU^{-1} = S.$$

The starting-point in the theory is mainly a matter of taste. Some authors prefer to start with the commutation relation (9.2) and work back to the invariance condition for the amplitude. We shall use the system which appears to be most convenient for a given problem.

The restriction (9.2) also applies to the associated R, M and T operators of Chapter 7. Furthermore, if a specific form of the S-matrix operator has been constructed from an interaction Lagrangian or Hamiltonian density, the same requirements will be placed upon them (this argument can, of course, be reversed)

$$\mathscr{L}' = \mathscr{L} = U\mathscr{L}U^{-1} \qquad [\mathscr{L}, U] = 0 \tag{9.3}$$

$$\mathscr{H}' = \mathscr{H} = U\mathscr{H}U^{-1} \qquad [\mathscr{H}, U] = 0.$$

Much of this chapter will be concerned with examining the implications of the relations given above.

9.2. SYMMETRY PROPERTIES AND REFLECTIONS

9.2(a). *Introduction*

In this section we shall examine how the transition amplitude (7.6)

$$S_{fi} = \langle f|\, S\, |i\rangle$$

behaves under the discrete transformations P (space reflection), C (charge conjugation) and T (time reversal).

The first two transformations are both unitary, and so from equations (9.1) and (9.2) we can write

$$[U, S] = 0, \qquad U \equiv C, P \tag{9.4}$$

$$\langle f'|\, S\, |i'\rangle = \langle f|\, S\, |i\rangle$$

if the interaction is invariant under the transformation U. In the case of the charge conjugation operation, the operator C changes particles to antiparticles, and since the transition probability is proportional to $|\langle f|\, S\, |i\rangle|^2$ the invariance under C implies that transitions should proceed at the same rate for particles and antiparticles, for example the cross-sections for $\pi^- p$ and $\pi^+ \bar{p}$ scattering should be the same.

We next consider time reversal, which is an antilinear transformation (see remark following (5.165)). For time reversal invariance (§ 5.6(a)), the transformations may be written as follows:

$$T\,|i\rangle = \langle i'|, \qquad \langle f|\, T^{-1} = |f'\rangle \tag{9.5}$$

$$[T, S] = 0, \qquad T^{-1}ST = S$$

$$\langle f|\, S\, |i\rangle = \langle f|\, T^{-1}ST\, |i\rangle = \langle i'|\, S\, |f'\rangle. \tag{9.6}$$

According to § 5.6, if the momentum and spin states were originally given by

$$|i\rangle \equiv |\mathbf{p}_i, s_i\rangle, \qquad \langle f| \equiv \langle \mathbf{p}_f, s_f|$$

then

$$T\,|i\rangle = \langle i'| \equiv \langle -\mathbf{p}_i, -s_i| \tag{9.7}$$

$$\langle f|\, T^{-1} = |f'\rangle \equiv |-\mathbf{p}_f, -s_f\rangle.$$

Equation (9.6) is called the reciprocal theorem; it is not the same as the principle of detailed balancing which states

$$|\langle f|\, S\, |i\rangle|^2 = |\langle i|\, S\, |f\rangle|^2. \tag{9.8}$$

The principle of detailed balance holds under certain circumstances; these can be derived from the reciprocal theorem.

(1) If the interaction is weak then the principle of detailed balance holds to good approximation. The S-matrix can be approximated by (8.33)

$$\langle f| S |i\rangle = -2\pi i \langle f| H' |i\rangle \, \delta(E_f - E_i)$$

where we have used time dependent perturbation theory. The transition probability is then given by (8.37)

$$\lambda = 2\pi \, |\langle f| H' |i\rangle|^2 \, \delta(E_f - E_i).$$

But H' is Hermitian and so by (4.17)

$$\langle f| H' |i\rangle = \langle f| H'^{\dagger} |i\rangle = (\langle i| H' |f\rangle)^*$$

therefore

$$|\langle f| H' |i\rangle|^2 = |\langle i| H' |f\rangle|^2. \tag{9.9}$$

(2) If the S-matrix operator is invariant under space reflection

$$P^{-1}SP = S$$

then the reciprocal theorem can be extended

$$\begin{aligned}\langle f| S |i\rangle &= \langle f| T^{-1}P^{-1}SPT |i\rangle \\ &= \langle i'| P^{-1}SP |f'\rangle \\ &= \langle \mathbf{p}_i, -s_i| S |\mathbf{p}_f, -s_f\rangle.\end{aligned} \tag{9.10}$$

If a summation over spin directions is made, then

$$\sum_{\text{spins}} |\langle \mathbf{p}_i, -s_i| S |\mathbf{p}_f, -s_f\rangle|^2 = \sum_{\text{spins}} |\langle \mathbf{p}_f, s_f| S |\mathbf{p}_i, s_i\rangle|^2. \tag{9.11}$$

This is known as the *principle of semi-detailed balance.*

(3) If S is invariant under spatial rotations and reflections a theorem of detailed balancing may be proved for a subsystem of S specified by the angular momentum quantum number j. Consider initial and final states denoted by

$$|i\rangle = |\alpha jm\rangle, \qquad \langle f| = \langle \beta jm|$$

where α, β contain all the information about the states apart from angular momentum, then the condition $S = T^{-1}ST$ yields (5.164)

$$\langle \beta jm| S |\alpha jm\rangle = \langle \beta jm| T^{-1}ST |\alpha jm\rangle = |C_{jm}|^2 \langle \alpha j - m| S |\beta j - m\rangle. \tag{9.12}$$

But if S is invariant under spatial rotation the change m to $-m$ leaves the system unaltered, and so for the channel j

$$\langle \beta jm| S |\alpha jm\rangle = \langle \alpha jm| S |\beta jm\rangle. \tag{9.13}$$

It is apparent from the above conditions that the principle of detailed balance holds in virtually all circumstances. The most important application of this principle has been in the determination of the spin of the π-meson (§ 2.4(b)) for which equation (9.11) is relevant.

9.2(b). *Construction of interaction terms*

The terms describing the interaction between fields can often be constructed from a Lagrangian density. The method has its applications even when Lagrangians are not specifically used in the S-matrix (as in dispersion relations), since it is often useful to consider an effective Lagrangian for a system in order to examine its transformation properties.

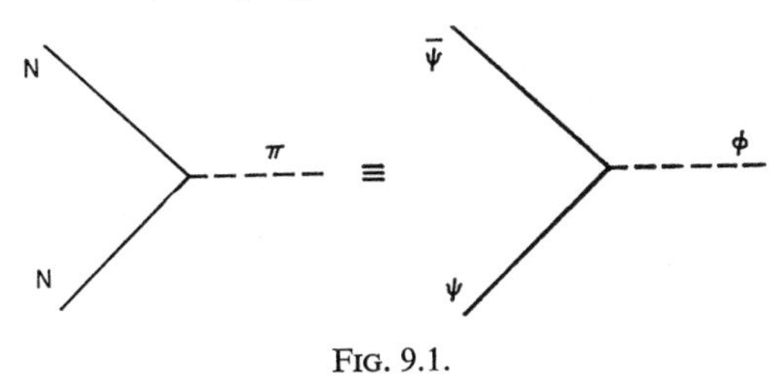

FIG. 9.1.

As we have stated previously (§ 4.2) the Lagrangian density is invariant under proper Lorentz transformations and must therefore be either a scalar or a pseudoscalar term. The general rules for its construction have been given in § 4.2 for non-interacting fields. The same principles can be used for interactions, and in § 8.3 the interaction term between electrons and photons was constructed according to these rules. In order to clarify the discussion of the symmetry properties of weak and strong interactions we shall again consider specific examples – the pion–nucleon interaction and the four-fermion interaction.

(1) The pion–nucleon interaction. We shall consider the interaction given in Fig. 9.1. The presence of a pion and two nucleons requires a combination of the Klein–Gordon field† φ with two Dirac bilinear covariants $\bar{\psi}$ and ψ. If we restrict ourselves to interactions which are linear in $\bar{\psi}$, ψ and φ and do not contain derivatives, we have two possible choices:

$$\mathscr{L}_{\text{int}} = g\bar{\psi}\psi\varphi = \mathscr{L}_s \tag{9.14}$$

$$\mathscr{L}_{\text{int}} = ig\bar{\psi}\gamma_5\psi\varphi = \mathscr{L}_{ps} \tag{9.15}$$

where the factor i has been introduced so that for a real coupling strength g the operator $\mathscr{L}_{\text{int}}$ is Hermitian, and so corresponds to an energy density. Strictly speaking the bilinear covariants should be antisymmetrised (§ 4.5(d)), but this is not essential for our present purposes.

† For simplicity, we shall assume for the moment that φ is Hermitian

Now let us see the effect of the pariy operation on these terms. From equations (5.96), (5.127) and Table 5.5 (p. 214) we may write

$$P\varphi P^{-1} = \xi_{P\pi}\varphi \tag{9.16}$$

$$P\overline{\psi}\psi P^{-1} = \overline{\psi}\psi$$

$$P\overline{\psi}\gamma_5\psi P^{-1} = -\overline{\psi}\gamma_5\psi$$

where $\xi_{P\pi}$ represents the intrinsic parity of the pion, and we have a assumed that both nucleons possess the same parity and so $|\xi_{PN}|^2 = 1$. Thus we find

$$P\mathscr{L}_s P^{-1} = \xi_{P\pi}\mathscr{L}_s, \qquad P\mathscr{L}_{ps}P^{-1} = -\xi_{P\pi}\mathscr{L}_{ps}$$

but we know that the pion has odd intrinsic parity $\xi_{P\pi} = -1$ (§ 9.5(b)), and so we may conclude that equation (9.15) is a suitable Lagrangian for an interaction which conserves parity.

The interaction (9.15) is called a *ps–ps* interaction (pseudoscalar Klein–Gordon field with pseudoscalar coupling). An alternative interaction term can be constructed if derivatives of fields are used; it is known as the *ps–pv* interaction (pseudoscalar pion with pseudovector coupling)

$$\mathscr{L}_{\text{int}} = i\frac{f}{m}\,\overline{\psi}\gamma_5\gamma_\lambda\psi\frac{\partial\varphi}{\partial x_\lambda} = \mathscr{L}_{pv} \tag{9.17}$$

where the factor $1/m$ has been introduced so that f and g possess the same dimensions ($\mathscr{L}$ has dimensions of erg cm^{-3}, φ of dyne$^{1/2}$ (§ 4.3(f)) and ψ of cm$^{-3/2}$ (§ 4.5(a)); thus g has the dimensions of electrical charge, dyne$^{1/2}$ cm). It is customary to choose the pion mass for m. Although the *ps–ps* and *ps–pv* interactions are fundamentally different, nevertheless one may show that they lead to similar results in the lowest order of the perturbation expansion provided that

$$\frac{f}{m} = \frac{g}{2M} \tag{9.18}$$

where M is the nucleon mass (Schweber, Bethe and de Hoffman, 1955).

We note, for later use, that in the nonrelativistic limit equation (9.17) assumes the form

$$\mathscr{L}_{\text{int}} = \frac{f}{m}\,\psi^\dagger\boldsymbol{\sigma}\psi\cdot\nabla\varphi. \tag{9.19}$$

(2) The four-fermion interaction. An example of this interaction is β-decay

$$n \to p + e^- + \overline{\nu}$$

which we can write as

$$\nu + n \to p + e^-$$

since the creation of a Dirac particle is equivalent to the destruction of its antiparticle. This form of interaction is more convenient for calculation and leaves the answers unaltered. The process can be illustrated diagramatically as Fig. 9.2, where two pairs of bilinear Dirac fields have been inserted because we are considering four fermions. The interaction Lagrangian can again be constructed from

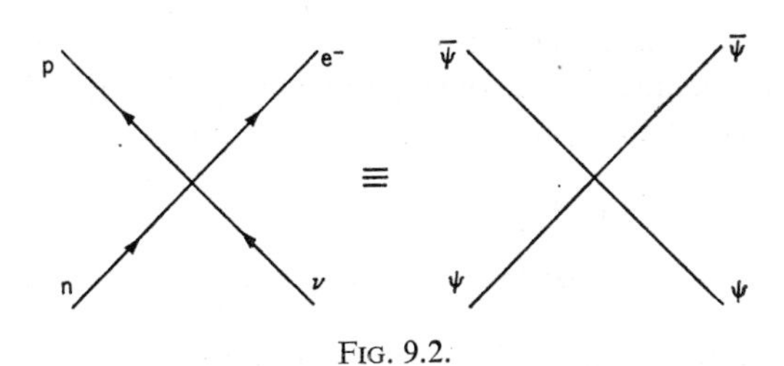

FIG. 9.2.

the bilinear covariants; the only forms which do not contain derivatives of fields and are invariant under proper Lorentz transformations are given by

$$\mathscr{L}_{\text{int}} = \sum_i C_i(\bar{\psi}_a \Gamma_i \psi_b)(\bar{\psi}_c \Gamma_i \psi_b) + \sum_i C_i'(\bar{\psi}_a \Gamma_i \psi_b)(\bar{\psi}_c \Gamma_i \gamma_5 \psi_d) + \text{h.c.}$$

$$= \mathscr{L} + \mathscr{L}' + \text{h.c.} \qquad (i = S, V, T, A, P) \tag{9.20}$$

where h.c. refers to the Hermitian conjugate expression, and the subscripts a, b, c and d have been included for purposes of identification in later algebraic manipulations.

The dimensions of C in c.g.s. units follow immediately from (9.20). They are erg cm^3, since $\mathscr{L}$ and ψ have dimensions of erg cm^{-3} and cm$^{-3/2}$ respectively (§ 4.5(a)).

With the aid of Table 5.5 (p. 214) it is apparent that the terms $\mathscr{L}$ and $\mathscr{L}'$ of equation (9.20) transform as follows under space reflections:

$$P\mathscr{L}P^{-1} = \mathscr{L}, \qquad P\mathscr{L}'P^{-1} = -\mathscr{L}'. \tag{9.21}$$

where we have set the product of intrinsic parities equal to unity. Thus $\mathscr{L}$ and $\mathscr{L}'$ represent, respectively, components which conserve and do not conserve parity in the Lagrangian. Since parity conservation fails in weak interactions, both terms must be present in the Lagrangian.

More elaborate expressions involving the derivatives of fields can be inserted in the Lagrangian for β-decay: they are found to disagree with the experimental data and so are ignored.

(3) Other forms of interaction. The number of examples of Lagrangians describing the interaction of fields can be multiplied almost indefinitely. Lists of typical Lagrangians (which preserve parity) may be found in papers by Chan (1962) and Matthews (1961).

9.2(c). *Reflections and strong interactions*

We have shown that the pseudoscalar coupling (9.15)

$$\mathscr{L}_{\text{int}} = ig\overline{\psi}\gamma_5\psi\varphi = \mathscr{L}_{ps}$$

is invariant under spatial reflections if the field φ has odd intrinsic parity. Let us now examine its behaviour for the charge conjugation and time reversal operations. If we use (5.157) and Table 5.6 (p. 231) we find

$$\begin{aligned} C\mathscr{L}_{ps}C^{-1} &= ig\xi_{C\pi}\overline{\psi}\gamma_5\psi\varphi \\ T\mathscr{L}_{ps}T^{-1} &= -ig\xi_{T\pi}\overline{\psi}\gamma_5\psi\varphi \end{aligned} \tag{9.22}$$

where we have assumed again that $|\xi_N|^2 = 1$ for the nucleons. Thus if $\mathscr{L}_{ps}$ is to be invariant under the operations C and T we may conclude that

$$\xi_{C\pi} = +1, \qquad \xi_{T\pi} = -1. \tag{9.23}$$

The above relations were obtained with uncharged pseudoscalar fields. If we assumed the fields were charged, a suitable Hermitian Lagrangian would be

$$\begin{aligned} \mathscr{L}_{ps} &= ig\overline{\psi}\gamma_5\psi\varphi + \text{h.c.} \\ &= ig\overline{\psi}\gamma_5\psi\varphi + ig^*\overline{\psi}\gamma_5\psi\varphi^\dagger \end{aligned} \tag{9.24}$$

and by use of (5.133), (5.134) and (5.171) we find

$$\begin{aligned} C\mathscr{L}_{ps}C^{-1} &= ig\xi^*_{C\pi}\overline{\psi}\gamma_5\psi\varphi^\dagger + ig^*\xi_{C\pi}\overline{\psi}\gamma_5\psi\varphi \\ T\mathscr{L}_{ps}T^{-1} &= -ig\xi_{T\pi}\overline{\psi}\gamma_5\psi\varphi^\dagger - ig^*\xi^*_{T\pi}\overline{\psi}\gamma_5\psi\varphi \end{aligned} \tag{9.25}$$

and it is apparent that $\mathscr{L}_{ps}$ is invariant under the operations C and T if

$$g = g^*, \qquad \xi^*_{C\pi} = \xi_{C\pi} = 1, \qquad \xi^*_{T\pi} = \xi_{T\pi} = -1, \tag{9.26}$$

that is, the coupling strength g should be real.

We may note also that under a strong reflection (5.192)

$$R_S = CTP$$

we find

$$\begin{aligned} R_S\mathscr{L}_{ps}R_S^{-1} &= CTPig\overline{\psi}\gamma_5\psi\varphi P^{-1}T^{-1}C^{-1} \\ &= -CTig\xi_{P\pi}\overline{\psi}\gamma_5\psi\varphi T^{-1}C^{-1} \\ &= Cig\xi_{P\pi}\xi_{T\pi}\overline{\psi}\gamma_5\psi\varphi^\dagger C^{-1} \\ &= ig\xi_{P\pi}\xi_{T\pi}\xi_{C\pi}\overline{\psi}\gamma_5\psi\varphi \\ &= ig\overline{\psi}\gamma_5\psi\varphi \end{aligned} \tag{9.27}$$

where (for ease of writing) we have considered only the first term in (9.24). An identical result can be achieved for the second term. The same result could have been obtained if the operations C, T, P were applied in any other order.

We note, finally, the properties of the Lagrangian for the electromagnetic interaction (8.40)

$$\mathscr{L}_{\text{int}} = j_\mu A_\mu = ie\overline{\psi}\gamma_\mu \psi A_\mu$$

under strong reflections. Using the methods indicated above and the properties listed in Chapter 5, one may easily show that

$$\xi_{P\mu} = \xi_{T\mu} = \xi_{C\mu} = -1 \qquad (\mu = 1, 2, 3) \tag{9.28}$$

$$\xi_{P\mu} = \xi_{T\mu} = -\xi_{C\mu} = +1 \quad (\mu = 4).$$

Since we may always work in the Coulomb gauge with $A_4 = 0$, these results imply that the intrinsic parities of the photon are negative.

9.2(d). *Experimental limits on P, T, C invariance in strong and electromagnetic interactions*

The Lagrangians we have discussed in this chapter can be used to construct transition amplitudes. Let us suppose that the amplitude for a given reaction can be written as

$$M_{fi} = \langle f| \, M_C + FM_V \, |i\rangle \tag{9.29}$$

where the subscripts C and V refer respectively to the components which conserve and violate, say, parity, and F represents the coefficient of admixture of the violating term. Then the probability of a transition which violates a given law is proportional to $|F|^2$, whilst the expectation value of the interference term (a pseudoscalar in the case of parity violation) is proportional to $(2\text{Re}\, F)/(1 + |F|^2)$.

The concept of parity as a good quantum number for a system was introduced by Wigner in 1927 as a generalisation of Laporte's rule for atomic spectra (1924). For our present purposes we may paraphrase this rule to state that transitions between atomic states are forbidden if they fail to conserve parity. Searches for spectral lines which are forbidden by this rule have proved unsuccessful and so an upper limit

$$|F|^2 \leqq |j_1(kr)|^2 \sim (kr)^2 \sim 10^{-6}$$

can be assigned for parity failure (Lee and Yang, 1956a). Here k and r represent the wave number of the photon and the atomic radius respectively. We have also used the expansion for spherical Bessel functions in the situation where $kr \ll 1$ (Appendix, A. 7, p. 704).

A similar approach has been used by Tanner (1957) to test parity violation in strong interactions. He bombarded the nucleus F^{19} with protons whose energies were suitable for forming the compound nucleus Ne^{20} in a state with spin and parity 1^+. A search was then made for transitions from this state to the ground state of O^{16} by the emission of α-particles (Fig. 9.3). Since the α-particle has zero spin, the transition can only occur if the conservation of parity (or angular momentum) is violated. Tanner was able to set a limit of $|F|^2 \lesssim 10^{-7}$. The method has been extended by other workers, and a limit of $|F|^2 \lesssim 7.10^{-13}$ on parity violation in strong interactions has been set by Donovan, Alburger and Wilkinson (1961) from an examination of the decay properties of a 2^- level in O^{16} (the authors also quote a limit of $\leqq 2 \times 10^{-13}$ for the nonconservation of angular momentum).

We now consider the second method of testing the invariance laws – by measuring the size of the interference term in equation (9.29). Consider an experiment in which two particles scatter and assume that one or both have in-

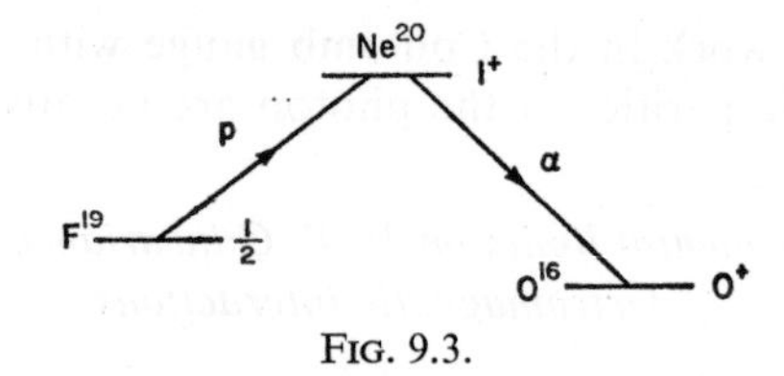

FIG. 9.3.

trinsic spin; then a measurement is made of the degree of polarisation of one of the particles along its direction of motion. The matrix element describing the interaction is then of the form

$$M_{fi} = a + b\frac{\boldsymbol{\sigma}\cdot\mathbf{p}}{|\mathbf{p}|} = a\left(1 + \frac{b}{a}\frac{\boldsymbol{\sigma}\cdot\mathbf{p}}{|\mathbf{p}|}\right)a \equiv \left(1 + F\frac{\boldsymbol{\sigma}\cdot\mathbf{p}}{|\mathbf{p}|}\right) \tag{9.30}$$

where a and b are scalar functions of the experimental parameters. Under an inversion of space coordinates $\mathbf{p}$ but not $\boldsymbol{\sigma}$ changes sign, and thus the second term in the matrix element does not conserve parity. The transition probability is proportional to

$$|M_{fi}|^2 = |a|^2 + |b|^2 + 2\,\mathrm{Re}\,ab\frac{\boldsymbol{\sigma}\cdot\mathbf{p}}{|\mathbf{p}|} \tag{9.31}$$

and so the degree of longitudinal polarisation, and hence of parity violation, is proportional to

$$2\,\mathrm{Re}\frac{ab}{|a|^2 + |b|^2} \sim 2\,\mathrm{Re}\,F \tag{9.32}$$

if $b \ll a$. A measurement of F in nucleon-nucleon scattering has been made by Jones, Murphy and O'Neill (1958). Protons of 380 MeV were used to bombard a beryllium target, and the longitudinal polarisation of the emergent neutrons was examined. The neutrons chosen for examination possessed energies greater than 350 MeV and travelled at 0^0 with respect to the proton beam. The spin of the eutrons was then rotated in a magnetic field so that any longitudinal polarisation vas converted to transverse polarisation. The magnitude of this polarisation was hen determined by an 'up–down' scattering in a hydrogen target (§ 6.2(f)). The neasured asymmetry indicated that $|F|^2 \leqq 4 \times 10^{-6}$.

Tests of time reversal invariance in strong interactions may be made by examining certain of the scattering properties of polarised nucleons. For example ve stated in § 6.2(f) that the following relation (6.73)

$$\text{tr } M\sigma M^\dagger = \text{tr } MM^\dagger\sigma$$

olds in the scattering of polarised nucleons providing that the reaction is invariant under space and time reversal. Now the matrix elements given above can e related to physical quantities in the following way.

(1) The polarisation acquired by unpolarised protons in scattering from an npolarised target at an angle θ is

$$\mathbf{P}(\theta) = P(\theta)\,\mathbf{n} = \frac{\text{tr } MM^\dagger\sigma}{\text{tr } MM^\dagger}.$$

(2) if protons with 100 per cent polarisation strike an unpolarised target, then he left–right asymmetry in the plane perpendicular to the direction of polariation is given by

$$e(\theta) = \frac{\text{tr } M\sigma M^\dagger}{\text{tr } MM^\dagger};$$

Ience for P and T invariance we expect

$$P(\theta) = e(\theta).$$

A measurement of these quantities has been made by Abashian and Hafner 1958), using protons with a kinetic energy of 200 MeV. They obtained the result

$$P(\theta) - e(\theta) = -0{\cdot}014 \pm 0{\cdot}014$$

t $\theta = 30°$.

A sensitive limit on the failure of both P and T invariance in electromagnetic nteractions is set by the absence of a dipole moment in the neutron (Lee and Yang, 1957(b)). If an electrically neutral particle has both a magnetic and electric lipole moment then the Hamiltonian for their interaction with an electronagnetic field will be of the form

$$H_{\text{int}} = \mu_m \sigma \cdot \mathbf{H} + \mu_e \sigma \cdot \mathbf{E} \tag{9.33}$$

n the nonrelativistic limit (compare § 3.3(f) for a pure magnetic interaction). Iere μ_m and μ_e represent the magnitudes of the intrinsic magnetic and electric dipole moments respectively. Now under reversals of time and space we can write

$$P\boldsymbol{\sigma}P^{-1} = \boldsymbol{\sigma}, \qquad T\boldsymbol{\sigma}T^{-1} = -\boldsymbol{\sigma} \tag{9.34}$$

$$P\mathbf{H}P^{-1} = \mathbf{H}, \qquad T\mathbf{H}T^{-1} = -\mathbf{H}$$

$$P\mathbf{E}P^{-1} = -\mathbf{E}, \qquad T\mathbf{E}T^{-1} = \mathbf{E}$$

where we have used (9.28) and the relations $\mathbf{H} = \nabla \times \mathbf{A}$ and $\mathbf{E} = -\partial\mathbf{A}/\partial t$. Thus the magnetic interaction is invariant under both space and time reversals

$$P(\boldsymbol{\sigma}\cdot\mathbf{H})P^{-1} = \boldsymbol{\sigma}\cdot\mathbf{H}, \qquad T(\boldsymbol{\sigma}\cdot\mathbf{H})T^{-1} = \boldsymbol{\sigma}\cdot\mathbf{H} \tag{9.35}$$

but the electric interaction changes sign

$$P(\boldsymbol{\sigma}\cdot\mathbf{E})P^{-1} = -\boldsymbol{\sigma}\cdot\mathbf{E}, \qquad T(\boldsymbol{\sigma}\cdot\mathbf{E})T^{-1} = -\boldsymbol{\sigma}\cdot\mathbf{E}. \tag{9.36}$$

Thus electric dipole interactions cannot occur unless both P and T invariance breaks down in electrodynamics. If the electric dipole moment existed it would have a magnitude given roughly by

$$\mu_e \sim edF \sim e \times 10^{-13}\, F \quad \text{cm}$$

where d is the neutron size. An upper limit on the magnitude of the electric dipole moment of the neutron has been obtained by Smith, Purcell, and Ramsey (1957)

$$\mu_e \leqq e\, 0{\cdot}5 \times 10^{-20} \text{ cm}$$

so that $|F|^2 \leqq 10^{-13}$.

A less sensitive limit has been set by examining the electric dipole moment of the muon; a value $\mu_e \leqq e\, 0\cdot 3 \times 10^{-17}$ cm was obtained (Charpak, *et al.*, 1961 b).

Experimental tests on the limits of charge conjugation invariance in strong and electromagnetic interactions are difficult because of the problems of obtaining sufficiently intense beams of antiparticles. However, the experimental observations of invariance under the operations P and T and under the combined operation CPT suggests that charge conjugation is preserved in strong and electromagnetic interactions.

Certain interesting properties are associated with the concept of charge conjugation invariance. One concerns the behaviour of the annihilation processes for electron–positron and proton–antiproton pairs; we shall discuss these processes in §§ 11.2(d) and 14.4(b) respectively. Another property worth considering is the number of photons appearing in the decay of the π^0 meson. In certain of the problems examined earlier we have assumed that the π^0 meson decays into two photons; there is good experimental evidence for this transition since the photons are observed to possess an energy $\frac{1}{2}m_{\pi^0}$. However, at rare intervals decay into larger numbers of photons should occur, for example four photons should appear in $\sim(1/137)^2$ of the decay events. The appearance of three photons is forbidden, however, if the decay process is assumed to be invariant under charge conjugation.

This statement may be proved with the aid of Furry's theorem (1937), which in modern language states that diagrams which contain a closed loop and an odd number of photon vertices contribute nothing to the matrix element. In particular this leads to the statement that diagrams containing an odd number

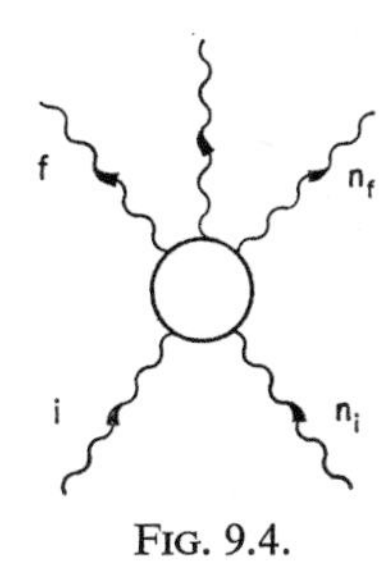

FIG. 9.4.

of external photon lines vanish. Consider Fig. 9.4, which represents some process with n_i and n_f photons in the initial and final states respectively, and

$$n_i + n_f = N.$$

Now the intrinsic charge parity of the photon is odd (equations (5.139) and (9.28)), and so the intrinsic charge parities of the initial and final states are

$$\xi_{Ci} = (-1)^{n_i}, \qquad \xi_{Cf} = (-1)^{n_f} \tag{9.37}$$

and hence

$$C\,|i\rangle = (-1)^{n_i}\,|i\rangle, \qquad C\,|f\rangle = (-1)^{n_f}\,|f\rangle \tag{9.38}$$

since the photon is a self-conjugate particle (§ 5.5(c)).

But charge conjugation invariance implies that

$$\langle f|\,S\,|i\rangle = \langle f|\,C^{-1}SC\,|i\rangle = (-1)^{n_i+n_f}\,\langle f|\,S\,|i\rangle = (-1)^N\,\langle f|\,S\,|i\rangle \tag{9.39}$$

and so it is clear that N should be even. Thus the matrix element associated with Fig. 9.4 vanishes if N is odd, which is Furry's theorem.

Now in the case of the decay

$$\pi^0 \rightarrow \gamma + \gamma + \gamma$$

we find

$$\langle 3\gamma|\,S\,|\pi^0\rangle = \langle 3\gamma|\,C^{-1}SC\,|\pi^0\rangle = (-1)^3\,\langle 3\gamma|\,S\,|\pi^0\rangle$$

since the π^0 meson is a self-conjugate particle with intrinsic charge parity +1 (§ 5.5(c)). Thus the decay $\pi^0 \rightarrow 3\gamma$ should be forbidden. No experimental search has been made for this process.

9.2(e). *Invariance principles and weak interactions*

The weak interaction Lagrangian (9.20)

$$\mathscr{L}_{\text{int}} = \sum_i C_i(\bar{\psi}_a \Gamma_i \psi_b)(\bar{\psi}_c \Gamma_i \psi_d) + \sum_i C_i'(\bar{\psi}_a \Gamma_i \psi_b)(\bar{\psi}_c \Gamma_i \gamma_5 \psi_d) + \text{h.c.}$$

for the β-decay process may be subjected to the transformations P, T and C in the same way as the strong interactions were treated in § 9.2(c). In all the transformations an adjustable phase factor of the form $\xi_a^* \xi_b \xi_c^* \xi_d$ will appear. This term may be set equal to one and ignored. The invariance requirements for the coupling constants can then be summarised in Table 9.1.

TABLE 9.1

	P	T	C	$R_s = CTP$
C_i	C_i	C_i^*	C_i^*	C_i
C_i'	$-C_i'$	$C_i'^*$	$-C_i'^*$	C_i'

It can be seen from this table that if the interaction is invariant under spatial reflection then

$$C_i' = -C_i'$$

a condition which can only be satisfied if $C_i' = 0$. Thus the second group of terms in the interaction Lagrangian fail to conserve parity.

The table also shows that for invariance under time reversal

$$C_i = C_i^*, \qquad C_i' = C_i'^* \tag{9.40}$$

so that the coupling constant must be real for time reversal invariance regardless of whether parity is conserved. Furthermore real coupling constants imply invariance under the product CP.

The realisation that parity conservation failed in weak interactions arose from attempts to explain the properties of the K-mesons. By 1956 strong evidence was accumulating that the same parent particle was responsible for the decay processes

$$K_{\pi 2} \equiv \theta \to \pi^+ + \pi^0$$

$$K_{\pi 3} \equiv \tau \to \pi^+ + \pi^- + \pi^+.$$

In the Table 9.2 we cite data quoted by Lee and Yang (1957b).

A problem existed, however, as the particles θ and τ apparently had different spatial parity. Consider the θ-particle and assume it has a spin s; the two pions

TABLE 9.2

Decay mode	Mass (m_e)	Lifetime (sec)	Fractional abundance
$\theta \rightarrow \pi^+ + \pi^0$	$966{\cdot}7 \pm 2{\cdot}0$	$1{\cdot}21 \pm 0{\cdot}02 \times 10^{-8}$	29%
$\tau \rightarrow \pi^+ + \pi^- + \pi^+$	$966{\cdot}3 \pm 2{\cdot}1$	$1{\cdot}19 \pm 0{\cdot}05 \times 10^{-8}$	5·6%

must possess relative orbital angular momentum $L(= s)$ and so the parity of the θ-particle is given by

$$\xi_{P\theta} = (\xi_{P\pi})^2(-1)^L = (-1)^2\,(-1)^L = (-1)^s \tag{9.41}$$

(compare § 5.4(b), especially (5.101)). Thus the spin and parity of the θ-particle can be 0^+, 1^-, 2^+, ...

The configurations of the three pions arising in the decay of the τ-meson were analysed by Dalitz (see § 9.5(c)); he concluded that the pions were in the possible states 0^-, $1^\pm$, $2^\pm$, ... and that the most likely state was 0^-. Thus if the spin and parity are conserved in the decay process these quantum numbers are also those for the τ-particle. The assignment 0^- is clearly incompatible with the possible assignments for the θ-particle, and yet the masses and lifetimes of the two particles are apparently the same.

The reconciliation of this apparently contradictory evidence was made by Lee and Yang (1956a), when they put forward their famous suggestion of the failure of parity conservation in weak interactions. They also suggested that their hypothesis should be tested by looking for features which did not conserve parity in other weak interactions, notably factors of the type $\boldsymbol{\sigma} \cdot \mathbf{p}$. We have already shown that the occurrence of a term of this type violates the conservation of parity (compare (9.30)). Following the suggestion of Lee and Yang non-conservation of parity was observed in β-decay by Wu *et al.* (1957) and in μ-decay by Garwin, Lederman and Weinrich (1957) and by Friedman and Telegdi (1957). Essentially the same feature was measured in both decay processess – the correlation between a direction of spin and the momentum of the outgoing decay particle – a $\boldsymbol{\sigma} \cdot \mathbf{p}$ term. The experiments revealed maximum violations of spatial parity in weak interactions. This conclusion has since been confirmed in many other weak decay processes.

Charge conjugation invariance also fails in weak interactions. The most readily available source of particles and antiparticles is the muon decay process

$$\mu^+ \rightarrow e^+ + \nu + \bar{\nu}$$

$$\mu^- \rightarrow e^- + \bar{\nu} + \nu.$$

Now because of the nonconservation of parity in this interaction the electrons will be polarised (§ 12.3). Let us assume that the positrons are right handed (this is the correct assumption, but the helicity is immaterial for our present considerations), and that we can write the S-matrix element for μ^+-decay as

$$\langle e_R^+ \nu\bar{\nu}|\, S \,|\mu^+\rangle = \langle e_R^+ \nu\bar{\nu}|\, C^{-1}SC \,|\mu^+\rangle = \xi_C \langle e_R^- \bar{\nu}\nu|\, S \,|\mu^-\rangle \qquad (9.42)$$

if charge conjugation invariance holds. The phase factor ξ_C can again be set equal to 1. This is not important, however; the important point is that under charge conjugation the mechanical properties of a particle remain unchanged (§ 5.5), and therefore, if charge conjugation holds, the electron from μ^--decay should also be right handed. Experiments by Culligan, Frank and Holt (1959) and by Macq, Crowe and Haddock (1958) have shown precisely the opposite. Both experiments measured the helicities of the electrons and positrons by observing the total transmission of their Bremsstrahlung photons in magnetised iron. The transmission coefficient is dependent upon the helicity of the photons, which in turn depends on that of the electrons and positrons (Dyson and McVoy, 1957). The experimental results were compatible with right-handed positrons and left-handed electrons each with 100 per cent polarisation, that is a maximum violation of the principle of charge conjugation. Similar effects have been observed in the polarisation of the muons from the decay $\pi \rightarrow \mu + \nu$ (§ 12.3) and in nuclear β-decay (§12.2(c)).

The invariance properties of weak interactions under time reversal involve more complex arguments than those for P and C invariance. Briefly the technique used is to search for evidence that the coupling terms C contain imaginary components (Jackson, Treiman and Wyld, 1957). Invariance under time reversal requires that C should be real (Table 9.1 and equation (9.40)). The available evidence shows no indication of a general failure of the invariance of time reversal in weak interactions (see, for example, § 12.2(d) but note § 12.9(d)).

9.2(f). *The CTP theorem*

Table 9.1 and equation (9.27) illustrate special examples of a very general theorem – the *CTP* theorem. The theorem was evolved in slightly different forms by a number of workers, notably Schwinger (1953) and Pauli (1955).

The theorem states that any local Lagrangian theory, which is invariant under proper Lorentz transformations, is invariant with respect to the operation $R_s = CTP$ (taken in any order), although the theory may not be separately invariant under the individual operations C, T and P.

Thus *CTP* invariance does not introduce any new restrictions or selection rules for an interaction. It does, however, lead to the following two important corollaries.

(1) If a process is not invariant under one of the operations C, T or P, it is not invariant under one of the remaining two.

(2) If a process is invariant under one of the transformations P, T or C it is also invariant under the product of the other two, even if the individual transformations of this pair are noninvariant.

The first corollary seems to be observed in weak interactions. Although C and invariance fails, T invariance occurs (within rather poor experimental limits). As stated above, Table 9.1 and equation (9.27) represent special cases of the TP theorem. The tables were constructed from Lagrangians which were scalar · pseudoscalar and therefore invariant under proper Lorentz transformations proof that a pseudoscalar quantity is invariant under a proper Lorentz trans-rmation is given in § 3.3(p), equation (3.203)). Another assumption is im-icit in our discussion of the transformation properties of the Lagrangians in 9.2(c) and (e); it is that boson (fermion) fields must be properly symmetrised ntisymmetrised). The necessity for symmetrisation or antisymmetrisation for ıantised fields was discussed in § 5.5(b), and it was shown to arise from the quirement that the quantised field operators possess commutation (or anti-ommutation) properties. This requirement in turn implies a spin–statistics re-tionship, since the boson fields we have discussed were associated with par-cles of spin zero or one, whilst the Dirac field had spin $\frac{1}{2}$. We have also noted in § 4.3(c) and 4.5(b) that whereas commutation relations were associated with oson fields, anticommutation relations were required for fermion fields in order obtain Hamiltonians with positive definite eigenvalues. This requirement so leads to the Pauli exclusion principle (4.210).

A rigorous proof of the spin–statistics relation has been given by Jost (1957), e also Pauli (1958).

9.2(g). *Some consequences of the spin–statistics relationship*

The spin–statistics relationship leads to useful rules concerning the behaviour f pairs of identical particles, which we shall have occasion to use later. Con-der a set of identical particles, then we have no way of distinguishing between em if they obey the rules of quantum mechanics, since the process of 'tagging' f one the particles would lead to a violation of the uncertainty principle (in rder to 'catch' it we would have to predict its path exactly – this is impossible).

We are thus led to the principle of indistinguishability for identical particles hich is of fundamental importance for the quantum mechanical investigations f systems of elementary particles. Consider first a system of two identical articles to which we assign the generalised quantum states ξ_1 and ξ_2

$$|\Psi(\xi_1, \xi_2)\rangle \equiv |\xi_1 \xi_2\rangle. \tag{9.43}$$

ince the particles are identical the system obtained upon exchanging the par-icles must be physically equivalent to the first – a phase factor may appear in the ransformation but this is unobservable (§ 4.1(b)). Thus we can write

$$|\xi_1 \xi_2\rangle = e^{i\alpha} |\xi_2 \xi_1\rangle. \tag{9.44}$$

f the exchange is repeated a second time the phase factor becomes $e^{2i\alpha}$, but the system must now be in its original state and so

$$e^{2i\alpha} = 1, \qquad e^{i\alpha} = \pm 1. \tag{9.4}$$

Therefore we find

$$|\xi_1\xi_2\rangle = \pm |\xi_2\xi_1\rangle. \tag{9.46}$$

Systems which do or do not change sign upon exchange of particles are said to be *antisymmetric* or *symmetric* respectively. The principle may be extended to systems containing any number of particles.

Equation (9.46) may be satisfied by writing $|\xi_1\xi_2\rangle$ in the following forms (the factor $\sqrt{2}$ is included for purposes of normalisation)

$$|\xi_1\xi_2\rangle = \frac{|\xi_1\rangle\,|\xi_2\rangle + |\xi_2\rangle\,|\xi_1\rangle}{\sqrt{2}} \tag{9.47}$$

$$|\xi_1\xi_2\rangle = \frac{|\xi_1\rangle\,|\xi_2\rangle - |\xi_2\rangle\,|\xi_1\rangle}{\sqrt{2}}. \tag{9.48}$$

It is apparent that (9.47) does not exchange sign upon exchange of particles whilst (9.48) does; furthermore if $|\xi_1\rangle = |\xi_2\rangle$ equation (9.48) vanishes. The last result is precisely that expected if $|\xi_1\xi_2\rangle$ represents two fermions—the Pauli exclusion (4.210) requires that no two fermions should exist in the same state. We may therefore conclude that particles obeying Fermi–Dirac statistics can be described by antisymmetrical state functions. Similarly, equation (9.47) is satisfactory for particles obeying Bose–Einstein statistics.

The principles described above for an assembly of two particles can easily be extended to more than this number. The equation (9.48) can also be written as

$$|\xi_1\xi_2\rangle = \frac{1}{\sqrt{2}} \det \begin{vmatrix} |\xi_1\rangle & |\xi_2\rangle \\ |\xi_1\rangle & |\xi_2\rangle \end{vmatrix}. \tag{9.49}$$

A determinant form is particularly suitable for describing assemblies containing many fermions (Dirac, 1947).

The conditions (9.47) and (9.48) impose certain restrictions on the possible states in which bosons and fermions may exist. Consider, for example, two spinless bosons in their mutual c-system. The interchange of these particles is equivalent to a reflection of space coordinates, but by (5.101) the effect of space reflections on states of two bosons is given by the relation

$$P\,|kLM\rangle = (-1)^L\,|kLM\rangle.$$

(We assume the two particles are identical and therefore $\xi_P^2 = 1$.) Thus equation (9.47) becomes

$$|\xi_1\xi_2\rangle = \frac{|kLM\rangle + (-1)^L\,|kLM\rangle}{\sqrt{2}} \tag{9.50}$$

vhich vanishes if L (the relative orbital angular momentum quantum number) is odd. Thus systems which decay into two identical spinless bosons must have ven spin.

The requirement of overall antisymmetry for fermions enables a useful restriction to be placed on the possible states in which two identical fermions, vith spin $\frac{1}{2}$, can exist. We may write the overall state function describing the ystem as a product of an orbital angular momentum function and a spin function. The behaviour of the spin function may be immediately ascertained from 9.47) and (9.48), namely

$$|\chi_1\chi_2\rangle = \frac{|\chi_1\rangle\,|\chi_2\rangle \pm |\chi_2\rangle\,|\chi_1\rangle}{\sqrt{2}} \tag{9.51}$$

vhere the χ terms represent the spin states. It is apparent from this equation that parallel spin states ($\chi_1 = \chi_2$) are symmetric and vice versa. Now the behaviour of the orbital function is given by (9.50), and so we may conclude that for overall antisymmetry of a two-fermion system we require

$$\text{even } L \text{ and } s = 0 \quad \text{or} \quad \text{odd } L \text{ and } s = 1 \tag{9.52}$$

vhere s represents the total spin of the system. Therefore we may summarise the antisymmetry requirements of the system as follows:

$$\begin{aligned} &\text{exchange function for } L && = (-1)^L \\ &\text{exchange function for } s && = (-1)^{s+1} \\ &\text{overall antisymmetry requirement} && = (-1)^{L+s+1} = -1. \end{aligned} \tag{9.53}$$

This rule could have decided upon almost intuitively. It shows that states of parallel spin (that is both particles have the same s_z quantum number) are only possible if L is odd. Thus if we use a modified spectroscopic notation

$$^{2s+1}L_j^{\xi_P}$$

the properly antisymmetrised states for two identical fermions are

$$^1S_0^+,\quad {}^3P_{0,1,2}^-,\quad {}^1D_2^+,\quad {}^3F_{2,3,4},\quad \ldots \tag{9.54}$$

9.2(h). *The masses and lifetimes of charge conjugate particles*

The CPT theorem may be used to show that the masses and lifetimes of particles and their antiparticles are identical (Lee, Yang, and Oehme, 1957; Lee and Yang 1957b)). This statement appears to be confirmed by the experimental data. If the charge conjugation operation is a symmetry transformation, then the equivalence of properties of particles and antiparticles would be expected from the invariance of the Hamiltonian under C

$$CHC^{-1} = H.$$

In weak interactions we know, however, that charge conjugation is violated, so that the equality of masses and lifetimes for particles and antiparticles is no longer immediately obvious. The equality is retained, however, as a consequence of the CPT theorem.

We first consider masses. Let the mass m of a particle a be represented as an eigenvalue of the equation

$$H\,|a\rangle = m\,|a\rangle \tag{9.55}$$

where the Hamiltonian H represents a sum of both the strong and weak internal interactions which contribute to the observed mass of the particle

$$H = H_S + H_W.$$

An equivalent expression to (9.55) is

$$\langle a|\,H\,|a\rangle = m\,\langle a\,|\,a\rangle. \tag{9.56}$$

If H is invariant under CTP, then

$$H = P^{-1}T^{-1}C^{-1}HCTP$$

and so

$$\langle a|\,P^{-1}T^{-1}C^{-1}\,H\,CTP\,|a\rangle = |\xi_S|^2\,\langle\bar{a}|\,H\,|\bar{a}\rangle = \langle\bar{a}|\,H\,|\bar{a}\rangle$$

where we have used (5.199) and denoted the antiparticle state by $\bar{a}$. Similarly the right-hand side of (9.56) can be written as

$$m\,\langle a\,|\,a\rangle = m\,\langle a|\,P^{-1}T^{-1}C^{-1}CTP\,|a\rangle = m\,\langle\bar{a}\,|\,\bar{a}\rangle$$

and so

$$\langle\bar{a}|\,H\,|\bar{a}\rangle = m\,\langle\bar{a}\,|\,\bar{a}\rangle \tag{9.57}$$

$$m_a = m_{\bar{a}}.$$

If the particle has spin, then the spin of the antiparticle points in the opposite direction (see remark following (5.199)). This feature is immaterial for a free particle; the condition of Lorentz invariance required for the CTP theorem covers rotational symmetry, and hence the original spin direction can always be restored (see equation (9.61) below).

We next consider the decay of charge conjugate particles

$$|i\rangle \rightarrow |f\rangle, \qquad |\bar{i}\rangle \rightarrow |\bar{f}\rangle.$$

If we assume that the decay process is a weak one, only the first term in the S-

atrix need be considered (8.31)

$$S = \hat{1} - i \int_{-\infty}^{+\infty} dt\, H_W(t) \tag{9.58}$$

o that

$$\langle f|\, S\, |i\rangle \propto \langle f|\, H_W\, |i\rangle .$$

Ve now assume that H_W may be split into parity conserving and nonconserving arts

$$H_W = H_1 + H_2 \tag{9.59}$$

$$PH_1P^{-1} = H_1, \qquad PH_2P^{-1} = -H_2$$

hen

$$\langle f|\, S\, |i\rangle \propto \langle f|\, H_1 + H_2\, |i\rangle .$$

he proof of the equality of lifetimes for particle and antiparticle states may e made by making use of the Hermitian property of the Hamiltonian (§ 4.1) nd the *CPT* theorem

$$\begin{aligned} \langle f|\, H_1\, |i\rangle^* &= \langle i|\, H_1\, |f\rangle \\ &= \langle i|\, C^{-1}T^{-1}P^{-1}H_1PTC\, |f\rangle \\ &= \langle \bar{i}|\, T^{-1}H_1T\, |\bar{f}\rangle . \end{aligned} \tag{9.60}$$

The action of the T operator on state vectors has been discussed in § 5.6(a) see also (9.7)). If the state $|\bar{i}\rangle$ is written as $|\alpha jm\rangle$ where α contains all the other nformation about the state apart from angular momentum, then by (5.164)

$$\langle \bar{i}|\, T^{-1} = \langle \alpha jm|\, T^{-1} = C^*_{jm}\, |\alpha j, -m\rangle$$

imilarly

$$T\, |\bar{f}\rangle = T\, |\beta jm\rangle = C_{jm}\, \langle \beta j, -m| \qquad |C_{jm}|^2 = 1 .$$

hus the matrix element $\langle i|\, T^{-1}H_1T\, |f\rangle$ becomes

$$|C_{jm}|^2\, \langle \beta j, -m|\, H_1\, |\alpha j, -m\rangle = \langle \beta j, -m|\, H_1\, |\alpha j, -m\rangle .$$

ut if the system is invariant under rotations (that is proper Lorentz transormations), then

$$\langle \beta j, -m|\, H_1\, |\alpha j, -m\rangle = \langle \beta jm|\, H_1\, |\alpha jm\rangle = \langle \bar{f}|\, H_1\, |\bar{i}\rangle . \tag{9.61}$$

hus we have shown that

$$\langle f|\, H_1\, |i\rangle^* = \langle \bar{f}|\, H_1\, |\bar{i}\rangle . \tag{9.62}$$

Similarly, because of the second of equations (9.59),

$$\langle f|\, H_2\, |i\rangle^* = -\langle \bar{f}|\, H_2\, |\bar{i}\rangle. \tag{9.63}$$

The lifetime of the decaying particle can be written as

$$\frac{1}{\tau} \propto \sum_{i,f} |\langle f|\, H_1 + H_2\, |i\rangle|^2 \tag{9.64}$$

but since τ is a scalar quantity the interference terms, which are pseudoscalars must vanish (in other words since we are summing over all final states, terms of the type $\boldsymbol{\sigma} \cdot \mathbf{p}$ disappear because we consider all directions). Thus equation (9.64) becomes

$$\frac{1}{\tau} \propto \sum_{i,f} (|\langle f|\, H_1\, |i\rangle|^2 + |\langle f|\, H_2\, |i\rangle|^2)$$

and similarly for the charge conjugate state

$$\frac{1}{\bar{\tau}} \propto \sum_{\bar{i},\bar{f}} (|\langle \bar{f}|\, H_1\, |\bar{i}\rangle|^2 + |\langle \bar{f}|\, H_2\, |\bar{i}\rangle|^2).$$

Thus by equations (9.62) and (9.63)

$$\tau = \bar{\tau} \tag{9.65}$$

to lowest order in perturbation theory.

The theorem may be extended to show that the branching ratios in the decay of charge conjugate particles should also be the same.

9.3. THE BEHAVIOUR OF INTERACTIONS UNDER CONTINUOUS TRANSFORMATIONS

9.3(a). *Introduction*

The invariance properties of the transition amplitude

$$\langle f|\, S\, |i\rangle$$

under continuous transformations lead to many useful conservation laws. Consider the action of a unitary transformation (§ 4.1 (e))

$$U = \hat{1} - iG\delta\lambda \qquad (\delta\lambda \to 0) \tag{9.66}$$

on the S-matrix operator (compare (4.33))

$$S' = USU^{-1} = S - i[G, S]\delta\lambda. \tag{9.67}$$

If the S-operator is invariant under this transformation then

$$[G, S] = 0. \tag{9.68}$$

A similar expression was given in § 7.3(b), and it was then shown that invariance under displacements and rotations led to conservation of linear and angular momenta respectively. The remainder of this section will be devoted to exploiting the relation (9.68) in special spaces.

9.3(b). *Gauge invariance and the electromagnetic interaction*

In §§ 3.4(f.2) and 4.4(a) gauge transformations of the first and second kind were introduced. We shall now show that they are related by considering the basic electromagnetic interaction between electrons and photons.

The total Lagrangian for the interacting system can be given as†

$$\begin{aligned}\mathscr{L} &= \mathscr{L}_e + \mathscr{L}_\gamma + \mathscr{L}_{\text{int}} \\ &= -\bar{\psi}\left(\gamma_\mu \frac{\partial}{\partial x_\mu} + m\right)\psi - \tfrac{1}{4}\left(\frac{\partial A_\mu}{\partial x_\nu} - \frac{\partial A_\nu}{\partial x_\mu}\right)^2 + ie\bar{\psi}\gamma_\mu\psi A_\mu.\end{aligned} \tag{9.69}$$

This equation can be rewritten as

$$\mathscr{L} = -\bar{\psi}\gamma_\mu\left(\frac{\partial}{\partial x_\mu} - ieA_\mu\right)\psi - m\bar{\psi}\psi - \tfrac{1}{4}\left(\frac{\partial A_\mu}{\partial x_\nu} - \frac{\partial A_\nu}{\partial x_\mu}\right)^2. \tag{9.70}$$

We now consider a gauge transformation of the second kind (4.148)

$$A_\mu(x) \to A'_\mu(x) = A_\mu(x) + \frac{\partial \chi(x)}{\partial x_\mu}. \tag{9.71}$$

Under this transformation $(\partial A_\mu/\partial x_\nu - \partial A_\nu/\partial x_\mu)^2$ stays invariant, and so the Lagrangian (9.70) becomes

$$\mathscr{L}' = -\bar{\psi}'\gamma_\mu\left(\frac{\partial}{\partial x_\mu} - ieA_\mu - ie\frac{\partial \chi(x)}{\partial x_\mu}\right)\psi' - m\bar{\psi}'\psi' - \tfrac{1}{4}\left(\frac{\partial A_\mu}{\partial x_\nu} - \frac{\partial A_\nu}{\partial x_\mu}\right)^2.$$

This equation stays invariant under (9.71) if we postulate that at the same time

$$\bar{\psi} \to \bar{\psi}' = \bar{\psi}e^{-i\alpha(x)}, \qquad \psi \to \psi' = \psi e^{i\alpha(x)} \tag{9.72}$$

then

$$\frac{\partial \psi'}{\partial x_\mu} = e^{i\alpha(x)}\frac{\partial \psi}{\partial x_\mu} + i\psi e^{i\alpha(x)}\frac{\partial \alpha(x)}{\partial x_\mu}$$

† We have used equation (4.162) for $\mathscr{L}_\gamma$; strictly speaking we should have used (4.161) (compare § 8.3(a)) the gauge invariance for $\mathscr{L}_\gamma$ is then less obvious (but still applicable).

and therefore

$$\mathscr{L}' = -i\bar{\psi}\gamma_\mu\psi\frac{\partial}{\partial x_\mu}[\alpha(x) - e\chi(x)] - \bar{\psi}\gamma_\mu\left(\frac{\partial}{\partial x_\mu} - ieA_\mu\right)\psi$$

$$- m\bar{\psi}\psi - \tfrac{1}{4}\left(\frac{\partial A_\mu}{\partial x_\nu} - \frac{\partial A_\nu}{\partial x_\mu}\right)^2. \quad (9.73)$$

Thus we find $\mathscr{L} = \mathscr{L}'$ if

$$\alpha(x) = e\chi(x) \quad (9.74)$$

but the transformations (9.72) represent gauge transformations of the first kind. Thus we have shown that the requirement of gauge invariance of the second kind leads to that of the first kind and therefore to charge conservation.

It should be noted that the interaction term in the Lagrangian given in (9.69) is the simplest that can be formed. More complicated expressions, which are Lorentz and gauge invariant, are also possible, for example

$$\mathscr{L}_{\text{int}} = i\mu\bar{\psi}(\gamma_\mu\gamma_\nu - \gamma_\nu\gamma_\mu)\psi f_{\mu\nu} \quad (9.75)$$

where μ has the dimensions of a dipole moment and $f_{\mu\nu}$ is the electromagnetic field tensor (§ 4.4(b)) in operator form. This interaction was introduced by Pauli (1941), and is sometimes called the Pauli dipole moment interaction since it represents the interaction of a dipole moment with a field strength. The presence of this term has not been found necessary in order to explain the behaviour of pure electromagnetic processes.

9.3(c). *The principle of minimal electromagnetic interaction*

The fact that all electromagnetic interactions can be adequately described by the product of a current and a potential (9.69) is known as the *principle of minimal electromagnetic interaction* (Gell-Mann, 1956). The principle implies that the differential operator $\partial/\partial x_\mu$ acting on ψ in the total Lagrangian for the free particles is replaced by $(\partial/\partial x_\mu - ieA_\mu)$. This effect can be seen by comparing (9.69) and (9.70); an equivalent relation exists for the $\bar{\psi}$ field (see, for example equation (8.44)). We may summarise the rules as

$$\frac{\partial}{\partial x_\mu} \to \frac{\partial}{\partial x_\mu} - ieA_\mu \quad \text{for } \psi \text{ field} \quad (9.76)$$

$$\frac{\partial}{\partial x_\mu} \to \frac{\partial}{\partial x_\mu} + ieA_\mu \quad \text{for } \bar{\psi} \text{ field}$$

This result can be applied to any other charged field.

9.3(d). *The conservation of fermions*

Since fermions possess half integral spin, and orbital angular momentum occurs in integral units of $\hbar$, total angular momentum can only be conserved in any process if the fermions occur in multiples of two. The word 'fermion' means both particle and antiparticle in the present context. Further restrictions can be deduced; certain reactions involving fermions occur whilst others do not. For example the threshold of the production of antiprotons in proton–proton collisions occurs when the incident protons possess a kinetic energy 5·6 GeV in the laboratory system. This figure implies that the production process is of the type

$$p + p \rightarrow p + p + p + \bar{p}.$$

If the process

$$p + p \rightarrow p + \bar{p} + \pi^+ + \pi^+$$

occurred, the threshold would have been 0·6 GeV. The second process satisfies the condition enunciated above and also conserves charge, yet it does not occur. Another example may be quoted; the Λ-hyperon decays as

$$\Lambda \rightarrow p + \pi^-$$

but not

$$\Lambda \rightarrow \bar{p} + \pi^+.$$

On the other hand, the collision of an antiproton with a proton occasionally yields two Λ-hyperons one of which decays into $\bar{p} + \pi^+$. This process can be interpreted as

$$\bar{p} + p \rightarrow \bar{\Lambda} + \Lambda \rightarrow \bar{p} + \pi^+ + p + \pi^-.$$

The nucleons and hyperons are collectively called baryons, and the above reactions, taken in conjunction with many others, permit the deduction that the *number of baryons minus the number of antibaryons in any process is conserved.* The law was suggested by Wigner (1949), and appears to hold for strong, electromagnetic and weak interactions. The law has been tested by attempting to observe the decay of the proton (Giamati and Reines, 1962). The result obtained indicated that the lifetime of the proton against decay into a charged particle with kinetic energy greater than 100 MeV was $\tau_p > 1{\cdot}5 \times 10^{26}$ years. Even if we accepted this figure as a genuine decay rate, the weakness of the interaction can be judged by comparing it with the lifetime range of 10^{-6} to 10^{-10} sec for typical weak interactions.

The law of the conservation of baryons bears a strong formal resemblance to that of electrical charge. A 'baryonic charge' of magnitude $B = 1$ can be assigned to nucleons and hyperons whilst $B = -1$ is assigned to antinucleons and antihyperons. The conservation law may then be expressed by postulating a gauge invariance of the first kind

$$\psi \rightarrow \psi' = \psi e^{iB\lambda}, \qquad \bar{\psi} \rightarrow \bar{\psi}' = \bar{\psi} e^{-iB\lambda} \tag{9.77}$$

for all baryon fields (compare the previous section), thus leading to a conserved baryon current. The term λ in equation (9.77) is an arbitrary variable.

A further conservation law also appears to exist for fermions—it refers to the *leptons* which are particles specifically associated with weak interactions. The leptons are the electron, muon and neutrino and their antiparticles. The conservation law is similar to that for baryons, namely the *number of leptons minus the number of antileptons in any weak interaction is a constant*. This law can also be expressed in terms of a leptonic charge $l = \pm 1$ and a gauge invariance of the first kind

$$\psi \to \psi' = \psi e^{il\lambda}, \qquad \bar{\psi} \to \bar{\psi}' = \bar{\psi} e^{-il\lambda}. \tag{9.78}$$

Let us examine a classification scheme for the leptons in β-decay. It is customary to regard the neutrino appearing in β-decay as an antineutrino

$$n \to p + e^- + \bar{\nu}.$$

Thus it is immediately obvious that the electron must be of opposite leptonic charge to the antineutrino. The argument may be extended in the same way to positron decay and so we may classify the leptons and antileptons as follows:

$$l = -1 \text{ (say)} \qquad \nu e^-$$

$$l = +1 \qquad \bar{\nu} e^+.$$

The choice of -1 or $+1$ is, of course, arbitrary. Since the neutrino appears to be massless, the two-component theory can be used to argue that opposite helicities should be assigned to the neutrino and antineutrino (§ 3.3(l)). Experiment (§ 2.4(d)) shows that the neutrino is left-handed, and so we may write

$$l = -1, \qquad \nu_L e^- \tag{9.79}$$

$$l = +1, \qquad \bar{\nu}_R e^+.$$

The conservation of leptonic charge can be examined by studying the reactions induced by neutrinos. The conservation law implies that whilst the inverse reaction to β-decay

$$\bar{\nu} + p \to n + e^+$$

should occur, the equivalent process with a neutron replacing a proton should fail

$$\bar{\nu} + n \nrightarrow p + e^-.$$

The first reaction has been observed to occur with a cross-section of $\sim 10 \times 10^{-45}$ cm^2 with antineutrinos from a reactor (§ 1.2). A search for the second process by Davis (1955) in the reaction

$$\bar{\nu} + \mathrm{Cl}^{37} \rightarrow \mathrm{A}^{37} + e^-$$

has led to an upper limit on the cross-section of $\sim 0{\cdot}9 \times 10^{-45}$ cm^2.

The experiments using high energy neutrinos have also given strong evidence for leptonic conservation in the processes which involve muons. If we make the following designation for antineutrinos and neutrinos†

$$\pi^- \rightarrow \mu^- + \bar{\nu}$$

$$\pi^+ \rightarrow \mu^+ + \nu$$

in analogy with β-decay, then the following processes are observed to be allowed and forbidden (Siena conference report of CERN neutrino experiment (1964))

$$\nu + n \rightarrow p + \mu^-$$

$$\nu + p \nleftrightarrow n + \mu^+$$

$$\bar{\nu} + n \nleftrightarrow p + \mu^-$$

$$\bar{\nu} + p \rightarrow n + \mu^+.$$

As in the case of β-decay there is strong experimental evidence that the muonic neutrino is left handed and the antineutrino right handed (the evidence is discussed in Chapter 12). We shall therefore assign leptonic charges to neutrinos, electrons and muons in the following manner:

$$l = -1, \qquad \nu_L e^- \mu^- \tag{9.80}$$

$$l = +1, \qquad \bar{\nu}_R e^+ \mu^+.$$

A further important conservation law has been demonstrated by a study of the reactions induced by high energy neutrinos. Arguments, which we shall discuss in § 12.11 and 12.12 have led to the conclusion that the neutrinos associated with muons and electrons are intrinsically different. Experiments with high energy neutrinos from pion decay (§ 1.2) have shown the existence of a law of conservation of muonic number (§ 12.12).

We return, finally, to the statement made at the beginning of this section that fermions occur in pairs in any process. Consider the process of β-decay

$$n \rightarrow p + e^- + \bar{\nu}.$$

If we adopt the convention that the operator $a_r^\dagger(\mathbf{p})$ (4.212) creates particles whilst $b^\dagger(\mathbf{p})$ creates antiparticles, then the appropriate destruction and creation opera-

† We shall use the symbol ν generically for both electronic and muonic neutrinos, unless their distinction happens to be important.

tors for the above process and their related field operators are

$$\begin{array}{cccc} n \rightarrow & p\ + & e^-\ + & \bar{\nu} \\ \downarrow & \downarrow & \downarrow & \downarrow \\ a & a^\dagger & a^\dagger & b^\dagger \\ \downarrow & \downarrow & \downarrow & \downarrow \\ \psi_n & \bar{\psi}_p & \bar{\psi}_{e^-} & \psi_{\bar{\nu}} \end{array} \tag{9.81}$$

Thus we have two baryon and two lepton operators, each occurring in the combination $\bar{\psi}\psi$, and so the requisite gauge invariance conditions are always fulfilled. All interactions involving fermions in fact require the combination $\bar{\psi}\psi$ never $\psi\psi$ or $\bar{\psi}\bar{\psi}$.

9.4. ISOSPIN AND STRANGENESS

9.4(a). *Charge independence in strong interactions*

Abundant experimental evidence exists to show that nuclear forces are *charge independent*. By this statement we mean that if a neutron is replaced by a proton in any system, then, providing the particles are in the same state of angular momentum, the system is unaltered. This point is well illustrated in the level structure given in Fig. 9.5 for the nuclei B^{12}, C^{12}, N^{12}. It can be seen that many levels correspond closely to each other in the three nuclei.

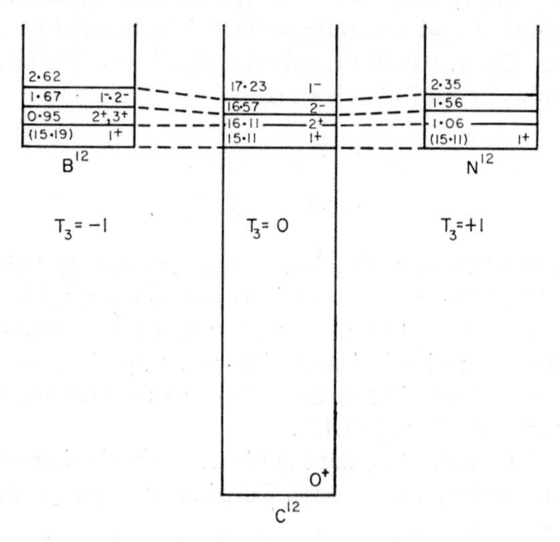

FIG. 9.5. Suspected $T = 1$ levels in B^{12}, C^{12} and N^{12}

It should be noted that the principle of charge independence does not imply that the cross-sections for *n–p* and *p–p* scattering are the same. The neutron and proton can interact in states which are forbidden to two protons by the Pauli exclusion principle.

The charge independence of nuclear forces can be conveniently formulated by introducing the concept of an isospin which acts in an isospin space (Cassen and Condon, 1936).† The latter concept was first introduced by Heisenberg (1932) as a method for describing the two charge states of the nucleon, in the same way as the spin states of the electron are described. The nucleon was assigned an isospin of $\frac{1}{2}$ and the proton and neutron states corresponded to 'spin up' and 'spin down' respectively. Spin up and down was taken along the 3-axis in isospin space (also called the z-axis). Similarly the equivalent levels in the nuclei B^{12}, C^{12}, N^{12} can be regarded as evidence for the existence of an isostate with spin $T = 1$, and components $-1, 0, +1$ along the 3-axis. The components may be identified by the relation

$$Z = \frac{A}{2} + T_3 \tag{9.82}$$

where Z and A refer to atomic number and weight respectively.

The idea of isospin can be extended to other strongly interacting particles; thus the three charge states of the pion can be regarded as eigenstates of an operator T_3 with eigenvalues $\pm 1, 0$

$$T_3 |\pi^+\rangle = |\pi^+\rangle, \qquad T_3 |\pi^-\rangle = -|\pi^-\rangle \tag{9.83}$$
$$T_3 |\pi^0\rangle = 0.$$

The existence of three states implies that

$$2T + 1 = 3, \qquad T = 1 \tag{9.84}$$

for the pion. Isospin assignments for the other strongly interacting particles are given in A. 9 (Appendixes, p. 715).

The analogy between isospin and angular momentum can be carried further by postulating invariance for strong interactions under rotations in isospin space. Under these conditions $\mathbf{T}^2$ and T_3 should be conserved (compare § 5.3). This requirement leads to many interesting relations which will be discussed in later sections. We shall mention one example here, since it is immediately obvious. Consider the hypothetical reaction

$$\begin{array}{ccccc} & d + & d \rightarrow & He^4 + & \pi^0 \\ T & 0 & 0 & 0 & 1 \end{array}$$

Neither the deuteron nor the He^4 nucleus have isobars and therefore they have $T = 0$. The pion has an isospin of one, however, and so the above reaction should be forbidden if isospin is conserved. Searches for the reaction have been made by Akimov, Savchenko and Soroko (1960) and by Poirier and Pripstein (1963). Their results indicate that the cross-section is at least 100 times smaller than that expected if isospin failed.

† This is one of three possible appellations; isospin is also called isotopic spin and isobaric spin—all three choices appear to be about equally popular.

Whilst isospin (or equivalently charge independence) appears to be conserved in strong interactions, electromagnetic interactions are obviously charge dependent since the photon is coupled through the electric charge. The superposition of electromagnetic effects on strong interactions therefore tends to destroy the conservation of T. Since T_3 is linked with charge, however, it is conserved; in other words T_3 commutes with the charge operator Q, for example

$$T_3 Q\,|\pi\rangle = QT_3\,|\pi\rangle \tag{9.85}$$

$$[T_3, Q] = 0.$$

The difference in mass between the various charge states of the strongly interacting particles is normally attributed to a violation of charge independence by electromagnetic interactions. This point is illustrated in the Table 9.3, where m refers to the average mass of the pair considered. It is apparent that the differences are of the order of the fine structure constant $\sim 1/137$.

TABLE 9.3

Particles	$\Delta m/m$
$\pi^{\pm} - \pi^0$	$3{\cdot}3 \times 10^{-2}$
$K^- - K^0$	$0{\cdot}8 \times 10^{-2}$
$n - p$	$0{\cdot}1 \times 10^{-2}$
$\Sigma^+ - \Sigma_-$	$0{\cdot}6 \times 10^{-2}$
$\Sigma^- - \Sigma^0$	$0{\cdot}4 \times 10^{-2}$

9.4(b). *Transformations in isospin space*

The formal structure of the angular momentum operator and the charge operator bear a striking resemblance to each other. Consider the operators for the charged scalar field (4.118)

$$\varphi = \frac{1}{\sqrt{2}}(\varphi_1 - i\varphi_2)$$

$$\varphi^\dagger = \frac{1}{\sqrt{2}}(\varphi_1 + i\varphi_2)$$

where φ_1 and φ_2 are both Hermitian. We shall denote the uncharged (Hermitian) scalar field by φ_3. The charge conservation of this field arises from the condition of gauge invariance under the transformations

$$\varphi \to \varphi' = \varphi e^{i\alpha}, \qquad \varphi^\dagger \to \varphi^{\dagger\prime} = \varphi^\dagger e^{-i\alpha} \tag{9.86}$$

(compare §§ 3.4(f.2) and 5.2).

This transformation leads to the conditions

$$\varphi_1 \to \varphi_1' = \frac{1}{\sqrt{2}}\,[\varphi(\cos\alpha + i\sin\alpha) + \varphi^\dagger(\cos\alpha - i\sin\alpha)]$$

$$= \frac{1}{\sqrt{2}}\,[(\varphi + \varphi^\dagger)\cos\alpha + i(\varphi - \varphi^\dagger)\sin\alpha]$$

$$= \varphi_1\cos\alpha + \varphi_2\sin\alpha$$

$$\varphi_2 \to \varphi_2' = -\varphi_1\sin\alpha + \varphi_2\cos\alpha$$

$$\varphi_3 \to \varphi_3' = \varphi_3 . \tag{9.87}$$

Thus the gauge transformation can be regarded as a rotation of the components φ_1 and φ_2 of a vector $\boldsymbol{\varphi}$ in some internal space of the field, the rotation being about the third axis φ_3.

Now we have already seen in § 5.2 that the generator of the gauge transformation is the charge operator for the field. If we use equation (4.138) we may write this operator as

$$Q = \frac{1}{i}\int d\mathbf{x}\, j_4$$

$$= \frac{e}{i}\int d\mathbf{x}\left(\frac{\partial\varphi_1}{\partial x_4}\,\varphi_2 - \frac{\partial\varphi_2}{\partial x_4}\,\varphi_1\right)$$

$$= -e\int d\mathbf{x}(\pi_1\varphi_2 - \pi_2\varphi_1) \tag{9.88}$$

where

$$\pi_i = \frac{\partial\varphi_i}{\partial t} \quad (i = 1, 2, 3). \tag{9.89}$$

The equation (9.88) has already appeared in a slightly different form. If we compare it with the third component of the spin operator for the photon field (5.52)

$$S_3 = -\int d\mathbf{x}(\pi_1 A_2 - \pi_2 A_1)$$

it can be seen that, apart from the electric charge and the trivial substitution of A for φ, the two expressions are identical. Thus the charge operator possesses similar properties to the third component of the spin operator of the vector field.

Thus we may introduce a special space in which the (spatial) scalar field can be regarded as a vector. This is called the isospin space and the φ-field is sometimes called an isovector field.† The gauge transformation (9.87) then represents

† Strictly speaking it is a pseudovector since the presence of the γ_5 term in the interaction $\mathcal{L}_{ps} = ig\bar{\psi}\gamma_5\psi\varphi$ (9.15) causes the nucleon part to behave as a pseudovector for reflections in isospin space.

a rotation of φ in this space. Furthermore, the charge operator Q can be regarded as the third component of a generator of rotations in the space. We will denote the generator by the symbol $\mathbf{T}$ and call it the isospin operator, where by analogy with (5.53) we may write

$$\mathbf{T} = -\int d\mathbf{x}\, \boldsymbol{\pi} \times \boldsymbol{\varphi} = \int d\mathbf{x}\, \boldsymbol{\varphi} \times \boldsymbol{\pi}. \tag{9.90}$$

It is a straightforward matter to show that the components of $\mathbf{T}$ obey the usual commutation laws for angular momentum operators (compare (5.54))

$$[T_i, T_j] = iT_k$$

where i, j and k are taken cyclically, and

$$T_i = -\int d\mathbf{x}(\boldsymbol{\pi} \times \boldsymbol{\varphi})_i = -\int d\mathbf{x}(\pi_j\varphi_k - \pi_k\varphi_j) \tag{9.91}$$

This equation can be written in matrix form as

$$T_i = \int d\mathbf{x}\, \frac{\partial\varphi^{(\dagger)}}{\partial x_\mu}\, t_i\varphi\delta_{\mu 4} = -\int d\mathbf{x}\varphi^{(\dagger)} t_i\, \frac{\partial\varphi}{\partial x_\mu}\, \delta_{\mu 4} \tag{9.92}$$

where

$$T_1 \equiv t_1 = \begin{pmatrix} 0 & 0 & 0 \\ 0 & 0 & -i \\ 0 & i & 0 \end{pmatrix} \qquad T_2 \equiv t_2 = \begin{pmatrix} 0 & 0 & i \\ 0 & 0 & 0 \\ -i & 0 & 0 \end{pmatrix} \tag{9.93}$$

$$T_3 \equiv t_3 = \begin{pmatrix} 0 & -i & 0 \\ i & 0 & 0 \\ 0 & 0 & 0 \end{pmatrix} \qquad \varphi = \begin{pmatrix} \varphi_1 \\ \varphi_2 \\ \varphi_3 \end{pmatrix}$$

$$\varphi^{(\dagger)} = (\varphi_1\, \varphi_2\, \varphi_3).$$

The matrix elements of T_i represent elements between two isospin states and are obtained by a full integration of field operators as in § 5.3(d). They can also be represented as†

$$(T_i)_{\beta\alpha} = i\varepsilon_{\beta i\alpha} \tag{9.94}$$

where the subscripts β and α refer to isospin states and the symbol $\varepsilon_{\beta i\alpha}$ behaves like

$$\varepsilon_{\beta i\alpha} = \begin{cases} +1 & \text{if } \beta i\alpha \text{ is an even permutation of 123,} \\ -1 & \text{if } \beta i\alpha \text{ is an odd permutation of 123,} \\ 0 & \text{if any two indices are identical.} \end{cases}$$

† Equations (9.93) and (9.94) have been cast in this form for later use; they can always be written in a more familiar form, for example, using the methods of equation (9.126) and (9.127) we find

$$< \pi^- | T_3 | \pi^- > = \tfrac{1}{2} < (\pi_1 + i\pi_2) | T_3 | (\pi_1 - i\pi_2) > = -1.$$

(compare A.4, Appendixes, p. 700). We may note for later use that the current density (4.138) and charge operators can be written as

$$j_\mu = ie\frac{\partial \varphi^{(\dagger)}}{\partial x_\mu} t_3\varphi = e\left(\frac{\partial \varphi_1}{\partial x_\mu}\varphi_2 - \frac{\partial \varphi_2}{\partial x_\mu}\varphi_1\right) \tag{9.95}$$

$$\frac{Q}{e} = \frac{1}{ie}\int d\mathbf{x}\, j_4 = \int d\mathbf{x}\frac{\partial \varphi^{(\dagger)}}{\partial x_\mu} t_3\, \varphi\delta_{\mu 4} = T_3$$

as expected (9.88).

We next consider nucleons. The nucleon has only two charge states, +1 and 0, and so the formalism developed above no longer applies. We introduce a spinor field with eight components

$$\psi_N = \begin{pmatrix}\psi_p\\ \psi_n\end{pmatrix} \tag{9.96}$$

where ψ_p and ψ_n are the ordinary four-component spinor fields for proton and neutron respectively.

We now wish to relate the operators T_3 and Q in a useful way. Consider the operators for charge and baryon conservation (compare equation (4.218) and the remark following (4.225)). They are given by

$$\begin{aligned} Q &= e\int d\mathbf{x}\, \bar{\psi}_p\gamma_4\psi_p = e\int d\mathbf{x}\, \psi_p^\dagger\psi_p \\ &= e\sum_{\mathbf{p},r}[N_r^+(\mathbf{p}) - N_r^-(\mathbf{p})]_{\text{protons}} \\ &= e(N_p - N_{\bar{p}}) \end{aligned} \tag{9.97}$$

and

$$B = \int d\mathbf{x}\, \bar{\psi}_N\gamma_4\psi_N = \int d\mathbf{x}\, \psi_N^\dagger\psi_N = N_p + N_n - N_{\bar{p}} - N_{\bar{n}} \tag{9.98}$$

where the occupation number operator

$$N_p = \sum_{\mathbf{p},r} N_r(\mathbf{p})$$

refers to all the proton states, and the field operators have not been written in their fully antisymmetrised form. The operators could also have been written as

$$Q = e\int d\mathbf{x}\, \psi_N^\dagger\begin{pmatrix}1 & 0\\ 0 & 0\end{pmatrix}\psi_N \tag{9.99}$$

$$B = \int d\mathbf{x}\, \psi_N^\dagger\begin{pmatrix}1 & 0\\ 0 & 1\end{pmatrix}\psi_N.$$

Now in nuclear physics T_3 is written as

$$T_3 = Z - \frac{A}{2}$$

(compare (9.82)). Thus a suitable operator for the nucleons can be written as

$$T_3 = \frac{Q}{e} - \frac{B}{2} = \tfrac{1}{2}\int d\mathbf{x}\, \psi_N^\dagger \tau_3 \psi_N. \tag{9.100}$$

The operator T_3 is linearly related to B and Q and so its eigenvalues are conserved quantities. It is apparent from equations (9.99) that τ_3 is mathematically equivalent to the Pauli spin operator σ_3

$$\tau_3 = \begin{pmatrix} 1 & 0 \\ 0 & -1 \end{pmatrix}. \tag{9.101}$$

This operator has the following property (compare (3.142)):

$$\tau_3 \psi_N = \begin{pmatrix} 1 & 0 \\ 0 & -1 \end{pmatrix} \begin{pmatrix} \psi_p \\ \psi_n \end{pmatrix} = \begin{pmatrix} \psi_p \\ -\psi_n \end{pmatrix} \tag{9.102}$$

so that if we introduce the terms

$$\chi_p = \begin{pmatrix} 1 \\ 0 \end{pmatrix}, \qquad \chi_n = \begin{pmatrix} 0 \\ 1 \end{pmatrix} \tag{9.103}$$

then

$$\tau_3 \chi_p = \chi_p, \qquad \tau_3 \chi_n = -\chi_n. \tag{9.104}$$

Thus ψ_N can be seen to be equivalent to the Pauli two component spinor (3.142) and τ_3 is the spin projection operator for this state. The terms ψ_N, χ_p and χ_n are therefore called *isospinors*, since they have the same properties in isospace as spinors have in ordinary space.

It should be noted that equation (9.100) can be written as

$$T_3 = \tfrac{1}{2}(N_p - N_n - N_{\bar{p}} + N_{\bar{n}}) \tag{9.105}$$

where we have used (9.97) and (9.98), therefore

$$T_3\,|p\rangle = \tfrac{1}{2}\,|p\rangle, \qquad T_3\,|\bar{p}\rangle = -\tfrac{1}{2}\,|\bar{p}\rangle \tag{9.106}$$

$$T_3\,|n\rangle = -\tfrac{1}{2}\,|n\rangle, \quad T_3\,|\bar{n}\rangle = \tfrac{1}{2}\,|\bar{n}\rangle.$$

Thus the antiproton isospin is 'down' relative to the proton, and that of the antineutron is 'up' relative to the neutron.

The remaining components of $\mathbf{T}$ can be defined from the mathematical equivalence of $\boldsymbol{\tau}$ and the Pauli spin operator $\boldsymbol{\sigma}$ (9.101) and lead to the vector

$$\mathbf{T} = \tfrac{1}{2}\int d\mathbf{x}\, \psi_N^\dagger \boldsymbol{\tau} \psi_N \tag{9.107}$$

$$\tau_1 = \begin{pmatrix} 0 & 1 \\ 1 & 0 \end{pmatrix}, \qquad \tau_2 = \begin{pmatrix} 0 & -i \\ i & 0 \end{pmatrix}, \qquad \tau_3 = \begin{pmatrix} 1 & 0 \\ 0 & -1 \end{pmatrix}.$$

This operator is the generator of rotations for the nucleon field in isospin space.

The formal manipulations associated with these rotations are in complete analogy with those discussed in §§ 5.3(a) and 5.4(d) for rotations in ordinary space, and so they will not be repeated here.

Two other operators are often encountered; they are represented as

$$\tau_+ = \tfrac{1}{2}(\tau_1 + i\tau_2) \qquad \tau_- = \tfrac{1}{2}(\tau_1 - i\tau_2) \tag{9.108}$$
$$= \begin{pmatrix} 0 & 1 \\ 0 & 0 \end{pmatrix} \qquad\qquad = \begin{pmatrix} 0 & 0 \\ 1 & 0 \end{pmatrix}$$

They transform neutron isospinors to proton isospinors and vice versa

$$\tau_+\chi_p = 0, \qquad \tau_+\chi_n = \chi_p \tag{9.109}$$
$$\tau_-\chi_p = \chi_n, \qquad \tau_-\chi_n = 0$$

and project out neutron and proton states from ψ_N

$$\tau_+\psi_N = \begin{pmatrix} 0 & 1 \\ 0 & 0 \end{pmatrix}\begin{pmatrix} \psi_p \\ \psi_n \end{pmatrix} = \begin{pmatrix} \psi_n \\ 0 \end{pmatrix} \tag{9.110}$$
$$\tau_-\psi_N = \begin{pmatrix} 0 & 0 \\ 1 & 0 \end{pmatrix}\begin{pmatrix} \psi_p \\ \psi_n \end{pmatrix} = \begin{pmatrix} 0 \\ \psi_p \end{pmatrix}.$$

9.4(c). *The pion–nucleon interaction* (1)

It is instructive to examine the implications of isospin for a charge independent interaction. In equation (9.15) the Lorentz invariant Lagrangian for the interaction of pseudoscalar and nucleon fields was given as

$$\mathscr{L}_{\text{int}} = ig\bar{\psi}\gamma_5\psi\varphi.$$

This equation represents the interaction for single component pion and nucleon fields. If we wish to include charge states, and at the same time keep the interaction invariant under rotations in isospin space, we must modify the above equation as follows:

$$\mathscr{L}_{\text{int}} = ig\bar{\psi}_N\gamma_5(\boldsymbol{\tau}\cdot\boldsymbol{\varphi})\,\psi_N. \tag{9.111}$$

Since a scalar product has been formed from two vectors the Lagrangian obviously remains invariant under rotations in isospin space. Now let us examine the implications of this expression for the charge states; we may expand it as

$$\begin{aligned}
\mathscr{L}_{\text{int}} &= ig\bar{\psi}_N\gamma_5(\tau_1\varphi_1 + \tau_2\varphi_2 + \tau_3\varphi_3)\,\psi_N \\
&= ig\bar{\psi}_N\gamma_5(\sqrt{2}\,\tau_+\varphi + \sqrt{2}\,\tau_-\varphi^\dagger + \tau_3\varphi_3)\,\psi_N \\
&= i\sqrt{2}\,g(\bar{\psi}_p\bar{\psi}_n)\gamma_5\begin{pmatrix} \psi_n \\ 0 \end{pmatrix}\varphi + i\sqrt{2}\,g\,(\bar{\psi}_p\bar{\psi}_n)\gamma_5\begin{pmatrix} 0 \\ \psi_p \end{pmatrix}\varphi^\dagger + ig(\bar{\psi}_p\bar{\psi}_n)\gamma_5\begin{pmatrix} \psi_p \\ -\psi_n \end{pmatrix}\varphi_3 \\
&= i\sqrt{2}\,g\bar{\psi}_p\gamma_5\psi_n\varphi + i\sqrt{2}\,g\bar{\psi}_n\gamma_5\psi_p\varphi^\dagger + ig\bar{\psi}_p\gamma_5\psi_p\varphi_3 - ig\bar{\psi}_n\gamma_5\psi_n\varphi_3
\end{aligned} \tag{9.112}$$

where we have used equations (9.102) and (9.110) and also the relation

$$\tau_1\varphi_1 + \tau_2\varphi_2 = \sqrt{2}\,(\tau_+\varphi + \tau_-\varphi^\dagger) \tag{9.113}$$

(compare (4.118) and (9.108)). Now the operator $\varphi = (\varphi_1 - i\varphi_2)/\sqrt{2}$ creates π^- mesons and absorbs π^+ mesons, whilst $\varphi^\dagger$ causes the reverse processes. Consequently the first term in (9.112) represents the following processes:

$$\begin{aligned} n &\to p + \pi^-, \qquad \pi^+ + n \to p \\ \bar{p} &\to \bar{n} + \pi^-, \qquad \pi^+ + \bar{p} \to \bar{n}. \end{aligned} \tag{9.114}$$

Thus if we consider the basic nucleon–nucleon interaction and assume that, apart from the charged states, all the other factors are the same in the interaction, then we may construct Fig. 9.6 with the help of (9.112).

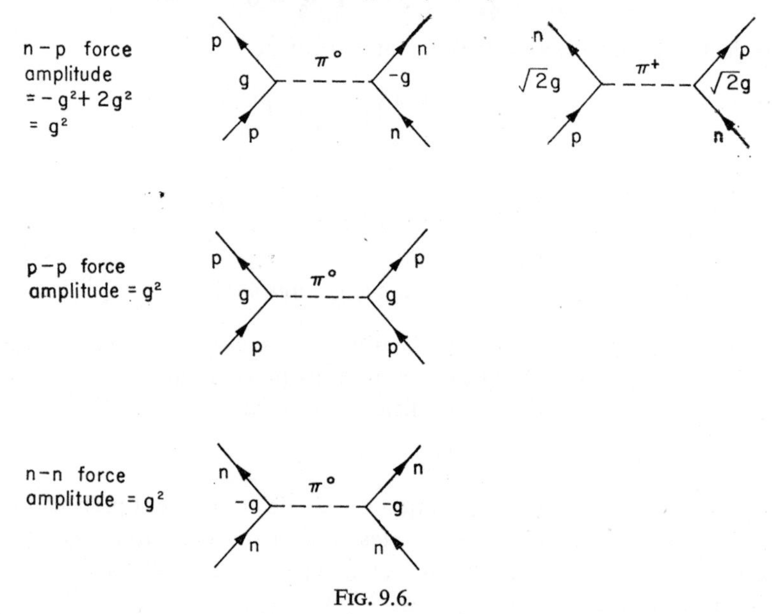

FIG. 9.6.

Thus the amplitudes of all three interactions are the same (providing all other physical conditions are equivalent) and nucleon–nucleon forces are charge independent. We have twisted the original argument to some extent. Kemmer (1938) showed that the charge independence of nuclear forces indicated that pions must exist in three charge states long before they had been identified by experiments.

9.4(d). *The pion–nucleon interaction* (2)

We shall now extend the principle of charge independence to pion–nucleon scattering. We shall work in the nucleon isospin space and label the meson isospin indices as α and β each of which takes on the values 1, 2 and 3. All other indices will be suppressed. The isospin part of the interaction contains two terms $\tau_\alpha\varphi_\beta$ and $\tau_\alpha\varphi_\beta$ (compare (9.111) and Fig. 14.4, for example), and so the scattering amplitude between isospin states β and α can be written as

$$T_{\beta\alpha} = T'\tau_\beta\tau_\alpha + T''\tau_\alpha\tau_\beta. \tag{9.115}$$

Since

$$\tau_\beta\tau_\alpha = \tfrac{1}{2}(\tau_\beta\tau_\alpha + \tau_\alpha\tau_\beta) + \tfrac{1}{2}(\tau_\beta\tau_\alpha - \tau_\alpha\tau_\beta) = \delta_{\beta\alpha} + \tfrac{1}{2}[\tau_\beta, \tau_\alpha] \tag{9.116}$$

we can rewrite (9.115) as

$$T_{\beta\alpha} = T_1\delta_{\beta\alpha} + \tfrac{1}{2}T_2[\tau_\beta, \tau_\alpha] \tag{9.117}$$

where T_1 and T_2 represent amplitudes for no isospin flip and isospin flip respectively.

We now wish to relate T_1 and T_2 to the amplitudes for the isospin channels and in turn to systems with specified electrical charges. Since the pion and the nucleon have isospin 1 and $\frac{1}{2}$ respectively, systems with total isospin $\frac{3}{2}$ and $\frac{1}{2}$ can be formed. We shall define the scattering amplitudes $T_{3/2}$ and $T_{1/2}$ for these states by the relation

$$T_{\beta\alpha} = T_{1/2}(P_{1/2})_{\beta\alpha} + T_{3/2}(P_{3/2})_{\beta\alpha} \tag{9.118}$$

where $P_{1/2}$ and $P_{3/2}$ are projection operators for the two states. We have shown in A.7 (Appendixes, p. 704) that the above equation leads to the relation

$$T_{\beta\alpha} = \tfrac{1}{3}(2T_{3/2} + T_{1/2})\,\delta_{\beta\alpha} + \tfrac{1}{6}(T_{1/2} - T_{3/2})\,[\tau_\beta, \tau_\alpha] \tag{9.119}$$

and so a comparison with (9.117) shows that

$$\begin{aligned} T_1 &= \tfrac{1}{3}(2T_{3/2} + T_{1/2}) \\ T_2 &= \tfrac{1}{3}(T_{1/2} - T_{3/2}) \end{aligned} \tag{9.120}$$

and

$$\begin{aligned} T_{3/2} &= T_1 - T_2 \\ T_{1/2} &= T_1 + 2T_2. \end{aligned} \tag{9.121}$$

Now let us examine the action of the operators $\tau_\alpha\tau_\beta$ on the isospin states. If we use equation (A.7.29) we can write

$$\tau_\beta\tau_\alpha = \delta_{\beta\alpha} - \langle\beta|\,\boldsymbol{\tau}\cdot\mathbf{T}\,|\alpha\rangle \tag{9.122}$$

where $\boldsymbol{\tau}\cdot\mathbf{T}$ has eigenvalues $+1$ and -2 for isospin states $\frac{3}{2}$ and $\frac{1}{2}$ respectively (A.7.23). Hence $\tau_\beta\tau_\alpha$ has eigenvalues

$$\begin{aligned} &0 \text{ for isospin } \tfrac{3}{2} \text{ states,} \\ &3 \text{ for isospin } \tfrac{1}{2} \text{ states,} \end{aligned} \tag{9.123}$$

and since

$$\tau_\beta\tau_\alpha + \tau_\alpha\tau_\beta = 2\delta_{\beta\alpha} \tag{9.124}$$

we may conclude that $\tau_\alpha \tau_\beta$ yields eigenvalues

$$2 \text{ for isospin } \tfrac{3}{2} \text{ states,} \qquad (9.125)$$
$$-1 \text{ for isospin } \tfrac{1}{2} \text{ states.}$$

Finally, let us consider charged particle states. Since the field $(\varphi_1 + i\varphi_2)/\sqrt{2}$ creates π^+ mesons (Table 4.1 (p. 138)), we can write

$$|\pi^+\rangle \equiv \frac{1}{\sqrt{2}} |(\pi_1 + i\pi_2)\rangle, \qquad |\pi^-\rangle \equiv \frac{1}{\sqrt{2}} |(\pi_1 - i\pi_2)\rangle. \qquad (9.126)$$

The elastic scattering of π^- mesons can therefore be represented as

$$\begin{aligned}\langle \pi^- p | T | \pi^- p\rangle &= \tfrac{1}{2} \langle (\pi_1 + i\pi_2) p | T | (\pi_1 - i\pi_2)\, p\rangle \\ &= \tfrac{1}{2} \langle p | T_{11} + T_{22} + iT_{21} - iT_{12} | p\rangle \\ &= \tfrac{1}{2} \langle p | T_1 + T_1 + \tfrac{1}{2} iT_2[\tau_2, \tau_1] - \tfrac{1}{2} iT_2[\tau_1, \tau_2] | p\rangle \\ &= \langle p | T_1 + T_2 \tau_3 | p\rangle = \langle p | T_1 + T_2 | p\rangle \\ &= T_1 + T_2 = T^- \end{aligned} \qquad (9.127)$$

where we have used (9.117). A similar treatment can be made for $\pi^+ p$ elastic scattering and for charge exchange scattering. In the former case a simpler treatment can be made; $\pi^+ p$ elastic scattering is in a pure isospin $\frac{3}{2}$ state and so from (9.121)

$$T^+ = T_{3/2} = T_1 - T_2.$$

Thus we find

$$T_1 = \tfrac{1}{2}(T^- + T^+) \qquad (9.128)$$
$$T_2 = \tfrac{1}{2}(T^- - T^+)$$

and with the aid of (9.120)

$$T^+ = T_{3/2} \qquad (9.129)$$
$$T^- = \tfrac{1}{3} T_{3/2} + \tfrac{2}{3} T_{1/2}.$$

Since the cross-section for any process is proportional to $|T|^2$, equations (9.129) imply that the cross-section for elastic $\pi^+ p$ scattering should be nine times greater than that for elastic $\pi^- p$ collisions in the corresponding angular momentum channel, if the $T_{1/2}$ amplitude is negligible. The experimental data confirm this ratio.

In order to provide a complete description of the scattering it is necessary to include nuclear spin variables. Let k and k' represent the initial and final four-momenta for the pions and p and p' the same for the nucleons, then conservation of energy and momentum can be expressed as

$$k + p = k' + p'$$

and we may express the scattering amplitude as

$$T_{\beta\alpha} = \bar{u}_{p'} O_{\beta\alpha} u_p \tag{9.130}$$

where $O_{\beta\alpha}$ is a Lorentz scalar and a 4×4 matrix in the spinor space of the nucleons. $O_{\beta\alpha}$ is therefore a function of the four-momentum variables k, k', p and p' and the γ-matrices taken in scalar combinations

$$O_{\beta\alpha} = f_S + f_V \gamma_\lambda + f_T \gamma_\lambda \gamma_\mu + f_A \gamma_\lambda \gamma_5 + f_P \gamma_5$$

where the functions f transform like scalars, vectors, tensors, axial vectors and pseudoscalars respectively. A suitable expression is given by

$$O_{\beta\alpha} = A_{\beta\alpha} - i\gamma \left(\frac{k + k'}{2}\right) B_{\beta\alpha} \tag{9.131}$$

where A and B are scalars and the factor 2 and the minus sign are introduced for later convenience.

The choice of this function is determined by arguments similar to those developed in § 11.4(c) for electron-nucleon scattering, for example the functions $p + p'$ and $p - p' = k' - k$ are not independent variables, because if we insert them in the Dirac equations (3.146)

$$(i\gamma p + M)\, u = 0$$

$$\bar{u}\, (i\gamma p + M) = 0$$

we find

$$i\bar{u}_{p'} \gamma (p + p')\, u_p = i\bar{u}_{p'} \gamma p u_p + i\bar{u}_{p'} \gamma p' u_p = -2M \bar{u}_{p'} u_p$$

$$i\bar{u}_{p'} \gamma (p - p') u_p = 0.$$

When we include the isospin terms (9.117) the full expression for the invariant scattering amplitude becomes

$$T_{fi} = T_{\beta\alpha}$$

$$= \bar{u}_{p'} \left\{ \left[A_1 - i\gamma \left(\frac{k + k'}{2}\right) B_1 \right] \delta_{\beta\alpha} + \tfrac{1}{2} \left[A_2 - i\gamma \left(\frac{k + k'}{2}\right) B_2 \right] [\tau_\beta, \tau_\alpha] \right\} u_p . \tag{9.132}$$

In the case of forward scattering $k = k'$, and so T_{fi} becomes a function of energy and isospin only.

Invariant expressions can be constructed for other processes under the assumption of charge independence; that for pion–pion scattering is

$$T_{\gamma\alpha,\delta\beta} = \tfrac{1}{3}\delta_{\alpha\beta}\delta_{\gamma\delta}(T_0 - T_2) + \tfrac{1}{2}\delta_{\alpha\gamma}\delta_{\beta\delta}(T_1 + T_2) - \tfrac{1}{2}\delta_{\alpha\delta}\delta_{\beta\gamma}(T_1 - T_2) \quad (9.133)$$

where the subscripts 0, 1 and 2 refer to the isospin channels (see, for example, Gasiorowicz, 1960).

9.4(e). *Vector addition coefficients and isospin*

The similarity of the isospin and angular momentum operators implies that isospin states can be combined in exactly the same manner as angular momentum states. Thus we can use all the relevant results of § 5.3. We note one useful relation at this point. The raising and lowering operators of equation (5.46) can be written as

$$T_- |T, T_3\rangle = \sqrt{[T(T+1) - T_3(T_3 - 1)]}\, |T, T_3 - 1\rangle \quad (9.134)$$

$$T_+ |T, T_3\rangle = \sqrt{[T(T+1) - T_3(T_3 + 1)]}\, |T, T_3 + 1\rangle$$

in isospin notation. Other results will be introduced in the relevant sections.

9.4(f). *The principle of extended symmetry*

In § 9.2(g) the spin–statistics relation was discussed and it was shown that systems of two identical fermions, for example protons, must be described by antisymmmetrical state functions, whilst systems of two identical bosons require symmetrical state functions.

The concept of isospin implies that the classification of symmetrical and antisymmetrical systems can be extended to particles which do not necessarily have the same charge.

Let us consider two nucleons. They each have isospin $\frac{1}{2}$, the proton being represented by a component $+\frac{1}{2}$ along the 3-axis and the neutron by $-\frac{1}{2}$ along the same axis. Using Clebsch–Gordan coefficients (A.7 (Appendixes, p. 704)) we can construct the states shown in Table 9.4.

TABLE 9.4

T_3	$T = 1$	$T = 0$
1	$(\frac{1}{2})(\frac{1}{2})$	
0	$\frac{1}{\sqrt{2}}[(\frac{1}{2})(-\frac{1}{2}) + (-\frac{1}{2})(\frac{1}{2})]$	$\frac{1}{\sqrt{2}}[(\frac{1}{2})(-\frac{1}{2}) - (-\frac{1}{2})(\frac{1}{2})]$
-1	$(-\frac{1}{2})(-\frac{1}{2})$	
	Symmetrical	Antisymmetrical

It is evident that the states $T_3 = +1$ and $T_3 = -1$ represent two protons and two neutrons respectively, and both are symmetrical in isospin space. The states $T_3 = 0$ can be symmetrical or antisymmetrical. For overall antisymmetry in the two nucleon system the space or spin parts of the $T = 1$ state must be antisymmetric, whilst they must be symmetric for the $T = 0$ state. The last condition

TABLE 9.5

Particles	T	T_3	Spectroscopic classification for space and spin components
pp	1	1	$^1S_0^+ \; ^1D_2^+ \ldots$
np	1	0	
nn	1	-1	$^3P_{0,1,2}^- \; ^3F_{2,3,4}^- \ldots$
np	0	0	$^3S_1^+ \; ^3D_{1,2,3}^+$ $^1P_1^- \; ^1F_3^-$

may be fulfilled if both space and spin parts are symmetric or both are antisymmetric. The possible configurations for two nucleons are summarised in Table 9.5.

We next consider the two-pion states. Since the isospin of the pion is one, we can construct states with $T = 0$, 1 and 2 and $T_3 = 2, 1, 0, -1$ and -2. Use of the Clebsch–Gordan coefficients (A.7 (Appendixes, p. 704)) yields the combinations shown in Table 9.6.

TABLE 9.6

T_3	$T = 2$	$T = 1$	$T = 0$
2	$(+1)(+1)$		
1	$\frac{1}{\sqrt{2}}[(+1)(0)+(0)(+1)]$	$\frac{1}{\sqrt{2}}[(+1)(0)-(0)(+1)]$	
0	$\frac{1}{\sqrt{6}}[(+1)(-1) + (-1)(+1)+2(0)(0)]$	$\frac{1}{\sqrt{2}}[(+1)(-1) - (-1)(+1)]$	$\frac{1}{\sqrt{3}}[(+1)(-1) + (-1)(+1)-(0)(0)]$
-1	$\frac{1}{\sqrt{2}}[(-1)(0)+(0)(-1)]$	$\frac{1}{\sqrt{2}}[(0)(-1)-(-1)(0)]$	
-2	$(-1)(-1)$		
	Symmetrical	Antisymmetrical	Symmetrical

Thus it can be seen that the states containing two identical particles are symmetric in isospin space, but the others have mixed symmetries. Overall symmetry in space and isospin in two-pion systems can be maintained if the combinations shown in Table 9.7 occur.

TABLE 9.7

Particles	T	T_3	Relative orbital angular momentum L
$\pi^+ \pi^+$	2	2	
$\pi^0 \pi^0$	0, 2	0	0, 2, 4, ...
$\pi^- \pi^-$	2	−2	
$\pi^+ \pi^0$	2	1	
$\pi^+ \pi^-$	2	0	0, 2, 4, ...
$\pi^- \pi^0$	2	−1	
$\pi^+ \pi^0$	1	1	
$\pi^+ \pi^-$	1	0	1, 3, 5, ...
$\pi^- \pi^0$	1	−1	

Thus states of odd isospin require odd orbital angular momentum to maintain overall symmetry. The antisymmetry of the $T=1$ state could have been decided upon intuitively by noting the close correspondence of the formal properties of isospin and angular momentum operators (compare § 9.2(g)).

9.4(g). *G conjugation and isospin space*

An interesting property, which we shall have occasion to use later, is the G parity of a system (Lee and Yang, 1956b). The operation of G conjugation is defined as the product of a rotation through π about the 2-axis in isospin space†

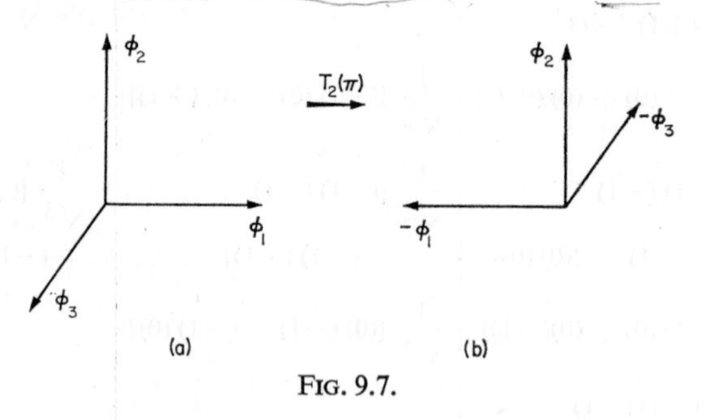

FIG. 9.7.

† The operation of rotation through π about the T_2 axis is sometimes called R conjugation

$$R = T_2(\pi).$$

and the charge conjugation operator

$$G = T_2(\pi)C. \tag{9.135}$$

Since all strong interactions are charge independent (if electromagnetic effects can be ignored) and invariant under charge conjugation, they are also G invariant.

Consider the components of the pion system in isospin space given in Fig. 9.7(a). A rotation through π about the 2-axis induces the following changes:

$$\varphi_1 \xrightarrow{T_2(\pi)} -\varphi_1$$

$$\varphi_2 \longrightarrow +\varphi_2$$

$$\varphi_3 \longrightarrow -\varphi_3$$

as can be seen in Fig. 9.7(b). Now consider charge conjugation; the fields φ and $\varphi^\dagger$ (4.118),

$$\varphi = \frac{1}{\sqrt{2}}(\varphi_1 - i\varphi_2)$$

$$\varphi^\dagger = \frac{1}{\sqrt{2}}(\varphi_1 + i\varphi_2)$$

transform as $\varphi \to \varphi^\dagger$ and $\varphi^\dagger \to \varphi$ under charge conjugation (compare (5.133) and (5.134), where we have set $\xi_C = \xi_C^* = +1$ (9.26)). Thus the components φ_1, φ_2 and φ_3 transform as

$$\varphi_1 \xrightarrow{C} \varphi_1$$

$$\varphi_2 \longrightarrow -\varphi_2$$

$$\varphi_3 \longrightarrow \varphi_3$$

under charge conjugation, and so the operation of G conjugation yields

$$(\varphi_1, \varphi_2, \varphi_3) \xrightarrow{G} (-\varphi_1, -\varphi_2, -\varphi_3). \tag{9.136}$$

Thus the pion has negative G parity and for a system containing n pions, we can represent the intrinsic G parity as

$$G = (-1)^n. \tag{9.137}$$

One consequence of this relation is an extension of Furry's theorem (§ 9.2(d)). Using the argument given in that section, one may easily show that the matrix elements, associated with graphs with an odd number of meson lines emerging (Fig. 9.8), must equal zero. Thus a process of the type $\pi^+ + \pi^- \to \pi^0 + \pi^0 + \pi^0$ is absolutely forbidden, if G conjugation holds.

Now let us consider neutral mesons of zero strangeness ($S = B = Q = 0$). A simple relation exists between G, T and ξ_C (the intrinsic charge parity) for these particles. We first note that the particles are self-conjugate (§ 5.5(c)), that is that they are in eigenstates of C with $\xi_C = \pm 1$. Because of the formal correspondence

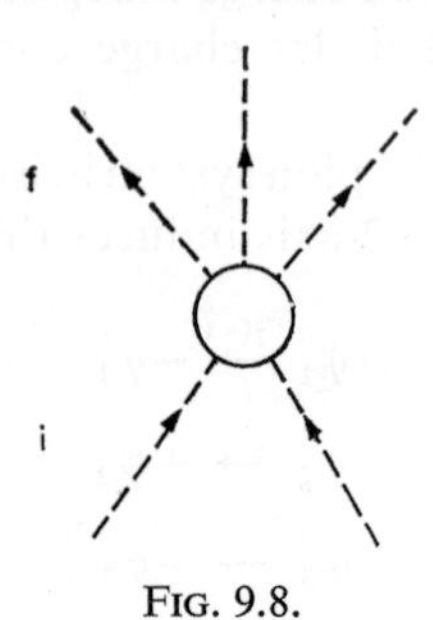

FIG. 9.8.

between isospin and angular momentum, we may represent the isospin wave function as $Y_T^0(\cos\theta)$, where the superscript arises through the relation (1.22)

$$\frac{Q}{e} = T_3 + \frac{B}{2} + \frac{S}{2}$$

and $\cos\theta$ represents the projection of the isospin vector along the 3-axis. Now upon a rotation of π about the 2-axis $\cos\theta \to \cos(\theta + \pi)$, and so the neutral mesons have the isospin symmetry property $(-1)^T$. Thus we find

$$G = \xi_C(-1)^T. \tag{9.138}$$

This result may be extended to a pair of charged bosons with the overall quantum numbers $S = B = Q = 0$. We then have $\xi_C = (-1)^L$ (compare (5.137)), and so

$$G = (-1)^{L+T}. \tag{9.139}$$

The action of the G conjugation operator on the spinor field can be readily evaluated. The field ψ transforms in isospin space in the same way as in ordinary space

$$\psi \to \psi' = S\psi = e^{i\frac{1}{2}\tau_k\varepsilon}\psi = \left(\cos\frac{\varepsilon}{2} + i\,\tau_k \sin\frac{\varepsilon}{2}\right)\psi \tag{9.140}$$

(compare (5.113) and (5.114)), where ε represents a rotation in the ij plane about the axis k. Thus for a rotation $\varepsilon = \pi$ about the 2-axis, we find

$$\psi_N = \begin{pmatrix}\psi_p\\ \psi_n\end{pmatrix} \xrightarrow{T_2(\pi)} \begin{pmatrix}\psi_n\\ -\psi_p\end{pmatrix} \tag{9.141}$$

since

$$i\tau_2 = \begin{pmatrix} 0 & 1 \\ -1 & 0 \end{pmatrix}$$

Finally, charge conjugation (5.148) leads to

$$\begin{pmatrix} \psi_n \\ -\psi_p \end{pmatrix} \xrightarrow{C} \begin{pmatrix} \gamma_2\gamma_4\bar{\psi}_n^T \\ -\gamma_2\gamma_4\bar{\psi}_p^T \end{pmatrix}. \tag{9.142}$$

9.4(h). *Hypercharge and strangeness*

Since the photon field is coupled directly to the electrical charge, electromagnetic interactions violate T but not T_3 invariance as we have already noted in § 9.4(a). A specific example of the relationship between the charge operator and the T_3 operator was given in equation (9.100) for the nucleon field.

$$T_3 = \frac{Q}{e} - \frac{B}{2}$$

or

$$\frac{Q}{e} = T_3 + \frac{B}{2}.$$

The relation between Q and T can be given in a more general fashion for all strongly interacting particles by introducing a *hypercharge operator* (d'Espagnat and Prentki, 1956)

$$\frac{Q}{e} = T_3 + \frac{Y}{2} \tag{9.143}$$

where Y is the centre of charge or average charge of any multiplet. Formally, the hypercharge operator and its eigenvalues can be constructed by postulating the invariance of interactions under a hypercharge gauge transformation for the fields (compare (9.77))

$$\eta \rightarrow \eta' = \eta e^{iY\lambda}, \qquad \eta^\dagger \rightarrow \eta^{\dagger\prime} = \eta^\dagger e^{-iY\lambda} \tag{9.144}$$

It is apparent from (9.143) that since Q and T_3 are both conserved in strong and electromagnetic interactions, Y should also be conserved in these interactions. Furthermore equation (9.143) shows us that if we have some particle state $|a\rangle$, and

$$Y\,|a\rangle = y\,|a\rangle \tag{9.145}$$

where y is the eigenvalue of Y, then

$$Y\,|\bar{a}\rangle = -y\,|\bar{a}\rangle \tag{9.146}$$

where $|\bar{a}\rangle$ is the corresponding antiparticle state to $|a\rangle$. Unless the distinction is important we shall normally use the symbol Y for both operator and eigenvalue. Using (9.143), values for Y for the particles appearing in Table 1.4 (p. 19), and their corresponding antiparticles, are listed in Table 9.8.

TABLE 9.8

Y	$B=+1$	$B=0$	$B=-1$
$+1$	N	K	$\overline{\Xi}$
0	Σ, Λ	π	$\overline{\Sigma}, \overline{\Lambda}$
-1	Ξ	$\overline{K}$	$\overline{N}$

The difference between Y and B is called the *strangeness* S of the system (§ 1.4(c))

$$S = Y - B \tag{9.147}$$

hence

$$\frac{Q}{e} = T_3 + \frac{B}{2} + \frac{S}{2}. \tag{9.148}$$

Chronologically the strangeness operator was introduced by Gell-Mann (1956) and Nishijima (1955) before the hypercharge operator. The scheme involving Y was suggested as a more symmetrical arrangement of the elementary particles.

A comparison of Tables 9.8 and 1.4 shows that the classification of particles using the hypercharge quantum number is more economical in numbers than one involving strangeness. The unitary symmetry scheme (§ 13.5(a)) also favours the use of hypercharge.

It is apparent from (9.148) that, since Q, T_3 and B are conserved in strong and electromagnetic interactions, S must also be conserved. In weak interactions involving strongly interacting particles, however, S, T_3 and Y are not conserved. Consider, for example,

$$\Lambda \rightarrow p + \pi^- \tag{9.149}$$

$$\begin{array}{cccc} T_3 & 0 & \frac{1}{2} & -1 \\ S & -1 & 0 & 0 \\ Y & 0 & 1 & 0. \end{array}$$

If further examples are considered the following rule is found for the weak decays into strongly interacting particles:

$$\Delta S = \pm 1, \qquad \Delta Y = \pm 1. \tag{9.150}$$

9.5. THE INVARIANCE LAWS AND THE PROPERTIES OF THE BOSONS

9.5(a). *Introduction*

The symmetry laws have two complementary applications:
(1) imposing limitations on the possible form for a given reaction;
(2) exploitation of the laws to yield the intrinsic properties of the particles.
This section will be concerned with the second application.

9.5(b). *The properties of the pions*

Mention has already been made in this chapter (§ 9.2(a)) and earlier (§ 2.4) of the use of detailed balancing to show that the π^+ meson has zero spin. We now consider some further properties.

We examine firstly the parity of the π^- meson. When π^- mesons are brought to rest in liquid hydrogen or deuterium the following reactions are observed (Panofsky, Aamodt and Hadley, 1951):

$$\pi^- + p \rightarrow n + \gamma \tag{a}$$

$$n + \pi^0$$

$$\pi^- + d \rightarrow n + n \tag{b}$$

$$n + n + \gamma.$$

The notable feature about these transitions is the absence of π^0 mesons from the reaction in deuterium, whereas in hydrogen about one-half of the reactions produce π^0 mesons. This result may be understood if the π^- meson possesses odd parity.

Let us consider the initial state of the reaction (b); the deuteron and pion have spin and parity 1^+ and $0^{\xi_{P\pi}}$ respectively, and the pion is captured from an s-shell about the deuterium nucleus,† therefore we may conclude that the spin and parity of the initial state is $j^{\xi_{P\pi}} = 1^{\xi_{P\pi}}$. Now let us consider the final state of the process

$$\pi^- + d \rightarrow n + n.$$

This contains two identical fermions and must be antisymmetrical. If we consult equation (9.54) we find that the only possible two neutron state with $j = 1$ is given as ${}^{2s+1}L_j^{\xi_P} = {}^3P_1^-$. Hence we may conclude that $\xi_{P\pi} = -1$, that is the π^- meson has odd parity.

† Calculations by Brueckner, Serber and Watson (1951) have shown that capture from a p-shell is much slower than capture from an s-shell.

A similar parity assignment may be made for the π^0 meson. Two arguments have been made, one concerns the low cross-section in the process $p + p \to p + p + \pi^0$ at threshold and will be given in § 14.3(a); the other is based on the decay process $\pi^0 \to \gamma + \gamma$.

Let us assume that the π^0 meson has zero spin (this point is discussed below) then the only suitable transition amplitude is of the form

$$a\mathbf{e}_1 \cdot \mathbf{e}_2 \varphi_{P\pi} + b(\mathbf{e}_1 \times \mathbf{e}_2) \cdot \mathbf{k} \varphi_{P\pi} \tag{9.151}$$

where a and b are scalar quantities, $\mathbf{e}_1$ and $\mathbf{e}_2$ electric polarisation vectors for the two photons, $\varphi_{P\pi}$ is a parity function for the π^0 meson (a scalar or pseudoscalar function) and $\mathbf{k}$ is the momentum of one of the photons (the momentum of the other photon is not an independent variable since it is directly related to $\mathbf{k}$ by conservation of momentum). Equation (9.151) represents the only expression which is both invariant under proper Lorentz transformations and symmetric under the exchange of the two photons as required by Bose–Einstein statistics. However, if the system is subjected to spatial reflections $(\mathbf{e}_1 \times \mathbf{e}_2) \cdot \mathbf{k}$ changes sign. Thus if parity is assumed to be preserved in the decay a must equal zero for a pseudoscalar pion, and $b = 0$ for a scalar pion. Now $\mathbf{e}_1 \cdot \mathbf{e}_2$ and $\mathbf{e}_1 \times \mathbf{e}_2$ are zero when the electric polarisation vectors are at 90° and 0° respectively. Thus we expect that the planes of polarisation are parallel or perpendicular respectively if the π^0 meson is scalar or pseudoscalar.

The direct measurement of the planes of polarisation of both photons has not been achieved since there is no known way of measuring the plane polarisation of high energy photons with high efficiency. However, the rare decay of the π^0 meson into two electron–positron pairs has been used to demonstrate that the π^0 particle is indeed pseudoscalar (Plano *et al.*, 1959). The correlation between the photon polarisations appears in this process as a correlation between the planes of the electron pairs. It is not difficult to see that the plane of the electron–positron pair lies preferentially in the plane of the electric vector, that is $\mathbf{e}$. The detailed theory of the process has been given by Kroll and Wada (1955).

We consider, finally, the spin of the π^0 meson. The best experimental evidence that it is zero is that it is *consistent* with all the data on the scattering and production of π^0 mesons near threshold. No experiment exists, however, which offers direct proof of this statement. The invariance laws show, however, that the spin of the neutral pion cannot be one. Consider the decay process

$$\pi^0 \to \gamma + \gamma$$

and let the two γ-rays emerge parallel and antiparallel to the axis of quantisation (the z-axis). Since the photon spins point parallel or antiparallel to their direction of motion the z-component of angular momentum can only take on the value

$$S_z = 0, 2. \tag{9.152}$$

Now if the spin of the π^0 meson is one, the only possible value for S_z is 0; this state can be represented by the angular momentum function $Y_1^0(\theta, \varphi)$ where $\theta = 0$ is along the $+z$-axis. Now the function $Y_1^0(\theta, \varphi)$ can be written as

$$Y_1^0 = \sqrt{\left(\frac{3}{4\pi}\right)} \cos \theta$$

(A.7 (Appendixes, p. 704)) and a rotation of this function through 180° causes its sign to change. However, a rotation through 180° is equivalent to interchanging the two photons, and since they are bosons their overall wave function must remain unchanged. Thus a particle of spin one cannot decay into two photons.

9.5(c). *Decay processes yielding three pions*

In this section we shall examine a method which was first developed by Dalitz (1953, 1954) for determining the properties of the K^- meson. This was done by examining the process

$$\tau \equiv K \rightarrow \pi^+ + \pi^+ + \pi^-.$$

The technique has since been applied to the pion resonance states.

We first note that since three spinless particles are involved in the final state, then two independent scalars are necessary to specify the kinematics of the decay process.

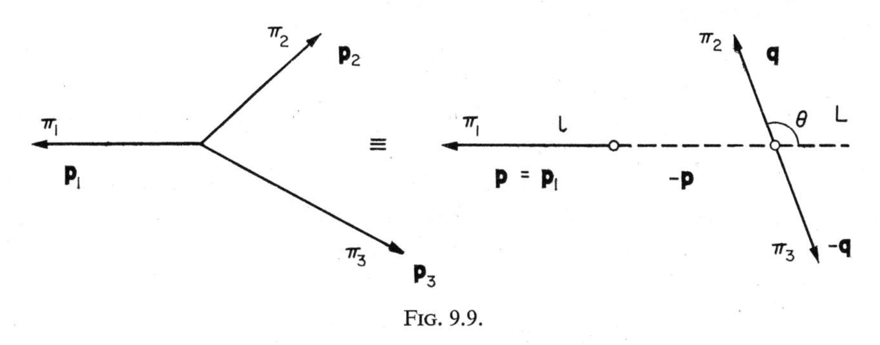

FIG. 9.9.

It is convenient to regard the three-pion system as a pion plus a dipion, each recoiling with momentum $\mathbf{p}$ and with the two mesons composing the dipion having momentum $\mathbf{q}$ and $-\mathbf{q}$ in their c-system (Fig. 9.9). It is a simple matter to show that

$$q = |\mathbf{q}| = \tfrac{1}{2}(M^2 - 3m_\pi^2 - 2M\omega_p)^{1/2}$$

where M is the mass of the kaon and

$$\omega_p^2 = m_\pi^2 + \mathbf{p}^2, \qquad |\mathbf{p}| = p$$

and so p and θ are chosen as independent variables.

Let L and l represent the angular momenta associated with the vectors **q** and respectively, then conservation of angular momentum gives

$$\mathbf{j} = \mathbf{L} + \mathbf{l} \tag{9.153}$$

$$|L - l| \leqq j \leqq L + l$$

where j is the total angular momentum of the three-pion system (and the spin o the kaon). It is convenient for our present problem to choose the two π^+ meson as the components of the dipion. Since they are identical bosons their wave func tion must be symmetric upon interchange of particles, that is reversal of **q**, an so the parity and hence the L-value of the dipion system must be even (compar Table 9.7). Now the intrinsic parity of the pion is -1, and so the overall parit of the three-pion system is given by

$$(\xi_{P\pi})^3 (-1)^l = (-1)^{l+1}. \tag{9.154}$$

If parity was conserved in weak interactions, this value would also represent th intrinsic parity of the kaon. The failure of parity conservation in weak inter actions means that we cannot specify the kaon parity from a decay process.

Now let us consider the matrix element T_{fi} for the decay process. Two spher ical wave functions will appear in this expression corresponding to the π^- an dipion systems. We may use vector addition coefficients to combine these func tions, and so we expect T_{fi} to be of the form

$$T_{fi} \equiv T_{fi}^m(\mathbf{p}, \mathbf{q}) \propto \mathscr{Y}_j^m f_{Ll}(p^2, q^2) \tag{9.155}$$

$$\propto \sum_{L,l,m_L,m_l} C_{j\ L\ l}^{m\ m_L\ m_l} Y_l^{m_l}(\theta_p)\, Y_L^{m_L}(\theta_q)\, f_{Ll}(p^2, q^2) \tag{9.156}$$

(compare (5.70)), where m is the magnetic quantum number associated with j and $f_{Ll}(p^2, q^2)$ is a product of spherical Bessel functions. If we choose the direc tion **p** as axis of quantisation, then $\cos\theta_p = 1$, $m_l = 0$ and $Y_l^{m_l}(\theta_p) =$ constant and so the above equation reduces to

$$T_{fi}^m(\mathbf{p}, \mathbf{q}) \propto \sum_{L,l} C_{j\ L\ l}^{m\ m\ 0} Y_L^m(\theta)\, f_{Ll}(p^2, q^2) \tag{9.157}$$

where θ is the angle between **p** and **q** (Fig. 9.9).

Now let us consider the term f_{Ll}

$$f_{Ll}(p^2, q^2) = j_l(pr)\, j_L(qr)\,. \tag{9.158}$$

Since the Q-value of the process $K \to 3\pi$ is 75 MeV the maximum kinetic energ of a pion is 50 MeV. Furthermore a suitable radius of interaction for the systen is $r \sim 1/m_\pi \sim 1/140\ \text{MeV}^{-1}$. Hence we may approximate f by

$$f_{Ll}(p^2, q^2) \sim \frac{p^l}{(2l+1)!!}\ \frac{q^L}{(2L+1)!!}\ \frac{1}{(140)^{l+L}} \tag{9.159}$$

A.7 (Appendixes, p. 704)), and it is not difficult to see that only the lowest values of L and l, consistent with the overall configuration, contribute to the sum in (9.157). Let us consider the first three configurations; they may be constructed from the conditions (9.153) and (9.154), and are as follows:

Configuration	L	l
0^-	0	0
1^+	0	1
1^-	2	2

(the configuration 0^+ is forbidden). Inspection of equations (9.157) and (9.159) then shows that $|T_{fi}|^2$ should be flat for the 0^- configuration, be zero at $p = 0$ and have a maximum value at $p = p_{\max}$ for 1^+ and have zeros at $p = 0$ and $p = p_{\max}$ for 1^- (the condition $p = p_{\max}$ corresponds to $q = 0$).

In addition to the matrix element T_{fi} we must also consider the phase space factor. If we use equations (7.40) and (7.36) the transition probability for a decay into three particles can be written as

$$d\lambda_{fi} \propto \overline{\sum_i} \sum_f |T_{fi}|^2 \frac{d\mathbf{p}_1}{2\omega_1} \frac{d\mathbf{p}_2}{2\omega_2} \frac{d\mathbf{p}_3}{2\omega_3} \delta(p_i - p_f) \tag{9.160}$$

where the energies and momenta refer to the kaon rest system. We may use the δ-function (compare (7.68)) to reduce this equation to one which involves two independent scalars

$$\begin{aligned} d\lambda_{fi} &\propto \overline{\sum_i} \sum_\alpha |T_{fi}|^2 \, d\omega_1 \, d\omega_2 \\ &= \overline{\sum_i} \sum_\alpha |T_{fi}|^2 \, d\omega_p \, d\omega_2 \end{aligned}$$

since $d\omega_1 = d\omega_p$ from Fig. 9.9. Now it is a simple matter to show that

$$\cos\theta = \frac{(\omega_2 - \omega_3)}{p} \frac{\omega_q}{q}$$

and so

$$d\omega_p \, d\omega_2 = d\omega_p \, d(\cos\theta) \frac{pq}{\omega_q} \propto d\omega_p \, d(\cos\theta) \frac{pq}{(M^2 + m_\pi^2 - 2M\omega_p)^{1/2}}. \tag{9.161}$$

The experimental data have been analysed both in terms of p and θ (see for example Dalitz, 1957) and are clearly consistent with the assignment of angular momentum and parity 0^- for the three-pion configuration. Therefore the kaon has zero spin.

The data can be displayed in another way which is also due to Dalitz (1957). He notes that all points inside an equilateral triangle have the property that the sum of the perpendicular distances from the three sides is a constant equal to the height of the triangle. Thus if an equilateral triangle of height Q is drawn, then each observed decay can be represented by a point (P of Fig. 9.10) whose perpendicular distances from the sides represent the kinetic energies of the three pions. The simultaneous requirement of the conservation of linear momentum and energy then restricts the points to within a circle if all three particles are nonrelativistic.

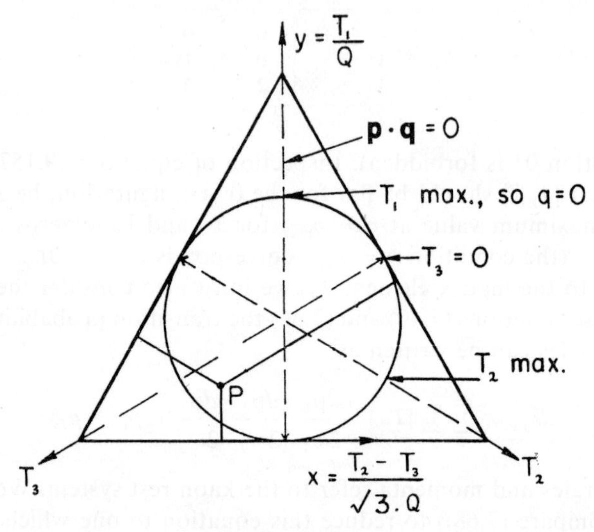

FIG. 9.10.

Deviations from this shape are caused by relativistic effects (Fabri, 1954), but these do not change the argument. It is convenient to introduce coordinates given by

$$x = \frac{T_2 - T_3}{\sqrt{3Q}}, \qquad y = \frac{T_1}{Q} \tag{9.162}$$

where T_1, T_2 and T_3 are the kinetic energies of the π_1, π_2 and π_3 mesons respectively; then

$$d\omega_1\, d\omega_2 = dT_1\, dT_2 \propto dx\, dy$$

since x and y are related to T_1 and T_2 by a linear transformation with constant Jacobian. Thus $|T_{fi}|^2$ is related directly to the population of points in Fig. 9.10. Since the two π^+ mesons are identical $|T_{fi}|^2$ is an even function of x and so the circle can be folded across the y-axis to give a semicircle. Arguments similar to the ones we have made previously can then be made about the population of points in the semicircle, for example the bottom and top of the semicircle correspond to $p = 0$ and $q = 0$ and so the population of dots in these regions would be expected to be $\sim p^l$ and $\sim q^L$ respectively (9.159). Experimentally a uniform opulation of dots is found throughout the permitted region of the plot; this esult suggests $l = L = 0$ and therefore a 0^- configuration.

We finally consider a decay by a strong interaction. In this situation we must onsider the conservation of isospin and related properties. As an example we hall consider the decay of an ω-particle (§ 13.3(b))

$$\omega \rightarrow \pi^+ + \pi^- + \pi^0.$$

n this case it is convenient to regard any pair of the mesons as the dipion.

The ω-meson only occurs in the electrically neutral state and has quantum umbers $B = S = Q = 0$. Therefore it has $T_3 = T = 0$. Now the strong decay o three pions indicates $G = -1$ by virtue of equation (9.137), and if we use the elation $G = \xi_C(-1)^T$ of (9.138) we may conclude that $\xi_C = -1$. Now the ntrinsic charge conjugation parity of the pion is even (§ 9.2(c)), therefore the ipion must be in a state with odd L (compare (5.137)) in order to yield an overll value of $\xi_C = -1$ for the ω-meson.

Now let us consider the isospin function. Since we have the overall requirenent of $T = 0$ and yet the pions possess isospin 1, we may conclude that the ipion is in a state with $T = 1$ which cancels the $T = 1$ value of the third pion. Consultation of Table 9.6 shows that a $T = 1$ system for two pions is antisymnetric, and so the isospin function for the three-pion system is also antisymmetric. Since Bose–Einstein statistics require overall symmetry we may therefore conclude that the spatial part of the wave function for the three-pion system must e antisymmetric.

If we assume various spin and parity values for the ω-particle, specific requirenents are set upon the properties of the three-pion system by the conservation aws. These properties in turn determine the density distribution in the Dalitz lot (Stevenson *et al.*, 1962). In Table 9.9 we display these consequences for hree-spin and parity assignments (Rosenfeld, 1963). In this table 'Space' refers o the spatial transformation properties of the wave function $\psi_{3\pi}$.

TABLE 9.9

ω-meson	$\psi_{3\pi}(\mathbf{p}, \mathbf{q}, L, l)$						
$j^{\xi_P G}$	Space	C and L	l	Leading terms	$\rightarrow$	Antisymmetrical form	Zeroes on Dalitz plot
0^{--}	$0^+(S)$	odd	odd	$\mathbf{q}\cdot\mathbf{p}$	$\rightarrow$	$(\omega_1 - \omega_2)(\omega_2 - \omega_3)(\omega_3 - \omega_1)$	medians
1^{+-}	$1^-(V)$	odd	even	$\mathbf{q}$	$\rightarrow$	$\omega_3(\mathbf{p}_1 - \mathbf{p}_2) + \omega_1(\mathbf{p}_2 - \mathbf{p}_3) + \omega_2(\mathbf{p}_3 - \mathbf{p}_1)$	$q = 0$ and centre
1^{--}	$1^+(A)$	odd	odd	$\mathbf{q} \times \mathbf{p}$	$\rightarrow$	$(\mathbf{p}_1 \times \mathbf{p}_2) + (\mathbf{p}_2 \times \mathbf{p}_3) + (\mathbf{p}_3 \times \mathbf{p}_1)$	boundary

This table has been constructed by inspection. Consider the first row as a example. The assignment $j^{\xi_P} = 0^-$ to the ω-particle implies an overall pseudo scalar wave function for the three pions; the odd intrinsic parity of the pion therefore implies that the spatial component transforms like a scalar. Since w have already shown earlier that L is odd, the parity requirement then forces l t be odd. The simplest possibility is $L = l = 1$, and so the simplest scalar functio

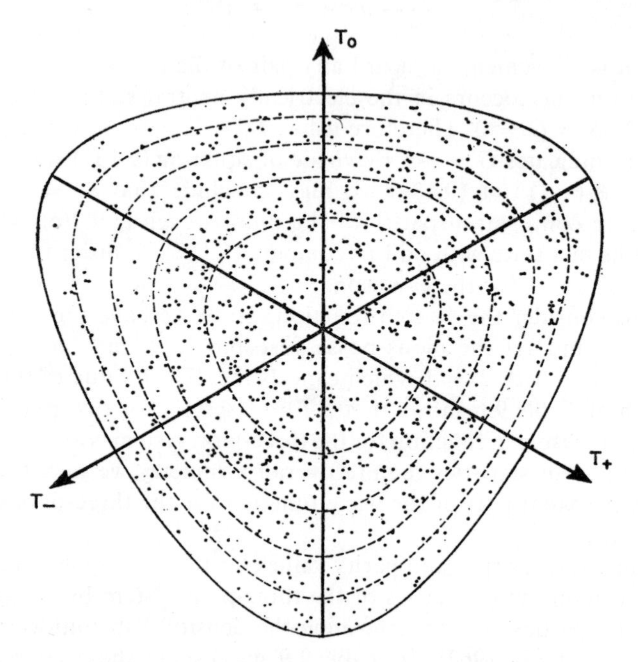

FIG. 9.11. Collected data for Dalitz plot for ω-decay (Puppi, 1962).

is $\mathbf{p} \cdot \mathbf{q} = pq \cos \theta$ (compare (9.157) and (9.159)). This vanishes when $\cos \theta = 0$ that is when any pair of pions have equal energy (Fig. 9.9). This condition occur along the medians of the Dalitz plot. The antisymmetrical form follows naturally In Fig. 9.11 we display the collected data of a number of groups (Puppi, 1962) they are clearly consistent with an assignment $j^{\xi_P} = 1^-$ for the ω-particle. Argu ments similar to those given above have been presented by Rosenfeld (1963) fo other systems decaying to three pions. A thorough discussion of the decay o unstable particles into three pions has been given by Zemach (1964).

CHAPTER 10

THE INTERACTION OF FIELDS V DISPERSION RELATIONS AND RELATED TOPICS

10.1. INTRODUCTION

In Chapter 8 the techniques of perturbation theory were used to evaluate the -matrix. The method involved development parameters which appeared in in-reasing powers in successive terms of the perturbation expansion. The para-eters possess the magnitudes

$$CM^2 \sim 10^{-5} \qquad \text{for weak interactions}$$

$$\frac{e^2}{4\pi} \sim \frac{1}{137} \qquad \text{for electromagnetic interactions}$$

$$\frac{g^2}{4\pi} \sim 10 \qquad \text{for strong interactions.}$$

Thus successive terms in the S-matrix expansion for strong interactions diverge apidly. Nevertheless, a surprisingly large amount of information about strong nteractions may be obtained from perturbation theory as we shall see later.

Considerable effort has been made to overcome the problems of a theory of trong interactions, and the most successful approach has been by the method of ispersion relations. The techniques used in these relations exploit very general hysical principles and the analytic structure of the S-matrix. The method can rovide exact relationships between certain components of the S-matrix, but does ot attempt to produce detailed pictures of the physical mechanisms as in per-urbation theory. Nevertheless, a substructure of field theoretical ideas is fre-uently present in a dispersion relation. Some physicists (see, for example, Lan-au, 1959) have suggested that the substructure can be abandoned, and that the nalytic structure of the S-matrix alone is adequate for constructing the scatter-ng amplitudes. The method has less direct physical appeal. Counter arguments n theoretical grounds have also been made against the proposal (see, for ex-mple, Mandelstam, 1962).

In this chapter we shall make a mainly historical approach to the subject and lay emphasis on the applications of the method, rather than approach the more difficult problem of the analyticity of the S-matrix.

10.2. BASIC FORMS FOR DISPERSION RELATIONS

10.2(a). *Mathematical preliminaries*

The mathematical basis of the dispersion relations is essentially the same in all problems. It is the theory of the functions of a complex variable, and, in particular the Cauchy theorem for complex integration. We begin by stating a number of mathematical theorems concerning the properties of complex functions.† Proofs will not always be given.

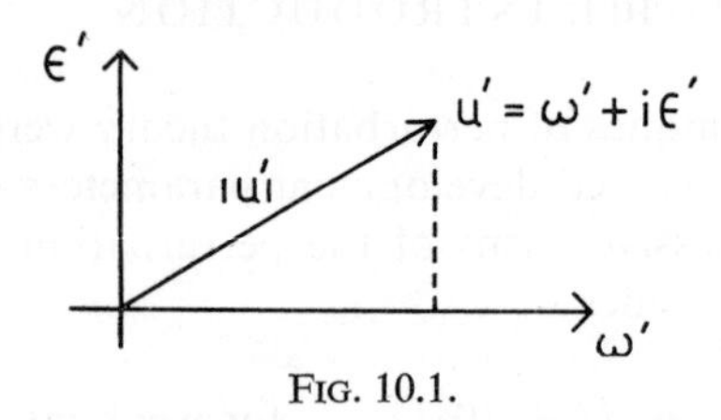

FIG. 10.1.

Consider a function $F(\omega')$ of a real variable ω'; it is often possible to extend its range of usefulness by employing a function $F(u')$ of a complex variable $u' = \omega' + i\varepsilon'$ (Fig. 10.1) which reduces to $F(\omega')$ on the real axis

$$\lim_{\varepsilon' \to 0} F(u') = F(\omega'). \tag{10.1}$$

The functions used in the study of dispersion relations are called *analytic functions*. They may be defined through their properties under integration; thus if a function $F(u')$ is *analytic* at all points within and on the domain of a closed curve C (Fig. 10.2) it satisfies the relation

$$\int_C du' F(u') = 0. \tag{10.2}$$

This formula is often called Cauchy's integral theorem, and the region enclosed by C is called the domain of analyticity.

The functions may possess points at which they are *singular*; for example the function

$$\varphi(u') = \frac{1}{u' - u} \tag{10.3}$$

† We follow the excellent introduction given by Corinaldesi (1959). A formal treatment of the topic of complex variables may be found in Morse and Feshbach (1953).

is singular at $u' = u$, and a *pole* or *singularity* is said to exist at that point. If the function $\varphi(u')$ is integrated along a closed curve that does not contain u (Fig. 10.3), the result is the same as in (10.2)

$$\int_C du' \varphi(u') = 0. \tag{10.4}$$

If, however, the curve contains u (Fig. 10.4) the integration yields

$$\int_C du' \varphi(u') = 2\pi i. \tag{10.5}$$

The product

$$F(u')\,\varphi(u') = \frac{F(u')}{u' - u} \tag{10.6}$$

is not analytic if the contour encloses u, even if $F(u')$ is analytic; the only ex-

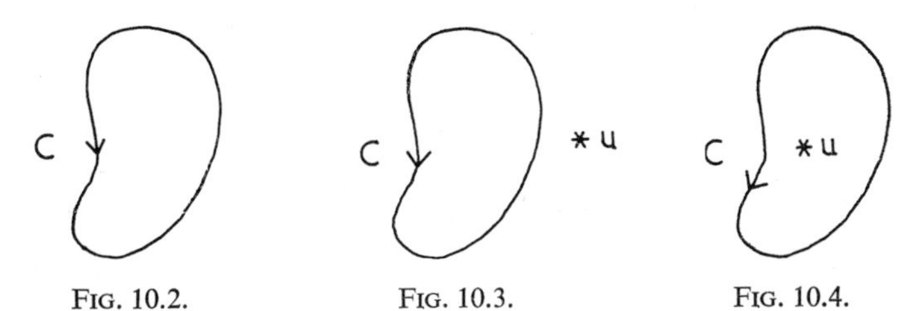

FIG. 10.2. FIG. 10.3. FIG. 10.4.

ception is in the trivial situation $F(u') = 0$. The product term possesses the following properties on integration:

$$\frac{1}{2\pi i} \int_C du' \frac{F(u')}{u' - u} = \begin{cases} F(u) & (u \text{ lies inside } C) \\ 0 & (u \text{ lies outside } C) \end{cases}. \tag{10.7}$$

This is the *Cauchy integral theorem*; in the situation when u lies on the contour of integration the integral becomes

$$\frac{1}{\pi i} P \int du' \frac{F(u')}{u' - u} = F(u). \tag{10.8}$$

In this equation the symbol P denotes the principal of the integral; it represents Cauchy's prescription for defining integrals of the type $\int dx\, F(x)$, where $F(x) \to \infty$ at a value $x = c$, by excluding from the contour a small interval about c

$$P \int_a^b dx \frac{F(x)}{x - c} = \lim_{\varrho \to 0} \left[\int_a^{c-\varrho} + \int_{c+\varrho}^b dx \frac{F(x)}{x - c} \right] \tag{10.9}$$

where c lies between a and b.

Equation (10.8) may also be obtained by adding the two contours of Fig. 10.5

$$F(u) = \frac{1}{2\pi i} \int_{C_2} du' \frac{F(u')}{u' - u}$$

$$0 = \frac{1}{2\pi i} \int_{C_1} du' \frac{F(u')}{u' - u}$$

and allowing them to run infinitesimally close together.

A considerable simplification of the above equations takes place if the function $F(u')$ falls off more quickly than $1/|u'|$ at large distances and if u lies near

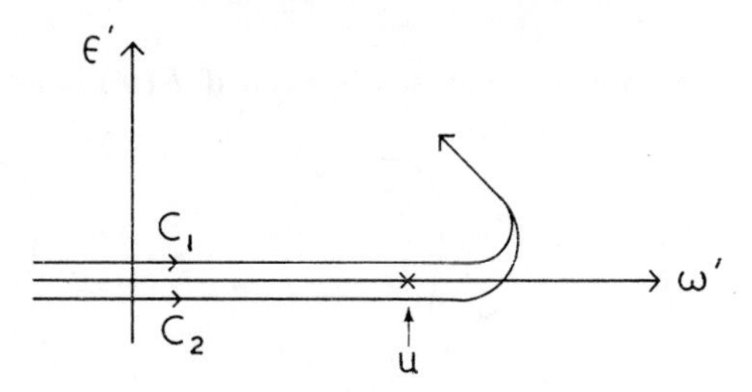

FIG. 10.5.

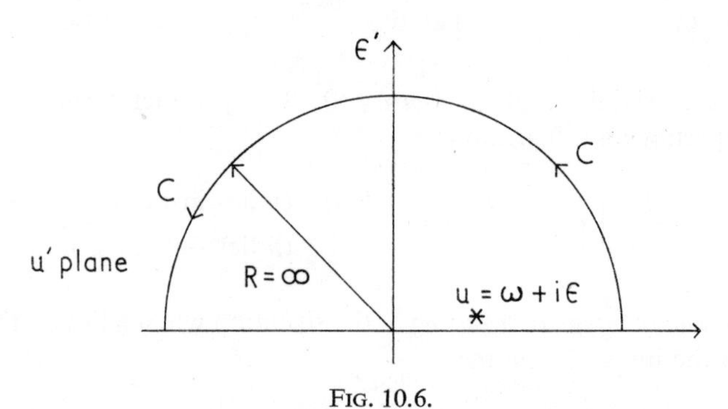

FIG. 10.6.

the real axis. Let the pole u lie at $\omega + i\varepsilon$ (Fig. 10.6). If the contour of integration is chosen to be the real axis and a semicircle of infinite radius in the upper half plane, we may write equation (10.8) as

$$F(u) = F(\omega + i\varepsilon) = \frac{1}{2\pi i} \int_{-\infty}^{+\infty} d\omega' \frac{F(\omega')}{\omega' - \omega - i\varepsilon} + C' \tag{10.10}$$

where C' represents the rest of the contour integration around the semicircle of infinite radius. Let us assume that in this region $F(u') \sim 1/|u'|$ and $|u'| \sim R$, then the integrand is $\sim 1/R^2$ and the contour integration gives

$$C' \sim \pi R \frac{1}{R^2} = \frac{\pi}{R} \sim 0$$

for $R \to \infty$. Thus equation (10.10) becomes

$$F(u) = F(\omega + i\varepsilon) = \frac{1}{2\pi i} \int_{-\infty}^{+\infty} d\omega' \frac{F(\omega')}{\omega' - \omega - i\varepsilon}. \tag{10.11}$$

Now consider the corresponding equation when u lies outside the semicircle at $u = \omega - i\varepsilon$ (ε positive); equation (10.7) then gives us

$$0 = \frac{1}{2\pi i} \int_{-\infty}^{+\infty} d\omega' \frac{F(\omega')}{\omega' - \omega + i\varepsilon} \tag{10.12}$$

and if we take the complex conjugate of this equation we find

$$0 = \frac{1}{2\pi i} \int_{-\infty}^{+\infty} d\omega' \frac{F^*(\omega')}{\omega' - \omega - i\varepsilon}. \tag{10.13}$$

If we subtract this equation from (10.11) and let $\varepsilon \to 0^+$ we obtain the relation

$$\lim_{\varepsilon \to 0+} F(\omega + i\varepsilon) = F(\omega) = \lim_{\varepsilon \to 0+} \frac{1}{\pi} \int_{-\infty}^{+\infty} d\omega' \frac{\text{Im } F(\omega')}{\omega' - \omega - i\varepsilon}. \tag{10.14}$$

This is one of the basic forms of the dispersion relations; it is normally written as

$$F(\omega) = \frac{1}{\pi} \int_{-\infty}^{+\infty} d\omega' \frac{\text{Im } F(\omega')}{\omega' - \omega - i\varepsilon} \tag{10.15}$$

where the limit is understood; sometimes the term $i\varepsilon$ is also dropped.

Let us finally consider the meaning of the denominator in the above expression. Although we write $\varepsilon \to 0^+$, we cannot simply put $\varepsilon = 0$ since it was essential for our arguments that the point $\omega + i\varepsilon$ lay inside (or outside) contours. Moreover, if $\varepsilon = 0$ problems arise at $\omega' = \omega$ unless the integral is suitably defined. Let us therefore re-examine (10.14)

$$F(\omega) = \lim_{\varepsilon \to 0+} F(\omega + i\varepsilon) = \lim_{\varepsilon \to 0+} \frac{1}{2\pi i} \int_{-\infty}^{+\infty} d\omega' \frac{F(\omega')}{\omega' - \omega - i\varepsilon}. \tag{10.16}$$

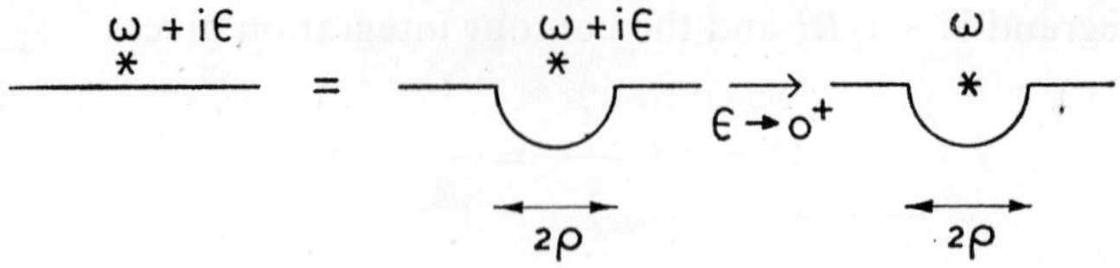

FIG. 10.7.

In the limit $\varepsilon \to 0^+$ we may retain $\omega + i\varepsilon$ inside the contour by deforming the curve as shown in Fig. 10.7, then if we write $\omega' = \omega + \varrho e^{i\theta}$ in the region of the deformation we obtain

$$\lim_{\varepsilon \to 0^+} \int_{-\infty}^{+\infty} d\omega' \frac{F(\omega')}{\omega' - \omega - i\varepsilon}$$

$$= \left[\left[\int_{-\infty}^{\omega - \varrho} + \int_{\omega + \varrho}^{\infty} \right] d\omega' \frac{F(\omega')}{\omega' - \omega} + \int_{\pi}^{2\pi} \frac{\varrho d(e^{i\theta})}{\varrho e^{i\theta}} F(\omega + \varrho e^{i\theta}) \right]$$

$$= P \int_{-\infty}^{+\infty} d\omega' \frac{F(\omega')}{\omega - \omega'} + i\pi F(\omega) \qquad \varrho \to 0.$$

If we now insert this result into (10.16) we find

$$F(\omega) = \frac{1}{\pi i} P \int_{-\infty}^{+\infty} d\omega' \frac{F(\omega')}{\omega' - \omega}. \tag{10.17}$$

The result obtained above shows that we can write, formally

$$\frac{1}{\omega' - \omega - i\varepsilon} = P \frac{1}{\omega' - \omega} + i\pi\delta(\omega' - \omega).$$

It is not difficult to extend the argument to yield the general result

$$\frac{1}{(\omega' - \omega) \pm i\varepsilon} = P \frac{1}{\omega' - \omega} \mp i\pi\delta(\omega' - \omega). \tag{10.18}$$

10.2(b). *Basic forms for the dispersion relations*

One of the most frequently used forms of the dispersion relation is that given in (10.15)

$$F(\omega) = \frac{1}{\pi} \int_{-\infty}^{+\infty} d\omega' \frac{\operatorname{Im} F(\omega')}{\omega' - \omega - i\varepsilon}.$$

It is also convenient to have relations between the real (*dispersive*) and imaginary (*absorptive*) parts of $F(\omega)$. The names in brackets derive from the original use of dispersion relations in optical problems, hence

$$F(\omega) = \text{Re}\,F(\omega) + i\,\text{Im}\,F(\omega) \tag{10.19}$$

$$= D(\omega) + iA(\omega). \tag{10.20}$$

If these equations are inserted in (10.17), we find that

$$\text{Re}\,F(\omega) = \frac{1}{\pi} P \int_{-\infty}^{+\infty} d\omega' \frac{\text{Im}\,F(\omega')}{\omega' - \omega} \tag{10.21}$$

$$\text{Im}\,F(\omega) = \frac{-1}{\pi} P \int_{-\infty}^{+\infty} d\omega' \frac{\text{Re}\,F(\omega')}{\omega' - \omega}.$$

Equation (10.15) and the first of the above equations are alternative forms of the same expression, since with the aid of equation (10.18) we can write

$$F(\omega) = \frac{1}{\pi} \int_{-\infty}^{+\infty} d\omega' \frac{\text{Im}\,F(\omega')}{\omega' - \omega - i\varepsilon}$$

$$= \frac{1}{\pi} P \int_{-\infty}^{+\infty} \frac{\text{Im}\,F(\omega')}{\omega' - \omega} + i\,\text{Im}\,F(\omega).$$

Equations (10.15) and (10.21) are known as *Hilbert transforms*. It must be emphasised that the derivation of these formulae involved taking contours in the upper half plane, and that we stipulated that $F(\omega')$ is analytic in this region.

It is sometimes necessary to extend the contour into the lower half plane. This may be done by the process of analytic continuation – that is if the function $F(\omega)$ is known in a certain region then its behaviour there may be used to predict its properties in another region.

Let us return to equation (10.7)

$$F(u) = \frac{1}{2\pi i} \int_C du' \frac{F(u')}{u' - u} \quad (u \text{ inside } C).$$

Consider the contour shown in Fig. 10.8 and assume

$$\text{Im}\,F(u') = 0 \tag{10.22}$$

for $-\infty < \omega' < \omega_c$ along the real ω' axis. If we make the usual assumptions about the vanishing of the functions at infinity we find

$$
\begin{aligned}
F(u) &= \frac{1}{2\pi i} \int_C du' \frac{F(u')}{u' - u} \\
&= \frac{1}{2\pi i} \lim_{\delta \to 0+} \left[\int_{\omega_c}^{\infty} d\omega' \frac{F(\omega' + i\delta)}{\omega' - u} + \int_{\infty}^{\omega_c} d\omega' \frac{F(\omega' - i\delta)}{\omega' - u} \right] \\
&= \frac{1}{2\pi i} \lim_{\delta \to 0+} \int_{\omega_c}^{\infty} d\omega' \frac{F(\omega' + i\delta) - F^*(\omega + i\delta)}{\omega' - u} \\
&= \frac{1}{\pi} \int_{\omega_c}^{\infty} d\omega' \frac{\operatorname{Im} F(\omega')}{\omega' - u}.
\end{aligned}
\tag{10.23}
$$

In Fig. 10.8 the point (ω_c) at which Im $F(\omega')$ vanishes is known as the *branch point*; the contours extending from ω_c to infinity are known as *branch lines or*

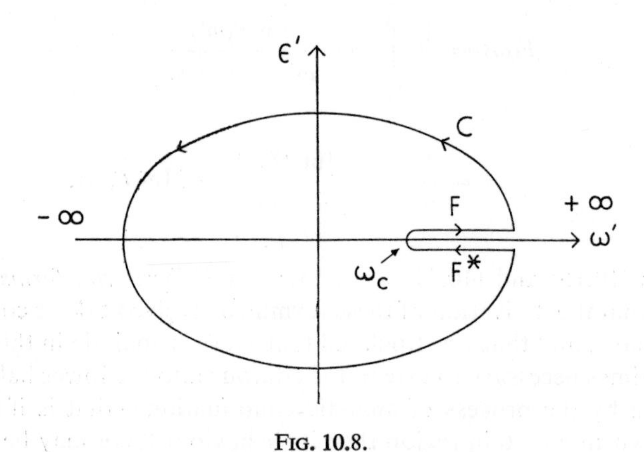

FIG. 10.8.

cuts. Thus the analytic function $F(u')$ has a discontinuity of magnitude $2i$ Im $F(\omega')$ across the cut, and elsewhere along the real ω' axis $F(\omega')$ is real. If we make the usual assumption that the pole is located at $u = \omega + i\varepsilon$ equation (10.23) becomes

$$
F(\omega) = \frac{1}{\pi} \int_{\substack{\omega_c \\ \varepsilon \to 0+}}^{\infty} d\omega \frac{\operatorname{Im} F(\omega')}{\omega' - \omega - i\varepsilon}. \tag{10.24}
$$

In problems involving elementary particles one often finds that isolated points may occur along the real axis where the imaginary part of F does not vanish. For example, let us consider Fig. 10.9 and assume

$$\begin{aligned} \operatorname{Im} F(\omega') &= 0 & -\infty < \omega' < \omega_R \\ &= F(\omega_R) & \omega' = \omega_R \\ &= 0 & \omega_R < \omega' < \omega_c \\ &\neq 0 & \omega_c < \omega' < \infty. \end{aligned}$$

The region surrounding ω_R must then be excluded from the contour of integration. Formally this may be done by deforming the contour to exclude ω_R, say by making a cross-cut from the contour C to ω_R in Fig. 10.9, and then by

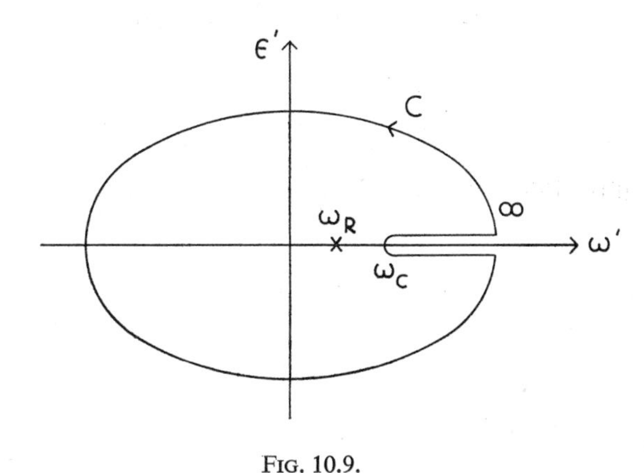

FIG. 10.9.

letting the lines of the cut run extremely close together. Since thes elines are traversed in opposite directions the integral along them then vanishes, so that the contour integral C is unaltered but ω_R is isolated. In practice an adequate procedure is to determine the existence of the isolated points, and then add their contribution to the function under investigation. Thus Fig. 10.9 implies that we should modify equation (10.24) in the following manner:

$$F(\omega) = \frac{F(\omega_R)}{\omega_R - \omega - i\varepsilon} + \frac{1}{\pi}\int_{\omega_c}^{\infty} d\omega' \frac{\operatorname{Im} F(\omega')}{\omega' - \omega - i\varepsilon} \tag{10.25}$$

$$\varepsilon \to 0^+.$$

The first expression on the right-hand side is often called a *pole term*. In § 10.3(e) we shall evaluate explicit expressions with the form given above.

10.2(c). *Causality and the dispersion relations*

An important feature of the use of dispersion relations is in the application of the principle of causality. The causality requirement in a scattering problem may be stated as follows. If an incident particle (or wave packet) reaches a scattering centre at a certain time, then if c is the velocity of light the scattered particle should not appear at a distance r from the centre until at least a time r/c has elapsed.

Let us assume that we have a system into which we can send a signal, and that this system then delivers an output signal. We shall designate the signals as

$$\text{input} \equiv \text{in}\,(t_1) \qquad \text{output} \equiv \text{out}\,(t_2). \tag{10.26}$$

We assume that the signals are related linearly, so that

$$\text{out}\,(t_2) = \int_{-\infty}^{+\infty} dt_1 g(t_1, t_2)\,\text{in}\,(t_1). \tag{10.27}$$

If they are also related causally we must assume that

$$g(t_1, t_2) = 0 \quad \text{for} \quad t_2 - t_1 < 0.$$

Furthermore, if the internal properties of the system are constant in time, we can state that

$$\text{out}\,(t_2) = \int_{-\infty}^{+\infty} dt_1 g(t_2 - t_1)\,\text{in}\,(t_1) \tag{10.28}$$

and

$$g(t_2 - t_1) = 0 \quad \text{for} \quad t_2 - t_1 < 0. \tag{10.29}$$

If we write $t = t_2 - t_1$, then this becomes

$$g(t) = 0 \quad \text{for} \quad t < 0. \tag{10.30}$$

The Fourier transform of this function therefore can be written as

$$F(\omega) = \int_{-\infty}^{+\infty} dt g(t)\, e^{i\omega t} = \int_{0}^{\infty} dt g(t)\, e^{i\omega t} \tag{10.31}$$

or

$$g(t) = \int_{-\infty}^{+\infty} d\omega\, e^{-i\omega t} F(\omega). \tag{10.32}$$

The relationship between causality and dispersion relations can be shown to arise from a very general theorem developed by Titchmarsh (1948). In the context of dispersion relations the theorem can be modified to state that a function $F(\omega)$ with any one of the following properties:

(1) satisfies Hilbert transforms (dispersion relations),
(2) has a Fourier transform which vanishes for $t < 0$,
(3) is analytic in the upper half plane,

automatically has the other two properties. A particularly clear proof of Titchmarsh's theorem in a suitable style for dispersion relations has been given by Hagedorn (1961). We shall repeat part of the proof by showing that if $F(\omega)$ satisfies a dispersion relation then $g(t)$ vanishes for $t < 0$. We have already shown in the previous section that $F(\omega)$ must be analytic in the upper half plane in order to yield the dispersion relation.

We assume $F(\omega)$ satisfies (10.17)

$$F(\omega) = \frac{1}{\pi i} P \int_{-\infty}^{+\infty} d\omega' \frac{F(\omega')}{\omega' - \omega}$$

then by (10.32)

$$g(t) = \int_{-\infty}^{+\infty} d\omega \, e^{-i\omega t} F(\omega) = \int_{-\infty}^{+\infty} d\omega \, e^{-i\omega t} \frac{1}{\pi i} P \int_{-\infty}^{+\infty} d\omega' \frac{F(\omega')}{\omega' - \omega}. \quad (10.33)$$

If we assume that the integrations can be interchanged, then

$$g(t) = \int_{-\infty}^{+\infty} d\omega' F(\omega') \left(\frac{-1}{\pi i}\right) P \int_{-\infty}^{+\infty} d\omega \frac{e^{-i\omega t}}{\omega - \omega'} \quad (10.34)$$

but the second integral has the following mathematical property

$$\frac{-1}{\pi i} P \int_{-\infty}^{+\infty} d\omega \frac{e^{-i\omega t}}{\omega - \omega'} = \begin{cases} e^{-i\omega' t} & \text{for} \quad t > 0 \\ -e^{-i\omega' t} & \text{for} \quad t < 0 \end{cases} \quad (10.35)$$

(compare (A.6.6), Appendixes, p. 704) hence

$$g(t) = \int_{-\infty}^{+\infty} d\omega' e^{-i\omega' t} F(\omega') = g(t) \quad \text{for} \quad t > 0 \quad (10.36)$$

$$g(t) = -\int_{-\infty}^{+\infty} d\omega' e^{-i\omega' t} F(\omega') = -g(t) \quad \text{for} \quad t < 0$$

and so $g(t) = 0$ for $t < 0$.

In most examples in particle physics the Titchmarsh theorem cannot be directly applied, but nevertheless it states clearly a principle which is represented in all dispersion relations with a single variable.

The causal form of $g(t)$ may also be expressed by introducing a function $G(t)$, which we define to be equal to $g(t)$ for $t > 0$ but is otherwise arbitrary. Then we can write

$$g(t) = G(t)\,\theta(t) = G(t)\,\tfrac{1}{2}[1 + \varepsilon(t)] \quad (10.37)$$

where

$$\varepsilon(t) = \frac{t}{|t|} = \begin{cases} +1 & \text{for} \quad t > 0 \\ -1 & \text{for} \quad t < 0 \end{cases} \quad (10.38)$$

and also

$$F(\omega) = \tfrac{1}{2} \int_{-\infty}^{+\infty} dtG(t)\,[1 + \varepsilon(t)]\, e^{i\omega t}. \tag{10.39}$$

We may then identify the dispersive and absorptive parts of $F(\omega)$ with the following expressions (compare (10.19))

$$F(\omega) = \text{Re } F(\omega) + i \text{ Im } F(\omega)$$

$$\text{Re } F(\omega) = \tfrac{1}{2} \int_{-\infty}^{+\infty} dtG(t)\, \varepsilon(t)\, e^{i\omega t} \tag{10.40}$$

$$i \text{ Im } F(\omega) = \tfrac{1}{2} \int_{-\infty}^{+\infty} dtG(t)\, e^{i\omega t}.$$

The identification may be demonstrated by using the following equation (compare (A.6.6))

$$\varepsilon(t)\, e^{i\omega t} = \frac{1}{\pi i} P \int_{-\infty}^{+\infty} d\omega' \frac{e^{i\omega' t}}{\omega' - \omega}. \tag{10.41}$$

Then from equation (10.40) we have

$$\text{Re } F(\omega) = \frac{1}{2\pi i} \int_{-\infty}^{+\infty} dtG(t) P \int_{-\infty}^{+\infty} d\omega' \frac{e^{i\omega' t}}{\omega' - \omega}$$

and if it is assumed that the order of integration can be interchanged, this equation becomes

$$\text{Re } F(\omega) = \frac{1}{2\pi i} P \int_{-\infty}^{+\infty} d\omega' \frac{1}{\omega' - \omega} \int_{-\infty}^{+\infty} dt\, G(t)\, e^{i\omega' t}$$

$$= \frac{1}{\pi} P \int_{-\infty}^{+\infty} d\omega' \frac{\text{Im } F(\omega')}{\omega' - \omega} \tag{10.42}$$

where we have used the second of equations (10.40). Equation (10.42) is identical with the first Hilbert transform in (10.21), and so the identification is justified. The inverse relation may also be shown to hold.

10.2(d) *Dispersion relations in classical physics – the Kramers–Kronig relation*

The main points in the theory of dispersion relations can be illustrated by considering an electromagnetic wave propagating in a dispersive medium. The classical relationship between the displacement (**D**), electric field (**E**) and polarisation (**P**) is given by

$$\mathbf{D} = \varepsilon \mathbf{E} = \mathbf{E} + 4\pi \mathbf{P} \tag{10.43}$$

where ε is the dielectric constant. Upon rearranging terms and recalling that we may replace ε by the refractive index n^2 we find that

$$\mathbf{P} = \frac{\varepsilon - 1}{4\pi} \mathbf{E} = \frac{n^2 - 1}{4\pi} \mathbf{E}. \tag{10.44}$$

Now this relationship is obviously causal since the polarisation must be zero until the disturbing electric field reaches the medium. The function $(n^2 - 1)/4\pi$ can be regarded as equivalent to $g(t)$ in (10.28). It is customary to assume that n is close to unity, and so can write the Fourier transform (compare (10.31)) as

$$F(\omega) \equiv \frac{n^2(\omega) - 1}{4\pi} \sim 2 \frac{[n(\omega) - 1]}{4\pi}. \tag{10.45}$$

Now

$$n(\omega) - 1 = \mathrm{Re}\, n(\omega) - 1 + i \,\mathrm{Im}\, n(\omega) \tag{10.46}$$

and by using this equation and (10.21) we obtain the following dispersion relation

$$\mathrm{Re}\, n(\omega) - 1 = \frac{1}{\pi} P \int_{-\infty}^{+\infty} d\omega' \frac{\mathrm{Im}\, n(\omega')}{\omega' - \omega}. \tag{10.47}$$

Thus we have obtained a dispersion relation relating the real (dispersive) and imaginary (absorptive) parts of the refractive index at any frequency *without making any specific assumptions about the physical processes.* It is instructive to consider a specific atomic model and to offer further evidence that the dispersion relation is justified. Consider a system with N electrons per cm^3 vibrating with a characteristic frequency ω_0, and apply to the system an external field $\mathbf{E}(\omega)\, e^{-i\omega t}$. Then the equation of motion is

$$\frac{d^2\mathbf{x}}{dt^2} + \Gamma \frac{d\mathbf{x}}{dt} + \omega_0^2 \mathbf{x} = \frac{e}{m} \mathbf{E}(\omega)\, e^{-i\omega t} \tag{10.48}$$

where Γ is a damping term. The polarisation of the system is then given by

$$\mathbf{P}(\omega) = Ne\mathbf{x} = \frac{Ne^2}{m} \frac{1}{(\omega_0^2 - \omega^2) - i\omega\Gamma} \mathbf{E}(\omega) = \frac{n^2 - 1}{4\pi} \mathbf{E} \tag{10.49}$$

where we have used (10.44). The term Γ can be seen to be the width of the resonance. The approximation $n(\omega) \sim 1$ then gives (compare (10.45))

$$\frac{n(\omega) - 1}{4\pi} = \frac{Ne^2}{2m} \frac{1}{(\omega_0^2 - \omega^2) - i\omega\Gamma}. \tag{10.50}$$

It is apparent that dispersion relations can be applied to $n(\omega)$ since:

(1) the equation represents a function in the complex ω plane with poles in the lower half plane only at positions given by

$$(\omega_0^2 - \omega^2) - i\omega\Gamma = 0$$

$$\omega = -i\frac{\Gamma}{2} \pm \sqrt{\left(\omega_0^2 - \frac{\Gamma^2}{4}\right)}. \tag{10.51}$$

(2) as $|\omega| \to \infty$

$$|n(\omega) - 1| \to \frac{1}{|\omega^2|}. \tag{10.52}$$

The first condition implies that we require contours in the upper half plane, and the second implies that we need only consider that part of the contour which lies along the real axis (compare Fig. 10.6). Thus the relation (10.47)

$$\text{Re}\, n(\omega) - 1 = \frac{1}{\pi} P \int_{-\infty}^{+\infty} d\omega' \frac{\text{Im}\, n(\omega')}{\omega' - \omega}$$

is justified.

The integral contains both positive and negative frequencies. Now the region $\omega' < 0$ is an unphysical one and not accessible to experiment; however, an inspection of (10.50)

$$\frac{n(\omega) - 1}{4\pi} = \frac{Ne^2}{2m} \frac{1}{(\omega_0^2 - \omega^2) - i\omega\Gamma}$$

shows that

$$n^*(\omega) = n(-\omega) \tag{10.53}$$

and so

$$\text{Re}\, n(\omega) = \text{Re}\, n(-\omega), \qquad \text{Im}\, n(\omega) = -\text{Im}\, n(-\omega). \tag{10.54}$$

This relation is known as a *crossing relation;* many types of crossing relations linking physical and unphysical regions exist in dispersion relations. If equation (10.54) is inserted into (10.47) we obtain the Kramers–Kronig relation

$$\text{Re}\, n(\omega) - 1 = \frac{2}{\pi} P \int_0^{+\infty} d\omega'\omega' \frac{\text{Im}\, n(\omega')}{\omega'^2 - \omega^2}. \tag{10.55}$$

This equation possesses historical importance as it is the first published dispersion relation for a physical problem (Kramers, 1927; Kronig, 1926).

In particle and nuclear physics the forward scattering amplitude

$$f(\omega) = f(\omega, \theta = 0) \tag{10.56}$$

is a more useful concept than refractive index (θ is the angle of scattering). The two functions are related as follows

$$n(\omega) = 1 + \frac{2\pi}{k^2} Nf(\omega) = 1 + \frac{2\pi}{\omega^2} Nf(\omega) \quad \text{for light} \tag{10.57}$$

(see, for example, Fermi, 1950). In the above expression N represents the number of scattering centres per cm³. The imaginary part of $f(\omega)$ is related to the total cross-section by the optical theorem, as we have shown in (6.12)

$$\sigma_T(\omega) = \frac{4\pi}{k} \operatorname{Im} f(\omega) = \frac{4\pi}{\omega} \operatorname{Im} f(\omega). \tag{10.58}$$

Thus the Kramers–Kronig relation (10.55) becomes

$$\begin{aligned} \operatorname{Re} f(\omega) &= \frac{2\omega^2}{\pi} P \int_0^{+\infty} d\omega' \frac{\operatorname{Im} f(\omega')}{\omega'(\omega'^2 - \omega^2)} \\ &= \frac{\omega^2}{2\pi^2} P \int_0^{\infty} d\omega' \frac{\sigma_T(\omega')}{\omega'^2 - \omega^2}. \end{aligned} \tag{10.59}$$

Thus if we know the total scattering cross-section at all frequencies the forward scattering amplitude can be evaluated at any required frequency.

10.2(e). *Difference formulae and convergence*

In equation (10.53) the crossing relation

$$n^*(\omega) = n(-\omega)$$

was obtained. This result was specific to the problem in hand; in general any function $F(\omega)$ can assume the following values:

$$\begin{aligned} F^*(\omega) &= F(-\omega) \\ F^*(\omega) &= -F(-\omega). \end{aligned} \tag{10.60}$$

Both relations are forms of the crossing relation and can be used to avoid the unphysical regions of negative frequencies. The two relations imply that

$$\begin{aligned} &\operatorname{Re} F(\omega) = \operatorname{Re} F(-\omega), \ \operatorname{Im} F(\omega) = -\operatorname{Im} F(-\omega) \quad \text{for} \quad F^*(\omega) = F(-\omega) \\ &\operatorname{Re} F(\omega) = -\operatorname{Re} F(-\omega), \ \operatorname{Im} F(\omega) = \operatorname{Im} F(-\omega) \quad \text{for} \quad F^*(\omega) = -F(-\omega) \end{aligned} \tag{10.61}$$

and so

$$\operatorname{Re} F(\omega) = \frac{1}{\pi} P \int_{-\infty}^{+\infty} d\omega' \frac{\operatorname{Im} F(\omega')}{\omega' - \omega} = \frac{2}{\pi} P \int_{0}^{\infty} d\omega' \omega' \frac{\operatorname{Im} F(\omega')}{\omega'^2 - \omega^2}$$

$$\text{for} \quad F^*(\omega) = F(-\omega),$$

$$= -\frac{2\omega}{\pi} P \int_{0}^{\infty} d\omega' \frac{\operatorname{Im} F(\omega')}{\omega'^2 - \omega^2} \quad \text{for} \quad F^*(\omega) = -F(-\omega). \tag{10.62}$$

It can be seen that the second of these equations is more convergent than the first, since the first equation includes the factor ω' in the numerator. In § 10.2(a) it was stated that the function $F(\omega')$ should fall away faster than $1/|\omega'|$ at large distances (see discussion relevant to equations (10.10) and (10.11)). If $F(\omega')$ does not fulfil this requirement, greater convergence may be achieved by using a subtraction technique. For functions of the type $F^*(\omega) = F(-\omega)$ this can be done by subtracting $\operatorname{Re} F(\alpha)$ at some known point α

$$\operatorname{Re} F(\omega) - \operatorname{Re} F(\alpha) = \frac{2}{\pi} P \int_{0}^{\infty} d\omega' \omega' \operatorname{Im} F(\omega') \left[\frac{1}{\omega'^2 - \omega^2} - \frac{1}{\omega'^2 - \alpha^2} \right]$$

or

$$\frac{\operatorname{Re} F(\omega) - \operatorname{Re} F(\alpha)}{\omega^2 - \alpha^2} = \frac{2}{\pi} P \int_{0}^{\infty} d\omega' \omega' \frac{\operatorname{Im} F(\omega')}{(\omega'^2 - \omega^2)(\omega'^2 - \alpha^2)}. \tag{10.63}$$

The second function in equation (10.60)

$$F^*(\omega) = -F(-\omega)$$

may be treated similarly

$$\alpha \operatorname{Re} F(\omega) - \omega \operatorname{Re} F(\alpha) = -\frac{2}{\pi} P \int_{0}^{\infty} d\omega' \operatorname{Im} F(\omega') \left[\frac{\alpha\omega}{\omega'^2 - \omega^2} - \frac{\alpha\omega}{\omega'^2 - \alpha^2} \right]$$

or

$$\frac{\alpha \operatorname{Re} F(\omega) - \omega \operatorname{Re} F(\alpha)}{\omega\alpha(\omega^2 - \alpha^2)} = -\frac{2}{\pi} P \int_{0}^{\infty} d\omega' \frac{\operatorname{Im} F(\omega')}{(\omega'^2 - \omega^2)(\omega'^2 - \alpha^2)}. \tag{10.64}$$

The functions $F(\alpha)$ preferably should be both known and simple. It should be noted that subtraction has yielded a more complicated integral for evaluation.

10.3. DISPERSION RELATIONS FOR FORWARD SCATTERING

10.3(a). *Introduction*

In this section we shall examine the forward scattering of scalar and pseudo-scalar particles of finite mass†. In doing so we shall exploit the principles of:

(1) Lorentz invariance,
(2) causality,
(3) unitarity,
(4) crossing symmetry.

The formalism developed in § 8.9 is particularly suitable for the application of the principle of causality, since causal conditions have been built into it. We therefore start with the scattering amplitude (8.174)

$$T_{fi} = -N_p N_{p'} \int_{-\infty}^{+\infty} d^4z e^{-i\frac{(k+k')}{2}z} \langle p'| R\left[j\left(\frac{z}{2}\right)j\left(-\frac{z}{2}\right)\right]|p\rangle$$

and simplify it for forward scattering. We shall work in the rest frame of one of the particles and write

$$p = p' = (0, iM) \qquad (10.65)$$

$$k = k' = (\mathbf{k}, i\omega)$$

hence

$$T_{fi} = -N_p^2 \int_{-\infty}^{+\infty} d^4z e^{-ikz} \langle p| R\left[j\left(\frac{z}{2}\right)j\left(-\frac{z}{2}\right)\right]|p\rangle$$

$$= -N_p^2 \int_{-\infty}^{+\infty} d^4z e^{-i(\mathbf{k}\cdot\mathbf{z}-\omega t)} \langle p| R\left[j\left(\frac{z}{2}\right)j\left(-\frac{z}{2}\right)\right]|p\rangle. \qquad (10.66)$$

10.3(b). *Crossing symmetry and other properties*

Let us now examine some of the properties of (10.66). First, it is a scalar function since it was constructed from scalar fields and contains no derivatives. Now the only scalar variables we can associate with a scattering amplitude are energy and scattering angle, and since we have fixed the latter we may conclude that T_{fi} is a function of energy alone

$$T_{fi} = T(\omega).$$

† We follow the treatment given by Jackson (1961).

A similar conclusion concerning forward scattering was reached in § 9.4(d) in connection with pion–nucleon scattering.

Now let us examine the property of crossing symmetry, we note that

$$T^*(-\omega) = -N_p^2 \int_{-\infty}^{+\infty} d^4z\, e^{i(\mathbf{k}\cdot\mathbf{z}+\omega t)} (\langle p|\, R\,|p\rangle)^*$$

but

$$\int_{-\infty}^{+\infty} d\mathbf{z}\, e^{i\mathbf{k}\cdot\mathbf{z}} = \int_{-\infty}^{+\infty} d\mathbf{z}\, e^{-i\mathbf{k}\cdot\mathbf{z}}$$

and

$$(\langle p|\, R\,|p\rangle)^* = \langle p|\, R^\dagger\,|p\rangle = \langle p|\, R\,|p\rangle$$

as a result of (4.17) and the Hermitian property of the retarded commutator R (§ 8.9(b)). Thus we find

$$\begin{aligned} T^*(-\omega) &= -N_p^2 \int_{-\infty}^{+\infty} d^4z\, e^{-i(\mathbf{k}\cdot\mathbf{z}-\omega t)} \langle p|\, R\left[j\left(\frac{z}{2}\right) j\left(-\frac{z}{2}\right)\right]|p\rangle \\ &= T(\omega) \end{aligned}$$

and so

$$T(-\omega) = T^*(\omega). \tag{10.67}$$

Next we split T into dispersive and absorptive parts, as in the classical case (§ 10.2(c)). This can be done by considering the definition of R (8.157) and then splitting $\theta(t)$ as in (10.37)

$$\theta(t) = \tfrac{1}{2}[1 + \varepsilon(t)].$$

Then we may write

$$T(\omega) = D(\omega) + iA(\omega)$$

where

$$D(\omega) = i\,\frac{N_p^2}{2} \int_{-\infty}^{+\infty} d^4z\, e^{-ikz} \langle p|\, \varepsilon(t)\left[j\left(\frac{z}{2}\right), j\left(-\frac{z}{2}\right)\right]|p\rangle \tag{10.68}$$

$$A(\omega) = \frac{N_p^2}{2} \int_{-\infty}^{+\infty} d^4z\, e^{-ikz} \langle p|\left[j\left(\frac{z}{2}\right), j\left(-\frac{z}{2}\right)\right]|p\rangle.$$

The functions D and A can be easily shown to be real; consider A^*

$$A^*(\omega) = \frac{N_p^2}{2} \int_{-\infty}^{+\infty} d^4z\, e^{ikz} \langle p|\left[j\left(-\frac{z}{2}\right), j\left(\frac{z}{2}\right)\right]|p\rangle$$

where we have used (4.17) and (4.21). Since we are summing over all values of z, we may change z to $-z$, yielding

$$A^*(\omega) = \frac{N_p^2}{2} \int_{-\infty}^{+\infty} d^4 z e^{-ikz} \langle p | j\left(\frac{z}{2}\right), j\left(-\frac{z}{2}\right)] | p \rangle = A(\omega).$$

Similarly

$$D^*(\omega) = D(\omega).$$

Equation (10.67) then implies that

$$D(\omega) = D(-\omega), \qquad A(\omega) = -A(-\omega). \tag{10.69}$$

10.3(c). *The analytic properties of $T(\omega)$*

The matrix element $\langle p | R [j(\frac{1}{2}z) j(-\frac{1}{2}z)] | p \rangle$ does not contain any direction in the state vectors and is therefore spherically symmetric. We can therefore integrate

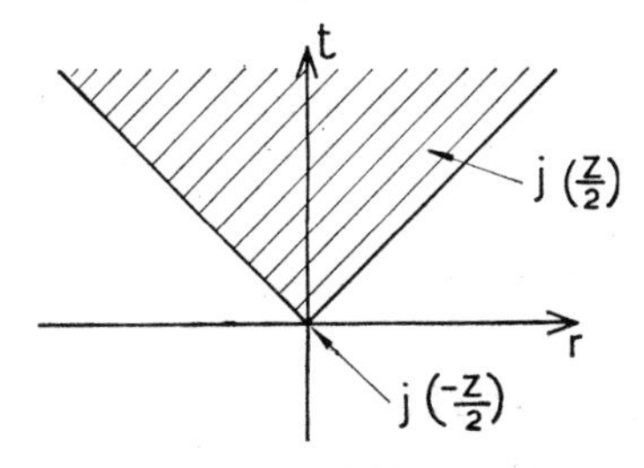

FIG. 10.10.

over all directions of $\mathbf{z}$ in the causal amplitude (10.66)

$$\begin{aligned} T &= -N_p^2 \int_{-\infty}^{+\infty} d^4 z \, e^{-ikz} \langle p | R \left[j\left(\frac{z}{2}\right) j\left(-\frac{z}{2}\right)\right] | p \rangle \\ &= N_p^2 \, 2\pi \int_{+1}^{-1} d(\cos\theta) \int_0^{\infty} dr \, r^2 \, e^{-ikr\cos\theta} \int_{-\infty}^{+\infty} dt \, e^{i\omega t} \langle p | R | p \rangle \\ &= -4\pi N_p^2 \int_0^{\infty} dr \, r \frac{\sin kr}{k} \int_{-\infty}^{+\infty} dt \, e^{i\omega t} \langle p | R | p \rangle \end{aligned} \tag{10.70}$$

where we have written

$$|\mathbf{k}| = k, \qquad |\mathbf{z}| = r.$$

The presence of the retarded commutator in (10.70) implies that T vanishes unless $j(\frac{1}{2}z)$ is in the forward light cone with respect to $j(-\frac{1}{2}z)$ (Fig. 10.10). The definition for z

$$z = x - y$$

(compare (8.169)) thus implies that T is a causal function, and we may replace the lower limit $-\infty$ in the time integral by r†

$$T(\omega) = -4\pi N_p^2 \int_0^\infty dr\, r \frac{\sin kr}{k} \int_r^\infty dt\, e^{i\omega t} \langle p| R\left[j\left(\frac{z}{2}\right) j\left(-\frac{z}{2}\right)\right] |p\rangle . \tag{10.71}$$

We now introduce a function $T(\omega, r)$ defined as

$$T(\omega, r) = -4\pi N_p^2 r \frac{\sin kr}{k} \int_r^{+\infty} dt\, e^{i\omega t} \langle p| R |p\rangle \tag{10.72}$$

so that

$$T(\omega) = \int_0^\infty dr\, T(\omega, r). \tag{10.73}$$

This function satisfies the conditions required for an analytic function, first it is causal, secondly it vanishes at $t = \infty$, since

$$j\left(\frac{z}{2}\right) j\left(-\frac{z}{2}\right) \underset{t\to\infty}{\longrightarrow} j_{\text{out}}\left(\frac{z}{2}\right) j_{\text{in}}\left(-\frac{z}{2}\right)$$

and by equations (8.145) and (8.114)

$$(\Box^2 - m^2)\, \varphi_{\substack{\text{in}\\ \text{out}}} = -j_{\substack{\text{in}\\ \text{out}}} = 0. \tag{10.74}$$

Furthermore, as $\omega \to \infty$, T will be proportional to $1/|\omega|$, and so it satisfies the requirement given in (10.10) and (10.11). Finally, if the function $T(\omega, r)$ is examined in the complex ω-plane

$$T(\omega, r) \to T(\omega + i\varepsilon, r), \quad \varepsilon \to 0+ \tag{10.75}$$

it is apparent that for $\varepsilon > 0$, T is analytic in the upper half plane since its dependence on ω is contained in the function

$$\frac{\sin kr}{k} e^{i\omega t} \equiv \frac{\sin \sqrt{(\omega^2 - m^2)}r}{\sqrt{(\omega^2 - m^2)}} e^{i\omega t} e^{-\varepsilon t} \tag{10.76}$$

and $t \geqq r$. It is evident that this function contains no poles in ω, furthermore, it vanishes as $t \to \infty$.

The double condition that $T(\omega, r)$ vanishes at infinity and is analytic in the upper half plane implies that we may write a dispersion relation for $T(\omega, r)$. We

† Strictly speaking we should use $r - \varrho$ where ϱ is positive and small, since δ-functions can arise along the light cone. These may be ignored—for a detailed treatment see Hagedorn (1961).

choose the form (10.15)

$$T(\omega, r) = \frac{1}{\pi}\int_{-\infty}^{+\infty} d\omega' \frac{A(\omega', r)}{\omega' - \omega - i\varepsilon} \tag{10.77}$$

where $A(\omega, r)$ is defined from (10.68)

$$A(\omega) = \int_0^\infty dr\, A(\omega, r)$$

$$A(\omega, r) = 4\pi N_p^2 r \frac{\sin kr}{2k} \int_r^\infty dt e^{i\omega t} \langle p| \left[j\left(\frac{z}{2}\right), j\left(-\frac{z}{2}\right)\right] |p\rangle. \tag{10.78}$$

In order to obtain a dispersion relation for $T(\omega)$ we must integrate over r. This operation presents no difficulty on the left-hand side of (10.77), and so we obtain

$$T(\omega) = \frac{1}{\pi}\int_0^\infty dr \int_{-\infty}^{+\infty} d\omega' \frac{A(\omega', r)}{\omega' - \omega - i\varepsilon}. \tag{10.79}$$

This dispersion relation is not yet in a standard form since it is necessary to interchange the order of integration on the right-hand side in order to obtain $A(\omega')$. The interchange can be carried out without trouble when $m = 0$ or when $\omega^2 > m^2$ (see equation (10.76)), since the function $\sin kr/k$ behaves in a regular fashion. However, if $\omega^2 < m^2$ then $k = \sqrt{(\omega^2 - m^2)}$ is imaginary and $\sin kr \to \sinh |k|\, r$ and the integral diverges for large r. Originally this problem was ignored in setting up dispersion relations for forward scattering. Later it was recognised and various techniques were evolved for dealing with this region (for a detailed discussion of these methods see the article by Goldberger, 1960). The result of the more detailed examination showed that the original interchange of integration over the r and ω variables could be justified. We will therefore ignore the technical problems associated with the interchange and write

$$T(\omega) = U(\omega) + \frac{1}{\pi}\left(\int_{-\infty}^{-m} + \int_{m}^{-\infty}\right) d\omega' \frac{A(\omega')}{\omega' - \omega - i\varepsilon} \tag{10.80}$$

where

$$U(\omega) = \frac{1}{\pi}\int_{-m}^{+m} d\omega' \frac{A(\omega')}{\omega' - \omega - i\varepsilon}. \tag{10.81}$$

10.3(d). *Physical and unphysical regions in the forward scattering dispersion relation*

The dispersion relation given in (10.80) possesses interesting properties. The region of physical scattering is when $\omega' > m$; in addition there is an unphysical region $m > \omega' > -\infty$. The region $-m < \omega' < -\infty$ can be reached by a crossing relation (compare (10.67) and § 10.2(e)). We are finally left with the region $-m < \omega' < m$ which is defined by the integral $U(\omega)$ (10.81). If $A(\omega')$

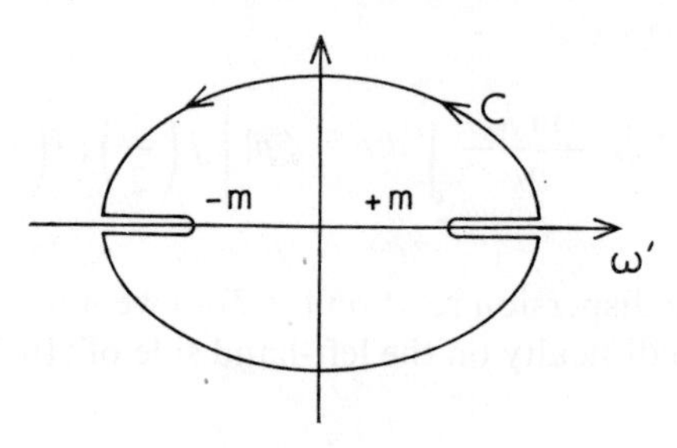

FIG. 10.11.

vanishes in this region, then we have branch points at $\omega'^2 = m^2$ and we may use the contour given in Fig. 10.11. The contour is similar in principle to that of Fig. 10.8, and the associated integral is the second term on the right-hand side of (10.80) (compare (10.24)).

We now consider the problems associated with $U(\omega)$ (10.81). A cursory inspection of this equation might suggest that $A(\omega')$ would vanish for $\omega' < m$ since $A(\omega')$ is linearly related to the total cross-section through the optical theorem. However, points can arise in this unphysical region where the symmetry laws are formally satisfied. We shall show that these points give rise to pole or Born terms in the scattering amplitudes. In order to demonstrate this feature we shall use the unitarity principle (§ 7.5(c)) and diagrams; let us give the blob on the left-hand side of Fig. 10.12 a structure by assuming the existence of intermediate states and stretching the single blob into two blobs (we shall consider only forward scattering so that $k = k'$, $p = p'$).

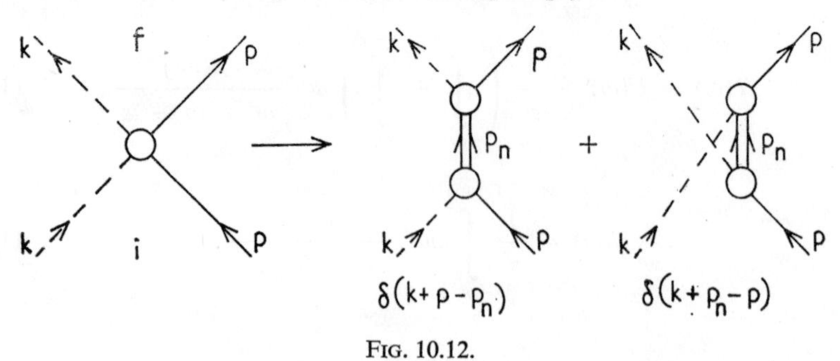

FIG. 10.12.

We shall represent the intermediate states by the vectors $|n\rangle$; their presence in the scattering amplitude may be provided for by inserting a complete set of states $\sum_n |n\rangle\langle n| = \hat{1}$ (4.15). The complex part of the scattering amplitude, $A(\omega)$, (10.68) then becomes

$$A(\omega) = \frac{N_p^2}{2}\int_{-\infty}^{+\infty} d^4z\, e^{-ikz}\langle p|\left[j\left(\frac{z}{2}\right), j\left(-\frac{z}{2}\right)\right]|p\rangle$$

$$= \frac{N_p^2}{2}\int_{-\infty}^{+\infty} d^4z\, e^{-ikz}\langle p|\, j\left(\frac{z}{2}\right) j\left(-\frac{z}{2}\right)|p\rangle - \langle p|\, j\left(-\frac{z}{2}\right) j\left(\frac{z}{2}\right)|p\rangle$$

$$= \frac{N_p^2}{2}\sum_n\int_{-\infty}^{+\infty} d^4z\, e^{-ikz}\left[\langle p|\, j\left(\frac{z}{2}\right)|n\rangle\langle n|\, j\left(-\frac{z}{2}\right)|p\rangle\right.$$

$$\left. - \langle p|\, j\left(-\frac{z}{2}\right)|n\rangle\langle n|\, j\left(\frac{z}{2}\right)|p\rangle\right]. \tag{10.82}$$

If we now make use of translational invariance (8.141) we find

$$j\left(\frac{z}{2}\right) = e^{-i(Pz/2)}\, j(0)\, e^{i(Pz/2)} \tag{10.83}$$

$$j\left(-\frac{z}{2}\right) = e^{i(Pz/2)}\, j(0)\, e^{-i(Pz/2)}$$

and equation (10.82) becomes

$$A(\omega) = \frac{N_p^2}{2}\sum_n\int_{-\infty}^{+\infty} d^4z\, e^{-ikz}\,[e^{-i(p-p_n)z}\langle p|\, j(0)\,|n\rangle\langle n|\, j(0)\,|p\rangle$$

$$- e^{i(p-p_n)z}\langle p|\, j(0)\,|n\rangle\langle n|\, j(0)\,|p\rangle]$$

$$= N_p^2\, 8\pi^4 \sum_n |\langle p|\, j(0)\,|n\rangle|^2\,[\delta(k+p-p_n) - \delta(k+p_n-p)]. \tag{10.84}$$

The two δ-functions can obviously be associated with the two diagrams on the right-hand side of Fig. 10.12. It can be seen that the ingoing particle can be destroyed at either blob. A similar situation arises in perturbation theory (compare Fig. 8.20 (p. 322)).

As in perturbation theory, it is customary to assume that only known particles are associated with the intermediate states and that the usual symmetry properties are preserved at the blobs. In addition more than one particle can be associated with the intermediate state.

The δ-functions in (10.84) can be written as

$$\delta(k + p - p_n) = \delta(\mathbf{k} + \mathbf{p} - \mathbf{p}_n)\, \delta(\omega + \omega_p - \omega_n)$$
$$= \delta(\mathbf{k} - \mathbf{p}_n)\, \delta(\omega + M - \omega_n) \tag{10.85}$$
$$\delta(k + p_n - p) = \delta(\mathbf{k} + \mathbf{p}_n - \mathbf{p})\, \delta(\omega + \omega_n - \omega_p)$$
$$= \delta(\mathbf{k} + \mathbf{p}_n)\, \delta(\omega + \omega_n - M) \tag{10.86}$$

since we are considering forward scattering and working in the reference frame (10.65)

$$p = (0, iM)$$

where M is the mass of the target particle.

The form of the scattering amplitude (10.84) is thus determined by

$$\sum_n |\langle p|\, j(0)\, |n\rangle|^2$$

and the δ-functions. Let us now consider some specific examples.

(1) Pseudoscalar meson–meson scattering. In order to be more specific let us say that the mesons are pions, and that $M = m$ in (10.85) and (10.86). Consider the first δ-function for energy, we may write its argument as

$$\omega = \omega_n - m. \tag{10.87}$$

We now consider the possible states $|n\rangle$. That with lowest mass is the one meson state, and so $\omega_n \geqq m$. Thus ω is always greater than zero, and the first δ-function will contribute to an integral over ω for all $\omega > 0$. Similarly the second δ-function contributes to all $\omega < 0$. Now let us consider the term $\langle p|\, j(0)\, |\pi\rangle$. The intermediate state has zero angular momentum since $\mathbf{p} = 0$ and j is a scalar under rotations. But both $|p\rangle$ and $|\pi\rangle$ have odd intrinsic parity and $j(0)$ is pseudoscalar. Thus if we require parity invariance $\langle p|\, j(0)\, |\pi\rangle$ must vanish. The same objection does not arise for the state $\langle p|\, j(0)\, |2\pi\rangle$; in this case $\omega_n \geqq 2m$ and hence ω must always be greater than m in (10·87). The function $U(\omega)$ therefore vanishes in (10.82), and a dispersion relation for forward scattering in the pion–pion interaction can, in principle, be established.

$$T(\omega) = \frac{1}{\pi}\left(\int_{-\infty}^{-m} + \int_{m}^{\infty}\right) d\omega'\, \frac{A(\omega')}{\omega' - \omega - i\varepsilon}.$$

(2) Scalar meson–meson scattering. The parity argument is of no use for scalar meson–meson scattering and so the nonphysical region $-m < \omega < m$ has to be investigated.

(3) Meson–nucleon scattering. In this case the state $|p\rangle$ is a nucleon and if we assume the intermediate states conserve baryon number, then $|n\rangle$ must be of the

form

$$|n\rangle = 1 \text{ nucleon} \tag{10.88}$$

$$1 \text{ nucleon } 1 \text{ pion}$$

$$1 \text{ nucleon } 2 \text{ pions}$$

$$\cdots\cdots\cdots\cdots\cdots\cdots$$

$$2 \text{ nucleons } 1 \text{ antinucleon}$$

$$\cdots\cdots\cdots\cdots\cdots\cdots\cdots\cdots$$

where again, for convenience, we have assumed the meson to be a pion. Equation (10.87) now becomes

$$\omega = \omega_n - M \tag{10.89}$$

where $\omega_n \geqq M$, and it is again apparent that $U(\omega)$ can contribute in the region $\omega^2 \leqq m^2$.

The regions with $\omega^2 \leqq m^2$ correspond to complex momenta and lead to the term sinh $|k|\,r$ as we have seen in the previous section. Since the problem associated with this term can be circumvented, as we have already mentioned, we shall ignore it in future discussions.

10.3(e). *The unphysical region and the scattering of scalar particles*

We return to equations (10.84) and consider a single particle in the intermediate state. The summation over n in this equation can be replaced by an integral (7.28)

$$\sum_n \to \frac{1}{(2\pi)^3}\int d\mathbf{p}_n \tag{10.90}$$

where we have set $V = 1$ since it would have cancelled. The integration of (10.84) over $\mathbf{p}_n$ then gives

$$A(\omega) = N_p^2\,\frac{8\pi^4}{8\pi^3}\,|\langle p|\,j(0)\,|n\rangle|^2\,[\delta(\omega + M - \omega_n) - \delta(\omega + \omega_n - M)]. \tag{10.91}$$

Consider the argument of the first δ-function and recall that it is associated with (10.85).

$$\delta(k + p - p_n) = \delta(\mathbf{k} - \mathbf{p}_n)\,\delta(\omega + M - \omega_n). \tag{10.92}$$

We may write

$$\omega^2 = \mathbf{k}^2 + m^2 = \mathbf{p}_n^2 + m^2 = \omega_n^2 - m_n^2 + m^2 \tag{10.93}$$

$$= (\omega_n - M)^2 = \omega_n^2 - 2\omega_n M + M^2 \tag{10.94}$$

hence

$$\omega_n = \frac{M^2 + m_n^2 - m^2}{2M} \tag{10.95}$$

and so

$$M - \omega_n = \omega_R = M - \frac{M^2 + m_n^2 - m^2}{2M}$$
$$= \frac{M^2 - m_n^2 + m^2}{2M}. \tag{10.96}$$

Let us assume that the scattering process involves identical scalar particles, and that the target meson, the scattered meson and propagator all have the same mass, then equation (10.96) gives

$$\omega_R = \frac{m}{2}, \quad \text{for} \quad m = M = m_n. \tag{10.97}$$

Now consider the energy δ-function in (10.85); it will appear in a function of the form

$$\int d\omega\, f(\omega)\, \delta(\omega + M - \omega_n) = \int dx\, f(\omega)\, \delta(x)\, \frac{\partial \omega}{\partial x} = \left([f(\omega)] \Big/ \left|\frac{\partial x}{\partial \omega}\right|\right)_{x=0}$$

where we have used the last of equations (A. 6.4) (Appendixes, p. 703); now by (10.93)

$$\frac{\partial x}{\partial \omega} = \frac{\partial}{\partial \omega}(\omega + M - \omega_n) = 1 - \frac{\partial \omega_n}{\partial \omega} = 1 - \frac{\omega}{\omega_n}$$
$$= 1 - \frac{\omega_n - M}{\omega_n} = \frac{M}{\omega_n}$$

hence

$$\int d\omega\, f(\omega)\, \delta(\omega + M - \omega_n) = \frac{\omega_n}{M} f(-\omega_R) \equiv \int d\omega\, f(\omega)\, \frac{\omega_n}{M}\, \delta(\omega + \omega_R)$$

so that

$$\delta(\omega + M - \omega_n) \to \frac{\omega_n}{M}\, \delta(\omega + \omega_R) \tag{10.98}$$

and similarly equation (10.86) gives

$$\delta(\omega + \omega_n - M) \to \frac{\omega_n}{M}\, \delta(\omega - \omega_R). \tag{10.99}$$

Thus the contribution to the scattering amplitude from the single particle intermediate state vanishes everywhere except at $\omega = \pm\omega_R = \pm\frac{1}{2}m$ in the unphysical region, and equation (10.91) can be written

$$A(\omega) = 2\pi\omega_n |\langle p|\, j(0)\, |n\rangle|^2\, [\delta(\omega + \omega_R) - \delta(\omega - \omega_R)] \tag{10.100}$$

where we substituted $N_p^2 = 2M$ (compare (7.36)). The above expression is the sole contributor to the term $U(\omega)$ in (10.81). As we have explained in the argunent following (10.87), the two meson contribution to the intermediate state mplies that $\omega_n \geqq 2m$, hence $\omega \geqq m$ and so this contribution extends over the continuum.

Now let us consider the term $|\langle p| j(0) |n\rangle|^2$ in (10.100). This expression must be a scalar quantity, and must also be a function of the momentum four-vectors p and n. The only Lorentz invariant term which we can construct is as follows:

$$|\langle p| j(0) |n\rangle|^2 = \frac{|F(p-n)^2|^2}{4\omega_p\omega_n} = \frac{|F(-m^2)|^2}{4\omega_p\omega_n} \tag{10.101}$$

where we have factored out the term $4\omega_p\omega_n$ for later convenience. For reasons which will become apparent later we shall denote $|F(-m^2)|^2$ by g^2. If we now collect terms in $A(\omega)$ from (10.100) we obtain

$$\begin{aligned} A(\omega) &= 2\pi\omega_n \frac{g^2}{4\omega_p\omega_n} [\delta(\omega + \omega_R) - \delta(\omega - \omega_R)] \\ &= g^2 \frac{\pi}{2m} [\delta(\omega + \omega_R) - \delta(\omega - \omega_R)] \end{aligned}$$

where we have set $\omega_p = m$, since we are considering $m = M$ and forward scattering in the laboratory system.

Thus the integral $U(\omega)$ in the unphysical region of the dispersion relation (10.81) has become

$$\begin{aligned} U(\omega) &= \frac{1}{\pi} \int_{-m}^{m} d\omega' A(\omega') \\ &= \frac{g^2}{2m} \int_{-m}^{m} d\omega' \left[\frac{\delta(\omega' + \omega_R)}{\omega' - \omega - i\varepsilon} - \frac{\delta(\omega' - \omega_R)}{\omega' - \omega - i\varepsilon}\right] \\ &= \frac{g^2}{2m} \left[\frac{1}{\omega - \omega_R} - \frac{1}{\omega + \omega_R}\right] \qquad \varepsilon \to 0 \end{aligned} \tag{10.102}$$

and so from equation (10.80) we obtain the following dispersion relation for scalar meson–meson scattering:

$$\begin{aligned} T(\omega) &= \frac{g^2}{2m} \left[\frac{1}{\omega - \omega_R} - \frac{1}{\omega + \omega_R}\right] + \frac{2}{\pi} \int_m^\infty d\omega' \frac{\omega' A(\omega')}{\omega'^2 - \omega^2 - i\varepsilon} \\ &= \frac{g^2}{2m} \left[\frac{1}{\omega - \frac{1}{2}m} - \frac{1}{\omega + \frac{1}{2}m}\right] + \frac{2}{\pi} \int_m^\infty d\omega' \frac{\omega' A(\omega')}{\omega'^2 - \omega^2 - i\varepsilon} \\ &= \frac{g^2}{2} \frac{1}{\omega^2 - \frac{1}{4}m^2} + \frac{2}{\pi} \int_m^\infty d\omega' \frac{\omega' A(\omega')}{\omega'^2 - \omega^2 - i\varepsilon} \end{aligned} \tag{10.103}$$

where we have used the crossing relation (10.67), $T^*(-\omega) = T(\omega)$ and the first of equations (10.62). It can be seen that the scattering amplitude $T(\omega)$ is represented by a function which is analytic in the upper half plane except for poles at $\pm\frac{1}{2}m$.

The first term on the right-hand side of (10.103) is called the *Born, pole* or *bound state term* by various authors. We show now that it has the form expected from lowest order perturbation theory. Let us consider the middle diagram of Fig. 10.12 and treat it as a Feynman diagram with coupling constants g at the vertices. The effective Lagrangian associated with the interaction of identical scalar particles is

$$\mathscr{L}_{\text{int}} = g\varphi^3 \tag{10.104}$$

and if we make judicious adjustments to the rules given in § 8.6(b), in accord with the scalar nature of the particles and the kinematic conditions of equation (10.65), and then use equation (7.21), we obtain a matrix element

$$M_{fi} = g^2 \frac{1}{2\omega} \frac{1}{2M} \frac{1}{p_n^2 + m_n^2} = \frac{1}{2\omega} \frac{1}{2M} T_{fi}$$

The last equation can be obtained from (7.34), hence

$$T_{fi} = g^2 \frac{1}{p_n^2 + m_n^2}. \tag{10.105}$$

Now from the middle diagram of Fig. 10.12

$$p_n = k + p$$

and

$$\begin{aligned} p_n^2 + m_n^2 &= (k + p)^2 + m_n^2 \\ &= \mathbf{k}^2 - (\omega + M)^2 + m_n^2 \\ &= \omega^2 - m^2 - \omega^2 - 2\omega M - M^2 + m_n^2 \end{aligned}$$

since we are using the conditions (10.65). For scalar meson–meson scattering we can write $m = m_n = M$, and so

$$T_{fi} = \frac{g^2}{-2\omega m - m^2} = \frac{-g^2}{2m(\omega + \frac{1}{2}m)}. \tag{10.106}$$

An inspection (10.103) and earlier equations shows that an expression of similar form was obtained for the Born term from the same diagram. The reasons for introducing the factor 4 in (10.101) and for writing $F[(p - n)^2]$ as g^2 are now apparent. It should be noted that the g^2 obtained from lowest order perturbation theory (10.106) refers to bare particle fields and unrenormalised coupling constants, whereas the g^2 of equation (10.103) was obtained with real particle fields and represents a renormalised coupling strength.

10.3(f). *Dispersion relations for the forward scattering of the pion–nucleon system*

The dispersion relations we have just derived can be used to construct equa-ions for the forward scattering of nonscalar systems. As an example we shall consider pion–nucleon scattering (Goldberger, Miyazawa and Oehme, 1955).

Firstly, isospin and spin states must be introduced. We shall consider the scattering of a pion with isospin component α and a nucleon with a parameter λ (all states apart from four-momentum) to final states of β and λ' respectively

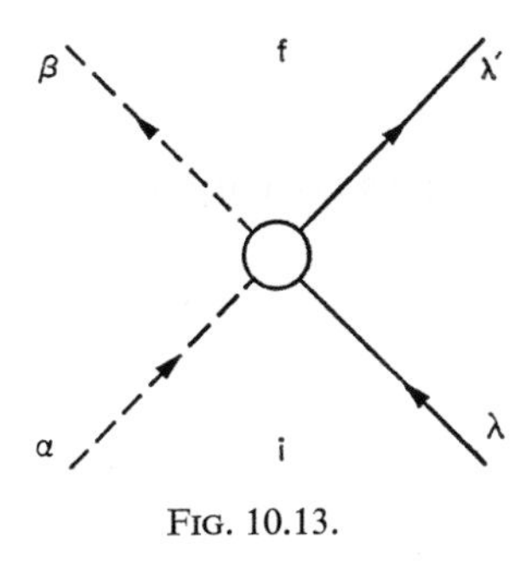

FIG. 10.13.

(Fig. 10.13). The nucleon variables can be suppressed (compare § 9.4(d)), and so an appropriate scattering amplitude can be obtained by modifying (10.66)

$$T_{\beta\alpha}(\omega) = -N_p^2 \int_{-\infty}^{+\infty} d^4z\, e^{-ikz} \langle p|\, R\left[j_\beta\left(\frac{z}{2}\right) j_\alpha\left(-\frac{z}{2}\right)\right] |p\rangle. \quad (10.107)$$

The crossing symmetry relation (10.67) becomes

$$T^\dagger_{\beta\alpha}(-\omega) = T_{\beta\alpha}(\omega) \quad (10.108)$$
$$T_{\beta\alpha}(-\omega) = T^\dagger_{\beta\alpha}(\omega)$$

where the adjoint is taken with respect to the nucleon variables. The isospin dependence of $T_{\beta\alpha}$ can be written as

$$T_{\beta\alpha} = T_1 \delta_{\beta\alpha} + \tfrac{1}{2} T_2 [\tau_\beta, \tau_\alpha] \quad (10.109)$$

(compare (9.117)); ordinary spin flip does not occur in forward scattering and so the above equation is adequate for our purpose. In this equation T_1 and T_2 correspond to no isospin flip and isospin flip amplitudes respectively. From (10.108) they behave as

$$T_1(-\omega) = T_1^*(\omega), \qquad T_2(-\omega) = -T_2^*(\omega). \quad (10.110)$$

The dispersion relations for T_1 and T_2 can be constructed in the same manner as in the previous sections†. We must also note that the same analytic problems arise in the unphysical region. We shall again ignore them; the analogous equation to (10.91) for $A(\omega)$ is

$$A(\omega) = N_p^2 \pi \sum_{n\,\text{spins}} [\langle p| j_\beta(0) |n\rangle \langle n| j_\alpha(0) |p\rangle \, \delta(\omega + M - E_n)$$

$$- \langle p| j_\alpha(0) |n'\rangle \langle n'| j_\beta(0) |p\rangle \, \delta(\omega + E_n - M)] \quad (10.111)$$

where we have replaced ω_n by E_n. Since we are dealing with forward scattering we may write

$$p = 0, iM, \qquad N_p^2 = 1 \quad (10.112)$$

where M is the nucleon mass, and since baryon number is conserved in Fig. 10.12.

$$m_n = M$$

It is not difficult to show that the δ-functions now become

$$\delta(\omega + M - E_n) = \frac{E_n}{M} \, \delta(\omega + \omega_R) \quad (10.113)$$

$$\delta(\omega + E_n - M) = \frac{E_n}{M} \, \delta(\omega - \omega_R)$$

(compare § 10.3(e)), where

$$\omega_R = \frac{m^2}{2M} \quad (10.114)$$

and m is the pion mass.

Thus equation (10.111) becomes

$$A(\omega) = \frac{\pi E_n}{M} \sum_{n\,\text{spins}} [\langle p| j_\beta(0) |n\rangle \langle n| j_\alpha(0) |p\rangle \, \delta(\omega + \omega_R)$$

$$- \langle p| j_\alpha(0)| n'\rangle \langle n'| j_\beta(0) |p\rangle \, \delta(\omega - \omega_R)]. \quad (10.115)$$

We next consider the current elements. The general structure of the terms can again be established with the aid of perturbation theory. A charge independent Lagrangian for the basic pion–nucleon interaction was given in (9.111)

$$\mathscr{L} = ig_0 \overline{\psi}_N \gamma_5 (\boldsymbol{\tau} \cdot \boldsymbol{\varphi}) \psi_N = ig_0 \overline{\psi}_N \gamma_5 \tau_\alpha \varphi_\alpha \psi_N \quad (10.116)$$

where we have now written g_0 for g to indicate that we are considering point

† The procedure which we are describing represents a standard method for applying dispersion relations to elementary particle physics. Firstly the scattering amplitude is constructed in terms of a set of invariant scalar functions (T_1 and T_2 in the present case), and then dispersion relations are evaluated for these functions. The results are finally related to physical conditions, for example charge or isospin states.

articles and g_0 is an unrenormalised coupling constant. This Lagrangian gives
he following equation of motion (compare § 8.3(a)):

$$(\Box^2 - m^2)\,\varphi_\alpha = -ig_0\overline{\psi}_N\gamma_5\tau_\alpha\psi_N = -j_\alpha(x). \tag{10.117}$$

hus, with the aid of Table 8.1, a typical matrix element in perturbation theory
ooks like

$$\langle p|\, j_\beta(0)\, |n\rangle_B = ig_0\,\frac{M}{\sqrt{(E_pE_n)}}\,\bar{u}(\mathbf{p})\,\gamma_5\tau_\beta u(\mathbf{p}_n) \tag{10.118}$$

here the subscript B signifies the Born term. Now let us consider the struc-
ure of the current elements in (10.115); from invariance arguments we require
urrent elements with the transformation properties of pseudocalars and so we
an write (compare (10.101))

$$\langle p|\, j_\beta(0)\, |n\rangle = i\,\frac{M}{\sqrt{(E_pE_n)}}\,F[(p - p_n)^2]\,\bar{u}(\mathbf{p})\,\gamma_5\tau_\beta u(\mathbf{p}_n) \tag{10.119}$$

nd

$$F[(p - p_n)^2] = F(-m^2).$$

imilarly,

$$\begin{aligned}\langle n|\, j_\alpha(0)\, |p\rangle &= \langle p|\, j_\alpha(0)\, |n\rangle^* \\ &= i\,\frac{M}{\sqrt{(E_pE_n)}}\,F^*(-m^2)\,\bar{u}(\mathbf{p}_n)\,\gamma_5\tau_\alpha u(\mathbf{p})\end{aligned} \tag{10.120}$$

y (4.17), and so we again may define a coupling constant

$$g^2 = |F(-m^2)|^2$$

s in the previous section, and again remark that g^2 represents the renormalised
oupling constant.

Thus the expression (10.115) for $A(\omega)$ in the unphysical region becomes

$$\begin{aligned}A(\omega) = \pi g^2 \sum_{n \text{ spins}} \{&[\bar{u}(\mathbf{p})\,\gamma_5\tau_\alpha u(\mathbf{p}_{n'})\,\bar{u}(\mathbf{p}_{n'})\,\gamma_5\tau_\beta u(\mathbf{p})]\,\delta(\omega - \omega_R) \\ &- [\bar{u}(\mathbf{p})\,\gamma_5\tau_\beta u(\mathbf{p}_n)\,\bar{u}(\mathbf{p}_n)\,\gamma_5\tau_\alpha u(\mathbf{p})]\,\delta(\omega + \omega_R)\}.\end{aligned} \tag{10.121}$$

he summation over the intermediate spin states can be carried out by the
nethod of (3.168), and, after using (10.86), we finally obtain

$$\begin{aligned}A(\omega) &= \pi g^2\,\frac{\omega_R}{2M}\,[\tau_\alpha\tau_\beta\delta(\omega - \omega_R) - \tau_\beta\tau_\alpha(\omega + \omega_R)] \\ &= \pi g^2\,\frac{\omega_R}{2M}\,\{\delta_{\alpha\beta}[\delta(\omega - \omega_R) - \delta(\omega + \omega_R)] \\ &\qquad + \tfrac{1}{2}[\tau_\beta, \tau_\alpha]\,[-\delta(\omega - \omega_R) - \delta(\omega + \omega_R)]\} \\ &= A_1(\omega)\,\delta_{\alpha\beta} + \tfrac{1}{2}[\tau_\beta, \tau_\alpha]\,A_2(\omega).\end{aligned} \tag{10.122}$$

We are now in a position to write dispersion relations for T_1 and T_2. We adapt equation (10.80) to read

$$T_j(\omega) = U_j(\omega) + \frac{1}{\pi}\left(\int_{-\infty}^{-m} + \int_{m}^{\infty}\right) d\omega' \frac{A_j(\omega')}{\omega' - \omega - i\varepsilon} \qquad (j = 1, 2)$$

and, by use of (10.122), the crossing relations (10.110) and equations (10.102) and (10.103), we find

$$T_1(\omega) = \frac{-g^2\omega_R^2}{M(\omega^2 - \omega_R^2)} + \frac{2}{\pi}\int_m^\infty d\omega' \frac{\omega' A_1(\omega')}{\omega'^2 - \omega^2 - i\varepsilon} \tag{10.123}$$

$$T_2(\omega) = \frac{g^2\omega_R\omega}{M(\omega^2 - \omega_R^2)} + \frac{2\omega}{\pi}\int_m^\infty d\omega' \frac{A_2(\omega')}{\omega'^2 - \omega^2 - i\varepsilon}$$

where (10.114)

$$\omega_R = \frac{m^2}{2M}$$

and m and M represent pion and nucleon masses respectively.

The amplitudes T_1, T_2, A_1 and A_2 are related to the scattering amplitudes for charged particles by equation (9.128)

$$T_1 = \tfrac{1}{2}(T^- + T^+), \qquad T_2 = \tfrac{1}{2}(T^- - T^+) \tag{10.124}$$

$$A_1 = \tfrac{1}{2}(A^- + A^+), \qquad A_2 = \tfrac{1}{2}(A^- - A^+)$$

where the superscripts refer to $\pi^- p$ and $\pi^+ p$ elastic scattering respectively.

We may use (7.34), (7.71) and the optical theorem to relate the amplitudes A to the total cross-sections and T to the forward scattering amplitudes $f(0)$

$$f_L(0) = \frac{1}{4\pi} T_{ii}^L \tag{10.125}$$

$$\sigma_T = \frac{A}{k_L}$$

where appropriate terms have been introduced for the normalisation factors N, and the symbol L has been introduced to remind the reader that we are working in the laboratory system. Thus upon collecting terms we obtain

$$f_L^-(\omega) + f_L^+(\omega) = -\frac{2g^2}{4\pi}\left(\frac{m}{2M}\right)^2 \frac{m^2}{M(\omega^2 - \omega_R^2)} + \frac{1}{2\pi^2}\int_m^\infty d\omega' \frac{k_L'\omega_L'[\sigma_T^-(\omega') + \sigma_T^+(\omega')]}{\omega'^2 - \omega^2 - i\varepsilon} \tag{10.126}$$

$$f_L^-(\omega) - f_L^+(\omega) = \frac{g^2}{4\pi}\left(\frac{m}{2M}\right)^2 \frac{4\omega}{\omega^2 - \omega_R^2}$$

$$+ \frac{\omega}{2\pi^2}\int_m^\infty d\omega' \frac{k_L'[\sigma_T^-(\omega') - \sigma_T^+(\omega')]}{\omega'^2 - \omega^2 - i\varepsilon}. \tag{10.127}$$

t should be noted that the replacement

$$\frac{f^2}{4\pi} = \frac{g^2}{4\pi}\left(\frac{m}{2M}\right)^2 \tag{10.128}$$

s often made.

In addition to the relations given above, the real and imaginary parts of the cattering amplitude are often related as in (10.21). For example, we may convert 10.127) to the form

$$\operatorname{Re} f_L^-(\omega) - \operatorname{Re} f_L^+(\omega) = \frac{f^2}{4\pi}\frac{4\omega}{\omega^2 - \omega_R^2} + \frac{\omega}{2\pi^2} P\int_m^\infty d\omega' \frac{k_L'[\sigma_T^-(\omega') - \sigma_T^+(\omega')]}{\omega'^2 - \omega^2} \tag{10.129}$$

Both this equation and (10.127) suffer from convergence difficulties at high nergies unless $\sigma_T^-(\omega') - \sigma_T^+(\omega') \to 0$ as $\omega' \to \infty$. We shall raise this point again n § 14.6. The equations are therefore sometimes given in subtracted form (compare § 10.2(e)); the subtraction is normally made at $\omega = m$.

10.3(g). *An application of a dispersion relation*

If we assume that the integral of (10.129) converges at high energies (§ 14.6), hen this equation can be rewritten in a form suitable for determining the coupling onstant $f^2/4\pi$. This was done by Haber-Schaim (1956) and represents one of the irst applications of dispersion relations to pion–nucleon scattering.

Consider equation (10.129); since

$$\omega_R^2 = \left(\frac{m^2}{2M}\right)^2 = 5{\cdot}5 \times 10^{-3}\, m^2$$

t can be neglected to good approximation. Furthermore, we may write part of he integrand as

$$\frac{1}{\omega'^2 - \omega^2} = \frac{1}{\omega'^2} + \frac{\omega^2}{\omega'^2(\omega'^2 - \omega^2)}.$$

f these two facts are incorporated in (10.129) and we multiply throughout by

$\frac{1}{2}\omega$ the equation becomes

$$\frac{1}{2}\omega[\operatorname{Re} f_L^-(\omega) - \operatorname{Re} f_L^+(\omega)] - \frac{\omega^4}{4\pi^2} P \int_m^\infty d\omega' \frac{k_L'}{\omega'^2(\omega'^2 - \omega^2)} [\sigma_T^-(\omega') - \sigma_T^+(\omega')]$$

$$= 2\frac{f^2}{4\pi} + \frac{\omega^2}{4\pi^2} \int_m^\infty d\omega' \frac{k_L'}{\omega'^2} [\sigma_T^-(\omega') - \sigma_T^+(\omega')]. \tag{10.130}$$

The left-hand side of this equation contains factors which may be determined by experiment, whilst the right-hand side is proportional to ω^2. If we call the left-hand side $\varphi(\omega)$ and plot it as a function of ω^2, and then extrapolate to $\omega^2 = 0$ we expect

$$\varphi(0) = 2\frac{f^2}{4\pi}.$$

The extrapolation of Haber-Schaim yielded

$$\frac{f^2}{4\pi} = 0{\cdot}082 \pm 0{\cdot}015 \quad \text{or} \quad \frac{g^2}{4\pi} = 15 \pm 3$$

for the pion–nucleon coupling strength. Later determinations have verified this result.

10.4. THE MANDELSTAM REPRESENTATION

10.4(a). *Introduction*

The dispersion relations for forward scattering, discussed in the previous section, give a good introduction to what can and cannot be done with this technique. It can be seen that the method can be used to relate components of the scattering amplitude, but does not yield the amplitude itself, in the sense of the way that the amplitude appears in perturbation theory.

Historically the dispersion relations for forward scattering were followed by those at fixed momentum transfer (Gell-Mann, Goldberger, Nambu and Oehme unpublished; Salam, 1956; Capps and Takeda, 1956). Instead of discussing them in detail we shall now pass to the Mandelstam representation and later indicate how the dispersion relations for fixed momentum transfer can arise from it.

It should be noted that we shall confine our discussion to scalar particles, that is particles without both spin and isospin.

10.4(b). *Kinematic features*

The scattering amplitude for a process containing two incoming and two outgoing scalar particles can always be expressed in terms of two scalar quantities, for example the energy and scattering angle in the *c*-system.

Let us therefore consider the scattering of scalar particles and denote the ingoing and outgoing four momenta by p_1, p_2 and $-p_3, -p_4$ respectively, for example, in terms of our previous notation

$$k \equiv p_1, \quad p \equiv p_2, \quad k' \equiv -p_3, \quad p' \equiv -p_4 \tag{10.131}$$

so that the conventional scattering diagram transforms as in Fig. 10.14.

Thus p_1, p_2, p_3 and p_4 correspond to four ingoing particles. This convention does effect our previous arguments, since an ingoing antiparticle and outgoing

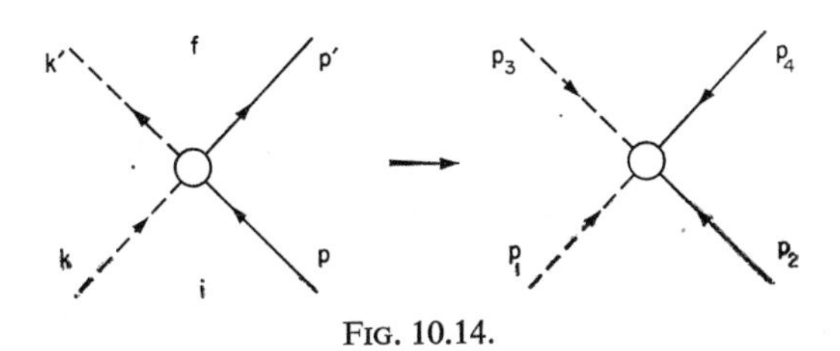

FIG. 10.14.

particle are equivalent processes. The conservation of energy and momentum can now be stated as

$$p_1 + p_2 + p_3 + p_4 = 0 \tag{10.132}$$

where

$$p_i^2 = -m_i^2 \quad (i = 1, 2, 3, 4)$$

and m_i denotes the mass of particle i. The only independent scalar variables that can be formed with the p_i terms are the set of six scalar products $p_i p_j$. However, the four conditions contained in (10.132) reduce the number of independent variables to two, as we have already stated above. However, rather than working with energy and scattering angle in the c-system, we shall use energy and four-momentum transfer. The latter quantity is more convenient since it is a Lorentz invariant. The variables may be introduced through three scalars defined by

$$\begin{aligned} u = s_1 &= -(p_1 + p_4)^2 = -(p_2 + p_3)^2 \\ t = s_2 &= -(p_2 + p_4)^2 = -(p_1 + p_3)^2 \\ s = s_3 &= -(p_3 + p_4)^2 = -(p_1 + p_2)^2 \end{aligned} \tag{10.133}$$

where we have given the two sets of symbols which appear most frequently in the literature.

It is a straightforward matter to show that

$$s + t + u = m_1^2 + m_2^2 + m_3^2 + m_4^2 = \sum_i m_i^2 \tag{10.134}$$

and so only two of the variables in the triplet are independent. If we choose p_1 and p_2 as the initial particles, then we can represent the four momenta in the c-system as

$$p_1 = \mathbf{p}_1, \quad iE_1 \equiv p_c^i, iE_1 \tag{10.135}$$
$$p_2 = \mathbf{p}_2, \quad iE_2 \equiv -p_c^i, iE_2$$
$$p_3 = -\mathbf{p}_3, \quad -iE_3 \equiv -p_c^f, -iE_3$$
$$p_4 = -\mathbf{p}_4, \quad -iE_4 \equiv p_c^f, -iE_4$$

and

$$s_3 = s = -(p_1 + p_2)^2 = (E_1 + E_2)^2 - (\mathbf{p}_1 + \mathbf{p}_2)^2$$
$$= (E_1 + E_2)^2 = E_c^2 \tag{10.136}$$
$$s_2 = t = -(p_1 + p_3)^2 = (E_1 - E_3)' - (\mathbf{p}_1 - \mathbf{p}_3)^2. \tag{10.137}$$

Thus s represents the square of the total energy in the centre of momentum system, and t minus the square of the four-momentum transfer, *when p_1 and p_2 represent the ingoing particles*. We shall examine the significance of s, t and u when we consider other pairs as ingoing particles in later sections.

In the case of elastic scattering

$$t = -(\mathbf{p}_1 - \mathbf{p}_3)^2 = -2p_c^2(1 - \cos\theta). \tag{10.138}$$

We may also note that if one particle, say particle 2, is stationary in the laboratory system and if we denote the energy of particle 1 in this system by ω, then

$$s_3 = s = (\omega + m_2)^2 - (\omega^2 - m_1^2) = 2\omega m_2 + m_1^2 + m_2^2. \tag{10.139}$$

10.4(c). *Physical processes and the variables s, t, u*

Consider the diagram shown in Fig. 10.15. We wish to associate physical processes with it on the assumption that two particles are ingoing and two are

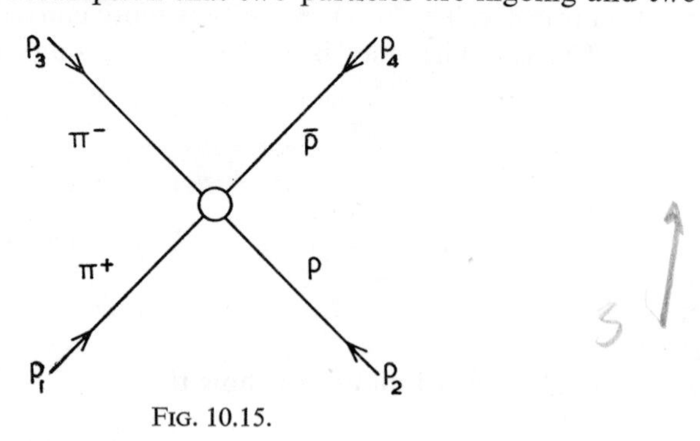

FIG. 10.15.

outgoing. The process will be obviously dependent upon which particles we choose as ingoing and which as outgoing, for example if we choose p_1 and p_2 to represent ingoing particles the processes is $\pi^+ + p \rightarrow \pi^+ + p$, on the other hand, ingoing p_2 and p_3 represent $\pi^- + p \to \pi^- + p$. In fact it is not difficult to see that the diagram represents six processes, which may be grouped in three pairs, each pair corresponding to a given process and its *TCP* reflection.

If complications due to spin and isospin are ignored, we shall assert that a single analytic function of any of the two variables s_1, s_2 and s_3 describes all six processes. This statement constitutes the most general form of the substitution law, and it represents one of the basic assumptions in the Mandelstam representation. If graphical methods of calculation are used (compare § 8.6(d)), it is not difficult to demonstrate its plausibility.

It is customary to consider each process with its *TCP* reversed process, and to refer to this pair as a single channel. We shall denote the particles associated with the four momenta p_1, p_2, p_3 and p_4 by the numbers 1, 2, 3 and 4 respectively, and represent the reactions as follows:

	Channel	Reaction		
u	I	$1 + 4 \to \bar{2} + \bar{3}$ and	$2 + 3 \to \bar{1} + \bar{4}$	(10.140)
t	II	$2 + 4 \to \bar{1} + \bar{3}$	$1 + 3 \to \bar{2} + \bar{4}$	
s	III	$3 + 4 \to \bar{1} + \bar{2}$	$1 + 2 \to \bar{3} + \bar{4}$	

where we have again used two popular alternative notations to indicate the channels (compare (10.133)).

Now let us examine the channels II(t) and III(s) a little more closely. We have already seen in the previous section that in channel III the parameters $s_3(s)$ and $s_2(t)$ represent the square of the energy in the c-system and minus the square of the four-momentum transfer respectively. In the c-system for channel II we can write

$$t = s_2 = -(p_1 + p_3)^2 = (E_1 + E_3)^2 - (\mathbf{p}_1 + \mathbf{p}_3)^2 \tag{10.141}$$
$$= (E_1 + E_3)^2 = E_c^2 = E_t^2$$
$$s = s_3 = -(p_1 + p_2)^2 = (E_1 - E_2)^2 - (\mathbf{p}_1 - \mathbf{p}_2)^2$$

and if the process is an elastic scattering

$$s = s_3 = -2p_c^2(1 - \cos\theta)$$
$$= -2p_t^2(1 - \cos\theta_t) \tag{10.142}$$

where we have used the subscript t to indicate the fact that we are working in the t-channel. It is apparent from equations (10.136), (10.137) and (10.141) that the variables s_2 and s_3 change roles in the two channels. In a similar manner one may show that in channel I (or u), s_1 (or u) represents the total energy in the c-system for reaction I.

We now consider the physical and non-physical domains for the three channels. It is convenient to introduce triangular coordinates (Fig. 10.16) (see, for example,

Kibble, 1960) where all lengths inside an equilateral triangle are positive, and all lengths outside are negative giving the condition

$$s_1 + s_2 + s_3 = \text{const} = \sum_i m_i^2$$

when s_1, s_2 and s_3 are plotted as perpendiculars to the axes.

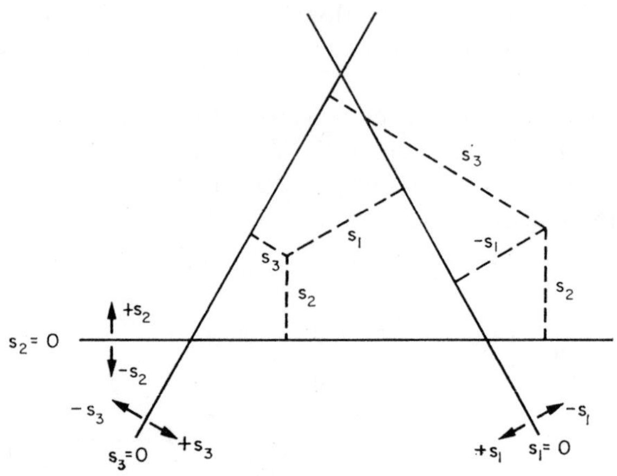

FIG. 10.16.

For convenience of expression we shall introduce particles

$$\pi^+ \equiv 1, \quad p \equiv 2, \quad \pi^- \equiv 3, \quad \bar{p} \equiv 4$$

and assume that they are all scalar. The three channels in (10.140) therefore become:

Channel		Reaction
u	I	$\pi^- + p \to \pi^- + p$
t	II	$\pi^- + \pi^+ \to \bar{p} + p$
s	III	$\pi^+ + p \to \pi^+ + p.$

Equations (10.136) and (10.137) then tell us that in channel III the physical limits are set by the following relations (M and m are the nucleon and pion masses respectively).

$$s_3 = E_c^2 \geqq (M + m)^2 \tag{10.143}$$

$$s_2 = -2p_c^2(1 - \cos\theta) = -4p_c^2 \sin^2 \theta/2 \leqq 0$$

$$s_1 = (E_1 - E_4)^2 - (\mathbf{p}_1 - \mathbf{p}_4)^2 \leqq (M - m)^2.$$

Limits on s_1, s_2 and s_3 may be deduced in a similar manner for channels II and III, and we may draw up the following table.

TABLE 10.1

Process	Physical region		
	s_1	s_2	s_3
I $\pi^- + p \to \pi^- + p$	$\geqq (M+m)^2$	$\leqq 0$	$\leqq (M-m)^2$
II $\pi^+ + \pi^- \to \bar{p} + p$	$\leqq 0$	$\geqq 4M^2$	$\leqq 0$
III $\pi^+ + p \to \pi^+ + p$	$\leqq (M-m)^2$	$\leqq 0$	$\geqq (M+m)^2$

The limits on the physical regions are sketched in Fig. 10.17.

It should be noted that if the mass of one of the particles exceeds the other three, then a fourth physical region corresponding to the decay process then

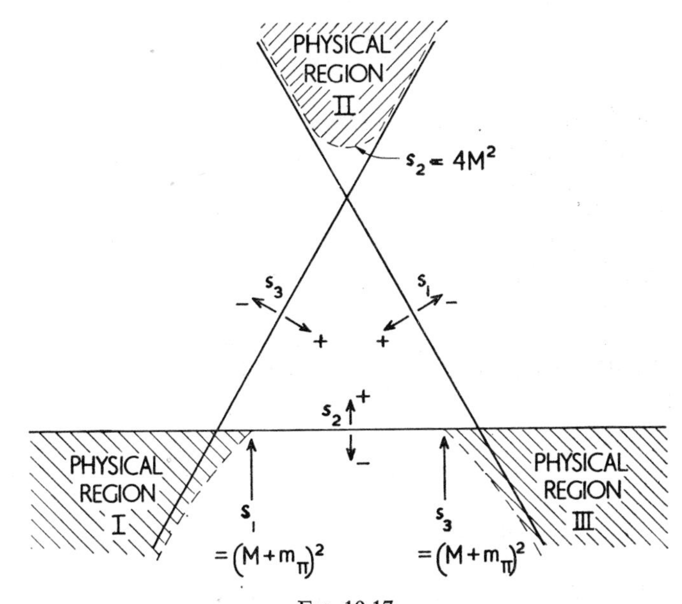

FIG. 10.17.

appears inside the triangle. We also remark that although Fig. 10.17 possesses some symmetry, this need not be true in general. The symmetry of Fig. 10.17 arises because only two masses are involved.

10.4(d). *The Mandelstam conjecture*

Mandelstam (1958a, 1959) introduced an expression for an invariant scattering amplitude as a function of the variables s_1, s_2 and s_3. It is based upon the assumption that:

(1) the six reactions occurring in the three channels I, II and III can be al described by the same analytic function $T(s_1, s_2, s_3)$,

(2) the amplitude is analytic in the entire complex plane, apart from cuts and poles which lie along the real axes.

Mandelstam conjectured that the part of the scattering amplitude which was dependent upon two independent variables possessed the following form:

$$T(s_1, s_2, s_3) = \frac{1}{\pi^2} \int ds_1' \, ds_2' \frac{\varrho_{12}(s_1', s_2')}{(s_1' - s_1)(s_2' - s_2)} + \frac{1}{\pi^2} \int ds_2' \, ds_3' \frac{\varrho_{23}(s_2', s_3')}{(s_2' - s_2)(s_3' - s_3)} + \frac{1}{\pi^2} \int ds_3' \, ds_1' \frac{\varrho_{31}(s_3', s_1')}{(s_3' - s_3)(s_1' - s_1)} \tag{10.144}$$

where the terms ϱ are real functions of the corresponding variables. In addition the amplitude may contain a part which is dependent upon only one of the variables s_1, s_2 and s_3. This component can be expressed in terms of one-dimensional dispersion integrals and possible pole terms; their form is similar to that which we have given for forward scattering (10.103).

A rigorous proof of the Mandelstam representation has never been given (Mandelstam, 1962). In our following treatment we intend using plausibility arguments only, and so we will ignore possible poles and limits for cuts.

The effective dispersion relation used in our discussion of forward scattering was of the form

$$T(\omega) = \frac{1}{\pi} \int d\omega' \frac{\operatorname{Im} T(\omega')}{\omega' - \omega - i\varepsilon}.$$

We shall now consider this relation in channel III. From equation (10.139) we can write $s_3 = 2\omega m_2 + m_1^2 + m_2^2$; also we shall let $\varepsilon \to 0$ in the above expression but only on the understanding that we are still dealing with complex variables. The equation can therefore be written as

$$T_{\mathrm{III}}(s_3) = \frac{1}{\pi} \int ds_3' \frac{\operatorname{Im} T_{\mathrm{III}}(s_3')}{s_3' - s_3}. \tag{10.145}$$

Now this is a dispersion relation for a special case of fixed momentum transfer namely $s_2 = 0$. It would seem plausible that a similar relationship exists for any value of fixed momentum transfer,

$$T_{\mathrm{III}}^{s_2}(s_3) = \frac{1}{\pi} \int ds_3' \frac{\operatorname{Im} T_{\mathrm{III}}^{s_2}(s_3')}{s_3' - s_3} \tag{10.146}$$

where the superscript indicates fixed momentum transfer. We note in passing that dispersion relations for fixed momentum transfer have in fact been proved for only limited ranges of s_2.†

Now let us also treat s_2 as a variable and assume that a dispersion relation can exist for Im T_{III} just as for T_{III}. A suitable form might be expected to be

$$\text{Im}\, T_{III}(s_2, s_3') = \frac{1}{\pi} \int ds_2' \frac{\varrho(s_2', s_3')}{s_2' - s_2} \tag{10.147}$$

and if we insert this expression into (10.146) we find

$$T_{\text{III}}(s_3, s_2) = \frac{1}{\pi^2} \int ds_3'\, ds_2' \frac{\varrho(s_2', s_3')}{(s_3' - s_3)(s_2' - s_2)}. \tag{10.148}$$

It can be seen that the above relation is similar in form to equation (10.144). In the general relation we might expect symmetry in all three variables, and so we write

$$T(s_1, s_2, s_3) = \frac{1}{\pi^2} \int ds_1'\, ds_2' \frac{\varrho_{12}(s_1', s_2')}{(s_1' - s_1)(s_2' - s_2)}$$
$$+ \frac{1}{\pi^2} \int ds_2'\, ds_3' \frac{\varrho_{23}(s_2', s_3')}{(s_2' - s_2)(s_3' - s_3)} + \frac{1}{\pi^2} \int ds_3'\, ds_1' \frac{\varrho_{31}(s_3', s_1')}{(s_3' - s_3)(s_1' - s_1)}.$$

The critical point in setting up this relation is the introduction of (10.147). Mandelstam evaluated Im $T_{III}(s_2, s_3')$ explicitly with the aid of fourth-order perturbation theory and showed that it was a boundary value of a function which satisfied a dispersion relation in this approximation. Subsequent work with more complex diagrams also appears to justify the assumption.

The symbols ϱ are called spectral functions. By applying unitarity, Mandelstam showed that these functions vanished except in certain regions which are indicated schematically by the curves C in Fig. 10.18. The asymptotic limits for these functions are determined by the masses of the particles which participate in the process.

If we choose to keep one of the parameters s_1, s_2 and s_3 fixed, the procedure we have outlined above can be reversed to yield single variable dispersion relations for s_1, s_2 and s_3 with the form of (10.146). In addition dispersion relations can be obtained for other quantities, for example partial wave dispersion relations (Chew, 1961).

† Proofs of dispersion relations for fixed momentum transfer have been given by a number of authors, see, for example, Lehmann (1959). They are far from simple and have only yielded results over a limited range of momentum transfer, for example

$$0 < -s_2 < \frac{32}{3} \frac{2M + m}{2M - m} m^2$$

in pion–nucleon scattering.

Applications of the double dispersion relations have yielded only modest results. In practice the techniques of detailed calculation are formidable. Most of the work has been concentrated on the pion–pion interaction and has led to divergence problems which have not been satisfactorily solved (see, for example, Mandelstam (1962) for a commentary).

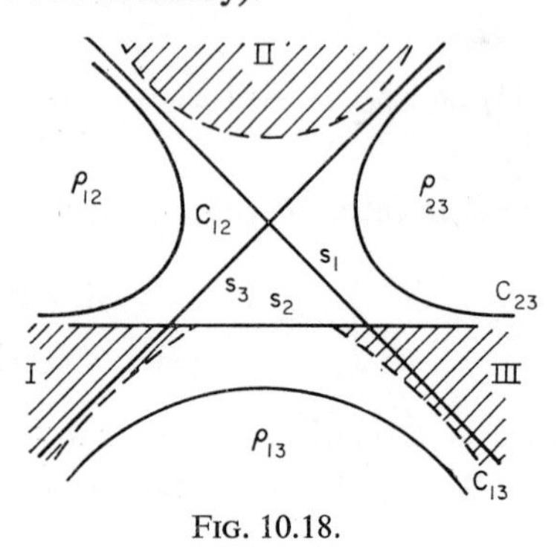

FIG. 10.18.

10.4(e). *Pole terms and 'polology'*

In addition to the dispersion integrals the scattering amplitude in the Mandelstam representation can contain pole terms. Their existence and position depends on the intermediate particle states. Consider the diagram shown in Fig. 10.19

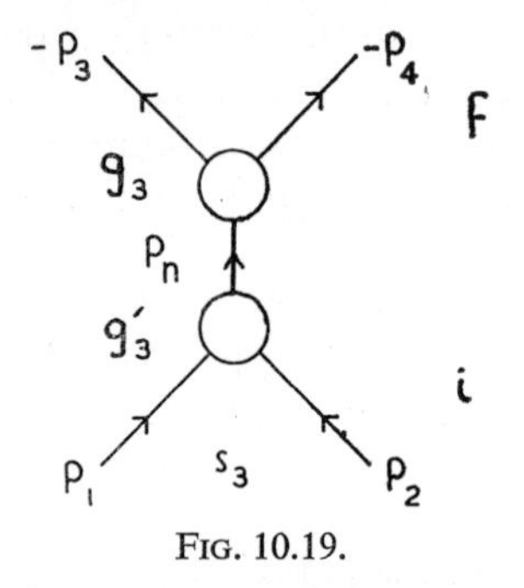

FIG. 10.19.

representing, say, a single particle intermediate state in channel III. If we denote its contribution to the transition amplitude as $T_{fi}(1)$, then

$$T_{fi}(1) = \frac{1}{\pi} \int_{-\infty}^{+\infty} ds_3' \frac{\operatorname{Im} T_{fi}(1)}{s_3' - s}. \qquad (10.149)$$

We may use the unitarity relation (7.106) to write $\operatorname{Im} T_{fi}(1)$ as

$$\operatorname{Im} T_{fi}(1) = \frac{1}{2} \sum_n \frac{(2\pi)^4}{N_n^2} \delta(p_i - p_n) T_{fn} T_{ni}^*.$$

We will assume that the intermediate particle is scalar so that $N_n^2 = 2\omega_n$ and so by equations (7.107) and (7.52)

$$\sum_n \frac{1}{N_n^2} = \frac{1}{(2\pi)^3} \int \frac{d\mathbf{p}_n}{2\omega_n} = \frac{1}{(2\pi)^3} \int d^4p_n\, \delta(p_n^2 + m_n^2)\, \theta(\omega_n)$$

hence

$$\begin{aligned} \operatorname{Im} T_{fi} &= \pi \int d^4p_n\, \delta(p_i - p_n)\, \delta(p_n^2 + m_n^2)\, \theta(\omega_n) T_{fn} T_{ni}^* \\ &= \pi T_{fn} T_{ni}^*\, \delta(p_i^2 + m_n^2) \\ &= \pi T_{fn} T_{ni}^*\, \delta(m_n^2 - s_3). \end{aligned}$$

If we define

$$(T_{fn})_{s_3 = m_n^2} = g_3, \qquad (T_{ni}^*)_{s_3 = m_n^2} = g_3'$$

then equation (10.149) gives us

$$\begin{aligned} T_{fi}(1) &= \frac{1}{\pi} \int_{-\infty}^{+\infty} ds_3' \frac{\pi g_3 g_3'}{s_3' - s_3}\, \delta(m_n^2 - s_3') \\ &= \frac{g_3 g_3'}{m_n^2 - s_3} = \frac{g_3 g_3'}{s_{3R} - s_3} \end{aligned} \tag{10.150}$$

where for later convenience we have written m_n^2 as s_{3R}. An inspection of § 10.3(e) shows that the above procedure is an alternative method for obtaining the second pole term in (10.102) (compare also equation (10.105)); the first term in equation (10.102) arises in channel I (u).

This formulation can be extended to other channels (if they possess appropriate singularities), and the full Mandelstam equation can be written as

$$\begin{aligned} T(s_1, s_2, s_3) = \sum_{i=1}^{3} \frac{g_i g_i'}{s_{iR} - s_i} + \sum_{i=1}^{3} \frac{1}{\pi} \int ds_i' \frac{\varrho_i(s_i')}{s_i' - s_i} \\ + \sum_{i,j} \frac{1}{\pi^2} \int ds_i'\, ds_j' \frac{\varrho_{ij}(s_i', s_j')}{(s_i' - s_i)(s_i' - s_j)}. \end{aligned} \tag{10.151}$$

The study of the pole terms and their residues† has yielded many of the most important quantitative results of the theory of strong interactions. The study is often called 'polology'. Chew (1959) has listed three facets of 'polology' that deserve attention:

(1) The existence and position of poles can be predicted on the basis of particle masses and intrinsic quantum numbers.

† For our purposes, the term residue means everything in the pole term apart from the denominator ($s_{iR} - s_i$).

(2) The residues of poles in different amplitudes or in different regions of the same amplitude are often simply related. In particular, coupling constants can usually be defined in terms of residues.

(3) Poles with large residues can dominate the behaviour of the scattering amplitude in their immediate neighbourhood.

There is an obvious corollary to (1), namely that the suspected existence of a pole can be used to predict the existence and properties of hitherto unknown particles (see § 11.4(f)).

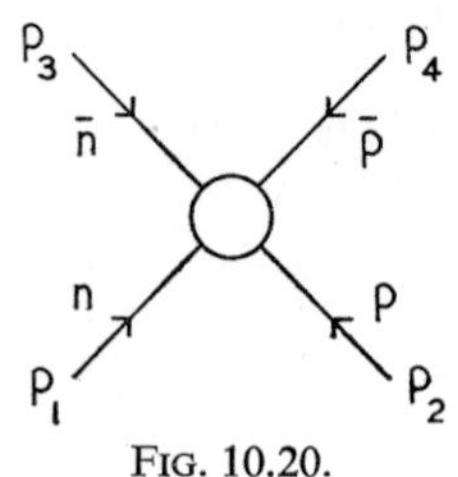

FIG. 10.20.

The location of the pole is most easily made with the aid of diagrams and perturbation theory. Consider, for example, nucleon–nucleon scattering (Fig. 10.20). The associated diagram has the following channels:

$$\begin{aligned} &\text{I } (u) \quad p + \bar{n} \to p + \bar{n} \\ &\text{II } (t) \quad n + \bar{n} \to p + \bar{p} \\ &\text{III } (s) \quad n + p \to n + p. \end{aligned} \tag{10.152}$$

In channel s the lowest bound state is that of the deuteron, so that $s_R = m_d^2$, whereas in channel t the isobaric number is 0, and so the lowest bound state is the pion with $t_R = m^2$. Since the deuteron is so much heavier than the pion we should therefore expect the pole term in the t channel to dominate the scattering amplitude at moderate energies. It should be noted that the photon could satisfy the symmetry requirements, but it is neglected in favour of the pion because of the relative weakness of the electromagnetic interaction.

The effective diagrams associated with the pion pole are given in Fig. 10.21, where diagrams (a) and (b) correspond to ordinary and charge exchange scat-

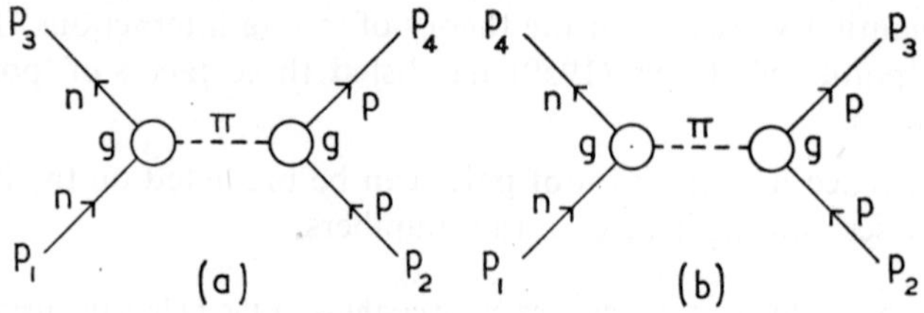

FIG. 10.21.

tering respectively. Since

$$\mathbf{p}_1 = -\mathbf{p}_2, \quad \mathbf{p}_3 = -\mathbf{p}_4, \quad E_1 = E_2 = E_3 = E_4$$

in the c-system, we may use equation (10.105), or (10.150) and (10.151), to show that the pole terms possess denominators which are given by

(a) $$p_n^2 + m^2 = (p_1 - p_3)^2 + m^2 = 2p_c^2(1 - \cos\theta) + m^2$$

(b) $$\begin{aligned} p_n^2 + m^2 &= (p_1 - p_4)^2 + m^2 \\ &= (\mathbf{p}_1 - \mathbf{p}_4)^2 + m^2 \\ &= (\mathbf{p}_1 + \mathbf{p}_3)^2 + m^2 = 2p_c^2(1 + \cos\theta) + m^2 \end{aligned}$$

where m is the pion mass. Thus the poles appear at the unphysical scattering angles

$$\cos\theta = \pm\left(1 + \frac{m^2}{2p_c^2}\right) \quad \begin{matrix} + \text{ diagram } (a) \\ - \text{ diagram (b).} \end{matrix}$$

Now the numerators of the Born terms are $g^2/4\pi$ (we ignore charge and spin complications), where g is the renormalised pion–nucleon coupling constant, and at high energies

$$\cos\theta = \pm\left(1 + \frac{m^2}{2p_c^2}\right) \to \pm 1 \quad \text{as} \quad p_c^2 \to \infty$$

that is the poles are nearly in the physical region, and they can dominate the scattering amplitude near $\theta = 0$ or π. It was therefore suggested by Chew (1958) that this behaviour could be turned to advantage to yield the coupling constants for the pion–nucleon interaction from neutron–proton scattering data. Measurements have been made by a number of groups (see, for example, Ashmore *et al.*, 1962) and yield values of $(g^2/4\pi) \sim 14$.

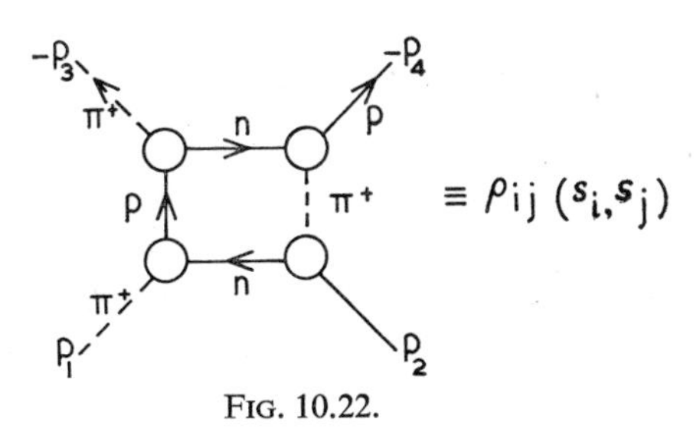

FIG. 10.22.

We mention, finally, that the one pion exchange term represents a force with range $\sim 1/m$. If the problem of n–p scattering is examined using the Born approximation and a Yukawa potential of the type $g^2(e^{-mr}/r)$ then a scattering amplitude

$$\frac{-2mg^2}{2p_c^2(1-\cos\theta)+m^2} \tag{10.153}$$

is obtained (compare § 6.3(b) and (6.89)). Thus, in so far as it is permissible to think in terms of potentials, the one pion exchange term represents the longest range part of the nucleon–nucleon potential. The exchange of two pions would represent a force with range $1/2m$.

The double integrals in (10.151) arise from diagrams of the type shown in Fig. 10.22 in which two or more particles are present in the intermediate states in all three channels. It is evident that diagrams of this type cannot give rise to simple pole terms in s_1, s_2, or s_3.

10.5. REGGE POLES AND TRAJECTORIES

10.5(a). *Introduction*

The combination of graph and pole techniques which we have described in this chapter works reasonably well for interactions involving single variables and simple poles. For example the one pion exchange in nucleon–nucleon scattering (§ 10.4(e)) yields a pole term in the t-channel of the form

$$\frac{-g^2}{2p_c^2(1+\cos\theta)+m^2} = \frac{g^2}{t-m^2}\quad\frac{g^2}{t-t_R}$$

which behaves in the manner expected from the known data on pion–nucleon scattering (m is the pion mass).

However, the boundary conditions associated with these methods do not always lead to physically sensible conclusions. Let us assume that we have a reaction involving four spinless particles in the s-channel $a+b\to d+e$ and that a particle of spin l gives a pole in the t-channel. The contribution of the pole term might then be expected to be of the form

$$\frac{g^2P_l(x_t)}{t-t_R}$$

if field theoretical arguments are a reliable guide to the construction of pole terms. In this expression $P_l(x_t)$ is a Legendre polynomial and x_t represents the cosine of the scattering angle in the t-channel, $x_t=\cos\theta_t$. Now x_t is related to the energy variable s in the s-channel (10.141), and if we assume for simplicity that all four particles have equal masses, then for large s and small t we can write

$$x_t = 1+\frac{2s}{t-4m^2} \sim -\frac{s}{2m^2}. \tag{10.154}$$

If x_t is large $P_l(x_t)$ will therefore behave like s^l. Thus the presence of the pole at an unphysical value of the momentum transfer t leads to an energy dependence of the form s^l in the s-channel for large values of s. This behaviour obviously leads to divergence difficulties for spins $l > 1$ as $s \to \infty$.

The problem of how to treat systems with spin greater than one is not restricted to polology, but is present in all forms of field theory.

A solution to this problem was proposed by Chew, Frautschi and Mandelstam (1962; see also Chew and Frautschi, 1961). The argument proposed by these workers was based upon the work of Regge (1959, 1960), who examined the nonrelativistic Schrödinger equation for a particle in a potential well of the Yukawa type with the objective of providing a nonperturbative proof of the Mandelstam representation for potential scattering. Chew *et al.* have conjectured that the results of this work are also valid for relativistic forms of the scattering amplitude in special circumstances, which we shall describe in § 10.5(d).

10.5(b). *Regge poles*

In equation (6.20) the scattering amplitude for a spinless interaction was given as

$$f(\theta) = \sum_{l=0}^{\infty} (2l + 1) \frac{\sin \delta_l e^{i\delta_l}}{k} P_l(\cos \theta)$$

where the detailed form of the interaction is associated with the phase shifts δ_l. The scattering amplitude f is proportional to our amplitude T_{fi} (see, for example, (7.95)), and so if we work in the t-channel we may introduce an amplitude

$$T(x_t, t) = \sum_{l=0}^{\infty} P_l(x_t) A(l, t) \tag{10.155}$$

where $A(l, t)$ contains the phase shift terms and kinematic factors. Now if l is treated as a complex variable the scattering amplitude can be analytically continued in the complex l-plane with the help of radial solutions of the Schrödinger equation with complex l values. For convenience of writing we shall replace l by α when considering it as a complex variable. The expansion (10.155) can be rewritten in the form of a contour integral by a method due to Watson and Sommerfeld

$$T(x_t, t) = \frac{1}{2\pi i} \oint d\alpha \, \pi \frac{P_\alpha(-x_t) A(\alpha, t)}{\sin \pi\alpha} \tag{10.156}$$

where the contour just surrounds the Re α axis (see solid line of Fig. 10.23). The residues from the terms $\pi/\sin\pi\alpha$ yield $(-1)^l$, which may easily be verified by writing $\alpha = l + \delta$ where $\delta \to 0$, and $(-1)^l P_l(x_t) = P_l(-x_t)$. Thus it can be seen that the integral leads back to the summation of (10.155).

Regge showed that, for a wide range of Yukawa potentials, the contour of (10.156) could be distorted to the vertical line $\alpha = -\frac{1}{2} - i\,\infty$ to $= -\frac{1}{2} + i\,\infty$ (Fig. 10.23) without encountering any singularities other than the poles in

$A(\alpha, t)$. The ith pole occurs at a position $\alpha_i(t)$ in the complex α-plane; its position varies continuously with t. Thus the amplitude becomes

$$T(x_t, t) = \frac{-1}{2i} \int_{-\frac{1}{2}-i\infty}^{-\frac{1}{2}+i\infty} d\alpha \frac{P_\alpha(-x_t)A(\alpha, t)}{\sin \pi\alpha} + \sum_i \beta_i(t) \frac{P_{\alpha_i}(-x_t)}{\sin \pi\alpha_i} \tag{10.157}$$

where $\beta_i(t)$ are residues at the poles $\alpha_i(t)$. The poles $\alpha_i(t)$ are called *Regge poles.*

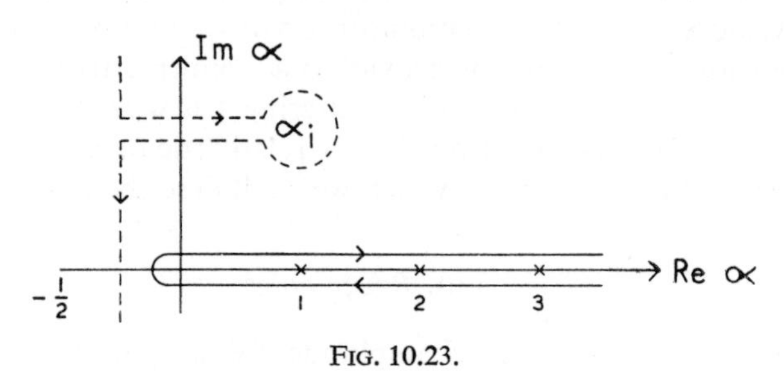

FIG. 10.23.

An inspection of equation (10.157) shows that the divergence difficulties described in § 10.5(a) have now been avoided. If we consider $T(x_t, t)$ in the unphysical limit $x_t \to +\infty$, that is infinitely large momentum transfer, then

$$P_\alpha(-x_t) \propto x_t^\alpha$$

and the integral in (10.157) can be shown to go to zero at least as fast as $x_t^{-1/2}$. Thus for high momentum transfers, $T(x_t, t)$ is dominated by the pole terms

$$T(x_t, t) \propto \sum_i \frac{\beta_i(t)\, x_t^{\alpha_i}}{\sin \pi\alpha_i} \quad (x_t \to \infty). \tag{10.158}$$

Furthermore it is clear that the summation will tend to be dominated by the pole, say α_j, which is furthest to the right in the complex α-plane. Thus the asymptotic form of the scattering amplitude for high momentum transfer is

$$T(x_t, t) \sim \frac{\gamma_j(t) s^{\alpha_j}}{\sin \pi\alpha_j} \quad (x_t \to \infty) \tag{10.159}$$

where we have used (10.154) ($x_t \propto s$ for large s) and have absorbed proportionality factors in the function $\gamma_j(t)$.

10.5(c). *Resonances and Regge trajectories*

We have stated in the previous section that the poles α_i are functions of energy in the t-channel, that is

$$\alpha_i = \alpha_i(t).$$

Let us now examine the behaviour of the pole contribution as the real part of α passes through some integral value, say l, at $t = t_R$ and assume that the imaginary part of α is small at that point. Then in the neighbourhood of t_R we can write

$$\alpha(t) \sim l + \alpha'(t - t_R) + iI_\alpha \tag{10.160}$$

where

$$\alpha' = \left(\frac{d}{dt}\,\mathrm{Re}\,\alpha\right)_{t_R}, \quad I_\alpha = \mathrm{Im}\,\alpha$$

and we have dropped the subscripts i for convenience of writing. Using equation (10.160) we may write

$$\sin \pi\alpha \sim (-1)^l \pi\alpha' \left(t - t_R + i\,\frac{I_\alpha}{\alpha'}\right)$$

and so the contribution of a single Regge pole term in (10.157) to the scattering amplitude is of the Breit–Wigner form

$$\frac{\beta(t_R)P_l(x_t)}{\pi\alpha'\left(t - t_R + i\,\dfrac{I_\alpha}{\alpha'}\right)} \tag{10.161}$$

and

$$\frac{I_\alpha}{\alpha'} \equiv \frac{\Gamma}{2}$$

where Γ is a resonance width.

The detailed behaviour of the pole term (10.161) obviously depends upon α' and I_α. Regge was able to show that α possesses the following properties (m_1 and m_3 refer to the masses of the initial particles in the t-channel – compare (10.140))

$$t < (m_1 + m_3)^2 \quad \mathrm{Re}\,\alpha(t) \geqq -\tfrac{1}{2}, \qquad I_\alpha = \mathrm{Im}\,\alpha(t) = 0 \tag{10.162}$$

$$\frac{d}{dt}\,\mathrm{Re}\,\alpha(t) = \alpha'(t) > 0$$

$$t > (m_1 + m_3)^2 \quad \mathrm{Re}\,\alpha(t) \geqq -\tfrac{1}{2}, \qquad I_\alpha > 0.$$

Weaker conditions exist for $\alpha'(t)$ for $t > (m_1 + m_3)^2$; initially $\alpha'(t)$ must be positive, but later on may go negative, since calculations using Born series suggest that $\mathrm{Re}\,\alpha(t) \leqq -\tfrac{1}{2}$ at large t values (Chew, Frautschi and Mandelstam, 1962).

The behaviour of the real and imaginary components of $\alpha(t)$ are indicated schematically in Fig. 10.24; the paths are called *Regge trajectories*. For $t < (m_1 + m_3)^2$ bound states can occur if integral values of $\mathrm{Re}\,\alpha(t)$ are reached, and for $t > (m_1 + m_3)^2$ resonances of width Γ can occur. The number of bound states is dependent upon the strength of the potential; for weak potentials $\mathrm{Re}\,\alpha(t)$ may not reach even $l = 0$ for $t < (m_1 + m_3)^2$, and there are no bound states.

It can be seen that the behaviour of the pole as a function of energy leads to the conclusion that a single pole can cause resonances in states with different l values at different energies. This condition arises naturally in an interaction through a potential, since several resonances, with different l values, may be produced by the same potential. The properties of the resonances are then related because of their common origin.

The description given above must be modified if exchange potentials are present as well as direct potentials. Nuclear potentials generally depend upon the exchange character of the wave function, or, even for distinguishable particles, upon the orbital parity. There are therefore two potentials present in nuclear

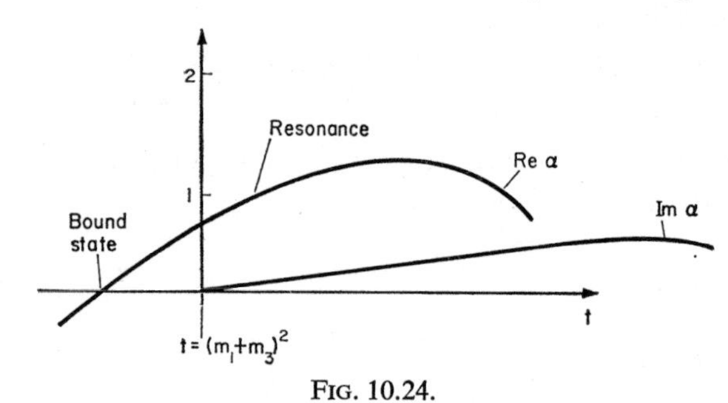

FIG. 10.24.

problems, one operative for odd parity states and the other for even parity states. Thus two trajectories must occur, and, because parity and orbital angular momentum are related to each other, it follows that one trajectory must be associated with even l and the other with odd l. Thus the pole terms appear in intervals $\Delta l = 2$ along each trajectory. Moreover, even if we deal with spinning particles, so that the operative quantum number is j rather than l, we must still distinguish the parity. Thus poles associated with j will also appear in steps of two.

The ordinary and exchange character of the forces may be introduced by multiplying the pole terms by an appropriate factor depending upon whether the systems are symmetric or antisymmetric upon the exchange of their coordinates (Frautschi, Gell-Mann and Zachariasen, 1962). The pole terms then assume the form

$$\sum_i \frac{\beta_i(t)}{\sin \pi\alpha_i} \tfrac{1}{2}[P_{\alpha_i}(-x_t) \pm P_{\alpha_i}(x_t)] \tag{10.163}$$

where the Regge terms corresponding to physical states with even (odd) l values take the $+$ $(-)$ sign. They are called states with *positive* (*negative*) *signature*. Thus we now have two trajectories corresponding to the direct and exchange potentials and the intervals between bound states (or resonances) are $\Delta l = 2$.

An equivalent expression to (10.163) which has a more useful form for later applications is

$$\sum_i \frac{\beta_i(t)}{\sin \pi\alpha_i} P_{\alpha_i}(-x_t)\left[\frac{1 \pm e^{i\pi\alpha_i(t)}}{2}\right] \tag{10.164}$$

where the choice of phase sign in the exponential is arbitrary (Frautschi, Gell-Mann and Zachariasen, 1962).

10.5(d). *The scattering amplitude at high energies*

The main application of the techniques associated with Regge poles has been made in the field of ultra high energy physics. This has been done by using crossing symmetry (§ 10.4(c)) to relate regions in the S-matrix which are functions of the same variables s and t, and which correspond to low energy and high momentum transfer in one channel (the t-channel) and high energy and low momentum transfer in a second one (the s-channel). Now the theory of Regge was developed for the nonrelativistic Schrödinger equation and in nonrelativistic theory there is no crossing symmetry to relate the t-channel with high energy behaviour, and so we are forced to conjecture that equation (10.157) is also valid in the relativistic theory. This conjecture is supported by non-perturbative calculations based upon field theory (see, for example, Amati, Fubini and Stanghellini, 1962).

Let us now consider forward scattering in the s-channel at high energies, that is according to equations (10.138) and (10.139)

$$s \to \infty, \qquad t \to 0^-.$$

In developing our treatment we shall require the optical theorem. In equation (7.109) this was given as

$$2 \operatorname{Im} T_{ii} = p_c E_c n_i^2 \sigma_T$$

where n_i^2 is a normalisation factor which we may set equal to 4 if we assume, for simplicity, that all our particles are spinless (compare § 7.5(a) and (7.75)). In the relativistic limit we may also write $E_c = 2p_c$, and so the above equation becomes

$$\operatorname{Im} T_{ii} = \tfrac{1}{2} p_c E_c 4\sigma_T = E_c^2 \sigma_T = s\sigma_T. \tag{10.165}$$

It is also convenient to work with differential cross-sections per element of t rather than per steradian. Using the relation $t = -2p_c^2(1 - \cos\theta)$ in the s-channel (compare (10.138)), it is not difficult to show that

$$\frac{d\sigma}{dt} \equiv \frac{4\pi}{s} \frac{d\sigma}{d\Omega}. \tag{10.166}$$

We return to the scattering amplitude. In § 10.5(b) we stated that the amplitude is dominated by the pole which is furthest to the right in the complex α-plane. The appropriate expression is then given by (10.159); this relation may be combined with the exchange term of (10.164) to yield the expression

$$T_{fi} = \frac{\gamma_j(t)\, s^{\alpha_j}}{\sin \pi\alpha_j} \left[\frac{1 \pm e^{i\pi\alpha_j(t)}}{2}\right]. \tag{10.167}$$

Now forward scattering corresponds to T_{ii} and $t = 0$, hence the total cross-section is given by

$$\sigma_T = \text{Im}\left\{\frac{\gamma_j(0)}{\sin \pi\alpha_j} \frac{[1 \pm e^{i\pi\alpha_j(0)}]}{2}\right\} s^{\alpha_j(0)-1} \tag{10.168}$$

where $\alpha_j(0)$ represents the pole which is furthest to the right in the α-plane at $t = 0$; we shall denote this pole by $\alpha_P(0)$. The experimental data (§ 14.6) indicates that the total cross-sections for strong interactions are nearly independent of energy at high energies. Equation (10.168) yields this result if

$$\alpha_j(0) = \alpha_P(0) = 1. \tag{10.169}$$

The subscript P is introduced because this pole can also be associated with the Pomeranchuk theorems (§§ 10.5(f) and 14.6), namely that in the high energy limit the total cross-sections are independent of isospin channel number and of whether particles or antiparticles participate in the reaction. These conditions may be satisfied if the Regge pole associated with the Pomeranchuk limits has quantum numbers $B = S = T = 0$. These quantum numbers are those for the vacuum, and so the pole is called the *vacuum* or *Pomeranchuk pole*. Since it cannot lead to exchange terms, we take the plus sign in equations (10.167) and (10.168) and obtain

$$T_{fi} \sim \frac{i}{2}\gamma_P(t)\, s^{\alpha_P(t)} \qquad (t \to 0^-) \tag{10.170}$$

$$T_{ii} \sim \frac{i}{2}\gamma_P(0)\, s$$

$$\sigma_T = \frac{1}{s}\,\text{Im}\, T_{ii} = \tfrac{1}{2}\gamma_P(0).$$

Thus the scattering amplitude contains only an imaginary component at high energies – a result which is in accord with § 6.1 (c).

The condition $\alpha_P(0) = 1$, which was introduced by an empirical argument, also has a theoretical basis since it can be shown that $\alpha_P(0) \leqq 1$ (Froissart, 1961; Martin, 1962). The argument of Martin is particularly simple, and is based upon the limitations imposed by unitarity on the imaginary part of the scattering amplitude.

Equations (10.170) may be used to make a prediction concerning the behaviour of the elastic cross-section at small momentum transfers. Using equations (7.74), (10.166) and (10.170), and assuming that we are dealing with spinless particles, we may show that

$$\left(\frac{d\sigma}{dt}\right) \equiv \frac{4\pi}{s}\frac{d\sigma}{d\Omega} = \frac{4\pi}{s}\frac{1}{64\pi^2 s}\left|\frac{i}{2}\gamma_P(t)s^{\alpha_P(t)}\right|^2$$

$$\sim \text{constant}\; s^{2(\alpha_P(t)-1)} \tag{10.171}$$

since $\gamma_P(t)$ is a slowly varying function of t. Now we can expand $\alpha_P(t)$ in the neighbourhood of $t = 0$ as

$$\alpha_P(t) = \alpha_P(0) + \alpha'_P(0)\, t \sim 1 + \alpha'_P(0)\, t$$

and we have stated earlier (§ 10.5(c)) that $\alpha' > 0$ for $t < 0$, hence for increasing momentum transfers, that is decreasing t, we expect $\alpha_P(t) < 1$. This conclusion is in accord with Froissart's theorem which requires $\alpha_P(t) \leqq 1$ for $t \leqq 0$. Thus we can rewrite (10.171) as

$$\left(\frac{d\sigma}{dt}\right) \propto s^{-2\alpha'_P(0)|t|} = e^{-2\alpha'_P|t| \log s} \tag{10.172}$$

$$\sigma_{sc} \propto \int_{-\infty}^{0} dt\, e^{-2\alpha'_P|t| \log s} \sim \frac{\text{const}}{\log s} \tag{10.173}$$

and so we would expect the elastic scattering cross-section to fall for fixed momentum transfer (t) as energy (s) increases, or in other words the width of the diffraction peak should decrease as energy increases. This prediction differs from that of the 'black body' theory of § 6.1 (c), which states that the width of the diffraction peak should be constant. The shrinkage has been observed in elastic proton–proton scattering (Fig. 10.25) but not in pion–nucleon scattering (Fig. 10.26) (Foley *et al.*, 1963). We shall return to this point in § 14.6.

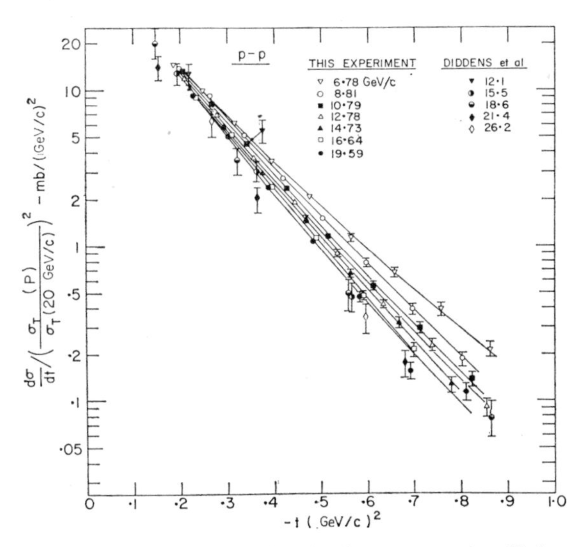

FIG. 10.25. Differential cross-sections for elastic *p–p* scattering (Foley *et al.*, 1963).

If the assumption is made that the vacuum pole does dominate the scattering at high energies then equation (10.171) can be used to trace the trajectory of the vacuum pole. This has been done by Foley *et al.*, (1963; see also Diddens, 1962a).

The results are displayed in Fig. 10.27. It can be seen that only the proton data give both $(d\alpha/dt) > 0$ and $\alpha(t) \to 1$ as $t \to 0$ as expected. The analysis of the data in this form is essentially an average over all contributing poles, but the value $\alpha(t) \to 1$ as $t \to 0$ suggests that the vacuum pole dominates.

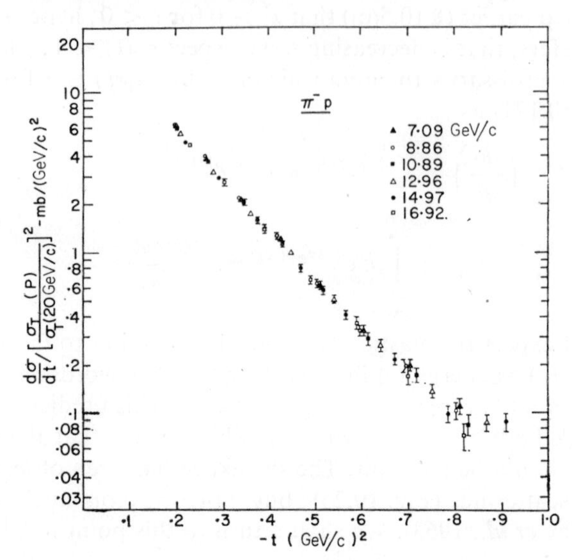

FIG. 10.26. Differential cross-sections for elastic π^-–p scattering (Foley *et al.*, 1963).

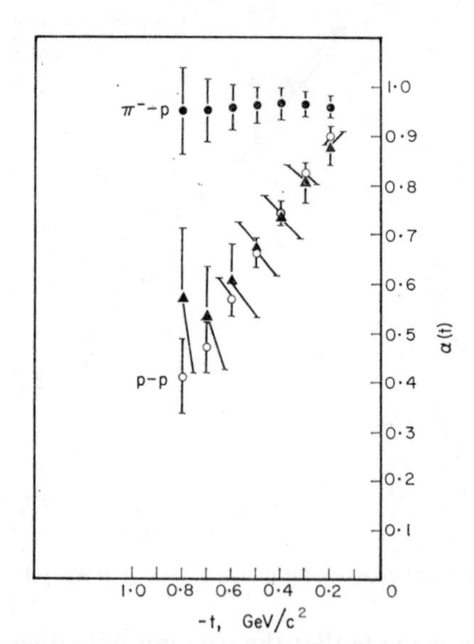

FIG. 10.27. $\alpha(t)$ vs t (Foley *et al.*, 1963) for π^-–p and p–p scattering. The open circles are for all data; the triangles represent results obtained using incident protons with momentum greater than 10 GeV/c.

10.5(e). *Rules for constructing the transition amplitude*

In the previous section we have examined a scattering amplitude which was dominated by a single trajectory – the vacuum trajectory with quantum numbers $B = S = T = 0$. Since the Regge poles occur in the complex angular momentum plane, we may associate other quantum numbers with the trajectories without modifying our previous conclusions. In turn we may associate families of particles with a given trajectory, each family having the same quantum numbers apart from their spin j which changes by $\Delta j = 2$ in going from 'particle' to 'particle' (compare (10.163)). The argument has been carried further by Chew and Frautschi (1962) who assert that there are no truly elementary particles, but instead particles represent a series of poles along Regge trajectories in the complex angular momentum plane (§ 13.5(c)).

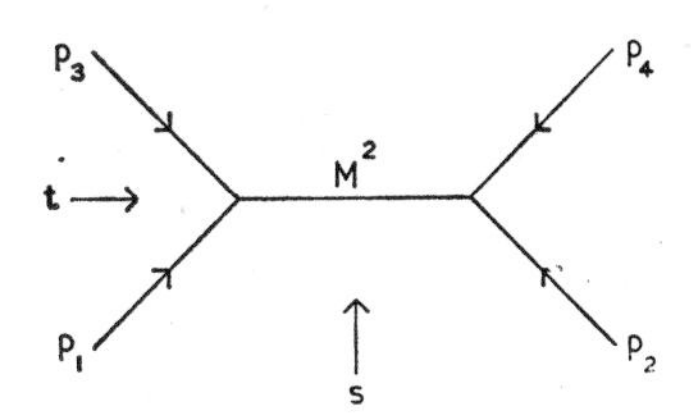

FIG. 10.28.

In our present discussion we may ignore the last problem, and concentrate upon a set of rules for constructing transition amplitudes which have been conjectured by Frautschi, Gell-Mann and Zachariasen (1962).

They may be summarised as follows.

(1) Construct a set of linearly independent invariant scattering amplitudes $T_k(s, t)$; for example in pion–nucleon scattering four amplitudes arise corresponding to the two isospin states with and without spin flip.

(2) Consider the reaction in the t-channel (Fig. 10.28) and take any set of values of the conserved quantum numbers except j, the total angular momentum. Then, as a function of j construct the contribution to the amplitudes T_k of a hypothetical exchanged particle with these quantum numbers. This step may be regarded as an intermediate one since we shall modify it in rule 3. As we have indicated above, the connection between pole and particle is tenuous, nevertheless the particle concept is a convenient one if only to relate the present argument to particle 'polology' (compare § 10.4(e)). The 'mass' of the particle is denoted by M, where $M = M(t)$. For each T_k the contribution of the pole terms will be a sum of terms containing Legendre functions of x_t (or derivatives thereof) with indices depending on j. For large s the pole terms assume the asymptotic form

$$\frac{c_k s^{(j-\nu_k)}}{t - M^2} \tag{10.174}$$

since $x_t \propto s$ (10.154), and constraints may exist on the amplitudes c_k (see below). A comparison of this relation with (10.161) shows that

$$M^2 \equiv t_R - i\frac{I_\alpha}{\alpha'}. \tag{10.175}$$

(3) Write $j = \alpha$ for integral spin in the t-channel or $j = \alpha + \frac{1}{2}$ for half integral spin and continue to complex α. The particle poles are replaced by Regge poles with the asymptotic behaviour

$$\frac{\gamma_k(t)}{2\sin\pi\alpha}[1 \pm e^{i\pi\alpha}]f(s,\alpha) \tag{10.176}$$

where the function $f(s, \alpha)$ depends on the properties of the Legendre function; commonly encountered terms are

$$\begin{array}{ccc} P_\alpha(x_t) & \underset{s\to\infty}{\longrightarrow} & s^\alpha \\ P'_\alpha(x_t) & & \alpha s^{\alpha-1} \\ x_t P'_\alpha(x_t) & & \alpha s^\alpha . \end{array}$$

The + and − signatures are chosen as explained in § 10.5(c), namely +(−) sign indicates no pole for odd (even) integral values of α.

As an example of the use of the above rules consider π–π scattering. Rule (1) indicates that we require three amplitudes $T_k(s, t)$ where $k = 0, 1, 2$ corresponding to the three isospin channels. Now let us consider pole terms apart from the vacuum pole (which is present in every reaction). The crossed reaction in the t-channel is still the π–π interaction and a prominent feature of this reaction is the ϱ resonance at 750 MeV with quantum numbers $T = 1$, $j^{\xi_P} = 1^-$, $G = +1$ (§ 13.3(a)). We shall assume that the ϱ-system is a physical manifestation of a Regge term with quantum numbers $T = 1$, $\xi_P = -1$, $G = +1$, $j = 1, 3, 5, \ldots$ The exchange of a particle with these quantum numbers causes the following contributions to (T_0, T_1, T_2)

$$(-2, -1, 1)\frac{P_j(x_t)c}{t - M^2}$$

according to rule (2), the figures $(-2, -1, 1)$ may be understood with the aid of (A.7.24, p. 712).

Rule (3) then gives us the Regge term

$$(-2, -1, 1)\frac{\gamma(t)}{2\sin\pi\alpha_\varrho(t)}[1 - e^{i\pi\alpha_\varrho}]\, s^{\alpha_\varrho} \tag{10.177}$$

for large x_t; the negative signature is chosen because of the requirements $j = 1, 3, 5, \ldots$

The example of pion–nucleon scattering is also treated by Frautschi, Gell-Mann and Zachariasen (1962). In this case the t-channel is the process $\pi + \pi \to N + \overline{N}$ and so the trajectory associated with the ϱ-meson can again appear.

10.5(f). *The factoring principle*

An interesting property of the Regge pole amplitude has been suggested by a number of workers (see, for example, Gribov and Pomeranchuk, 1962; Gell-Mann, 1962a). This is the factoring principle by which a given amplitude may be factored into a product of terms which may be related to other amplitudes. We shall develope the result by means of a plausibility argument.

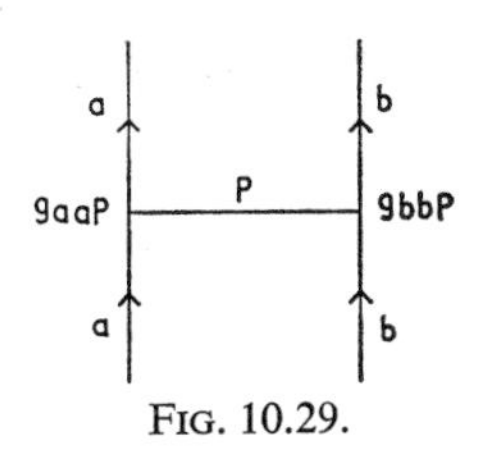

FIG. 10.29.

Consider the forward scattering of two particles a and b in the s-channel in the limit $s \to \infty$. In this limit we may expect the Pomeranchuk pole to dominate, and so we may represent the scattering by Fig. 10.29. If we regard the process as the exchange of the Pomeranchuk pole one may argue that the process is analogous to the conventional Feynman diagrams and so the term $\gamma_P(t)$ in the Regge pole amplitude may be written as

$$\gamma_P(0) \propto g_{aaP} g_{bbP} \tag{10.178}$$

where g_{aaP} represents the coupling strength of the particles aa to P. Thus we may represent the imaginary part of the forward scattering amplitude as

$$\text{Im } T_{ii} \equiv \text{Im } T_{ab} \propto g_{aaP} g_{bbP} s$$

(compare (10.170)), and so the total cross-section is given by

$$\sigma_{Tab} = \frac{1}{s} \text{Im } T_{ii} \propto g_{aaP} g_{bbP}. \tag{10.179}$$

Now consider processes involving the particles aa, ab and bb in the high energy limit, each reaction is dominated by the same vacuum pole and so each total cross-section is proportional to $\gamma_P(0)$. Therefore we may represent the total cross-sections as

$$\sigma_{aa} \propto g_{aaP} g_{aaP} \tag{10.180}$$

$$\sigma_{bb} \propto g_{bbP} g_{bbP}$$

$$\sigma_{ab} \propto g_{aaP} g_{bbP}$$

where we have dropped the subscript T for convenience of writing. The above relations therefore give us

$$(\sigma_{ab})^2 = \sigma_{aa}\sigma_{bb} \tag{10.181}$$

so that if, say, $a = \pi$, $b = N$

$$\sigma_{\pi N}^2 = \sigma_{\pi\pi}\sigma_{NN}$$

similarly

$$\sigma_{KN}^2 = \sigma_{KK}\sigma_{NN}.$$

Thus using the experimental data $\sigma_{NN} \sim 40$ mb, $\sigma_{\pi N} \sim 25$ mb and $\sigma_{KN} \sim 22$ mb, we obtain $\sigma_{\pi\pi} \sim 16$ mb and $\sigma_{KK} \sim 12$ mb.

The factoring principle also leads to the Pomeranchuk theorem, which we shall discuss in § 14.6. This theorem states that at very high energies ($s \to \infty$) the cross sections for particle–particle and particle–antiparticle scattering should be the same. If we consider equation (10.181) and substitute the antiparticle symbol $\bar{a}$ for a, we find that

$$(\sigma_{\bar{a}b})^2 = \sigma_{\bar{a}\bar{a}}\sigma_{bb}$$

but $\sigma_{aa} = \sigma_{\bar{a}\bar{a}}$ by charge conjugation invariance (§ 9.2(a)) and so we obtain the Pomeranchuk theorem

$$\sigma_{ab} = \sigma_{\bar{a}b}. \tag{10.182}$$

Field theoretical justification of the factoring principle has been provided by Amati, Fubini and Stanghellini (1962) by use of ladder diagrams.

CHAPTER 11

ELECTROMAGNETIC INTERACTIONS

11.1. COMPTON SCATTERING

11.1(a). *Kinematics*

We are now in a position to use the techniques developed in Chapters 6–10 to make calculations on measured physical processes. We start by considering the Compton scattering of photons from electrons (Compton, 1923); later the scattering of photons from protons will be examined.

Consider a photon with four-momentum k elastically scattering from an electron with momentum p, yielding final momentum states of k' and p' respectively. Conservation of energy–momentum may be expressed as

$$k + p = k' + p'. \tag{11.1}$$

We shall evaluate the cross-sections in the coordinate system of the initial electron – effectively the laboratory system, and so we can represent the four momenta as

$$k = \mathbf{k}, i\omega, \qquad p = \mathbf{p}, iE = 0, im \tag{11.2}$$

$$k' = \mathbf{k}', i\omega', \qquad p' = \mathbf{p}', iE'.$$

If we use the relation (3.24)

$$p^2 = p'^2 = -m^2, \qquad k^2 = k'^2 = 0$$

equation (11.1) can be re-written as follows:

$$p' = k + p - k'$$

$$p'^2 = -m^2 = k^2 + p^2 + k'^2 + 2kp - 2kk' - 2pk'$$

$$= -m^2 + 2kp - 2kk' - 2pk'$$

hence

$$kk' = kp - pk'. \tag{11.3}$$

The scalar products can be written as

$$kk' = \mathbf{k}\cdot\mathbf{k}' - \omega\omega' = \omega\omega'(\cos\theta - 1) \qquad (11.4)$$

$$kp = -\omega m$$

$$pk' = -m\omega'$$

where we have used equations (3.17) and (11.2), and the fact that the photon is massless

$$|\mathbf{k}| = \omega, \qquad |\mathbf{k}'| = \omega'.$$

The angle θ in (11.4) is the angle of scattering of the photon.

If equations (11.4) are inserted in (11.3) the Compton scattering formula is obtained

$$\omega\omega'(\cos\theta - 1) = -\omega m + \omega' m$$

$$\omega'[\omega(\cos\theta - 1) - m] = -\omega m$$

$$\omega' = \frac{\omega m}{m + \omega(1 - \cos\theta)} = \frac{\omega}{1 + (\omega/m)(1 - \cos\theta)}. \qquad (11.5)$$

We next consider the kinematic factors in the cross-section. If we compare the present problem with the process $a + b \to d + e$ considered in § 7.4 (c), we can write

$$k + p \to k' + p' \qquad (11.6)$$

$$a + b \to d + e.$$

The differential cross-section for a general scattering process was given in (7.71)

$$\frac{d\sigma}{d\Omega} = \frac{V^4}{(2\pi)^2}\frac{1}{v_{ab}}\overline{\sum_i}\sum_\alpha |M_{fi}|^2 \frac{E_d E_e |\mathbf{p}_d|^3}{p_d^2 E_i - \mathbf{p}_i\cdot\mathbf{p}_d E_d}$$

where v_{ab} represents the relative velocity of the particles. In the present situation b is at rest and a is a photon and so (in $c = 1$ units)

$$v_{ab} = 1. \qquad (11.7)$$

Equations (11.2) and (11.6) allow us to make the following identifications:

$$E_d \equiv \omega', \qquad E_e \equiv E', \qquad p_d \equiv \mathbf{k}', \quad |\mathbf{p}_d| \equiv \omega' \qquad (11.8)$$

$$E_i \equiv \omega + m, \qquad \mathbf{p}_i \equiv \mathbf{k}, \qquad |\mathbf{p}_i| \equiv \omega$$

hence

$$\frac{E_d E_e \,|\mathbf{p}_d|^3}{\mathbf{p}_d^2 E_i - \mathbf{p}_i \cdot \mathbf{p}_d E_d} = \frac{\omega' E' \omega'^3}{\omega'^2(\omega + m) - \omega\omega'^2 \cos\theta}$$

$$= \frac{\omega'^2 E'}{m + \omega(1 - \cos\theta)} = \frac{\omega'^3 E'}{\omega m} \tag{11.9}$$

where we have used (11.5).

Thus the scattering cross-section becomes

$$\frac{d\sigma}{d\Omega} = \frac{V^4}{(2\pi)^2} \overline{\sum_i} \sum_\alpha |M_{fi}|^2 \frac{\omega'^3 E'}{\omega m}. \tag{11.10}$$

Now the matrix element M_{fi} for Compton scattering in lowest order perturbation theory can be obtained from (7.21) and (8.93)

$$M_{fi} = \langle p'k' |\, M_2(\gamma e^-) \,| pk \rangle$$

$$= \frac{e^2}{V^2}\left(\frac{m^2}{4EE'\omega'\omega}\right)^{\frac{1}{2}} \bar{u}_{r'}(\mathbf{p}')\, O u_r(\mathbf{p})$$

$$= \frac{e^2}{V^2}\left(\frac{m^2}{4EE'\omega'\omega}\right)^{\frac{1}{2}} T_{r'r} \tag{11.11}$$

where we have used (7.76). Thus equation (11.10) reduces to

$$\frac{d\sigma}{d\Omega} = \frac{V^4}{(2\pi)^2} \frac{e^4}{4V^4} \overline{\sum_i} \sum_\alpha \frac{m^2}{EE'\omega'\omega} |T_{r'r}|^2 \frac{\omega'^3 E'}{\omega m}$$

$$= \left(\frac{e^2}{4\pi}\right)^2 \frac{\omega'^2}{\omega^2} \overline{\sum_i} \sum_\alpha |T_{r'r}|^2 \tag{11.12}$$

since $E = m$ (11.2).

11.1.(b) *The sum over polarisation states*

The summation of equation (11.12) over the different states of polarisation is straightforward but tedious. The full matrix element $T_{r'r}$ was given in (8.93)

$$T_{r'r} = \bar{u}_{r'}(\mathbf{p}')\, O u_r(\mathbf{p}) = \bar{u}_{r'}(\mathbf{p}') \left[\gamma_\mu e_\mu^{(\lambda)} \frac{i\gamma(p - k') - m}{(p - k')^2 + m^2} \gamma_\nu e_\nu^{(\lambda')} \right.$$

$$\left. + \gamma_\nu e_\nu^{(\lambda')} \frac{i\gamma(p + k) - m}{(p + k)^2 + m^2} \gamma_\mu e_\mu^{(\lambda)} \right] u_r(\mathbf{p}). \tag{11.13}$$

Certain simplifications can be introduced. In A.5.9 (Appendixes, p. 703) we have noted that the scalar product of a γ-matrix with a four-vector has the following property:

$$\gamma e \gamma p = -\gamma p \gamma e + 2ep. \tag{11.14}$$

Now in the present problem $\mathbf{p} = 0$, and we can use the Coulomb gauge condition $e_4^{(\lambda)} = e_4^{(\lambda')} = 0$, hence $ep = 0$ and so the above equation reduces to

$$\gamma e \gamma p = -\gamma p \gamma e. \tag{11.15}$$

Now consider the right-hand numerator of equation (11.13); this operates on the spinor $u_r(\mathbf{p})$, and using the above result we can write

$$[i\gamma(p+k) - m]\,\gamma e u = -\gamma e(i\gamma p + m)\,u + i\gamma k \gamma e u = i\gamma k \gamma e u$$

since the spinor form of the Dirac equation obeys the relation (3.129)

$$(i\gamma p + m)\,u = 0.$$

The result we have obtained can be written in a different form for later convenience; using (11.14) we may write

$$i\gamma k \gamma e u = -i\gamma e \gamma k u + 2keu = -i\gamma e \gamma k u \tag{11.16}$$

since $ek = 0$ because of gauge invariance (§ 4.4(a)).

We can apply similar techniques to those given above to the left-hand numerator in (11.13) with similar results. Inserting these results into the matrix element and using (11.4), we find

$$T_{r'r} = \bar{u}_{r'}(\mathbf{p}')\,O u_r(\mathbf{p})$$
$$= \bar{u}_r(\mathbf{p}')\left[\frac{i\gamma_\mu e_\mu^{(\lambda)}\gamma_\nu e_\nu^{(\lambda')}\gamma k'}{2m\omega'} + \frac{i\gamma_\nu e_\nu^{(\lambda')}\gamma_\mu e_\mu^{(\lambda)}\gamma k}{2m\omega}\right]u_r(\mathbf{p}) \tag{11.17}$$

At this point we shall introduce a simplified notation†

$$\gamma_\mu e_\mu^{(\lambda)} = \hat{e}, \qquad \gamma_\nu e_\nu^{(\lambda')} = \hat{e}' \tag{11.18}$$
$$\gamma k = \hat{k}, \qquad \gamma p = \hat{p}$$

so that we can represent O as

$$O = \frac{1}{2m}\left(i\hat{e}\hat{e}'\,\frac{\hat{k}'}{\omega'} + i\hat{e}'\hat{e}\,\frac{\hat{k}}{\omega}\right) \tag{11.19}$$

The summation of the matrix element over spinor states can now be formulated with the aid of equations (7.80) and (7.81)

$$\overline{\sum_i}\sum_\alpha |T_{r'r}|^2 = \tfrac{1}{2}\sum_i\sum_\alpha|\,\bar{u}_{r'}(\mathbf{p}')\,O u_r(\mathbf{p})|^2 = \tfrac{1}{2}\mathrm{tr}\,[O\Lambda^+(\mathbf{p})\,\tilde{O}\Lambda^+(\mathbf{p}')]$$
$$= \frac{1}{32m^4}\,\mathrm{tr}\left[\left(i\hat{e}\hat{e}'\,\frac{\hat{k}'}{\omega'} + i\hat{e}'\hat{e}\,\frac{\hat{k}}{\omega}\right)(i\hat{p} - m)\right.$$
$$\left.\left(i\,\frac{\hat{k}'}{\omega'}\,\hat{e}'\hat{e} + i\,\frac{\hat{k}}{\omega}\,\hat{e}\hat{e}'\right)(i\hat{p}' - m)\right] \tag{11.20}$$

† It should be noted that the scalar product of γ with a four-vector is frequently written in the following form: $\gamma_\lambda A_\lambda = \gamma A = A\!\!\!/$.

here we have used (7.78), for example

$$(i\hat{e}\hat{e}'\hat{k}')^{\sim} = \gamma_4(i\hat{e}\hat{e}'\hat{k}')^{\dagger}\gamma_4 = i\hat{k}'\hat{e}'\hat{e}$$

nd have reversed the signs of the $(m - i\hat{p})$ terms; this procedure is permissible nce they occur twice.

At this stage a sum over the photon polarisation vectors can be made in the anner of § 7.4(d). We shall retain them, however, as it is instructive to first xamine Compton scattering for a polarised photon beam and then sum over all irections of photon spin.

We therefore evaluate (11.20) by first making use of (11.14)

$$i\hat{e}\hat{e}'\frac{\hat{k}'}{\omega'} + i\hat{e}'\hat{e}\frac{\hat{k}}{\omega} = ie\hat{e}'\frac{\hat{k}'}{\omega'} - i\hat{e}\hat{e}'\frac{\hat{k}}{\omega} + 2iee'\frac{\hat{k}}{\omega} = i\hat{e}\hat{e}'\hat{a} + 2iee'\frac{\hat{k}}{\omega} \tag{11.21}$$

here we have introduced the term

$$a = \frac{k'}{\omega'} - \frac{k}{\omega} \tag{11.22}$$

hich has the following properties (compare 11.1))

$$pa = 0 \tag{11.23}$$

$$a^2 = -2\frac{kk'}{\omega'\omega} = \frac{(k'-k)^2}{\omega'\omega} = \frac{(p-p')^2}{\omega'\omega} = \frac{-2m^2+2mE'}{\omega'\omega}$$

$$= \frac{-2m^2+2m(m+\omega-\omega')}{\omega'\omega} = \frac{2m}{\omega'\omega}(\omega'-\omega).$$

With the aid of the technique used in (11.21), equation (11.20) becomes

$$\overline{\sum_i}\sum_\alpha |T_{r'r}|^2$$

$$= \frac{1}{32m^4}\,\mathrm{tr}\left[\left(i\hat{e}\hat{e}'\hat{a} + 2iee'\frac{\hat{k}}{\omega}\right)(i\hat{p}-m)\left(i\hat{a}\hat{e}'\hat{e} + 2iee'\frac{\hat{k}}{\omega}\right)(i\hat{p}'-m)\right]$$

$$= \frac{1}{32m^4}\,\mathrm{tr}\,(A+B+C+D) \tag{11.24}$$

here the terms A, B, C and D will be identified in the following equations. /e now evaluate the traces A, B, C and D with the aid of A.5 (Appendixes, p. 701).

$$\mathrm{tr}A = \mathrm{tr}\,[i\hat{e}\hat{e}'\hat{a}(i\hat{p}-m)\,i\hat{a}\hat{e}'\hat{e}(i\hat{p}'-m)]$$

$$= -\mathrm{tr}\,[(i\hat{p}+m)\,i\hat{e}\hat{e}'\hat{a}i\hat{a}\hat{e}'\hat{e}]\,[i(\hat{p}+\hat{k}-\hat{k}')-m]$$

$$= a^2 \operatorname{tr} [(i\hat{p} + m)] [i(\hat{p} + \hat{k} - \hat{k}') - m]$$

$$= -a^2 \operatorname{tr} [\hat{p}(\hat{k} - \hat{k}')]$$

$$= -4a^2 p(k - k')$$

$$= \frac{8m^2}{\omega\omega'} (\omega - \omega')^2 \qquad (11.25$$

where we have used equations (11.14), (11.23), (A.5.6) (p. 702) and (11.4), and a adaptation of equation (A.5.9) (p. 703)

$$\hat{A}\hat{A} = A^2 . \qquad (11.26$$

In developing equations for traces B, C, and D the same techniques can b used as for trace A, together with the facts that

$$\hat{k}\hat{k} = \hat{k}'\hat{k}' = 0 , \qquad \hat{p}\hat{p} = -m^2$$

and that the trace of an odd number of γ-matrices is zero (§ 3.3 (d)). Trace B ca be written as

$$\operatorname{tr} B = 4(ee')^2 \operatorname{tr} \left[i \frac{\hat{k}}{\omega} (i\hat{p} - m) \, i \frac{\hat{k}}{\omega} (i\hat{p}' - m) \right]$$

$$= -4 \frac{(ee')^2}{\omega^2} \operatorname{tr} [\hat{k}(i\hat{p} - m) \, \hat{k}(i\hat{p}' - m)]$$

$$= -4i \frac{(ee')^2}{\omega^2} \operatorname{tr} [\hat{k}\hat{p}\hat{k}] [i(\hat{p} + \hat{k} - \hat{k}')]$$

$$= 4 \frac{(ee')^2}{\omega^2} \operatorname{tr} [\hat{k}\hat{p}\hat{k}(\hat{p} - \hat{k}')]$$

$$= 4 \frac{(ee')^2}{\omega^2} \operatorname{tr} [\hat{k}(- \hat{k}\hat{p} + 2kp) (\hat{p} - \hat{k}')]$$

$$= 8 \frac{(ee')^2}{\omega^2} kp \operatorname{tr} [\hat{k}(\hat{p} - \hat{k}')]$$

$$= 32 \frac{(ee')^2}{\omega^2} (kp)^2 \left(1 - \frac{kk'}{kp}\right) = 32(ee')^2 m^2 \left(1 - \frac{kk'}{kp}\right) \qquad (11.27$$

It may be shown without difficulty that the expressions which we designat as trace C and trace D in equation (11.24) are the same. Thus we may write

$$\operatorname{tr} C + \operatorname{tr} D$$

$$= 4(ee') \operatorname{tr} \left[i\hat{e}\hat{e}'\hat{a}(i\hat{p} - m) \, i \frac{\hat{k}}{\omega} (i\hat{p}' - m) \right]$$

$$= -4 \frac{ee'}{\omega} \operatorname{tr}[(i\hat{p} + m)\hat{e}\hat{e}'\hat{a}\hat{k}i\hat{k}']$$

$$= 4 \frac{ee'}{\omega} \operatorname{tr}[\hat{p}\hat{e}\hat{e}'\hat{a}\hat{k}k']$$

$$= 4 \frac{ee'}{\omega} \operatorname{tr}\left[\hat{p}\hat{e}\hat{e}'\left(\frac{\hat{k}'}{\omega'} - \frac{\hat{k}}{\omega}\right)\hat{k}\hat{k}'\right]$$

$$= 4 \frac{ee'}{\omega\omega'} \operatorname{tr}[\hat{p}\hat{e}\hat{e}'\hat{k}'(-\hat{k}'\hat{k} + 2kk')]$$

$$= 8 \frac{ee'}{\omega\omega'} kk' \operatorname{tr}[\hat{p}\hat{e}\hat{e}'\hat{\boldsymbol{k}}']$$

$$= 8 \frac{ee'}{\omega\omega'} kk' [4(pe)(e'k') - 4(pe')(ek') + 4(pk')(ee')]$$

$$= 32 \frac{(ee')^2}{\omega\omega'} kk'pk' = 32 \frac{(ee')^2 m^2}{(pk)(pk')} (kk')(pk')$$

$$= 32(ee')^2 m^2 \frac{kk'}{pk}. \tag{11.28}$$

Thus equation (11.24) becomes

$$\overline{\sum_i} \sum_\alpha |T_{r'r}|^2 = \frac{1}{32m^4} \operatorname{tr}(A + B + C + D)$$

$$= \frac{1}{32m^4}\left[8m^2 \frac{(\omega - \omega')^2}{\omega\omega'} + 32(ee')^2 m^2\left(1 - \frac{kk'}{kp}\right) + 32(ee')^2 m^2 \frac{kk'}{kp}\right]$$

$$= \frac{1}{4m^2}\left[\frac{(\omega - \omega')^2}{\omega\omega'} + 4(ee')^2\right]. \tag{11.29}$$

This equation may be inserted in (11.12) to yield the following expression for the differential cross-section for Compton scattering

$$\frac{d\sigma}{d\Omega} = \left(\frac{e^2}{4\pi}\right)^2 \frac{\omega'^2}{\omega^2} \frac{1}{4m^2}\left[\frac{(\omega - \omega')^2}{\omega\omega'} + 4(ee')^2\right]$$

$$= \frac{1}{4m^2}\left(\frac{e^2}{4\pi}\right)^2 \frac{\omega'^2}{\omega^2}\left[\frac{\omega}{\omega'} + \frac{\omega'}{\omega} - 2 + 4(ee')^2\right]. \tag{11.30}$$

The term ω' may be eliminated from this equation with the aid of (11.
yielding

$$\frac{d\sigma}{d\Omega}=\frac{1}{4m^2}\left(\frac{e^2}{4\pi}\right)^2\frac{1}{[1+(\omega/m)(1-\cos\theta)]^2}\left[\frac{(\omega^2/m^2)(1-\cos\theta)^2}{1+(\omega/m)(1-\cos\theta)}+4(ee')^2\right] \quad (11.31)$$

This is the Klein–Nishina formula for the Compton effect (Klein and Nishina 1929; Tamm, 1930). It is instructive to write this equation in c.g.s units rather than $\hbar = c = 1$ units. Since e^2 has the dimensions of energy × length, one may easily show that a factor $1/c^4$ must be inserted into the above formula since cross-sections have the dimensions of area. Thus equation (11.31) may be written as

$$\frac{d\sigma}{d\Omega}=\frac{1}{4}r_0^2\frac{1}{[1+(\omega/m)(1-\cos\theta)]^2}\left[\frac{(\omega^2/m^2)(1-\cos\theta)^2}{1+(\omega/m)(1-\cos\theta)}+4(ee')^2\right] \quad (11.32)$$

where

$$r_0=\frac{e^2}{4\pi}\frac{1}{m}\equiv\frac{e^2}{4\pi\hbar c}\frac{\hbar}{mc}=\alpha\lambda_c=2{\cdot}818\times10^{-13}\text{ cm} \quad (11.33)$$

is called the *Thomson radius*. The term α is the fine structure constant and λ_c the Compton wavelength. In the rest of this chapter we shall use $\hbar = c = 1$ units in this system

$$\alpha=\frac{e^2}{4\pi},\qquad r_0=\frac{\alpha}{m}.$$

It should be noted that (11.32) represents the differential cross-section for scattering for a certain relative orientation of the polarisation vectors e and e'. If we wish to consider unpolarised radiation the states e and e' must be summed and averaged. This process will be carried out in the next section.

11.1(c) *The properties of the scattered photon beam*

In the nonrelativistic limit $\omega/m \to 0$, equation (11.32), reduces to

$$\frac{d\sigma}{d\Omega}=\frac{r_0^2}{4}4(ee')^2=r_0^2(ee')^2 \quad (11.34)$$

Now the term $(ee')^2$ can be written as

$$(ee')^2=(\mathbf{e}\cdot\mathbf{e}')^2=\cos^2\chi \quad (11.35)$$

since we chose the gauge condition $e_4 = 0$; χ represents the angle between the polarisation vectors (Fig. 11.1). These vectors point along the direction of the electric component of the electromagnetic field (see discussion of § 4.4(a)). The nonrelativistic formula for the scattering is therefore given by

$$\frac{d\sigma}{d\Omega} = r_0^2 \cos^2 \chi \tag{11.36}$$

which is the classical Thomson formula for the scattering of polarised low energy radiation by a static charge.

It is worth pausing to consider the implications of (11.36). We shall use the elementary principle, appearing in textbooks on optics, that in examining the state of polarised light it is only necessary to consider parallel and perpendicular states of mutual polarisation, since all other states may be resolved into them. We shall therefore take an arbitrary direction for the polarisation **e** of the incident photons and then examine the following two states for a scattering angle θ.

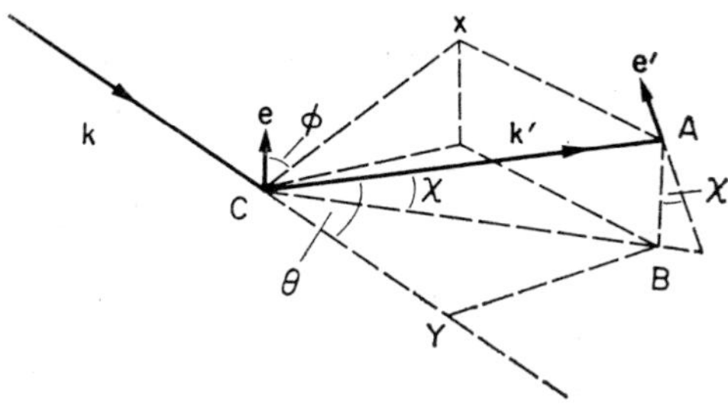

FIG. 11.1.

(1) **e** and **e**′ in the same plane (the **e**, **k**′ plane)—Fig. 11.1. In this diagram **e** and AB are parallel and so the angle between AB and **e**′ equals χ. The angle between the **e**, **k** and **k**,**k**′ planes is given by φ hence

$$\sin \chi = \frac{AB}{AC}$$

$$AB = CX \cos \varphi, \qquad AC \sin \theta = AY$$

$$\sin \chi = \sin \theta \cos \varphi \quad \text{since} \quad CX = AY$$

therefore

$$\cos^2 \chi = 1 - \sin^2 \theta \cos^2 \varphi.$$

(2) **e**′ perpendicular to **e** then $\cos \chi = 0$.

We therefore have two different cross-sections for the two orientations of the polarisation. If polarisation is not detected in the final system these cross-sections must be added:

$$\frac{d\sigma_1}{d\Omega} = r_0^2(1 - \sin^2 \theta \cos^2 \varphi), \qquad \frac{d\sigma_2}{d\Omega} = 0 \tag{11.37}$$

$$\frac{d\sigma}{d\Omega} = \sum_{\mathbf{e}'} r_0^2 \cos^2 \chi = \frac{d\sigma_1}{d\Omega} + \frac{d\sigma_2}{d\Omega} = r_0^2(1 - \sin^2 \theta \cos^2 \varphi).$$

It is apparent therefore that $d\sigma_1/d\Omega$ reaches a maximum when **e** is perpendicular to the **k**, **k**′ plane or $\mathbf{e}\cdot(\hat{\mathbf{k}}\times\hat{\mathbf{k}}') = 1$, where $\hat{\mathbf{k}}$ and $\hat{\mathbf{k}}'$ represent unit vectors in the direction of **k** and **k**′ respectively. Now let us consider the situation when **e** lies first perpendicularly and then parallel to the **k**, **k**′ plane (Fig. 11.2).

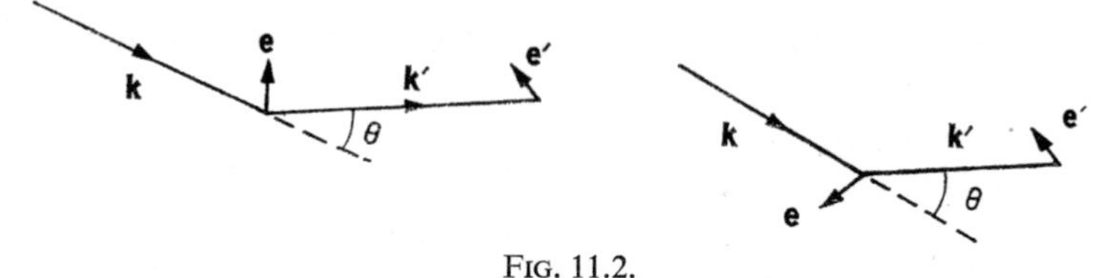

FIG. 11.2.

If we denote the partial cross-sections for **e**′ to point perpendicular or parallel to the **k**, **k**′ plane by $d\sigma_\perp$ and $d\sigma_{\parallel}$ respectively we obtain the following results:

$$\left.\begin{aligned} \mathbf{e}\cdot(\hat{\mathbf{k}}\times\hat{\mathbf{k}}') &= 1 \\ \cos^2\varphi &= 0 \end{aligned}\right\} \quad \left\{\begin{aligned} \frac{d\sigma_\perp}{d\Omega} &= r_0^2 \\ \frac{d\sigma_\parallel}{d\Omega} &= 0 \end{aligned}\right. \tag{11.38}$$

$$\left.\begin{aligned} \mathbf{e}\cdot(\hat{\mathbf{k}}\times\hat{\mathbf{k}}') &= 0 \\ \cos^2\varphi &= 1 \end{aligned}\right\} \quad \left\{\begin{aligned} \frac{d\sigma_\perp}{d\Omega} &= 0 \\ \frac{d\sigma_\parallel}{d\Omega} &= r_0^2\cos^2\theta \end{aligned}\right.$$

It is apparent from this crude discussion that the scattered photon beam can be polarised even if the incident beam is unpolarised, and that the direction of polarisation points perpendicularly to the scattering plane. Furthermore, the polarisation vanishes at the scattering angles $\theta = 0°$ and $180°$ has a maximum at $\theta = 90°$. A more extensive discussion of polarisation effects in Compton scattering may be found in the article by McMaster (1961).

In most experiments on Compton scattering angular distributions only are considered, and no attempt is made to detect polarisation. We must therefore average over initial conditions and sum over the final ones; hence from (11.37)

$$\frac{d\sigma}{d\Omega} = \overline{\sum_e}\frac{d\sigma_1}{d\Omega} + \overline{\sum_e}\frac{d\sigma_2}{d\Omega} = \overline{\sum_e} r_0^2(1-\sin^2\theta\cos^2\varphi).$$

Now the average value of (cosine)^2 is $\frac{1}{2}$, and so the nonrelativistic scattering cross-section for unpolarised conditions becomes†

$$\frac{d\sigma}{d\Omega} = r_0^2(1-\tfrac{1}{2}\sin^2\theta) = \frac{r_0^2}{2}(1+\cos^2\theta). \tag{11.39}$$

† Alternatively equation (11.39) can be obtained by summing the partial cross-sections in (11.38) and then dividing by two, since we considered two independent initial states.

Upon integrating over the solid angle we obtain the classical Thomson scattering cross-section

$$\sigma_T = \frac{8\pi}{3} r_0^2 = \frac{8\pi}{3} \left(\frac{\alpha}{m}\right)^2 = 0{\cdot}665 \times 10^{-24}\ \text{cm}^2. \tag{11.40}$$

We return, finally, to the relativistic cross-section (11.32). It is apparent from our discussion between (11.35) and (11.39) that if we wish to consider the scattering of unpolarised photons we must make the summation of $d\sigma_1/d\Omega$ and $d\sigma_2/d\Omega$ and average over φ. Thus the Klein–Nishina formula (11.32) becomes

$$\begin{aligned}\frac{d\sigma}{d\Omega} &= \overline{\sum_e}\left[\frac{d\sigma_1}{d\Omega} + \frac{d\sigma_2}{d\Omega}\right] \\ &= \frac{r_0^2}{2} \frac{1}{[1 + (\omega/m)(1 - \cos\theta)]^2}\left[\frac{(\omega^2/m^2)(1 - \cos^2\theta)^2}{1 + (\omega/m)(1 - \cos\theta)} + 1 + \cos^2\theta\right].\end{aligned} \tag{11.41}$$

11.1(d). *Comparison with experiment*

The formulae relating to the polarisation of photons by Compton scattering have never been subjected to a comprehensive experimental test. The correctness of the formula has been assumed and exploited in the construction of γ-ray polarimeters for work in low energy nuclear physics—the assignment of spins and parities to nuclear levels. The data accumulated in this manner agrees with that obtained by other techniques.

Indirect tests have also been made by double Compton scattering experiments (Hoover, Faust and Dohne, 1952). The photons were polarised in the first scattering (see remarks following (11.38)), and the azimuthal distribution of the photons was examined in the second scattering. The agreement between theory and experiment checked within 5 per cent.

Considerable work has been done on the angular distributions observed when unpolarised photons are scattered by electrons. Figure 11.3 shows theoretical curves for the ratio of the differential cross-sections at angles θ and zero, as a function of θ, for three different photon energies. The experimental points at $\varepsilon = \omega/m = 0{\cdot}173$ are due to Friedrich and Goldhaber (1927); those at $\varepsilon = 2{\cdot}5$ are due to Hofstadter and McIntyre (1949).

Numerous measurements have been on the total cross-sections for the absorption of photons. These normally take the form of transmission measurements

$$N_x = N_0 e^{-x/\lambda}$$

where N_x represents the number of photons remaining after traversing a thickness x of an absorbing material, and λ the mean free path for the photons. The term λ is a function of the cross-sections for the photoelectric effect, Compton scattering and pair production. If the absorbers used are of low atomic weight, and if the energies of the photons are of the order of a few MeV, then Compton scattering predominates. Measurements by Colgate (1952) using photons with energies between 0·4 and 17·6 MeV agreed with the Klein–Nishina formula to within 0·2 to 1 per cent.

At high photon energies the Compton effect tends to be masked by the pair production process. Measurements by Emigh (1952) using a cloud chamber

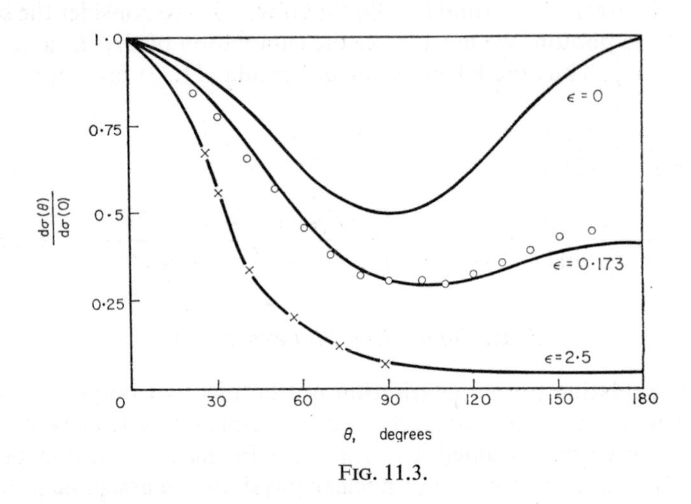

FIG. 11.3.

indicate that the Klein–Nishina formula is satisfactory up to energies ~ 200 MeV.

An excellent account of the experimental work, relating to the Compton scattering of photons by electrons, may be found in an article by Evans (1958).

11.1(e). *Compton scattering by protons*

The Compton scattering of photons by protons is a more difficult problem than scattering by electrons, both experimentally and theoretically.

Firstly, the cross-section for the process is small; in the nonrelativistic limit it is of the order of the Thomson cross-section (11.40) with the proton mass M substituted for m

$$\sigma_T = \frac{8\pi}{3} r_0^2 \left(\frac{m}{M}\right)^2 = \frac{8\pi}{3} \frac{\alpha^2}{M^2} \tag{11.42}$$

$$\sim 20 \times 10^{-32} \text{ cm}^2$$

and is therefore difficult to detect. Secondly, the background due to the process

$$\gamma + p \to p + \pi^0 \to p + \gamma + \gamma$$

tends to swamp the Compton effect at photon energies greater than 140 MeV.

If we consider the theoretical problem, the process we have discussed for electrons makes only one of many contributions to the elastic scattering of γ-rays by protons. The proton does not behave like a point Dirac particle – this is obvious from the value of its magnetic moment. This anomalous behaviour is normally attributed to the meson cloud surrounding the nucleon core, and so mesonic effects are expected to contribute to the Compton scattering.

For high photon energies the following processes might be expected to contribute to the amplitude for the elastic scattering of photons by protons:

(1) photon scattering by a point Dirac particle

(2) excitation of the nucleon isobaric states

$$\gamma + p \rightarrow N^* \rightarrow p + \gamma, \tag{11.43}$$

(3) photon scattering through a virtual pion state (Fig. 11.4).

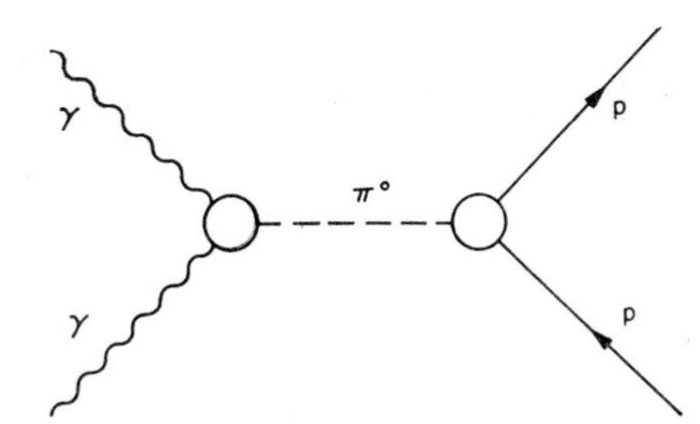

FIG. 11.4.

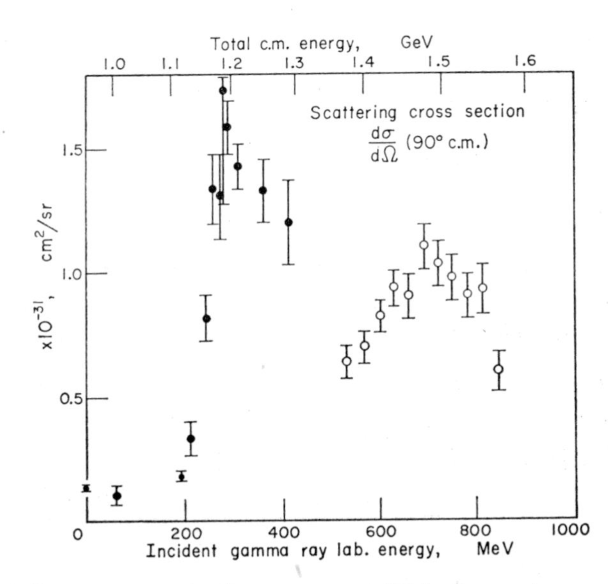

FIG. 11.5. Compton scattering from protons at 90° in the c-system as a function of photon energy.

The last mechanism is interesting since, in principle, it offers a method for measuring the lifetime of the meson (Jacob and Mathews, 1960). Unfortunately, the scattering amplitude does not appear to be sensitive to this term.

In Fig. 11.5 we display collected experimental data on Compton scattering from protons at 90° in the *c*-system as a function of energy (Stiening *et al.*, 1963). The peaks in the figure are due to the N_{33}^* and N_{13}^* states (§ 13.2). It can be seen that as the energy of the photon tends to zero, the cross approaches the expected value for scattering by a point Dirac particle (compare (11.39)).

$$\frac{d\sigma}{d\Omega} = \frac{1}{2} r_0^2 \left(\frac{m}{M}\right)^2 = \frac{1}{2} \frac{\alpha^2}{m^2} \left(\frac{m}{M}\right)^2 = \frac{1}{2} \frac{\alpha^2}{M^2} \sim 10^{-32}\ \text{cm}^2. \quad (11.44)$$

11.2. THE ELECTRON–ELECTRON INTERACTION

The above title covers a number of interactions:

(1) elastic scattering of electrons by electrons, Møller scattering (1932),
(2) elastic scattering of electrons by positrons, Bhabha scattering (1936),
(3) annihilation of positrons by electrons yielding photons.

Since electromagnetic processes are invariant under charge conjugation, the elastic scattering of positrons by positrons will be the same as Møller scattering. Process (3) is of especial interest since the electron and positron can form a bound state before their mutual annihilation. This is known as *positronium*.

Certain aspects of the three processes will be discussed in this section in their lowest order approximations. It should be borne in mind that in practice the process can be more complicated than we have indicated, for example in addition to elastic scattering, the scattering of electrons by electrons can involve the emission of photons (this is called a *radiative effect*).

11.2(a). *Møller scattering*

We shall represent the scattering by the four momenta

$$p_1 + p_2 = p_1' + p_2', \quad a + b \to d + e. \quad (11.45)$$

The relevant diagrams and *S*-matrix element for this process were given in (§ 8.6(c)). The matrix element M_{fi} is then obtained from equation (7.21)

$$\begin{aligned} M_{fi} &= \langle p_1' p_2' |\, M_2(e^-e^-)\, | p_1 p_2 \rangle \\ &= -\frac{e^2}{V^2}\left(\frac{m^4}{E_1'E_2'E_1E_2}\right)^{\frac{1}{2}} \left[\frac{\bar{u}_{r'}(\mathbf{p}_1')\gamma_\nu u_r(\mathbf{p}_1)\, \bar{u}_{s'}(\mathbf{p}_2')\, \gamma_\nu u_s(\mathbf{p}_2)}{(p_1 - p_1')^2} \right. \\ &\qquad \left. - \frac{\bar{u}_{s'}(\mathbf{p}_2')\gamma_\nu u_r(\mathbf{p}_1)\, \bar{u}_{r'}(\mathbf{p}_1')\, \gamma_\nu u_s(\mathbf{p}_2)}{(p_1 - p_2')^2}\right] \\ &= -\frac{e^2}{V^2}\left(\frac{m^4}{E_1'E_2'E_1E_2}\right)^{\frac{1}{2}} T_{s'r'sr}. \end{aligned} \quad (11.46)$$

In this problem it is convenient to work in the c-system, and so we may write

$$\mathbf{p}_1 = -\mathbf{p}_2, \qquad \mathbf{p}_1' = -\mathbf{p}_2' \tag{11.47}$$

$$|\mathbf{p}_1| = |\mathbf{p}_2| = |\mathbf{p}_1'| = |\mathbf{p}_2'| = p_c$$

$$\mathbf{p}_1 \cdot \mathbf{p}_1' = p_c^2 \cos\theta$$

$$E_1 = E_2 = E_1' = E_2' = E, \qquad 2E = E_c$$

$$p_1 p_1' = p_c^2 \cos\theta - E^2, \qquad p_1 p_2 = -p_c^2 - E^2$$

$$p_1 p_2' = -p_c^2 \cos\theta - E^2$$

where θ represents the angle of scattering.

The differential scattering cross-section in the c-system for any process was given in (7.72).

$$\begin{aligned}\frac{d\sigma}{d\Omega} &= \frac{V^4}{(2\pi)^2}\frac{E_1E_2E_1'E_2'}{E_c^2}\frac{p_c^f}{p_c^i}\overline{\sum_i}\sum_\alpha |M_{fi}|^2 \\ &= \frac{V^4}{(2\pi)^2}\frac{e^4}{V^4}\frac{E_1E_2E_1'E_2'}{E_c^2}\frac{m^4}{E_1E_2E_1'E_2'}\overline{\sum_i}\sum_\alpha |T_{s'r'sr}|^2 \\ &= \left(\frac{e^2}{4\pi}\right)^2\frac{m^4}{E^2}\overline{\sum_i}\sum_\alpha |T_{s'r'sr}|^2 \end{aligned} \tag{11.48}$$

where the term $T_{s'r'sr}$ represents

$$\frac{\bar{u}_{r'}(\mathbf{p}_1')\gamma_\nu u_r(\mathbf{p}_1)\,\bar{u}_{s'}(\mathbf{p}_2')\,\gamma_\nu u_s(\mathbf{p}_2)}{(p_1 - p_1')^2} - \frac{\bar{u}_{s'}(\mathbf{p}_2')\,\gamma_\nu u_r(\mathbf{p}_1)\,\bar{u}_{r'}(\mathbf{p}_2')\,\gamma_\nu u_s(\mathbf{p}_2)}{(p_1 - p_2')^2}. \tag{11.49}$$

The averaging of T over initial spinor states and the summation over final states can be carried out with the aid of (7.84). The result becomes

$$\overline{\sum_i}\sum_\alpha |T_{s'r'sr}|^2 = \frac{1}{4}\left[\frac{A}{(p_1 - p_1')^4} + \frac{B}{(p_1 - p_2')^4} - \frac{2C}{(p_1 - p_1')^2\,(p_1 - p_2')^2}\right] \tag{11.50}$$

$$\begin{aligned} A &= \mathrm{tr}\,[\gamma_\mu \Lambda^+(\mathbf{p}_1)\,\tilde{\gamma}_\nu \Lambda^+(\mathbf{p}_1')]\,\mathrm{tr}\,[\gamma_\mu \Lambda^+(\mathbf{p}_2)\,\tilde{\gamma}_\nu \Lambda^+(\mathbf{p}_2')] \\ &= \frac{1}{16m^4}\,\mathrm{tr}\,[\gamma_\mu(m - i\hat{p}_1)\,\gamma_\nu(m - i\hat{p}_1')]\,\mathrm{tr}\,[\gamma_\mu(m - i\hat{p}_2)\,\gamma_\nu(m - i\hat{p}_2')] \end{aligned} \tag{11.51}$$

$$B = \frac{1}{16m^4} \operatorname{tr} [\gamma_\mu(m - i\hat{p}_1)\, \gamma_\nu(m - i\hat{p}_2')] \operatorname{tr} [\gamma_\mu(m - ip_2)\, \gamma_\nu(m - i\hat{p}_1')]$$

$$C = \frac{1}{16m^4} \operatorname{tr} [\gamma_\mu(m - i\hat{p}_1)\, \gamma_\nu(m - i\hat{p}_2')\, \gamma_\mu(m - i\hat{p}_2)\, \gamma_\nu(m - i\hat{p}_1')].$$

In developing these equations we have used the fact that although

$$\tilde{\gamma}_k = \gamma_4 \gamma_k^\dagger \gamma_4 = -\gamma_k \quad (k = 1, 2, 3) \tag{11.52}$$

$$\tilde{\gamma}_4 = \gamma_4$$

each γ-term appears an even number of times and so the product is always positive.

As an illustration of the technique used for evaluating the matrix element we will consider the term containing A in (11.50)

$$\frac{1}{64m^4(p_1 - p_1')^4} \operatorname{tr} [\gamma_\mu(m - i\hat{p}_1)\, \gamma_\nu\, (m - i\hat{p}_1')] \times$$
$$\times \operatorname{tr} [\gamma_\mu(m - i\hat{p}_2)\, \gamma_\nu(m - i\hat{p}_2')]. \tag{11.53}$$

The denominator may be evaluated with the aid of (11.47)

$$(p_1 - p_1')^4 = [(\mathbf{p}_1 - \mathbf{p}_1')^2 - (E_1 - E_1')^2]^2$$
$$= (\mathbf{p}_1 - \mathbf{p}_1')^4 = 16p_c^4 \sin^4 \tfrac{1}{2}\theta. \tag{11.54}$$

Similarly the factor $(p_1 - p_2')^2$ appearing in the other denominators in (11.50) can be written as

$$(p_1 - p_2')^2 = (\mathbf{p}_1 - \mathbf{p}_2')^2 = (\mathbf{p}_1 + \mathbf{p}_1')^2 = 2p_c^2(1 + \cos\theta) = 4p_c^2 \cos^2 \tfrac{1}{2}\,\theta. \tag{11.55}$$

Hence equation (11.50) becomes

$$\overline{\sum_i} \sum_\alpha |T_{s'r'sr}|^2$$
$$= \frac{1}{64m^4} \cdot \frac{1}{16p_c^4} \left[\frac{A'}{\sin^4 \frac{1}{2}\theta} + \frac{B'}{\cos^4 \frac{1}{2}\theta} - \frac{2C'}{\sin^2 \frac{1}{2}\theta \cos^2 \frac{1}{2}\theta} \right] \tag{11.56}$$

where A', B' and C' are the trace terms defined in equations (11.51). Upon inserting this expression in equation (11.48) the differential scattering cross-section becomes

$$\frac{d\sigma}{d\Omega} = \left(\frac{e^2}{4\pi}\right)^2 \frac{m^4}{E^2 64m^4 16p_c^2} \left[\frac{A'}{\sin^4 \frac{1}{2}\theta} + \frac{B'}{\cos^4 \frac{1}{2}\theta} - \frac{2C'}{\sin^2 \frac{1}{2}\theta \cos^2 \frac{1}{2}\theta} \right]. \tag{11.57}$$

The trace terms A', B' and C' can be evaluated by the methods of A.5 (Appendixes, p. 701). Consider first the term

$$\operatorname{tr}\,[\gamma_\mu(m - i\hat{p})\,\gamma_\nu(m - i\hat{p}')] \tag{11.58}$$

where for simplicity we have dropped the subscripts on the p's.

The traces of terms containing odd numbers of γ-matrices vanish, and so we need to examine terms containing two and four γ-matrices. The following properties for the traces are listed in A.5 (Appendixes, p. 701):

$$\operatorname{tr}\,\gamma_\mu\gamma_\nu = 4\delta_{\mu\nu}$$

$$\operatorname{tr}\,\gamma_\mu\gamma_\lambda\gamma_\nu\gamma_\varrho = 4\delta_{\mu\lambda}\delta_{\nu\varrho} - 4\delta_{\mu\nu}\delta_{\lambda\varrho} + 4\delta_{\mu\varrho}\delta_{\lambda\nu}.$$

Therefore we can write

$$\begin{aligned}
&\operatorname{tr}\,[\gamma_\mu(m - i\hat{p})\,\gamma_\nu(m - i\hat{p}')] \\
&\quad = \operatorname{tr}\,\gamma_\mu\gamma_\nu m^2 - \operatorname{tr}\,\gamma_\mu\gamma_\lambda p_\lambda\gamma_\nu\gamma_\varrho p'_\varrho \\
&\quad = 4\delta_{\mu\nu}m^2 - 4\delta_{\mu\lambda}p_\lambda\delta_{\nu\varrho}p'_\varrho + 4\delta_{\mu\nu}p_\lambda\delta_{\lambda\varrho}p'_\varrho - 4\delta_{\mu\varrho}p_\lambda\delta_{\lambda\nu}p'_\varrho \\
&\quad = 4\delta_{\mu\nu}(m^2 + pp') - 4(p_\mu p'_\nu + p_\nu p'_\mu)
\end{aligned} \tag{11.59}$$

and so A' becomes

$$\begin{aligned}
A' &= \operatorname{tr}\,[\gamma_\mu(m - i\hat{p}_1)\,\gamma_\nu(m - i\hat{p}'_1)]\operatorname{tr}\,[\gamma_\mu(m - i\hat{p}_2)\,\gamma_\nu(m - i\hat{p}'_2)] \\
&= 16\,\{[\delta_{\mu\nu}(m^2 + p_1p'_1) - (p_{1\mu}p'_{1\nu} + p_{1\nu}p'_{1\mu})] \times \\
&\qquad \times [\delta_{\mu\nu}(m^2 + p_2p'_2) - (p_{2\mu}p'_{2\nu} + p_{2\nu}p'_{2\mu})]\} \\
&= 32[2m^2(m^2 + p_1p'_1) + (p_1p_2)^2 + (p_1p'_2)^2]
\end{aligned} \tag{11.60}$$

where we have used the Einstein summation convention (§ 3.2(b)), for example $\delta_{\mu\nu}\delta_{\mu\nu} = 4$, and the fact that the conservation of four-momentum

$$p_1 + p_2 = p'_1 + p'_2$$

can be made to yield the relations

$$p_1p_2 = p'_1p'_2, \qquad p_1p'_1 = p_2p'_2, \qquad p_1p'_2 = p'_1p_2.$$

With some labour the remaining trace terms can be evaluated; they yield the expressions

$$B' = 32[2m^2(m^2 + p_1p'_2) + (p_1p'_2)^2 + (p_1p'_1)^2] \tag{11.61}$$

$$C' = -32[m^2(m^2 + p_1p_2 + p_1p'_1 + p_1p'_2) + (p_1p_2)^2].$$

The full expression for the cross-section may be obtained by inserting A', B' and C' into (11.57). It can be seen that it is formidable. If the terms $p_1, p_2 \ldots$ are fully evaluated using (11.47) the following formula can be obtained:

$$\frac{d\sigma}{d\Omega} = \frac{1}{4}\left(\frac{e^2}{4\pi}\right)^2 \frac{1}{m^2} \frac{m^2}{E^2 p_c^4} \left[\frac{4(E^2 + p_c^2)^2}{\sin^4\theta} - \frac{(8E^4 - 4E^2m^2 - m^4)}{\sin^2\theta} + p_c^4\right]. \tag{11.62}$$

However, if we consider the nonrelativistic limit a more recognisable expression can be obtained. In this limit we can write

$$E \to m, \qquad |\mathbf{p}| \to 0$$

$$p_1 p_2 \to -m^2, \qquad p_1 p_1' \to -m^2, \qquad p_1 p_2' \to -m^2$$

hence

$$A' \to 64m^4, \quad B' \to 64m^4, \quad C' \to 32m^4$$

and so the scattering cross-section (11.57) becomes

$$\frac{d\sigma}{d\Omega} = \left(\frac{e^2}{4\pi}\right)^2 \frac{m^2}{16p_c^4}\left[\frac{1}{\sin^4\frac{1}{2}\theta} + \frac{1}{\cos^4\frac{1}{2}\theta} - \frac{1}{\sin^2\frac{1}{2}\theta\cos^2\frac{1}{2}\theta}\right]. \tag{11.63}$$

If this equation is transferred from the c-system to the L-system, then

$$p_L = 2p_c, \qquad \theta_L = \frac{\theta}{2}$$

in the nonrelativistic limit (p_L represents the momentum of the incident electron), and (11.63) can be written as

$$\frac{d\sigma}{d\Omega_L} = \left(\frac{e^2}{4\pi}\right)^2 \frac{4m^2\cos\theta_L}{p_L^4}\left[\frac{1}{\sin^4\theta_L} + \frac{1}{\cos^4\theta_L} - \frac{1}{\sin^2\theta_L\cos^2\theta_L}\right] \tag{11.64}$$

with the aid of A.8 (Appendixes, p. 713). The first term in this expression is the classical Rutherford scattering formula, whilst the second and third terms are the exchange factors which arise from the fact that the particles are identical.

11.2(b). *Bhabha scattering*

The relevant diagrams for Møller and Bhabha scattering in momentum space are compared in Fig. 11.6. In these graphs $p \equiv \mathbf{p}, E_-$ and $q \equiv \mathbf{q}, E_+$ refer to the momentum four-vectors of electron and positron respectively.

It can be seen that there is a close correspondence between the upper diagrams. The lower diagram for Bhabha scattering corresponds to the virtual annihilation and creation of a pair, whilst that for Møller scattering arises from the Pauli exclusion principle.

The two processes follow the same kinematic laws since the masses of electron and positron are the same. The topological equivalence of the graphs in Fig. 11.6 implies, however, an even stronger relationship between the two processes. The

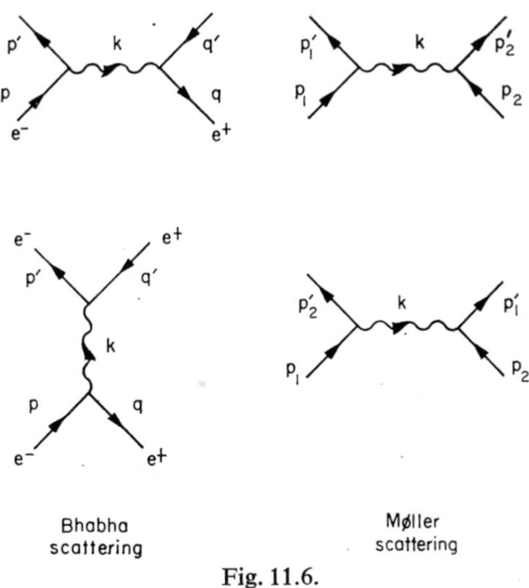

Fig. 11.6.

substitution law of § 8.6(d) states that matrix elements corresponding to Fig. 11.6 are equivalent if we make the following alterations:

$$p_1 \to p \quad p_1' \to p', \quad p_2 \to -q' \quad p_2' \to -q \tag{11.65}$$

hence

$$(p_1 - p_1') \to (p - p'), \quad (p_1 - p_2') \to (p + q). \tag{11.66}$$

The calculation of the matrix elements for Bhabha scattering can therefore be easily performed, since all we have to do is to take the results from Møller scattering and apply substitution law. We represent the relevant kinematical factors as

$$\mathbf{p} = -\mathbf{q}, \qquad \mathbf{p}' = -\mathbf{q}' \tag{11.67}$$

$$|\mathbf{p}| = |\mathbf{q}| = |\mathbf{p}'| = |\mathbf{q}'| = p_c$$

$$\mathbf{p} \cdot \mathbf{p}' = p_c^2 \cos\theta$$

$$E_+ = E_- = E_+' = E_-' = E.$$

The cross-section may then be written as (compare (11.48))

$$\frac{d\sigma}{d\Omega} = \left(\frac{e^2}{4\pi}\right)^2 \frac{m^4}{E^2} \overline{\sum_i \sum_\alpha} |T_{r's'rs}|^2 \tag{11.68}$$

where by using equations (11.50) and (11.66) we find that

$$\overline{\sum_i} \sum_\alpha |T_{r's'rs}|^2$$

$$= \frac{1}{4}\left[\frac{A}{(p_1 - p_1')^4} + \frac{B}{(p_1 - p_2')^4} - \frac{2C}{(p_1 - p_1')^2 (p_1 - p_2')^2}\right]$$

$$\rightarrow \frac{1}{4}\left[\frac{A}{(p - p')^4} + \frac{B}{(p + q)^4} - \frac{2C}{(p - p')^2 (p + q)^2}\right]$$

$$= \frac{1}{64m^4}\left[\frac{A'}{16p_c^4 \sin^4 \frac{1}{2}\theta} + \frac{B'}{E_c^4} + \frac{2C'}{4p_c^2 E_c^2 \sin^2 \frac{1}{2}\theta}\right] \tag{11.69}$$

where we have used (11.67) and we have extracted the factor $16m^4$ from A, B and C (compare (11.51)). Thus the cross-section becomes

$$\frac{d\sigma}{d\Omega} = \left(\frac{e^2}{4\pi}\right)^2 \frac{m^4}{E^2} \frac{1}{64m^4}\left[\frac{A'}{16p_c^4 \sin^4 \frac{1}{2}\theta} + \frac{B'}{E_c^4} + \frac{2C'}{4p_c^2 E_c^2 \sin^2 \frac{1}{2}\theta}\right]. \tag{11.70}$$

If this equation is compared with (11.57) a strong resemblance can be seen between the first terms in each expression. In fact they are identical; we may show this by inserting the kinematic factors into A' (11.60). For Møller scattering this becomes

$$A' = 32[2\mathrm{m}^2(m^2 + p_1p_1') + (p_1p_2)^2 + (p_1p_2')^2]$$

$$= 32[2m^2(m^2 + p_c^2 \cos\theta - E^2) + (p_c^2 + E^2)^2 + (p_c^2 \cos\theta + E)^2] \tag{11.71}$$

where we have used (11.47); the application of the substitution law (11.65) gives the corresponding term for Bhabha scattering

$$A' = 32[2m^2(m^2 + p_1p_1') + (p_1p_2)^2 + (p_1p_2')^2]$$

$$\rightarrow 32[2m^2(m^2 + pp') + (pq')^2 + (pq)^2]$$

$$= 32[2m^2(m^2 + p_c^2 \cos\theta - E^2) + (p_c^2 \cos\theta + E)^2 + (p_c^2 + E^2)^2]. \tag{11.72}$$

Thus the A' terms are identical in the two expressions. The exchange and annihilation factors (B') and the interference terms (C') differ, however. The latter can be written as follows for Bhabha scattering:

$$B' = 32[2m^2(m^2 - pq) + (pq)^2 + (pp')^2] \tag{11.73}$$

$$C' = -32[m^2(m^2 - pq' + pp' - pq) + (pq')^2].$$

In the nonrelativistic limit $A' \to 64m^4$ and for small scattering angles the Bhabha scattering cross-section (11.70) becomes

$$\frac{d\sigma}{d\Omega} = \left(\frac{e^2}{4\pi}\right)^2 \frac{m^4}{m^2} \frac{64m^4}{64m^4} \frac{1}{16p_c^4 \sin^4 \frac{1}{2}\theta} = \left(\frac{e^2}{4\pi}\right)^2 \frac{m^2}{16p_c^4 \sin^4 \frac{1}{2}\theta}. \tag{11.74}$$

Thus for small angles the Møller and Bhabha scattering cross-sections are the same in the nonrelativistic limit.

All the equations given above for Møller and Bhabha scattering assume that no attempt is made to detect polarisation, and that the incident beams are unpolarised. Polarisation effects for these processes are discussed by McMaster (1961).

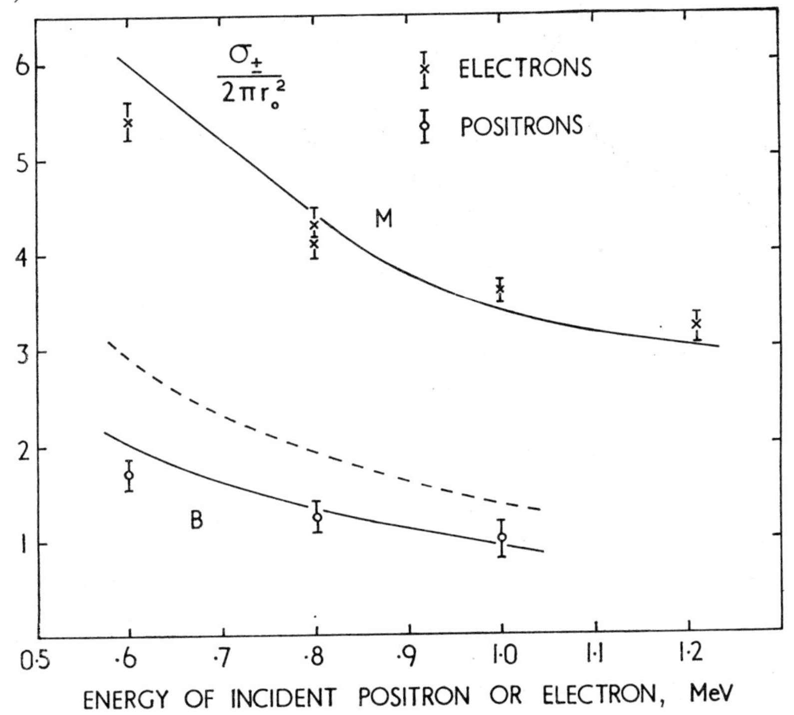

FIG. 11.7. The scattering of electrons and positrons from electrons (Ashkin, Page and Woodward, 1954). The letters M and B relate to the Møller and Bhabha scattering formulae respectively; the dashed curve represents the Bhabha cross-section without the annihilation term.

The validity of the scattering formulae has been checked by Ashkin, Page and Woodward (1954). Instead of measuring angular distributions they detected the differential cross-sections as a function of the energy transferred to the target particle. The relevant formulae may be easily obtained from (11.57) and (11.70) and are quoted by the authors. In some of the experiments the differential cross-section was measured for the transfer of half the energy of the incident particle to the target particle as a function of incident energy. The results obtained are shown in Fig. 11.7 for incident electrons striking electrons and positrons striking electrons. It can be seen that the agreement with theoretical formulae is excellent, and that the effect of including the exchange and annihilation diagrams respectively in the theory for electron–electron and positron–electron scattering is considerable.

11.2(c). *The annihilation of free electron–positron pairs*

In addition to Bhabha scattering the collision of an electron–positron pair can lead to their mutual annihilation. This process can lead to 2, 3 or more photons.† The production of each additional photon implies that a further coupling term e

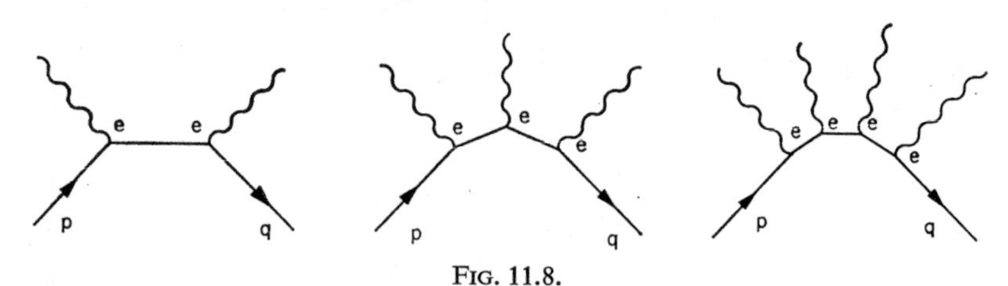

FIG. 11.8.

appears in the matrix element (Fig. 11.8), and therefore a term $\alpha = e^2/4\pi \sim 1/137$ in the transition probability.

Thus the rate of production of three or more photons in the annihilation process will only be about $\sim 1/100$ of that for two photons, and so can be ignored in most circumstances. We shall return to three-photon processes, however, when we consider positronium in the next section.

We shall represent the annihilation process by the four vectors

$$p + q = k_1 + k_2 \tag{11.75}$$

$$a + b \rightarrow d + e$$

and first consider a situation which is experimentally accessible, namely the annihilation of moving positrons by stationary (or virtually stationary) electrons. We shall therefore work in the rest system of the electron ($p = 0, im$) and if we represent the four-momentum of the positron by $q = \mathbf{q}, iE_+$, conservation of four momentum yields

$$k_2^2 = 0 = (p + q - k_1)^2 = -2m^2 + 2pq - 2pk_1 - 2qk_1$$

$$0 = -m^2 - mE_+ + m\omega_1 - \mathbf{q} \cdot \mathbf{k}_1 + E_+\omega_1 \tag{11.76}$$

† The production of a single photon is excluded by energy–momentum conservation in the collision of free particles; single photon production can occur when the electron is tightly bound to the nucleus.

but

$$E_+ + m = E_i, \qquad \mathbf{q} = \mathbf{p}_i$$

where E_i represents the total energy and $\mathbf{p}_i$ the total momentum. Therefore equation (11.76) can be written as

$$\omega_1 E_i - \mathbf{p}_i \cdot \mathbf{k}_1 = mE_i. \tag{11.77}$$

We shall make use of this equation later. Another equation which we shall have occasion to use later is obtained by writing

$$q^2 = -m^2 = (k_1 + k_2 - p)^2 = 2k_1k_2 - 2k_1p - 2k_2p - m^2$$

but

$$k_1p + k_2p = -m(\omega_1 + \omega_2) = -mE_i \tag{11.78}$$

$$k_1k_2 = \omega_1\omega_2(\cos\varphi - 1)$$

hence

$$mE_i = \omega_1\omega_2(1 - \cos\varphi) \tag{11.79}$$

where φ represents the angle between $\mathbf{k}_1$ and $\mathbf{k}_2$.

Now the cross-section for any process of the type $a + b \to d + e$ was given in (7.71)

$$\frac{d\sigma}{d\Omega} = \frac{V^4}{(2\pi)^2}\frac{1}{v_{ab}}\overline{\sum_i}\sum_\alpha |M_{fi}|^2 \frac{E_dE_e|\mathbf{p}_d|^3}{|\mathbf{p}_d|^2E_i - \mathbf{p}_i\cdot\mathbf{p}_dE_d}.$$

If we make the necessary identifications in equation (11.75) we can write

$$E_d \equiv \omega_1, \quad E_e \equiv \omega_2, \quad \mathbf{p}_d \equiv \mathbf{k}_1, \quad |\mathbf{p}_d| \equiv \omega_1, \quad v_{ab} = \frac{|\mathbf{q}|}{E_+} \tag{11.80}$$

since we consider the electron to be at rest (7.45). Thus the phase space term becomes

$$\frac{\omega_1\omega_2\omega_1^3}{\omega_1^2E_i - \mathbf{p}_i\cdot\mathbf{k}_1\omega_1} = \frac{\omega_1^3\omega_2}{\omega_1E_i - \mathbf{p}_i\cdot\mathbf{k}_1} = \frac{\omega_1^3\omega_2}{mE_i} \tag{11.81}$$

where we have used (11.77). The diagrams associated with the matrix element for annihilation of the electron–positron pair into two photons are given in Fig. 11.9. Using the rules given in § 8.6(b) (p. 325) and (7.21) it is evident that we can write the matrix element as

$$M_{fi} = \langle k_1k_2|\, M_2(e^+e^-)\,|pq\rangle = \frac{e^2}{V^2}\left(\frac{m^2}{4mE_+\omega_1\omega_2}\right)^{\frac{1}{2}} T_{rs} \tag{11.82}$$

since the electron is stationary, $p = (0, im)$. Using equations (11.80–11.82) we therefore find the cross-section to be

$$\frac{d\sigma}{d\Omega} = \frac{V^4}{(2\pi)^2}\frac{E_+}{|\mathbf{q}|}\overline{\sum_i}\sum_\alpha \frac{e^4}{V^4}\frac{m}{4E_+\omega_1\omega_2}|T_{rs}|^2\frac{\omega_1^3\omega_2}{mE_i}$$

$$= \left(\frac{e^2}{4\pi}\right)^2\frac{\omega_1^2}{|\mathbf{q}|E_i}\overline{\sum_i}\sum_\alpha |T_{rs}|^2. \tag{11.83}$$

We next consider the structure of T_{rs}. The diagrams which contribute in lowest order perturbation theory to the annihilation into two photons are shown in Fig. 11.9. Underneath them the diagrams which contribute in lowest order to

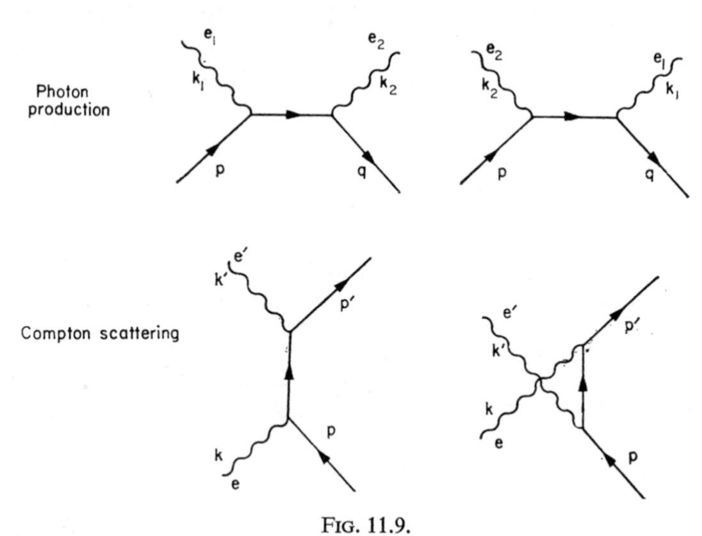

FIG. 11.9.

Compton scattering are displayed. It can be seen that both contain two photons and two electrons (we regard the positron as an anti-electron). It is therefore possible to use the matrix elements associated with one set of diagrams to calculate the matrix elements of the other process by means of the substitution law. We therefore consider our calculations for Compton scattering with the following substitutions (compare Table 8.2 (p. 330):

$$p \to p, \quad p' \to -q, \quad e \to e_1, \quad e' \to e_2,$$
$$k \to -k_1, \quad k' \to k_2. \tag{11.84}$$

Now in equation (11.24) we wrote

$$\overline{\sum_i}\sum_\alpha |T_{r'r}|^2 = \frac{1}{32m^4}\operatorname{tr}(A + B + C + D)$$

and then evaluated the traces of A, B, C and D. Trace A was given in (11.25) as

$$\operatorname{tr} A = -4a^2 p(k - k') = \frac{8m^2}{\omega\omega'}(\omega' - \omega).$$

This equation is unsuitable for our present purposes as we need to make substitutions for four vectors. The problem can be overcome by writing the denominators of a as Lorentz scalars

$$a = -m\left(\frac{k'}{pk'} - \frac{k}{pk}\right) \tag{11.85}$$

which still yields (11.22)

$$a = \frac{k'}{\omega'} - \frac{k}{\omega}$$

for $p = (0, im)$. Using the definition (11.85) we find that

$$a^2 = -m^2 \frac{2kk'}{(pk')(pk)}$$

and hence

$$\operatorname{tr} A = 8m^2 \frac{kk'}{pk'} - 8m^2 \frac{kk'}{pk}. \tag{11.86}$$

Therefore we can re-write (11.29) as

$$\overline{\sum_i}\sum_\alpha |T_{r'r}|^2 = \frac{1}{32m^4}\operatorname{tr}(A + B + C + D)$$

$$= \frac{1}{32m^4}\left[8m^2\frac{kk'}{pk'} - 8m^2\frac{kk'}{pk} + 32(ee')^2 m^2\left(1 - \frac{kk'}{pk}\right) + 32(ee')^2 m^2 \frac{kk'}{pk}\right]$$

$$= \frac{1}{4m^2}\left[\frac{kk'}{pk'} - \frac{kk'}{pk} + 4(ee')^2\right]. \tag{11.87}$$

We are now in a position to apply the substitution law. In addition to making the substitutions listed in (11.84), we must multiply the expression by $\frac{1}{2}$ since we now average over two initial electron spin states rather than one as in Compton scattering. We must also multiply (11.87) by a factor -1 as one electron becomes a positron (8.95). Hence, if the initial and final systems are unpolarised, equation (11.87) becomes

$$\overline{\sum_i}\sum_\alpha |T_{rs}|^2 = \frac{-1}{8m^2}\sum_{e_1 e_2}\left[-\frac{k_1 k_2}{pk_2} - \frac{k_1 k_2}{pk_1} + 4(e_1 e_2)^2\right]$$

$$= \frac{1}{8m^2}\sum_{e_1 e_2}\left[\frac{k_1 k_2}{pk_2} + \frac{k_1 k_2}{pk_1} - 4(e_1 e_2)^2\right]$$

$$= \frac{1}{8m^2} \sum_{e_1 e_2} \left[\frac{\mathbf{k}_1 \cdot \mathbf{k}_2 - \omega_1 \omega_2}{-m\omega_2} + \frac{\mathbf{k}_1 \cdot \mathbf{k}_2 - \omega_1 \omega_2}{-m\omega_1} - 4(e_1 e_2)^2 \right]$$

$$= \frac{1}{8m^2} \sum_{e_1 e_2} \left[\frac{\omega_1 \omega_2}{m\omega_2} (1 - \cos\varphi) + \frac{\omega_1 \omega_2}{m\omega_1} (1 - \cos\varphi) - 4(e_1 e_2)^2 \right]$$

$$= \frac{1}{8m^2} \sum_{e_1 e_2} \left[\frac{E_i}{m} (1 - \cos\varphi) - 4(e_1 e_2)^2 \right] \tag{11.88}$$

where we have used conservation of energy to write $\omega_1 + \omega_2 = E_i$.

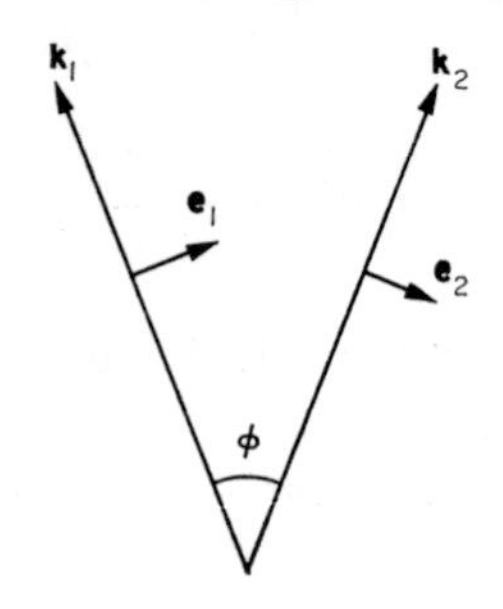

FIG. 11.10.

We finally consider the sum over states of polarisation. We use the same method as in § 11.1(c), but in this case the properties are simpler. As in (11.35) we can write

$$(e_1 e_2)^2 = (\mathbf{e}_1 \cdot \mathbf{e}_2)^2 = \cos^2\chi \tag{11.89}$$

where χ is the angle between $\mathbf{e}_1$ and $\mathbf{e}_2$. We shall work in the $\mathbf{k}_1\mathbf{k}_2$ plane, and we have four possible states (the final state is illustrated in Fig. 11.10)

$$\mathbf{e}_1 \perp \mathbf{e}_2 \perp, \quad \mathbf{e}_1 \perp \mathbf{e}_2 \parallel, \quad \mathbf{e}_1 \parallel \mathbf{e}_2 \perp, \quad \mathbf{e}_1 \parallel \mathbf{e}_2 \parallel$$

$$\cos^2\chi = 1, \quad \cos^2\chi = 0, \quad \cos^2\chi = 0, \quad \cos^2\chi = \cos^2\varphi$$

where the sign $\mathbf{e}_1 \perp$ implies that $\mathbf{e}_1$ is perpendicular to the $\mathbf{k}_1\mathbf{k}_2$ plane and so on. Thus the expression in square brackets in (11.88) reads

$$\mathbf{e}_1 \perp \mathbf{e}_2 \perp \quad \left[\frac{E_i}{m} (1 - \cos\varphi) - 4 \right] \tag{11.90}$$

$$\mathbf{e}_1 \perp \mathbf{e}_2 \parallel \quad \left[\frac{E_i}{m} (1 - \cos\varphi) \right]$$

$$\mathbf{e}_1 \parallel \mathbf{e}_2 \perp \quad \left[\frac{E_i}{m} (1 - \cos\varphi) \right]$$

$$\mathbf{e}_1 \parallel \mathbf{e}_2 \parallel \quad \left[\frac{E_i}{m} (1 - \cos\varphi) - 4\cos^2\varphi \right]$$

and so

$$\overline{\sum_i}\sum_\alpha |T_{rs}|^2 = \frac{1}{2m^2}\left[\frac{E_i}{m}(1-\cos\varphi) - 1 - \cos^2\varphi\right] \tag{11.91}$$

and the annihilation cross-section (11.83) becomes (with the aid of (11.79))

$$\begin{aligned}\frac{d\sigma}{d\Omega} &= \frac{1}{2m^2}\left(\frac{e^2}{4\pi}\right)^2 \frac{\omega_1^2}{|\mathbf{q}|\,E_i}\left[\frac{E_i}{m}(1-\cos\varphi) - 1 - \cos^2\varphi\right]\\ &= \frac{1}{2m^2}\left(\frac{e^2}{4\pi}\right)^2 \frac{m}{|\mathbf{q}|}\frac{\omega_1}{\omega_2}\left[\frac{E_i}{m} - \frac{1+\cos^2\varphi}{1-\cos\varphi}\right].\end{aligned} \tag{11.92}$$

In the nonrelativistic limit

$$E_i \to 2m, \qquad \omega_1 \to m, \qquad \omega_2 \to m, \qquad \varphi \to \pi$$

since the two photons must travel in opposite directions in order to conserve both energy and momentum ($|\mathbf{q}| \to 0$); in this situation (11.92) reduces to

$$\frac{d\sigma}{d\Omega} = \frac{1}{2m^2}\left(\frac{e^2}{4\pi}\right)^2 \frac{m}{|\mathbf{q}|}\left(2 - \frac{2}{2}\right) = \frac{1}{2m^2}\left(\frac{e^2}{4\pi}\right)^2 \frac{m}{|\mathbf{q}|}. \tag{11.93}$$

Thus the photons are emitted isotropically; the total cross-section is then given by

$$\sigma_T = \frac{1}{2}\,4\pi\,\frac{1}{2m^2}\left(\frac{e^2}{4\pi}\right)^2 \frac{m}{|\mathbf{q}|} = \frac{\pi}{m^2}\left(\frac{e^2}{4\pi}\right)^2 \frac{m}{|\mathbf{q}|} = \pi r_0^2\,\frac{m}{|\mathbf{q}|} \tag{11.94}$$

when the factor $\frac{1}{2}$ appears because the differential cross-section refers to *any* photon, and therefore we count all final states twice in integrating over 4π. The Thomson radius r_0 has also been inserted in (11.94).

The factors ω_1, ω_2 and φ can be eliminated from full expression for the cross-section (11.92) by using (11.77) and (11.79); we then obtain

$$\begin{aligned}\frac{d\sigma}{d\Omega} = \frac{1}{2m^2}\left(\frac{e^2}{4\pi}\right)^2 \frac{m^2}{qE_+ - q^2\cos\theta_1}\Bigg[1 - \frac{E_+}{m} &+ \frac{(E_i - q\cos\theta_1)^2}{E_+E_i - E_iq\cos\theta_1}\\ &- 2\,\frac{(E_iE_+ - E_iq\cos\theta_1)}{(E_i - q\cos\theta_1)^2}\Bigg]\end{aligned} \tag{11.95}$$

where $q = |\mathbf{q}|$ and θ_1 is the angle between $\mathbf{k}_1$ and $\mathbf{q}$. As we have seen above this distribution is isotropic in the nonrelativistic limit $|\mathbf{q}| \to 0$. At high energies $E_+ \gg m$ it peaks strongly in the forward direction because of the conservation of momentum.

If the above equation is integrated and the factor $\frac{1}{2}$ inserted, as in (11.94), the total cross-section is obtained (Dirac, 1930)

$$\sigma_T = \pi r_0^2\,\frac{m}{E_i}\left[\frac{E_+^2 + 4mE_+ + m^2}{q^2}\log\frac{(E_+ + q)}{m} - \frac{(E_+ + 3m)}{q}\right]. \tag{11.96}$$

When $E_+ \gg m$ this equation tends to the following form:

$$\sigma_T \to \pi r_0^2 \frac{m}{E_i}\left(\log \frac{2E_+}{m} - 1\right). \tag{11.97}$$

Both the differential and total cross-sections for the positron–electron annihilation process have been measured, and found to agree with experiment. The differential cross-sections have been measured at low energies (Kendall and Deutsch, 1956; Seward, Hatcher and Fultz, 1961), and the total cross-sections at higher energies. Some results for σ_T are given in Table 11.1; the data for positron energies up to 200 MeV is that of Colgate and Gilbert (1952), and the high energy results are those of Fabiani *et al.* (1961).

TABLE 11.1

Energy of positrons (in MeV)	Measured σ_T (mb)	Calculated σ_T (mb)
50	11·0 ± 2·5	10·8
100	6·3 ± 1·2	6·35
200	{ 2·6 ± 1·0 3·7 ± 0·6 }	3.53
1940	0·521 ± 0·015	0·523
5800	0·197 ± 0·006	0·199
7710	0·156 ± 0·006	0·154
9640	0·127 ± 0·007	0·126

11.2(d). *Positronium*

The annihilation cross-section for positrons at low velocities was given in (11.94)

$$\sigma_T = \pi r_0^2 \frac{m}{|\mathbf{q}|} = \frac{\pi r_0^2}{\beta} \tag{11.98}$$

where β represents the relative velocity of the electron and positron. Thus the rate of annihilation of positrons in a medium containing n electrons per cm^3 is given by

$$\lambda = \sigma_T n\beta = \pi r_0^2 n \equiv \pi r_0^2 cn. \tag{11.99}$$

If the medium is a gas we can write n as

$$n = 6 \times 10^{24} \varrho \frac{Z}{A} \frac{P}{P_0}$$

where ϱ is the density of the gas at standard conditions and P/P_0 the pressure in atmospheres. For most light gases $\varrho \sim 10^{-3}$ gm cm^{-3} and $Z/A \sim \frac{1}{2}$ and so $n \sim 3 \times 10^{21} P/P_0$, hence the annihilation rate becomes

$$\lambda \sim \pi 9 \times 10^{-26} \times 3 \times 10^{10} \times 3 \times 10^{21} \frac{P}{P_0}$$

$$\sim 10^7 \frac{P}{P_0} \text{ sec}^{-1}$$

where we have inserted the Thomson radius (11.33). The measured dependence of λ upon pressure is illustrated in the lines A and B of Fig. 11.11. The data is that of Deutsch (1951). His work showed that the decay of positrons in gases involved

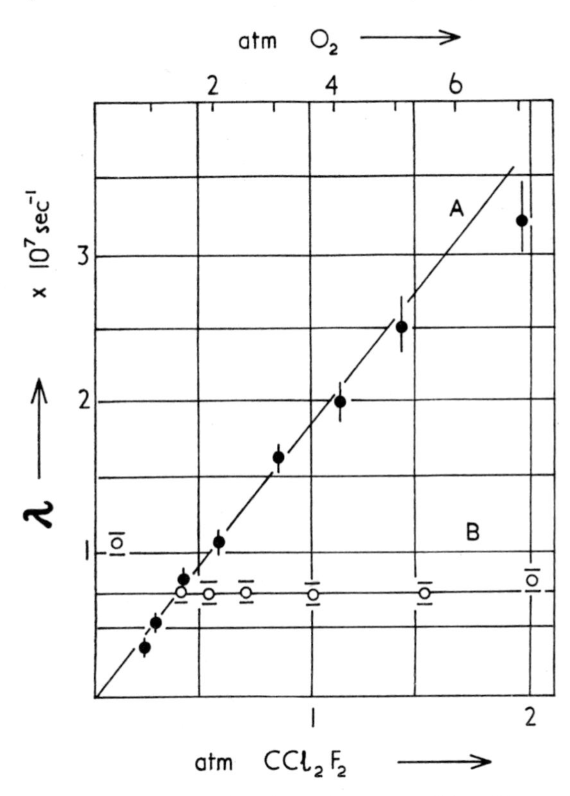

FIG. 11.11. Disappearance rate of positrons in oxygen (A) and freon (B) as a function of pressure (Deutsch, 1951).

three separate periods. One was too short to measure, a second was of order 10^{-7} sec and varied inversely as the pressure and the third was practically independent of the pressure with a lifetime of $\sim 1{\cdot}4 \times 10^{-7}$ sec. The relative occurrence of the three components, and the lifetime of the pressure dependent component was found to depend critically on the nature of the gas used. Figure 11.11 shows the lines for oxygen (A) and freon (B); it can be seen that the second component is strong in oxygen.

The decay modes which are independent of pressure are due to the formation and decay of *positronium* – the bound state of an electron and positron. This state is equivalent, in many ways, to the hydrogen atom with the proton replaced by a positron. The energy levels of the hydrogen atom are given by

$$E_n \sim -\frac{1}{2}\left(\frac{e^2}{4\pi}\right)^2 \frac{\mu}{n^2} = -\frac{1}{2}\alpha^2 \frac{\mu}{n^2} \tag{11.100}$$

where

$$\mu = \frac{m_p m_e}{m_p + m_e} \tag{11.101}$$

is the reduced mass which equals $\frac{1}{2}m$ for the electron–positron system, and n is the principal quantum number. Upon inserting numbers in (11.100) positronium is found to have an ionisation potential of 6·8 eV.

The conservation laws, discussed in Chapters 5 and 9, impose an interesting condition on the decay mechanism of positronium. The ground state ($L = 0$) of the electron–positron system can be formed with spins pointing parallel or antiparallel to each other. The two states have the properties shown in Table 11.2.

TABLE 11.2

Direction of Spins	Notation	Statistical weight
↑ ↑	$^3S_1$ triplet orthopositronium	$\frac{3}{4}$
↓ ↑	$^1S_0$ singlet parapositronium	$\frac{1}{4}$

Now both systems should be antisymmetric with respect to the exchange of particles. A similar situation was examined in § 9.2(g) for two *identical* fermions; it was then concluded that the overall requirement for antisymmetry was given by the equation (9.53)

$$(-1)^{L+s+1} = -1.$$

In the case of positronium we must modify this equation to include the intrinsic charge parity ξ_C of the particle–antiparticle system, hence

$$(-1)^{L+s+1}\,\xi_C = -1 \tag{11.102}$$

$$\xi_C = (-1)^{L+s}. \tag{11.103}$$

Now if positronium decays into n photons the charge parity of the final state is given by $(-1)^n$, since the intrinsic charge parity of the photon is -1 (equation (9.28)), hence invariance of the decay process under charge conjugation implies that

$$(-1)^{L+s} = (-1)^n. \tag{11.104}$$

This condition yields the following result for $L = 0$:

$$\begin{aligned} &{}^3S_1 \text{ state } s = 1 \qquad \text{requires } n \text{ odd} \\ &{}^1S_0 \text{ state } s = 0 \qquad\qquad\qquad n \text{ even.} \end{aligned} \tag{11.105}$$

In practice this result may be stated as

$$\begin{aligned} &{}^3S \text{ positronium} \to 3\gamma \\ &{}^1S \text{ positronium} \to 2\gamma \end{aligned} \tag{11.106}$$

since the next favourable states would be smaller by a factor $(1/137)^2 \sim 10^{-4}$.

The decay time for singlet positronium may be easily calculated by modifying the expression for the annihilation rate of free electron–positron pairs (11.99)

$$\lambda = \pi r_0^2 n.$$

We can modify this expression to

$$\lambda_{2\gamma} = \frac{1}{\tau_{2\gamma}} = 4\pi r_0^2\, |\psi|^2 \tag{11.107}$$

where $|\psi|^2$ represents the density of the wave function at the position of the particles, and the factor 4 appears because the annihilation in flight was averaged over the four possible relative spin orientations of the incident particles, whereas positronium decays from a specific state.

The density function can be written as

$$|\psi|^2 = |f(0)|^2$$

where $f(0)$ represents the radial wave function $f(\mathbf{r})$ of the positronium atom at position 0. We may obtain this function straightaway from the textbook treatments of the theory of the hydrogen atom in the ground state

$$f(0) = \frac{1}{(\pi a^3)^{\frac{1}{2}}} = \left(\frac{\alpha^3 m^3}{8\pi}\right)^{\frac{1}{2}} \tag{11.108}$$

since the ground state of positronium has a radius twice as large as that of the Bohr orbit of the hydrogen atom

$$a = 2a_H = \frac{2}{\alpha m}.$$

The lifetime of parapositronium in its ground state is therefore given by the relationship

$$\lambda_{2\gamma} = \frac{1}{\tau_{2\gamma}} = 4\pi r_0^2 \frac{\alpha^3 m^3}{8\pi} = 4\pi \frac{\alpha^2}{m^2} \frac{\alpha^3 m^3}{8\pi}$$

or

$$\tau_{2\gamma} = \frac{2}{\alpha^5 m} = 1{\cdot}25 \times 10^{-10} \text{ sec} \tag{11.109}$$

where we have replaced r_0 by α/m (11.33).

The three quantum annihilation may be calculated in a similar fashion. The calculation for the annihilation in flight is similar in principle, but more complicated in detail, to that used for the two quantum annihilation. The calculation was made by Ore and Powell (1949), and in the nonrelativistic limit, yields the following expression for the total cross-section:

$$\sigma_T = \frac{4}{3} \frac{\alpha r_0^2}{\beta} (\pi^2 - 9) \tag{11.110}$$

hence

$$\lambda_{3\gamma} = \frac{1}{\tau_{3\gamma}} = \frac{4}{3} \frac{4}{3} \frac{\alpha^3}{m^2} (\pi^2 - 9) \frac{\alpha^3 m^3}{8\pi}$$

or

$$\tau_{3\gamma} = \frac{9\pi}{2(\pi^2 - 9)} \frac{1}{\alpha^6 m} = 1{\cdot}39 \times 10^{-7} \text{ sec.} \tag{11.111}$$

The above data is in excellent agreement with the work of Deutsch. If the pressure independent periods in the annihilation of positrons in gases is interpreted as the decay of positronium, then the very short period can be recognised as coming from the decay of the singlet state and the long period from the triplet state. In fact upon extrapolating line B in Fig. 11.11 to zero pressure (thus

$\mathbf{k}_R$ $-\mathbf{k}_R$ $\mathbf{k}_L$ $-\mathbf{k}_L$

FIG. 11.12.

eliminating the effect of collisions), Deutsch found a lifetime of $(1{\cdot}47 \pm 0{\cdot}15)$ 10^{-7} sec, which compares favourably with the calculated figure of $1{\cdot}39 \times 10^{-7}$ sec.

Much experimental work has been done upon the properties of positronium. Review articles on the topic have been written by Deutsch (1953) and by Simons (1958). We conclude by mentioning two interesting properties.

First, let us consider the polarisation of the two outgoing photons from the decay of singlet positronium. The total angular momentum of the system is zero and so the spins of both photons must lie parallel or antiparallel to their direction of motion as in Fig. 11.12. Now the intrinsic parity of the ground state of positronium is negative, since a Dirac particle and antiparticle have opposite intrinsic parity (see (5.125) and the following discussion). Thus the parity of the two-photon system must also be negative. It was shown in § 5.4(c) that the linear combination (5.109)

$$|\mathbf{k}R;\ -\mathbf{k}R\rangle - |\mathbf{k}L;\ -\mathbf{k}L\rangle \tag{11.112}$$

is an eigenstate of the parity operator P with eigenvalue -1; using this combination it is not difficult to show that outgoing photons must have their planes of polarisation at right angles (see § 9.5(b) for an alternative argument – singlet positronium has the same spin and parity as the π° meson).

A further interesting property of positronium is that it does not reveal a first order Zeeman splitting in any of its states. This may be shown by either classical or quantum mechanical arguments; the first method is clearer. Since the masses of the two particles are equal and the signs of their charges are opposite, the orbital motion cannot provide a net circulating current and so $g_L = 0$ (g is the Landé factor). In addition there is no contribution from the spin states; the singlet states have no preferred spin directions and so the expectation value of the magnetic moment is zero, and although the spins point together in the triplet states the magnetic moments of the two particles are in opposition. Thus simple Zeeman splitting cannot occur; a magnetic moment can be induced in the system, however, giving rise to second order Zeeman splitting (Berestetskii, 1949; Ferrell, 1951).

11.3. DIVERGENCES AND THE SELF-ENERGY OF PARTICLES

11.3(a). *Introduction*

The work described in §§ 11.1 and 11.2 indicates good agreement between experiment and theory if one considers only the lowest order terms in the iteration solution of the S-matrix. However, if the theory is an exact one, it should be possible to go on to consider more sophisticated problems and processes. For

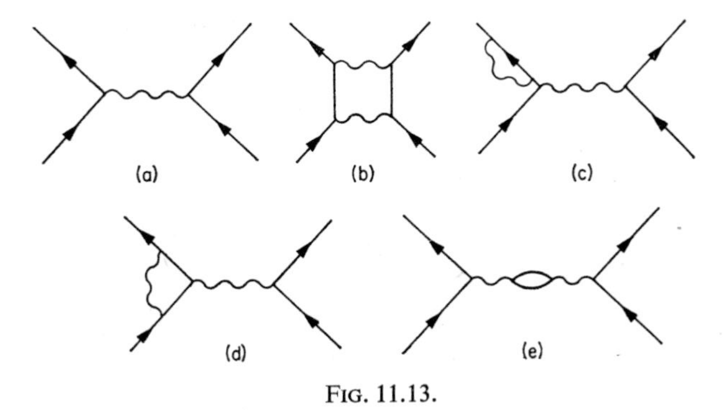

FIG. 11.13.

example, in addition to the lowest order diagram in Møller scattering (Fig 11.13(a)), more complex diagrams would appear as corrections in the next order of approximation (diagrams (b) to (e)). All the diagrams lead to the same final states as (a) and therefore contribute to any relevant cross-section. Now each of these diagrams has four vertices and so a factor e^4 appears in the matrix element. We would therefore expect the following contributions to appear in the transition probability:

$$\text{diagram (a)} \qquad e^4 \equiv \alpha^2 \sim \left(\frac{1}{137}\right)^2 \tag{11.113}$$

$$\text{diagrams (b–e)} \qquad e^8 \equiv \alpha^4 \sim \left(\frac{1}{137}\right)^4$$

$$\text{interference terms} \quad e^6 \equiv \alpha^3 \sim \left(\frac{1}{137}\right)^3.$$

Thus one might expect the higher order terms to make little contribution to any transition probability. However, the diagrams (b–e) lead to divergent integrals and therefore nonsensical contributions.

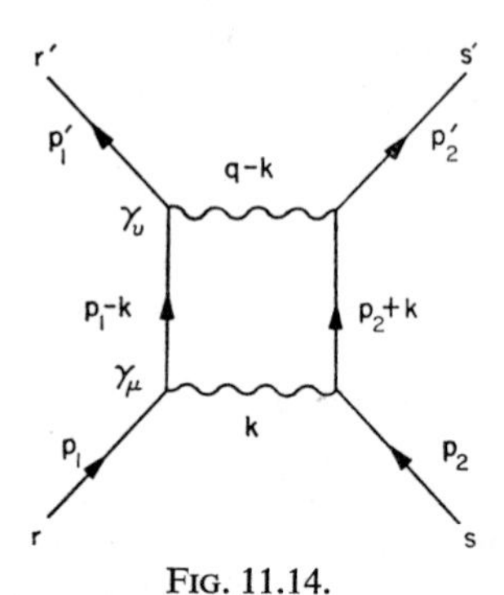

FIG. 11.14.

Let us consider Fig. 11.13(b) in more detail (Fig. 11.14). Using the rules given in § 8.6(b) and dropping off factors which are irrelevant to our present discussion, we can write the matrix element as

$$T \propto \int d^4k \left[\bar{u}_{s'}\gamma_\nu \frac{i\gamma(p_2+k) - m}{(p_2+k)^2 + m^2}\gamma_\mu u_s\right] \frac{1}{k^2}\frac{1}{(q-k)^2} \times$$
$$\times \left[\bar{u}_{r'}\gamma_\nu \frac{i\gamma(p_1-k) - m}{(p_1-k)^2 + m^2}\gamma_\mu u_r\right]. \tag{11.114}$$

The integral $\int d^4k$ appears in the matrix element since we must sum over all possible intermediate states, and the external kinematics only allow us to define the overall four-momentum transfer q.

Using the relation $p^2 = -m^2$ we can reduce the denominator in (11.114) to

$$(2p_2k + k^2)\, k^2(q - k)^2\, (-2p_1k + k^2). \tag{11.115}$$

Therefore in the limits $k \to \infty$ and $k \to 0$ the matrix element behaves as follows:

$$k \to \infty \quad T \to \int d^4k \cdot \frac{k^2}{k^8} \equiv \int \frac{dk}{k^3} \tag{11.116}$$

$$k \to 0 \quad T \to \int \frac{d^4k}{k^4} \equiv \int \frac{dk}{k}. \tag{11.117}$$

Thus we have a logarithmic divergence at the low frequency limit; this is known as *infrared divergence*. The existence of this problem has been known for many years and its solution was first provided by Bloch and Nordsieck (1937). The procedure used is to consider the diagrams leading to infrared divergences along

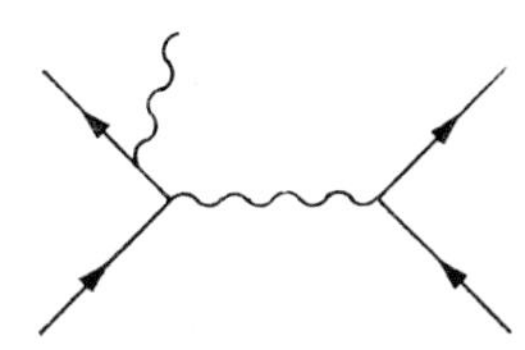

FIG. 11.15.

with diagrams of the type shown in Fig. 11.15. This process is known as a *radiative process*. Now any apparatus which is used to measure elastic scattering possesses a finite energy resolution ΔE, and is therefore incapable of distinguishing between a particle of energy E and one of energy $E + \Delta E$. Thus if one or more photons with total energy less than ΔE are liberated in the scattering process, they would not be detected. This effect must be included in any exact calculation, and when diagrams of the type Figs. 11.14 and 11.15 are considered in conjunction it is found that the infrared divergence may be eliminated (a similar cancellation can be shown to occur in all orders of perturbation theory, see, for example, Jauch and Rohrlich, 1955).

The problem of radiative corrections in electron scattering experiments is an important one and considerable work has been done upon it (see § 11.4(e) for references).

11.3(b). *Ultraviolet divergences*

The problem of divergences or infinities appearing in field theories is an old one. It first arose in classical field theory in considering the self-energy of an electron. This is given as $\sim e^2/a$ but how big is a? If we assume the electron to be a point charge then the energy is infinite; if, on the other hand, we assume a finite extension the field will tend to blow the charge apart.

The divergences appearing in quantum electrodynamics are of three main types; these are listed below.

(1) Divergences associated with vacuum fluctuations; these are of the type shown in Fig. 11.16. They cause the elements in the S-matrix to be multiplied by a phase factor $M \to Me^{i\delta}$, where δ is infinite. Since we always consider $|M|^2$ the contribution is of no importance.

(2) Infrared divergencies (§ 11.3(a)).

(3) Ultraviolet divergencies. These are represented by the basic diagrams shown in Fig. 11.17; process (a) is known as an *electron self-energy correction*, (b)

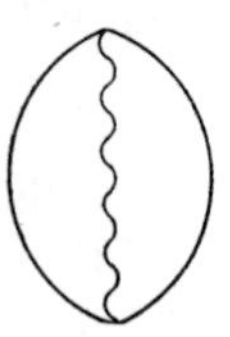

FIG. 11.16.

is called *vacuum polarisation* since the occurrence of a pair implies that the vacuum can be polarised by an external field, and (c) is known as a *vertex correction.*

The problems associated with (3) are considerable, and their solution is far from simple. The procedure used in handling them is to split the matrix elements

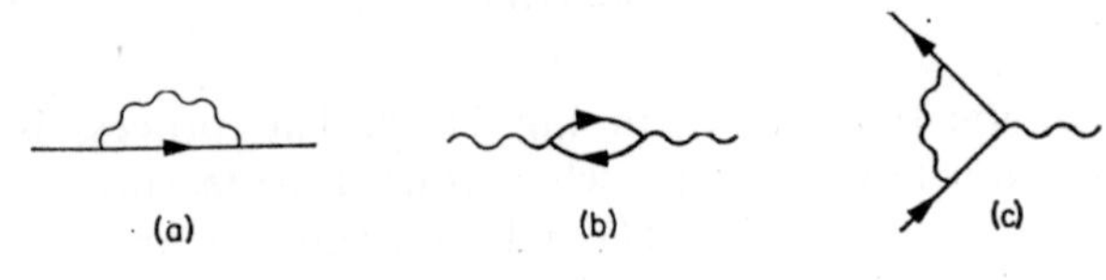

FIG. 11.17.

into two parts containing the finite and divergent terms and the latter are then eliminated by a redefinition of mass and charge. This procedure is known as *mass and charge renormalisation.*

The renormalisation programme has lead to considerable successes in explaining the electromagnetic level shifts in atomic systems and the anomalous magnetic moments of the charged leptons. Nevertheless, it should be noted that the renormalisation process has not really solved the basic problem of why the divergencies occur. They appear to be inextricably associated with the assumption that interactions can be described by a local interaction Lagrangian.

11.3(c). *The self-energy of an electron*

As an example of the problem of treating a serious divergence we shall examine the self-energy of an electron. The treatment of vacuum polarisation and radiative corrections can be found in a number of books—see for example Feynman (1961).

Consider Fig. 11.13(c); this shows the emission and reabsorption of an virtual photon by an electron during a scattering process. In fact there is nothing to prevent the electron from emitting and reabsorbing virtual photons in an infinite number of ways during its passage between two points (Fig. 11.18).

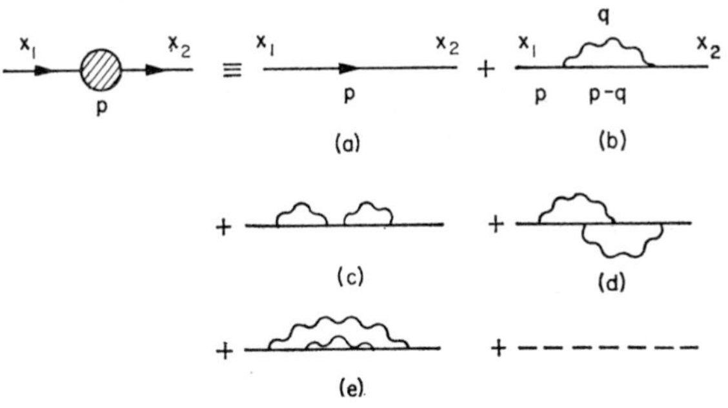

FIG. 11.18.

Let us therefore consider the two simplest diagrams for an electron propagator; the matrix elements are given by Table 8.1 (p. 326):

diagram (a)

$$\frac{i\gamma p - m}{p^2 + m^2} = \frac{-1}{i\gamma p + m} \tag{11.118}$$

diagram (b)

$$e^2 \int d^4q \, \frac{i\gamma p - m}{p^2 + m^2} \gamma_\mu \frac{i\gamma(p-q) - m}{(p-q)^2 + m^2} \gamma_\mu \frac{1}{q^2} \frac{i\gamma p - m}{p^2 + m^2} \tag{11.119}$$

where we have used (4.287) for diagram (a), and have summed over intermediate states for diagram (b). We have also inserted the term e^2 in (b) to remind us of the relative coupling strength. Equation (11.119) can be re-written as

$$\frac{i\gamma p - m}{p^2 + m^2} \, C \, \frac{i\gamma p - m}{p^2 + m^2} = \frac{1}{i\gamma p + m} \, C \, \frac{1}{i\gamma p + m} \tag{11.120}$$

where

$$C = e^2 \int \frac{d^4q}{q^2} \gamma_\mu \frac{i\gamma(p-q) - m}{(p-q)^2 + m^2} \gamma_\mu$$

$$= -e^2 \int \frac{d^4q}{q^2} \gamma_\mu \frac{1}{i\gamma(p-q) + m} \gamma_\mu \tag{11.121}$$

$$= i\gamma p A(p^2) - B(p^2) \tag{11.122}$$

where A and B are two Lorentz invariant functions.

Now if we have any two operators D and C in the form $1/(D - C)$ we may expand the expression as

$$\frac{1}{D - C} = \frac{1}{D} + \frac{1}{D} C \frac{1}{D} + \frac{1}{D} C \frac{1}{D} C \frac{1}{D} + \cdots \qquad (11.123)$$

and so

$$\frac{-1}{i\gamma p + m - C} = -1\left[\frac{1}{i\gamma p + m} + \frac{1}{i\gamma p + m} C \frac{1}{i\gamma p + m} + \frac{1}{i\gamma p + m} C \frac{1}{i\gamma p + m} C \frac{1}{i\gamma p + m} + \cdots\right]. \qquad (11.124)$$

An inspection of this expression shows that the first term in the expansion corresponds to diagram (a), the second to diagram (b) and the third to diagram (c) of Fig. 11.18. The first two statements follow from (11.118) and (11.119), and the verification of the third is quite straightforward. It should be noted, however, that the term

$$\frac{1}{i\gamma p + m} C \frac{1}{i\gamma p + m} C \frac{1}{i\gamma p + m} \qquad (11.125)$$

does not represent the diagrams (d) and (e) of Fig. 11.18 if C is in the form given in (11.122). Diagrams (d) and (e) can be represented in the general expansion (11.124) by adding correction terms of order e^4 to C.

To a fair approximation, therefore, the expression (11.124)

$$\begin{aligned}\frac{-1}{i\gamma p + m - C} &= \frac{-1}{i\gamma p + m - iA\gamma p + B} \\ &= \frac{-1}{i\gamma p(1 - A) + (m + B)} \\ &= \frac{i\gamma p(1 - A) - (m + B)}{p^2(1 - A)^2 + (m + B)^2} \qquad (11.126)\end{aligned}$$

represents a 'bare' particle (Fig. 11.18(a)) plus the associated radiative processes. Now if this expression is also intended to represent a physical particle with a measurable mass m_{ex}, then it should have poles at $p^2 = -m_{ex}^2$. Thus (11.126) should have a pole at

$$p^2 = -\frac{(m + B)^2}{(1 - A)^2} = -m_{ex}^2 \qquad (11.127)$$

or

$$m_{ex} = \frac{m + B}{1 - A} = m\left(1 + \frac{B}{m}\right)(1 + A) = m + \delta m$$

ince B and A are both proportional to e^2 and we are considering approx-mations to this order only, thus

$$\delta m = B(m^2) + mA(m^2) \tag{11.128}$$

o a good approximation.

Thus the action of the radiation field on a bare electron of mass m changes ts mass by an amount δm. The quantity δm is unobservable and divergent. The ast statement can be easily verified by examining (11.122)

$$\begin{aligned} C = i\gamma p A - B &= e^2 \int \frac{d^4 q}{q^2} \gamma_\mu \frac{i\gamma(p-q) - m}{(p-q)^2 + m^2} \gamma_\mu \\ &= -2e^2 \int \frac{d^4 q}{q^2} \frac{i\gamma(p-q) + 2m}{(p-q)^2 + m^2} \end{aligned} \tag{11.129}$$

vhere we have used the relations (A.5, Appendixes, p. 701)

$$\gamma_\mu \gamma_\mu = 4, \qquad \gamma_\mu \gamma A \gamma_\mu = -2\gamma A.$$

The magnitudes of A and B can be picked out from this relation and we find

$$\delta m = 2e^2 \int \frac{d^4 q}{q^2} \frac{m}{(p-q)^2 + m^2} = 2e^2 m \int \frac{d^4 q}{q^4} \quad q \to \infty \tag{11.130}$$

vhere the factor $i\gamma q$ has disappeared by symmetry; for large q this integral bviously diverges. The existence of this problem in various forms has been nown for thirty years; one way of treating it was suggested by Feynman (1949). A relativistic convergence factor can be introduced into (11.130)

$$C(q)^2 = \frac{\Lambda^2}{q^2 + \Lambda^2} = \frac{1}{1 + \dfrac{q^2}{\Lambda^2}} \tag{11.131}$$

hen

$$\begin{aligned} \frac{1}{q^2} \frac{\Lambda^2}{q^2 + \Lambda^2} &= \frac{1}{q^2} - \frac{1}{q^2 + \Lambda^2} \\ &\sim \frac{1}{q^2} \qquad q^2 \ll \Lambda^2 \\ &\sim 0 \qquad q^2 \gg \Lambda^2 \end{aligned} \tag{11.132}$$

vhere the assumption is made that $\Lambda \gg m$. The presence of the term $1/(q^2 + \Lambda^2)$ uggests the presence of a vector boson of mass Λ (see, however, the remark at he end of this section); $C(q^2)$ is sometimes called the *regulator*, and Λ the *regu-*

lator mass. The evaluation of (11.130) is not simple even with the regulator present, however it is apparent that the lower and upper limits will look like e^2ma and $e^2mb \log \Lambda/m$ respectively where $a \sim b \sim 1$. An exact computation (see, for example, Feynman, 1961) yields

$$\frac{\delta m}{m} = \frac{\alpha}{2\pi}\left(3 \log \frac{\Lambda}{m} + \frac{3}{4}\right). \tag{11.133}$$

The procedure outlined above is of little significance for free particles. Consider the diagram (Fig. 11.19); the symbols $\bar{u}$ and u are the free particle spinors

FIG. 11.19.

and the shaded region the radiative effects. We can use (11.124) to represent this diagram as

$$\begin{aligned}\bar{u}\frac{1}{i\gamma p + m - C}u &= \bar{u}\frac{1}{i\gamma p + m_{\text{ex}} - \delta m - C}u \\ &= \bar{u}\frac{1}{i\gamma p + m_{\text{ex}}}\left[1 + \frac{\delta m}{i\gamma p + m_{\text{ex}}} + \frac{i\gamma pA - B}{i\gamma p + m_{\text{ex}}}\right]u = \bar{u}\frac{1}{i\gamma p + m_{\text{ex}}}u\end{aligned} \tag{11.134}$$

where we have used (11.122) and the spinor equation (3.146) $i\gamma pu = -mu$. When the electron is not free, however, the self-energy and δm terms do not cancel exactly. It is then necessary to adopt the following rules:

(1) insert a regulator $\Lambda^2/(q^2 + \Lambda^2)$ for each propagator $1/q^2$,
(2) substitute $m_{\text{ex}} - \delta m$ for each m, where m_{ex} is the particle's measured mass,
(3) expand δm if necessary, certain terms involving $\log \Lambda/m$ then cancel, and in the remainder let $\Lambda \to \infty$.

It should be noted that the programme of mass renormalisation has led to a matrix element which is no longer unitary. The function C (11.121) contains the factor e^2/q^2; the introduction of the regulator changes this term in the following manner (compare (11.132)):

$$\frac{e^2}{q^2} \to \frac{e^2}{q^2} - \frac{e^2}{q^2 + \Lambda^2}. \tag{11.135}$$

Thus the vector boson Λ is coupled by a term ie and the Lagrangian is no longer Hermitian. The modified propagator can be shown to violate conservation of probability to order $(m/\Lambda)^2$.

11.3(d). *Divergences and dispersion relations*

An alternative approach to the problem of divergences in quantum electrodynamics is provided by dispersion relations (Drell and Zachariasen, 1958). These have not produced any new results in quantum electrodynamics; however, they permit calculations of higher approximations in an easier manner than by ordinary field theory (although the detailed treatment is still far from simple). Also concepts like 'bare' and 'dressed' particles may be dispensed with.

Let us assume we examine a scattering problem and the amplitude is of the form $T(t)$, where t represents minus the square of the four-momentum transfer. Then we can write a dispersion relation in the form

$$T(t) = \frac{1}{\pi} \int dt' \frac{\operatorname{Im} T(t')}{t' - t}$$

and if $\operatorname{Im} T \to 0$ as $t' \to \infty$, then no problem exists; if, however, $\operatorname{Im} T$ remains constant or increases slowly, then the integral has a logarithmic divergence at its upper limit; it can be made to converge by making a subtraction (§ 10.2(e)) at some convenient point $T(t_0)$ (which has a known, finite value)

$$T(t_0) = \frac{1}{\pi} \int dt' \frac{\operatorname{Im} T(t')}{t' - t_0}$$

then

$$T(t) - T(t_0) = \frac{t - t_0}{\pi} \int dt' \frac{\operatorname{Im} T(t')}{(t' - t)(t' - t_0)} \qquad (11.136)$$

and the integral behaves like $\int dt'/t'^2$ at large values for t'. Since the divergences which appear in quantum electrodynamics are logarithmic in character, it is apparent that the above procedure may be used to examine the higher order terms in electromagnetic interactions.

11.3(e). *Consequences of the higher terms and the limits of validity of quantum electrodynamics*

The calculation of higher order terms with the aid of the renormalisation programme has shown remarkable agreement between experiment and theory. One of the most important results has been associated with the Lamb shift in the hydrogen atom.

If the energy levels of the hydrogen atom are calculated using the Dirac equation, then no difference is found between the positions of $2\,S_{1/2}$ and $2\,P_{1/2}$ levels (we use nL_j notation). However, experiments have shown a difference in level positions corresponding to a frequency of $1057{\cdot}77 \pm 0{\cdot}10$ Mc/s (for references see Lamb, 1951).

Calculations† of the higher order terms, using renormalisation theory, have shown that the major contributions to the level shift arise from the following effects:

(1) electron self-energy	1011 Mc/s
(2) vacuum polarisation	−27 Mc/s
(3) vertex corrections	68 Mc/s.

It can be seen that these terms account for most of the level shift. Exact calculations of the level shift have been made for hydrogen and other atoms, and a comparison between experiment and theory is made in Table 11.3 (Layzer, 1960).

TABLE 11.3

Atom	H	D	He
Theory	1057·70 ± 0·15	1058·96 ± 0·16	14046·3 ± 3·0
Experiment	1057·77 ± 0·10	1059·00 ± 0·10	14040·2 ± 4·5

The agreement between the experimental and theoretical values for the Lamb shift shown in the table is impressive. Similar striking agreement may be found in other problems, for example the g values for the charged leptons.

In § 3.3(f) the magnetic moment of a Dirac particle was calculated in its most simplest form, and a value $g = 2$ was obtained. The experimental values for the

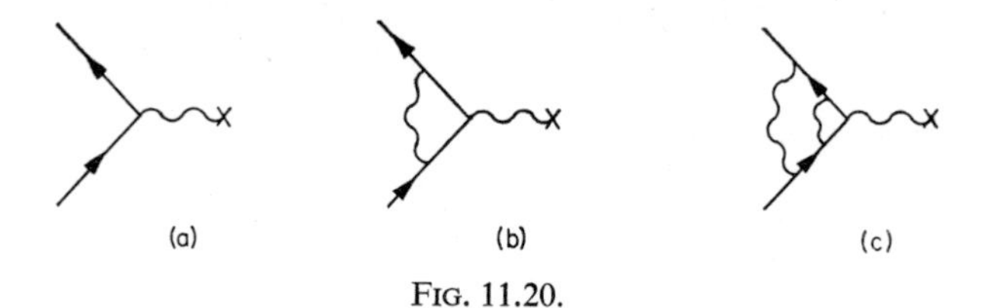

FIG. 11.20.

electron and muon are slightly larger than this figure (§ 2.5(e)). The discrepancy is not surprising since in effect the calculation was made for a process of the type shown in Fig. 11.20(a); more complex processes as shown in diagrams (b) and (c) should also be included and these can be expected to modify the magnetic moment μ to the form

$$\mu = g\,\frac{e}{2m}\,s(1 + A\alpha + B\alpha^2 + \cdots) = \frac{e}{2m}\,(1 + a) \qquad (11.137)$$

† Details of the calculation may be found in Jauch and Rohrlich (1955). An illuminating semiclassical argument (concerning the term associated with the electron self-energy) has been given by Welton (1948).

where we have set

$$s = \tfrac{1}{2}, \quad g = 2(1 + a)$$

$$a = A\alpha + B\alpha^2 + \cdots. \tag{11.138}$$

The coefficient A was first calculated by Schwinger (1948a) and found to be $1/2\pi$. The value for B is different for electrons and muons (see, for example, Petermann, 1958); the difference arises from the contribution associated with the

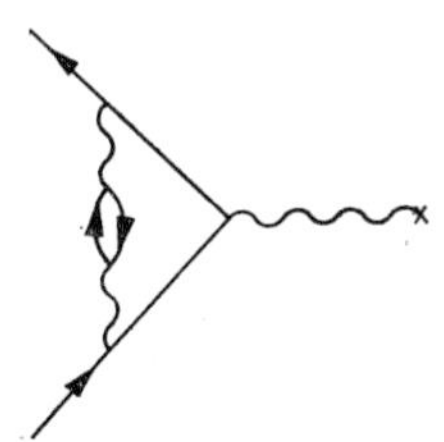

FIG. 11.21.

diagram shown in Fig. 11.21. The theoretical values of a are then as follows (if terms higher than B are neglected):

$$a = \frac{\alpha}{2\pi} + 0{\cdot}75\,\frac{\alpha^2}{\pi^2} = 0{\cdot}0011654 \text{ muons} \tag{11.139}$$

$$= \frac{\alpha}{2\pi} - 0{\cdot}328\,\frac{\alpha^2}{\pi^2} = 0{\cdot}0011596 \text{ electrons.}$$

Experimental values for a were quoted in § 2.5(e); the comparison between experiment and theory shows that

$$a_{\text{exp}} = (0{\cdot}9974 \pm 0{\cdot}0042)\, a_{\text{th}} \quad \text{muons}$$

$$a_{\text{exp}} = (1{\cdot}0011 \pm 0{\cdot}0020)\, a_{\text{th}} \quad \text{electrons.}$$

Thus within the limits of experimental errors no discrepancy has yet appeared between experiment and theory.

The agreement appears all the more remarkable when it is recalled that what is really being tested are the radiative corrections which remain when the divergences have been removed by the process of renormalisation. Furthermore, after the divergent terms have been removed one must assume that the remaining factors in the S-matrix lead to a convergent series in powers of α. It has been shown by Dyson (1952) that this assumption is not necessarily correct.

The question therefore arises as to whether the theory of quantum electrodynamics eventually fails, and as to how to put quantitative limits on this failure. Let us therefore first recall how the theory was constructed. We started from a basic interaction Hamiltonian density $\mathscr{H}_I = -j_\mu(x)\, A_\mu(x)$ (8.51), and from it the S-matrix was constructed as a power series in $\mathscr{H}_I$. The existence of divergences in the higher order processes then led to the introduction of convergence factors and the renormalisation programme.

A convergence factor was introduced for the photon in (11.131); it was of the form

$$C(q^2) = \frac{\Lambda}{q^2 + \Lambda^2}$$

so that the photon propagator $1/q^2$ became

$$\frac{1}{q^2}\,\frac{\Lambda^2}{q^2 + \Lambda^2}.$$

Now the convergence factor can be regarded as a form factor for the photon (Drell, 1958) giving the photon a "size" of $1/\Lambda$ (the Compton wavelength of the regulator mass divided by 2π). It also modifies Coulomb's law from

$$V(r) \propto \frac{1}{r}$$

to

$$V(r) \propto \frac{1}{r}\,(1 - e^{-\Lambda r}) \tag{11.140}$$

(compare remark after (10.153)).

Thus Λ represents a measure for the distance at which quantum electrodynamics breaks down, since the basic interaction no longer takes place at a point signified by $j_\mu(x)A_\mu(x)$ but is measured over a distance Λ. The idea has been extended by Drell to other propagators than the photon.

The most suitable places for estimating Λ are in processes which involve photons and leptons only, since we lack detailed knowledge about the pion clouds which surround strongly interacting particles. The form of (11.131) also makes it plain that in order to detect Λ the values of q^2 should be large since

$$\frac{\Lambda^2}{q^2 + \Lambda^2} \to 1 \quad \text{as} \quad q^2 \to 0.$$

Let us consider, for example, Møller scattering (Fig. 11.22). The invariant four-momentum transfer q is given by

$$\begin{aligned} q^2 &= (p_1 - p_1')^2 \\ &= (\mathbf{p}_1 - \mathbf{p}_1')^2 - (E_1 - E_1')^2 \\ &= 2p_c^2(1 - \cos\theta) \leqq 4p_c^2. \end{aligned} \tag{11.141}$$

Now if an electron with energy E_L in the laboratory system strikes a stationary electron, the total energy in the c-system is $\sim \sqrt{(2m_e E_L)}$. Thus if $E_L = 1$ GeV

$$p_c \sim \tfrac{1}{2}\sqrt{(2mE_L)} \sim 15 \text{ MeV}$$

and $q \sim 30$ MeV. Thus an experiment of this type would probe a distance o $\sim 1/30 \text{ MeV}^{-1} \sim 6\times 10^{-13}$ cm.

If, however, we could have two electron beams each of 500 MeV, and they could be made to collide head-on, then the total energy in the c-system would be 1 GeV. Hence we would find $q \sim 1$ GeV, and much shorter distances could be probed. Considerations of this type provide the motivation for the construction of electron storage rings for use with electron synchrotrons.

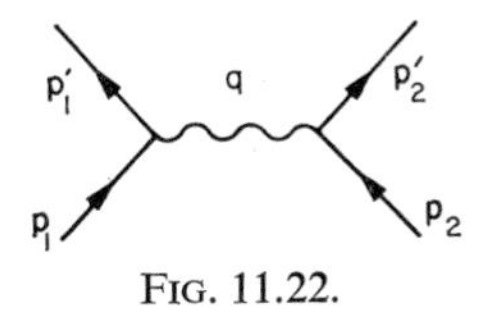

FIG. 11.22.

It should be noted that experiments of this nature cannot examine the photon form factor in isolation, since the electron may also have a shape. We shall show in § 11.4(f) that the form factor for a charge can be written as

$$G_E(q^2) \sim (1 - \tfrac{1}{6}q^2a^2) \sim (1 + \tfrac{1}{6}q^2a^2)^{-1}$$

where a^2 may be regarded as a charge distribution parameter for the electron. We can also write the photon convergence factor in the form

$$C(q^2) = \frac{\Lambda^2}{q^2 + \Lambda^2} = \left(1 + \frac{q^2}{\Lambda^2}\right)^{-1}. \tag{11.142}$$

Now each photon propagator is connected with two vertices, thus the propagator $1/q^2$ in the Møller scattering amplitude is modified as follows:

$$\frac{1}{q^2} \to \frac{1}{q^2}C(q^2)\,G_E^2 \to \frac{1}{q^2}\left(1 + \frac{q^2}{\Lambda^2}\right)^{-1}(1 + \tfrac{1}{3}q^2a^2)^{-1}$$

$$\sim \frac{1}{q^2}\left(1 + \frac{q^2}{\Lambda_{e\gamma}^2}\right)^{-1} \tag{11.143}$$

where

$$\frac{1}{\Lambda_{e\gamma}^2} = \frac{1}{\Lambda^2} + \frac{1}{3}a^2.$$

The measurement of the anomalous magnetic moment of the muon (see § 2.5(e)) also provides a sensitive test of the limit for Λ. The diagrams associated with the anomalous moment are complicated, and so the cut-off can be inferred in a number of ways. For example the anomalous moment can be expressed as a dispersion integral (Drell and Zachariasen, 1958), and the integral is cut off at an upper limit Λ instead of at infinity (Berestetskii, 1961). Another method is to cut off the photon propagator at the regulator mass (Berestetskii, Krokhin and Klebnikov, 1956); this causes the factor $\alpha/2\pi$ in (11.139) to change as follows:†

$$\frac{\alpha}{2\pi} \to \frac{\alpha}{2\pi}\left(1 - \frac{2}{3}\frac{m_\mu^2}{\Lambda^2}\right).$$

All the cut-off procedures yield essentially the same result–that the comparison of experimental and theoretical values for the magnetic moment of the muon implies that quantum electrodynamics holds at least down to distances $\sim 10^{-14}$ cm. This result is equivalent to a propagator mass greater than 1 GeV.

11.4. THE ELECTROMAGNETIC STRUCTURE OF THE NUCLEON

11.4(a). *Introduction*

The production of beams of electrons of high energy in recent years has enabled a study of the electromagnetic structure of the nucleon to be made. Since the nucleon is a strongly acting particle one might guess that its basic structure consisted of a core of size $1/M \sim 2\times 10^{-14}$ cm (M is the nucleon mass) surrounded by a pion cloud extending over a distance $1/m \sim 10^{-13}$ cm. Now an electron with an energy of 1 GeV has a reduced wavelength of $\sim 2\times 10^{-14}$ cm, and so it can be used as a sensitive probe for examining the electromagnetic structure of the nucleon, since it has no strong interactions and its weak interactions are of negligible strength.

11.4(b). *Form factors*

In § 3.3(f) the interaction of a point particle with an electromagnetic field was examined. We assumed that the particle could be described by the Dirac equation, and that it had a point charge e and mass m. The equation for the interaction was shown to lead in the nonrelativistic limit to a magnetic moment (3.124)

$$\boldsymbol{\mu} = \frac{e}{2m}\boldsymbol{\sigma}.$$

As we have seen in Table 3.2 (p. 65), this relation is moderately satisfactory for electrons and muons, but totally incorrect for baryons. In order to take

† It can be seen from this relation that the measurement of the anomalous magnetic moment for the muon provides a far more sensitive test of electromagnetic breakdowns than is possible with the electron.

account of the anomalous part of the magnetic moment, the basic equation for the interaction (3.119)

$$\gamma_\mu \frac{\partial}{\partial x_\mu} \psi + m\psi - ie\gamma_\mu A_\mu \psi = 0 \tag{11.144}$$

was modified by Pauli (1941) to the following form:

$$\gamma_\mu \frac{\partial}{\partial x_\mu} \psi + m\psi - ie\gamma_\mu A_\mu \psi - i\mu_a \gamma_\mu \gamma_\nu \left(\frac{\partial A_\mu}{\partial x_\nu} - \frac{\partial A_\nu}{\partial x_\mu} \right)$$

$$= \gamma_\mu \frac{\partial}{\partial x_\mu} \psi + m\psi - ie\gamma_\mu A_\mu \psi - i\mu_a \sigma_{\mu\nu} q_\nu A_\mu = 0 \tag{11.145}$$

where μ_a represents the anomalous part of the magnetic moment and we have written $-(i/2)(\gamma_\mu \gamma_\nu - \gamma_\nu \gamma_\mu) = \sigma_{\mu\nu}$ (3.110). In the nonrelativistic limit this equation can be written as

$$i \frac{\partial \psi}{\partial t} = \left[m + \frac{1}{2m} \left(\mathbf{p} - e\mathbf{A} \right)^2 + e\varphi + \frac{\mu_a}{2m} \Box^2 \varphi - \left(\mu_a + \frac{e}{2m} \right) \hat{\boldsymbol{\sigma}} \cdot \mathbf{H} \right] \tag{11.146}$$

where φ is the scalar potential $\varphi = iA_4$ and $\mathbf{H}$ is the magnetic field (compare (3.123)).

The term in square brackets in the above expression can be regarded as the total Hamiltonian for the system. The term $e\varphi$ represents the electric interaction

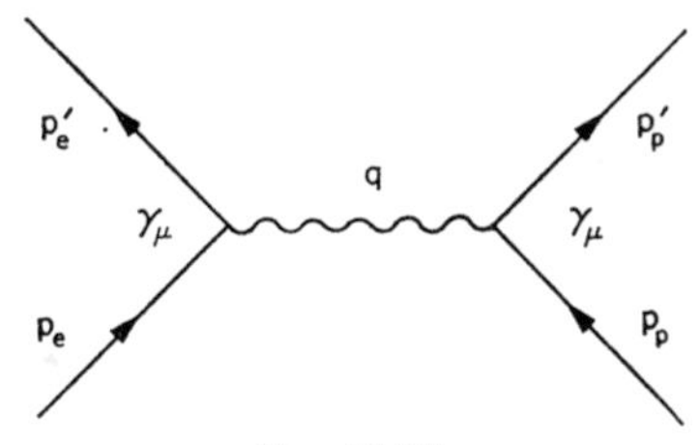

FIG. 11.23.

energy and the final term describes the magnetic interaction. The most interesting feature of (11.146) is that an uncharged particle with anomalous magnetic moment can interact electrically. This occurs through the term $(\mu_a/2m)\,\Box^2 \varphi$, since (see equation (4.150))

$$\Box^2 \varphi = -\varrho$$

where ϱ is the external charge density. Because of the occurrence of this term a neutron can interact with a charged particle both electrically and magnetically.

Now let us examine electron–proton scattering. If we considered the interaction as one between point particles then we could start with the Hamiltonian density operator (8.51)

$$\mathscr{H}_I = -j_\mu(x)\, A_\mu(x) = -ie\bar{\psi}(x)\, \gamma_\mu \psi(x)\, A_\mu(x)$$

and obtain a matrix element in lowest order perturbation theory which corresponds to Fig. 11.23. The matrix element can be written down immediately from the equivalent diagram for Møller scattering (Fig. 8.22 and equations (11.46) and (11.49))

$$M_{fi} = \frac{e^2}{V^2}\left(\frac{m_e^2 m_p^2}{E_e' E_p' E_e E_p}\right)^{\frac{1}{2}} T_{e'p'ep} \tag{11.147}$$

$$T_{e'p'ep} = \bar{u}_{e'}\gamma_\mu u_e \frac{1}{q^2} \bar{u}_{p'}\gamma_\mu u_p \tag{11.148}$$

$$q = p_e - p_e' = p_p' - p_p \tag{11.149}$$

where we have allowed for the difference in sign of charge of the electron and proton. Only the first spinor function in (11.49) appears in (11.147) because the electron and proton are different particles. We now wish to modify this equation to take into account:

(1) the nonlocality of the interaction because of the nucleon structure,

(2) the anomalous part of the magnetic moment as in (11.145).†

The first modification may be introduced by rewriting the Hamiltonian as $j_\mu(x)\, G(x - x')\, A_\mu(x')$, where $G(x - x')$ is a *shape* or *form* factor describing the

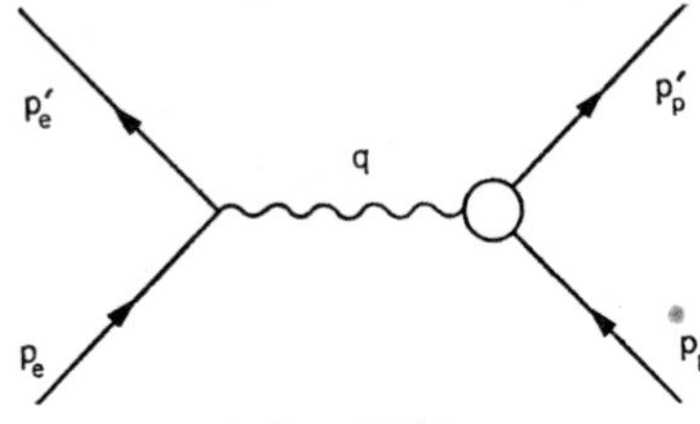

FIG. 11.24.

distribution of charge over the proton. In practice it is more convenient to work with invariant momentum transfers rather than spatial functions, and so we introduce a Fourier transform of $G(x - x')$

$$G(q^2) = \int d^4y e^{-iqy}\, G(y) \tag{11.150}$$

where $y = (x - x')$. Thus $G(q^2)$ represents a form factor in momentum space, and we represent its presence by modifying the proton vertex from a point to a blob (Fig. 11.24).

† It should be noted that the categories (1) and (2) are not mutually exclusive; the division has been made for later convenience.

The second modification may be made by adding on a form factor for the anomalous magnetic moment to the interaction. Its manner of introduction becomes obvious if equations (11.144) and (11.145) are considered in conjunction with the rules for graphs given in § 8.6(b).

We therefore modify (11.148) to the following form:

$$T_{e'p'ep} = \frac{1}{e}\,\bar{u}_{e'}\gamma_\mu u_e\,\frac{1}{q^2}\,\bar{u}_{p'}[\gamma_\mu G_1(q^2) + \sigma_{\mu\nu}q_\nu G_2(q^2)]\,u_p \tag{11.151}$$

where the factor $1/e$ has been introduced for later convenience.

11.4(c). *Invariance properties and form factors*

Equation (11.151) could have been obtained by considering the most general form of matrix element permitted by the requirements of Lorentz and gauge invariance. We shall repeat the derivation of this equation as we wish to use the same techniques in other situations.

The term $\bar{u}_{p'}\gamma_\mu u_p$ in (11.148) has the transformation properties of a four-vector. This statement is perhaps obvious because of the presence of the single γ_μ matrix (compare § 3.3(p)). It should also be recalled that $\bar{u}_{p'}\gamma_\gamma u_p$ was constructed from the proton current four-vector.

We shall therefore assume that the most general form for the proton current is given by

$$\bar{u}_{p'}O_\mu u_p \tag{11.152}$$

where O_μ is any expression (or group of expressions) which has the transformation properties of a four-vector and is also a 4×4 matrix in the spin space of the proton. It is a function of the four-vectors p_p and p_p' (Fig. 11.24) and we can construct from them a scalar variable $q^2 = (p_p' - p_p)^2$. Rather than use p_p and p_p' separately, we shall use them in the linear combinations

$$q = p_p' - p_p, \qquad P = p_p' + p_p. \tag{11.153}$$

These variables must be combined with the γ-matrices in order to give O_μ the form of a 4×4 matrix in spinor space. From the combinations (§ 3.3 (p))

$$\Gamma_i = 1,\quad \gamma_\mu,\quad \sigma_{\mu\nu},\quad i\gamma_\mu\gamma_5,\quad \gamma_5$$

the last two can be rejected because O_μ is not a pseudovector. Thus the most general form for O_μ is given by

$$O_\mu = a(q^2)\,q_\mu + b(q^2)\,P_\mu + c(q^2)\,\gamma_\mu + d(q^2)\,\sigma_{\mu\nu}q_\nu + e(q^2)\,\sigma_{\mu\nu}P_\nu \tag{11.154}$$

where a, b, c, d and e are scalar functions of q^2. If equations (3.146) for free spinors are exploited, it is not difficult to show that

$$\bar{u}_{p'}\sigma_{\mu\nu}q_\nu u_p = -i\bar{u}_{p'}P_\mu u_p - 2m_p\bar{u}_{p'}\gamma_\mu u_p \tag{11.155}$$

$$\bar{u}_{p'}\sigma_{\mu\nu}P_\nu u_p = -i\bar{u}_{p'}q_\mu u_p.$$

Hence O_μ can be reduced to

$$O_\mu = a'(q^2)\, q_\mu + c'(q^2)\, \gamma_\mu + d'(q^2)\, \sigma_{\mu\nu}q_\nu \tag{11.156}$$

where a', c' and d' represent appropriate linear combinations of a, b, c, d and e.

Equation (11.156) can be reduced still further by considering gauge invariance. We have shown in § 3.4(f.2) that gauge invariance leads to the continuity equation (3.242)

$$\frac{\partial j_\mu}{\partial x_\mu} = 0.$$

However, if we work in momentum space we may change $j_\mu(x)$ by a Fourier transform into

$$j_\mu(x) = \int d^4q e^{iqx} j_\mu(q^2)$$

and so the continuity equation gives us

$$\frac{\partial j_\mu(x)}{\partial x_\mu} = i\int d^4q e^{iqx}\, q_\mu j_\mu(q^2) = 0$$

$$q_\mu j_\mu(q^2) = 0. \tag{11.157}$$

This result implies that we can also write

$$\bar{u}_{p'}q_\mu O_\mu u_p = 0 \tag{11.158}$$

hence

$$\bar{u}_p[a'(q^2)\, q_\mu^2 + c'(q^2)\, \gamma_\mu q_\mu + d'(q^2)\, \sigma_{\mu\nu}q_\nu q_\mu]\, u_p = 0.$$

The last term is zero, since $\sigma_{\mu\nu} = -\sigma_{\nu\mu}$, and the second can be written as

$$\bar{u}_{p'}\gamma_\mu q_\mu u_p = \bar{u}_{p'}\gamma_\mu(p'_p - p_p)_\mu\, u_p$$

$$= i(m_p - m_p)\, \bar{u}_{p'}u_p = 0. \tag{11.159}$$

Gauge invariance therefore requires that

$$a'(q^2)\, q^2 = 0.$$

Since the virtual photon in Fig. 11.24 is off the mass shell, $q^2 \neq 0$ and so $a'(q^2) = 0$. Thus we have reduced (11.156) to

$$\bar{u}_{p'}O_\mu u_p = \bar{u}_{p'}[c'(q^2)\, \gamma_\mu + d'(q^2)\, \sigma_{\mu\nu}q_\nu]\, u_p \tag{11.160}$$

and if this equation is compared to the function between proton spinors in (11.151) it is apparent that

$$c'(q^2) \equiv G_1(q^2), \qquad d'(q^2) \equiv G_2(q^2)$$

and so (11.160) becomes

$$\bar{u}_{p'} O_\mu u_p = \bar{u}_{p'}[\gamma_\mu G_1(q^2) + \sigma_{\mu\nu} q_\nu G_2(q^2)]\, u_p. \tag{11.161}$$

It is convenient to have the form factors in dimensionless form for later work. Now in the limit $q^2 \to 0$, it is apparent from our discussion in the previous section that

$$G_1(q^2) = e, \quad G_2(q^2) = \mu_a = 1{\cdot}79\,\frac{e}{2m_p}, \quad q^2 \to 0. \tag{11.162}$$

We will therefore write

$$G_1(q^2) = eF_1(q^2), \quad G_2(q^2) = \frac{e}{2m_p}\,F_2(q^2) \tag{11.163}$$

$$F_1(0) = 1, \qquad F_2(0) = 1{\cdot}79$$

and equation (11.161) becomes

$$\bar{u}_{p'} O_\mu u_p = e\bar{u}_{p'}\left[\gamma_\mu F_1(q^2) + \sigma_{\mu\nu} q_\nu \frac{1}{2m_p}\,F_2(q^2)\right] u_p. \tag{11.164}$$

This equation may be inserted in the scattering matrix (11.147) with the aid of (11.151)

$$M_{fi} = \frac{e^2}{V^2}\left(\frac{m_e^2 m_p^2}{E_e' E_p' E_e E_p}\right)^{\frac{1}{2}} T_{e'p'ep}$$

$$= \frac{e^2}{V}\left(\frac{m_e^2 m_p^2}{E_e' E_p' E_e E_p}\right)^{\frac{1}{2}} \bar{u}_{e'}\gamma_\mu u\,\frac{1}{q^2}\,\bar{u}_{p'}\left[\gamma_\mu F_1(q^2) + \sigma_{\mu\nu} q_\nu \frac{1}{2m_p}\,F_2(q^2)\right] u_p. \tag{11.165}$$

This equation is sometimes written as an interaction between two 'currents'; the proton current between the states p and p' is given by

$$\langle p'|\, j_\mu \,|p\rangle = i\bar{u}_{p'} O_\mu u_p$$

$$= ie\bar{u}_{p'}\left[\gamma_\mu F_1(q^2) + \sigma_{\mu\nu} q_\nu \frac{1}{2m_p}\,F_2(q^2)\right] u_p \tag{11.166}$$

so that

$$M_{fi} = -\frac{ie}{V^2}\left(\frac{m_e^2 m_p^2}{E_e' E_p' E_e E_p}\right)^{\frac{1}{2}} \bar{u}_{e'}\gamma_\mu u_e\,\frac{1}{q^2}\,\langle p'|\, j_\mu \,|p\rangle. \tag{11.167}$$

The cross-section resulting from the amplitude (11.165) can be evaluated by the methods of § 11.2(a). This was first done by Rosenbluth† (1950) who obtained the following formula (the *Rosenbluth formula*) for the scattering cross-section in the laboratory system:

$$\frac{d\sigma}{d\Omega} = \sigma_M(\theta)\left\{F_1^2 + \frac{q^2}{4m_p^2}\left[2(F_1 + F_2)^2 \tan^2 \tfrac{1}{2}\theta + F_2^2\right]\right\} \tag{11.168}$$

where $\sigma_M(\theta)$ represents the Mott scattering cross-section

$$\sigma_M(\theta) = \left(\frac{e^2}{4\pi}\right)^2 \frac{1}{4E_e^2} \frac{\cos^2 \frac{1}{2}\theta}{\sin^4 \frac{1}{2}\theta} \frac{1}{1 + (2E_e/m_p)\sin^2 \frac{1}{2}\theta}. \tag{11.169}$$

It is the cross-section which would be obtained for Møller or Bhabha scattering, if exchange and annihilation diagrams were ignored and the appropriate kinematic conditions were inserted. The term q^2 is given by

$$q^2 = (p_e - p_e')^2 = \frac{4E_e^2 \sin^2 \frac{1}{2}\theta}{1 + (2E_e/m_p)\sin^2 \frac{1}{2}\theta}. \tag{11.170}$$

In evaluating these expressions the approximation $m_e^2 \to 0$ has been used.

An inspection of equation (11.168) shows that in the limit $q^2 \to 0$, the scattering cross-section reduces to

$$\frac{d\sigma}{d\Omega} = \sigma_M(\theta)\, F_1(0) = \sigma_M(\theta). \tag{11.171}$$

Thus in the limit $q^2 \to 0$ the electron should 'see' only the charge on the proton. The experimental observations of low energy electron scattering from protons and nuclei are in accord with $\sigma_M(\theta)$. A similar situation exists in Compton scattering. Figure 11.5 shows that at low energies the incoming photon 'sees' only the charge on the proton, and so the measured cross-section tends asymptotically to that for classical Thomson scattering as the energy of the photon tends to zero.

11.4(d). *Form factors for the proton and neutron*

Form factors can be defined for the neutron in an analogous manner to (11.166). The neutron has zero charge but has a magnetic moment, and therefore must have some electromagnetic structure. We therefore represent the neutron 'current' by the equation

$$\langle n'|\, j\, |n\rangle = ie\bar{u}_{n'}\left[\gamma_\mu F_{1n}(q^2) + \sigma_{\mu\nu} q_\nu \frac{1}{2m_n} F_{2n}(q^2)\right] u_n \tag{11.172}$$

† The original Rosenbluth formula was calculated for point Dirac particles with anomalous magnetic moments.

where $q = p_n' - p_n$. The static limits of F_1 and F_2 are given by

$$F_{1n}(0) = 0 \qquad F_{2n}(0) = -1{\cdot}91. \tag{11.173}$$

(The neutron magnetic moment is $-1{\cdot}91$ magnetons.)

In our discussion later in this section it will be more convenient to use isospin notation rather than refer to scattering from a neutron or proton. The expressions (11.166) and (11.172) for the proton and neutron currents can then be combined in the following manner:

$$\langle N'|j_\mu|N\rangle = ie\bar{u}_{N'}\left\{\gamma_\mu\left[F_1^S(q^2) + \tau_3 F_1^V(q^2)\right] + \sigma_{\mu\nu}q_\nu \frac{1}{2M}\left[F_2^S(q^2) + \tau_3 F_2^V(q^2)\right]\right\}u_N \tag{11.174}$$

where M is the nucleon mass, $\bar{u}_{N'}$ and u_N represent the nucleon spinors for the final and initial states respectively, and (9.102)

$$\tau_3 u_N = \tau_3\begin{pmatrix} u_p \\ u_n\end{pmatrix} = \begin{pmatrix} u_p \\ -u_n\end{pmatrix}.$$

The superscripts define isotopic scalar (S) and vector (V) components, for example F_2^V represents the isotopic vector component of the magnetic form factor. The isotopic form factors are therefore related to those for protons and neutrons as follows

$$F_{1p} = F_1^S + F_1^V, \qquad F_{1n} = F_1^S - F_1^V \tag{11.175}$$

$$F_{2p} = F_2^S + F_2^V, \qquad F_{2n} = F_2^S - F_2^V.$$

It is apparent from (11.163) and (11.173) that the isotopic components assume the following form at $q = 0$

$$F_1^S(0) = F_1^V(0) = \tfrac{1}{2} \tag{11.176}$$

$$F_2^S(0) = \frac{F_{2p}^{(0)} + F_{2n}^{(0)}}{2} = -0{\cdot}06$$

$$F_2^V(0) = \frac{F_{2p}^{(0)} - F_{2n}^{(0)}}{2} = 1{\cdot}85$$

11.4(e). *The evaluation of the form factors from the experimental data*

Considerable experimental work has been done on the scattering of electrons from protons and deuterons. The pioneering work was performed at Stanford under Hofstadter. At the present time much work is also done at Cambridge, Cornell and Orsay.

The observed scattering cross-sections deviate enormously from those expected if the nucleons were point particles. This fact is illustrated in Fig. 11.25 which has been taken from the paper of Olson, Schopper and Wilson (1961). The experimental data represent the scattering of electrons from protons at a fixed angle, and the curve labelled point cross-section represents the Rosenbluth formula with $F_{1p} = 1$, $F_{2p} = 1{\cdot}79$ for all values of q^2 (compare (11.163)).

The evaluation of the form factors from the experimental cross-sections is fairly straightforward for protons. The measured cross-sections first have to be subjected to a radiative correction ($\sim$ 15 per cent) to allow for the loss of electrons by Bremsstrahlung processes (Schwinger, 1949; Tsai, 1961; Krass, 1962). The data is then ready for analysis. The case of the neutron is not as simple, since free neutrons are not available as targets. Deuterons therefore have to be used. A typical measurement on a deuteron target is shown in Fig. 11.26, together

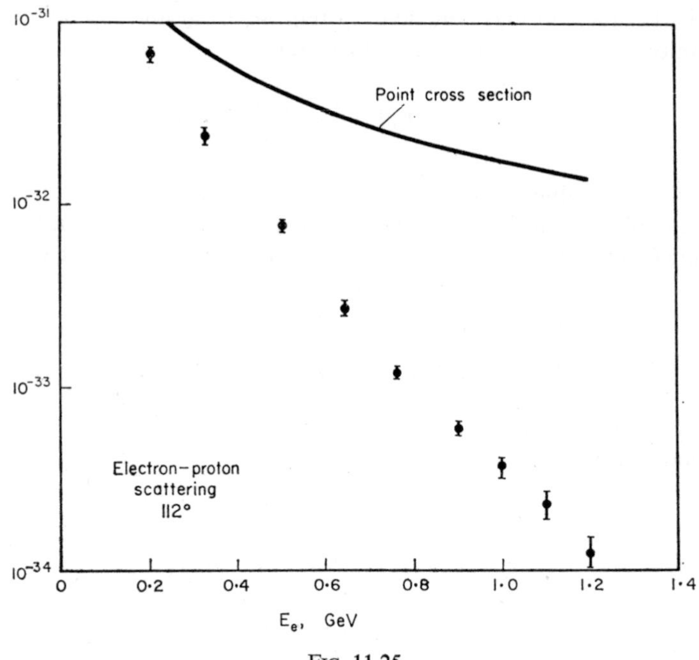

FIG. 11.25.

with the data from a hydrogen run superposed (Hofstadter, de Vries and Herman, 1961). The deuteron data represent that from the process

$$e^- + d \rightarrow n + p + e^-.$$

In addition, an elastic scattering peak for $e^- + d \rightarrow e^- + d$ occurs at a higher energy. The scattering from the deuteron can be seen to correspond to the process of quasi-elastic scattering from a single nucleon with internal motion in the deuteron nucleus. If the problem is treated in this manner, then to a fair approximation the neutron scattering cross-section is given by

$$\frac{d\sigma_n}{d\Omega_{e'}} = \int dE_{e'} \frac{d^2\sigma_d}{dE_{e'}d\Omega_{e'}} - \frac{d\sigma_p}{d\Omega_{e'}}. \tag{11.177}$$

The first term on the left-hand side can be evaluated either by direct measurement, or by measuring the cross-section at its peak and then employing a correc-

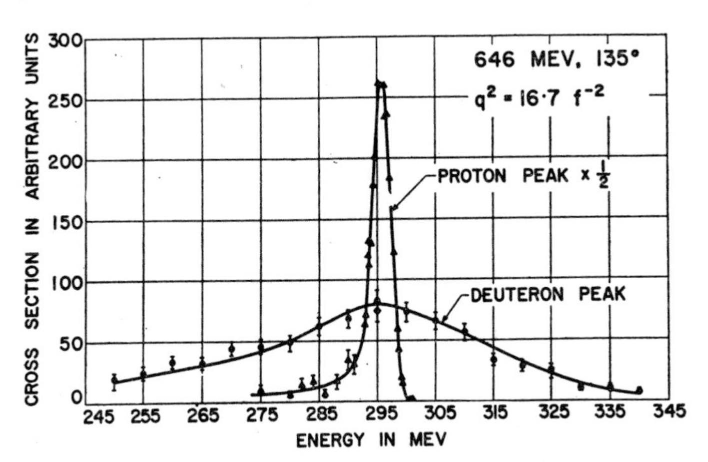

FIG. 11.26.

tion factor which is based on the impulse approximation (Goldberg, 1958; Durand 1961).

The experimental data can be analysed in many ways. The data can be used to give F_1 and F_2 directly, or it can be used to give linear combinations of F_1 and F_2 with somewhat different physical interpretations†. The latter terms were suggested by Ernst, Sachs and Wali (1960; see also Sachs, 1962; Barnes, 1962), and are given by

$$G_M(q^2) = F_1(q^2) + F_2(q^2) \tag{11.178}$$

$$G_E(q^2) = F_1(q^2) - \frac{q^2}{4M^2} F_2(q^2).$$

It is apparent that in the limit $q^2 \to 0$

$$G_M(0) = \mu$$

$$G_E(0) = \begin{cases} 1 \text{ protons} \\ 0 \text{ neutrons} \end{cases}$$

† If the form factors F_1 and F_2 are used, then cross terms F_1F_2 appear in the cross-section and the analysis of data is not quite as straightforward.

where μ represents the magnetic moment (in magnetons). G_M and G_E are sometimes called the magnetic and electric form factors respectively. Scalar and vector form factors may be constructed from these terms as in (11.175). If the nucleon current (11.166) is reconstructed in terms of the form factors G_M and G_E,

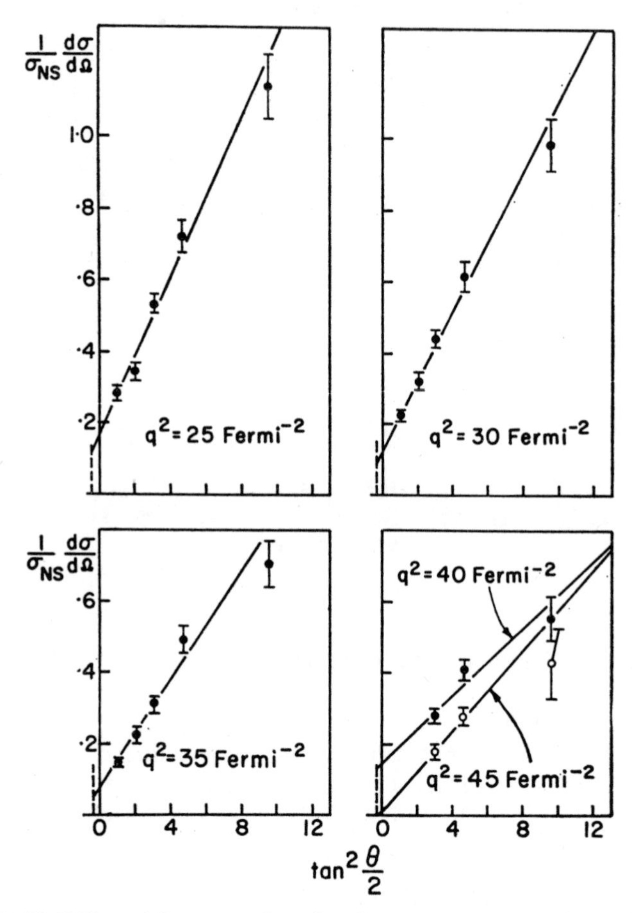

FIG. 11.27. Differential cross-sections for electron-proton scattering at fixed momentum transfers (Berkelman, *et al.*, 1963). The term σ_{NS} is the same as σ_M in the text.

then one finds that it has been split into a spin independent term involving G_E, and a spin dependent term involving G_M.

If equations (11.178) are inserted in (11.168) the expansion for the cross-section is modified to

$$\frac{d\sigma}{d\Omega} = \sigma_M(\theta)\left[\frac{1}{1+(q^2/4M^2)}\left(G_E^2 + \frac{q^2}{4M^2}G_M^2\right) + \frac{2q^2}{4M^2}G_M^2 \tan^2 \tfrac{1}{2}\theta\right]. \tag{11.179}$$

Thus the form factors for a fixed value of q^2 may be obtained by plotting $(1/\sigma_M)(d\sigma/d\Omega)$ against $\tan^2 \frac{1}{2}\theta$. It is apparent from the above equation that straight lines should be obtained; a deviation from a straight line would indicate failure of the Rosenbluth formula. In Fig. 11.27 we display data for electron-proton scattering from an experiment performed at Cornell (Berkelman *et al.*, 1963). It can be seen that the results are consistent with the Rosenbluth formula (if in fact deviations do occur they are expected in a region which is not easily accessible by experiment – Berkelman *et al.*, 1963). Equation (11.179) then shows

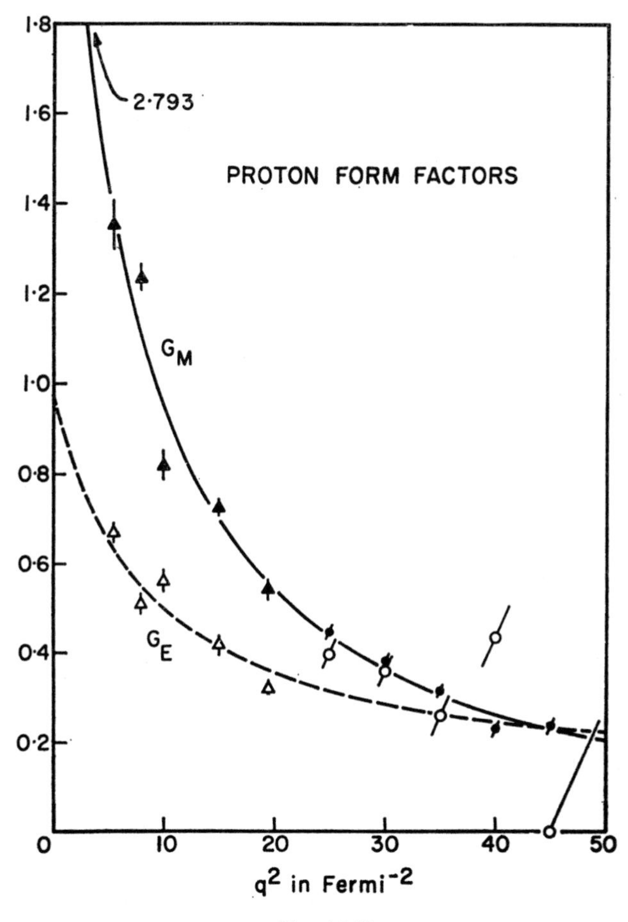

FIG. 11.28.

us that the slopes of the lines in Fig. 11.27 gives G_M^2 whilst the intercept at $\tan^2 \frac{1}{2}\theta = 0$ leads to G_E^2. Similar analyses may be made for the neutron data. With one exception (G_E for the neutron) the signs of the terms may be obtained by recalling that $G_E(0)$ and $G_M(0)$ have known values, and that $G(q^2)$ is expected to be a smooth function of q^2. Values of $G_M(q^2)$ and $G_E(q^2)$ for the data displayed in Fig. 11.27 are given in Fig. 11.28. At low momentum transfers ($q^2 < 20\,f^{-2}$) the data indicates that

$$G_E \sim \frac{G_M}{\mu}$$

for protons. Analysis of the neutron data (see Hand *et al.*, 1963 for detailed references) also indicates that $G_{Mp} \sim G_{Mn}\mu_p/\mu_n$. The term G_{En} is small and some uncertainty exists regarding its sign.

11.4(f). *Interpretation of the form factors*

The experimental data can be analysed to yield an electromagnetic 'shape' for the nucleon although the physical meaning of the result is ambiguous especially for high momentum transfers (Wilson, 1961).

If we work in the c-system we can write $q = \mathbf{q}$ since

$$q = (\mathbf{p}_e - \mathbf{p}_e'), \quad i(E_e - E_e')$$

and $E_e = E_e'$; we may then modify (11.150) to define spatial charge and moment densities

$$G_i(\mathbf{q}^2) = \int d\mathbf{r} e^{-i\mathbf{q}\cdot\mathbf{r}} \varrho_i(r) \qquad (i = E, M) \tag{11.180}$$

where $r = |\mathbf{r}|$. For small values of $\mathbf{q}$ we may expand the exponential in the above equation to yield

$$\begin{aligned} G_i(\mathbf{q}^2) &= \int d\mathbf{r}\varrho_i(r) - \tfrac{1}{2}\int d\mathbf{r}(\mathbf{q}\cdot\mathbf{r})^2 \varrho_i(r) \\ &= G_i(0)\,(1 - \tfrac{1}{6}\,\mathbf{q}^2 a_i^2) \end{aligned} \tag{11.181}$$

where we have used the fact that

$$\int d\mathbf{r} i(\mathbf{q}\cdot\mathbf{r})\,\varrho_i(r) = 0$$

and have introduced the following definitions

$$G_i(0) = \int d\mathbf{r}\varrho_i(r) \tag{11.182}$$

$$a_i^2 = \frac{\int d\mathbf{r} r^2 \varrho_i(r)}{\int d\mathbf{r}\varrho_i(r)} \equiv -6\frac{\partial}{\partial \mathbf{q}^2} G_i(\mathbf{q}^2)\cdot\frac{1}{G_i(0)}. \tag{11.183}$$

It is clear that a_i corresponds to a root mean square radius, but it must be emphasized that this definition has been made in a particular Lorentz frame in which the proton is not stationary and therefore it is not necessarily a physical radius.† At low energies, however, the difference between quantities expressed in the c-system and the proton's coordinate frame are small and the relation

† Ernst, Sachs and Wali (1960) have suggested that shape analyses should be made in the Breit reference frame.

(11.183) has some physical meaning. In addition we can use the covariance property, $q^2 = \mathbf{q}^2$, and so equation (11.181) can be written as

$$G_i(q^2) = G_i(0)\,(1 - \tfrac{1}{6}\,q^2 a_i^2). \tag{11.184}$$

The experimental data may then be used to show that

$$a_{Ep} \sim a_{Mp} \sim a_{Mn} \sim 0{\cdot}9f$$

$$a_{En} \sim 0.$$

These are roughly the radii to be expected if the nucleon core is surrounded by a pion cloud. An extensive discussion of the fitting of form factors to experimental data may be found in an article by Hand *et al.* (1963).

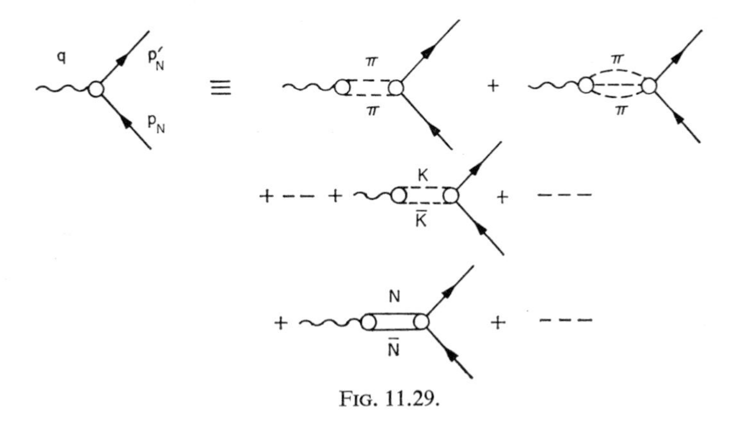

FIG. 11.29.

The main approach to a physical understanding of the form factors has been based upon dispersion relations. In this treatment the form factors are regarded as analytic functions in the plane of complex momentum transfer q^2.

The construction of the dispersion relations may be carried out as in Chapter 10. The details will not be repeated here (they may be found, for example, in the work of Federbush, Goldberger and Treiman, 1958). It will be recalled, however, that the unitarity requirement permits the imaginary part of the scattering amplitude to be determined by a sum over the intermediate states (Fig. 11.29). Furthermore, since the kinematic variables appear in the denominators in the pole terms, the intermediate states with the lightest masses are of the greatest importance. We shall therefore concentrate on the two-pion and three-pion states (conservation of angular momentum excludes a one-pion state).†

† It is evident from Fig. 11.29 that in examining electron–nucleon scattering, we are tackling the wrong problem first. The more logical process would be to examine electron–pion scattering, and then move on to the more complex problem of electron–nucleon scattering. This ideal situation cannot be realised in practice and so we must infer the effect of the interactions given in Fig. 11.29 from the available data in electron–nucleon scattering.

The quantum numbers for the pion states will be first examined. Since the photon has spin 1, the states must possess angular momentum $j = 1$. The two- and three-pion states may then be separated into an isotopic scalar and a vector contribution by the operation of G conjugation (charge conjugation times a rotation by π about the 2-axis in isospin space— see § 9.4(g)). If a state contains n pions then G conjugation yields the result

$$G\,|n\pi\rangle = (-1)^n\,|n\pi\rangle. \tag{11.185}$$

Now let us examine the effect of G upon an electric current; we may split a current operator into an isotopic scalar and a third component of an isotopic vector (§ 12.6(a))

$$j_\mu = j_\mu^S + j_\mu^{(3)}$$

and since j_μ is odd under charge conjugation

$$Gj_\mu^S\,G^{-1} = -j_\mu^S, \qquad Gj_\mu^{(3)}\,G^{-1} = j^{(3)}. \tag{11.186}$$

Now the isotopic scalar current corresponds to $T = 0$, whilst the vector current is associated with $T = 1$. Thus we may conclude

$$\begin{aligned} |3\pi\rangle &\equiv j = 1, \qquad & T &= 0 \\ |2\pi\rangle &\equiv j = 1, \qquad & T &= 1. \end{aligned} \tag{11.187}$$

The analysis of the nucleon form factors by the methods of dispersion relations is usually made with the Mandelstam parameter t replacing $-q^2$ (compare § 10.4(b))

$$t = -(p_e - p_e')^2 = -(p_N' - p_N)^2 = -q^2. \tag{11.188}$$

In the c-system t represents minus the square of the momentum transfer. The form factors can then be expressed as $F(t)$ or $G(t)$. Consider the isovector form factors which are associated with the two-pion contribution (Fig. 11.30); the kinematics of this vertex show that if

$$\begin{aligned} t &\geqq -(k_\pi + k_\pi')^2 \geqq -(-2m_\pi^2 + 2\mathbf{k}_\pi \cdot \mathbf{k}_\pi' - 2\omega_\pi\omega_\pi') \\ &\geqq 4m_\pi^2 \end{aligned} \tag{11.189}$$

then the photon has sufficient energy to create a pion pair. Thus for $t \geqq 4m_\pi^2$ the function $G^V(t)$ must be complex since it will contain an absorptive term. Thus a branch cut occurs for the function $G^V(t)$ in the complex t-plane stretching from $+4m_\pi^2$ to infinity. This conclusion may be contrasted with the experimental range of t for electron–proton scattering, which covers only negative values of t between 0 and $-\infty$ as shown in Fig. 11.30. The contour of integration is therefore as indicated in Fig. 11.30, and if the usual assumptions are made about the vanishing of functions at infinity, then the dispersion relation for the isovector form factors becomes (compare (10.24) with $\varepsilon \to 0$)

$$G_i^V(t) = \frac{1}{\pi} \int_{4m_\pi^2}^{\infty} dt' \, \frac{\text{Im } G_i^V(t')}{t' - t} \qquad i = E, M. \tag{11.190}$$

A similar relation holds for the isoscalar form factor with a lower limit of $9m_\pi^2$ replacing $4m_\pi^2$.

The first calculations for $G(t)$ were made with the assumption that no pion–pion interaction occurred. The results did not agree with experiment. It was then shown by Frazer and Fulco (1959, 1960) that the assumption of a strong

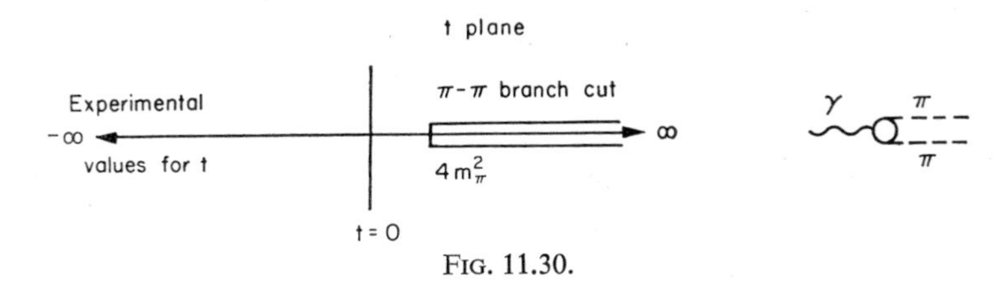

FIG. 11.30.

pion–pion interaction in the intermediate state removed the disagreement. The details of their calculation will not be repeated here, but the main physical argument may be sketched with the aid of Fig. 11.31. The spectral function Im $G(t)$ is essentially a weighting factor. If no pion–pion interaction occurs then Im $G(t)$ is a smoothly varying function of t as shown in Fig. 11.31(a). If, on the other

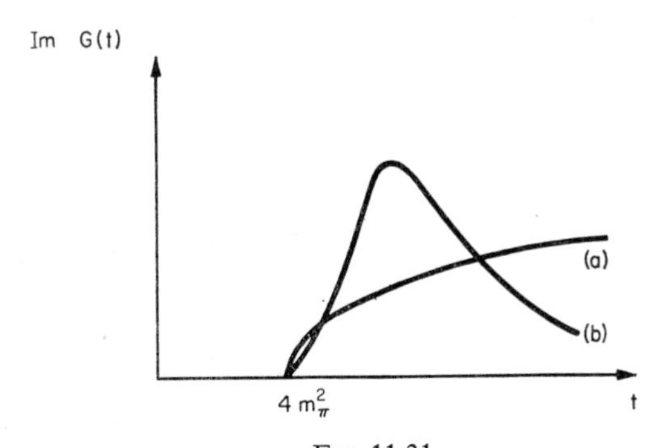

FIG. 11.31.

hand, there is a strong pion–pion interaction then Im $G(t)$ will behave as shown in Fig. 11.31(b). Frazer and Fulco found that the form factors fell too slowly in value for decreasing t unless a strong resonant state existed near the threshold. They estimated the position of the resonance to be at $t \sim 11m_\pi^2$.

The argument of Frazer and Fulco has been extended and simplified by many workers (see, for example, Bergia *et al.*, 1961; Matthews, 1961). Let us assume that the interaction is transmitted with the aid of an isotopic vector' particle' of mass m_V, a scalar particle of mass m_S and more massive complex systems, mass m_D (Fig. 11.32). It should be noted that we are now calling a resonant state a 'particle'. The distinction is not entirely academic, but the problems associated with it can be ignored in most circumstances.

Let us ignore the complication introduced by the contributions from the E, M and S, and V channels for the moment. If we assume that the mass (m_S or m_V) of

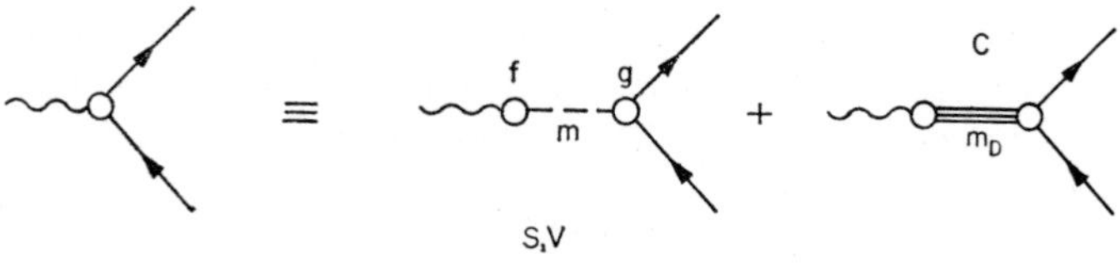

FIG. 11.32.

the particle lies close to the cut, we may expect its pole term to make a significant contribution to the scattering amplitude. This term may be calculated by the methods of § 10.4(e) (see in particular equation (10.150)), and we may therefore write

$$G(t) = \frac{fg}{m^2 - t} + C = \frac{a}{1 - (t/m^2)} + C \tag{11.191}$$

where the symbols f and g represent the vertex functions and C represents the contributions from intermediate states of higher energy and from any substractions which may be necessary. The value of C may be fixed by fitting $G(t)$ at $t = 0$ to the known values of the charge and magnetic moment, for example for the scalar form factor $G_E^S(0) = \frac{1}{2}$ and so

$$G_E^S(0) = \tfrac{1}{2} = a_S + C_S \tag{11.192}$$

$$C_S = \tfrac{1}{2} - a_S \tag{11.193}$$

hence

$$G_E^S(t) = \tfrac{1}{2} - a_S + \frac{a_S}{1 - (t/m_S^2)} \tag{11.194}$$

and similarly

$$G_E^V(t) = \tfrac{1}{2} - a_V + \frac{a_V}{1 - (t/m_V^2)} \tag{11.195}$$

$$G_M^S(t) = 0{\cdot}440 - b_S + \frac{b_S}{1 - (t/m_S^2)}$$

$$G_M^V(t) = 2{\cdot}353 - b_V + \frac{b_V}{1 - (t/m_V^2)}.$$

The experimental data may be analysed in terms of these expressions and yield masses for the scalar and vector particles which vary between 500 and 600 MeV (for references see Bishop, 1962). These figures are somewhat lower than the observed masses for systems with the appropriate quantum numbers, namely 750 and 780 MeV for the ϱ-meson ($T = 1$) and ω-meson ($T = 0$) respectively. The discrepancy could arise from (1) further pion resonances with appropriate quantum numbers (Table A.9.3, p. 717), (2) oversimplification of the theory and (3) experimental errors, especially in the neutron data.

CHAPTER 12

THE WEAK INTERACTIONS

12.1. INTRODUCTION

Most of our knowledge concerning the weak interactions has been obtained from studies of the decay processes of the elementary particles. In Table A.9.4 (p. 718) the lifetimes and decay modes of particles undergoing weak decays are listed. In the table the symbol ν has been used generically for both neutrino and antineutrino. In addition to the processes listed in Table A.9.4, studies of the weak interactions have been made by examining the muon capture reaction $\mu^- + p \rightarrow n + \nu$ and the neutrino induced reactions.

The weak interactions may be conveniently classified depending on whether leptons (ν, e, μ) occur in the individual processes, and whether changes in strangeness are involved. The classifications are given in Table 12.1.

TABLE 12.1

Groups	Subgroups	Examples
Leptonic	pure leptonic	$\mu \rightarrow e + \nu + \nu$
	$\Delta S = 0$	$n \rightarrow p + e^- + \nu$, $\mu^- + p \rightarrow n + \nu$
	$\lvert\Delta S\rvert = 1$	$K \rightarrow \mu + \nu$, $\Lambda \rightarrow p + e^- + \nu$
Nonleptonic	$\lvert\Delta S\rvert = 1$	$K \rightarrow \pi + \pi$, $\Lambda^0 \rightarrow p + \pi^-$, $\Xi^- \rightarrow \Lambda^0 + \pi^-$

Two points of interest emerge from the table. First, the leptons always occur in the pairs ($e\nu$) or ($\mu\nu$), or both as in muon decay; secondly, nonleptonic decays always involve a change in strangeness of one unit.

Only one example of a pure leptonic transition is quoted in Table 12.1; in fact, apart from muon decay, all easily accessible weak interactions involve strongly interacting particles. This occurrence makes exact calculations extremely difficult. Consider, for example, the β-decay of the neutron (Fig. 12.1(a)). In addition to the basic point interaction given in Fig. 12.1(b), higher order effects can occur such as the vertex correction shown in Fig. 12.1(c).

The vertex correction is a strong interaction with all the usual attendant problems. Nevertheless, the simple point interaction given in Fig. 12.1(b) can yield valuable information. We shall therefore start by examining the nuclear β-decay process in this order of approximation. This process has been more intensively

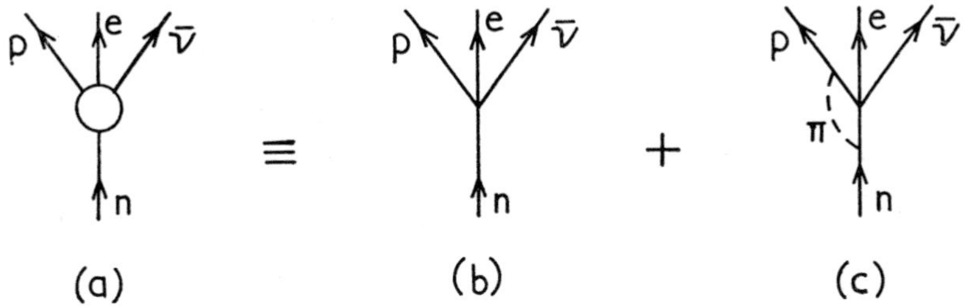

FIG. 12.1.

studied than any other, and a comparison of the experimental and theoretical data has enabled severe restrictions to be placed on the form of the interaction. We shall then use the information obtained from β-decay to examine other processes, and show that a large amount of experimental information can be explained by remarkably few parameters.

12.2. THE β-DECAY OF THE NEUTRON

12.2(a). *Nuclear β-decay – selection rules*

Ideally the β-decay of the neutron should be examined as an isolated process

$$n \rightarrow p + e^- + \bar{\nu}$$

but most of the important experimental information on this process has been gathered from observations on the β-decay of more complex nuclei. The β-decay of a nucleus of atomic weight A and atomic number Z can be represented as

$$(A, Z) \rightarrow (A, Z + 1) + e^- + \bar{\nu} \tag{12.1}$$

and the associated process of positron decay as

$$(A, Z) \rightarrow (A, Z - 1) + e^+ + \nu$$

where $\bar{\nu}$ represents the antineutrino and ν the neutrino (the association of the symbols $\bar{\nu}$ and ν with electron and positron decay respectively is a matter of definition).

We shall start by examining the selection rules for nuclear β-decay. The spatial part of the wave function of the outgoing particles is of the form e^{ipr}; now most β-decay processes involve energies of less than 1 MeV, and a useful size for r is the nuclear radius, which is

$$r \sim \frac{A^{1/3}}{m_\pi} \sim \frac{A^{1/3}}{140}\, MeV^{-1}$$

hence

$$pr \sim \frac{1}{100}, \qquad e^{ipr} \sim 1. \tag{12.2}$$

This result implies that the outgoing particles prefer to take away zero orbital angular momentum, since e^{ipr} can be expanded into a series of spherical harmonics containing Bessel functions which behave like $(pr)^l$ for small values of pr (A. 7.6). When the β-decay process involves $l = 0$ it is called an *allowed transition*, if one unit of orbital angular momentum is removed it is called *a first forbidden transition* and two units represent a *second forbidden transition*. The lifetimes for the forbidden transitions are considerably longer than for allowed transitions.

Now let us consider the allowed transitions in more detail. The electron–neutrino system can be emitted with spins parallel in a triplet state or with spins antiparallel (singlet state). The requirement of overall conservation of angular momentum in the decay process then leads to the selection rules (Table 12.2) for the nuclear spin.

TABLE 12.2

Lepton state	Change in nuclear spin	Terminology
Singlet	$\Delta j = 0$	Fermi transition
Triplet	$\Delta j = 0, 1$ not $0 \to 0$	Gamow–Teller transition

In both transitions no change in nuclear parity occurs. Examples of both types of transition are well known. The processes

$$O^{14} \to N^{14*} + e^+ + \nu \quad \text{Fermi}$$

spin 0 0

$$B^{12} \to C^{12} + e^- + \bar{\nu} \quad \text{Gamow–Teller}$$

spin 1 0

are unambiguous; on the other hand,

$$n \rightarrow p + e^- + \bar{\nu}$$

$$\text{spin } \tfrac{1}{2} \qquad \tfrac{1}{2}$$

is a mixed transition.

12.2(b). *Parity independent forms of the matrix element for β-decay*

The first successful theory of the nuclear β-decay process was put forward by Fermi (1934). He visualised the β-decay of a complex nucleus as basically the decay of one of its constituent neutrons into a proton (or vice versa)

$$n \rightarrow p + e^- + \bar{\nu}.$$

Fermi then constructed his theory by making an analogy with the theory of the generation of electromagnetic radiation. The latter is proportional to the electric current four-vector $j_\mu = ie\bar{\psi}\gamma_\mu\psi$, and Fermi assumed that nucleons generated β-radiation in proportion to the 'current' associated with the neutron to proton transformation $\bar{\psi}_p\gamma_\mu\psi_n$. Now the interaction amplitude must be scalar, and hence a second four-vector must be associated with the 'current'. The Hamiltonian for the interaction was therefore written as

$$\mathscr{H}_{\text{int}} = C(\bar{\psi}_p\gamma_\mu\psi_n)(\bar{\psi}_e\gamma_\mu\psi_\nu) + \text{h.c.} \tag{12.3}$$

where C is a coupling constant defining the strength of the interaction. It can be seen that the process

$$n \rightarrow p + e^- + \bar{\nu}$$

has been represented as

$$\nu + n \rightarrow p + e^-$$

as the problem is easier to handle in this more symmetric form.

Now the Hamiltonian (12.3) represents an energy density. There is no reason why it should not be constructed from any of the other five covariant densities associated with the Dirac equation (Konopinski and Uhlenbeck, 1935), and so equation (12.3) can be re-written as

$$\mathscr{H}_{\text{int}} = \sum_i C_i(\bar{\psi}_p\Gamma_i\psi_n)(\bar{\psi}_e\Gamma_i\psi_\nu) + \text{h.c.} \tag{12.4}$$

where the terms Γ_i are products of γ-matrices and transform as S, V, T, A and P (scalars, vectors, tensors, axial vectors and pseudoscalars – compare § 3.3(p)); for the moment we assume that $\mathscr{H}_{\text{int}}$ must be a scalar under spatial reflections.

The process represented by (12.4) must be embedded in a complex nucleus, and allowance made for the multiplicity of the nucleons and their spatial extent.

The transition effectively may be represented by Fig. 12.2 which has a matrix element†

$$M_{fi} = \langle N^A_{Z+1} e^- \bar{\nu} | \; M \; | N^A_Z \rangle$$

$$= C_i \langle N^A_{Z+1} | \; M' O_i \; | N^A_Z \rangle \sqrt{\frac{m_e}{E_e}} \sqrt{\frac{m_\nu}{E_\nu}} \, \bar{u}_e \Gamma_i u_\nu \tag{12.5}$$

where we have factored out the lepton spinors by using the Feymann rules associated with Fig. 12.2 (Table 8.1 (p. 326)). We have set the normalisation factor $V = 1$ and have used the polite fiction that the neutrino has a small but

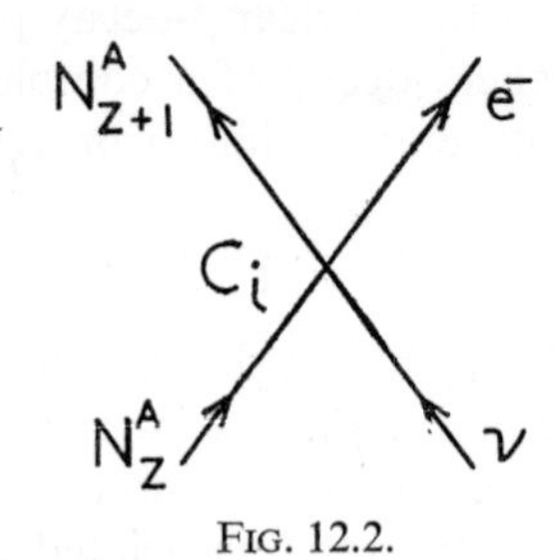

FIG. 12.2.

finite mass; this complication later vanishes. The functions O_i transform like S, V, T, A, P. The matrix element is of the form

$$\langle N^A_{Z+1} | \; M' O_i \; | N^A_Z \rangle \equiv \langle N^A_{Z+1} | \sum_{j=1}^{A} \tau_j^+ O_{ij} \; | N^A_Z \rangle \tag{12.6}$$

where τ_j^+ represents the operator which transforms the nucleon j from a neutron to a proton (9.109).

The matrix elements in (12.6) can be considerably simplified by using non-relativistic approximations. This is justifiable since the energy release in nuclear β-decay is small compared with the nucleon mass. Thus we factor the spinors into their large and small components (3.133)

$$u = \begin{pmatrix} u_L \\ u_s \end{pmatrix}, \qquad u_s \sim 0 \tag{12.7}$$

and consider terms of the type $\bar{u} O_i u$ in 2×2 notation (3.103). It is not difficult to show that only the following factors need be retained in the matrices O_i

$$O_S \cdot \Gamma_S \rightarrow \hat{1} \cdot \hat{1} \tag{12.8.}$$

$$O_V \cdot \Gamma_V \rightarrow \hat{1} \cdot \gamma_4$$

† The transformation amplitude is normally written in Hamiltonian form and the transitions calculated in lowest order time-dependent perturbation theory, as this form is simplest for handling the *nuclear* part of the problem. We are interested in the particle aspects and the lepton part in particular, and so equation (12.5) is adequate.

$$O_T \cdot \Gamma_T \to \boldsymbol{\sigma} \cdot \hat{\boldsymbol{\sigma}}$$

$$O_A \cdot \Gamma_A \to \boldsymbol{\sigma} \cdot \hat{\boldsymbol{\sigma}}\gamma_4$$

$$O_P \cdot \Gamma_P \to 0$$

where the first factor is associated with the nucleons and the second with the leptons; $\boldsymbol{\sigma}$ represents the 2×2 Pauli spin matrix (3.104), and $\hat{\boldsymbol{\sigma}}$ the 4×4 matrix (3.110).

It is apparent from (12.8) that the scalar and vector terms are incapable of causing nuclear spin changes and they must therefore be associated with the Fermi transitions of Table 12.3. The tensor and axial vector terms can induce spin changes, however, and are therefore associated with Gamow–Teller transitions. This property can be readily demonstrated by rewriting the scalar product $\boldsymbol{\sigma} \cdot \hat{\boldsymbol{\sigma}}$ in terms of raising and lowering operators (compare (9.112) or (12.48)) and $\sigma_3\hat{\sigma}_3$; use of the properties listed in equations (5.45) and (5.46) then lead to the Gamow-Teller selection rules.

The nuclear matrix elements are therefore often given in one of the following notations:

$$M_F \equiv \langle \hat{1} \rangle \equiv \int \hat{1} \equiv \langle N^A_{Z+1} | \sum_j \tau^+_j \, | N^A_Z \rangle \tag{12.9}$$

$$M_{GT} \equiv \langle \boldsymbol{\sigma} \rangle \equiv \int \boldsymbol{\sigma} \equiv \langle N^A_{Z+1} | \sum_j \tau^+_j \boldsymbol{\sigma}_j \, | N^A_Z \rangle . \tag{12.10}$$

and we shall represent the complete matrix element for β-decay as

$$M_{fi} = M_F \frac{1}{N_l} \left[\bar{u}_e (C_S + C_V \gamma_4) \, u_\nu \right] + M_{GT} \frac{1}{N_l} \left[\bar{u}_e \hat{\boldsymbol{\sigma}} (C_T + C_A \gamma_4) \, u_\nu \right] \tag{12.11}$$

where N_l represents the normalisation terms for the leptons in (12.5). This equation can be inserted in the standard expression for transition probability (7.39) and evaluated for any required process.

Let us consider, for example, a pure Fermi transition ($M_{GT} = 0$), and examine the distribution in electron energy and the angular correlation between the electron and neutrino. We will ignore kinematic terms for the moment; the transition probability is then proportional to

$$\bar{\sum_i} \sum_\alpha |M_{fi}|^2 = \bar{\sum_i} \sum_\alpha |M_F|^2 \frac{m_e m_\nu}{E_e E_\nu} \, |\bar{u}_e (C_S + C_V \gamma_4) \, u_\nu|^2 \tag{12.12}$$

when we sum over spin states. The summation over spinor states may be carried out by adapting (7.81)

$$\sum_{\text{spins}} |\bar{u} O u|^2 = \text{tr} \left[O \frac{m - i\gamma p}{2m} \gamma_4 O^\dagger \gamma_4 \frac{m - i\gamma p}{2m} \right] . \tag{12.13}$$

We now use this relation in (12.12) and assume the coupling constants to be real, since invariance under time reversal requires real coupling constants (compare (9.40) and § 12.2(d)). The spinor factor becomes

$$\frac{m_e m_\nu}{E_e E_\nu} \operatorname{tr}\left[(C_S + C_V\gamma_4)\frac{m_e - i\gamma p_e}{2m_e}\gamma_4(C_S + C_V\gamma_4)^\dagger\gamma_4\frac{m_\nu - i\gamma p_\nu}{2m_\nu}\right]$$

$$= \frac{-i}{4E_eE_\nu}\operatorname{tr}[(C_S + C_V\gamma_4)(m_e - i\gamma p_e)(C_S + C_V\gamma_4)\gamma p_\nu]_{m_\nu=0}$$

$$= \frac{-1}{4E_eE_\nu}\operatorname{tr}[(C_S + C_V\gamma_4)\gamma p_e(C_S + C_V\gamma_4)\gamma p_\nu]$$

$$\frac{-im_e}{4E_eE_\nu}\operatorname{tr}[(C_S + C_V\gamma_4)(C_S + C_V\gamma_4)\gamma p_\nu]$$

$$= \frac{-C_S^2}{4E_eE_\nu}\operatorname{tr}\gamma p_e\gamma p_\nu - \frac{C_V^2}{4E_eE_\nu}\operatorname{tr}(-\boldsymbol{\gamma}\cdot\mathbf{p}_e + \gamma_4 p_{e4})\gamma p_\nu - \frac{iC_SC_Vm_e}{2E_eE_\nu}\operatorname{tr}\gamma_4\gamma p_\nu$$

$$= C_S^2\left(1 - \frac{\mathbf{p}_e\cdot\mathbf{p}_\nu}{E_eE_\nu}\right) + C_V^2\left(1 + \frac{\mathbf{p}_e\cdot\mathbf{p}_\nu}{E_eE_\nu}\right) + 2C_SC_V\frac{m_e}{E_e} \tag{12.14}$$

where we have used the following relations (A.5, Appendixes, p. 701):

$$\operatorname{tr}\hat{1} = 4, \qquad \operatorname{tr}\gamma_\lambda\gamma_\mu = 4\delta_{\lambda\mu}, \quad \operatorname{tr}\gamma = 0. \tag{12.15}$$

Thus equation (12.12) becomes

$$\overline{\sum_i}\sum_\alpha |M_{fi}|^2 = |M_F|^2\left[C_S^2 + C_V^2 - (C_S^2 - C_V^2)\frac{\mathbf{p}_e\cdot\mathbf{p}_\nu}{E_eE_\nu} + 2C_SC_V\frac{m_e}{E_e}\right]. \tag{12.16}$$

It should be noted that the last term inside the brackets is positive only for electrons; if we had considered positron emission a minus sign would have occurred by virtue of (3.168) and (3.160).

The kinematics of the β-decay process have already been examined in § 8.7(c) where in effect we set $M_{fi} = C$. Using (8.108) and subsequent equations and (12.16), the transition probability becomes for electrons (−) and positrons (+)

$$\frac{d\lambda_\mp}{dE_e d\Omega_e d\Omega_\nu} = \frac{M_F|^2}{(2\pi)^5}p_eE_e(\varDelta - E_e)^2 \times$$

$$\times\left[(C_S^2 + C_V^2) - (C_S^2 - C_V^2)\frac{\mathbf{p}_e\cdot\mathbf{p}_\nu}{E_eE_\nu} \pm 2C_SC_V\frac{m_e}{E_e}\right] \tag{12.17}$$

where E_e, p_e and $\varDelta$ refer to electron (positron) energy, momentum and the total energy released in the decay respectively.

A similar expression could be constructed for a pure Gamow–Teller transition;† the answer is similar except for the coupling constants and the electron–neutrino correlation term which converts in the following manner:

$$-(C_S^2 - C_V^2)\frac{\mathbf{p}_e \cdot \mathbf{p}_\nu}{E_e E_\nu} \to +\tfrac{1}{3}(C_T^2 - C_A^2)\frac{\mathbf{p}_e \cdot \mathbf{p}_\nu}{E_e E_\nu}. \tag{12.18}$$

The energy spectrum of the emitted electrons can be obtained by integrating (12.17) and its equivalent for the Gamow–Teller transition over angles (compare (8.109)). In the integration the correlation term vanishes; the result for a mixed transition by an unpolarised nucleus becomes

$$\frac{d\lambda_\mp}{dE_e} = \frac{p_e E_e}{2\pi^3}(\Delta - E_e)^2 \Big[(C_S^2 + C_V^2)|M_F|^2 + (C_T + C_A)^2 |M_{GT}|^2 \pm \frac{2m_e}{E_e}(C_S C_V |M_F|^2 + C_T C_A |M_{GT}|^2)\Big]. \tag{12.19}$$

It can be seen that this equation contains no cross terms of the type $M_F M_{GT}$; this is because

$$M_F M_{GT} \equiv \langle \hat{1} \rangle \langle \sigma \rangle$$

(compare (12.9) and (12.10)), which vanishes upon averaging over the spin directions of the unpolarised nuclei. Equation (12.19) must be modified if the coulomb interaction between the outgoing electron and the final nucleus is considered; the modification is slight for light nuclei.

Let us finally consider the information we can obtain about the coupling constants from experiment. The existence of pure Fermi and Gamow–Teller transitions implies that neither of the following hypotheses are true:

$$C_S = C_V = 0, \qquad C_T = C_A = 0. \tag{12.20}$$

Furthermore, detailed examination of the spectral shapes in pure Fermi and Gamow–Teller transitions has yielded the following results:

$$\frac{C_S C_V}{C_S^2 + C_V^2} = 0{\cdot}00 \pm 0{\cdot}15, \qquad \frac{C_T C_A}{C_T^2 + C_A^2} = 0{\cdot}00 \pm 0{\cdot}02 \tag{12.21}$$

(Data quoted by Michel, 1957.)

Thus there appears to be only two effective couplings in the β-decay matrix element. Differing types of experiment have provided independent evidence that these couplings are V and A. In this section we shall describe the evidence obtained from experiments on electron–neutrino correlation. Data associated with the failure of parity conservation will be examined in the following section.

† The problem is a little harder; it is worked out in the excellent review article by Smorodinskii on the β-decay process (1959a).

Consider equation (12.16); we may write the correlation term as

$$\frac{\mathbf{p}_e \cdot \mathbf{p}_\nu}{E_e E_\nu} = v_e \cos \theta_{e\nu}$$

and, for a pure Fermi interaction, the transition probability becomes

$$|M_{fi}|^2 \propto C_S^2(1 - v_e \cos \theta_{e\nu}) \qquad (C_V = 0) \tag{12.22}$$

$$|M_{fi}|^2 \propto C_V^2(1 + v_e \cos \theta_{e\nu}) \qquad (C_S = 0)$$

in the two extreme limits. Similarly, by (12.18) a pure Gamow–Teller transition predicts

$$|M_{fi}|^2 \propto C_T^2(1 + \tfrac{1}{3}v_e \cos \theta_{e\nu}) \qquad (C_A = 0) \tag{12.23}$$

$$|M_{fi}|^2 \propto C_A^2(1 - \tfrac{1}{3}v_e \cos \theta_{e\nu}) \qquad (C_T = 0).$$

The correlation experiments (Robson, 1955; Allen, 1959) measure the direction of emission of the electron (or positron) and the recoiling nucleus. The direction

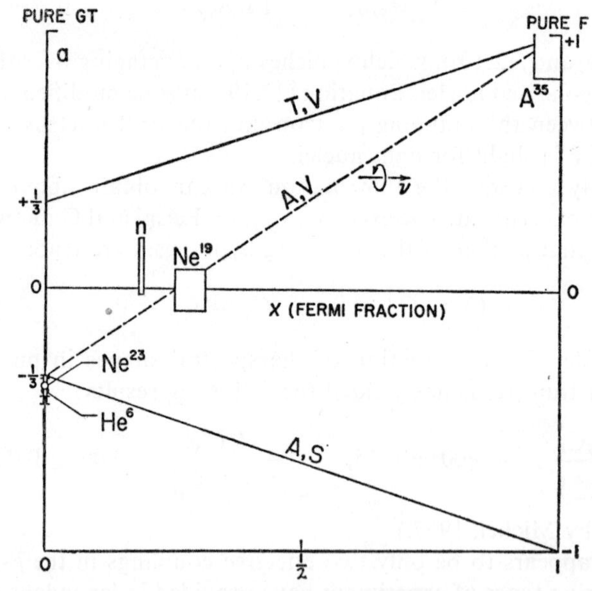

FIG. 12.3. Data on the correlation coefficient for electrons and neutrinos in β-decay (Konopinski, 1959). The letters S, V, T and A refer to scalar, vector, tensor and axial vector couplings respectively.

of the neutrino can then be inferred. The results are displayed in Fig. 12.3 (Konopinski, 1959), and provide clear evidence for a pure V, A coupling. The function

x represents

$$x = \frac{|M_F|^2}{|M_F|^2 + |M_{GT}|^2}$$

that is, the proportion of the Fermi contribution in a mixed transition.

The knowledge that $C_S = C_T = 0$ in nuclear β-decay permits us to simplify the expression for the energy spectrum (12.19) to

$$\frac{d\lambda}{dE_e} = \frac{p_e E_e}{2\pi^3} (\Delta - E_e)^2 \, [C_V^2 \, |M_F|^2 + C_A^2 \, |M_{GT}|^2] \tag{12.24}$$

The integration of this expression over energy (compare § 8.7(c)) then yields the mean lifetime τ for the decay process

$$\lambda = \frac{1}{\tau} = \frac{1}{2\pi^3} [C_V^2 |M_F|^2 + C_A^2 \, |M_{GT}|^2] \int_{m_e}^{\Delta} dE_e p_e E_e (\Delta - E_e)^2. \tag{12.25}$$

The integral is normally called the f-value for the decay, and it is clear that $(f\tau)^{-1}$ is proportional to the coupling strength of the interactions.

Careful measurements of the $f\tau$ values have been made for many nuclei. The most important are those for the pure Fermi transitions and the neutron. The Fermi transitions yield†

$$O^{14} \; ft = 3066 \pm 10 \text{ sec}$$

$$Al^{26} \, ft = 3015 \pm 12 \text{ sec}$$

$$Cl^{34} \, ft = 3055 \pm 20 \text{ sec}$$

$$Co^{54} \, ft = 2966 \pm 18 \text{ sec}.$$

The result for O^{14} is due to Bardin *et al.* (1962), and the data for the remaining nuclei represent the work of Freeman *et al.* (1962, 1964). It can be seen that the ft values appear to be independent of the nature of the nucleus. The raw data listed above has to be subjected to corrections for electronic screening, nuclear form factors and radiative effects; a figure $ft = 3126$ sec is then obtained for O^{14}. From this figure the following values have been deduced for the vector coupling strength:

$$C_V = (1{\cdot}4025 \pm 0{\cdot}0022) \, 10^{-49} \text{ erg-cm}^3 \equiv (1{\cdot}0031 \pm 0{\cdot}0016) \frac{10^{-5}}{m_p^2} \tag{12.26}$$

where m_p is the mass of the proton.

Now let us consider the decay of the neutron. The measurements of Sosnovskii *et al.*, (1959) have yielded a half life of 11·7 ± 0·3 min and a ft value 1180 ± 35 sec. This figure may be used to deduce C_A/C_V. Now the neutron decay is a mixed transition. A specific neutron only has one final state available when it emits

† The figures are normally quoted using half lives rather than mean lives; we have followed this convention by using t to represent half life.

Fermi radiation – that state is a proton with the same spin orientation. Now consider Gamow–Teller radiation; from (12.10) we can write $|M_{GT}|^2$ as

$$|M_{GT}|^2 \equiv |\langle \boldsymbol{\sigma} \rangle|^2 = |\langle \sigma_x \rangle|^2 + |\langle \sigma_y \rangle|^2 + |\langle \sigma_z \rangle|^2 = 3|\langle \hat{1} \rangle|^2$$

since if we consider lifetimes we sum over all possible states of polarisation and then cross terms must vanish. Thus there is a weighting factor 3 for Gamow–Teller radiation by the neutron, and so

$$\frac{1}{ft} = C_V^2 + 3C_A^2.$$

Now O^{14} is the $T_3 = +1$ component of the isotopic triplet C^{14}, N^{14*}, O^{14} and so $|M_F|^2$ has a weighting factor of two for O^{14}, since the decay can come from two equivalent protons in the O^{14} nucleus†, thus

$$\frac{ft_0}{ft_n} = \frac{C_V^2 + 3C_A^2}{2C_V^2} = \frac{3066 \pm 10}{1180 \pm 35} \tag{12.27}$$

$$\frac{C_A^2}{C_V^2} = 1{\cdot}40 \pm 0{\cdot}06, \qquad \left|\frac{C_A}{C_V}\right| = 1{\cdot}18 \pm 0{\cdot}03$$

where the ft values have not been subjected to radiative corrections. A safer figure for $|C_A/C_V|$ is probably $1{\cdot}2 \pm 0{\cdot}05$ (Blin-Stoyle, 1961) in view of the uncertainty in the radiative and other corrections.

12.2(c). *Parity failure and the matrix element for β-decay*

The proposal of Lee and Yang that parity was not conserved in weak interactions, and its rapid proof in β-decay and μ-decay, led to an intense attack on the β-decay problem. In their original paper Lee and Yang (1956a) proposed that the Hamiltonian for β-decay should be modified to include parity nonconserving terms in the lepton part

$$C_i \Gamma_i \to C_i \Gamma_i + C_i' \Gamma_i \gamma_5 \tag{12.28}$$

Now let us consider the lepton spinor part of the β-decay matrix element $C_i \bar{u}_e \Gamma_i u_\nu$. The modification (12.28) then enables us to write this term as

$$C_i \bar{u}_e \Gamma_i u_\nu + C_i' \bar{u}_e \Gamma_i \gamma_5 u_\nu = \frac{(C_i + C_i')}{2} \bar{u}_e \Gamma_i (1 + \gamma_5) u_\nu + \frac{(C_i - C_i')}{2} \bar{u}_e \Gamma_i (1 - \gamma_5) u_\nu . \tag{12.29}$$

† More formally one could say that O^{14} and N^{14*} represent the $T_3 = +1$ and 0 components respectively of an isotopic triplet, hence by (9.134)

$$T_- |T, T_3\rangle = \sqrt{[T(T+1) - T_3(T_3 - 1)]}\, |T, T_3 - 1\rangle$$
$$T_- |1, 1\rangle = \sqrt{2}\, |1, 0\rangle .$$

If the mass of the neutrino is zero the terms $(1 + \gamma_5)\, u_\nu$ and $(1 - \gamma_5)\, u_\nu$ lead to longitudinally polarised particles with spin pointing respectively antiparallel and parallel to their direction of motion (§ 3.3(1)). Lee and Yang (1957a and b), and independently Landau (1957) and Salam (1957), pointed out that if only one of the polarisation states existed in nature, maximum violation of parity would occur in weak interactions. This is the basic statement of the two-component theory of the neutrino. The experimental data which we shall discuss in this chapter supports the concept of maximum violation, and so one of the terms in (12.29) should be dropped. The experiment of Goldhaber, Grodzins and Sunyar (1958; see § 2.4(d)) has shown that the spin and momentum of the neutrino are antiparallel† (that is, it is a left-handed particle), and so we set

$$C_i = C_i' = \frac{G_i}{\sqrt{2}} \tag{12.30}$$

in equation (12.29). The introduction of the factor $\sqrt{2}$ is for convenience; with this modification the magnitude of G_i remains the same as C_i in parity independent expressions.

Thus the basic interaction Hamiltonian (12.4) can be re-written as

$$\mathscr{H}_{\text{int}} = \sum_i \frac{G_i}{\sqrt{2}} (\overline{\psi}_p \Gamma_i \psi_n)\, [\overline{\psi}_e \Gamma_i (1 + \gamma_5)\, \psi_\nu] \tag{12.31}$$

where for convenience of writing we shall omit the expression "+ h.c.". This expression possesses an interesting property. If the electron operator ψ_e is written as

$$\psi_e = \tfrac{1}{2}(1 + \gamma_5)\psi_e + \tfrac{1}{2}(1 - \gamma_5)\psi_e \tag{12.32}$$

then

$$\psi_e^\dagger = \tfrac{1}{2}\psi_e^\dagger(1 + \gamma_5) + \tfrac{1}{2}\psi_e^\dagger(1 - \gamma_5)$$

since we are using the Hermitian form of the γ-matrices (3.103). The lepton part of (12.31) now becomes

$$\begin{aligned}
\overline{\psi}_e \Gamma_i (1 + \gamma_5)\, \psi_\nu &= \psi_e^\dagger \gamma_4 \Gamma_i (1 + \gamma_5)\, \psi_\nu \\
&= \tfrac{1}{2}\psi_e^\dagger (1 + \gamma_5)\, \gamma_4 \Gamma_i (1 + \gamma_5)\, \psi_\nu + \tfrac{1}{2}\psi_e^\dagger (1 - \gamma_5)\, \gamma_4 \Gamma_i (1 + \gamma_5)\, \psi_\nu \\
&= \tfrac{1}{2}\overline{\psi}_e (1 - \gamma_5)\, \Gamma_i (1 + \gamma_5)\, \psi_\nu + \tfrac{1}{2}\overline{\psi}_e (1 + \gamma_5)\, \Gamma_i (1 + \gamma_5)\, \psi_\nu
\end{aligned} \tag{12.33}$$

where we have used the anticommutation properties of the γ-matrices (3.93). A further application of these properties shows that the two expressions in (12.33)

† The experimental result was 68 per cent polarisation for the neutrinos. Sufficient depolarisation mechanisms could occur in the aparatus for Goldhaber *et. al.* to conclude that the true figure was 100 per cent polarisation.

assume the following forms:

$$\tfrac{1}{2}\bar{\psi}_e(1-\gamma_5)\,\Gamma_i(1+\gamma_5)\,\psi_\nu = \begin{cases} 0 & (i = S, T, P) \\ \bar{\psi}_e\Gamma_i(1+\gamma_5)\,\psi_\nu & (i = V, A) \end{cases} \tag{12.34}$$

$$\tfrac{1}{2}\bar{\psi}_e(1+\gamma_5)\,\Gamma_i(1+\gamma_5)\,\psi_\nu = \begin{cases} \bar{\psi}_e\Gamma_i(1+\gamma_5)\,\psi_\nu & (i = S, T, P) \\ 0 & (i = V, A). \end{cases}$$

We have seen in the previous section that the β-decay coupling is V and A, hence we should expect the electrons from β-decay to possess left-handed polarisation since the first term in (12.34) is associated with the projection operator $(1+\gamma_5)$ in (12.32).

If the electron were massless or if $m_e/E_e \to 0$ this conclusion would imply 100 per cent polarisation for the electron. Neither of these statements is correct, and so the electron polarisation is less than 100 per cent. The magnitude of this polarisation may be easily calculated by the two-component theory. We first note that the insertion of (12.34) into the Hamiltonian (12.31) leads to a considerable simplification

$$\mathscr{H}_{\text{int}} = \frac{G_V}{\sqrt{2}}(\bar{\psi}_p\gamma_\lambda\psi_n)\,[\bar{\psi}_e\gamma_\lambda(1+\gamma_5)\,\psi_\nu]$$

$$+\frac{G_A}{\sqrt{2}}(\bar{\psi}_p i\gamma_\lambda\gamma_5\psi_n)\,[\bar{\psi}_e i\gamma_\lambda\gamma_5(1+\gamma_5)\,\psi_\nu] \tag{12.35}$$

$$= \frac{G_V}{\sqrt{2}}\left[\bar{\psi}_p\gamma_\lambda\left(1-\frac{G_A}{G_V}\gamma_5\right)\psi_n\right][\bar{\psi}_e\gamma_\lambda(1+\gamma_5)\,\psi_\nu] \tag{12.36}$$

where we have used the relation

$$\gamma_5(1+\gamma_5) = 1+\gamma_5.$$

Thus the lepton term is the same for both V and A couplings, and one therefore expects the polarisation of the leptons to be the same in both Fermi and Gamow–Teller transitions, whether pure or mixed. Let us now calculate the magnitude of the polarisation. We assume the initial nuclei to be unpolarised, and hence from (12.32) the rate of transition will be proportional to

$$\sum_{\text{spins}} |\bar{\psi}_e\gamma_\lambda(1+\gamma_5)\,\psi_\nu|^2 \propto \sum_{\text{spins}} |\bar{u}_e\gamma_\lambda(1+\gamma_5)\,u_\nu|^2$$

$$\equiv \tfrac{1}{4}\sum_{\text{spins}} |[(1+\gamma_5)\,u_e]^\dagger\,\gamma_4\gamma_\lambda(1+\gamma_5)\,u_\nu|^2. \tag{12.37}$$

Let us examine the electron term $(1+\gamma_5)\,u_e$; using equations (3.103) and

(3.136) we find

$$(1 + \gamma_5)\, u_e = \begin{pmatrix} 1 & -1 \\ -1 & 1 \end{pmatrix} \begin{pmatrix} u_L \\ \dfrac{\boldsymbol{\sigma} \cdot \mathbf{p}}{E + m}\, u_L \end{pmatrix} = \begin{bmatrix} \left(1 - \dfrac{\boldsymbol{\sigma} \cdot \mathbf{p}}{E + m}\right) u_L \\ -\left(1 - \dfrac{\boldsymbol{\sigma} \cdot \mathbf{p}}{E + m}\right) u_L \end{bmatrix} \tag{12.38}$$

where u_L represents the large spinor component.

Now an electron with 100 per cent polarisation parallel or antiparallel to its momentum can be represented by the functions

$$\begin{aligned} &\left(1 + \frac{\boldsymbol{\sigma} \cdot \mathbf{p}}{|\mathbf{p}|}\right) u_L \quad \text{parallel} \\ &\left(1 - \frac{\boldsymbol{\sigma} \cdot \mathbf{p}}{|\mathbf{p}|}\right) u_L \quad \text{antiparallel.} \end{aligned} \tag{12.39}$$

Therefore if we split the operator appearing in (12.38) into

$$\left(1 - \frac{\boldsymbol{\sigma} \cdot \mathbf{p}}{E + m}\right) = a\left(1 + \frac{\boldsymbol{\sigma} \cdot \mathbf{p}}{|\mathbf{p}|}\right) + b\left(1 - \frac{\boldsymbol{\sigma} \cdot \mathbf{p}}{|\mathbf{p}|}\right) \tag{12.40}$$

where

$$a = \tfrac{1}{2}\left(1 - \frac{|\mathbf{p}|}{E + m}\right), \qquad b = \tfrac{1}{2}\left(1 + \frac{|\mathbf{p}|}{E + m}\right)$$

then the polarisation along the direction of the electron's momentum is given by

$$\begin{aligned} P = \frac{|a|^2 - |b|^2}{|a|^2 + |b|^2} &= \frac{(-2|\mathbf{p}|)/(E + m) - (2|\mathbf{p}|)/(E + m)}{2 + (2|\mathbf{p}|^2)/[(E + m)^2]} \\ &= -\frac{|\mathbf{p}|}{E} = -v_e. \end{aligned} \tag{12.41}$$

Thus the polarisation of the electron is proportional to its velocity, and in the limit $v_e \rightarrow 1$, 100 per cent polarisation is expected. The above conclusion may be extended to positron emission. In this situation we make the replacements

$$\begin{aligned} &e^- \rightarrow e^+ \quad \text{emission} \\ &\nu \rightarrow \bar{\nu} \quad \text{absorption} \end{aligned}$$

In addition the two-component theory for the neutrino is noninvariant under charge conjugation and requires that we make the replacement

$$(1 + \gamma_5) \rightarrow (1 - \gamma_5).$$

The same replacement occurs in front of the positron spinor and leads to a polarisation $+v_e$, hence (in $c = 1$ units)

$$P_- = -v_e, \quad P_+ = +v_e. \tag{12.42}$$

Frauenfelder *et al.*, (1957) first demonstrated the existence of longitudinal polarisation in β-decay. They used electrons from Co^{60} (Fig. 12.4). The longitudinal polarisation was first converted to transverse polarisation by passing the electrons through an electrostatic deflector which changes the direction of momentum but not the spin. The electrons were then subjected to Coulomb scattering

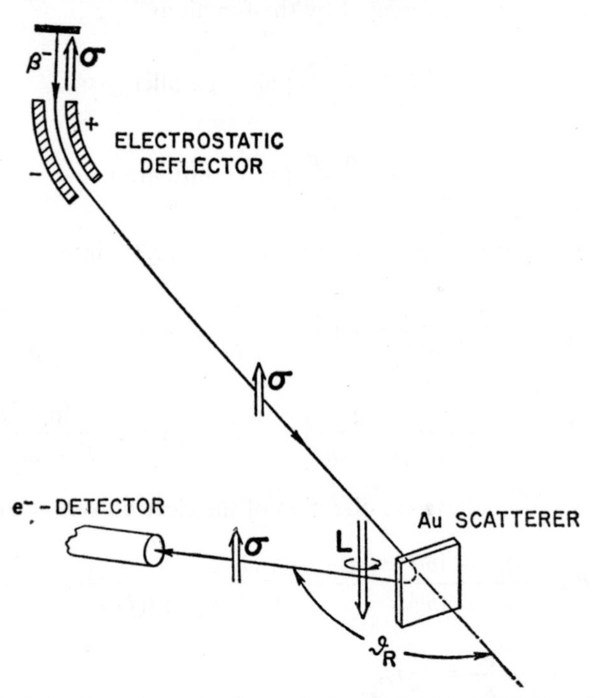

FIG. 12.4. Principle of experiment to detect the longitudinal polarisation of electrons in β-decay (Konopinski, 1959).

in a gold foil, and right–left asymmetries were sought. If the electrons were polarised an asymmetry was expected to arise from spin–orbit coupling. This can be easily seen if we consider the situation semi-classically. Since $\mathbf{L} = \mathbf{r} \times \mathbf{p}$ scatterings to the right and left are associated with orbital angular momentum pointing downwards and upwards respectively (Fig. 12.4). The spin–orbit coupling then leads to terms of the type $\sigma \cdot \mathbf{L}$ with different signs in the Hamiltonian for the scattering and hence a right–left asymmetry occurs. Frauenfelder and his colleagues observed negative polarisation, although the magnitude was smaller than v/c; this is not surprising since depolarisation mechanisms are usually present in such experiments.

The measurements of electron and positron polarisation have now been made by many groups (see *Proceedings Rehovoth Conference on Nuclear Structure*, North Holland, 1958, for details). Some have used variants of the method described above, others have examined the polarisation of the Bremsstrahlung from the decay electrons. All the data appears to be consistent with a polarisation $-v_e$ and $+v_e$ for electrons and positrons respectively.

12.2(d). *The relative sign of the coupling terms V and A*

In the previous sections evidence has been summarised which shows that the coupling in nuclear β-decay is of the type V and A. The final question which we examine is the relative sign of these couplings. The answer was found by examining correlations in the decay of polarised neutrons.

Consider equation (12.35)

$$\mathscr{H}_{\text{int}} = \frac{G_V}{\sqrt{2}} (\overline{\psi}_p\gamma_\lambda\psi_n)\,[\overline{\psi}_e\gamma_\lambda(1+\gamma_5)\,\psi_\nu] + \frac{G_A}{\sqrt{2}} (\overline{\psi}_p i\gamma_\lambda\gamma_5\psi_n)\,[\overline{\psi}_e i\gamma_\lambda\gamma_5(1+\gamma_5)\,\psi_\nu];$$

if we use the relations

$$\gamma_5(1+\gamma_5) = 1+\gamma_5, \qquad \tfrac{1}{2}(1+\gamma_5)^2 = 1+\gamma_5$$

the equation becomes

$$\mathscr{H}_{\text{int}} = \frac{G_V}{2\sqrt{2}} (\overline{\psi}_p\gamma_\lambda\psi_n)\,[\psi_e^\dagger(1+\gamma_5)\,\gamma_4\gamma_\lambda(1+\gamma_5)\,\psi_\nu]$$
$$- \frac{G_A}{2\sqrt{2}} (\overline{\psi}_p\gamma_\lambda\gamma_5\psi_n)\,[\psi_e^\dagger(1+\gamma_5)\,\gamma_4\gamma_\lambda(1+\gamma_5)\,\psi_\nu]. \tag{12.43}$$

We now consider the spinors only, and use a nonrelativistic approximation for the nucleons

$$u_p = \begin{pmatrix} u_{pL} \\ u_{pS} \end{pmatrix} \sim \begin{pmatrix} p \\ 0 \end{pmatrix}, \qquad u_n \sim \begin{pmatrix} n \\ 0 \end{pmatrix}$$

and represent the lepton spinors as (compare (12.38))

$$(1+\gamma_5)\,u_e = \begin{bmatrix} \left(1 - \dfrac{\boldsymbol{\sigma}\cdot\mathbf{p}}{E+m}\right) u_{eL} \\ -\left(1 - \dfrac{\boldsymbol{\sigma}\cdot\mathbf{p}}{E+m}\right) u_{eL} \end{bmatrix} = \begin{pmatrix} e_L \\ -e_L \end{pmatrix} \tag{12.44}$$

$$(1+\gamma_5)\,u_\nu = \begin{pmatrix} \nu_L \\ -\nu_L \end{pmatrix}.$$

The subscript L in the final column will be used to refer to the left-handedness of the leptons, since their helicity is given by $-v/c$ (compare § 12.2(c)).

It the γ-matrices are written in 2×2 notation (3.103), the matrix element for the transition then becomes

$$M_{fi} = \frac{G_V}{\sqrt{2}} (p^\dagger n) (e_L^\dagger \nu_L) + \frac{G_A}{\sqrt{2}} (p^\dagger \boldsymbol{\sigma} n) (e_L^\dagger \boldsymbol{\sigma} \nu_L). \tag{12.45}$$

Now let us apply this matrix element to the decay of polarised neutrons. The direction of neutron spin is chosen as axis of quantisation and so the neutron spinor is $\begin{pmatrix}1\\0\end{pmatrix}$; now the term σ_3 cannot cause nucleon spin flip, since

$$\begin{aligned}\sum_{p \text{ spins}} p^\dagger \sigma_3 n &= (1\ 0) \begin{pmatrix}1 & 0\\0 & -1\end{pmatrix} \begin{pmatrix}1\\0\end{pmatrix} + (0\ 1) \begin{pmatrix}1 & 0\\0 & -1\end{pmatrix} \begin{pmatrix}1\\0\end{pmatrix} \\ &= (1\ 0) \begin{pmatrix}1\\0\end{pmatrix} + (0\ -1) \begin{pmatrix}1\\0\end{pmatrix} = (1\ 0) \begin{pmatrix}1\\0\end{pmatrix}.\end{aligned} \tag{12.46}$$

Therefore the terms in equation (12.45) can be regrouped as

$$\begin{aligned}M_{fi} = \frac{1}{\sqrt{2}} (p^\dagger n) [e_L^\dagger (G_V + G_A \sigma_3) \nu_L] + \frac{G_A}{\sqrt{2}} (p^\dagger \sigma_1 n) (e_L^\dagger \sigma_1 \nu_L) \\ + \frac{G_A}{\sqrt{2}} (p^\dagger \sigma_2 n) (e_L^\dagger \sigma_2 \nu_L).\end{aligned} \tag{12.47}$$

Now consider the last two terms; we can write

$$\begin{aligned}\sigma_1 \sigma_1 + \sigma_2 \sigma_2 &= 2 \left(\frac{\sigma_1 - i\sigma_2}{2}\right) \left(\frac{\sigma_1 + i\sigma_2}{2}\right) + 2 \left(\frac{\sigma_1 + i\sigma_2}{2}\right) \left(\frac{\sigma_1 - i\sigma_2}{2}\right) \\ &= 2\sigma_- \sigma_+ + 2\sigma_+ \sigma_-.\end{aligned} \tag{12.48}$$

But

$$\left(\frac{\sigma_1 + i\sigma_2}{2}\right) n = \sigma_+ n = \begin{pmatrix}0 & 1\\0 & 0\end{pmatrix} \begin{pmatrix}1\\0\end{pmatrix} = 0 \tag{12.49}$$

from equation (3.104). Therefore the matrix element (12.47) becomes

$$M_{fi} = \frac{1}{\sqrt{2}} (p^\dagger n) [e_L^\dagger (G_V + G_A \sigma_3) \nu_L] + \frac{2G_A}{\sqrt{2}} (p^\dagger \sigma_- n) (e_L^\dagger \sigma_+ \nu_L) \tag{12.50}$$

where the first and second terms represent the non-spin flip and spin flip contributions for the nucleons respectively.

Rather than evaluate this matrix element in detail, we shall consider possible orientations of the final particles which allow overall conservation of angular momentum, and which are consistent with negative helicity for the emitted electron and positive helicity for the antineutrino. (It is more convenient to make the physical argument considering the emission of antineutrinos rather than the absorption of neutrinos.) The possible orientations along the direction of neutron spin are then shown in Fig. 12.5; the light arrows represent momenta.

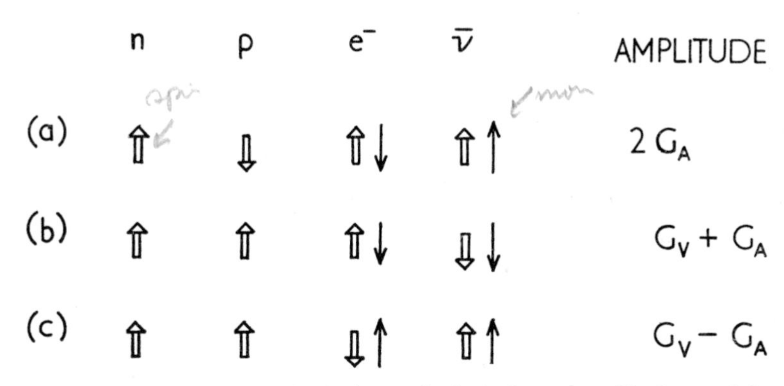

FIG. 12.5. Spins and momenta in the decay of polarised neutrons. The long and short arrows represent momenta and spins respectively.

The association of an amplitude with each process is straightforward; (a) follows directly from (12.50) and (b) and (c) arise from letting σ_3 operate on either $e_L^\dagger$ or ν_L – if the latter method is used one must remember that the antiparticle's spin points in the opposite direction to the particle's, for example

$$\sigma_3\begin{pmatrix}0\\1\end{pmatrix} = -\begin{pmatrix}0\\1\end{pmatrix} \equiv \sigma_3\,(\nu \Downarrow \text{ or } \bar{\nu} \Uparrow). \tag{12.51}$$

It can be seen that Fig. 12.5 leads to differing ratios for the numbers of electrons with momenta pointing up (parallel) and down with respect to the direction of the neutron spin. They are given in Table 12.3; in calculating the numerical coefficients we have used $G_A = 1{\cdot}2\,|G_V|$ from § 12.2(b).

TABLE 12.3

Particle	Up–down ratio	$\dfrac{G_A}{G_V} = +1{\cdot}2$	$\dfrac{G_A}{G_V} = -1{\cdot}2$
e^-	$\dfrac{\lvert G_V - G_A\rvert^2}{4\,\lvert G_A\rvert^2 + \lvert G_V + G_A\rvert^2}$	$\dfrac{0{\cdot}04}{10{\cdot}7}$	$\dfrac{4{\cdot}8}{5{\cdot}8}$
		$\sim 1 - \cos\theta$	$\sim 1 - 0{\cdot}1\cos\theta$
$\bar{\nu}$	$\dfrac{4\,\lvert G_A\rvert^2 + \lvert G_V - G_A\rvert^2}{\lvert G_V + G_A\rvert^2}$	$\dfrac{5{\cdot}8}{4{\cdot}8}$	$\dfrac{10{\cdot}6}{0{\cdot}04}$
		$\sim 1 + 0{\cdot}1\cos\theta$	$\sim 1 + \cos\theta$

The measurement of the correlation coefficients has been made by Burgy *et al.* (1958). Neutrons were polarised by reflecting them from magnetised cobalt mirrors. The up–down coefficients were determined by simultaneously detecting the decay proton and electron at suitable angles and hence inferring the direction of the antineutrino. The measurements were performed twice, with the neutron in opposite states of polarisation (this was achieved by reversing the direction of the magnetic field in the cobalt mirror). If we write the angular distributions of the decay products as $1 + a \cos \theta$, the measured values of the coefficients a were

$$a_e = -0{\cdot}09 \pm 0{\cdot}03, \qquad a_{\bar{\nu}} = 0{\cdot}88 \pm 0{\cdot}15. \tag{12.52}$$

Thus the results clearly show that V and A couplings are of opposite sign. The result can also be stated another way; if we write

$$G_A = G_V \Gamma e^{i\varphi} \tag{12.53}$$

where $\Gamma = 1{\cdot}2$, then the results obtained above indicate that $\varphi \sim \pi$, that is the relative phase of the coupling constants is real. This is the result to be expected if the β-decay matrix element is invariant under time reversal.

This concludes our examination of the nuclear β-decay process. The final form of the interaction Hamiltonian is obtained by inserting the above result in equation (12.36)

$$\mathscr{H}_{\text{int}} = \frac{G_V}{\sqrt{2}} [\bar{\psi}_p \gamma_\lambda (1 + \Gamma \gamma_5) \psi_n] [\bar{\psi}_e \gamma_\lambda (1 + \gamma_5) \psi_\nu] + \text{h.c.} \qquad (\Gamma = 1{\cdot}2) \tag{12.54}$$

where we have included the Hermitian conjugate term for completeness. A more complete analysis of the β-decay matrix element than that given here may be found in the paper of Jackson, Treiman and Wyld (1957).

12.3. MUON DECAY

The process of muon decay is unique; it is the only easily accessible weak process in which all participating particles are leptons, and it therefore offers an excellent opportunity for studying weak interactions without the complicating factors associated with virtual processes (small electromagnetic corrections do have to be made, however, in exact evaluations).

Muons are formed in pion decay by the process

$$\pi \to \mu + \nu$$

(the symbol ν is used generically at this point.) Now the pion is a spinless particle and the neutrino which appears in β-decay has negative helicity (§ 2.4(d)). If the assumption is made that the helicity of the neutrino appearing in β-decay is identical† with that in pion decay, then the simultaneous requirement of the

† We shall later present evidence that the two neutrinos are not the same particle. This leaves our present argument regarding helicities unaffected.

conservation of linear and angular momentum in π-decay permits us to draw up the schemes shown in Fig. 12.6.

The experiments of Alikhanov *et al.* (1960), and others (Backenstoss *et al.*, 1961; Bardon, Franzini and Lee, 1961) have shown the helicity of the negatively (positively) charged muons to be positive (negative) respectively,† thus we retain the schemes (a) and (d)

$$\pi^+ \rightarrow \mu^+ + \nu, \quad \pi^- \rightarrow \mu^- + \bar{\nu}. \tag{12.55}$$

We next consider the process of muon decay. The energy of the emitted electron is not unique and is known to extend from zero to about half the rest mass

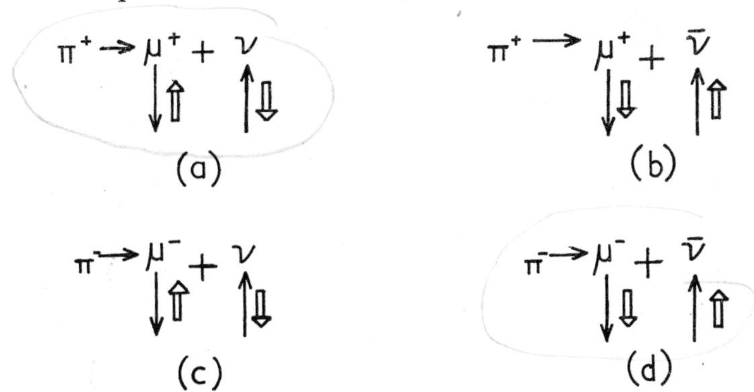

FIG. 12.6. Possible helicity assignments in pion decay. The long arrows represent momenta.

energy of the muon. The shape of the energy spectrum is consistent with a three-body decay process, and so the possible decay modes into an electron and two neutrinos are

$$\begin{aligned} \mu \rightarrow\ & e + \nu + \nu \\ & e + \bar{\nu} + \bar{\nu} \\ & e + \nu + \bar{\nu}. \end{aligned}$$

The first two possibilities can be eliminated by an argument which combines four pieces of experimental information:

(1) the helicities of the μ^+ and μ^- particles are negative and positive respectively,

(2) the electrons arising from μ^+ and μ^- decay have positive and negative helicity respectively; the experimental data is consistent with 100 per cent polarisation (Culligan, Frank and Holt, 1959; Macq, Crowe and Haddock, 1958).

(3) the energy spectrum of the electrons is large near its upper limit (Plano, 1960; Kruger, 1961).

(4) when muons are brought to rest in matter the angular distribution of both electrons and positrons is strongly peaked in the backward direction with

† The data is better for the negatively charged muons than for their positive counterparts.

respect to the momentum of the incoming muon (Garwin, Lederman and Weinrich, 1957; Friedman and Telegdi, 1957).

Now consider the emission of electrons with energy $E_0 = \frac{1}{2}m_\mu$ in the backward direction with respect to the incoming muon; conservation of energy and momentum requires both neutrinos to travel in the opposite direction, and so we

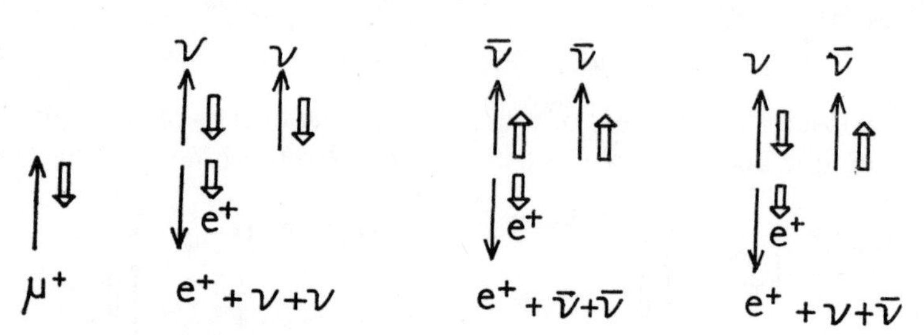

FIG. 12.7. Possible helicity assignments in muon decay. The long and short arrows indicate momenta and spins respectively. The arrow for momentum on the muon indicates its direction of motion before being brought to rest.

have the possible configurations shown in the diagrams of Fig. 12.7. An inspection of the diagrams reveals that only the process $\mu^+ \to e^+ + \bar{\nu} + \nu$ conserves angular momentum. In a similar manner the process of μ^- decay (Fig. 12.8) can be shown to satisfy the scheme

$$\mu^- \to e^- + \bar{\nu} + \nu.$$

Let us now consider a suitable Hamiltonian for the muon decay process. For ease of comparison with nuclear β-decay we shall examine the process $\mu^- \to e^- + \bar{\nu} + \nu$. In the case of β-decay the grouping of terms $(pn)\,(e^-\nu)$ in the Hamiltonian is

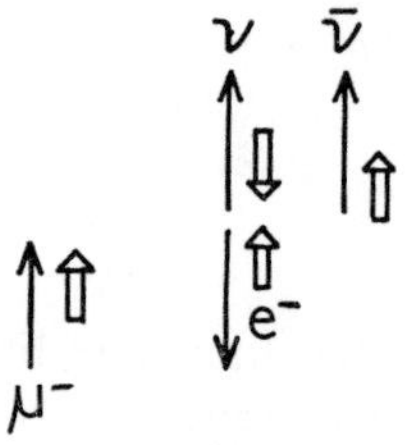

FIG. 12.8.

written almost instinctively since two baryons and two leptons occur. A natural grouping for the four leptons in muon decay is less obvious, but by analogy with the β-decay process we shall take the grouping† $(\nu\mu)\,(e\nu)$ corresponding to the process

$$\nu + \mu^- \to e^- + \nu.$$

† In fact the grouping is immaterial for the only two terms, V and A, which appear in the Hamiltonian.

Now there are distinct similarities in the process of β- and μ-decay for electrons of the same charge

$$\begin{aligned} \nu + n &\rightarrow e^- + p \\ \nu + \mu^- &\rightarrow e^- + \nu . \end{aligned} \tag{12.56}$$

In both processes neutrinos (antineutrinos) are absorbed (emitted), and in both processes the electron possesses negative helicity. We may therefore use the arguments of § 12.2(c) to eliminate all couplings except V and A modes, and obtain an appropriate Hamiltonian for μ-decay by suitably modifying (12.35)

$$\mathscr{H}_{\text{int}} = \frac{G_V}{\sqrt{2}} (\overline{\psi}_p \gamma_\lambda \psi_n) [\overline{\psi}_e \gamma_\lambda (1 + \gamma_5) \psi_\nu)] + \frac{G_A}{\sqrt{2}} (\overline{\psi}_p i \gamma_\lambda \gamma_5 \psi_n) [\overline{\psi}_e i \gamma_\lambda \gamma_5 (1 + \gamma_5) \psi_\nu].$$

An inspection of (12.56) shows that the required modifications are†

$$\psi_n \rightarrow \psi_\mu, \qquad \psi_p \rightarrow \tfrac{1}{2}(1 + \gamma_5) \psi_\nu \tag{12.57}$$

since we assume that the neutrinos are left-handed (3.177). When these modifications are inserted in $\mathscr{H}_{\text{int}}$, it becomes

$$\begin{aligned} \mathscr{H}_{\text{int}} &= \frac{G_V}{2\sqrt{2}} [\overline{\psi}_\nu (1 - \gamma_5) \gamma_\lambda \psi_\mu] [\overline{\psi}_e \gamma_\lambda (1 + \gamma_5) \psi_\nu] \\ &\quad - \frac{G_A}{2\sqrt{2}} [\overline{\psi}_\nu (1 - \gamma_5) \gamma_\lambda \gamma_5 \psi_\mu] [\overline{\psi}_e \gamma_\lambda \gamma_5 (1 + \gamma_5) \psi_\nu] \\ &= \frac{G_V - G_A}{2\sqrt{2}} [\overline{\psi}_\nu \gamma_\lambda (1 + \gamma_5) \psi_\mu] [\overline{\psi}_e \gamma_\lambda (1 + \gamma_5) \psi_\nu] \end{aligned} \tag{12.58}$$

where we have used the relations

$$\overline{\psi}_p \rightarrow \tfrac{1}{2} \overline{\psi}_\nu (1 - \gamma_5), \qquad \gamma_5 (1 + \gamma_5) = 1 + \gamma_5 .$$

Thus our assumption of left-handed neutrinos and electrons and right-handed antineutrinos has led to a form for the Hamiltonian for muon decay which is of the type $V - A$ (we are also assuming invariance under time reversal and hence real coupling constants).

We shall write the coupling constant in (12.58) as

$$G_\mu = \frac{G_V - G_A}{2}.$$

† In constructing the Hamiltonian for muon decay we are assuming that the mass of the muon neutrino in zero. The experimental limits are $m_{\nu_e} < 200$ eV (Sakurai 1958a), and $m_{\nu_\mu} < 2{\cdot}5$ MeV (Barkas et al., 1956). The choice of neutrinos with zero mass in the Hamiltonian is normally made for reasons of simplicity.

Thus equation (12.58) becomes

$$\mathscr{H}_{\text{int}} = \frac{G_\mu}{\sqrt{2}} [\bar{\psi}_\nu \gamma_\lambda (1 + \gamma_5) \psi_\mu] [\bar{\psi}_e \gamma_\lambda (1 + \gamma_5) \psi_\nu] + \text{h.c.} \tag{12.59}$$

where the Hermitian conjugate has been inserted for completeness. A comparison of this equation with (12.54) reveals a similar structure with $G_V \to G_\mu$, $\Gamma \to 1$ and appropriate alterations in particle subscripts; it yields the following matrix element:

$$M_{fi} = \frac{G_\mu}{\sqrt{2}} \frac{1}{N_l} [\bar{u}_\nu \gamma_\lambda (1 + \gamma_5) u_\mu] [\bar{u}_e \gamma_\lambda (1 + \gamma_5) u_\nu]. \tag{12.60}$$

The evaluation of (12.60) can be carried out with the standard techniques of Chapters 7 and 8. More elaborate calculations may be found in the papers of Kinoshita and Sirlin (1957) and Okun and Sehter (1958). The probability of producing electrons in the energy interval dE_e per steradian is given by

$$\frac{d\lambda}{dE_e\, d\Omega_e} = \frac{G_\mu^2}{4\pi^4} E_0 E_e^2 \left[E_0 - \tfrac{2}{3} E_e \pm \tfrac{1}{3} \frac{\boldsymbol{\sigma}_\mu \cdot \mathbf{p}_e}{|\mathbf{p}_e|} (2E - E_0) \right] \tag{12.61}$$

where $\boldsymbol{\sigma}_\mu$ represents the direction of the muon spin, the signs + and − are associated with μ^+ and μ^- respectively, and $E_0 = \frac{1}{2} m_\mu$ represents the maximum energy of the electron.† A double integration of equation (12.61) leads to the following expression for the muon lifetime:

$$\lambda = \frac{1}{\tau_\mu} = \frac{G_\mu^2 m_\mu^5}{192\pi^3}. \tag{12.62}$$

Recent measurements of the lifetime of the positive muon have yielded values which lie close to a mean of $\tau_\mu = (2{\cdot}210 \pm 0{\cdot}003)\, 10^{-6}$ sec (data assembled by Lundy, 1962), and upon inserting this figure into (12.62) a value is obtained for G_μ which is remarkably close to the vector coupling constant for nuclear β-decay (12.26). We quote from the data assembled by Bardin *et al.* (1962) in Table 12.4.

In this table we have used $G_V \equiv C_V$ of (12.26) and τ_V represents a lifetime calculated from (12.62) with G_V replacing G_μ

$$\tau_V^{-1} = \frac{G_V^2 m_\mu^5}{192\pi^3}. \tag{12.63}$$

The corrections in the table are for electronic screening, nuclear form factors and radiative effects in the case of β-decay, and radiative effects only in μ-decay. The data given in Table 12.4 shows a small but nevertheless significant difference in the coupling strengths for β- and μ-decay. One method of explaining this discrepancy is by assuming that the interaction is propagated by a charged

† Despite the change in signs of the asymmetry coefficients for positrons and electrons, both particles emerge preferentially backwards with respect to muon momentum because of the difference in helicity of the μ^+ and μ^- particles (Fig. 12.6(a) and (d)).

TABLE 12.4

Quantity	Raw data	Corrected data
$ft\ O^{14}$	3066 ± 10 sec	3126 sec
G_V	$(1{\cdot}0131 \pm 0{\cdot}0016)\dfrac{10^{-5}}{m_p^2}$	$1{\cdot}0031 \times \dfrac{10^{-5}}{m_p^2}$
τ_μ	$(2{\cdot}210 \pm 0{\cdot}003)\ 10^{-6}$ sec	$2{\cdot}201 \times 10^{-6}$ sec
G_μ	$(1{\cdot}0215 \pm 0{\cdot}0008)\dfrac{10^{-5}}{m_p^2}$	$1{\cdot}0237 \times \dfrac{10^{-5}}{m_p^2}$
$\dfrac{G_\mu - G_V}{G_\mu}$	$(0{\cdot}8 \pm 0{\cdot}2)\ 10^{-2}$	$2{\cdot}0 \times 10^{-2}$
$\dfrac{\tau_\mu - \tau_V}{\tau_\mu}$	$(-1{\cdot}6 \pm 0{\cdot}4)\ 10^{-2}$	-4×10^{-2}

boson of mass m_W as indicated in Fig. 12.9 (Feynman and Gell-Mann, 1958). This changes the lifetime τ_V to

$$\tau_V' \sim \tau_V \left[1 - 0{\cdot}6\left(\frac{m_\mu}{m_W}\right)^2\right]. \tag{12.64}$$

Thus a mass $m_w \sim 500$ MeV (§ 12.11) reduces τ_V by about 2·5 per cent. However, the high energy neutrino experiments (§ 12.12) indicate that the mass of the boson is at least 1.3 GeV. The discrepancy could also be caused by uncertainties

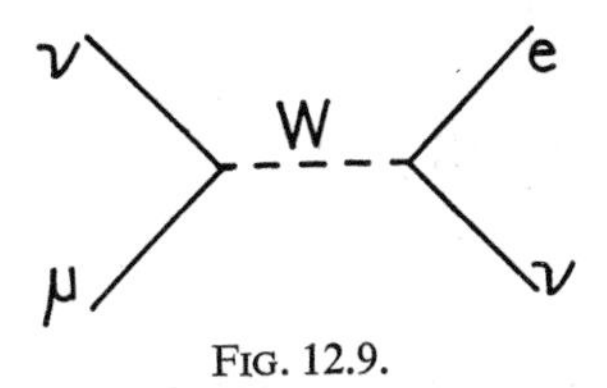

FIG. 12.9.

in the nuclear matrix element. Blin-Stoyle and Le Tourneux (1961) have pointed out that differences of the right magnitude could arise from isotopic spin impurities in the nuclear wave functions. This conclusion has been queried by Altmann and MacDonald (1962). A possible reason for the discrepancy within the framework of unitary symmetry has been suggested by Cabibbo (1963) (see remarks in § 12.9(a)).

We examine, finally, the differential forms of the electron spectrum. It is convenient to rewrite (12.61) in a simpler form by introducing the substitutions

$$x = \frac{E}{E_0}, \quad \cos\theta = \frac{\boldsymbol{\sigma} \cdot \mathbf{p}_e}{|\mathbf{p}_e|}$$

hence

$$\frac{d\lambda}{dx\,d\Omega} = \frac{1}{4\pi}\frac{1}{\tau_\mu}2x^2\,[3 - 2x \pm \cos\theta(2x - 1)]$$

$$= \frac{1}{4\pi}\frac{1}{\tau_\mu}2x^2(3 - 2x)\left[1 \pm \cos\theta\,\frac{(2x - 1)}{3 - 2x}\right]. \quad (12.65)$$

The integrated expressions for the energy spectrum and angular distributions are therefore

$$\frac{d\lambda}{dx} = \frac{2x^2}{\tau_\mu}(3 - 2x) \quad (12.66)$$

$$\frac{d\lambda}{d\Omega} = \frac{1}{4\pi}\frac{1}{\tau_\mu}\left(1 \pm \frac{1}{3}\cos\theta\right). \quad (12.67)$$

The shapes of the energy spectrum (12.66) and the asymmetry factor as a function of energy are given in Fig. 12.10. The results quoted below are in general agreement with these shapes.

The expressions given above can be derived using less restrictive assumptions than those employed in this section. For example, Kinoshita and Sirlin (1957)

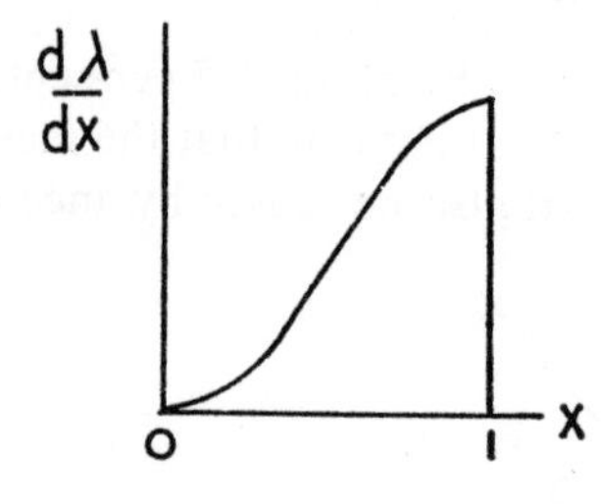

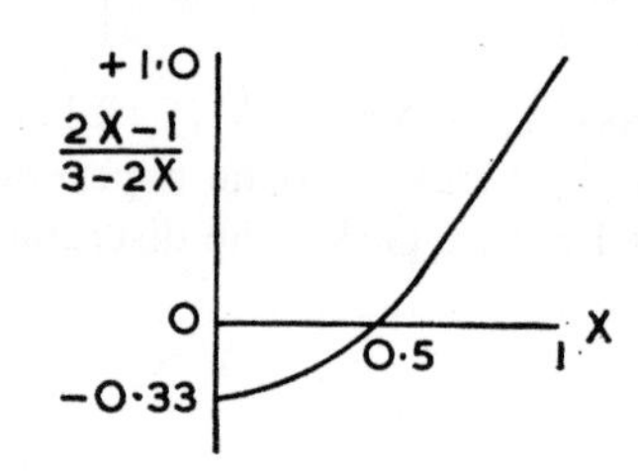

FIG. 12.10.

calculated the spectral shapes using a four-component neutrino theory and general parity violating terms. Their results can be given as

$$\frac{d\lambda}{dx\,d\Omega} = \frac{x^2}{\pi\tau_\mu}\{3(1 - x) + 2\varrho(\tfrac{4}{3}x - 1)$$

$$\mp \xi\cos\theta[1 - x + 2\delta(\tfrac{4}{3}x - 1)]\} \quad (12.68)$$

$$\frac{d\lambda}{dx} = \frac{4x^2}{\tau_\mu}[3(1 - x) + 2\varrho(\tfrac{4}{3}x - 1)] \quad (12.69)$$

where ϱ, ξ and δ are functions of the coupling constants. The function ϱ is known as the Michel ϱ-parameter (Michel, 1950, 1952); it is given by

$$\varrho = \frac{3(C_V^2 + C_V'^2 + C_A^2 + C_A'^2) + 6(C_T^2 + C_T'^2)}{(C_S^2 + C_S'^2) + (C_P^2 + C_P'^2) + 4(C_V^2 + C_V'^2 + C_A^2 + C_A'^2) + 6(C_T^2 + C_T'^2)} \tag{12.70}$$

(compare (12.28)) and it is apparent that it should equal 0·75 if

$$C_S = C_S' = C_T = C_T' = C_P = C_P' = 0.$$

In fact, a comparison of equations (12.68) and (12.69) with equations (12.65) and (12.66) shows that we should expect

$$\varrho = \delta = \tfrac{3}{4}, \qquad \xi = -1. \tag{12.71}$$

We list below some recent determinations of these parameters; the determination of ξ is complicated by the depolarisation of the muon which can occur in the detection system, we therefore give a lower limit, ξP, where P represents the polarisation retained by the muon.

ϱ = 0·741 ± 0·027 Dudziak, Sagane and Vedder (1959)

= 0·78 ± 0·025 Plano (1960)

= 0·774 ± 0·042 Kruger (1961)

δ = 0·78 ± 0·05 Plano (1960)

= 0·782 ± 0·031 Kruger (1961)

ξP = −0·97 ± 0·06 Ali-Zade *et al.* (1959)

= −0·94 ± 0·07 Plano (1960)

= −0·97 ± 0·05 Bardon, Berley and Lederman (1959)

= −0·87 ± 0·04 Lynch, Orear and Rosendorff (1960)

The data can be seen to be consistent with the figures given in (12.71) and suggests $C_V = C_V' = -C_A = -C_A'$.

It should be noted that if a vector boson propagates the reaction as in Fig. 12.9, then ϱ is expected to have the value

$$\varrho = 0{\cdot}75 + \frac{1}{3}\left(\frac{m_\mu}{m_W}\right)^2. \tag{12.72}$$

12.4. CONSERVATION OF LEPTONS

So far in this chapter we have examined three processes

$$N \to N + e + \nu$$

$$\pi \to \mu + \nu$$

$$\mu \to e + \nu + \nu$$

where the symbols have been used generically. All three processes have been found to be consistent with a law of conservation of lepton number; for further discussion of this point see § 9.3(d). If we define a neutrino as a lepton, then we may draw up a table of leptons and antileptons (Table 12.5).

TABLE 12.5

Lepton $l = +1$	Antilepton $l = -1$
ν	$\bar{\nu}$
e^-	e^+
μ^-	μ^+

It should be noted, however, that the handedness or helicity of the particles does not follow the same grouping:

$$\begin{aligned} &\text{left-handedness (negative helicity)} \quad && \nu,\ e^-,\ \mu^+ \\ &\text{right-handedness (positive helicity)} \quad && \bar{\nu},\ e^+,\ \mu^-. \end{aligned} \tag{12.73}$$

The data on the process $K \to \mu + \nu$ is also consistent with the above assignments.

12.5. STRUCTURAL TERMS IN WEAK INTERACTIONS

12.5(a). *Introduction*

We have treated the subjects of β- and μ-decay by assuming that the basic interaction was a local one between bare Dirac particles. This assumption is not entirely justifiable; as we shall show later, it can lead to infinite cross-sections for high energy processes induced by neutrinos. In addition the problem of strong virtual processes arises in β-decay, since two of the participating particles are strongly interacting. The same problem appears in all other weak interactions (apart from muon decay).

Even in this complex situation certain simplifications may be introduced. The conservation of leptons implies that they occur in pairs in weak interactions. Furthermore, leptons have no strong interactions, and so we may assume that the only virtual processes associated with them are the small electromagnetic corrections arising from their electric charge. It therefore seems reasonable to conjecture that, regardless of the detailed structure of a given interaction, the leptonic fields will couple locally at some point and can be factored out of the interaction term (Goldberger and Treiman, 1958b). Tests for this conjecture have been suggested by Pais (1962).

Let us consider, for example, a process involving two strongly interacting particles A and B and a lepton pair

$$A \to B + e^- + \bar{\nu}.$$

Then an appropriate matrix element for this interaction with lepton components factored out would be given by

$$M_{fi} = \sum_i \langle B| J_i |A\rangle \frac{1}{N_l} \bar{u}_e \Gamma_i (1 + \gamma_5) u_\nu \tag{12.74}$$

where the u terms are spinors and N_l is a normalisation factor for the leptons. The operators J_i are called *currents*, but may have any of the transformation properties S, V, T, A, P.

The matrix elements of the currents are determined by the strong interactions, and so the safest way of treating them is by exploiting their symmetry properties. We therefore commence by splitting the current into a part which conserves strangeness (j), and one which changes strangeness (s). To lowest order only one of them occurs in any process

$$J_i = j_i + s_i \tag{12.75}$$

for example j occurs in neutron β-decay and s in hyperon β-decay.

12.5(b). *The $\Delta S = 0$ leptonic transitions*

In the remainder of this section we shall discuss the strangeness conserving leptonic transitions with two strongly interacting particles. Examples are

$$n \to p + e^- + \bar{\nu}$$

$$\mu^- + p \to n + \nu$$

$$\Sigma^- \to \Lambda + e^- + \bar{\nu}$$

$$\Sigma^+ \to \Lambda + e^+ + \nu.$$

The success of perturbation theory in accounting for the main features of β-decay (§ 12.2) suggests that the currents can be constructed from bilinear sets of strongly interacting fields, for example in the β-decay of the neutron

$$j_i \equiv \frac{G_i}{\sqrt{2}} \bar{\psi}_p \Gamma_i \psi_n.$$

More complex terms could be included, for example

$$n \to p + \pi^- \to p + e^- + \bar{\nu}.$$

However, in what follows no specific fields will be assumed for j. The general structure of the matrix element must have the form (compare §§ 9.4(d) and 11.4(c))

$$\langle p| j_i |n\rangle = \frac{1}{N} \bar{u}_p O_i(p, n)\, u_n \tag{12.76}$$

where O_i represents Lorentz invariant terms formed from the four-momenta and γ-matrices with transformation properties appropriate to the label i, and N is a normalisation factor for the spinors of the form $\sqrt{(E/m)}$.

The structure of the O operators for β-decay can be established in the same manner as in §§ 9.4(d) and 11.4(c) (the vector term is in fact indentical with (11.161)). They are as follows:

$$\begin{aligned}
O_S &= G^S(q^2) \qquad\qquad (12.77)\\
O_V &= G_1^V(q^2)\, \gamma_\lambda + G_2^V(q^2)\, \sigma_{\lambda\nu} q_\nu\\
O_T &= G_1^T(q^2)\, \sigma_{\lambda\nu} + G_2^T(q^2)\, (q_\lambda \gamma_\nu - q_\nu \gamma_\lambda) + iG_3^T(q^2)\, (p_\lambda n_\nu - p_\nu n_\lambda)\\
O_A &= iG_1^A(q^2)\, \gamma_\lambda \gamma_5 + G_2^A(q^2)\, q_\lambda \gamma_5\\
O_P &= G^P(q^2)\, \gamma_5
\end{aligned}$$

where $q = n - p$ represents the four-momentum transfer. The terms G are coupling strengths or form factors; if the transition amplitude M_{fi} (12.74) is invariant under time reversal they must be real quantities (compare § 9.2(e)). The expressions given in (12.77) ensure that O is invariant under the G parity operation (§ 9.4(g)). Noninvariance under G parity would imply the existence of additional terms in the operators O_i, for example a factor $iG_3^V(q^2)\, q_\lambda$ in O_V (Weinberg, 1958).

Despite the complicated structure of (12.77), the actual situation encountered in β-decay is relatively simple. As we have seen in § 12.2(b) the scalar and tensor terms do not appear or are vanishly small. The matrix elements for the pseudoscalar case couples the large components of spinors with the small components and these vanish in the nonrelativistic limit; thus G^P is not detectable unless it is abnormally large compared to the other form factors. The experimental evidence indicates that G^P does not enjoy this privileged status. Thus we are left with the V and A terms; now the four-momentum transfer in β-decay is relatively small, and so only the terms

$$G_1^V(q^2)\,\gamma_\lambda, \qquad iG_1^A(q^2)\,\gamma_\lambda\gamma_5$$

make an appreciable contribution. The G terms are constant for all practical purposes over the restricted range of values for q^2 occuring in β-decay and so the classical methods of analysis for the β-decay process, which we have used in § 12.2, are justified if we assume the limiting values

$$G_1^V(q^2) \to \frac{G_V}{\sqrt{2}}, \qquad G_1^A(q^2) \to \frac{G_A}{\sqrt{2}} \quad (q^2 \to 0). \tag{12.78}$$

12.6. THE CONSERVED VECTOR CURRENT HYPOTHESIS

12.6(a). *Introduction*

In § § 12.2 and 12.3 it was shown that the basic coupling forms for β- and μ-decay were essentially V–A. The data may be summarised as follows:

$$\left.\begin{array}{c} G_V^\beta \cong G_V^\mu \\ G_A^\mu \cong -G_V^\mu \quad G_A^\beta \cong -1{\cdot}2G_V^\beta \end{array}\right\} \tag{12.79}$$

where the sign $\cong$ implies 'exactly or very closely equal to'.

The near equality of the vector coupling strength for β and μ decay is surprising since it might be expected that virtual strong interactions could seriously affect the structure of the np pair in β-decay, and hence modify the coupling strength. Or in more technical language we could say that the vector coupling does not appear to be renormalised.

An explanation for the lack of renormalisation has been made by Gershtein and Zeldovitch (1956) and independently by Feynman and Gell-Mann (1958).

FIG. 12.11.

Their idea is known as the conserved vector current hypothesis. It is based upon an analogy with electrodynamics, where it is well known that the strength of electromagnetic interactions (and therefore the electric charge) is unaffected by the presence of strong interactions.

Consider, for example, a proton current interacting with an electromagnetic field. Part of the time the proton will be dissociated into a neutron and π^+

(Fig. 12.11), and so when the photon cannot couple to the proton it couples to the π^+ so that the Langrangian for the electromagnetic interactions is given by

$$\mathscr{L}_{\text{int}}\,(\text{em}) = j_\mu A_\mu = (j_N + j_\pi)_\mu\, A_\mu. \tag{12.80}$$

The conservation of electric charge in this system is then given by the requirement

$$\frac{\partial j_\mu}{\partial x_\mu} = \frac{\partial}{\partial x_\mu}(j_N + j_\pi)_\mu = 0. \tag{12.81}$$

Now consider a weak interaction (Fig. 12.12) and assume that G_V is conserved in the same manner as e

$$\frac{G_V}{\sqrt{2}} \equiv e. \tag{12.82}$$

This may occur, by analogy with the electromagnetic case, if a weak vector current exists which satisfies the relation

$$\frac{\partial j^V_\mu}{\partial x_\mu} = 0. \tag{12.83}$$

Let us now examine the two currents and find what they have in common. With the aid of equations (9.95) and (9.102) the electromagnetic current can be written as

$$\begin{aligned}(j_N + j_\pi)_\mu &= ie\left[\overline{\psi}_p\gamma_\mu\psi_p - i\left(\frac{\partial\varphi_1}{\partial x_\mu}\varphi_2 - \frac{\partial\varphi_2}{\partial x_\mu}\varphi_1\right)\right]\\ &= ie\left[\overline{\psi}_N\gamma_\mu\left(\frac{1+\tau_3}{2}\right)\psi_N + \frac{\partial\varphi^{(\dagger)}}{\partial x_\mu}t_3\varphi\right]\\ &= ie\,[j^S + j^{(3)}]_\mu\end{aligned} \tag{12.84}$$

where j^S and $j^{(3)}$ represent an isoscalar current and a third component of an isovector current respectively. The condition for charge conservation (12.81) now becomes

$$\frac{\partial j^S_\mu}{\partial x_\mu} = \frac{\partial j^{(3)}_\mu}{\partial x_\mu} = 0 \tag{12.85}$$

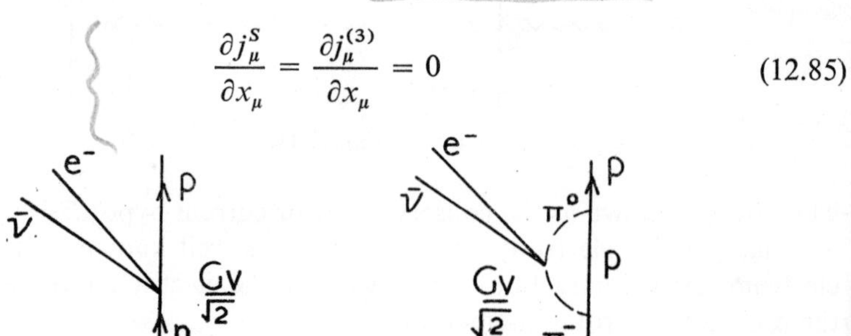

FIG. 12.12.

since both baryon number and the third component of isospin are conserved quantities (compare § 9.4 (b)). But the postulate of charge independence (or, equivalently, invariance under rotation in isospin space) implies that the whole isospin vector current must be conserved, hence

$$\frac{\partial \mathbf{j}_\mu}{\partial x_\mu} = 0 \tag{12.86}$$

where

$$\mathbf{j}_\mu = \tfrac{1}{2}\,\overline{\psi}_N \gamma_\mu \boldsymbol{\tau} \psi_N + \frac{\partial \varphi^{(\dagger)}}{\partial x_\mu}\,\mathbf{t}\varphi$$

Now the $j^{(1)} + ij^{(2)}$ component of this current will also be conserved, and it contains the necessary operators to effect the transformations appearing in Fig. 12.12 (compare (9.108))

$$\begin{aligned}[j^{(1)} + ij^{(2)}]_\mu &= \tfrac{1}{2}\,\overline{\psi}_N \gamma_\mu (\tau_1 + i\tau_2)\,\psi_N + \frac{\partial \varphi^{(\dagger)}}{\partial x_\mu}\,(t_1 + it_2)\,\varphi \\ &= \overline{\psi}_N \gamma_\mu \tau_+ \psi_N + \frac{\partial \varphi^{(\dagger)}}{\partial x_\mu}\,t_+\varphi \\ &= \psi_p \gamma_\mu \psi_n + \sqrt{2}\left(\frac{\partial \varphi_3}{\partial x_\mu}\,\varphi^\dagger - \frac{\partial \varphi^\dagger}{\partial x_\mu}\,\varphi_3\right)\end{aligned} \tag{12.87}$$

where $\varphi^\dagger = (\varphi_1 + i\varphi_2)/\sqrt{2}$ creates and destroys charged pions and is the same as the operator $\varphi^\dagger$ of § 4.3(i). Hence the hypothesis of a conserved vector current in weak interactions may be satisfied if that current behaves like the $x + iy$ component of the isovector current

$$\frac{\partial j^V_\mu}{\partial x_\mu} \equiv \frac{\partial}{\partial x_\mu}\,[j^{(1)} + ij^{(2)}]_\mu = 0 \tag{12.88}$$

$$j^V_\mu = \frac{G_V}{\sqrt{2}}\left[\overline{\psi}_p \gamma_\mu \psi_n + \sqrt{2}\left(\frac{\partial \varphi_3}{\partial x_\mu}\,\varphi^\dagger - \frac{\partial \varphi^\dagger}{\partial x_\mu}\,\varphi_3\right)\right].$$

The analogy between the weak and electromagnetic vector currents may be pushed further (Gell-Mann, 1958). The interaction terms in the two processes are:

$$\text{electromagnetic} \quad j_\mu A_\mu = ie[j^S + j^{(3)}]_\mu A_\mu \tag{12.89}$$

$$\text{weak} \quad \langle p|\, j^V_\mu\, |n\rangle\, \frac{1}{N_l}\,\bar{u}_e \gamma_\mu (1 + \gamma_5)\, u_\nu. \tag{12.90}$$

Let us examine the structural forms of the two currents. We have seen in § 11.4(d) that the isotopic vector part of the nucleon electromagnetic current has the following form:

$$\langle N'|\, j_\mu^V\, |N\rangle = ie\bar{u}_{N'}[F_1^V(q^2)\,\gamma_\mu + \frac{1}{2M}\,F_2^V(q^2)\,\sigma_{\mu\nu}q_\nu]\,\tau_3 u_N \tag{12.91}$$

and from (12.77) we may write the weak vector current as

$$\begin{aligned}\langle p|\, j_\mu^V\, |n\rangle &= \bar{u}_p[G_1^V(q^2)\,\gamma_\mu + G_2^V(q^2)\,\sigma_{\mu\nu}q_\nu]\,u_n \\ &= \frac{G_V}{\sqrt{2}}\,\bar{u}_p[a_V(q^2)\,\gamma_\mu + b_V(q^2)\,\sigma_{\mu\nu}q_\nu]\,u_n\end{aligned} \tag{12.92}$$

where

$$a_V(0) = 1$$

so that the matrix element is the same as in the classical theory of β-decay in the nonrelativistic limit. The limiting values of F_1 and F_2 were given in (11.176)

$$F_1^V(0) = \tfrac{1}{2}, \qquad F_2^V(0) = \frac{F_{2p}(0) - F_{2n}(0)}{2} = \frac{\mu_a - \mu_n}{2} = 1{\cdot}85 \tag{12.93}$$

where μ_n and μ_a represent the neutron and anomalous proton magnetic moments respectively. Hence, if the analogy between electromagnetic and weak vector currents is a real one,

$$b_V\,(0) = \frac{\mu_a - \mu_n}{2M}. \tag{12.94}$$

This contribution was called *weak magnetism* by Gell-Mann.

12.6(b). *Tests of the conserved vector current hypothesis*

Two main tests of the conserved vector current theory have been proposed. The first concerns the process appearing in Fig. 12.12

$$\pi^- \to \pi^0 + e^- + \bar{\nu}\,. \tag{12.95}$$

The conserved vector current theory implies that the strength of the interaction is $G_V/\sqrt{2}$, and the requisite matrix element can be calculated by a formal development of the field operators appearing in (12.87). This procedure is not necessary, however. We first note that if we ignore the difference in pion masses

$$\langle \pi^0|\, T_+\, |\pi^-\rangle = \sqrt{2}$$

since (9.134)

$$T_+|\,T, T_3\rangle = \sqrt{[T(T+1) - T_3(T_3+1)]}\,|T, T_3+1\rangle.$$

Secondly, since the energy release in the process is small the π° meson is virtually stationary and so only the $\mu = 4$ component of the meson transformation current need be considered. Thus we can write

$$\langle \bar{\nu} e^- \pi^0 | M | \pi^- \rangle = \frac{G_V}{\sqrt{2}} \sqrt{2}\, \delta_{\mu 4} \frac{1}{N_l} \bar{u}_e \gamma_\mu (1 + \gamma_5)\, u_\nu$$

$$= \frac{G_V}{N_l} \bar{u}_e \gamma_4 (1 + \gamma_5)\, u_\nu . \tag{12.96}$$

This matrix element may be evaluated using (12.13) and gives $2G_V^2$. Now there are three particles in the final state, hence if we use the phase space factor $(1/60\pi^3)\,\Delta^5$ derived for β-decay (§ 8.7(c)) the transition probability for the process $\pi^- \rightarrow \pi^0 + e^- + \bar{\nu}$ is given by

$$\lambda \sim \frac{2G_V^2}{60\pi^3} \Delta^5 \sim 0{\cdot}5 \text{ sec}^{-1} \tag{12.97}$$

where $\Delta = m_{\pi^-} - m_{\pi^0} \sim 4{\cdot}6$ MeV and we have ignored blocking factors. Now the decay process $\pi \rightarrow \mu + \nu$ has a measured lifetime of $\sim 2{\cdot}5 \times 10^{-8}$ sec, and so the branching ratio for $\pi^- \rightarrow \pi^0 + e^- + \bar{\nu}$ is given by

$$\frac{\lambda(\pi^- \rightarrow \pi^0 + e^- + \bar{\nu})}{\lambda(\pi \rightarrow \mu + \nu)} \sim \frac{0{\cdot}5}{0{\cdot}4 \times 10^8} \sim 10^{-8}. \tag{12.98}$$

The ratio is a precise one, and an exact evaluation predicts $(1{\cdot}06 \pm 0{\cdot}02)\, 10^{-8}$; now if we make no assumptions about conserved vector currents, the decay $\pi^- \rightarrow \pi^0 + e^- + \bar{\nu}$ can occur by the process shown in Fig. 12.13. This diagram

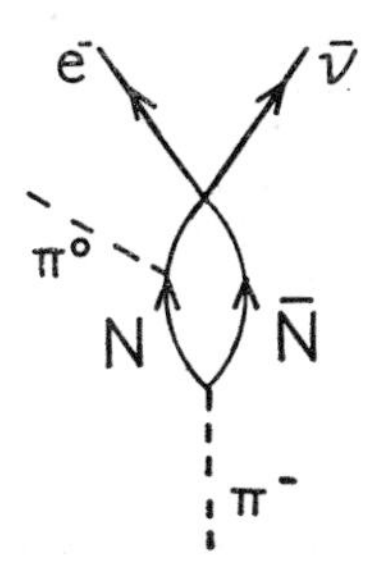

FIG. 12.13.

involves two strong interaction vertices, and hence all the problems associated with strong interactions arise when one tries to calculate the transition rate. A rough estimate can be made by pertubation theory and yields a branching ratio within the limits 10^{-7} to 10^{-9}.

Thus the test of the conserved vector current is essentially a negative one. A branching ratio of 10^{-8} does not necessarily prove the correctness of the theory, but a ratio not equal to 10^{-8} would rule it out. The measurement is a difficult one, but nevertheless the decay has been detected with the following results for the branching ratio:†

$$\left(1\cdot1 \pm \begin{matrix}1\cdot0\\0\cdot5\end{matrix}\right) 10^{-8} \quad \text{Dunaitsev } et\ al.\ (1962)$$

$$(1\cdot15 \pm 0\cdot22)\ 10^{-8} \quad \text{Depommier } et\ al.\ (1963)$$

$$(1\cdot7 \pm 0\cdot5)\ 10^{-8} \quad \text{Bacastow } et\ al.\ (1962).$$

Thus the experimental results are consistent with the conserved vector current theory.

A search for the weak magnetism term has also been made. It is based upon a test proposed by Gell-Mann (1958). Consider the isotopic triplet B^{12}, C^{12*}, N^{14} (Fig. 12.14).

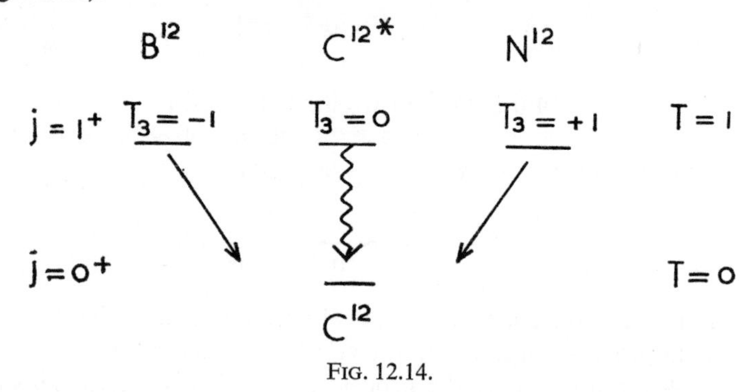

FIG. 12.14.

The transition $C^{12*} \to C^{12}$ proceeds by the emission of pure magnetic dipole radiation and the measured transition rate gives the size of the matrix element. If Coulomb corrections are ignored, the allowed Gamow–Teller transition would give the same matrix elements for the positron and the electron emission. However, a finite vector contribution also occurs as a first forbidden transition because of the large energy release. If the conserved vector current theory holds, the strength of the vector term is uniquely determined by the weak magnetism term (12.94) which in turn is related to the matrix element for the γ-radiation. The vector contribution leads to an interference term of opposite sign in the energy spectrum for the positrons and electrons, and if the ratio of intensities in the two spectra are examined as a function of energy (E) Gell-Mann showed that a term of the following type would be expected:

$$\frac{B^{12} \text{ spectrum}}{N^{12} \text{ spectrum}} = 1 + \frac{16}{3} \frac{a}{2M} E$$

where

$$a = 1 + \mu_a - \mu_n \sim 4\cdot7\,.$$

† The experiments actually measured the ratio for the process $\pi^+ \to \pi^0 + e^+ + \nu$, which, of course, has the same decay rate as that for $\pi^- \to \pi^0 + e^- + \bar{\nu}$.

Thus the correction should be $16 \times 4{\cdot}7/3 \times 2{\cdot}10^3$ per MeV $\sim 1{\cdot}2$ per cent per MeV. Calculations which include the Coulomb corrections for the nuclear wave functions have shown that the effect is still detectable. A measurement of the effect has yielded $(1{\cdot}19 \pm 0{\cdot}24)$ per cent per MeV (Lee, Mo and Wu, 1963), which is good agreement with the calculated value of $(1{\cdot}10 \pm 0{\cdot}17)$ per cent per MeV.

12.7. PION DECAY

The process of π–μ decay can be visualised as occurring through the virtual dissociation of the pion into a nucleon–antinucleon pair, which subsequently undergoes a process equivalent to muon capture. The lower vertex in this process (Fig. 12.15) is a strong interaction, and all the attendant problems of strong interactions arise when one attempts to evaluate it.

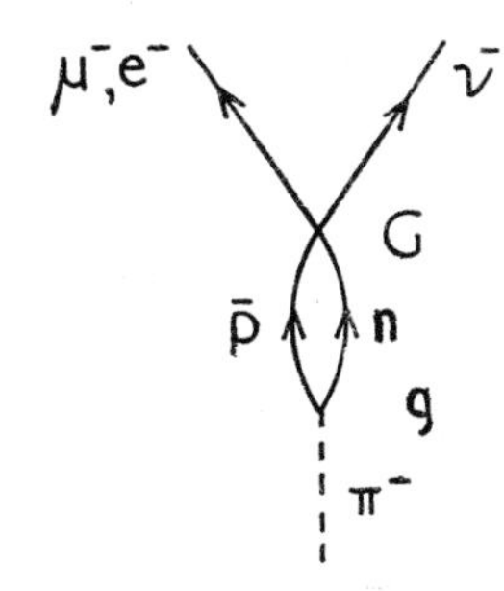

FIG. 12.15.

We start therefore with a simpler problem – the branching ratio for the processes $\pi \to e + \nu$ and $\pi \to \mu + \nu$. The process of π–e decay can be visualised in terms of Fig. 12.15 with an electron replacing the muon line. A good estimate of the branching ratio for the two processes can be made with perturbation theory since the lower vertex cancels (Ruderman and Finkelstein, 1949). An even easier evaluation may be made by applying invariance principles to the strong interaction currents.

Since the pion is a spinless pseudoscalar particle only the axial vector and pseudoscalar currents ($\partial\varphi/\partial x_\lambda$ and φ in field language) can contribute to the matrix element. The matrix element is therefore

$$M_{fi} = \langle 0|\, j_\lambda^A \,|\pi^-\rangle \frac{1}{N_l}\, \bar{u}_l i\gamma_\lambda(1 + \gamma_5)\, u_\nu + \langle 0|\, j^P \,|\pi^-\rangle \frac{1}{N_l}\, \bar{u}_l(1 + \gamma_5)\, u_\nu \tag{12.99}$$

where N_l is a normalisation factor for the leptons and the subscripts l on $\bar{u}_l$ refer to muon or electron. The currents must possess the following structures:

$$\langle 0|\, j_\lambda^A \,|\pi^-\rangle = -\frac{1}{\sqrt{2m_\pi}} f_A \frac{p_\lambda}{m_\pi}, \qquad \langle 0|\, j^P \,|\pi^-\rangle = \frac{1}{\sqrt{2m_\pi}} f_P \tag{12.100}$$

where $p_\lambda = (p_l + p_\nu)_\lambda$ and the mass m_π is included so that f_A and f_P should have the same dimensions; the factor $\sqrt{2m_\pi}$ ensures that the matrix element has a structure analogous to that obtained from field theory (compare Table 8.1). The form factors f may be regarded as coupling strengths. If we exploit the Dirac equations (3.146)

$$(i\gamma_\lambda p_\lambda + m)\, u = 0$$

$$\bar{u}(i\gamma_\lambda p_\lambda + m) = 0$$

and the fact that $m_\nu = 0$, M_{fi} may be reduced to

$$M_{fi} = \frac{1}{\sqrt{2m_\pi}} \left(f_A \frac{m_l}{m_\pi} + f_P \right) \frac{1}{N_l}\, \bar{u}_l(1 + \gamma_5)\, u_\nu \tag{12.101}$$

and with the aid of (7.41), (12.13), (7.59) and (A.8.2) (p. 713) the branching ratio becomes

$$R = \frac{\lambda(\pi \to e + \nu)}{\lambda(\pi \to \mu + \nu)} = \left| \frac{f_A m_e + f_P m_\pi}{f_A m_\mu + f_P m_\pi} \right|^2 \frac{(m_\pi^2 - m_e^2)^2}{(m_\pi^2 - m_\mu^2)^2}. \tag{12.102}$$

In the two extreme limits $f_A = 0$ and $f_P = 0$ the ratio takes on the following values:

$$f_A = 0, \qquad R = 5{\cdot}4; \qquad f_P = 0, \qquad R = 1{\cdot}36 \times 10^{-4}.$$

The experimental data may be summarised as follows:

$$R = (1{\cdot}22 \pm 0{\cdot}30)\, 10^{-4} \quad \text{Ashkin } et\ al. \text{ (1959)}$$

$$(1{\cdot}21 \pm 0{\cdot}07)\, 10^{-4} \quad \text{Anderson } et\ al. \text{ (1960)}$$

and obviously agrees with the assumption $f_P = 0$. Indeed, when electromagnetic corrections are made the theoretical figure becomes $1{\cdot}23 \times 10^{-4}$ (Kinoshita, 1959). The comparison between theory and experiment therefore offers striking proof of the absence of a pseudoscalar term in pion decay, and also of the interchangeability of muons and electrons in weak interactions.

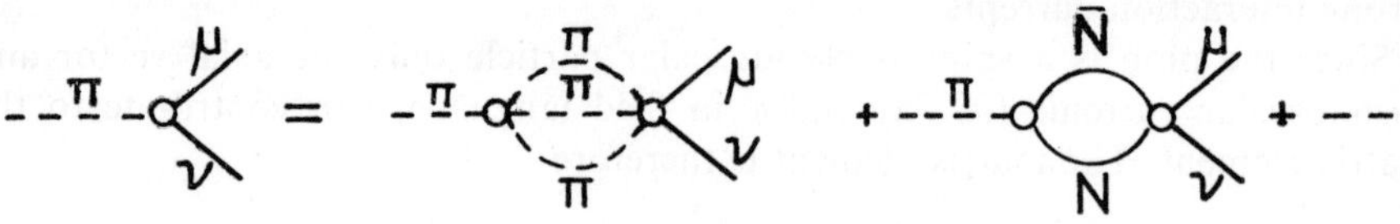

FIG. 12.16.

Let us next consider the problem of the absolute decay rate of the pion. The result obtained above implies that the matrix element for π–μ decay is

$$M_{fi} = \frac{1}{\sqrt{2m_\pi}}\, f_A \frac{m_\mu}{m_\pi} \frac{1}{N_l}\, \bar{u}(1 + \gamma_5)\, u_\nu. \tag{12.103}$$

A calculation of the decay rate has been made by Goldberger and Treiman (1958a); these workers treated f_A as an analytic function and evaluated it by means of a dispersion relation. This technique involved the summation over intermediate states consistent with the quantum numbers of the pion (Fig. 12.16) and led to the result

$$f_A = -Mm_\pi \frac{G_A}{g}$$

where M is the nucleon mass, G_A the axial vector coupling constant in β-decay and g the pion–nucleon coupling constant.

The value of f_A found from this result is in remarkable agreement with that calculated from the measured lifetime of the pion. The above equation has been reproduced in many other ways. One of the most interesting treatments is that of Nambu (1960), who argued that the smallness of the renormalisation effect on the vector coupling constant in β-decay ($G_A = -1{\cdot}2\, G_V$) implies that the axial vector current (12.77)

$$\begin{aligned} \langle p \,|\, j_\lambda^A \,|\, n\rangle &= \frac{1}{N}\, \bar{u}_p[iG_1^A(q^2)\,\gamma_\lambda\gamma_5 + G_2^A(q^2)\, q_\lambda\gamma_5]\, u_n \\ &= \frac{1}{N}\frac{G_A}{\sqrt{2}}\, \bar{u}_p[i\gamma_\lambda\gamma_5 + c(q^2)\, q_\lambda\gamma_5]\, u_n \end{aligned} \qquad (12.104)$$

is almost conserved. Exact conservation would occur if the expression in the brackets was

$$O_\lambda = i\gamma_\lambda\gamma_5 - 2\frac{M}{q^2}\, q_\lambda\gamma_5 \quad \text{since} \quad \bar{u}_p q_\lambda O_\lambda u_n = 0$$

(compare the argument associated with equation (11.157)), but O_λ leads to an impossibly large pseudoscalar term.

Nambu argued that, if a mass term is included, the denominator of the second term in O_λ is similar to that of a pion propagator in conventional field theory

$$-\frac{2M}{q^2} \sim -\frac{2M}{q^2 + m_\pi^2} = c(q^2).$$

The current j_λ^A is then approximately conserved for low q^2 and exactly conserved in the limit $q^2 \gg m_\pi^2$.

Now, if we recall Fig. 9.6, axial vector β-decay through an intermediate pion state can be represented in perturbation theory by a matrix element

$$\sqrt{2}\, g \frac{1}{N}\, \bar{u}_p\gamma_5 u_n \frac{1}{q^2 + m_\pi^2} f_A \frac{q_\lambda}{m_\pi} \text{(lepton terms)}.$$

Thus if we compare coupling terms with (12.104) we obtain the relation of Goldberger and Treiman

$$-\frac{G_A}{\sqrt{2}}\,2M = \sqrt{2}\,g\,\frac{f_A}{m_\pi}$$

$$f_A = -Mm_\pi\,\frac{G_A}{g} \tag{12.105}$$

$$M_{fi} = -\frac{1}{\sqrt{2m_\pi}}\,Mm_\pi\,\frac{G_A}{g}\,\frac{m_\mu}{m_\pi}\,\frac{1}{N_l}\,\bar{u}_\mu(1+\gamma_5)\,u_\nu.$$

Alternative methods have been used for deriving the formula, for example Gell-Mann and Levy (1960) have shown that if it is assumed that the divergence of the axial vector current $\partial j_\lambda^A/\partial x_\lambda$ is proportional to the pion field operator then the same result may be achieved. This assumption implies that the pion–nucleon coupling should be of the derived type.

12.8. MUON CAPTURE

12.8(a). *Structure of the matrix element*

There is a formal similarity between the processes of β-decay and muon capture

$$\nu + n \rightarrow p + e^-$$

$$\mu^- + p \rightarrow n + \nu$$

It can be seen that one is in effect the time reversed process of the other with muon substituted for electron. The invariance of the β-decay process under time reversal and the apparently interchangeable properties of muon and electron (apart from mass) therefore suggests that the same matrix element can be used in both processes. The difference in mass does imply, however, that momentum-dependent terms can be more important in muon capture than in β-decay. The matrix element can be written with the aid of equation (12.77) and (12.92) as

$$M_{fi} - \frac{G_V}{\sqrt{2}}\,\bar{u}_n[a_V(q^2)\,\gamma_\lambda + b_V(q^2)\,\sigma_{\lambda\nu}q_\nu]\,u_p\,\frac{1}{N}\,\bar{u}_\nu(1-\gamma_5)\,\gamma_\lambda u_\mu$$

$$+\frac{G_A}{\sqrt{2}}\,\bar{u}_n\,[ia_A(q^2)\,\gamma_\lambda\gamma_5 + b_A(q^2)\,q_\lambda\gamma_5)]\,u_p\,\frac{1}{N}\,\bar{u}_\nu(1-\gamma_5)\,i\gamma_\lambda\gamma_5 u_\mu \tag{12.106}$$

where

$$q = p - n = \nu - \mu$$

represents the four-momentum transfer, and N covers all normalisation terms.

An application of the Dirac equation as in (12.101) and conservation of energy–momentum allows us to rewrite M_{fi} as

$$M_{fi} = \frac{G_V}{N\sqrt{2}}\left[\bar{u}_n\gamma_\lambda\left(1 - \frac{G_A}{G_V}\gamma_5\right)u_p\right][\bar{u}_\nu(1-\gamma_5)\gamma_\lambda u_\nu]$$

$$+ \frac{G_V}{N\sqrt{2}} b_V(q^2)\,\bar{u}_n\sigma_{\lambda\nu}q_\nu u_p[\bar{u}_\nu(1-\gamma_5)\gamma_\lambda u_\mu]$$

$$- \frac{G_p}{N}\,\bar{u}_n\gamma_5 u_p[\bar{u}_\nu(1-\gamma_5)\gamma_5 u_\mu] \tag{12.107}$$

where we have set

$$a_V(q^2) = a_A(q^2) \sim 1, \qquad q^2 \sim m_\mu^2$$

$$G_P = \frac{G_A}{\sqrt{2}} b_A(q^2)\, m_\mu$$

since a_V and a_A are not expected to change significantly from the β-decay form factors $a_V(0)$, $a_A(0)$. If the conserved vector current theory is exploited, we can also write

$$b_V(q^2) \sim \frac{\mu_a - \mu_n}{2M}$$

(compare (12.94)). More exact calculations by Fujii and Primakoff (1959) have shown that the approximations are good ones.

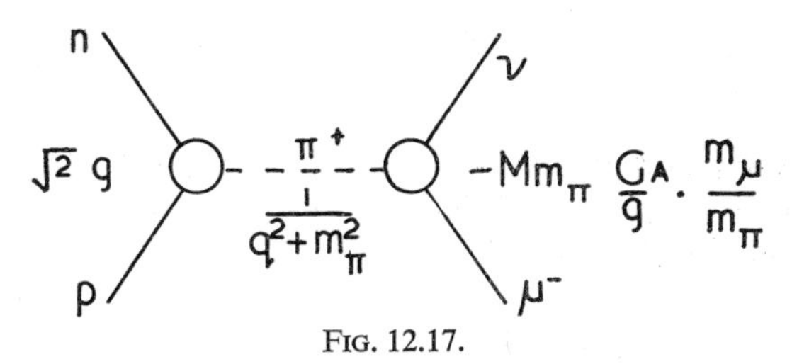

FIG. 12.17.

An inspection of (12.107) shows that the first term is the equivalent of the β-decay matrix element (12.36), when it is realised that the lepton term has now been written in reversed order

$$(1-\gamma_5)\gamma_\lambda = \gamma_\lambda(1+\gamma_5).$$

The second and third terms are recoil terms and are much more important in muon capture than in β-decay because of the larger four-momentum transfer. The third term has the form of a pseudoscalar and is known as the *induced pseudoscalar interaction.* A rough magnitude for the coupling strength of this term may be set by considering Fig. 12.17, where we have inserted the coupling strength for a strong interaction (9.112) and π–μ decay amplitude (12.105) at the two vertices. The matrix element corresponding to this diagram is

$$-\sqrt{2}g\,\frac{1}{q^2+m_\pi^2}\,Mm_\pi\,\frac{G_A}{g}\,\frac{m_\mu}{m_\pi}\,\frac{1}{N}\,\bar{u}_n\gamma_5 u_p[\bar{u}_\nu(1-\gamma_5)\,\gamma_5 u_\mu]$$

$$\sim -\sqrt{2}\,\frac{Mm_\mu}{m_\pi^2}\,\frac{G_A}{N}\,\bar{u}_n\gamma_5 u_p[\bar{u}_\nu(1-\gamma_5)\,\gamma_5 u_\mu]. \tag{12.108}$$

Thus the effective coupling strength for the induced pseudoscalar interaction is $G_P \sim 10G_A$ and with same sign as the axial vector term. A more elaborate calculation by Goldberger and Treiman (1958b) using dispersion relations has yielded a similar result, $G_P \sim 8G_A$.

The detailed theory of the muon capture process in hydrogen and more complex nuclei has been given by Primakoff (1959), using the matrix element (12.107). The recoil terms lead to important modifications of the figures predicted by a point interaction in the processes of muon capture. For example the muon can be captured by a proton in a triplet ($j = 1$) or singlet ($j = 0$) state of angular momentum; in the absence of recoil terms the ratio for capture in the two states is given by

$$\frac{\lambda_0}{\lambda_1} = \frac{\lambda(j=0)}{\lambda(j=1)} = \frac{(G_V - 3G_A)^2}{(G_V + G_A)^2} \sim 500 \tag{12.109}$$

where we have used the value $G_A = -1{\cdot}2\,G_V$ from β-decay. The calculation of Primakoff yields

$$\frac{\lambda_0}{\lambda_1} = \frac{636\ \text{sec}^{-1}}{13\ \text{sec}^{-1}} \sim 50.$$

For a further example we quote the angular distribution of the neutrons arising from polarised muon capture. This is represented by $1 + \alpha \cos\theta$, where θ is the angle between the direction of emission of the neutron and the muon spin, and

$$\alpha = \frac{G_V^2 - G_A^2}{G_V^2 + 3G_A^2} = -0{\cdot}08 \tag{12.110}$$

without recoil terms. The presence of the recoil terms raises this figure to $-0{\cdot}39$.

12.8(b). *Experimental data*

An experimental study of the basic process

$$\mu^- + p \rightarrow n + \nu$$

in liquid hydrogen is beset by two problems – firstly the ratio of muon capture to decay is about 10^{-3} and, secondly, molecular problems are introduced through the process

$$(p\mu) + (pe) \rightarrow (p\mu p) + e + 124 \text{ eV}$$

which occurs with a mean transition time $0{\cdot}7 \pm 0{\cdot}3\ \mu$sec (Bleser *et al.*, 1962a). The molecules are mainly in the ortho state, and Weinberg (1960) has shown that the absorption rate for muons in this state is given by the relation

$$\lambda_{p\mu p}^{(\text{theor})} = 1{\cdot}17[\eta\lambda_0 + (1 - \eta)\,\lambda_1]$$

where the factor 1·17 represents the enhanced probability density of the muon wave function in the $p\mu p$ molecule, λ_0 and λ_1 the capture rates for the atomic singlet and triplet states respectively, and η represents the probability that capture takes place from an effective singlet state in the molecule. Limits of $\frac{3}{8}$ to $\frac{3}{4}$ may be set upon η, and calculations by Halpern and Kroll (unpublished) indicate that η is close to the upper limit.

The capture rate has been measured in bubble chambers (Hildebrand, 1962; Bertolini *et al.*, 1962), and in counter experiments (Bleser *et al.*, 1962b). In both experiments the recoil neutron was detected; the first technique allows better identification of the neutron whilst the second allows time discrimination to be used for distinguishing direct atomic (μp) from molecular ($p\mu p$) capture. The results obtained together with the expected values are shown below.

Bubble chambers	425 ± 60 sec^{-1}	expected 583 sec^{-1}
Counter	515 ± 85 sec^{-1}	expected 560 sec^{-1}

The calculated figures are based upon the matrix element (12.107) and assume the coupling strengths for β-decay (apart from slight modifications due to form factors) and an induced pseudoscalar coupling equal to $8G_A$. These coupling strengths lead to $\lambda_0 = 636$ sec^{-1} and $\lambda_1 = 13$ sec^{-1} for a V–A theory (Primakoff, 1959), and hence to the values given above. If the coupling was of the form $V + A$, a capture rate of less than 200 sec^{-1} would be expected.

The results clearly support a V–A interaction. Independent evidence for this conclusion has been provided by Egorov *et al.*, (1961) and Culligan *et al.* (1961), who examined muon capture rates in hyperfine states in complex nuclei (compare (12.109)).

A method of examining the Gamow–Teller part of the capture process is provided by the reaction

$$\mu^- + C^{12} \rightarrow B^{12} + \nu$$

$$\text{spin} \qquad 0 \qquad 1$$

since considerable information about the nuclear matrix element is provided by the β-decay process

$$B^{12} \rightarrow C^{12} + e^- + \bar{\nu}\,.$$

The experimental data yields a capture rate of about 7×10^{-3} sec^{-1} and suggests $G_A^{(e)} = G_A^{(\mu)}$ within 20 per cent. In principle the method is capable of detecting the presence of the recoil terms, but the experimental evidence is too conflicting to allow conclusions to be reached at present (see Rubbia, 1961, for a critical survey).

We have previously mentioned (12.110) that the presence of the recoil terms cause a considerable modification of the parameter α, which represents the magnitude of the asymmetry in the angular distribution of the neutrons emitted following polarised muon capture. Most of this increase is caused by the induced pseudoscalar term G_P. Measurements by various groups (Table 12.6) have yielded asymmetry coefficients which are consistent with the theoretical prediction of $\alpha = -0{\cdot}4$, which is based upon the assumption $G_P \sim 8G_A$.

TABLE 12.6

Asymmetry $-\alpha$	Nucleus	Reference
$>0{\cdot}22 \pm 0{\cdot}07$	S^{32}	Astbury *et al.* (1962)
$>0{\cdot}24 \pm 0{\cdot}09$	S^{32}	Telegdi (1960)
$>0{\cdot}30 \pm 0{\cdot}08$	Mg^{24}	Telegdi (1960)
$\sim 0{\cdot}93 \pm 0{\cdot}33$	Ca^{40}	Evseev *et al.* (1962)

12.8(c). *Universality in the strangeness conserving weak interactions*

The evidence we have examined for the processes

$$n \rightarrow p + e^- + \bar{\nu}$$

$$\mu \rightarrow e + \nu + \bar{\nu}$$

$$\pi \rightarrow \mu + \nu$$

$$\mu^- + p \rightarrow n + \nu$$

is consistent with the following structure for strangeness conserving weak interactions:

(1) two-component theory of the neutrino with left-handed neutrinos,
(2) universality of the properties of muons and electrons apart from mass,
(3) lepton conservation,
(4) conserved vector coupling strength,
(5) time reversal invariance.

It is therefore tempting to construct a single interaction term which covers all the processes except $\pi \to \mu + \nu$; a suitable expression is

$$M_{\text{weak}} = \frac{G_V}{\sqrt{2}} j_\lambda j_\lambda^l + \text{h.c.} \tag{12.111}$$

where

$$j_\lambda = j_\lambda^V + i j_\lambda^A = \langle f | V_\lambda | i \rangle + i \langle f | A_\lambda | i \rangle \tag{12.112}$$

$$j_\lambda^l = \frac{1}{N_l} \bar{u}_e \gamma_\lambda (1 + \gamma_5) u_{\nu_e} + \frac{1}{N_l} \bar{u}_\mu \gamma_\lambda (1 + \gamma_5) u_{\nu_\mu}.$$

The operators V and A have the transformation properties of vectors and axial vectors respectively and can be defined with the aid of (12.77) and (12.100). It should be noted, however, that $G_V/\sqrt{2}$ has been factored out from these expressions.

An even more symmetric form has been conjectured by Feynman and Gell-Mann (1958); it is

$$M_{\text{weak}} = \frac{G_V}{\sqrt{2}} (j_\lambda + j_\lambda^l)^\dagger (j_\lambda + j_\lambda^l). \tag{12.113}$$

A basic interaction of this form permits weak interactions between nucleons. So far they have not been detected; for further discussion of this point see Blin-Stoyle (1960).

12.9. THE DECAY OF STRANGE PARTICLES–LEPTONIC MODES

12.9(a). *Introduction*

We have seen in the previous section that a set of comparatively simple assumptions can account for the observed properties of the strangeness conserving weak interactions. The decay of the strange particles is a more sophisticated problem, and an immediate question is what ideas can be retained in moving from the one problem to the other?

There are obvious analogies between certain processes in the strangeness conserving and nonconserving weak interactions, for example

$\Delta S = 0$	$\lvert\Delta S\rvert = 1$
$n \to p + e^- + \bar{\nu}$	$\Lambda \to p + e^- + \bar{\nu}$
$\pi \to \mu + \nu$	$K \to \mu + \nu.$

It is a simple matter to calculate decay rates for the processes on the right-hand side by using the matrix element given in (12.74), for example the process

$\Lambda \to p + e^- + \bar{\nu}$ could be represented as

$$M_{fi} = \sum_i \langle p | s_i | \Lambda \rangle \frac{1}{N_i} \bar{u}_e \Gamma_i (1 + \gamma_5) u_\nu. \tag{12.114}$$

If one assumes V, A terms only and takes the coupling strengths from β-decay in the limit $q^2 \to 0$ so that only nonrelativistic terms remain then a decay rate of $0{\cdot}6 \times 10^8$ sec^{-1} may be calculated. If we take the known lifetime of the Λ-hyperon, then a branching ratio

$$\frac{\Lambda \to p + e^- + \bar{\nu}}{\Lambda \to N + \pi} \sim \frac{0{\cdot}6 \times 10^8}{0{\cdot}4 \times 10^{10}} \sim 1{\cdot}5 \times 10^{-2}$$

is found. Similar ratios may be calculated for the other hyperon decays; a comparison with experimental data is given in Table 12.7 (Bingham, 1964).

TABLE 12.7

Decay Process	Branching ratio	
	Calculated	Experimental
$\Lambda \to p + e^- + \bar{\nu}$	$1{\cdot}5 \times 10^{-2}$	$(0{\cdot}85 \pm 0{\cdot}09)\ 10^{-3}$
$\Lambda \to p + \mu^- + \bar{\nu}$	$2{\cdot}5 \times 10^{-3}$	$(0{\cdot}13 \pm 0.06)\ 10^{-3}$
$\Sigma^- \to n + e^- + \bar{\nu}$	$5{\cdot}6 \times 10^{-2}$	$(1{\cdot}35 \pm 0{\cdot}4)\ 10^{-3}$
$\Sigma^- \to n + \mu^- + \bar{\nu}$	$2{\cdot}5 \times 10^{-2}$	$(0{\cdot}88 \pm 0{\cdot}3)\ 10^{-3}$

It can be seen that the observed branching ratios are much smaller than those expected from a conserved vector current theory indicating much smaller effective coupling strengths. The structural terms (§ 12.5(a)) could be more important in hyperon decay than in neutron decay since the energy release is much larger ($\sim$200 MeV), but it would be surprising if they could lead to order of magnitude effects. A similar effect has been observed in kaon decay. If we use the arguments of § 12.7 the measured decay rates for the processes $\pi \to \mu + \nu$ and $K \to \mu + \nu$ lead to effective coupling strengths f_π^2 and f_K^2 respectively. They are found to be in the ratio

$$\frac{f_\pi^2}{f_K^2} \sim 10.$$

The result is not surprising, since kaon decay could proceed through the leptonic decay of hyperons, whereas pion decay cannot (Fig.12.18).

An explanation of the low branching ratios within the framework of the unitary symmetry scheme (§ 13.5(a)) has been offered by Cabibbo (1963). He has suggested that the interaction describing the leptonic decay of strongly interacting particles is of the form (compare (12.111) and § 12.5)

$$\frac{G_V}{\sqrt{2}}\,[j_\lambda j_\lambda^l \cos\theta + s_\lambda j_\lambda^l \sin\theta]$$

where j_λ and s_λ represent strangeness conserving and nonconserving currents respectively. An angle $\theta = 0{\cdot}26$ radians then yields the correct branching ratio

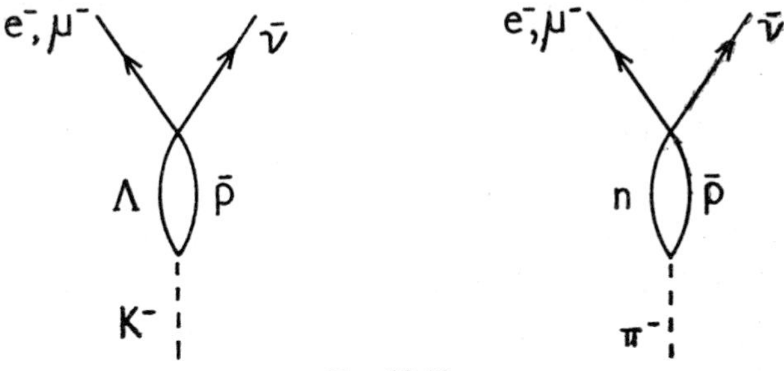

FIG. 12.18.

for the leptonic decay of strange particles, and simultaneously offers an explanation of the apparent discrepancy between the vector coupling strengths in β and μ-decay.

However, similarities are found in strangeness conserving and nonconserving weak interactions. For example the relative rates for hyperon decay to electrons and muons given in Table 12.7 are in accord with (12.114), thus illustrating the apparent interchangeability of muon and electron currents which we have previously noted. In addition the angular distribution of the decay electrons from the sequence $K \to \mu \to e$ has been measured about the direction of muon momentum, and the asymmetry coefficient is found to have the same amplitude and sign as in the $\pi \to \mu \to e$ sequence (Coombes *et al.*, 1957). This result suggests that the lepton current in kaon decay is the same as in pion decay, and also that the kaon is spinless.

12.9(b). *The process* $K^- \to \pi^0 + l^- + \bar{\nu}$.

Additional evidence about the kaon decay can be gained from the process

$$K^- \to \pi^0 + \begin{pmatrix} e^- \\ \mu^- \end{pmatrix} + \bar{\nu}$$

and its charge conjugate. The matrix element has the form

$$M_{fi} = \sum \langle\pi|\, s_i \,|K\rangle \frac{1}{N_l}\, \bar{u}_l \Gamma_i (1 + \gamma_5)\, u_\nu. \qquad (12.115)$$

The experimental data (§ 13.4(a)) strongly suggests that the kaon is pseudoscalar, and so only scalar, vector and tensor currents can contribute. Appropriate invariants can be constructed for the currents (Pais and Treiman, 1957) and lead to different predictions for the spectral shapes for pion momenta and pion-charged lepton angular correlations. The analytic shapes of the spectra for the $K \to \pi^0 + \bar{\nu} + e$ mode are especially simple because of terms of the type m_l/m_K in the matrix element. Measurements of the momentum spectra and pion–lepton correlation functions have been made by a number of groups (for a critical survey see Crawford, 1962). They have shown that the data is consistent with a pure vector coupling.

12.9(c). *Isospin changes in the leptonic decays of strange particles*

The leptonic decay of strange particles can, in principle, occur by the transitions $|\Delta S| = 0, 1, 2, \ldots$ The $\Delta S = 0$ transitions occur in processes of the type $\Sigma^- \to \Lambda + e^- + \nu$ and, like the β-decay of the neutron, involve an isospin change $|\Delta T| = 1$ for the strongly interacting particles. Leptonic decays involving $|\Delta S| = 2$ have not been observed, but in view of the small branching ratios for the leptonic decay modes of strange particles, the evidence is not at present conclusive that they cannot occur (see, however, the argument concerning $|\Delta S| = 2$ transitions given in § 12.9(d)).

The transitions involving $|\Delta S| = 1$ pose interesting questions concerning possible isospin changes. The rule $|\Delta S| = 1$ is linked with isospin changes through the relation (1.22)

$$\frac{Q}{e} = \frac{B}{2} + \frac{S}{2} + T_3 . \tag{12.116}$$

If we define the change in charge as $\Delta Q = (Q/e)_{\text{final}} - (Q/e)_{\text{initial}}$ then the above relation yields

$$\Delta Q = \frac{\Delta S}{2} + \Delta T_3 \tag{12.117}$$

and we then find

$$\frac{\Delta S}{\Delta Q} = 1 \to |\Delta T_3| = \tfrac{1}{2} \to |\Delta T| \geqq \tfrac{1}{2} \tag{12.118}$$

$$\frac{\Delta S}{\Delta Q} = -1 \to |\Delta T_3| = \tfrac{3}{2} \to |\Delta T| \geqq \tfrac{3}{2}.$$

Thus $|\Delta T| = \frac{1}{2}$ implies $\Delta S/\Delta Q = +1$ but not vice versa. This isospin change must occur by virtue of the observed process

$$\Lambda \to p + e^- + \bar{\nu}$$

$$T \quad 0 \quad \tfrac{1}{2}$$

Virtually all the observed leptonic decays of strange particles are consistent with the rule $\Delta S/\Delta Q = +1$ (compare Table 12.7); however, the processes $\Delta S/\Delta Q = -1$ may occur, for example

$$\Sigma^+ \rightarrow n + \mu^+ + \nu$$

$$K^0 \rightarrow \pi^+ + e^- + \bar{\nu}.$$

An example of the decay $\Sigma^+ \rightarrow n + \mu^+ + \nu$ has been reported by Barbaro-Galtieri *et al.* (1962) but the branching ratio appears to be small compared with equivalent leptonic decay modes involving $\Delta S/\Delta Q = +1$. Before discussing the process $K^0 \rightarrow \pi^+ + e^- + \bar{\nu}$, we shall examine K^0 decays in general.

12.9(d). *The neutral kaons*

In § 1.4(a) we have described how the postulate of associated production in strong interactions led to the prediction that there should be two neutral kaons – called K^0 and $\bar{K}^0$ with strangeness $+1$ and -1 respectively.

The cosmic ray work and early experiments with the Cosmotron showed the existence of a neutral particle which was produced in association with the hyperons as expected. Its presence was detected by its decay into $\pi^+ + \pi^-$ mesons with a lifetime of $\sim 10^{-10}$ sec. However, the rate of occurence of these particles was lower than that expected according to the associated production scheme, thus suggesting that further neutral decay modes existed.

A problem was also presented by the decay mode (Gell-Mann and Pais, 1955b). If the K^0 and $\bar{K}^0$ are particle and antiparticle, then the combined operation of parity and charge conjugation yields†

$$CP\,|K^0\rangle = |\bar{K}^0\rangle \tag{12.119}$$

but the decay product behaves like

$$CP\,|\pi^+\pi^-\rangle = |\pi^+\pi^-\rangle \tag{12.120}$$

hence the final state is an eigenstate of CP but the initial state is not.

In order to overcome these difficulties Gell-Mann and Pais postulated that the K^0 and $\bar{K}^0$ mesons could be regarded as mixtures of systems with eigenstates ± 1 under the CP operation

$$|K_1^0\rangle = \frac{|K^0\rangle + |\bar{K}^0\rangle}{\sqrt{2}}, \qquad CP\,|K_1^0\rangle = |K_1^0\rangle \tag{12.121}$$

$$|K_2^0\rangle = \frac{|K^0\rangle - |\bar{K}^0\rangle}{\sqrt{2}}, \qquad CP\,|K_2^0\rangle = -|K_2^0\rangle.$$

† The original argument was made using charge conjugation only, but was changed to the operation CP following the discovery of parity failure in weak interactions.

Since CP holds in weak interactions the decay mode $\pi^+ + \pi^-$ must be associated with K_1^0

$$K_1^0 \rightarrow \pi^+ + \pi^- \tag{12.122}$$

whilst the K_2^0 particle must decay into a pion and two leptons or three pions since†

$$CP\,|\pi\rangle = -|\pi\rangle . \tag{12.123}$$

By making a comparison with the decay rate for the comparable process $K^+ \rightarrow \mu^+ + \nu + \pi^0$, Gell-Mann and Pais further suggested that the lifetime for the K_2^0 decay might be a factor $\sim 10^3$ larger than for K_1^0.

A search for the K_2^0 meson was made by a number of groups, for example Landé *et al.* (1956) looked for decay processes in a cloud chamber located 6 m away (equivalent to 3×10^{-8} sec flight path) from a target bombarded by 3 GeV protons. Decay processes of the type $K^0 \rightarrow \pi^\pm + l^\mp + \nu$ were observed. By altering the distance between target and cloud chamber a lifetime of $\sim 8 \times 10^{-8}$ sec was estimated for the K_2^0 particle.

The existence of the $K^0 - \overline{K}^0$ doublet has presented many intriguing possibilities in studying the physics of strange particles. Thus suppose we generate a 'beam' of 1000 K^0 mesons (Gell-Mann and Rosenfeld, 1957) in the process

$$\pi^- + p \rightarrow \Lambda + K^0 .$$

These mesons have definite strangeness $S = +1$ and since

$$|K^0\rangle = \frac{|K_1^0\rangle + |K_2^0\rangle}{\sqrt{2}} \tag{12.124}$$

there are 500 K_1^0 and 500 K_2^0 mesons at time $t = 0$.‡ After a time $t \sim 10^{-9}$ sec virtually all the K_1^0 mesons will have decayed leaving a beam of ~ 500 K_2^0 particles, and it is apparent from (12.121) that the beam has now become equally divided between $S = \pm 1$ states. These states behave differently if the K_2^0 beam is then allowed to undergo strong interactions:

$$K^0 + p \rightarrow K^+ + n \quad (S = +1)$$

$$\overline{K}^0 + p \rightarrow \Sigma^+ + \pi^0 \quad (S = -1).$$

Thus the neutral K_1^0 and K_2^0 states do not possess a definite strangeness.

† Evidence has now been obtained for the existence of the process $K_2^0 \rightarrow \pi^+ + \pi^-$ with a branching ratio of $(2{\cdot}0 \pm 0{\cdot}4) \times 10^{-3}$ (Christenson *et al.*, 1964). This result implies a failure of CP (and even T) invariance in K_2^0 decay; it could have a more subtle interpretation.

‡ A check on this point has been made by the Michigan group using a xenon-filled bubble chamber (Glaser, 1959). They found $0{\cdot}53 \pm 0{\cdot}05$ of the Λ-hyperon production processes were accompanied by the K_1^0 decay modes, $K_1^0 \rightarrow \pi^0 + \pi^0$ and $K_1^0 \rightarrow \pi^+ + \pi^-$; the K_2^0 mesons left the chamber.

The variation in intensity of the K^0 and $\bar{K}^0$ components as a function of time has been exploited to yield a value for the mass difference of the neutral kaons. At a time t after production, the nonrelativistic wave function for the complex system can be written as

$$\psi(t) \equiv \frac{1}{\sqrt{2}}|K_1^0\rangle\, e^{[-t/(2\tau_1)]-im_1t} + \frac{1}{\sqrt{2}}|K_2^0\rangle\, e^{[-t/(2\tau_2)]-im_2t}$$

$$= \frac{|K^0\rangle + |\bar{K}^0\rangle}{2}\, e^{[-t/(2\tau_1)]-im_1t} + \frac{|K^0\rangle - |\bar{K}^0\rangle}{2}\, e^{[-t/(2\tau_2)]-im_2t} \quad (12.125)$$

where m_1 and m_2 represent the K_1^0 and K_2^0 masses respectively and τ_1 and τ_2 the lifetimes. The flux of this beam is given by $|\psi(t)|^2$ and for times $t \ll \tau_2$ the intensity of the K^0 and $\bar{K}^0$ components is given by

$$N(K^0) = \tfrac{1}{4}[1 + e^{(-t/\tau_1)} + 2(\cos \Delta m\, t)\, e^{[-t/(2\tau_1)]}] \quad (12.126)$$

$$N(\bar{K}^0) = \tfrac{1}{4}[1 + e^{(-t/\tau_1)} - 2(\cos \Delta m\, t)\, e^{[-t/(2\tau_1)]}]$$

where $\Delta m = m_2 - m_1$. The behaviour of these components as a function of time are sketched in Fig. 12.19. In the diagram with $\Delta m \gg 1/\tau_1$ we have assumed that the cosine function oscillates too rapidly for experimental detection.

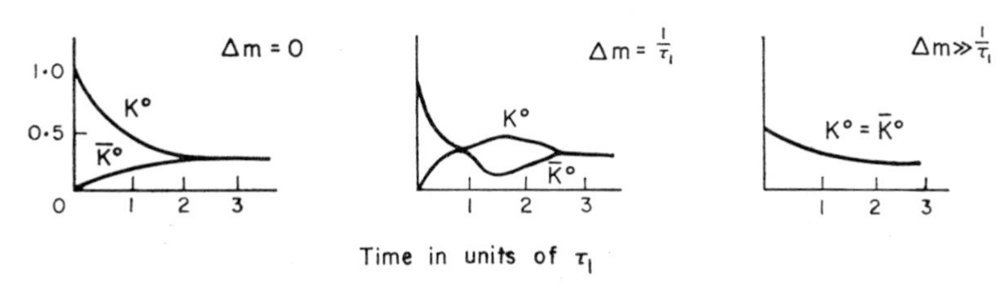

FIG. 12.19.

A combined group from Berkeley, Padua and Wisconsin (Camerini *et al.*, 1962) have measured the mass difference by exposing a propane-filled bubble chamber to a K^+ beam. Charge exchange collisions produced K^0 mesons in the chamber (the symbols N and N' represent complex nuclei)

$$K^+ + N \rightarrow N' + K^0 \quad (S = 1).$$

They then looked for the production of hyperons as a function of distance from the first reaction

$$\bar{K}^0 + N \rightarrow N' + \begin{cases} \Lambda \\ \Sigma \end{cases} \quad (S = -1)$$

and were thus able to deduce the fraction of $\bar{K}^0$ mesons present as a function of time and hence Δm. The experiment yielded $\Delta m = (1{\cdot}5 \pm 0{\cdot}2)\, 1/\tau_1$; this value has been confirmed by other groups.

A mass difference $\sim 1/\tau_1$ can be used to argue that transitions involving $\Delta S = 2$ should not occur (Okun and Pontecorvo, 1957). At nonrelativistic energies we can write

$$H\,|K_1^0\rangle = m_1\,|K_1^0\rangle, \qquad H\,|K_2^0\rangle = m_2\,|K_2^0\rangle \tag{12.127}$$

where H is the Hamiltonian operator and m_1 and m_2 the masses of the K_1^0 and K_2^0 mesons respectively. Using (12.121) we can therefore write

$$\begin{aligned}\langle K^0|\,H\,|\bar{K}^0\rangle &= \langle K^0|\,H|\left(\frac{|K_1^0\rangle - |K_2^0\rangle}{\sqrt{2}}\right)\\ &= \frac{m_1}{\sqrt{2}}\,\langle K^0\,|\,K_1^0\rangle - \frac{m_2}{\sqrt{2}}\,\langle K^0\,|\,K_2^0\rangle\\ &= \frac{m_1}{\sqrt{2}}\,\langle K^0|\left(\frac{|K^0\rangle + |\bar{K}^0\rangle}{\sqrt{2}}\right) - \frac{m_2}{\sqrt{2}}\,\langle K^0|\left(\frac{|K^0\rangle - |\bar{K}^0\rangle}{\sqrt{2}}\right)\\ &= \frac{\Delta m}{2}.\end{aligned} \tag{12.128}$$

Let us consider the mechanism by which the K^0 meson can transform to $\bar{K}^0$. Two possible methods are given in Fig. 12.20(a) and (b). Diagram (a) involves a

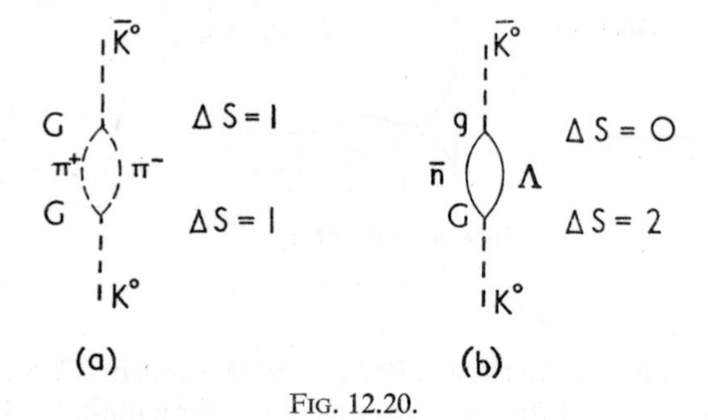

FIG. 12.20.

weak interaction (in the K_1^0 mode) at each vertex and so we may conclude that the transition amplitude is of order $1/\tau_1$

$$\frac{\Delta m}{2} = \langle K^0|\,H\,|\bar{K}^0\rangle \sim G^2 \sim \frac{1}{\tau_1}. \tag{12.129}$$

On the other hand, diagram (b) involves a weak interaction at one vertex and a strong one at the other so that

$$\frac{\Delta m}{2} = \langle K^0|\,H|\bar{K}^0\rangle \sim gG \sim \frac{g}{G}\frac{1}{\tau_1} \sim \frac{10^6}{\tau_1}. \tag{12.130}$$

This result is clearly incompatible with the experimental data.

12.9(e). *The decay process* $K^0 \to e^\pm + \pi^\mp + \nu$

We have mentioned earlier (§ 12.9(c)) that decay schemes involving the change $\Delta S = -\Delta Q$ represent a sensitive test for the occurrence of the change in isospin $|\Delta T| \geqq \frac{3}{2}$ in the leptonic decay of strange particles. Consider the decay processes

$$\left.\begin{array}{l} K^0 \to \pi^- + e^+ + \nu \\ \bar{K}^0 \to \pi^+ + e^- + \bar{\nu} \end{array}\right\} \frac{\Delta S}{\Delta Q} = +1, \qquad \left.\begin{array}{l} K^0 \to \pi^+ + e^- + \bar{\nu} \\ \bar{K}^0 \to \pi^- + e^+ + \nu \end{array}\right\} \frac{\Delta S}{\Delta Q} = -1. \tag{12.131}$$

If the $\Delta S = \Delta Q$ rule operates, then certain predictions can be made about the decays $K^0 \to e^+$ and $\bar{K}^0 \to e^-$. They are related to the change in structure of a K^0 beam as a function of time as presented in the previous section. The predictions are:

(1) the ratio

$$R(t) = \frac{N^+(t)}{N^-(t)} = \frac{\text{number of decays } e^+ + \pi^- + \nu}{\text{number of decays } e^- + \pi^+ + \bar{\nu}} = \frac{N(K^0)}{N(\bar{K}^0)}$$

where $N(K^0)$ and $N(\bar{K}^0)$ are the numbers of neutral kaons present as a function of time as given in equations (12.125), (12.126) and Fig. 12.19;

(2) $$N^-(t) + N^+(t) = \text{const}\,[e^{(-t/\tau_1)} + e^{(-t/\tau_2)}]$$

since we used the approximation $e^{(-t/\tau_2)} \sim 1$ in (12.126).

Experimental attempts to confirm these predictions have unfortunately led to inconclusive results (Siena conference report, 1964). Measurements on the branching ratios for the decay of K_2^0 mesons indicate, however, the existence of transitions with dominant $|\Delta T| = \frac{1}{2}$ amplitudes. Thus Luers *et al.* (1964) and Astier *et al.* (1961) have obtained the data listed in Table 12.8. The ratios may be predicted from the data for K^+ decay together with the assumption of $|\Delta T| = \frac{1}{2}$ or $|\Delta T| = \frac{3}{2}$ change in the decay processes (Dalitz, 1956; Behrends and Sirlin, 1961). The results clearly support the assumption of $\Delta T = \frac{1}{2}$ transitions.

TABLE 12.8

Decay mode	Luers *et al.* (%)	Astier *et al.* (%)	$\lvert\Delta T\rvert = \frac{1}{2}$ (%)	$\lvert\Delta T\rvert = \frac{3}{2}$ (%)
$\pi e \nu$	49 ± 5	34 ± 10	46 ± 4	30 ± 3
$\pi \mu \nu$	35 ± 7	48 ± 10	38 ± 4	26 ± 3
$\pi^+\pi^-\pi^0$	16 ± 2	19 ± 3	16 ± 1	44 ± 4

It should be noted that the substantial $K_2^0 \to 3\pi^0$ decay mode has been omitted from this data.

We may sum up §§ 12.9(c) and (e) by stating that the dominant modes for the leptonic decay of strange particles appear to occur as follows

$$|\Delta T| = 1 \quad \text{for} \quad \Delta S = 0 \tag{12.132}$$

$$\Delta S/\Delta Q = +1, \quad |\Delta T| = \tfrac{1}{2} \quad \text{for} \quad |\Delta S| = 1$$

12.10. THE DECAY OF STRANGE PARTICLES–NONLEPTONIC MODES

12.10(a). *Hyperon decay*

Let us consider the production and decay of a hyperon. The production of Λ and Σ hyperons occurs in reactions of the type

$$\pi + p \rightarrow K + Y$$

and, as we have seen in § 6.2, the interaction of spin 0 and spin $\frac{1}{2}$ systems can lead to polarisation of the spin $\frac{1}{2}$ particle in a direction perpendicular to the scattering plane, or in this case the $\pi - Y$ production plane (Fig. 12.2).

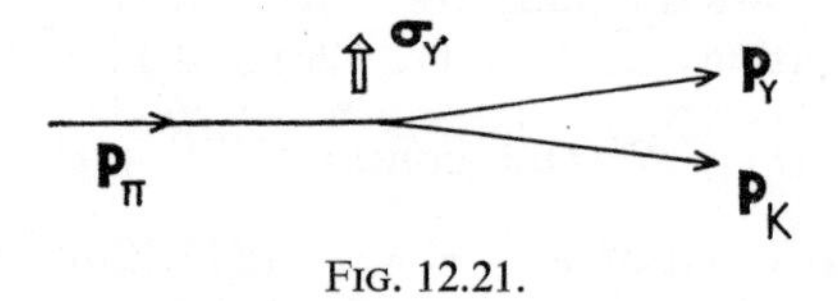

FIG. 12.21.

Let us now consider the decay process $Y \rightarrow N + \pi$. In the coordinate frame of the hyperon the decay amplitude can be characterised by the nucleon spin and the direction of the pion momentum (we shall restrict our remarks to the Λ and Σ hyperons). Since the hyperon possesses spin $\frac{1}{2}$ the decay particles can only be emitted in s and p-wave states if angular momentum is to be conserved; since the pion possesses odd intrinsic parity these states violate and preserve parity conservation respectively. The energy release in hyperon decay is comparatively low, and so it is adequate to work with Pauli spinors; the most general form of decay matrix is then given by

$$M = S\hat{1} + P\boldsymbol{\sigma}\cdot\hat{\mathbf{k}} \tag{12.133}$$

(compare § 6.2(e)), where S and P represent the s and p-wave amplitudes respectively, and $\hat{\mathbf{k}}$ is a unit vector in the direction of the momentum of the decay pion. Hence, if we call the hyperon polarisation $\langle\boldsymbol{\sigma}_Y\rangle = P_Y\mathbf{n}$ where $\mathbf{n}$ is a unit vector perpendicular to the production plane and use density matrix techniques (§ 6.2(b)), we can write the density matrix for the hyperon as

$$\varrho_Y = \tfrac{1}{2}I(\hat{1} + \langle\boldsymbol{\sigma}_Y\rangle\cdot\boldsymbol{\sigma}) = \tfrac{1}{2}I(\hat{1} + P_Y\mathbf{n}\cdot\boldsymbol{\sigma})$$

(compare (6.49)). Then by equations (6.57) and (6.63) we can represent the decay probability as

$$\frac{d\lambda}{d\Omega} = \frac{\operatorname{tr} M\varrho_Y M^\dagger}{\operatorname{tr} \varrho_Y} = (|S|^2 + |P|^2)(1 + \alpha P_Y \cos\theta) \tag{12.134}$$

where

$$\alpha = \frac{S^*P + P^*S}{|S|^2 + |P|^2} = \frac{2 \operatorname{Re} SP}{|S|^2 + |P|^2} \tag{12.135}$$

and θ represents the angle between the directions of the decay pion and the normal to the production plane

$$P_Y \mathbf{n} \cdot \hat{\mathbf{k}} = P_Y \cos\theta. \tag{12.136}$$

Further information concerning the hyperon decay process may be obtained from the polarisation $\langle \boldsymbol{\sigma}_N \rangle$ of the outgoing nucleon. This polarisation can be obtained from the expressions used in § 6.2 (compare (6.44) and (6.56)).

$$\langle \boldsymbol{\sigma}_N \rangle \operatorname{tr} \varrho_f = \operatorname{tr} \varrho_f \boldsymbol{\sigma} \tag{12.137}$$

$$\varrho_f = M\varrho_Y M^\dagger$$

Substitution for ϱ_Y and M and use of equations (6.63) and (6.61) then leads to the expression

$$(1 + \alpha P_Y \cos\theta) \langle \boldsymbol{\sigma}_N \rangle$$

$$= (\alpha + P_Y \cos\theta)\, \hat{\mathbf{k}} + \beta P_Y \hat{\mathbf{k}} \times \mathbf{n} + \gamma P_Y (\hat{\mathbf{k}} \times \mathbf{n}) \times \hat{\mathbf{k}} \tag{12.138}$$

where

$$\beta = \frac{2 \operatorname{Im} S^*P}{|S|^2 + |P|^2} \qquad \gamma = \frac{|S|^2 - |P|^2}{|S|^2 + |P|^2}. \tag{12.139}$$

As a result of parity violation in hyperon decay the nucleon possesses longitudinal polarisation even if the hyperon is unpolarised, since for $P_Y = 0$

$$\langle \boldsymbol{\sigma}_N \rangle = \alpha \hat{\mathbf{k}}.$$

The terms α, β and γ satisfy the relation

$$\alpha^2 + \beta^2 + \gamma^2 = 1.$$

It should be noted that the term β is a test of time reversal invariance and should be zero or very small.

The arguments given above may also be applied to the Ξ-hyperon if it possesses spin $\frac{1}{2}$; the term $\langle \boldsymbol{\sigma}_N \rangle$ in (12.138) would then refer to the Λ-hyperon.

The determination of the parameters α, β and γ has been made with the aid of the bubble and spark chamber techniques (see Crawford (1962) for detailed references). The principal properties of the nonleptonic decay modes of the hyperons are listed in Table 12.9.

TABLE 12.9

Decay mode	Branching fraction	Lifetime (10^{-10} sec)	Asymmetry parameters α	β	γ
$\Lambda \to p + \pi^-$	$0{\cdot}68 \pm 0{\cdot}02$	$2{\cdot}55 \pm 0{\cdot}1$	$-0{\cdot}62 \pm 0{\cdot}07$	$0{\cdot}17 \pm 0{\cdot}24$	$0{\cdot}78 \pm 0{\cdot}06$
$\Lambda \to n + \pi^0$	$0{\cdot}32 \pm 0{\cdot}02$		$-0{\cdot}62 \pm 0{\cdot}20$		$0{\cdot}8^{+0{\cdot}2}_{-0{\cdot}4}$
$\Sigma^+ \to p + \pi^0$	$0{\cdot}51 \pm 0{\cdot}02$	$0{\cdot}77 \pm 0{\cdot}04$	$+0{\cdot}78 \pm 0{\cdot}08$		
$\Sigma^+ \to n + \pi^+$	$0{\cdot}49 \pm 0{\cdot}02$		$+0{\cdot}03 \pm 0{\cdot}08$		
$\Sigma^- \to n + \pi^-$	$\sim 1{\cdot}00$	$1{\cdot}58 \pm 0{\cdot}06$	$+0{\cdot}1 \pm 0{\cdot}16$		
$\Xi^- \to \Lambda + \pi^-$	$\sim 0{\cdot}92$	$1{\cdot}75 \pm 0{\cdot}05$	$+0{\cdot}6 \pm 0{\cdot}1$	$-0{\cdot}68 \pm 0{\cdot}27$	$0{\cdot}6 \pm 0{\cdot}3$
$\Xi^0 \to \Lambda + \pi^0$	$\sim 1{\cdot}00$	$2{\cdot}80 \pm 0{\cdot}26$			

The near equality of the lifetimes suggests basically similar processes in all of the decays since the phase space factors are roughly the same. We may also note that $\alpha_\Lambda \sim -\alpha_\Xi$ (Rosen, 1962).

Let us now consider Σ-hyperon decay; we denote the decay amplitudes and asymmetry parameters as follows:

$$\begin{aligned} \Sigma^+ &\to p + \pi^0 \qquad N_0 \qquad \alpha_0 \\ &\quad\; n + \pi^+ \qquad N_+ \qquad \alpha_+ \\ \Sigma^- &\to n + \pi^- \qquad N_- \qquad \alpha_- . \end{aligned}$$

We shall examine the figures for α in conjunction with the decay rates λ in the three processes. From Table 12.9 we find

$$\lambda_0 = \frac{0{\cdot}51}{\tau_{\Sigma^+}} = \frac{0{\cdot}51}{0{\cdot}8 \times 10^{-10}} = 0{\cdot}64 \times 10^{10}\ \text{sec}^{-1}, \qquad p_0 = 189\ \text{MeV/c}$$

$$\lambda_+ = \frac{0{\cdot}49}{\tau_{\Sigma^+}} = \frac{0{\cdot}49}{0{\cdot}8 \times 10^{-10}} = 0{\cdot}61 \times 10^{10}\ \text{sec}^{-1}, \qquad p_+ = 185\ \text{MeV/c}$$

$$\lambda_- = \frac{1}{\tau_{\Sigma^-}} = \frac{1}{1{\cdot}6 \times 10^{-10}} = 0{\cdot}62 \times 10^{10}\ \text{sec}^{-1}, \qquad p_- = 192\ \text{MeV/c}.$$

The kinematic factors are about the same in each process (we have illustrated this point by quoting the momenta of the pions of appropriate charge); hence we may conclude from the above figures that

$$|N_0|^2 \sim |N_+|^2 \sim |N_-|^2 \tag{12.140}$$

where

$$|N|^2 = |S|^2 + |P|^2.$$

Combining the data from Table 12.9, (12.135) and (12.140) we may conclude that the decay processes $\Sigma^+ \to n + \pi^+$, $\Sigma^- \to n + \pi^-$ are virtually pure S or P, and that $\Sigma^+ \to p + \pi^0$ contains S and P contributions in roughly equal quantities.

Let us make these considerations a little more quantitative. In general the amplitudes N are complex numbers, $N = ne^{i\delta}$. The simultaneous requirements of unitarity of the S-matrix and time-reversal invariance lead to the requirement

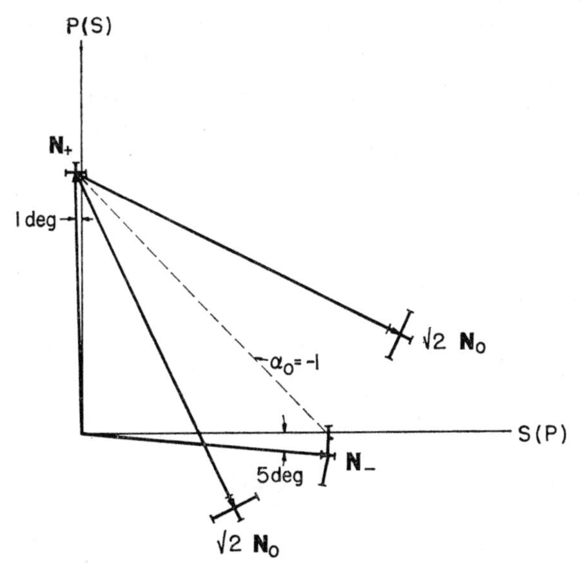

FIG. 12.22.

that the phase factors in N are pion–nucleon scattering phase shifts (compare § 7.5(b)). But since the Σ-hyperon has spin $\frac{1}{2}$, these phase shifts are small and may be neglected. Therefore to a good approximation the amplitudes N may be regarded as real numbers.

It is convenient to treat the amplitudes as vectors in an orthogonal S and P wave space, so that

$$\mathbf{N}^2 = S^2 + P^2 \tag{12.141}$$

and we may write the term α of equation (12.135) as

$$\alpha = \frac{2SP}{S^2 + P^2} = 2\nu \tag{12.142}$$

where we have introduced the reality condition discussed above, and ν represents the angle to the coordinate axes in SP space.

The results for N_0, N_+ and N_- obtained in this manner are displayed in Fig. 12.22 (Tripp, Watson and Ferro-Luzzi, 1962b). The data does not permit a decision between which are the S and P axes, consequently an ambiguity arises in the direction N_0 which depends on whether S/P is greater or less than 1.

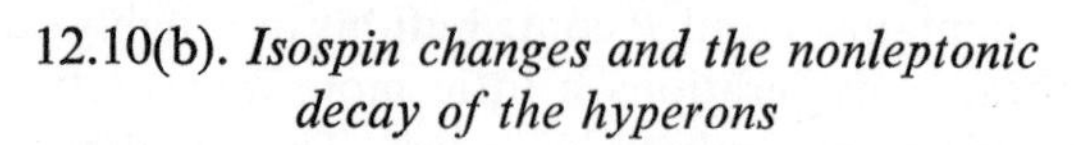

12.10(b). *Isospin changes and the nonleptonic decay of the hyperons*

As in the case of the leptonic decay modes, not all of the nonleptonic decay modes of the strange particles appear to be consistent with the rule $|\Delta T| = \frac{1}{2}$. The associated rule $\Delta S = \Delta Q$, which we discussed in § 12.9(c) does not have the same implications when all of the decay products possess isospin.

Consider Λ decay, the particles possess the following values of T and T_3:

	$\Lambda \to$	$p +$	π^-	$\Lambda \to$	$n +$	π^0
T	0	$\frac{1}{2}$	1	0	$\frac{1}{2}$	1
T_3	0	$\frac{1}{2}$	-1	0	$-\frac{1}{2}$	0.

Thus the decay can proceed by a change $|\Delta T_3| = \frac{1}{2}$ and $|\Delta T| = \frac{1}{2}$ or $\frac{3}{2}$. Using Clebsch–Gordan coefficients (Table 5.2) and denoting a final state of isospin T by $|T\rangle$, we find the following amplitudes for the two channels:

$$|p\pi^-\rangle = -\sqrt{\tfrac{2}{3}}\,|\tfrac{1}{2}\rangle + \sqrt{\tfrac{1}{3}}\,|\tfrac{3}{2}\rangle \tag{12.143}$$

$$|n\pi^0\rangle = \sqrt{\tfrac{1}{3}}\,|\tfrac{1}{2}\rangle + \sqrt{\tfrac{2}{3}}\,|\tfrac{3}{2}\rangle$$

so that

$$\frac{\Lambda \to p\pi^-}{\Lambda \to n\pi^0} = 2 \quad \text{for} \quad \Delta T = \tfrac{1}{2} \tag{12.144}$$

$$= \tfrac{1}{2} \qquad \Delta T = \tfrac{3}{2}.$$

It is apparent from Table 12.9 that the experimental ratio is 2, and hence a $\Delta T = \frac{1}{2}$ transition occurs. The ratio of 2 could arise, however, if the final states were of the form

$$|f\rangle = a_{1/2}\,|\tfrac{1}{2}\rangle + a_{3/2}\,|\tfrac{3}{2}\rangle$$

where the amplitudes a are related as follows

$$a_{3/2} = -2\sqrt{2}a_{1/2}$$

(Okubo, Marshak and Sudarshan, 1959).

We next consider the decay of the Σ-hyperon; this particle has isospin 1 and the possible states of the πN system are $\frac{1}{2}$ and $\frac{3}{2}$, and so a simple rule $\Delta T = \frac{1}{2}$ would have no restrictions. However, we must remember that an isospin operator possesses a 'direction' in isospin space, and so we assume the rule

$|\Delta \mathbf{T}| = \frac{1}{2}$.† The rule is conveniently visualised by introducing a *spurion* (Wentzel, 1956) with isospin $\frac{1}{2}$ so that

$$\mathbf{T}_i = \mathbf{T}_f + \mathbf{T}_s.$$

The spurion carries neither energy nor momentum, and so the dynamical features of the system are unchanged. The function of the spurion can be visualised as follows:

$$\begin{array}{cccccc} & \Sigma \rightarrow & N & + \ \pi & + & s \\ T & 1 & \frac{1}{2} & 1 & & \frac{1}{2} \\ & & \multicolumn{2}{c}{\underbrace{\qquad}_{\frac{1}{2} \text{ or } \frac{3}{2}}} & & \\ & & \multicolumn{4}{c}{\underbrace{\qquad\qquad}_{1}} \end{array}$$

Now the $T = \frac{1}{2}$ and $T = \frac{3}{2}$ channels can be related to the π–N system as follows (Table 5.2):

$$\mathscr{Y}^{1/2}_{1/2} = \sqrt{\tfrac{2}{3}}\, n\pi^+ - \sqrt{\tfrac{1}{3}}\, p\pi^0 \tag{12.145}$$

$$\mathscr{Y}^{1/2}_{3/2} = \sqrt{\tfrac{1}{3}}\, n\pi^+ + \sqrt{\tfrac{2}{3}}\, p\pi^0$$

$$\mathscr{Y}^{-3/2}_{3/2} = n\pi^-$$

and these must in turn be related to the spurion system by Clebsch–Gordan coefficients to yield the appropriate isospin functions for the hyperons. We shall denote the isospin functions for the Σ^+ and Σ^- hyperons by A^1_1 and A^{-1}_1 respectively, and the isospin function of the spurion by $\chi^{1/2}_{1/2}$. If a and b represent the $T_f = \frac{1}{2}$ and $\frac{3}{2}$ amplitudes respectively, then an application of the Clebsch–Gordan coefficients given in Table A.7.1 (p. 708) leads to the following result:

$$A^1_1 = (a\mathscr{Y}^{1/2}_{1/2} - \tfrac{1}{2}b\mathscr{Y}^{1/2}_{3/2})\,\chi^{1/2}_{1/2}$$

$$= \left\{\left[a\sqrt{\tfrac{2}{3}} - \tfrac{1}{2}b\sqrt{\tfrac{1}{3}}\right] n\pi^+ - \left[a\sqrt{\tfrac{1}{3}} + \tfrac{1}{2}b\sqrt{\tfrac{2}{3}}\right] p\pi^0\right\}\chi^{1/2}_{1/2}$$

$$A^{-1}_1 = -\frac{\sqrt{3}}{2}\, b\mathscr{Y}^{-3/2}_{3/2}\,\chi^{1/2}_{1/2} = -\frac{\sqrt{3}}{2}\, bn\pi^-\chi^{1/2}_{1/2}.$$

If we now dispense with the spurion the amplitudes for the three decay modes are given by

$$\Sigma^+ \rightarrow n + \pi^+ \quad N_+ = a\sqrt{\tfrac{2}{3}} - \tfrac{1}{2}b\sqrt{\tfrac{1}{3}} \tag{12.146}$$

$$\Sigma^+ \rightarrow p + \pi^0 \quad N_0 = -a\sqrt{\tfrac{1}{3}} - \tfrac{1}{2}b\sqrt{\tfrac{2}{3}}$$

$$\Sigma^- \rightarrow n + \pi^- \quad N_- = -\frac{\sqrt{3}}{2}\, b$$

† The same rule also applies to Λ-hyperon decay, but in that case there is no need to give it such a formal expression.

hence

$$N_+ + \sqrt{2}N_0 = N_-. \tag{12.147}$$

Now we have already seen (12.140) that the near equality of the decay rates for the three processes implies

$$|N_+|^2 \sim |N_0|^2 \sim |N_-|^2$$

and hence the requirement (12.147) implies that the three amplitudes form a right-angled isosceles triangle. The data displayed in Fig. 12.22 fails this test by between two and three standard deviations, thus indicating that the non-leptonic decay of Σ-hyperons is not necessarily a pure $|\Delta\mathbf{T}| = \frac{1}{2}$ transition.

12.10(c). *The nonleptonic decay of kaons and isospin changes*

Consider the decay of the kaon to two pions

$$K \to \pi + \pi$$

$$T \quad \tfrac{1}{2} \qquad 1 \qquad 1$$

The $|\Delta\mathbf{T}| = \frac{1}{2}$ rule then implies that only the final states $T_f = 0, 1$ are available. The latter state is forbidden by symmetry requirements; since the kaon has spin 0 the pions must emerge in an s-state which is spatially symmetric, but since the final state involves two bosons we require overall symmetrisation of the wave function and this is not possible for a $T = 1$ state (Table 9.6). We therefore conclude that only the $T = 0$ state is available.

Now let us examine K_1^0 decay to the $T = 0$ state, the appropriate isospin function is given in Table 9.6

$$\frac{1}{\sqrt{3}}(\pi_1^+\pi_2^- + \pi_1^-\pi_2^+ - \pi_1^0\pi_2^0)$$

from which we expect the following branching ratio:

$$\frac{K_1^0 \to \pi^0 + \pi^0}{(K_1^0 \to \pi^0 + \pi^0) + (K_1^0 \to \pi^+ + \pi^-)} = \frac{1}{3}. \tag{12.148}$$

The experimental results indicate a branching ratio of 0·31 ± 0·01 (from data assembled by Crawford (1962)), and obviously support the concept of a $|\Delta\mathbf{T}| = \frac{1}{2}$ transition.

Nevertheless, some admixture of $|\Delta\mathbf{T}| = \frac{3}{2}$ transitions could be present in the nonleptonic decay of kaons. The two pions in the process $K^+ \to \pi^+ + \pi^0$, for example, must be in a state with $T = 2$ since $T_3 = 1$ and $T = 1$ is forbidden to them. Thus the transition involves $|\Delta\mathbf{T}| = \frac{3}{2}$. The observed decay rate for the process $K^+ \to \pi^+ + \pi^0$ is $\sim 5 \times 10^{-2}$ times that for $K_1^0 \to \pi^+ + \pi^-$, and could be explained therefore by a $|\Delta\mathbf{T}| = \frac{3}{2}$ transition of small amplitude. Alternatively the decay could take place through a virtual electromagnetic transition, but a slowing factor $\alpha^2 \sim 10^{-4}$ might be expected for this mechanism.

12.10(d). *On the existence of a universal Fermi interaction*

The data on the weak interactions of strange particles is not inconsistent with a basic weak interaction of the V, A type, but so little is still known about the detailed behaviour of these processes that a stronger statement than the above one would be impossible.

Nevertheless, speculations have been made; Feynman and Gell-Mann (1958), for example, have suggested an extension of (12.113) to include strangeness violating currents†

$$(j_\lambda + s_\lambda + j_\lambda^l)^\dagger (j_\lambda + s_\lambda + j_\lambda^l)$$

so that nonleptonic decay would arise from terms of the type $s_\lambda^\dagger j_\lambda$. This simple structure is not easily reconcilable with the complicated isospin currents which appear to exist in the decay of strange particles; why, for example, is the nonleptonic decay of Λ-hyperons consistent with $|\Delta \mathbf{T}| = \frac{1}{2}$ whilst Σ-hyperons appear to require $|\Delta \mathbf{T}| = \frac{1}{2}$ and $\frac{3}{2}$ amplitudes? The discrepancy could be genuine, or due to experimental errors. In fact most of the present experimental data needs further refining, and until this has been achieved satisfactory comprehensive schemes for describing weak decay processes will probably not emerge.

12.11. THE INTERMEDIATE BOSON HYPOTHESIS

The (current) × (current) structure for the matrix element for the strangeness conserving weak interactions (12.113)

$$M_{\text{weak}} = \frac{G_V}{\sqrt{2}} (j_\lambda + j_\lambda^l)^\dagger (j_\lambda + j_\lambda^l)$$

does not permit a process of the type $\mu^- \to e^- + e^+ + e^-$ because the lepton currents are of the type $\bar{u}_l \gamma_\lambda (1 + \gamma_5) u_\nu$, where l represents electron or muon. The decay of a muon to three electrons appears to be a very rare process (Table 12.10), and yet it breaks none of the rules which appear to be associated with other weak interactions. Other examples of processes which do not appear to occur are $\mu^- + p \to n + e^-$ and $\mu \to e + \gamma + \gamma$; upper limits for their observation are quoted in Table 12.10.

TABLE 12.10

Process	Observed upper limit
$\mu \to e + e + e$	$2{\cdot}6 \times 10^{-7}$ ($\mu \to e + \nu + \bar{\nu}$) (a)
$\mu \to e + \gamma + \gamma$	5×10^{-6} ($\mu \to e + \nu + \bar{\nu}$) (b)
$\mu^- \to p \to n + e^-$	$2{\cdot}4 \times 10^{-7}$ ($\mu^- + p \to n + \nu$) (c)

(a) Alikhanov *et al.* (1962), (b) York, Kim and Kernan (1959), (c) Conversi *et al.* (1962)

† It should be noted that an interaction of this form leads to nonleptonic transitions involving $\Delta \mathbf{T}| = \frac{1}{2}$ and $\frac{3}{2}$ amplitudes; if the experiments should show that there is a $|\Delta \mathbf{T}| = \frac{1}{2}$ rule then the interaction given above is unacceptable.

Some explanation for the phenomenological structure of (12.113) would therefore be welcome. Several have been advanced; they may be classified as

(1) intermediate boson hypothesis,
(2) the existence of more than one type of neutrino,
(3) the hypothesis of a multiplicative conservation law in weak interactions.

Combinations of these theories have also been discussed. The multiplicative conservation law (Feinberg and Weinberg, 1961) is associated with the fact that invariance laws are of two types – additive (for example charge conservation) and multiplicative (for example parity conservation)†.

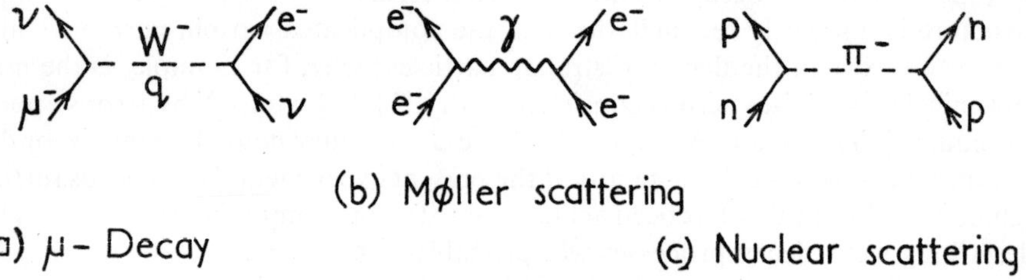

FIG. 12.23.

The intermediate boson hypothesis as first formulated (Feynman and Gell-Mann, 1958) assumed the existence of a charged boson ($W^{\pm}$) which acted as a propagator in weak interactions (Fig. 12.23(a)). It can be seen that the local coupling of four fields has been replaced by two vertices, and that the overall picture is similar to that encountered in electromagnetic and strong interactions. This is an attractive feature of the theory.

Let us examine some of the properties of these bosons. They may be listed as follows (Lee, 1961):

(1) Spin = 1 in order to reproduce the observed vector and axial vector character of the weak interactions.

(2) A mass m_W greater than that of the kaon in order to explain the absence of the process $K \rightarrow W + \gamma$.

(3) A charge $\pm e$, a magnetic moment $(e/2m_W)(1 + K)$ and a quadrupole moment $(-eK/m_W^2)$, where K represents the anomalous part of the magnetic moment in magnetons.

(4) Weak interactions with leptons of the following form:

$$\mathscr{L}_{\text{int}} = G_W W_\lambda \overline{\psi}_\lambda \gamma_\lambda (1 + \gamma_5) \psi_\nu + \text{h.c.} \tag{12.149}$$

where W_λ represents the boson field and G_W the coupling strength of the boson to the lepton fields.

† Additive and multiplicative conservation laws are associated with continuous and discrete transformations respectively (§ 5.1).

We may use the above properties to make further deductions about the behaviour of the bosons. The matrix element associated with Fig. 12.23(a) has the form

$$\frac{G_W^2}{N_l}\bar{u}_\nu\gamma_\lambda(1+\gamma_5)\,u_\mu\frac{1}{q^2+m_W^2}\left(\delta_{\lambda\varrho}+\frac{q_\lambda q_\varrho}{m_W^2}\right)\bar{u}_e\gamma_\varrho(1+\gamma_5)\,u_\nu \quad (12.150)$$

(compare (12.60)), where λ and ϱ represent the initial and final state polarisations respectively. Since $q^2 \ll m_W^2$ in the muon decay process (and in all the strangeness conserving weak decays), we can reduce the above expression to

$$\frac{G_W^2}{N_l m_W^2}\bar{u}_\nu\gamma_\lambda(1+\gamma_5)\,u_\mu\bar{u}_e\gamma_\lambda(1+\gamma_5)\,u_\nu. \quad (12.151)$$

A further comparison with (12.60) then shows that

$$G_W^2 \equiv \frac{G_\mu}{\sqrt{2}}m_W^2 = \frac{10^{-5}}{\sqrt{2}}\frac{m_W^2}{m_p^2} > 1\cdot 8\times 10^{-6} \quad (12.152)$$

where we have set $m_W > m_K$. G_W is therefore called the *semi-weak coupling strength*; its magnitude implies that the decay processes of the boson, say $W \to e + \nu$, should be fast when compared with weak interactions, but production cross-sections should be small compared with strong interactions; the relevant figures are

$$\lambda(W \to e + \nu) > 8\times 10^{16}\ \text{sec}^{-1}.$$

$$\text{production} \sim 10^{-5}\ \text{geometrical}$$

The matrix element (12.150) leads to specific predictions which are not easily tested by experiment (§ 12.3); they were given in (12.64) and (12.72)

$$\tau_V' = \tau_V\left[1 - 0\cdot 6\left(\frac{m_\mu}{m_W}\right)^2\right]$$

$$\varrho = \frac{3}{4} + \frac{1}{3}\left(\frac{m_\mu}{m_W}\right)^2$$

where τ_V is the muon lifetime without the presence of the boson and ϱ is the Michel ϱ-parameter (12.68).

One experiment which might have been expected to reveal the presence of the boson is the observation of the process $\mu \to e + \gamma$. The associated diagrams are given in Fig. 12.24, and order of magnitude approach suggests a matrix element of the form

$$M_{fi} \sim G_W\frac{F(q^2)}{m_W^2}G_W e \sim G_\mu F(q^2)\,e \quad (12.153)$$

where $F(q^2)$ is a form factor and e is the electronic charge. If one makes the reasonable guess $F(q^2) \sim 1$, then the ratio for the matrix elements $\mu \to e + \gamma$ and $\mu \to e + \nu + \bar{\nu}$ would be

$$\frac{M(\mu \to e + \gamma)}{M(\mu \to e + \nu + \bar{\nu})} \sim \frac{G_\mu e}{G_\mu} \sim e$$

so that a branching ratio $e^2 \sim \alpha \sim 10^{-2}$ might be expected. More elaborate calculations (Feinberg, 1958) lead to a divergent expression, but upon inserting a reasonable cut-off, a branching ratio of $\sim 10^{-4}$ is obtained. The experimental upper limit for this process is:

6×10^{-8} (Bartlett, Devons and Sachs, 1962),

2×10^{-7} (Frankel *et al.*, 1962).

Two reasons for this discrepancy have been advanced (whilst still retaining the boson concept). Firstly, since the decay $\mu \to e + \gamma$ involves an electromagnetic process its rate can depend on the detailed electromagnetic structure of the

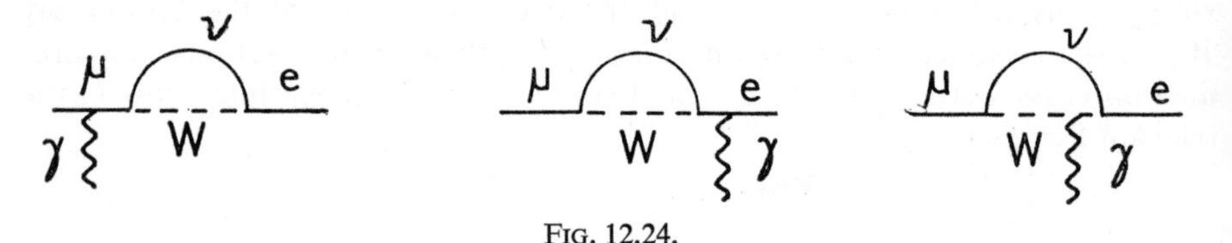

FIG. 12.24.

boson, a suitable choice of anomalous magnetic moment and quadrupole moment (Ebel and Ernst, 1960; Meyer and Salzman, 1959) enables the branching ratio to be suppressed. The alternative argument rests on the fact that the neutrinos occurring in different processes are not all the same type of particle. Therefore if it is assumed that the neutrinos associated with the muon and electron vertices are different, then processes of the type shown in Fig. 12.24 would be forbidden. The experiment of Danby *et al.* (1962) on the reactions induced by high energy neutrinos (§§ 1.2 and 12.12) supports the assumption that the neutrinos associated with β and μ decay are, indeed, different.

We note, finally, that so far in this section we have examined only decays which preserve strangeness. In order to introduce bosons in a consistent manner for the weak interactions of strange particles at least four more bosons would be required (Lee, 1962).

The present evidence for the existence of bosons will be described in the next section.

12.12. NEUTRINO INDUCED REACTIONS

In the previous section mention was made of the possible existence of vector bosons and more than one type of neutrino. Possible methods of testing these hypotheses and many others are provided by experiments in which neutrinos are used as primary particles (Lee and Yang, 1960; Pontecorvo, 1959).

The experimental problems are formidable. Consider the process

$$\bar{\nu} + p \rightarrow n + e^{+}$$

the cross-section in the c-system is given by (7.72)

$$\frac{d\sigma}{d\Omega} = \frac{V^4}{(2\pi)^2} \overline{\sum_i} \sum_\alpha \frac{E_{\bar{\nu}} E_p E_n E_{e^+}}{E_c^2} \frac{p_c^f}{p_c^i} |M_{fi}|^2 \tag{12.154}$$

where E_c is the total energy in the c-system. For antineutrino energies above a few MeV we can make the aproximations

$$E_{\bar{\nu}} \sim E_{e^+}, \qquad E_p \sim E_n, \qquad p_c^f \sim p_c^i$$

and the matrix element can be evaluated as

$$\overline{\sum_i} \sum_\alpha |M_{fi}|^2 \sim \frac{1}{V^4} (G_V^2 + 3G_A^2) \sim 4 \frac{G^2}{V^4}$$

for a $V - A$ interaction (compare § 12.2). Thus equation (12.154) leads to a total cross-section

$$\sigma_T \sim \frac{4\pi}{(2\pi)^2} 4G^2 \frac{E_{\bar{\nu}}^2 E_p^2}{E_c^2} = \frac{4}{\pi} \left(\frac{10^{-5}}{m_p^2}\right)^2 \frac{E_{\bar{\nu}}^2 E_p^2}{E_c^2}.$$

If we now turn to the laboratory system the total section remains the same. If we represent the energy of the neutrinos in this system as E_L, then

$$E_c^2 = m_p^2 + 2m_p E_L$$

and for moderate energies

$$E_{\bar{\nu}}^2 \sim E_L^2, \quad E_p^2 \sim m_p^2$$

hence

$$\sigma_T \sim \frac{4}{\pi} \left(\frac{10^{-5}}{m_p}\right)^2 \frac{E_L^2 m_p^2}{m_p^2 (m_p^2 + 2m_p E_L)} \sim \left(\frac{10^{-5}}{m_p}\right)^2 \left(\frac{E_L}{m_p}\right)^2 \frac{1}{1 + 2\, E_L/m_p}. \tag{12.155}$$

The proton Compton wavelength is given by $m_p^{-1} \sim 2 \times 10^{-14}$ cm and for $E_L \sim 1$ MeV the above relation yields a cross-section of $\sim 5 \times 10^{-44}$ cm^2. This is very small, but nevertheless experiments have been done by Cowan and Reines (§ 1.2) using the intense flux of antineutrinos from a reactor.

At high energies both neutrinos and antineutrinos are available through the decay process

$$\pi^+ \to \mu^+ + \nu$$

$$\pi^- \to \mu^- + \bar{\nu}$$

and although the intensity of the particles is not as great as from reactors, the lack of numbers is partially compensated by the rise of the cross-section with energy, for example at $E \sim 1\text{GeV}$ σ_T has risen to $\sim 2\times 10^{-38}$ cm². Obviously there is a limit to this rise in cross-section with energy and we shall examine this point later. Before doing so some possible applications of the neutrino induced reactions will be mentioned.

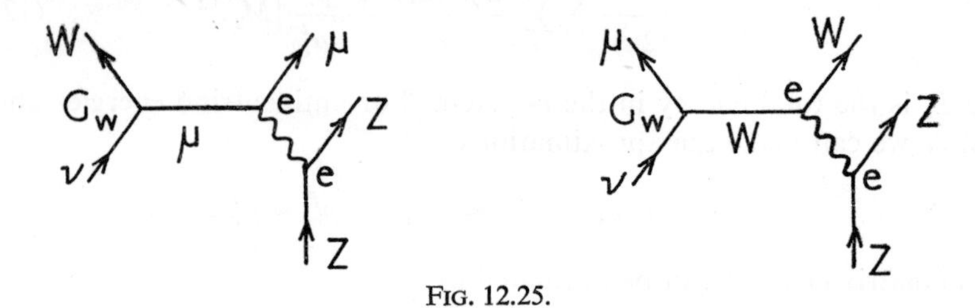

FIG. 12.25.

(1) The production of vector bosons. In the previous section we have stated that the coupling strength of the vector bosons to leptons is given by

$$G_W^2 = \frac{G\mu}{\sqrt{2}} m_W^2 \sim \frac{10^{-5}}{\sqrt{2}} \left(\frac{m_W}{m_p}\right)^2.$$

Since the vector coupling strengths are the same in β and μ decay, we may conclude that the vector boson couples to strongly interacting particles with roughly the same strength. Calculations using this semi-weak coupling strength indicate cross-sections $\sim 10^{-32}$ cm² for processes of the type

$$\pi^+ + p \to W^+ + p.$$

Since the lifetime of the bosons would be $\sim 10^{-17}$ sec their separate existence would not be observed – their presence could be revealed, however, by the processes indicated below.

$$\pi^+ + p \to W^+ + p \begin{cases} K^+ + \gamma + p & \text{apparent violation of strangeness} \\ \mu^+ + \nu + p & \text{apparent production of leptons.} \end{cases}$$

The problems of detecting these reactions against background under experimental conditions are formidable. The background problems in detecting the production of bosons are less difficult if a high energy neutrino beam can be used.

The bosons can be produced in processes of the type shown in Fig. 12.25; Z represents a target nucleus. Break up of the nucleus may occur

$$\nu + Z \begin{cases} W^+ + \mu^- + Z & \text{coherent} \\ W^+ + \mu^- + \text{star} & \text{incoherent}. \end{cases}$$

Let us examine the condition for coherent production; now the minimum recoil momentum transferred to the nucleus is given by

$$q_{\min} \sim \sqrt{(m_W^2 + \mathbf{k}_W^2)} - |\mathbf{k}_W| \sim \frac{m_W^2}{2|\mathbf{k}_W|} \sim \frac{m_W^2}{2E_\nu} \tag{12.156}$$

where the high energy approximations $|\mathbf{k}_W| \gg m_W$ and $|\mathbf{k}_W| \sim E_\nu$ have been used. At very high energies $q_{\min}$ can be much smaller than $1/R_Z$ where R_Z represents the nuclear radius. In these circumstances the nucleus can recoil as a whole and coherent production can occur with a cross-section proportional to

$$G_W^2 \alpha^2 Z^2 \sim \frac{10^{-5}}{m_p^2} 10^{-4} Z^2 \sim 10^{-37} Z^2.$$

If the nucleus breaks up in the reaction (incoherent production), a cross-section of order $G_W^2 \alpha^2 Z \sim 10^{-37} Z$ is expected. Detailed calculations have been made by Lee and Yang.

The high energy neutrino experiments have not yet yielded positive evidence for the existence of bosons. The data from the CERN experiment indicates that if the bosons exist they must have a mass of at least 1·3 GeV (Siena conference report, 1964).

(2) On the existence of different types of neutrino. In the previous section mention was made of the apparent absence of the decay process $\mu \to e + \gamma$.

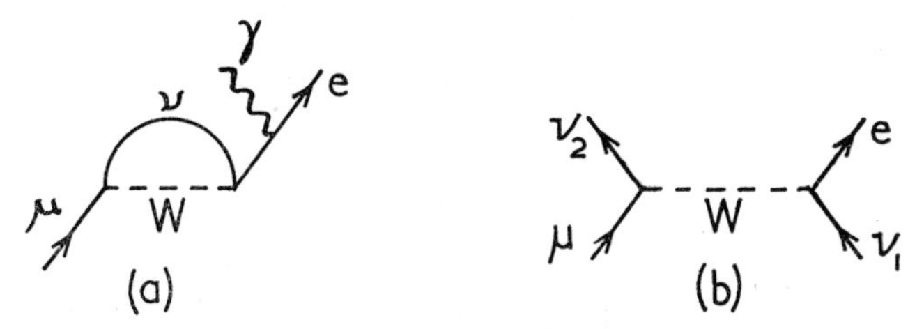

FIG. 12.26.

One explanation of the absence of this process is that the neutrinos associated with the muons and electrons are different. This is illustrated in Fig. 12.26(b); it is evident that if $\nu_1 \neq \nu_2$ the loop appearing in Fig. 12.26(a) cannot be constructed. The difference can be given quantitative form by introducing a second conservation law in weak interactions – *conservation of muonic number m*

$$\begin{aligned} m &= +1 \quad \text{for} \quad \mu^-, \nu_\mu \\ m &= -1 \qquad\qquad \mu^+, \bar{\nu}_\mu \\ m &= 0 \qquad\qquad \text{all others (including } e, \nu_e) \end{aligned} \tag{12.157}$$

where the subscripts μ and e indicate the neutrinos associated with muons and electrons respectively. The law is an additive one, and it is simultaneously operative with lepton conservation (§ 12.4); for example if l represents leptonic number, then

$$\begin{array}{ccccc} & \mu^- \rightarrow & e^- + & \nu_\mu + & \bar{\nu}_e \\ m & +1 & 0 & +1 & 0 \\ l & +1 & +1 & +1 & -1 \end{array}$$

$$\begin{array}{cccc} & \pi^+ \rightarrow & \mu^+ + & \nu_\mu \\ m & 0 & -1 & +1 \\ l & 0 & -1 & +1\,. \end{array}$$

Confirmation of the existence of the law can be sought by high energy neutrino experiments. The neutrinos arise in the decay process $\pi \rightarrow \mu + \nu$ and so $m = \pm 1$ depending on the pion charge; hence processes of the type

$$\nu_\mu + n \rightarrow p + \mu^-$$

would be expected, whereas

$$\nu_\mu + n \rightarrow p + e^-$$

would be forbidden. It should be noted, however, that electrons could appear if the cross-section for the boson production is high by the process

$$\begin{array}{c} \nu_\mu + n \rightarrow W^+ + \mu^- + n \\ \downarrow \\ e^+ + \nu_e\,. \end{array}$$

The observations of Danby *et al.* (1962) support the existence of the muonic number law (§ 1.2). Their results have been confirmed in experiments at CERN.

(3) The structure of weak interactions. The cross-section (12.155) predicts $\sigma_T \propto E_L^2$ for $E_L \ll m_p$ and $\sigma_T \propto E_L$ for $E_L \gg m_p$. This prediction is obviously an impossible one at high energies since it predicts infinite cross-sections and so breaks unitarity. In Fig. 12.27 we show the effect of abandoning a simple point interaction in favour of an interaction which uses the electromagnetic form factors of the nucleons (Lee and Yang, 1960). It should be noted that the weak form factors for the vector interaction only can be obtained from the electromagnetic

data (by application of the conserved vector current hypothesis). A detailed study of the neutrino–nucleon interaction can therefore lead to the weak interaction form factors, in the same manner as the electron–nucleon scattering leads to the electromagnetic form factors.

The preliminary data on the 'elastic' process $\nu_\mu + p \rightarrow n + \mu^-$ in the CERN neutrino experiment indicates the presence of form factors of about the expected

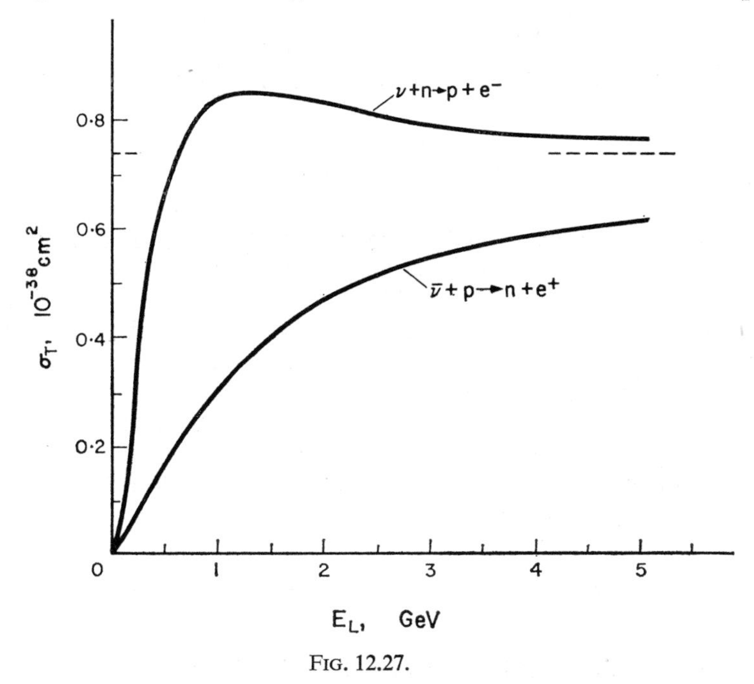

FIG. 12.27.

magnitude (Siena conference report, 1964). A notable feature in this experiment is the absence of neutral currents corresponding to $\nu_\mu + p \rightarrow p + \nu_\mu$. If weak interactions are assumed to be propagated by bosons, then this result argues against the existence of neutral vector bosons.

Many other interesting possibilities arising from high energy neutrino experiments are discussed in the article of Lee and Yang (1960).

CHAPTER 13

STRONG INTERACTIONS I RESONANCES AND STRANGE PARTICLES

13.1. INTRODUCTION

This chapter will be concerned with systems which decay by strong interactions, and with some properties of the strange particles which are not well known. Finally we shall discuss some possible classification schemes for the particles.

Many of the systems we shall examine possess short lifetimes and are called both resonances and particles. Recent discoveries have blurred the distinction between these words. The term resonance is normally reserved for scattering systems which go through a phase change of 90° (§ 6.1). The word particle is given to independent systems which get beyond the range of nuclear forces. Let us therefore examine a few numbers. The range of nuclear forces (r) is given roughly by the pion Compton wavelength

$$r \sim \lambda\!\!\!^{-}_{\pi} = \frac{1}{m_{\pi}} \equiv \frac{\hbar}{m_{\pi}c} \sim 10^{-13} \text{ cm} = 1 \text{ fermi}\dagger \tag{13.1}$$

so that a natural time unit for a strong interaction is given by

$$\tau = \frac{r}{c} = \frac{\hbar}{m_{\pi}c^2} \sim 10^{-23} \text{ sec}. \tag{13.2}$$

Now if a reaction takes place in a time τ, its natural width is given by

$$\Gamma = \frac{\hbar}{\tau} = m_{\pi}c^2 \sim 140 \text{ MeV}. \tag{13.3}$$

The nucleon isobars (§ 13.2) have widths which are comparable with this figure. In addition at least one of them (the $T = \frac{3}{2}$, $j = \frac{3}{2}$ state) is known to go through a phase change of 90° at the appropriate energy (Fig. 14.2). This state may therefore be regarded as a resonance in the classical sense. The ϱ-system (which decays $\varrho \to 2\pi$) also possesses the properties of a resonance, since $\Gamma \sim 100$ MeV and there is evidence that pion–pion scattering undergoes a phase change

† A more exact (and sometimes useful) relationship is 1 f ≡ 197 MeV/c ~ 200 MeV/c.

of 90° at the appropriate energy. On the other hand, the ω-system (which decays $\omega \to 3\pi$) has a width $\Gamma \sim 10$ MeV, which corresponds to a distance of travel of about 20 f before decay. Thus ω has a separate existence, and could be described as a particle.

However, the words resonance and particle now have a recognised and interchangeable meaning for these systems, and so we shall follow convention and use both terms.

The material reported in this chapter is naturally incomplete because of the nature of the subject; it is adequate, however, to indicate the main lines of analysis. A table showing the data at present available on the resonant systems is given in A.9 (Appendixes, p. 715).

13.2. THE NUCLEON ISOBARS

The resonant states reveal themselves as a peak either in a scattering cross-section plotted as a function of energy or in a plot of effective mass for the final

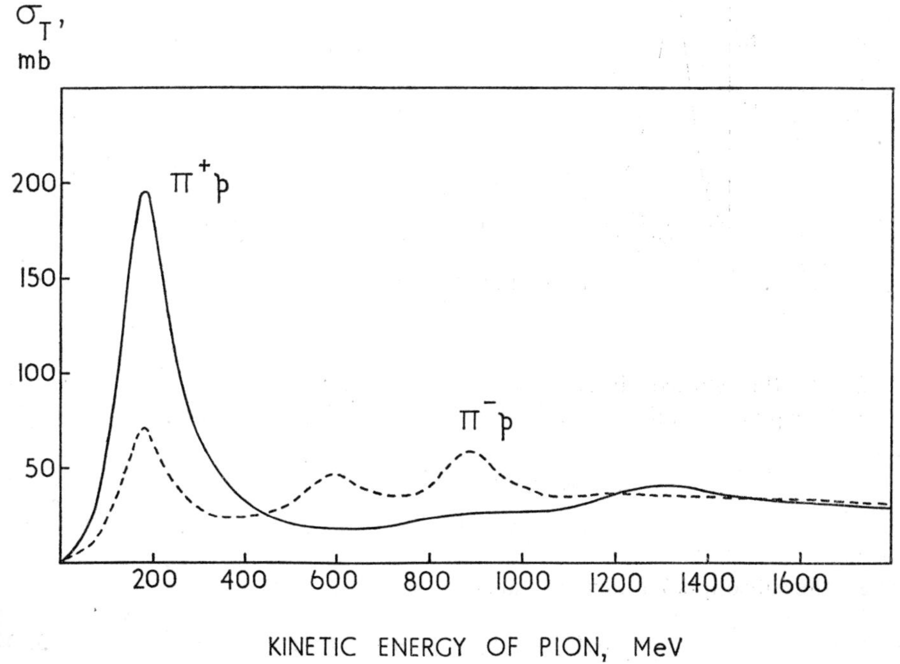

FIG. 13.1.

products of an interaction. The nucleon isobars belong to the former category and appear in the total cross sections for the scattering of pions from protons (Fig. 13.1).

The material assembled in this diagram comes from many groups; references may be found in the work of Klepikov, Mescheryakov and Sokolov (1960), and of Devlin, Moyer and Perez-Mendez (1962).

The data shows bumps suggestive of resonances. It is instructive to examine the material in isospin channels; using Table 5.2 (p. 200) we can write

$$|\pi^+p\rangle = |\tfrac{3}{2}\rangle \tag{13.4}$$

$$|\pi^-p\rangle = \frac{1}{\sqrt{3}}\left|\frac{3}{2}\right\rangle - \sqrt{\frac{2}{3}}\left|\frac{1}{2}\right\rangle$$

where the kets on the right-hand side refer to the total isospin channels (the hypothesis of charge independence implies that the amplitudes cannot depend on

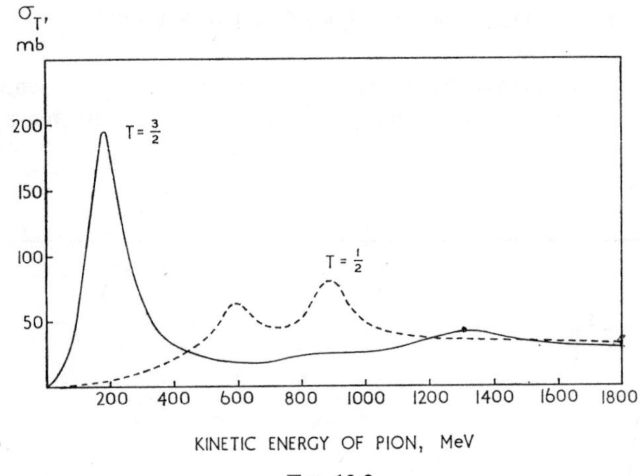

FIG. 13.2.

T_3). Since the two isospin states are orthogonal, there can be no interference after the squares of all the amplitudes have been added, and so we may conclude that

$$\sigma_T(\pi^+p) = \sigma_T(\tfrac{3}{2}) \tag{13.5}$$

$$\sigma_T(\pi^-p) = \tfrac{1}{3}\sigma_T(\tfrac{3}{2}) + \tfrac{2}{3}\sigma_T(\tfrac{1}{2})$$

where the subscripts T indicate total cross-sections, hence

$$\sigma_T(\tfrac{1}{2}) = \tfrac{3}{2}\sigma_T(\pi^-p) - \tfrac{1}{2}\sigma_T(\pi^+p). \tag{13.6}$$

The cross-sections in the two isospin channels are displayed in Fig. 13.2.

The curves strongly indicate the presence of resonances. The magnitudes of the cross-sections at the peaks and their positions are summarised in Table 13.1. In it T_π and M* refer to the kinetic energy of the pion in the laboratory system and total energy in the c-system respectively.

TABLE 13.1

Position of peak (MeV)		Cross section (mb)			
T_π	M^*	π^+p	π^-p	$T=\frac{3}{2}$	$T=\frac{1}{2}$
180	1238	195	67	195	3
605	1510	18	47	18	64
890	1680	20	59	20	80
1300	1900	41	36	41	34

Peaks in the cross-sections are also observed in the data for the photoproduction of pions. Total cross-section curves for the processes

$$\gamma + p \begin{matrix} \nearrow p + \pi^0 \\ \searrow n + \pi^+ \end{matrix}$$

are shown in Fig. 13.3, and the positions and heights of the peaks are tabulated

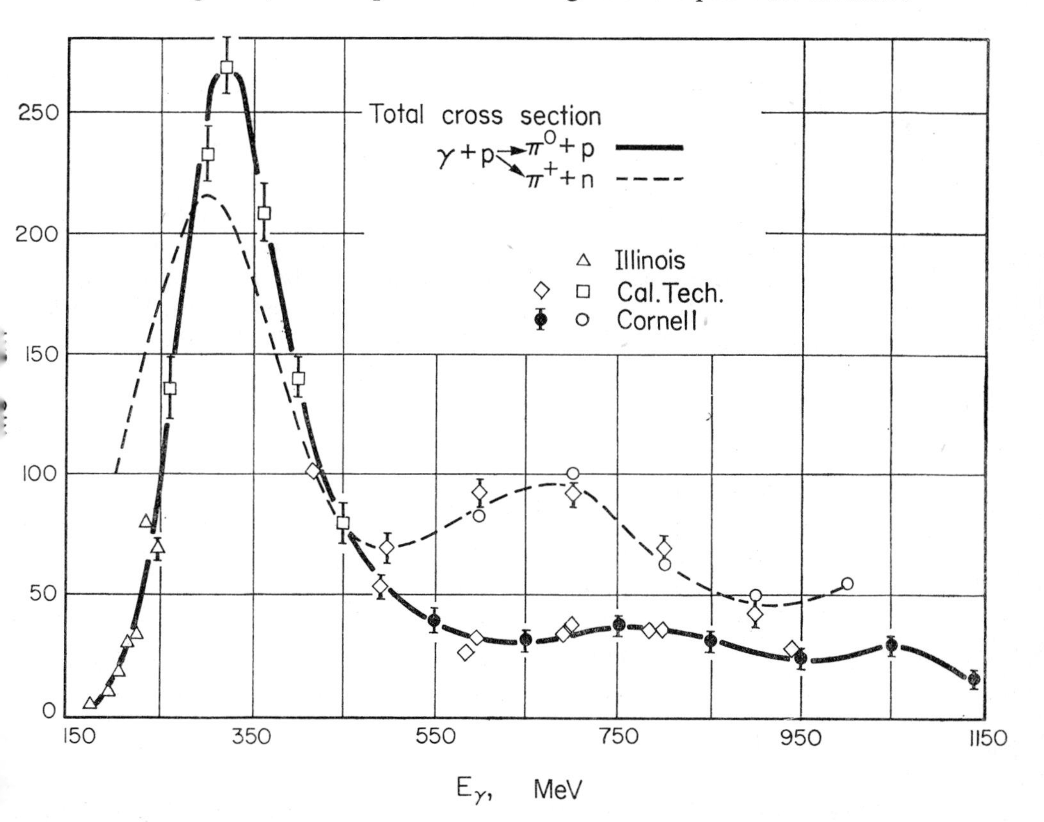

FIG. 13.3.

in Table 13.2. In the table E_γ and M^* refer to the photon energy in the laboratory system and total energy in the c-system respectively.

The difference in position of the peaks for the production of π^+ and π^0 mesons may be attributed to a photoelectric interference term (§ 14.1(d)). A comparison

TABLE 13.2

Position of peak (MeV)		Cross-section (μb)	Reaction products
E_γ	M^*		
310	1210	220	$\pi^+ n$
320	1220	270	$\pi^0 p$
700	1484	90	$\pi^+ n$
740	1507	40	$\pi^0 p$
1050	1697	25	$\pi^0 p$

with Table 13.1 shows that the peaks occur at the same energies for M^*. The peaks at $E_\gamma \sim 300$ and 700 MeV have also been observed in Compton scattering from protons (Fig. 11.5).

We shall label the systems with energies of 1238, 1510, 1680 and 1900 MeV as N_1^*, N_2^*, N_3^* and N_4^* respectively.† They obviously all share the quantum number $B = 1$, and the states N_1^* and N_4^* have isospin $T = \frac{3}{2}$ whilst N_2^* and N_3^* have $T = \frac{1}{2}$. The assignment of angular momentum numbers and parities can be made from the data on the angular distributions and polarisations. These will be examined in Chapter 14. The final results are quoted in Table 13.3.

TABLE 13.3

Isobar	M^* (MeV)	Width Γ (MeV)	T	L_j^{parity}
N_1^*	1238	90	$\frac{3}{2}$	$P_{3/2}^+$
N_2^*	1510	115	$\frac{1}{2}$	$D_{3/2}^-$?
N_3^*	1680	100	$\frac{1}{2}$	$F_{5/2}^+$?
N_4^*	1900	200	$\frac{3}{2}$	$F_{7/2}^+$, $G_{7/2}^-$?

† These systems are also often written as N_{2T2j}^* where T and j refer to isospin and spin states respectively.

13.3. THE PION SYSTEMS

In this section we shall examine some of the states, with quantum numbers

$$B = S = 0,$$

which decay to two or more pions.

13.3(a). *The ϱ-system*

The first clear evidence for the existence of the ϱ-state came from the experiment of Erwin *et al.* (1961); this involved a study of the reactions

$$\pi^- + p \rightarrow n + \pi^- + \pi^+$$
$$p + \pi^- + \pi^0$$

at a pion momentum of 1·9 GeV. The measured distribution of the total energy in the pion–pion *c*-system is displayed in Fig. 13.4. The events have been divided

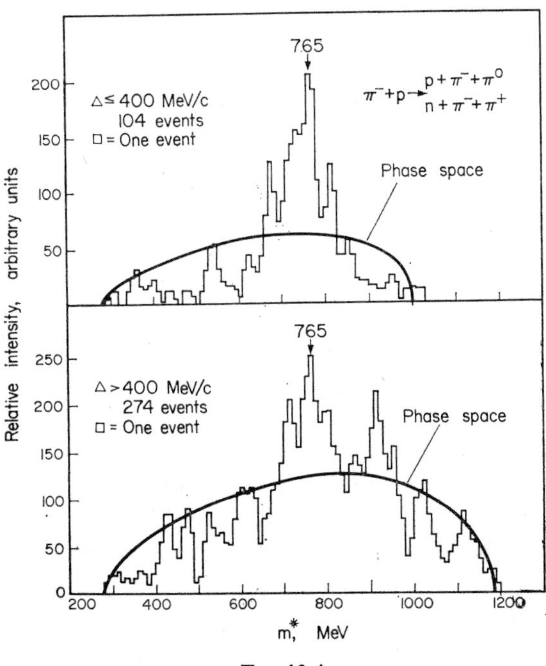

FIG. 13.4.

according to the momentum transferred to the nucleon (*Δ* in Fig. 13.4). Those which involve small momentum transfer show a prominent peak at $m^* \sim 750$ MeV with width $\Gamma \sim 100$ MeV.

The prominence of the peak in the events which involve low momentum transfer is probably due to the fact that these are peripheral collisions (that is collisions with large impact parameter) between the incident pion and a pion on the edge of the pion cloud about the nucleon (Fig. 14.24). At the appropriate π–π relative momentum the ϱ-state is then formed with little momentum transfer to the nucleon.

There is good evidence that this system has isospin $T = 1$. Thus Erwin *et al.* (1961) found a ratio of $(1{\cdot}7 \pm 0{\cdot}3)$ for the final states $(n\pi^-\pi^+/p\pi^-\pi^0)$; ratios of 2

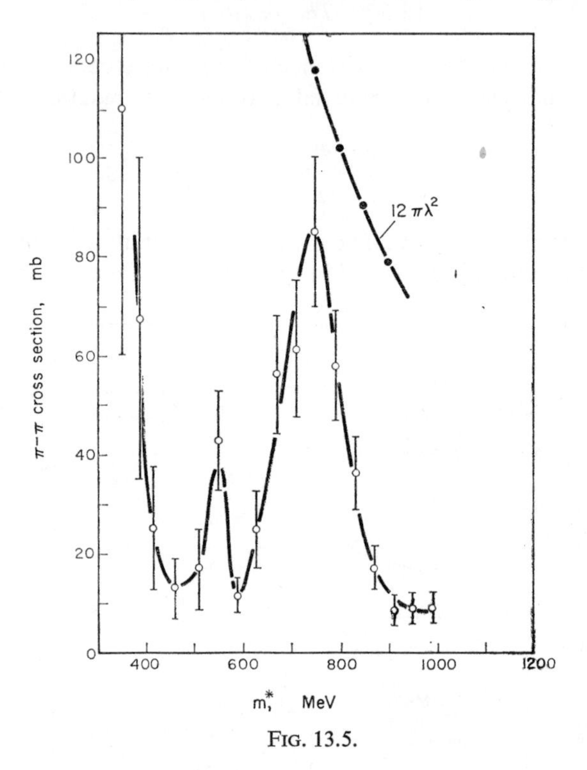

FIG. 13.5.

and 2/9 were expected if the assumption was made that the π–π scattering was dominated by states with $T = 1$ or 2 respectively. Independent evidence for $T = 1$ has been given by Stonehill *et al.* (1961) and by Anderson *et al.* (1961) who observed resonances in the $\pi^+\pi^0$ and $\pi^-\pi^0$ systems, but not in the charge states $\pi^\pm\pi^\pm(T = 2)$.

The occurrence of a state with $T = 1$ implies that the isospin part of the wave function is antisymmetric (Table 9.6 (p. 391)), therefore its spin j must be odd so that the overall system obeys Bose–Einstein statistics. The available evidence indicates $j = 1$; this result has been shown in two ways.

Firstly, if it is assumed that the production process involves mainly one pion exchange (Fig. 14.24), then the magnitude of the pion–pion cross-section can be estimated from the experimental data on the process $\pi + p \rightarrow 2\pi + N$ by using the following expression, derived by Chew and Low (1959)

$$\sigma_{\pi\pi}(\omega) = \frac{4\pi}{f^2} \frac{q^2}{(\omega^2 - 4m_\pi^2)^{1/2}\omega} \left\langle \frac{(\Delta^2 + m_\pi^2)m_\pi^2}{\Delta^2} \frac{d^2\sigma}{d\omega^2 d\Delta^2} \right\rangle \tag{13.7}$$

where Δ = four-momentum transfer to nucleon, ω = total energy in π–π c-system, q = laboratory momentum of incident pion, and $f^2/4\pi$ = pion–nucleon coupling constant $\sim 0{\cdot}08$.

The cross-section $\sigma_{\pi\pi}(\omega)$ has been obtained by a number of workers (Erwin *et. al.*, 1961; Pickup, Robinson and Salant, 1961). The data of the first group is shown in Fig. 13.5, where $\omega \equiv m^*$. It can be see that at $m^* = m_\varrho$ the cross-section $\sigma_{\pi\pi}$ is comparable with the geometric limit $12\pi\lambda\!\!\!^{-2}$, appropriate for a $j = 1$ resonance.

Independent evidence for spin $j = 1$ for the ϱ-system has been obtained by Carmony and Van de Walle (1962), who examined the angular distributions of the $\pi^+\pi^0$ system relative to the direction of the ϱ-meson. They found the distributions exhibited forward–backward asymmetries above and below the resonance energy, but in the region of the resonance a symmetric distribution $\sim 4 \cos^2 \theta + 1$ was obtained. This is again consistent with $j = 1$.

Now $j = 1$ for the pion–pion system implies $j = l = 1$ and therefore odd parity. We may therefore conclude that the system has $T = 1, j = 1^-$. These are the quantum numbers required for the system which was used to explain the behaviour of the nucleon isovector form factor in § 11.4(f).

We finally note that the observed width of the ϱ-decay enables us to make a rough estimate of the $\varrho\pi\pi$ coupling strength. A suitable Lagrangian for a vector meson with odd parity decaying to two pions is given by

$$\mathscr{L} = ig_{\varrho\pi\pi}\boldsymbol{\rho}_\mu \cdot \left(\boldsymbol{\pi} \times \frac{\partial \boldsymbol{\pi}}{\partial x_\mu}\right) \equiv ig_{\varrho\pi\pi}\varepsilon_{rst}\varrho_{\mu r}\left(\pi_s \frac{\overleftrightarrow{\partial}\pi_t}{\partial x_\mu}\right) \tag{13.8}$$

(compare Chan, 1962). In this expression ϱ and π refer to the ϱ and π-meson fields respectively, ε_{rst} is a tensor defined in A.4 (Appendixes, p. 700) and the symbol $\overleftrightarrow{\partial}$ was introduced in (8.120). It is not difficult to see that the Lagrangian leads to the following width for ϱ-decay:

$$\begin{aligned} \Gamma &= g_{\varrho\pi\pi}^2(2\pi)^4 \overline{\sum_i} \sum_\alpha \left| \frac{e_\mu(k_1 - k_2)_\mu}{\sqrt{(8m_\rho\omega_1\omega_2)}} \right|^2 \frac{4\pi}{(2\pi)^6} \omega_1\omega_2 \frac{|\mathbf{k}_\pi|}{m_\varrho} \\ &= \frac{g_{\varrho\pi\pi}^2}{\pi} \overline{\sum_i} \sum_\alpha \left| \frac{2\mathbf{e} \cdot \mathbf{k}_\pi}{\sqrt{(8m_\varrho\omega_\pi^2)}} \right|^2 \omega_\pi^2 \frac{|\mathbf{k}_\pi|}{m_\varrho} = \frac{g_{\varrho\pi\pi}^2}{4\pi} \frac{2}{3} \frac{|\mathbf{k}_\pi|^3}{m_\varrho^2} \end{aligned} \tag{13.9}$$

where the first line can be obtained by appropriate modifications of Table 8.1 (p. 326) together with equations (8.91), (7.39) and (7.59); the factor $e_\mu(k_1 - k_2)_\mu$ arises from the vector nature of the ϱ-meson (13.8), e_μ refers to its polarisation vector and k_1 and k_2 represent the four momenta of the two pions. The second line represents the simplifications which come from working in the c-system of the ϱ-particle. The factor 1/3 in the final line arises from the averaging over initial spin orientations. The momentum $|\mathbf{k}_\pi|$ is given by (A.8.2) (p. 713)

$$|\mathbf{k}_\pi| = \frac{1}{2m_\varrho}(m_\varrho^2 - 4m_\pi^2)^{1/2}m_\varrho \sim \frac{m_\varrho}{2}$$

hence

$$\Gamma \sim \frac{g_{\varrho\pi\pi}^2}{4\pi}\frac{1}{12}m_\varrho.$$

Since the measured value of Γ is ~ 100 MeV, we finally obtain

$$\frac{g_{\varrho\pi\pi}^2}{4\pi} \sim \frac{12\Gamma}{m_\varrho} \sim \frac{1200}{750} \sim 2. \tag{13.10}$$

13.3(b). *The ω-system*

The first evidence for a state decaying to three pions was obtained by Maglić *et al.* (1961) from a study of the final three pion states in the interaction

$$p + \bar{p} \to \pi^+ + \pi^+ + \pi^- + \pi^- + \pi^0.$$

It was found that if an effective mass distribution was plotted for states involving $\pi^+\pi^-\pi^0$, then a spike appeared on the smooth curve predicted by the consideration of phase space. No spike appeared, however, on the data for three pion states with net charge one or two. The peak was also observed by Xuong and Lynch (1961) in the antiproton annihilation with seven final meson states

$$p + \bar{p} \to 3\pi^+ \to 3\pi^- + \pi^0.$$

The clearest evidence for the ω-state was obtained, however, by Pevsner *et al.* (1961) from at a study of the reaction

$$\pi^+ + d \to p + p + \pi^+ + \pi^- + \pi^0$$

for pions with momentum 1·23 GeV/c. Their data is displayed in Fig. 13.6. It can be seen that well over half the reactions form the ω-state. Essentially the reaction is

$$\pi^+ + n \to p + \omega \to p + \pi^+ + \pi^- + \pi^0.$$

The measured width of the ω-resonance is ~ 10 MeV; thus the ω-particle possesses a relatively long lifetime if the ϱ-state is used as a standard of comparison. Part of this difference can be accounted for by the much smaller volume of phase space that is available for decay into three pions rather than two pions. A further factor could be that the ω-decay is mediated by a pair of heavy particles (for example a nucleon–antinucleon pair), whilst the ϱ-mesons are coupled directly to the pions.

No charged state of the ω-meson has ever been observed, and so it is reasonable to conjecture that $T = 0$ for the ω-system. A spin and parity assignment of 1^- has been made with the aid of a Dalitz plot (Stevenson *et al.*, 1962) – see

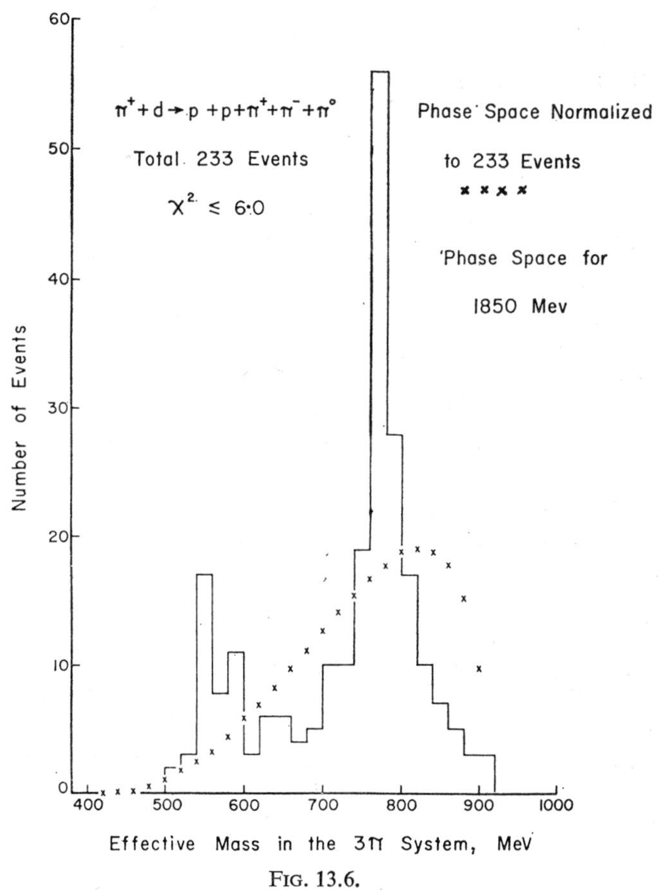

FIG. 13.6.

§ 9.5(c). Thus the quantum numbers of the ω-particle satisfy the requirements of the isotopic scalar system which appears as an intermediate state in the electron scattering experiments (§ 11.4(f)).

13.3(c). *The η-system*

Figure 13.6 shows a second peak in the region of 550 MeV. It has also been observed by Bastien *et al.* (1962b) in the reaction

$$K^- + p \rightarrow \Lambda + \pi^+ + \pi^- + \pi^0 .$$

The three-pion system is called the η-state; it has a width less than 10 MeV and so is long lived. It is believed to have zero isospin since reactions producing charged η-mesons have not been observed (Carmony, Rosenfeld and Van de Walle, 1962).

A notable feature of the decay of the η-particle is that a substantial fraction of the decays yield neutral particles only. This has been demonstrated by Bastien *et al.* (1962b), who plotted the effective mass spectrum of the neutral system in the reaction $K^- + p \to \Lambda +$ neutrals and obtained the ratio

$$\frac{\eta \to \text{neutrals}}{\eta \to \pi^+ + \pi^- + \pi^0} = 3{\cdot}2 \pm 1.$$

An examination of the neutral decay modes has been made by Chrétien *et al.* (1962), who concluded that about half of the neutral decay processes are $\eta \to 2\gamma$, and that their observations were not inconsistent with the mechanism $\eta \to 3\pi^0$ for the remainder. Now the process $\eta \to 2\gamma$ is electromagnetic, and since isospin is not conserved in the decay $\eta \to 3\pi^0$† this decay mode must also be electromagnetic. The expected width for electromagnetic processes would be of order 0·1–1·0 KeV.

The existence of the decay mode $\eta \to 2\gamma$ implies that the spin of the η-particle is 0 or 2 (see the argument regarding π^0 mesons in § 9.5(b)). However, a particle with even spin and parity could decay strongly to two pions, and since this decay mode is not observed, the parity of the η-particle must be odd. Now consider the G parity, the decay $\eta \to 2\gamma$ implies that the η-particle must possess even charge parity, and since $T = 0$ then $G = \xi_C(-1)^T = +1$ by (9.138). The same conclusion may have been reached by recalling that the G parity of a three-pion system is $(-1)^3$ by (9.137), and that the failure of isospin conservation automatically implies the failure of G conservation. The quantum numbers of the η-particle have also been examined by using a Dalitz plot (§ 9.5 (c)) for the process $\eta \to \pi^+ + \pi^- + \pi^0$, and again the assignment 0^-, even G parity was made (see, for example, Foelsche *et al.*, 1962).

13.4. STRANGE PARTICLES

13.4(a). *The parity of the nonresonant states*

Many of the intrinsic properties of the strange particles and their resonant systems are still unknown or have only been recently established. This statement is particularly true of the parity and so we start this section with a discussion of this problem.

The determination of the parity of the strange particles can only be made by examining their strong and electromagnetic interactions; the decay processes

† The isospin of the η-particle is zero, but that for a system of three neutral pions must be ≥ 1.

of the nonresonant systems cannot yield information because of the failure of parity conservation in weak interactions. Strange particles occur in pairs, however, in strong and electromagnetic processes, and so an absolute parity assignment is not possible for states with $S \neq 0$. We shall therefore use the convention that the parity of the Λ-hyperon is even, and that the parities of the other strange particles are given in relation to it.

We consider first the parity of the kaon. There is strong evidence that it is odd from the existence of the processes

$$K^- + He^4 \rightarrow {}_{\Lambda}He^4 + \pi^-$$

$$ {}_{\Lambda}H^4 \;\; + \pi^0$$

which have been observed in a helium bubble chamber (Block *et al.*, 1959). The symbols ${}_{\Lambda}He^4$ and ${}_{\Lambda}H^4$ indicate nuclei in which a neutron has been replaced by a Λ-hyperon (*hypernuclei*). About 3 per cent of all K^--mesons stopping in the helium yield the above reactions, thus indicating that there are no selection rules which strongly forbid the reaction. Dalitz and Liu (1959) (see also Dalitz, 1961) have produced strong arguments that the spin of the (${}_{\Lambda}He^4$, ${}_{\Lambda}H^4$) doublet is zero. Therefore since the pion possesses odd intrinsic parity, angular momentum and parity can only be preserved in the reactions listed above if the relative $K\Lambda$ parity is odd.

The above argument is independent of whether the K^--meson is captured from an s-state about the helium nucleus or from a higher orbit. Day and Snow (1959) have argued that the capture rate is predominantly from the s-state. If this argument is accepted then it can be readily shown that the angular distribution of the π^--mesons from the process

$$K^- + He^4 \rightarrow {}_{\Lambda}H^4 + \pi^0$$

$$\downarrow$$

$$He^4 + \pi^-$$

uniquely determines the K^- parity and the ${}_{\Lambda}H^4$ spin (Block, Lendinara and Monari, 1962). In particular, an isotropic angular distribution would be expected if the parity of the K^--meson is odd and ${}_{\Lambda}H^4$ has zero spin. The data of Block *et al.* support this conclusion.

Next let us consider the parity of the Σ-hyperon. Indirect experimental evidence from hypernuclei, which we shall mention in § 14.5(e), suggests that the relative $\Sigma\Lambda$ parity is even. A similar conclusion has been reached by Tripp, Watson and Ferro-Luzzi (1962a) by a more direct experimental method. These workers examined the angular distribution and polarisation of the Σ-hyperons in the process

$$K^- + p \rightarrow \Sigma + \pi .$$

Since the interaction is a strong one, the angular distribution and polarisation of the Σ-hyperons is determined by the initial angular momentum states and by the relative parity of the $K\Sigma$-system. The polarisation of the Σ-hyperon can be determined by examining the angular distribution of its decay products (compare § 12.10). The complete argument is somewhat complicated and will not be reproduced here. The measured data was found to be consistent with odd $K\Sigma$ parity.

The parity of the Ξ-particle has not been measured; proposals for measuring it by examining the capture process

$$\Xi + p \rightarrow \Lambda + \Lambda$$

have been made by Okun (1958) and by Treiman (1959).

13.4(b). *The strange particle resonance states*

In recent years several resonant states of the strange particles have been reported. The intrinsic properties and quantum numbers of the particles discussed in this section are summarised in Table 13.4; later data may be found in Table A.9.3 (p. 717).

TABLE 13.4

Type	Mass (MeV)	Width (MeV)	B	S	T	j	P
K^*	890	50	0	± 1	$\frac{1}{2}$	1	$-$
Y_0^*	1405	50	1	-1	0	$\frac{1}{2}$?	?
Y_0^{**}	1520	15	1	-1	0	$\frac{3}{2}$	$-$
Y_0^{***}	1815	70	1	-1	0	$\frac{5}{2}$	$+$
Y_1^*	1385	50	1	-1	1	$\frac{3}{2}$	$+$
Ξ^*	1532	7	1	-2	$\frac{1}{2}$	$\frac{3}{2}$	$+$

It can be seen from the table that some information remains to be acquired. The errors on masses are of the order of ~ 5 MeV. We consider first the K^* state; this was first observed by Alston *et al.* (1961a) during a study of K^-p reactions in a bubble chamber. The reactions

$$K^- + p \rightarrow p + \pi^0 + K^-$$

$$p + \pi^- + K^0$$

were frequently found to have arisen from the process $K^- + p \rightarrow p + K^{-*}$. The resonant system could be identified by examining the effective mass of the $K\pi$-system. A distribution from a later experiment (reported by Lynch, 1962) is given in Fig. 13.7. The state has since been observed in other processes

$$\pi^- + p \rightarrow \Sigma^- + K^{+*}$$

$$p + \bar{p} \rightarrow K^0 + \bar{K}^{0*}.$$

(See Gregory, 1962, for detailed references.)

Evidence obtained by Armenteros *et al.* (1962b) suggests a nonzero spin for the K^*-system (§ 14.4(b)). This conclusion has been confirmed by the work of Chinowsky *et al.* (1962), which assigned a spin and (relative) parity of 1^- to the

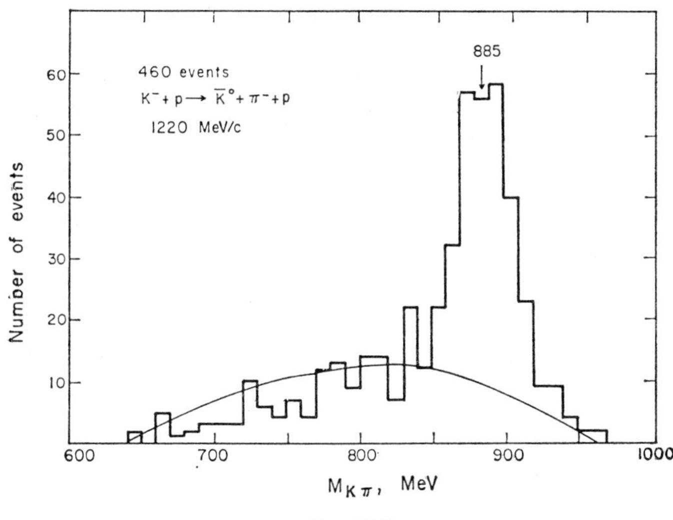

FIG. 13.7.

K^*-system by means of an Adair analysis (§ 2.4(e)) of the reaction $K^+p \rightarrow K^*N_1^*$.

The isospin of the K^*-system can be assigned from data on the branching ratio (Lynch, 1962)

$$\frac{K^{-*} \rightarrow \bar{K}^0 + \pi^-}{K^{-*} \rightarrow K^- + \pi^0} = 1{\cdot}4 \pm 0{\cdot}4.$$

This result indicates an isospin value of $\frac{1}{2}$ since values of 0·5 and 2 would be expected for $T = \frac{3}{2}$ and $\frac{1}{2}$ repectively.

Let us now consider the hyperon states. A state with isospin 0 (Y_0^*) and mass 1405 MeV was first reported by Alston *et al.* (1961b). These workers identified the process

$$Y_0^* \rightarrow \Sigma^\pm + \pi^\mp.$$

Its existence has since been confirmed by many groups (Gregory, 1962). We shall mention the properties of this state in § 13.5(b).

We next mention the state Y_0^{**}. It was first reported by Ferro-Luzzi, Tripp and Watson (1962), who observed a bump in the curves for K^-p scattering in the region of kaon momenta of 400 MeV/c

$$K^- + p \rightarrow Y_0^{**} \begin{cases} \nearrow \Lambda + \pi^+ + \pi^- \\ \searrow \overline{K}^0 + n . \end{cases}$$

The cross-sections for these processes are illustrated in Fig. 13.8. The resonance also appears in all three of the $\Sigma\pi$ channels but not in the $\Lambda\pi^0$ final state. Now the $\Lambda\pi^0$ system has isospin 1, whereas the observed decay modes have $T = 0$ or 1; it therefore seems reasonable to conjecture that $T = 0$ for the Y_0^{**}-state. This conclusion is supported by the absence of a bump in the K^+p cross-sections.

There is good evidence to show that the Y_0^{**}-system has spin $\frac{3}{2}$. Up to the region of the resonance the K^-p elastic scattering is s-wave (§ 14.5(d)), since

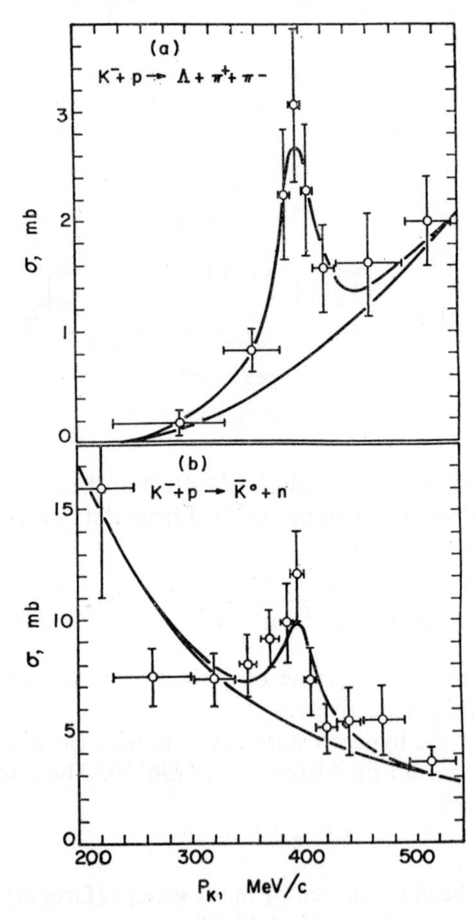

FIG. 13.8.

nearly isotropic distributions are found. In the region of the resonance a $\cos^2\theta$ term appears and then tends to disappear again at higher energies (Fig. 13.9). The only angular momentum state which can be added to s-wave scattering to produce a strong $\cos^2\theta$ term without introducing strong contributions from

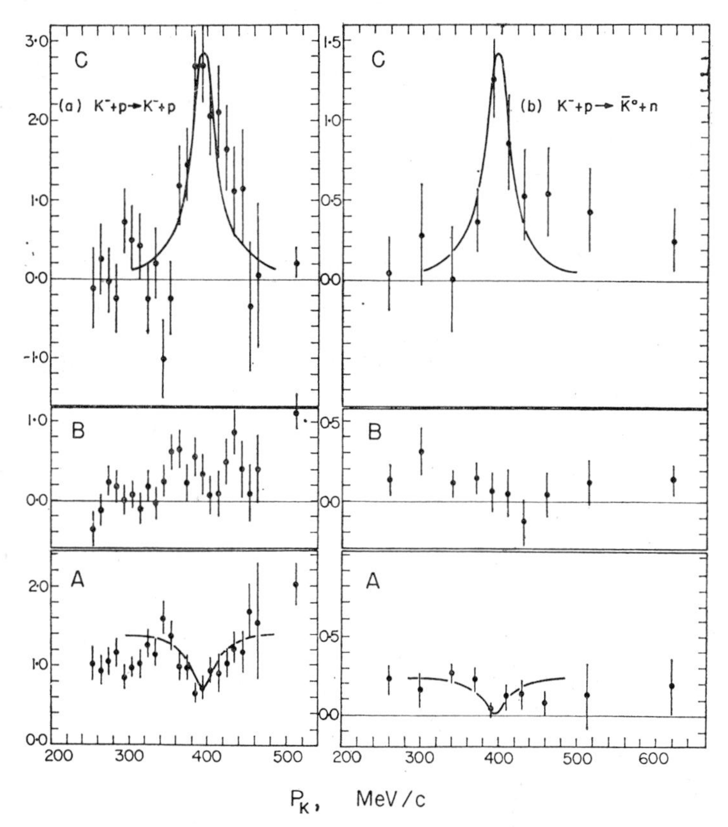

FIG. 13.9. Coefficients A, B and C for the angular distributions $A+B\cos\theta+C\cos^2\theta$ in the reactions $K^-+p\rightarrow K^-+p$ and $K^-+p\rightarrow \bar{K}^0+n$ (Ferro-Luzzi *et al.*, 1962).

other powers of $\cos\theta$ is the $D_{3/2}$ state. Ferro-Luzzi *et al.* (1962) have shown that the angular distributions in K^-p elastic and charge exchange scattering can be explained by a combination of the s-wave scattering amplitude extrapolated from the lower energy data, together with a d-wave amplitude of the Breit–Wigner form with a central value and width determined by the data on the total cross-sections (Fig. 13.8). The result is displayed by the solid lines in Fig. 13.9.

The presence of a $D_{3/2}$ state implies that the Y_0^{**}-system has even parity with respect to the K^-p-system.

The third hyperon resonance with isospin zero Y_0^{***} was detected by the work of Chamberlain *et al.* (1962). It appears as a peak in K^-p scattering at

1050 MeV/c but not in K^-n scattering (Fig. 13.10). The resonance is therefore in the $T = 0$ state. There is some evidence that it possesses spin $\frac{5}{2}$ (Beall *et al.*, 1962).

The hyperon resonance with $T = 1$, Y_1^*, was the first of the strange particle resonances to be detected (Alston *et al.*, 1960). It was observed in the interaction of negative kaons with protons at a kaon momentum of 1·15 GeV/c. The process

$$K^- + p \rightarrow \Lambda + \pi^- + \pi^+$$

was found to be almost entirely

$$K^- + p \begin{cases} \nearrow Y_1^{-*} + \pi^+ \\ \searrow Y_1^{+*} + \pi^- \end{cases}$$

$$Y_1^{\pm *} \rightarrow \Lambda + \pi^{\pm}.$$

Now the Λ-hyperon and the pion have $T = 0$ and 1 respectively, and so the Y_1^*, state must have $T = 1$. There is a strong indication that the Y_1^*, particle possesses

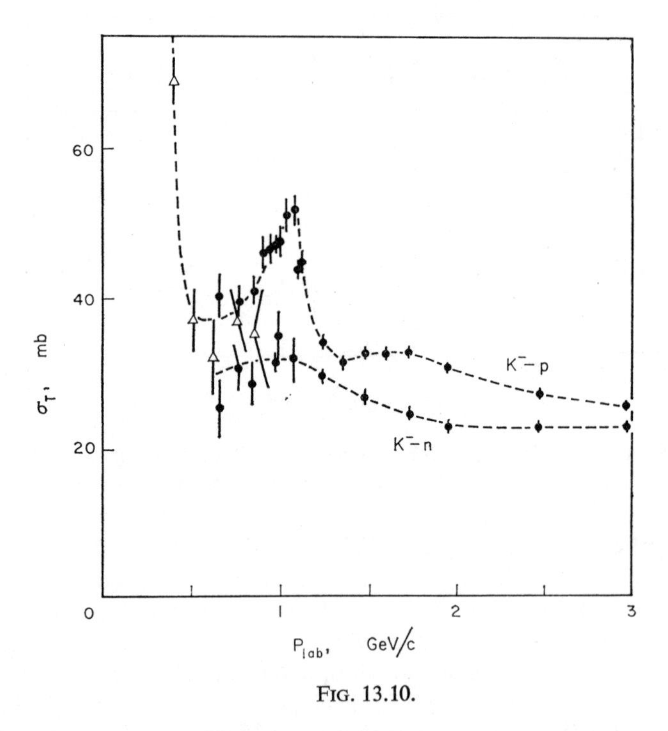

FIG. 13.10.

ordinary spin $\frac{3}{2}$; this assignment follows from an analysis of the up–down distributions of the hyperons relative to the plane of production of the Y_1^* state (Gregory, 1962).

Finally, an excited hyperon state with strangeness −2, the Ξ^* state, was found in the reaction

$$K^- + p \begin{cases} \nearrow \Xi^- + \pi^0 + K^+ \\ \searrow \Xi^- + \pi^+ + K^- \end{cases}$$

(Bertanza *et al.*, 1962; Pjerrou *et al.*, 1962). The branching ratio for the two reactions ($\Xi^-\pi^0K^+/\Xi^-\pi^+K^0 = 0{\cdot}27 \pm 0{\cdot}07$) is consistent with an isospin assignment $T = \frac{1}{2}$ for the Ξ^*-system.

13.5. INTERPRETATIONS

Numerous schemes to bring order to the empirical data on particle masses and quantum numbers have been suggested. They may be broadly classified as dynamical and empirical. The first classification covers such diverse topics as field theories, bound states and Regge pole concepts; whilst the second varies between empirical formulae and symmetry schemes.

In this section we consider some possible interpretations.

13.5(a) *Unitary Symmetry*

The most promising scheme which has been developed so far for classifying the elementary particles is that of unitary symmetry (Ikeda, Ogawa and Ohnuki, 1959; Gell-Mann 1961, 1962b; Ne'eman, 1961; Salam and Ward, 1961). If the strongly interacting particles are examined one is immediately struck by the large number of particles and by the remarkably restricted range of their quantum numbers, and so one might hope that regularities might appear in their classification. Let us first examine the mesons and baryons with spin and parity 0^- and $\frac{1}{2}^+$ respectively†. Now if hypercharge Y (9.143) is correlated with the third component of isospin T_3 for these particles we find the result shown in Table 13.5.

TABLE 13.5

Spin and Parity		0⁻					½⁺				
T_3		−1	−½	0	½	1	−1	−½	0	½	1
	1		K^0		K^+			n		p	
Y	0	π^-		π^0, η		π^+	Σ^-		Σ^0, Λ		Σ^+
	−1		K^-		$\bar{K}^0$			Ξ^-		Ξ^0	

† This statement holds within the context that the parities of strange particles are determined relative to the Λ-hyperon which is defined to be the same as the nucleon.

It is apparent from the table that there is a strong regularity in the two patterns. The data suggest that the groups of eight particles might form supermultiplets, although there are considerable differences in the masses of the particles within each group. The situation has been compared with that of the existence of isospin multiplets; the latter arise from the concept of charge independence in strong interactions (§ 9.4), and the differences in masses of the members of isospin multiplets are attributed to charge dependent electromagnetic interactions. By analogy it is presumed that a higher symmetry property exists for the strongly interacting particles and that the differences of mass within a supermultiplet arise from an unknown interaction which breaks the symmetry. Within this framework remarkable success has been achieved in accounting for differences of mass within a supermultiplet as we shall indicate later.

13.5(a.1) *Elements of group theory*

In previous chapters we have discussed the symmetry properties of particles and interactions in terms of concepts such as gauge invariance or invariance under rotations in various spaces. These symmetry operations have the same formal properties as the elements of more general mathematical systems which are known as groups, and the full exploitation of the symmetry properties is most conveniently realised by using the methods of group theory; for example, charge conservation can be related to a unitary transformation in one dimension (the U_1 group), whilst conservation of isospin follows from the properties of a two dimensional unitary group (SU_2). The existence of higher symmetries is normally discussed in the language of group theory, and so before proceeding further we shall mention its main points.

A group G comprises a set of abstract elements $a, b \ldots$, which can be either infinite or finite in number. These elements satisfy the following rules of composition:

(1) if a and b belong to the set† G, then so also does the product $ab = c$;
(2) a law of association

$$a(bc) = (ab)\,c;$$

(3) The set contains a unit element e so that

$$ae = ea = a;$$

(4) to every element in the set there exists one inverse element which is also a member of the set

$$aa^{-1} = a^{-1}a = e.$$

† The statement 'a belongs to the set G' is often represented as $a \in G$.

13.5(a.2) *Group theory and the unitary transformation*

The concept of unitary symmetry can be conveniently discussed in the language of angular momentum or isospin, namely that higher symmetry operators exist, the 'unitary spin operators', and that physical systems remain invariant under rotations in 'unitary spin space'. We have seen in Chapters 5 and 9 that rotations of a system in a real three-dimensional space lead to the concept of invariant spin operators, and that these rotations may be achieved with aid of unitary transformations, whose operators satisfy the condition

$$U^{-1}U = U^{\dagger}U = UU^{\dagger} = \hat{1} \tag{13.11}$$

and leave the Lagrangian for the system unchanged. The principle may be extended to a complex n dimensional space (denoted by L_n)†, and so the fields in this space transform as

$$\psi_a \to \psi'_a = \sum_{b=1}^{n} U_{ab}\psi_b \tag{13.12}$$

$$\psi_a^{\dagger} \to \psi_a'^{\dagger} = \sum_{b=1}^{n} \psi_b^{\dagger}U_{ba}^{*}.$$

It is apparent that n^2 parameters are required to specify U. The set of unitary matrices in the space L_n possesses the group property. If for example U_a and U_b are unitary matrices in L_n

$$U_a^{\dagger}U_a = U_aU_a^{\dagger} = U_b^{\dagger}U_b = U_bU_b^{\dagger} = \hat{1}$$

then the product U_aU_b also satisfies the unitarity condition

$$(U_aU_b)^{\dagger}(U_aU_b) = (U_aU_b)(U_aU_b)^{\dagger} = \hat{1}.$$

Thus the set contains $\hat{1}$ and since $U^{\dagger} = U^{-1}$ by definition, it also contains the reciprocal of its members. Thus the set of $n \times n$ unitary matrices in L_n constitute a group; it is called the n dimensional unitary group.

Let us consider a form for U; the unitarity condition implies that

$$|\det U| = 1, \quad \det U = e^{i\varphi}. \tag{13.13}$$

Since the choice of a particular value for φ does not disturb the unitarity condition, we can set $\varphi = 0$ so that

$$\det U = 1. \tag{13.14}$$

If U satisfies this condition it is said to be *unimodular*, the introduction of this restriction implies that we may reduce the number of parameters required to specify the unitary transformation from n^2 to $n^2 - 1$. Now if we write

$$U = e^{iH}$$

† In general a distinction should be drawn between the dimensions of the space and of the basis vectors in that space (as for ordinary spin). We shall refer to the latter as systems with x components.

then it is apparent that the unitarity condition implies that H should be Hermitian. The condition of unimodularity leads to the requirement that $\operatorname{tr} H = 0$; this relationship follows from the general algebraic property

$$\det (e^{H}) = e^{\operatorname{tr} H}.$$

It is often convenient to write H as

$$H = \sum_j \alpha_j F_j \equiv \alpha_j F_j$$

where the terms F_j are called the generators of the unitary group and α_j are a set of real continuous parameters† (for example, angles in the case of rotations.) If we consider infinitesimal transformations, then we can set $\alpha_j \to 0$ and the unitary operator becomes

$$U = e^{iH} = e^{i\alpha_j F_j} \to \hat{1} + i\alpha_j F_j. \tag{13.15}$$

The unimodularity condition then implies that the generators F_j should be traceless.

The unimodular transformation which we have introduced above defines the representation of the *special unitary group* SU_n, where n refers to the dimensionality of the space. Since it is a complex space, we may represent the transformation of the components of fields in it as

$$\psi'_a = \sum_{b=1}^{n} U_{ab}\psi_b \equiv U_{ab}\psi_b$$

$$\psi'^*_i = U^*_{ij}\psi^*_j.$$

We note that unitarity condition for U implies that

$$U^\dagger = U^{-1}, \quad U^\dagger_{ji} = U^*_{ij} = U^{-1}_{ji}.$$

In what follows a tensor notation (A.4, Appendixes p. 700) will be more convenient, and so we shall use this property for U and write

$$\psi'_a = U_{ab}\psi_b \tag{13.16}$$

$$\psi^*_i = \psi'^i = U^*_{ij}\psi^j = U^{-1}_{ji}\psi^j$$

where the lower line represents a contragradient tensor. Now let us consider the transformation of products of these terms. If we represent a system which transforms like $\psi_a\psi_b$ as ψ_{ab} then the general mixed tensor transforms as

$$\psi^{ij\ldots}_{ab\ldots} = U_{ad}U_{be}\ldots U^*_{il}U^*_{jm}\ldots\psi^{lm\ldots}_{de\ldots}. \tag{13.17}$$

These sets of tensor components can normally be split into subsets, each of which transforms only into itself under the general transformation. To illustrate this

† A group of transformations, which is characterised by a set of continuous parameters is called a *Lie group*.

point let us consider a more familiar example in a real three dimensional space, a pair of nucleons forms two subsystems in isospin space with isospin 0 and 1; each of these systems transforms only into itself under rotations in that space. If a system can be broken down in this fashion the product tensor is said to be reducible. It frequently happens that after an initial reduction the process may be repeated; when no further reductions can be made the system is said to be broken down into its *irreducible tensor products*.

In practice the reduction of tensor products of high rank is not a simple matter. In order to illustrate the process, however, we consider two simple examples.

(1) A tensor with two upper or lower indices may be split into a symmetric and an antisymmetric tensor (we show in A. 4, p. 700 that symmetric and antisymmetric parts do not mix under a transformation)

$$\begin{aligned}\psi_{ab} &= \tfrac{1}{2}(\psi_{ab} + \psi_{ba}) + \tfrac{1}{2}(\psi_{ab} - \psi_{ba}) \\ &= \psi_{ab,} + \psi_{a,b}.\end{aligned} \tag{13.18}$$

In an n dimensional space the symmetric part ψ_{ab}, has $\frac{1}{2}n(n+1)$ components and the antisymmetric part $\frac{1}{2}n(n-1)$ components. Equation (13.18) is often written in the notation

$$\psi_a \otimes \psi_b \sim \psi_{ab,} \oplus \psi_{a,b} \tag{13.19}$$

where the symbol $\sim$ means 'transforms like', and the symbols $\otimes$ and $\oplus$ indicate direct products and sums respectively; a specific example of a direct product is given in (§ 6.2(c)). In the case of a triple product the reduction process yields† a totally symmetric part, two parts of mixed symmetry and a totally antisymmetric part, which in a space with $n = 3$ have respectively 10, 8, 8 and 1 independent components

$$\psi_a \otimes \psi_b \otimes \psi_c \sim \psi_{abc,} \oplus \psi_{ab,c} \oplus \psi_{ac,b} \oplus \psi_{a,b,c}. \tag{13.20}$$

(2) Mixed tensors may be reduced with the aid of the process of contraction (the summing over identical upper and lower indices). Consider the operation

$$\psi_a'^i = U_{ab} U_{ij}^* \psi_b^j.$$

Then if we take $a = i$

$$\psi_i'^i = U_{ib} U_{ij}^* \psi_b^j = \delta_{bj} \psi_b^j = \psi_j^j. \tag{13.21}$$

Thus, as we might have expected from the properties of the unitary matrix, the trace term behaves like a scalar. Now let us examine the remaining terms which are in the form of a traceless tensor

$$\varphi_b^j = \psi_b^j - \frac{1}{n} \delta_{bj} \psi_c^c.$$

† Further details concerning the reduction processes may be found in the excellent article by Behrends, Dreitlen, Fronsdal and Lee (1962); the book by Hamermesh (1962) is also most useful.

Then

$$\begin{aligned}\varphi_a'^i &= U_{ab}U_{ij}^*\left(\psi_b^j - \frac{1}{n}\delta_{bj}\psi_c^c\right)\\ &= U_{ab}U_{ij}^*\psi_b^j - \frac{1}{n}U_{aj}U_{ij}^*\psi_c^c\\ &= \psi_a'^i - \frac{1}{n}\delta_{ai}\psi_a'^a\\ &= \varphi_a'^i.\end{aligned}$$

Thus the components of φ transform into each other, and so in the notation of equation (13.19) we can represent the reduction of the mixed tensor as

$$\psi_a \otimes \psi^i \sim \psi_a^i - \frac{1}{n}\delta_{ai}\psi_c^c \oplus \delta_{ai}\psi_c^c. \tag{13.22}$$

Further examples of the reduction process are considered in the article by Behrends, Dreitlen, Fronsdal and Lee (1962).

13.5(a.3) *The SU_2 group*

We first note that the SU_2 group (which can be regarded as a rotation in a complex two dimensional space) behaves like a rotation group in real three dimensional space R_3. For purposes of illustration we note that the conditions $UU^\dagger = \hat{1}$, $\det U = 1$ are given, for infinitesimal transformations, by

$$U = \begin{pmatrix} \alpha & \beta \\ -\beta^* & \alpha^* \end{pmatrix} \rightarrow \begin{pmatrix} 1 & 0 \\ 0 & 1 \end{pmatrix} + i\begin{pmatrix} a & b \\ b^* & -a \end{pmatrix} \tag{13.23}$$

where a is real. The expansion is equivalent to writing $U = e^{iH}$ with $H \rightarrow 0$ (compare (13.15)). Now consider the behaviour of the isospin rotation operator $e^{\frac{1}{2}i\tau_j\theta_j}$ where $\boldsymbol{\tau}$ represents the isospin operator (Pauli matrices—see § 9.4); for an infinitesimal angular rotation $\boldsymbol{\theta} \rightarrow 0$ we obtain

$$e^{\frac{1}{2}i\tau_j\theta_j} \rightarrow \begin{pmatrix} 1 & 0 \\ 0 & 1 \end{pmatrix} + \tfrac{1}{2}i\begin{pmatrix} \theta_3 & \theta_1 - i\theta_2 \\ \theta_1 + i\theta_2 & -\theta_3 \end{pmatrix}.$$

A comparison of this equation with (13.23) shows the equivalence of SU_2 and R_3 and we see further how to identify the parameters a and b in terms of rotation angles.

The process outlined above can be considered in a more general way, and also the reduction of products of state vectors can be undertaken (these properties are demonstrated in some text books on group theory, see for example Meijer and Bauer (1962)). The reduction process leads to the Clebsch–Gordan coefficients which we have introduced in Chapter 5.

Now let us examine how the SU_2 scheme can be used for classifying particles. Let us represent the nucleon and antinucleon doublet as

$$\psi = \begin{pmatrix} p \\ n \end{pmatrix} \qquad \psi^* = \begin{pmatrix} \bar{p} \\ \bar{n} \end{pmatrix}.$$

Then the outer product of these two terms can be written as a matrix

$$\psi\psi^\dagger = \begin{pmatrix} p \\ n \end{pmatrix} (\bar{p} \;\; \bar{n}) = \begin{pmatrix} \frac{1}{2}(p\bar{p} - n\bar{n}) & p\bar{n} \\ n\bar{p} & -\frac{1}{2}(p\bar{p} - n\bar{n}) \end{pmatrix} + \begin{pmatrix} \frac{1}{2}(p\bar{p} + n\bar{n}) & 0 \\ 0 & \frac{1}{2}(p\bar{p} + n\bar{n}) \end{pmatrix} \tag{13.24}$$

and it is apparent that the matrix elements can be identified with the irreducible tensor products of equation (13.22). For example the matrix elements $p\bar{p} + n\bar{n}$ correspond to the singlet system and so are scalars in isospin space. The elements $p\bar{p} + n\bar{n}$ in fact have the quantum numbers of the η meson – this statement should not be taken to imply that the meson is literally a bound state of nucleons and antinucleons, but merely that the system $p\bar{p} + n\bar{n}$ has the transformation properties of that particle.

In a similar manner the elements of the first matrix can be recognized as pions and assume the form†

$$\begin{pmatrix} \frac{\pi^0}{\sqrt{2}} & \pi^+ \\ \pi^- & -\frac{\pi^0}{\sqrt{2}} \end{pmatrix}.$$

They transform as the components of an isovector in isospin space, for example

$$\pi^0 \sim \psi^\dagger \tau_3 \psi \equiv p\bar{p} - n\bar{n}. \tag{13.25}$$

13.5(a.4) *The SU_3 group*

In order to incorporate strangeness within the framework of unitary symmetry we must add a third basic object to the neutron and proton. A three dimensional space is then required and so we must use the SU_3 or the U_3 groups.

An apparently obvious candidate for the third system is the Λ hyperon, and the properties of the p, n, Λ triplet in relation to the U_3 group‡ have been ex-

† The factor $\sqrt{2}$ appears as we have expressed π^0 as the $T_3 = 0$ component of an isotopic triplet (Table 9.4). The formal properties of the operators $T_1 \pm iT_2$ require that the single particle state for the antiproton should be written as $-|\bar{p}\rangle$; this is only important when relative phases of eigenstates are required.

‡ Prior to the work on unitary symmetry Sakata (1956) had proposed a model in which particles represented composites of the basic systems p, n, Λ; again we emphasize that this statement merely means composite systems which transform like combinations of p, n, Λ.

tensively studied by Ikeda, Ogawa and Ohnuki (1959). Unfortunately the predictions are in disagreement with experiment.

Various other schemes have been suggested, for example the eightfold way of Gell-Mann and Ne'eman is based upon SU_3 and a basic octed of particles. Gell-Mann (1964) and Zweig (1964), have shown independently that the eightfold way can be conveniently studied, and perhaps further delimited, by use of a basic triplet of spin $\frac{1}{2}$ 'particles' whose charges are multiples of $e/3$. The arguments of these workers give a useful introduction to the methods employed in obtaining irreducible tensor products, and so we shall examine them from that viewpoint. The 'particles' used in the scheme are called quarks by Gell-Mann and aces by Zweig; the properties of the particles and their antiparticles are given in Table 13.6.

TABLE 13.6

Symbol	B	Q	T_3	Y
u ψ_1	$\frac{1}{3}$	$\frac{2}{3}$	$\frac{1}{2}$	$\frac{1}{3}$
d ψ_2	$\frac{1}{3}$	$-\frac{1}{3}$	$-\frac{1}{2}$	$\frac{1}{3}$
s ψ_3	$\frac{1}{3}$	$-\frac{1}{3}$	0	$-\frac{2}{3}$
$\bar{u}$ ψ^1	$-\frac{1}{3}$	$-\frac{2}{3}$	$-\frac{1}{2}$	$-\frac{1}{3}$
$\bar{d}$ ψ^2	$-\frac{1}{3}$	$\frac{1}{3}$	$\frac{1}{2}$	$-\frac{1}{3}$
ψ^3	$-\frac{1}{3}$	$\frac{1}{3}$	0	$\frac{2}{3}$

Thus the terms u and d form an isospin doublet. Systems with the quantum numbers of mesons and baryons can now be constructed from these terms; for example mesons would appear as systems which transform as ψ_a^i whilst baryons can be formed from ψ_{abc} (or higher combinations).

Since the SU_3 operation corresponds to rotations in a complex three dimensional space, the decomposition of ψ_a^i for mesons yields an octet and a singlet according to equation (13.22). The construction of terms can be again carried out by considering outer products as in equation (13.24)

$$\psi\psi^\dagger = \begin{pmatrix} u\bar{u} & u\bar{d} & u\bar{s} \\ d\bar{u} & d\bar{d} & d\bar{s} \\ s\bar{u} & s\bar{d} & s\bar{s} \end{pmatrix} \tag{13.26}$$

$$= \begin{pmatrix} \dfrac{2u\bar{u} - d\bar{d} - s\bar{s}}{3} & u\bar{d} & u\bar{s} \\ d\bar{u} & \dfrac{2d\bar{d} - u\bar{u} - s\bar{s}}{3} & d\bar{s} \\ s\bar{u} & s\bar{d} & \dfrac{2s\bar{s} - u\bar{u} - d\bar{d}}{3} \end{pmatrix} + \frac{1}{3}\begin{pmatrix} \mathrm{tr} & 0 & 0 \\ 0 & \mathrm{tr} & 0 \\ 0 & 0 & \mathrm{tr} \end{pmatrix}$$

where

$$\text{tr} = (u\bar{u} + d\bar{d} + s\bar{s}).$$

We may now identify the matrix elements with known mesons. In all this work the assumption is made that the unitary operators commute with the operators for space-time transformations. Thus the spin and parity of each member of a multiplet should be the same, but these numbers can be different in different multiplets. Let us therefore consider possible members of a multiplet with spin and parity 0^-; we find that we may make the following identifications

$$u\bar{s} \sim K^+ \qquad s\bar{u} \sim K^-$$

$$d\bar{s} \sim K^0 \qquad s\bar{d} \sim \bar{K}^0.$$

Using Clebsch–Gordan coefficients (Tables 9.4 and 9.6) we also find (compare (13.24) and subsequent remarks)

$$u\bar{d} \sim \pi^+ \qquad d\bar{u} \sim \pi^- \qquad \frac{u\bar{u} - d\bar{d}}{\sqrt{2}} \sim \pi^0 \qquad \frac{u\bar{u} + d\bar{d} - 2s\bar{s}}{\sqrt{6}} \sim \eta.$$

In addition we require an isotopic singlet state, χ^0, which has not yet been identified. Thus the penultimate matrix in (13.26) yields the octet

$$\begin{pmatrix} \frac{\pi^0}{\sqrt{2}} + \frac{\eta}{\sqrt{6}} & \pi^+ & K^+ \\ \pi^- & -\frac{\pi^0}{\sqrt{2}} + \frac{\eta}{\sqrt{6}} & K^0 \\ K^- & \bar{K}^0 & -\frac{2}{\sqrt{6}}\eta \end{pmatrix}. \qquad (13.27)$$

An octet of vector mesons can be constructed in the same way with the same set of quantum numbers T_3 and Y. In the case of the vector mesons nine candidates are known, which is adequate for the octet plus a singlet. Unfortunately it is not entirely clear which is the candidate for the singlet state. The two candidates are the ω and φ mesons. Neither particle satisfies the mass formula which we shall discuss below; however their mean square mass value is in accord with this formula. It has been suggested (Sakurai, 1963) that since the ω and φ particles have the same quantum numbers, apart from those associated with unitary symmetry itself, then the physical ω and φ particles might represent linear combinations of pure ω_0 and φ_0 unitary states. There are grounds for regarding the ω

particle as the unitary singlet particle. The octet of vector mesons then becomes

$$\begin{array}{ccccc} & & K^{*0} & & K^{*+} \\ Y\uparrow & \varrho^- & & \begin{Bmatrix}\varrho^0\\ \varphi_0\end{Bmatrix} & & \varrho^+ \\ & & K^{*-} & & \bar{K}^{*0} \\ & & & \overrightarrow{T_3} & \end{array}$$

Baryon multiplets can be constructed by using the same principles as these developed above. The quantum numbers for u, d and s imply that we require systems which transform like $\psi_a\psi_b\psi_c$. The basic relation is then given in equation (13.20), and since a, b and c each have three elements, we can write

$$\psi_a \otimes \psi_b \otimes \psi_c \sim \psi_{abc}, \oplus \psi_{ab,c} \oplus \psi_{ac,b} \oplus \psi_{a,b,c}$$

$$[3] \otimes [3] \otimes [3] \sim [10] \oplus [8] \oplus [8] \oplus [1]$$

where [10] indicates an irreducible system with 10 components. The submultiplets of isospin and hypercharge associated with these systems together with their possible identifications are shown in Table 13.7. Particles which have not yet been discovered have been indicated by brackets. We have also included the systems $[\overline{10}]$ and [27] which result from the decomposition $[8] \otimes [8]$. Systems of the types $[\overline{10}]$ and [27] might be expected to be associated with the decay of excited baryon systems $B^* \to B + M$ (where B and M denote baryon and meson respectively), since both B and M belong to octets and the B^* systems could then result from the decomposition of $[8] \otimes [8]$†. It is noteworthy that the quantum numbers associated with the $[\overline{10}]$ and [27] systems imply the existence of resonances which could decay into a K^+ meson and a nucleon; none have been observed.

The prediction of the existence of the Ω particle in the [10] multiplet represents a great triumph for the unitary symmetry scheme. This particle should have strangeness -3 and charge $-e$, by virtue of the relations

$$Y = S + B$$

$$\frac{Q}{e} = T_3 + \frac{Y}{2}$$

Furthermore its mass may be predicted to be 1680 MeV by the formula which we shall discuss in § 13.5(a.5). A particle with this mass cannot decay by strong or

† Invariant amplitudes may be constructed for these processes, in much the same way as, say, the interaction of a pion and a nucleon can be split into amplitudes for the $T=\frac{1}{2}$ and $T=\frac{3}{2}$-systems. The equivalent of the Clebsch-Gordan coefficients associated with these amplitudes may be found in a number of papers (see for example de Swart, 1963).

TABLE 13.7

Basis	Y	T	Identification		
			Symbol	Mass (MeV)	j^{parity}
[1]	0	0	Y_0^*	1405	$\frac{1}{2}^-$ (?)
[8]	1 0 −1	$\frac{1}{2}$ 0, 1 $\frac{1}{2}$	N Λ, Σ Ξ	939 1115, 1192 1320	$\frac{1}{2}^+$
[8]	1 0 −1	$\frac{1}{2}$ 0, 1 $\frac{1}{2}$	N_2^* Y_0^{**}, Y_1^{**} (Ξ^{**})	1512 1520, 1660 (1810)	$\frac{3}{2}^-$
[10]	1 0 −1 −2	$\frac{3}{2}$ 1 $\frac{1}{2}$ 0	N_1^* Y_1^* Ξ^* Ω	1238 1385 1533 1686	$\frac{3}{2}^+$
[$\overline{10}$]	2 1 0 −1	0 $\frac{1}{2}$ 1 $\frac{3}{2}$			
[27]	2 1 0 −1 −2	1 $\frac{1}{2}, \frac{3}{2}$ 0, 1, 2 $\frac{1}{2}, \frac{3}{2}$ 1			

electromagnetic interactions (the $K\Xi$ threshold is at 1805 MeV), and so it can only suffer a weak decay into $\pi\Xi$, $\bar{K}\Sigma$ or $\bar{K}\Lambda$.

A particle with the properties of mass, charge and strangeness, predicted for the Ω has been identified by Barnes *et al.*, (1964); K^- mesons with momentum 5 GeV/c were passed through a hydrogen bubble chamber and the following reaction was identified

$$K^- + p \rightarrow \Omega^- + K^+ + K^0$$
$$\quad\quad \hookrightarrow \Xi^0 + \pi^-$$

The mass of the particle was measured to be 1686 ± 12 MeV.

13.5(a.5) *The mass formula*

Some of the most interesting features of the SU_3 scheme are concerned with the laws for the breakdown of unitary symmetry. For example, it has been shown both by Gell-Mann and Zweig (1964) that the following transformation properties of the strongly interacting particles in weak interactions:

(1) $|\Delta T| = 1$ for $\Delta S = 0$

(2) $|\Delta T| = \frac{1}{2}$, $\dfrac{\Delta S}{\Delta Q} = 1$ for $|\Delta S| = 1$ leptonic decays

follow naturally from the classification scheme given in Table 13.6. Another interesting property is the mass values of particles within a unitary multiplet; these may be expressed in an elegant mass formula developed by Okubo (1962).

Prior to deriving this formula let us discuss unitary spin. An infinitesimal unitary transformation was discussed in § 13.5(a.3) for a two dimensional space and it was shown that the generators were the three τ matrices. Let us now examine the situation for a three dimensional space

$$U = e^{iH} \to \hat{1} + iH = \hat{1} + i\alpha_j F_j.$$

The generators F_j may be regarded as the components of the unitary spin operator. In their simplest representation they are defined as $F_j = \lambda_j/2$ in analogy with $\tau_j/2$ in the case of isospin. Since we are working in three dimensional space we require $3^2 - 1$ traceless matrices for a complete representation of the terms F_j.

The form for the λ-matrices given by Gell-Mann (1962b) follows from the properties of the triplets listed in Table 13.6 (compare also (13.25) and the discussion of the isospin generators for the baryon field given in § 9.4(b)); for example

$$\pi^+ \sim \bar{d}u = \tfrac{1}{2}\psi^\dagger(\lambda_1 - i\lambda_2)\,\psi$$

$$\pi^- \sim \bar{u}d = \tfrac{1}{2}\psi^\dagger(\lambda_1 + i\lambda_2)\,\psi$$

if we write

$$\lambda_1 = \begin{pmatrix} 0 & 1 & 0 \\ 1 & 0 & 0 \\ 0 & 0 & 0 \end{pmatrix} \qquad \lambda_2 = \begin{pmatrix} 0 & -i & 0 \\ i & 0 & 0 \\ 0 & 0 & 0 \end{pmatrix}.$$

The remaining matrices can be constructed in a similar way:

$$\lambda_3 = \begin{pmatrix} 1 & 0 & 0 \\ 0 & -1 & 0 \\ 0 & 0 & 0 \end{pmatrix} \qquad \lambda_4 = \begin{pmatrix} 0 & 0 & 1 \\ 0 & 0 & 0 \\ 1 & 0 & 0 \end{pmatrix} \qquad \lambda_5 = \begin{pmatrix} 0 & 0 & -i \\ 0 & 0 & 0 \\ i & 0 & 0 \end{pmatrix}$$

$$\lambda_6 = \begin{pmatrix} 0 & 0 & 0 \\ 0 & 0 & 1 \\ 0 & 1 & 0 \end{pmatrix} \qquad \lambda_7 = \begin{pmatrix} 0 & 0 & 0 \\ 0 & 0 & -i \\ 0 & i & 0 \end{pmatrix} \qquad \lambda_8 = \frac{1}{\sqrt{3}}\begin{pmatrix} 1 & 0 & 0 \\ 0 & 1 & 0 \\ 0 & 0 & -2 \end{pmatrix}.$$

It is not difficult to show that F_1, F_2 and F_3 are identical with the three components of the isospin operator whilst F_8 represents $\sqrt{3}/2$ times the hypercharge operator. It is evident that F_3 and F_8 commute since we have chosen them to be diagonal. The matrices also satisfy the following relations (Gell-Mann, 1962b):

$$\operatorname{tr} F_i F_j = \tfrac{1}{4} \operatorname{tr} \lambda_i \lambda_j = \tfrac{1}{2}\delta_{ij} \tag{13.28}$$

$$[F_i, F_j] = if_{ijk}F_k$$

$$\{F_i, F_j\} = \tfrac{1}{3}\delta_{ij}\hat{1} + d_{ijk}F_k$$

where the terms f_{ijk} and d_{ijk} are real, totally antisymmetric and symmetric tensors respectively. The only nonzero d's are

$$d_{118} = d_{228} = d_{338} = -d_{888} = \frac{1}{\sqrt{3}} \tag{13.29}$$

$$d_{146} = d_{157} = -d_{247} = d_{256} = d_{344} = d_{355} = -d_{366} = -d_{377} = \frac{1}{2}$$

$$d_{448} = d_{558} = d_{668} = d_{778} = \frac{-1}{2\sqrt{3}}.$$

Now let us consider the mass formula; we shall use a method described by London (1963). If no interactions violated SU_3 symmetry all particles belonging to the same supermultiplet would have the same mass, spin, parity and baryon number. Let us assume that a relatively weak interaction exists which violates SU_3. We shall proceed in analogy with the violation of the SU_2 symmetry by electromagnetic interactions; in this case it is customary to assume that the degeneracy of the $2T + 1$ particles in an isospin multiplet arises from a dependence on T_3

$$M(T, T_3) = M(T) + \Delta M(T_3)$$

$$\Delta M = \langle T, T_3 | a + bT_3 + cT_3^2 | T, T_3 \rangle.$$

By analogy, in the violation of SU_3 symmetry it is assumed that ΔM is dependent upon the diagonal generator F_8 (the only other diagonal generator, F_3, should be responsible for electromagnetic mass differences). Two general terms can then be inserted in the expression for ΔM (we denote the state $|T, Y\rangle$ by $|\alpha\rangle$ and use the relation $\sum_j F_j^2 =$ constant; in the present representation $\sum_j F_j^2 = \frac{4}{3}$)

$$\Delta M = \langle \alpha | AF_8 + B \sum_{j,k} d_{8jk} F_j F_k | \alpha \rangle$$

$$= A \langle \alpha | F_8 | \alpha \rangle + B \langle \alpha | \frac{1}{\sqrt{3}} (F_1^2 + F_2^2 + F_3^2)$$

$$- \frac{1}{2\sqrt{3}} (F_4^2 + F_5^2 + F_6^2 + F_7^2) - \frac{1}{\sqrt{3}} F_8^2 | \alpha \rangle$$

$$= A\frac{\sqrt{3}}{2}Y + B\langle\alpha|\frac{1}{\sqrt{3}}\mathbf{T}^2 - \frac{1}{2\sqrt{3}}\left(\frac{4}{3} - \mathbf{T}^2 - \frac{3}{4}Y^2\right) - \frac{1}{\sqrt{3}}\frac{3}{4}Y^2|\alpha\rangle$$

$$= \frac{\sqrt{3}}{2}AY + \frac{\sqrt{3}}{2}B\left[T(T+1) - \frac{1}{4}Y^2\right] - \frac{2}{2\sqrt{3}}B$$

therefore

$$M = M_0\{1 + aY + b[T(T+1) - \tfrac{1}{4}Y^2]\}. \tag{13.30}$$

This formula leads to the following relation for the baryons in the [8] representation

$$3M_\Lambda + M_\Sigma = 2M_N + 2M_\Xi.$$

It is satisfied to better than 1 per cent accuracy. In general the values $a = \cdot 174$ and $b = \cdot 035$ appear to fit any known baryon representation whilst M_0 is dependent upon the representation. In the [10] multiplet the relation $Y = 2(T-1)$ holds (compare Table 13.7), and so the mass formula reduces to

$$M = M_0[1 + 2b + (a + \tfrac{3}{2}b)\,Y]$$

$$= M_0(c + dY). \tag{13.31}$$

From the known mass values of the N_1^* (1238 MeV), Y_1^* (1385) and Ξ^* (1533) particles, a mass of 1680 MeV may be deduced for the Ω particle; this figure is in excellent agreement with the value of 1686 ± 12 MeV reported by Barnes *et al.*, (1964).

The requirements of *CPT* invariance imply that $a = 0$ for the meson families. It has also been suggested that m^2 should replace m in formula (13.30) (Feynman, unpublished). The mass equation then reduces to

$$3m_0^2 + m_1^2 = 4m_{\frac{1}{2}}^2$$

where the subscripts refer to the isospin values, (the two $T = \frac{1}{2}$ isospin multiplets have equal masses since they are particle and antiparticle). The mesons in the 0^- octet satisfy the above relation to within 3 per cent, but the formula requires a meson with mass $m_0 = 920$ MeV in the 1^- octet. Neither the ω or φ meson has this mass, which lies roughly midway between them. As we mentioned earlier the explanation may lie in the suggestion that the physical ω and φ states may be linear combinations of pure unitary states ω_0 and φ_0.

The unitary symmetry scheme may also be used to make predictions about the electromagnetic mass difference of the particles. Let us assume that the unknown interaction which violates unitary symmetry is switched off leaving only the electromagnetic interaction as a means of violating the equality of the masses of all the particles within a multiplet. If we look at the baryons in Table 13.5 along axes of equal charge and denote the electromagnetic mass differences by δB, then we can write

$$\delta\Sigma^- = \delta\Xi^-$$

$$\delta n = \delta\Xi^0$$

$$\delta p = \delta\Sigma^+$$

hence

$$\begin{aligned} m_{\Xi^-} - m_{\Xi^0} &= \delta\Xi^- - \delta\Xi^0 \\ &= \delta\Sigma^- - \delta\Sigma^+ + \delta p - \delta n \\ &= m_{\Sigma^-} - m_{\Sigma^+} + m_p - m_n. \end{aligned}$$

The data on particle masses yield figures of 6·8 ± 1·6 and 7·0 ± 0·5 MeV for the left- and right-hand sides of this equation respectively.

The unitary symmetry scheme leads to many other interesting predictions concerning electromagnetic properties and processes (Coleman and Glashow, 1961).

13.5(b). *Dynamical models* (1). *Bound states*

The models involving bound states start from the assumption that certain systems or particles are elementary, and that other systems may be constructed from them as dynamical resonances. As an example we may quote the well-known $j = \frac{3}{2}^+$, $T = \frac{3}{2}$ resonance N_1^* for the pion–nucleon system. If the coupling between these particles is of the Yukawa type, then the interaction is most attractive in the $j = \frac{3}{2}^+$, $T = \frac{3}{2}$ state (§ 14.1(c)). The combination of p-wave centrifugal barrier and the attractive interaction at short distances then leads to a resonance in the pion–nucleon interaction if the coupling is sufficiently strong.

One of the possible systems which may have arisen from an interaction is the Y_0^* state. It has a mass of 1405 MeV which lies very close to the threshold of the $\bar{K}N$-system (~1430 MeV) and also has appropriate quantum numbers. The problem of K^-p scattering at low energies has been extensively studied, both experimentally (Humphrey and Ross, 1962; Watson *et. al.*, 1963) and theoretically (Dalitz and Tuan, 1960; Dalitz, 1962, 1963; Akiba and Capps, 1962). Some details of this work are given in § 14.5(d).

Let us consider s-wave $\bar{K}N$ scattering in the isospin channel $T = 0$ at low energies. If we use the notation of § 6.3(d) we may express the scattering function f in terms of either the complex scattering length A or phase shift δ

$$f = \frac{A}{1 - ikA} = \frac{\sin \delta e^{i\delta}}{k} \tag{13.32}$$

$$f^{-1} = \frac{1}{A} - ik = k \cot \delta - ik$$

where

$$A = a + ib = (k \cot \delta)^{-1}.$$

In addition to elastic $\bar{K}N$ scattering, the process $\bar{K}N \to \pi\Sigma$ can occur at all kaon energies, and so we shall describe the interaction with the aid of a matrix notation; the elements of the matrix will refer to the $\bar{K}N$ and $\pi\Sigma$ channels. We shall use the inverse scattering matrix of equation (7.112)

$$T_{fi}^{-1} = \text{Re}\, T_{fi}^{-1} - ip_c^f \delta_{fi} \frac{1}{2\pi E_c n_f^2}$$

where the symbols f and i refer respectively to the final and initial particle states. We may absorb the normalising terms with the matrix element and define a new relation

$$t_{fi}^{-1} = K_{fi}^{-1} - ik\delta_{fi} \tag{13.33}$$

where we have replaced p_c^f by k for later convenience. The elements of the K-matrix are normally defined in the following way (Dalitz, 1963):

$$\begin{array}{cc} \bar{K}N & \pi\Sigma \end{array}$$
$$\begin{pmatrix} \alpha & \beta \\ \beta & \gamma \end{pmatrix} \begin{array}{c} \bar{K}N \\ \pi\Sigma \end{array}$$

where the equality of the terms in the off-diagonal elements arises from the invariance properties of the K-matrix under time reversal.

If the requisite inversion of the matrices is made (compare (A. 3.12), p. 699), the following matrix elements are obtained (k and q denote respectively momenta in the c-system for KN and $\pi\Sigma$ scattering)

$$\langle \bar{K}N|\, t\, |\bar{K}N\rangle = \frac{A}{1 - ikA} \qquad A = a + ib = \alpha + \frac{iq\beta^2}{1 - iq\gamma} \tag{13.34}$$

$$\langle \pi\Sigma|\, t\, |\pi\Sigma\rangle = \frac{B}{1 - iqB} \qquad B = \gamma + \frac{ik\beta^2}{1 - ik\alpha}$$

$$\langle \pi\Sigma|\, t\, |\bar{K}N\rangle = \frac{\beta}{(1 - ikA)(1 - iq\gamma)}.$$

The elastic $\pi\Sigma$ scattering phase shift is then given by the second of equations (13.34)

$$q \cot \delta_{\pi\Sigma} = B^{-1}.$$

Now the Y_0^*-state (1405 MeV) is below the $\bar{K}N$ threshold, and so we must use an imaginary momentum $k = i\,|k|$ in the above equations

$$B = \gamma - \frac{|k|\,\beta^2}{1 + |k|\,\alpha}.$$

Thus a virtual bound state resonance occurs in the $\pi\Sigma$-system when $\delta_{\pi\Sigma} = 90°$, that is when

$$1 + |k|\,\alpha = 0 \tag{13.35}$$

and from the first of equations (13.34) α is given by

$$\alpha = a + bq\gamma.$$

The values of a, b and γ inferred from the scattering data (Humphrey and Ross, 1962; Watson *et al.*, 1963) are consistent with the value of α required for a $\pi\Sigma$ resonance at 1405 MeV, but the errors associated with these parameters are too large for any stronger statement to be made.

13.5(c). *Dynamical models* (2). *Regge poles and trajectories*

A totally different viewpoint to the two we have discussed in the previous sections has been proposed by Chew and Frautschi (1962; see also Chew, 1962). They assume that there are no hidden symmetries and that no particles are more fundamental than others; instead they conjecture that all particles arise from poles in the complex angular momentum plane (§ 10.5).

In Fig. 13.11 we display a plot of angular momentum versus mass squared ($\equiv t$) for known particles and resonances. Each point with the same 'internal' quantum numbers (B, S and T) is assumed to lie on a Regge trajectory, but because of the rule $\Delta j = 2$ only a few pairs of particles can lie along the same trajectory. These pairs have been connected by a straight line, but a linear relationship is not necessarily expected. Nevertheless, the uniformity of the slopes connecting suspected members of a trajectory is striking. The slopes are roughly given by

$$\frac{d}{dt}(\mathrm{Re}\,\alpha) \sim \frac{1}{M^2} \sim \frac{1}{1\,\mathrm{GeV}^2};$$

a similar slope has been found for the vacuum trajectory at $t = 0$ in high energy scattering (§ 10.5(d)).

An inspection of Fig. 13.11 shows that all particles lie to the right of the vacuum trajectory, and so the diagram satisfies the Froissart theorem (1961), which predicts $\alpha(t) \leqq 1$ for $t < 0$. This theorem is based upon the analytic and unitarity properties of the S-matrix. It should also be noted that systems with the simplest internal quantum numbers lie furthest to the left in Fig. 13.11. It has been suggested by Chew that this property might be connected with the *principle of maximal strength* (Chew and Frautschi, 1961) which postulates that strong interactions possess the maximum strength which is possible whilst still being consistent with the requirements of unitarity and the analyticity of the S-matrix. Chew has conjectured that in general the forces may be strongest in systems with the simplest quantum numbers; the conjecture is partially based on a study of the π–π system (Chew, Frautschi and Mandelstam, 1962) which showed that the long-range forces were most attractive for $T = 0$, less attractive for $T = 1$, and perhaps repulsive for $T = 2$. Therefore in systems with the quantum numbers of the vacuum ($T = B = S = G = 0$) all the partial contributors to the force add with the same sign, and the strongest possible interaction is obtained. The trajectories in Fig. 13.11 support this conjecture, since

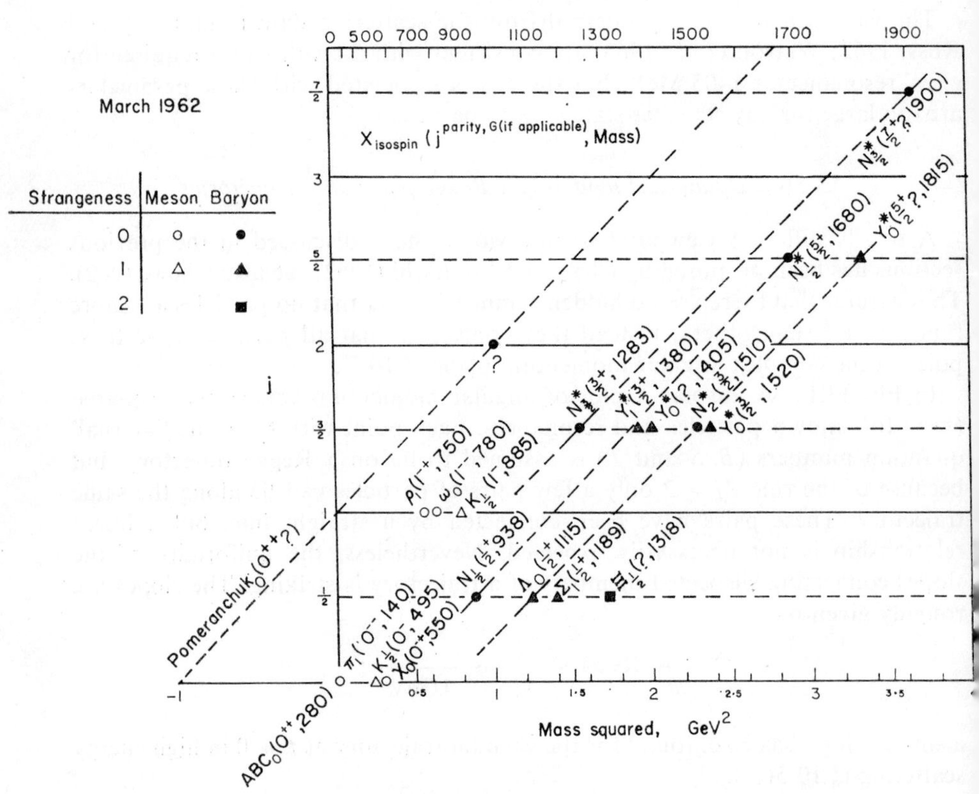

FIG. 13.11.

it can be seen that the height of the trajectory (and in particular $\alpha(t)$ at $t = 0$) is correlated with simplicity of quantum numbers.

Blankenbecker, Cook and Goldberger (1962) have raised the problem of why the Regge pole hypothesis should be restricted to strong interactions, and in particular whether the photon should be treated as a Regge pole. They have pointed out that the effective electromagnetic interaction must decrease at high energies, otherwise it would overtake the strong interactions if the latter are determined by Regge poles. At present, however, no experimental evidence exists to support the concept of a photon as a Regge pole.

CHAPTER 14

STRONG INTERACTIONS II REACTIONS

14.1. PION–NUCLEON SCATTERING AND THE PHOTOPRODUCTION OF PIONS

14.1 (a). *Introduction*

Considerable data exists on pion–nucleon scattering for energies in the *c*-system of less than 250 MeV; above this figure the experimental information becomes progressively worse with increasing energies. The data which may be acquired in pion–nucleon scattering may be summarised as follows:

(1) total cross-sections,

(2) total and differential cross-sections for elastic scattering,

(3) total and differential cross-sections for charge exchange scattering,

(4) polarisation of the recoil protons in (2) and (3),

(5) cross-sections and nucleon polarisation in the photoproduction of pions,

(6) inelastic channels.

At energies below the threshold for the production of a pion by the process $\pi + N \rightarrow \pi + \pi + N$, the total cross-sections for (2) and (3) must add to yield that for (1); the threshold occurs at a pion kinetic energy in the laboratory of ~ 170 MeV. The process of photoproduction may be regarded as radiative scattering in reverse, and obviously processes of the type

$$\gamma + p \begin{cases} \nearrow p + \pi^0 \\ \searrow n + \pi^+ \end{cases}$$

can provide valuable information which is not available in direct scattering experiments.

Data on the total cross-sections for pion–nucleon scattering and photoproduction have been given in Chapter 13. In this chapter we shall concentrate on differential cross-sections and polarisation.

14.1(b). *Partial wave analysis of the data below* 250 MeV *in pion–nucleon scattering*

At pion kinetic energies below 250 MeV only elastic and charge exchange scattering is of importance; for example

$$\pi^+ + p \to \pi^+ + p \qquad (\pi^+ \to \pi^+)$$

$$\pi^- + p \begin{cases} \pi^- + p & (\pi^- \to \pi^-) \\ \pi^0 + n & (\pi^- \to \pi^0). \end{cases}$$

Typical angular distributions in this energy range are displayed in Fig. 14.1.

The analysis of the scattering data is carried out by first making a phase shift analysis. Since the nucleon and pion are spin $\frac{1}{2}$ and 0 systems respectively, we

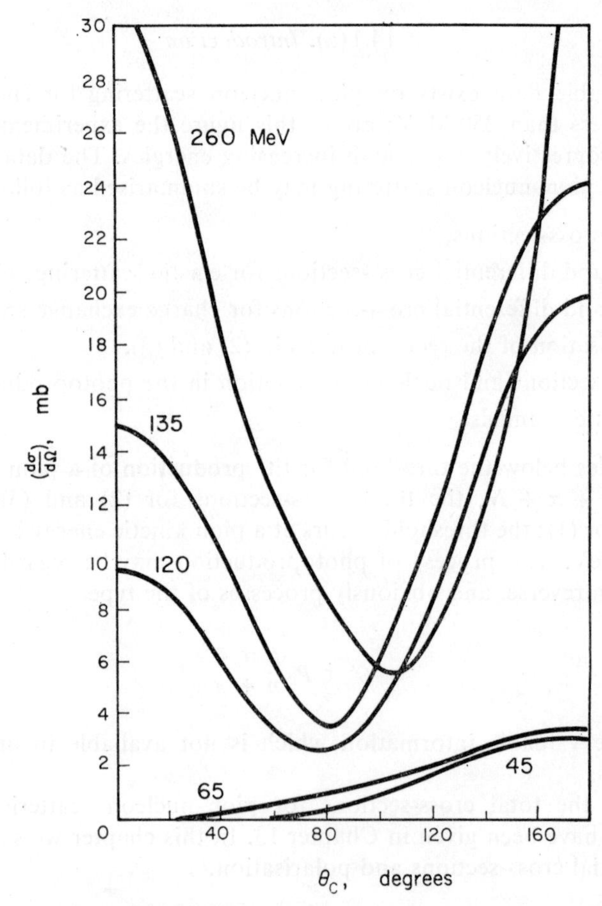

FIG. 14.1. Angular distributions, in the *c*-system, of positive pions scattered from protons at different kinetic energies (Yuan, 1956).

may use equation (6.62) to represent the scattering operator in nucleon spin space

$$M(\theta, \varphi) = g(\theta)\hat{1} + ih(\theta)\,\boldsymbol{\sigma}\cdot\mathbf{n}.$$

The differential cross-section is then given by

$$\frac{d\sigma_{sc}}{d\Omega} = |M|^2 = |g(\theta)|^2 + |h(\theta)|^2 \tag{14.1}$$

where $g(\theta)$ and $h(\theta)$ represent the nonspin–flip and spin–flip amplitudes respectively. They were given in (6.34)

$$g(\theta) = \sum_{l=0}^{\infty} [(l+1) f_{l+} + l f_{l-}]\, P_l^0(\cos\theta) \tag{14.2}$$

$$h(\theta) = -\sum_{l=0}^{\infty} (f_{l+} - f_{l-})\, P_l^1(\cos\theta)$$

where $f_{l\pm}$ is the scattering function for orbital angular momentum l and total angular momentum $j = l \pm \frac{1}{2}$. It was defined in (6.36) and (6.18)

$$f_{lj} = \frac{\eta_{lj} - 1}{2ik} = \frac{\varrho_{lj} e^{2i\delta_{lj}} - 1}{2ik} \tag{14.3}$$

$$\varrho_{lj} \leqq 1 \qquad \delta_{lj} \text{ real}$$

$$f_{lj} = \frac{1}{k} \sin\delta_{lj} e^{i\delta_{lj}} \qquad \text{for } \varrho_{lj} = 1$$

where k is the pion momentum in the c-system.

These terms must be subdivided further to allow for the isospin channels $T = \frac{3}{2}$ and $T = \frac{1}{2}$. The relevant substates can be constructed with the aid of Table 5.2 (p. 200):

$$|\pi^+ p\rangle = |\tfrac{3}{2}\rangle \tag{14.4}$$

$$|\pi^- p\rangle = \frac{1}{\sqrt{3}}\left|\frac{3}{2}\right\rangle - \sqrt{\frac{2}{3}}\left|\frac{1}{2}\right\rangle$$

$$|\pi^0 n\rangle = \sqrt{\frac{2}{3}}\left|\frac{3}{2}\right\rangle + \frac{1}{\sqrt{3}}\left|\frac{1}{2}\right\rangle.$$

Consider, for example, $\pi^- p$ scattering; if we represent the proton spin state by χ, the relation between incident and scattered waves is of the form (§ 6.2(d))

$$e^{ikz}\chi \rightarrow \frac{e^{ikr}}{r} f(\theta) = \frac{e^{ikr}}{r} M\chi$$

where we have ignored the charge states. Because of the isospin channels we may modify this relation in accord with (14.4)

$$e^{ikz}\chi\,|\pi^-p\rangle \to \frac{e^{ikr}}{r}\left[\frac{1}{\sqrt{3}}\left|\frac{3}{2}\right\rangle M^3 - \sqrt{\frac{2}{3}}\left|\frac{1}{2}\right\rangle M^1\right]\chi$$

where M^3 and M^1 represent the scattering matrices in ordinary spin space for the $T = \frac{3}{2}$ and $T = \frac{1}{2}$ channels respectively. Using (14.4) again to revert to the final charge states we find

$$e^{ikz}\chi\,|\pi^-p\rangle \to \frac{e^{ikr}}{r}\left\{\left[\frac{1}{3}M^3 + \frac{2}{3}M^1\right]\chi|\pi^-p\rangle + \left[\frac{\sqrt{2}}{3}(M^3 - M^1)\chi|\pi^0n\rangle\right]\right\}. \tag{14.5}$$

The same factors appear in the g and h functions, when we consider isospin channels, and so we may write

$$\begin{aligned} g^+(\theta) &= g^3(\theta) && \text{for } \pi^+p \to \pi^+p \qquad (14.6)\\ g^-(\theta) &= \tfrac{1}{3}g^3(\theta) + \tfrac{2}{3}g^1(\theta) && \pi^-p \to \pi^-p\\ g^0(\theta) &= \frac{\sqrt{2}}{3}[g^3(\theta) - g^1(\theta)] && \pi^-p \to \pi^0n \end{aligned}$$

where the superscripts on the left refer to the final pion charge and those on the right to twice the isospin; similar expressions can be constructed for $h(\theta)$. The notation may be extended to the scattering functions f_{lj} and phase shifts δ_{lj} by writing f_{lj}^{2T} and δ_{lj}^{2T} respectively.†

In practice only s and p waves need be considered at low energies, since the range of the strong interaction is $\sim 1/m_\pi$ and so only values of the orbital angular momentum up to $l \sim k/m_\pi$ can be expected to contribute to the interaction. The appropriate figures are displayed in Table 14.1, where T_π represents the kinetic energy of the pion in the laboratory system.

TABLE 14.1

T_π(MeV)	l
80	1
270	2
540	3
860	4

† The s and p scattering phase shifts are normally written in the following notation:

$$\delta^3_{s1/2} \equiv \delta_3 \text{ or } \alpha_3, \quad \delta^3_{p1/2} \equiv \delta_{31} \text{ or } \alpha_{31}, \quad \delta^3_{p3/2} \equiv \delta_{33} \text{ or } \alpha_{33}$$

$$\delta^1_{s1/2} \equiv \delta_1 \text{ or } \alpha_1, \quad \delta^1_{p1/2} \equiv \delta_{11} \text{ or } \alpha_{11}, \quad \delta^1_{p3/2} \equiv \delta_{13} \text{ or } \alpha_{13}.$$

If s and p waves only are considered the amplitudes $g(\theta)$ and $h(\theta)$ in (14.2) reduce to

$$g^{2T}(\theta) = f^{2T}_{s\,1/2} + (2f^{2T}_{p\,3/2} + f^{2T}_{p\,1/2}) \cos\theta \tag{14.7}$$

$$h^{2T}(\theta) = (f^{2T}_{p\,3/2} - f^{2T}_{p\,1/2}) \sin\theta \tag{14.8}$$

since

$$P^0_0(\cos\theta) = 1, \qquad P^0_1(\cos\theta) = \cos\theta\,,$$

$$P^1_1(\cos\theta) = -\sin\theta\,.$$

The cross-sections for pion scattering may then be obtained with the aid of equations (14.1) and (14.6), and yield

$$\frac{d\sigma_{sc}}{d\Omega}(\pi^+ \to \pi^+) = [|f^3_{s\,1/2} + (2f^3_{p\,3/2} + f^3_{p\,1/2}) \cos\theta|^2 + |(f^3_{p\,3/2} - f^3_{p\,1/2}) \sin\theta|^2]$$

$$\frac{d\sigma_{sc}}{d\Omega}(\pi^- \to \pi^-) = \tfrac{1}{9}\,[|f^3_{s\,1/2} + 2f^1_{s\,1/2} + (2f^3_{p\,3/2} + f^3_{p\,1/2} + 4f^1_{p\,3/2} + 2f^1_{p\,1/2}) \cos\theta|^2$$
$$+ |(f^3_{p\,3/2} - f^3_{p\,1/2} + 2f^1_{p\,3/2} - 2f^1_{p\,1/2}) \sin\theta|^2]$$

$$\frac{d\sigma_{sc}}{d\Omega}(\pi^- \to \pi^0) = \tfrac{2}{9}\,[|f^3_{s\,1/2} - f^1_{s\,1/2} + (2f^3_{p\,3/2} + f^3_{p\,1/2} - 2f^1_{p\,3/2} - f^1_{p\,1/2}) \cos\theta|^2$$
$$+ |(f^3_{p\,3/2} - f^3_{p\,1/2} - f^1_{p\,3/2} + f^1_{p\,1/2}) \sin\theta|^2]\,. \tag{14.9}$$

Thus the cross-sections have the general form

$$\frac{d\sigma_{sc}}{d\Omega} = A + B\cos\theta + C\cos^2\theta \tag{14.10}$$

in each of the three processes, and the problem of analysing the experimental data consists in obtaining six phase shifts from the nine parameters $\overset{\pm}{A^0}$, $\overset{\pm}{B^0}$ and $\overset{\pm}{C^0}$ (the superscripts refer to the notation of (14.6)). This is not easy as one phase shift tends to dominate the others (see below); in addition Coulomb scattering corrections have to be made at small scattering angles. Nevertheless, the magnitudes of the phase shifts as a function of energy are now reasonably well known. The procedure for analysis is to take experimental distributions and to determine sets of phase shifts which minimise the least squares relation

$$M = \sum_i \left(\frac{\Delta_i}{\varepsilon_i}\right)^2$$

where ε_i is the experimental error in the ith value and Δ_i the deviation of the calculated cross-section from the observed cross-section. Ambiguities in signs of the phase shifts have been eliminated by examining the Coulomb interference (destructive) at small angles (Orear, 1954; Ferrari *et al.* (1956)), and by use of dispersion relations (Davidon and Goldberger, 1956; Gilbert and Screaton, 1956).

Magnitudes of the phase shifts as a function of kinetic energy in the laboratory system, resulting from these analyses, are shown in Fig. 14.2. It can be seen that the $T = \frac{3}{2}, j = \frac{3}{2}$ phase shift dominates the data, and that a resonance ($\delta^3_{p3/2} = 90°$)

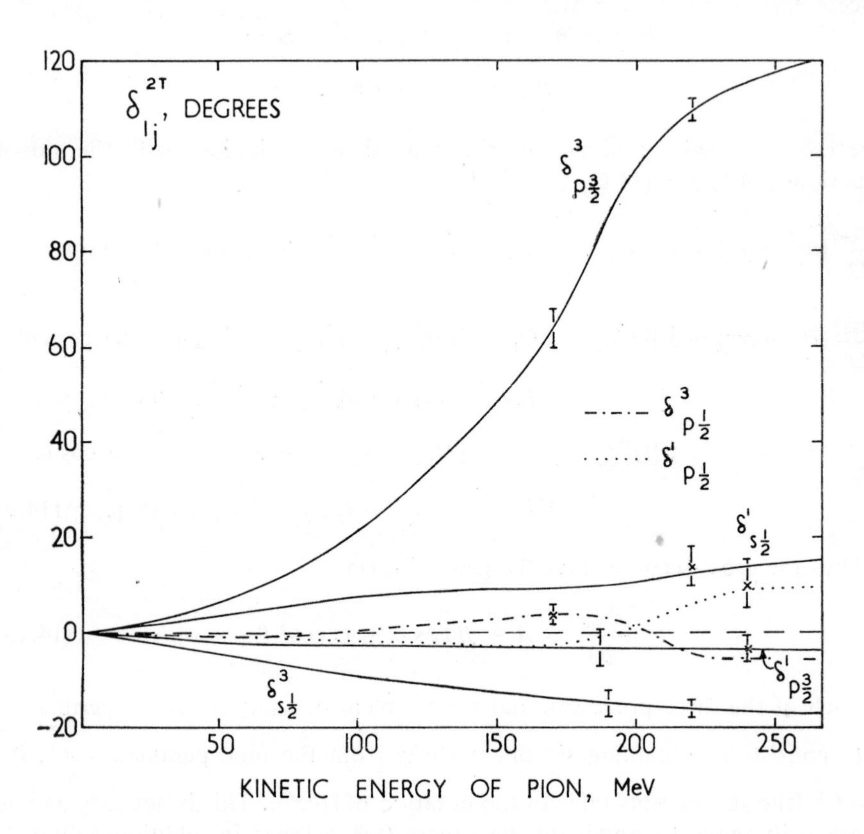

FIG. 14.2. Phase shifts for pion–nucleon scattering as a function of energy. The errors indicate the experimental accuracy achieved in the appropriate region. The values of the small p-wave phase shifts should be treated with caution, especially in the high energy region.

occurs at 180 MeV; this corresponds to the N_1^* nucleon isobaric state of § 13.2. We may therefore conclude that the resonant N_1^* state has the following quantum numbers:

$$B = 1, \quad T = \tfrac{3}{2}, \quad L_j^{\text{parity}} = p_{3/2}^{+}$$

(the parity sign + arises from the product of the p-wave contribution (odd) and the intrinsic parity function of the pion (odd)). If we ignore all terms apart from the $\frac{3}{2}, \frac{3}{2}^+$ contribution and write

$$f_{lj} = \frac{1}{k} \sin \delta_{lj} e^{i\delta_{lj}} \tag{14.11}$$

since $\varrho_{lj} = 1$ at these energies, the cross-sections in (14.9) assume the following forms in the region of the resonance:

$$\frac{d\sigma_{sc}}{d\Omega} = \frac{a}{k^2}(1 + 3\cos^2\theta)\sin^2\delta^3_{p\,3/2} \tag{14.12}$$

$$\sigma_T = a\frac{8\pi}{k^2}\sin^2\delta^3_{p\,3/2}$$

where

$$a = 1 \text{ for } \pi^+ \to \pi^+, \quad a = \tfrac{1}{9} \text{ for } \pi^- \to \pi^-, \quad a = \tfrac{2}{9} \text{ for } \pi^- \to \pi^0.$$

Thus the shape of the differential cross-sections should be roughly the same in the region of the resonance; (exact agreement is not expected because of the interference effects from the smaller phase shifts which have a greater effect on differential than on total cross-sections). It is also clear from the formulae given above that we should expect $\sigma_T \sim (8\pi/k^2)a$ in the region of the resonance. In Table 14.2 we compare the values given by this expression with those found experimentally at 180 MeV; the cross-sections are in millibarns.

TABLE 14.2

Total cross-section	$\pi^+ \to \pi^+$	$\pi^- \to \pi^-$	$\pi^- \to \pi^0$
$\frac{8\pi}{k^2}a$	202	23	45
experiment	195	22	45

Thus the cross-sections associated with the $\frac{3}{2}, \frac{3}{2}^+$ states reach the maximum value allowed by unitarity at this energy.

The dominance of the $\frac{3}{2}, \frac{3}{2}^+$ term in the region of the resonance implies that the smaller phase shifts are not well known. The situation improves below 100 MeV where the s-waves assume importance. At very low energies the s-waves predominate, and the scattering becomes increasingly isotropic as the energy is reduced. The phase shifts are small near zero momentum, and we may then use the approximation given in (6.96)

$$\sin \delta_l \sim \delta_l \propto k^{2l+1}.$$

We therefore introduce a parameter a^{2T} defined by the relation

$$\delta^{2T}_{lj} = a^{2T}_{lj}\eta^{2l+1} \tag{14.13}$$

where $\eta = k/m_\pi$. The experimental data yield the following values for a in radians (Woolcock, 1961):

$$a^3_{s\,1/2} = -0{\cdot}089 \pm 0{\cdot}004, \qquad a^1_{s\,1/2} = 0{\cdot}170 \pm 0{\cdot}005$$
$$a^3_{p\,1/2} = -0{\cdot}040 \pm 0{\cdot}004 \qquad a^1_{p\,1/2} = -0{\cdot}104 \pm 0{\cdot}006$$
$$a^3_{p\,3/2} = 0{\cdot}215 \pm 0{\cdot}004 \qquad a^1_{p\,3/2} = -0{\cdot}030 \pm 0{\cdot}005\,.$$

The behaviour of the s-wave phase shifts as a function of energy are displayed in Fig. 14.3 (Miyake, Kinsey and Knapp, 1962). It can be seen that they possess the expected linear dependence on momentum at low kinetic energies.

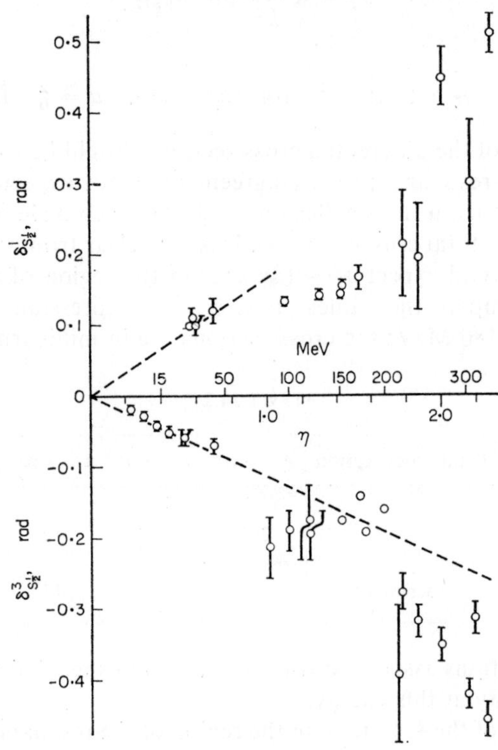

FIG. 14.3.

14.1(c). *Interpretation of the low energy scattering data*

In §9.2(b) two possible Lagrangians for the pion–nucleon interaction were introduced:

$$\mathscr{L}_{\text{int}} = ig\bar{\psi}\gamma_5\psi\varphi = \mathscr{L}_{ps} \tag{14.14}$$

$$\mathscr{L}_{\text{int}} = i\frac{f}{m}\bar{\psi}\gamma_5\gamma_\lambda\psi\frac{\partial\varphi}{\partial x_\lambda} = \mathscr{L}_{pv}\,.$$

Both terms are satisfactory as far as symmetry requirements are concerned. They also yield identical results in lowest order perturbation theory, providing $g/2M = f/m$ (where m and M represent pion and nucleon masses respectively); unfortunately, these results do not agree with experiment. For example, they predict that s-wave interactions should dominate in pion–nucleon scattering at low energies (compare § 8.7(b) and (8.101)).

If higher order expansions are considered, it is found that the pseudovector coupling does not satisfy the requirements for a renormalisable field theory; the pseudoscalar interaction can be made renormalisable by adding a meson–meson interaction term to it, so that the full Langrangian for the interaction becomes

$$\mathscr{L}_{\text{int}} = ig\bar{\psi}\gamma_5\psi\varphi + \lambda\varphi^4 \tag{14.15}$$

where λ is a coupling strength (Matthews and Salam, 1951). Unfortunately, this knowledge does not help the problem of how to make meaningful calculations in perturbation theory when the coupling parameter g is so large ($g^2/4\pi \sim 14$).

The first successful description of the main features of the pion–nucleon interaction was made by Chew and Low (Chew, 1954; Chew and Low, 1956). The theory of Chew was Hamiltonian in form and based upon the expression

$$H_{\text{int}} = \frac{f}{m}\int d\mathbf{x}\varrho(\mathbf{x})\,(\boldsymbol{\sigma}\cdot\nabla)\,(\boldsymbol{\tau}\cdot\boldsymbol{\varphi}). \tag{14.16}$$

In this equation f is a coupling strength, $\boldsymbol{\sigma}$ is the Pauli spin operator, $\boldsymbol{\tau}$ the nucleon isospin operator, $\boldsymbol{\varphi}$ is the pion field operator (a vector in isospace) and $\varrho(\mathbf{x})$ is the nucleon density function (the *extended source function*). The term m is the pion mass and gives f the dimensions of electric charge.

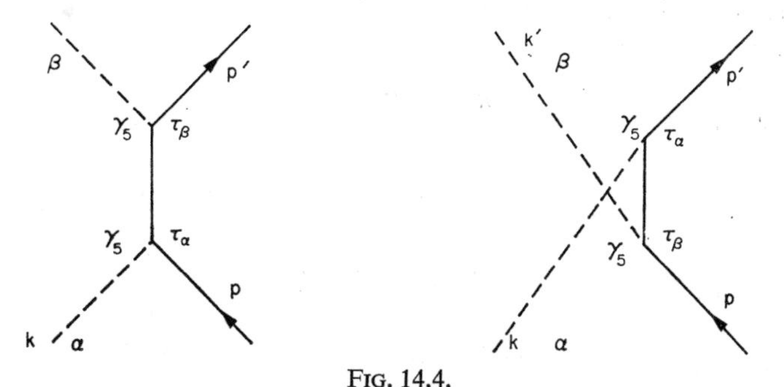

FIG. 14.4.

The form of the Hamiltonian is determined by the assumption that the interaction is linear in the meson field and that the nucleon source is infinite in mass so that it is not allowed to move. Therefore, the only states of the source are given by the orientation of the spin and the isospin. The symmetry requirements discussed in Chapter 9 then determine the final form of the Hamiltonian.

An attractive feature of equation (14.16) is that since φ behaves like r^l at short distances only the partial wave $l = 1$ has a finite derivative at the origin. Thus the theory predicts that p-waves should predominate at low energies.

Instead of developing the original theory,† we shall use the graph techniques of Chapter 8 and then allow the mass of the nucleon to tend to infinity. The relevant graphs in lowest order perturbation theory are shown in Fig. 14.4. If we denote the energy of the pion as ω, and the total energy in the centre of momentum system as E_c, then a comparison of equations (7.95), (7.75) and (8.98) suggests the following form for the scattering amplitude in lowest order perturbation theory:

$$f(\theta) = \frac{M}{4\pi E_c} T_{fi}$$

$$\begin{aligned} T_{fi} &= g^2 \bar{u}_{p'} \left[\tau_\beta \tau_\alpha \gamma_5 \frac{i\gamma q - M}{q^2 + M^2} \gamma_5 + \tau_\alpha \tau_\beta \gamma_5 \frac{i\gamma q' - M}{q'^2 + M^2} \gamma_5 \right] u_p \\ &= -g^2 \bar{u}_{p'} \left[\tau_\beta \tau_\alpha \frac{i\gamma q + M}{M^2 - s} + \tau_\alpha \tau_\beta \frac{i\gamma q' + M}{M^2 - u} \right] u_p \end{aligned} \tag{14.17}$$

where the first and second terms refer to Fig. 14.4(a) and (b) respectively, and

$$q = p + k, \qquad q' = p - k'$$

$$q^2 = -s, \qquad q'^2 = -u$$

$$p + k = p' + k'.$$

In the above equations, s and u represent the Mandelstam variables (§ 10.4(b)). Now consider the expression

$$\begin{aligned} i\gamma q + M &= i\gamma p + i\gamma \frac{\boldsymbol{k}}{2} + i\frac{\gamma}{2}(p' + k' - p) + M \\ &= \tfrac{1}{2}(i\gamma p + M) + \tfrac{1}{2} i\gamma (k + k') + \tfrac{1}{2}(i\gamma p' + M). \end{aligned}$$

If we use the spinor forms of the Dirac equation (3.146)

$$(i\gamma p + M) u = 0, \qquad \bar{u}(i\gamma p + M) = 0$$

we find that

$$\bar{u}_{p'} \frac{i\gamma q + M}{M^2 - s} u_p = \bar{u}_{p'} \frac{i\gamma\left(\dfrac{k + k'}{2}\right)}{M^2 - s} u_p$$

† In addition to the original paper of Chew (1954), an illuminating discussion of the method may be found in an article by Wick (1955).

and similarly

$$\bar{u}_{p'} \frac{i\gamma q' + M}{M^2 - u} u_p = -\bar{u}_{p'} \frac{i\gamma\left(\frac{k + k'}{2}\right)}{M^2 - u} u_p.$$

Thus equation (14.17) becomes

$$T_{fi} = -g^2 \bar{u}_{p'} i\gamma \left(\frac{k + k'}{2}\right)\left[\frac{\tau_\beta \tau_\alpha}{M^2 - s} - \frac{\tau_\alpha \tau_\beta}{M^2 - u}\right] u_p.$$

Now, if we use the relation (compare (9.116))

$$F_1 \tau_\beta \tau_\alpha + F_2 \tau_\alpha \tau_\beta = (F_1 + F_2)\, \delta_{\beta\alpha} + \tfrac{1}{2}(F_1 - F_2)\, [\tau_\beta, \tau_\alpha]$$

and compare T_{fi} with the invariant expression for pion–nucleon scattering given in (9.132)†

$$T_{fi} \equiv T_{\beta\alpha} = \bar{u}_{p'} \left\{\left[A^+ - i\gamma\left(\frac{k + k'}{2}\right) B^+\right] \delta_{\beta\alpha} + \tfrac{1}{2}\left[A^- - i\gamma\left(\frac{k + k'}{2}\right) B^-\right] [\tau_\beta, \tau_\alpha]\right\} u_p$$

then it is clear that the pole terms for the scattering amplitude are given by

$$A^\pm = 0 \qquad (14.18)$$

$$B^\pm = \frac{g^2}{M^2 - s} \mp \frac{g^2}{M^2 - u}.$$

Now let us consider the form of the elastic scattering amplitude

$$f(\theta) = \frac{M}{4\pi E_c} T_{fi}$$

in terms of the Lorentz invariant expression (compare (9.130) and (9.131))

$$T_{fi} = \bar{u}_{p'} \left[A - i\gamma\left(\frac{k + k'}{2}\right) B\right] u_p.$$

Since $\mathbf{k} = -\mathbf{p}$ and $\mathbf{k}' = -\mathbf{p}'$ in the c-system, we may write the four component spinors as

$$u_p = \frac{1}{\sqrt{[2M(E + M)]}} \begin{pmatrix} E + M \\ -\boldsymbol{\sigma} \cdot \mathbf{k} \end{pmatrix} \chi_i$$

$$\bar{u}_{p'} = u_{p'}^\dagger \gamma_4 = \frac{\chi_f^\dagger}{\sqrt{[2M(E + M)]}} (E + M, \boldsymbol{\sigma} \cdot \mathbf{k}')$$

† For convenience of notation we have replaced $A_{\frac{1}{2}}$ and $B_{\frac{1}{2}}$ in equation (9.132) by $A^\pm$ and $B^\pm$ respectively.

where χ represents a two-component Pauli spinor, and we have used equations (3.136) and (3.151). Upon using the explicit forms for the γ-matrices given in (3.103), we find

$$f(\theta) = \chi_f^\dagger \left\{ \frac{E + M}{2E_c} \left[\frac{A + (E_c - M) B}{4\pi} \right] + \frac{E - M}{2E_c} \left[\frac{-A + (E_c + M)B}{4\pi} \right] \boldsymbol{\sigma} \cdot \hat{\mathbf{k}}' \boldsymbol{\sigma} \cdot \hat{\mathbf{k}} \right\} \chi_i$$

$$= \chi_f^\dagger (f_1 + f_2 \boldsymbol{\sigma} \cdot \hat{\mathbf{k}}' \boldsymbol{\sigma} \cdot \hat{\mathbf{k}}) \chi_i \tag{14.19}$$

where $\hat{\mathbf{k}}$ and $\hat{\mathbf{k}}'$ are unit vectors in the directions $\mathbf{k}$ and $\mathbf{k}'$ respectively.

Scattering functions

$$f_{l\pm} = \frac{\sin \delta_{l\pm}}{k} e^{i\delta_{l\pm}}$$

can be projected out from this equation by means of the following operation (Chew *et al.*, 1957).

$$f_{l\pm} = \tfrac{1}{2} \int_{-1}^{+1} dz \, [f_1 P_l(z) + f_2 P_{l\pm 1}(z)]$$

where $z = \cos \theta$ and θ is the scattering angle. However, since we wish to consider low energy terms and p-waves only, the problem can be handled just as easily by straightforward expansion.

We first note that by using (9.121)

$$T_{3/2} = T^+ - T^-$$

$$T_{1/2} = T^+ + 2T^-$$

the scattering amplitude can be expressed in tems of isospin channels. Considering the channel with isospin $\frac{3}{2}$, we find from (14.18) that

$$A^3 = 0, \qquad B^3 = \frac{-2g^2}{M^2 - u}$$

where the superscript indicates twice the isospin number. Now let us consider the low energy limits. We can write

$$E_c \to M, \qquad E + M \to 2M$$

$$E - M \to \frac{k^2}{2M}, \qquad E_c - M \to \omega$$

$$\frac{1}{M^2 - u} = \frac{1}{M^2 + (p - k')^2} = \frac{1}{-2pk' - m^2}$$

$$= \frac{1}{-2\mathbf{p} \cdot \mathbf{k}' + 2E\omega - m^2} \to \frac{1}{2M\omega} \left(1 - \frac{k^2}{M\omega} \cos \theta \right)$$

since $\mathbf{p} = -\mathbf{k}$. Thus the amplitude with isospin $\frac{3}{2}$ in equation (14.19) becomes

$$f^3 = \frac{1}{4\pi}\chi_f^\dagger\left[\frac{-g^2\omega}{M\omega}\left(1 - \frac{k^2}{M\omega}\cos\theta\right) - \frac{k^2}{2M}\frac{g^2}{M\omega}\left(1 - \frac{k^2}{M\omega}\cos\theta\right)\boldsymbol{\sigma}\cdot\hat{\mathbf{k}}'\boldsymbol{\sigma}\cdot\hat{\mathbf{k}}\right]\chi_i$$

in the low energy limit. We have already shown in § 8.7(b) that the *s*-wave amplitude dominates this expression. However, we wish to examine the *p*-wave contributions only; we note from equation (6.95) that the *p*-wave contribution to the scattering amplitude should have a k^2 dependence in the low energy limit, and so the *p*-wave terms in the above equation give us

$$\begin{aligned} f_p^3 &= \frac{g^2}{4\pi}\frac{1}{2M^2}\frac{k^2}{\omega}\chi_f^\dagger[2\cos\theta - \boldsymbol{\sigma}\cdot\hat{\mathbf{k}}'\boldsymbol{\sigma}\cdot\hat{\mathbf{k}}]\chi_i \\ &= 2\frac{g^2}{4\pi}\left(\frac{m}{2M}\right)^2\frac{k^2}{m^2\omega}\chi_f^\dagger[2\cos\theta - (\cos\theta + i\boldsymbol{\sigma}\cdot\mathbf{n}'\sin\theta)]\chi_i \\ &= 2\frac{f^2}{4\pi}\frac{k^2}{m^2\omega}\chi_f^\dagger(\cos\theta - i\boldsymbol{\sigma}\cdot\mathbf{n}'\sin\theta)\chi_i \end{aligned}$$

where we have introduced the coupling constant f^2 (10.128) and used the relation

$$\boldsymbol{\sigma}\cdot\mathbf{A}\boldsymbol{\sigma}\cdot\mathbf{B} = \mathbf{A}\cdot\mathbf{B} + i\boldsymbol{\sigma}\cdot\mathbf{A}\times\mathbf{B}$$

and have written $\hat{\mathbf{k}}'\times\hat{\mathbf{k}} = \mathbf{n}'$. Now from equation (A.7.21) (p. 711) we can write the contribution of the lth wave in the partial wave scattering analysis as

$$\chi_f^\dagger\{(l+1)f_{l+} + lf_{l-}]P_l^0 + i(f_{l+} - f_{l-})P_l^1\boldsymbol{\sigma}\cdot\mathbf{n}'\}\chi_i.$$

Hence if we equate terms in this equation for $l = 1$ with those in f_p^3 we find

$$2f_{p\,3/2}^3 + f_{p\,1/2}^3 = 2\frac{f^2}{4\pi}\frac{k^2}{m^2\omega}$$

$$f_{p\,3/2}^3 - f_{p\,1/2}^3 = 2\frac{f^2}{4\pi}\frac{k^2}{m^2\omega}$$

and hence

$$f_{p\,3/2}^3 = \frac{4}{3}\frac{f^2}{4\pi}\frac{k^2}{m^2\omega}, \qquad f_{p\,1/2}^3 = -\frac{2}{3}\frac{f^2}{4\pi}\frac{k^2}{m^2\omega}.$$

A similar analysis can be carried out for the channels with isospin $\frac{1}{2}$ by use of (9.121); the final results in Born approximation may be written as

$$f_\alpha = \frac{\sin\delta_\alpha}{k}e^{i\delta_\alpha} = \frac{n_\alpha}{3}\frac{f^2}{4\pi}\frac{k^2}{m^2\omega} \tag{14.20}$$

where α represents the isospin (T) and angular momentum (j) channels and n_α is given in Table 14.3.

TABLE 14.3

j \ T	$\frac{3}{2}$	$\frac{1}{2}$
$\frac{3}{2}$	4	−2
$\frac{1}{2}$	−2	−8

These expressions are the same as those of the Chew theory apart from a source function $v(k^2)$ (the Fourier transform of $\varrho(\mathbf{x})$ in (14.16)); this function is normally written as $v(k^2) = 1$ in any case. At first sight the results given above appear to be unsatisfactory, since the strongest amplitude appears to be associated with the $T = \frac{1}{2}$, $j = \frac{1}{2}$ state. However, the sign associated with the $\frac{3}{2}, \frac{3}{2}$ amplitude indicates that the interaction is attractive in this state whilst the other states are repulsive (compare § 6.3(c)). There is a general tendency for the Born approximation to lead to underestimates of the amplitudes associated with attractive interactions, and the reverse for repulsive interactions. Thus the results obtained above can be regarded as a satisfactory first approximation.

The calculation may be improved by using the Born terms in a dispersion relation. We shall follow a method due to Feldman, Matthews and Salam (1960); see also Matthews (1961). The method employs dispersion relations for inverse scattering amplitudes (compare § 7.5(c)). Consider the inverse of the scattering function f_α

$$f_\alpha^{-1} = k \cot \delta_\alpha - ik \tag{14.21}$$

and assume f_α^{-1} is an analytic function for which we may construct a dispersion relation of the form

$$Bf^{-1} = \frac{1}{\pi} \oint d\omega' \frac{\operatorname{Im} Bf^{-1}}{\omega' - \omega - i\varepsilon}$$

where $\oint$ is an integral around a closed contour (compare § 10.2(a)), and B represents the Born approximation to the scattering function. If we assume similar analytic properties (including cuts) for f_α and f_α^{-1} we may write

$$Bf^{-1} = \frac{1}{\pi} \int_{\omega_c}^{\infty} d\omega' \frac{\operatorname{Im} Bf^{-1}}{\omega' - \omega - i\varepsilon} + C$$

(compare (10.24)), where C represents the contribution from the pole term B and the infinite circle of the contour. If we now evaluate the above relation at $\omega = \omega_B$ and assume that B dominates the scattering amplitude in the region of the pole, so that we may write $f \sim B$, then

$$1 = \frac{1}{\pi} \int_{\omega_c}^{\infty} d\omega' \frac{\text{Im } Bf^{-1}}{\omega' - \omega_B} + C$$

and upon subtraction we find

$$Bf^{-1} = 1 + \frac{\omega - \omega_B}{\pi} \int_{\omega_c}^{\infty} d\omega' \frac{\text{Im } Bf^{-1}}{(\omega' - \omega - i\varepsilon)(\omega' - \omega_B)}.$$

Now equation (14.21) gives us $\text{Im} f^{-1} = -k$, and it is not difficult to show that

$$\omega_B = \frac{m^2}{2M}$$

(compare (10.114)), hence we may neglect ω_B and write the dispersion relation as

$$\begin{aligned} Bf_\alpha^{-1} &= \frac{n_\alpha}{3} \frac{f^2}{4\pi} \frac{k^2}{m^2\omega} (k \cot \delta_\alpha - ik) \\ &= 1 - \frac{\omega}{\pi} \int d\omega' \frac{Bk'}{\omega'(\omega' - \omega - i\varepsilon)} \\ &= 1 - \frac{n_\alpha}{3m^2} \frac{f^2}{4\pi} \frac{\omega}{\pi} \int d\omega' \frac{k'^3}{\omega'^2(\omega' - \omega - i\varepsilon)} \end{aligned}$$

and so (by equation (10.18) and § 10.2(b))

$$\frac{n_\alpha}{3} \frac{f^2}{4\pi} \frac{k^3}{m^2\omega} \cot \delta_\alpha = 1 - \frac{n_\alpha}{3m^2} \frac{f^2}{4\pi} \frac{\omega P}{\pi} \int d\omega' \frac{k'^3}{\omega'^2(\omega' - \omega)}. \quad (14.22)$$

This equation is essentially the chew-low equation for pion-nucleon scattering (we have neglected crossing symmetry). The lower and upper limits of integration are given respectively by the pion mass and a cut-off value; the latter must be applied to avoid divergence problems at infinity. If the right-hand side vanishes, then cot δ_α passes through a resonance. The occurrence of a resonance is determined by a positive sign for n_α and its position by the cut off applied to the integral. The former condition is satisfied only for the $j = \frac{3}{2}$, $T = \frac{3}{2}$ resonance; if we replaced the integral (apart from ω) by the symbol ω_r^{-1} we can then write

$$\frac{4}{3} \frac{f^2}{4\pi} \frac{k^3}{m^2\omega} \cot \delta_{p\,3/2}^3 = 1 - \frac{\omega}{\omega_r}. \quad (14.23)$$

This equation was first derived by Chew and Low (1956); it is known as the effective range formula. Chew and Low showed that the accuracy of the formula may be improved by replacing ω by $\omega^* = \omega + (k^2/2M)$, where M is the nucleon mass; this amendment allows for nucleon recoil. In Fig. 14.5 a plot of

$$\frac{k^3}{m^2\omega^*} \cot \delta^3_{p\,3/2}$$

against ω^* is displayed (Barnes *et al.*, 1960). The data is given in units of $m = 1$; a resonance occurs for $\omega_r \equiv \omega_0^* = 2{\cdot}17\ m = 304$ MeV, and an extrapolation to

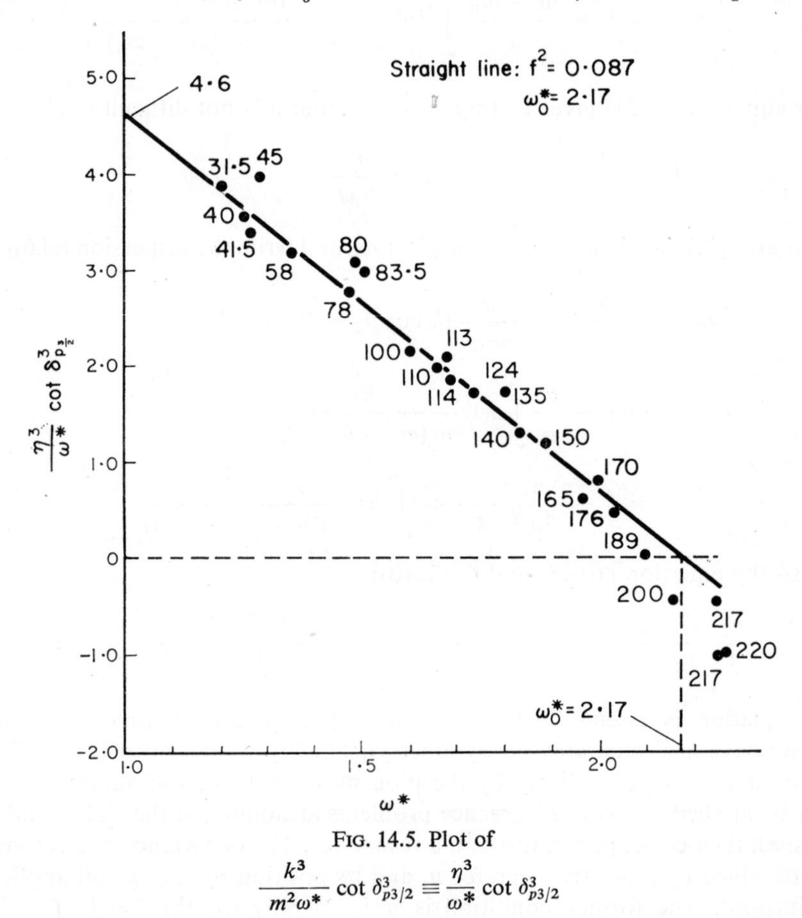

FIG. 14.5. Plot of

$$\frac{k^3}{m^2\omega^*} \cot \delta^3_{p3/2} \equiv \frac{\eta^3}{\omega^*} \cot \delta^3_{p3/2}$$

against ω^* in units of m (the pion mass) = 1 (Barnes *et al.*, 1960). The figures along the line refer to kinetic energies of the pions. The symbol f^2 is equivalent to $f^2/4\pi$ in the text.

$\omega^* = 1$ yields $f^2/4\pi = 0{\cdot}087$. This value for $f^2/4\pi$ is in accord with that found by other techniques.

A more refined version of the above theory may be produced by incorporating in the Born term the contribution from the $\frac{3}{2}$, $\frac{3}{2}$ resonance as a pole amplitude arising from a pseudoparticle (N_1^*) with appropriate properties (Amati and Fubini, 1962). The basic form of (14.23) remains unchanged, however, and so we shall not pursue the matter further. The effective range formula has been reproduced in other ways by application of the method of dispersion relations (Chew *et al.*, 1957).

A considerable improvement in the theoretical scattering amplitude for *s*-waves can be made by assuming the existence of an additional pole term which could

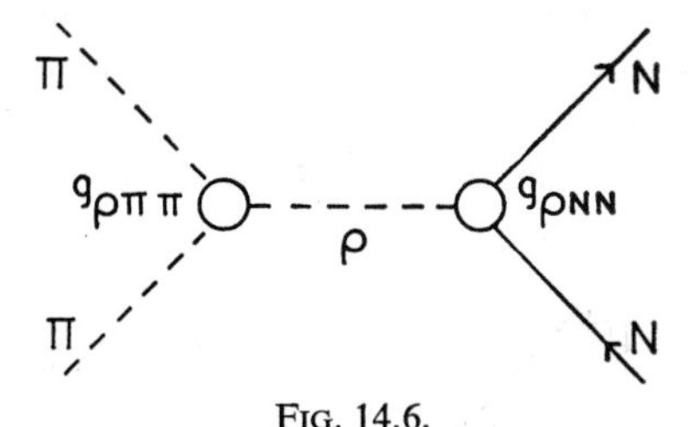

FIG. 14.6.

arise from the exchange of a ϱ-meson (Matthews, 1961; Sakurai, 1960). This process is illustrated in Fig. 14.6; it leads to a pole contribution

$$\frac{g_{\varrho\pi\pi}g_{\varrho NN}}{4\pi}\begin{pmatrix}2\\-1\end{pmatrix}\frac{2\omega}{m_\varrho^2+2\mathbf{k}^2} \tag{14.24}$$

(Matthews, 1961). The upper and lower numbers in the brackets refer to isospin states $\frac{1}{2}$ and $\frac{3}{2}$ respectively (compare (A.7.23), p. 711). If this term dominates the scattering function (by cancellations among the remaining contributions), we obtain

$$f_{s\,1/2}^{2T}=\frac{e^{i\delta}\sin\delta}{k}=a_{s\,1/2}^{2T}=\begin{pmatrix}2\\-1\end{pmatrix}\frac{g_{\varrho\pi\pi}g_{\varrho NN}}{4\pi}\frac{2m_\pi}{m_\varrho^2} \tag{14.25}$$

in the low energy limit. Here $a_{s\,1/2}^{2T}$ represents the *s*-wave scattering length (compare § 6.3(d)). Now

$$m_\varrho\sim 5{\cdot}5m_\pi,\qquad \frac{g_{\varrho\pi\pi}^2}{4\pi}\sim 2$$

where the coupling strength is obtained from the observed width for ϱ decay (13.10). If we assume $g_{\varrho\pi\pi}\sim g_{\varrho NN}$ (including sign), then

$$a_{s\,1/2}^{2T}\sim\begin{pmatrix}2\\-1\end{pmatrix}2\frac{2}{30m_\pi}=\begin{pmatrix}+0{\cdot}2\\-0{\cdot}1\end{pmatrix}\frac{1}{m_\pi}$$

compared with the experimental values

$$a_{s\,1/2}^{2T}=\begin{pmatrix}+0{\cdot}178\\-0{\cdot}087\end{pmatrix}\frac{1}{m_\pi}$$

(Hamilton and Woolcock, 1960).

The inclusion of the pole term due to a ϱ-meson also leads to some improvement in the comparison of theoretical and experimental data for the small p-phase shifts (Matthews, 1961). Its presence has virtually no effect on the $\frac{3}{2}, \frac{3}{2}$ state, however, and so equation (14.23) can be regarded as still valid.

14.1 (d). *The photoproduction of pions at low energies*

The analysis of the data on the photoproduction of pions is more complicated than that of pion–nucleon scattering. This is because different production mechanisms are available in π^+ and π^0-meson production. For example, the process shown in Fig. 14.7 (the 'photoelectric effect') can only arise when the pion possesses electric charge.

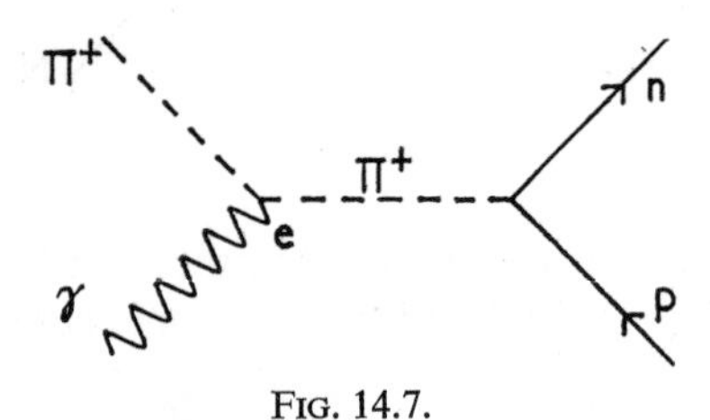

FIG. 14.7.

One manifestation of the different possible mechanisms for π^+ and π^0-meson photoproduction is in the difference in behaviour of the total cross-sections near the threshold (Fig. 14.8). The two cross-sections possess the energy dependence

$$\sigma_T(\pi^+) \propto (E - E_0)^{1/2}, \quad \sigma_T(\pi^0) \propto (E - E_0)^{2\cdot 2} \qquad (14.25)$$

(Bernardini, 1955), where E is the photon energy and E_0 is the threshold energy for pion production. The results are indicative of the production of π^+ and π^0 mesons in s and p-states respectively, since the cross-section is expected to behave like k^{2l+1} at threshold (see remark below), where k is the momentum of the outgoing pion, and since $k^2 \propto (E - E_0)$ near threshold

$$\sigma_T \propto (E - E_0)^{\frac{2l+1}{2}}. \qquad (14.26)$$

A comparison with equation (14.25) then shows that the assignments $l = 0$ and 1 are in accord with the data for π^+ and π^0 photoproduction respectively. The factor k^{2l+1} arises because the matrix element M_{fi} contains a term $j_l(kr)$ for the outgoing pion spherical wave which behaves like k^l at low momenta. Hence $|M_{fi}|^2 \propto k^{2l}$ and the density of states term produces an additional k (compare (7.59)) giving $\sigma_T \propto k^{2l+1}$.

The production of π^+ mesons in s-states near threshold is also apparent in the angular distributions which are nearly flat (see, for example, Walker and Burg, 1962).

It is apparent from the general shape of Fig. 14.8 and the data on the peaks of cross-sections for pion–nucleon scattering, given in the tables of § 13.2, that the photoproduction of both π^+ and π^0 mesons is dominated by the $N_1^*(\frac{3}{2}, \frac{3}{2}^+)$ resonance. This conclusion may be confirmed by examining the angular distributions for the photoproduction of pions, and in particular the data for π^0

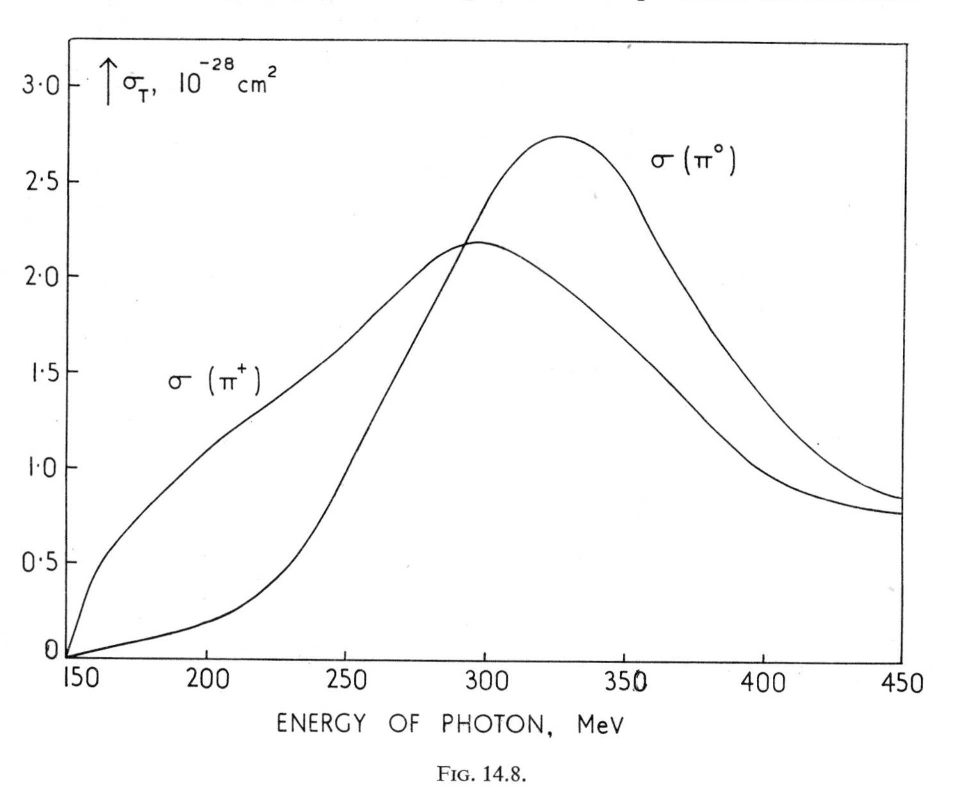

FIG. 14.8.

mesons (the production of π^+-mesons is more complicated because of electrical terms, as we have indicated earlier; a detailed discussion of these terms has been given by Moravcsik (1956, 1957)).

Below photon energies of ~450 MeV the angular distribution of the π° mesons can be analysed in the following manner:

$$\frac{d\sigma}{d\Omega} = A + B\cos\theta + C\cos^2\theta.$$

The coefficients A, B and C are shown in Fig. 14.9 (McDonald, Peterson and Corson, 1957). It can be seen that the B-term is small; this result implies that the s-wave contribution is small since the $\cos\theta$ term would arise from the interference of s and p-waves.

In the region of the resonance, the angular distribution is given by

$$\frac{d\sigma}{d\Omega} \sim (28 - 16 \cos^2 \theta)\, \mu b/\text{ster} \sim 5 \cdot 5\,(5 - 3 \cos^2 \theta)\, \mu b/\text{ster}\,.$$

This result is again indicative of an excited $p_{3/2}$ nucleon state; it may be excited by the absorption by the nucleon of magnetic dipole ($M1$) radiation (compare § 5.4(c)). Let us consider a pure $M1$ excitation, leading to a final state for the

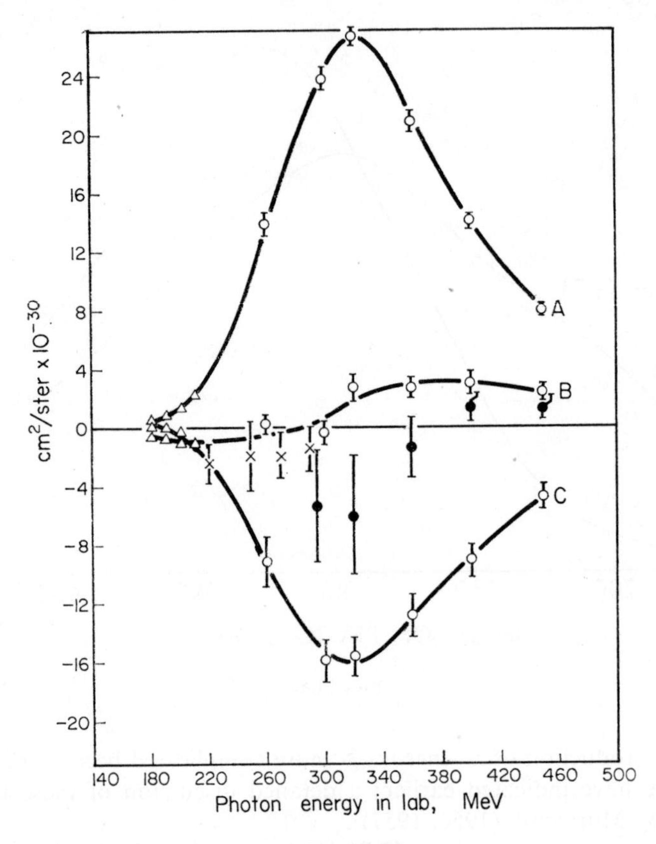

FIG. 14.9.

pion–nucleon system with angular momentum and parity $\frac{3}{2}^+$. A photon has angular momentum components $m_\gamma = \pm 1$ along its direction of motion (§ 5.3(f)), and $M1$ radiation represents total angular momentum $j = 1$ and even parity. We are then able to conserve angular momentum and parity by using the functions given in Table 14.4; in it we shall use the notation $\chi_{1/2}^{+1/2}$ and $\chi_{1/2}^{-1/2}$ for the nucleon spins and we shall make a double application of the Clebsch–Gordan coefficients from Table 5.2 (p. 200). It should be noted that since isospin

TABLE 14.4

Initial state	Final state
γp	$p\pi^0$
$\gamma\ M1\ (j = 1^+) \searrow$ $\nearrow j = \frac{3}{2}^+$ $p\ s = \frac{1}{2} \nearrow$	$\nearrow l = 1$ $j = \frac{3}{2}^+$ $\searrow s = \frac{1}{2}$

Initial state					Final state
			γp system		$p\pi^0$
m_γ	m_p	M	C	$\mathscr{Y}^M_{3/2}$	
1	$\frac{1}{2}$	$\frac{3}{2}$	1	$\mathscr{Y}^{3/2}_{3/2}$	$\mathscr{Y}^{3/2}_{3/2} = Y^1_1\chi^{1/2}_{1/2}$
1	$-\frac{1}{2}$	$\frac{1}{2}$	$\frac{1}{\sqrt{3}}$	$\mathscr{Y}^{1/2}_{3/2}$	$\frac{1}{\sqrt{3}}\mathscr{Y}^{1/2}_{3/2} = \frac{1}{\sqrt{3}}\left(\sqrt{\frac{2}{3}}\,Y^0_1\chi^{1/2} + \frac{1}{\sqrt{3}}\,Y^1_1\chi^{-1/2}_{1/2}\right)$

m_γ = photon angular momentum projection along axis of quantisation (the photon momentum vector); m_p = proton spin projection; M = total angular momentum projection; C = Clebsch–Gordan coefficient.

is not a good quantum number in electromagnetic processes, we do not have to consider isospin channels as in pion–nucleon scattering.

Strictly speaking we should also have included the states arising from $m_\gamma = -1$ in our analysis, but, since the final results are symmetrical for unpolarised systems, the inclusion of these states is an unneccessary complication of our present problem. Since

$$Y^1_1 = -\sqrt{\left(\frac{3}{8\pi}\right)}\sin\theta e^{i\theta}, \quad Y^0_1 = \sqrt{\left(\frac{3}{4\pi}\right)}\cos\theta \qquad \text{(A.7.10)}$$

the angular distribution is given by

$$\frac{d\sigma}{d\Omega} \propto \frac{3}{8\pi}\left[\sin^2\theta + \tfrac{1}{3}\left(\tfrac{2}{3}\times 2\cos^2\theta + \tfrac{1}{3}\sin^2\theta\right)\right] \propto 5 - 3\cos^2(\theta). \quad (14.27)$$

The angular distribution for other pure states may be evaluated in a similar manner and the results are summarised in Table 14.5 (Feld, 1953).

From this table it can be seen that an angular distribution $5 - 3\cos^2\theta$ can be associated with both the $\frac{3}{2}^+$ and $\frac{3}{2}^-$ states; however the latter involves d-waves and these can make only small contributions at the resonance (compare Table 14.1). Thus the measured angular distribution suggests that the resonance is dominated by a state with $j = \frac{3}{2}$ and even parity, arising from $M1$ absorption.

TABLE 14.5

Photon absorbed	Total j and parity	l_π	Pion angular distribution
$E1$ (electric dipole)	$\frac{1}{2}^-$	0	isotropic
	$\frac{3}{2}^-$	2	$5 - 3\cos^2\theta$
$M1$ (magnetic dipole)	$\frac{1}{2}^+$	1	isotropic
	$\frac{3}{2}^+$	1	$5 - 3\cos^2\theta$
$E2$ (electric quadrupole)	$\frac{3}{2}^+$	1	$1 + \cos^2\theta$
	$\frac{5}{2}^+$	3	$1 + 6\cos^2\theta - 5\cos^4\theta$

Table 14.5 involves transitions between pure states. In practice mixed states occur and so relative phases must also be considered. These phases may be found by exploiting the unitarity of the S-matrix operator and the invariance of the S-matrix element under time reversal, in order to establish a relationship between pion scattering and photoproduction (Watson, 1954). The result is found to be that the complex parts of the contribution from each multipole are related to the phase factors $e^{i\delta}$ in pion–nucleon scattering (δ is the phase shift). The argument has been outlined in § 7.5(b) (see (7.103) in particular). The technique has been extended by Peierls (1960) to include the situation when inelastic channels open in the scattering process.

We mention, finally, the interpretation of the photoproduction data in terms of the static model. We have already stated that the cross-sections for the production of π^+ and π^0-mesons differ considerably near threshold. Part of the difference can be explained by the photoelectric term (Fig. 14.7); another factor arises from the form for the static interaction. The Hamiltonian (14.16) contained the expression

$$(\boldsymbol{\sigma} \cdot \boldsymbol{\nabla})\,(\boldsymbol{\tau} \cdot \boldsymbol{\varphi}).$$

In the presence of the electromagnetic field this must be altered to preserve gauge invariance (§ 9.3(c))

$$\boldsymbol{\nabla} \to \boldsymbol{\nabla} + ie\mathbf{A}$$

where the operator $\mathbf{A}$ acts only on the charged pion field, and so a term

$$\sqrt{2}\, ie \frac{f}{m}\, \boldsymbol{\sigma} \cdot \mathbf{A}(\tau\varphi)_{\text{charged}} + \text{h.c.} \tag{14.28}$$

must be added to the Hamiltonian density; the factor $\sqrt{2}$ arises from the method of normalisation of the charged pion field operator (compare § 9.4(c)).

Now the presence of the gradient operator in (14.16) caused the pion–nucleon interaction to be in the p-state. The interaction associated with (14.28), however, can be represented diagrammatically as in Fig. 14.10; it is known as the *catastrophic effect* and gives rise to s-state production of pions, since there is no term $\nabla\varphi$ in (14.28) and φ behaves like r^l at short distances (compare remark after (14.16)). The problem of photoproduction can of course, be treated covariantly rather than in the static limit; an example of a covariant treatment may be found in the paper of Gourdin and Salin (1963).

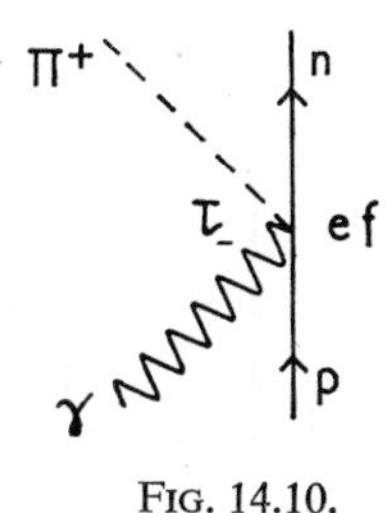

FIG. 14.10.

14.1 (e). *The higher nucleon resonances*

In this section we shall be concerned with the quantum numbers for the resonances N_2^*, N_3^*, and N_4^* given in Tables 13.1 and 13.2. We have remarked previously that they all share the baryon quantum number $B = 1$, and that the N_2^* and N_3^* states have $T = \frac{1}{2}$, whilst N_4^* has $T = \frac{3}{2}$. The assignment of spin and parity is more complicated than in the case of the N_1^* resonance since interference effects from many angular momentum states occur at the higher energies.

Let us first consider the isobar N_2^*. The angular distribution of the π^0-mesons in the process $\gamma + p \to p + \pi^0$ at $E_\gamma = 750$ MeV is roughly given by

$$\frac{d\sigma}{d\Omega} \propto 5 - 3\cos^2\theta \tag{14.29}$$

(Wilson, 1958; Peierls, 1960). According to Table 14.5 this distribution is compatible with a state with $j = \frac{3}{2}^-$ (electric dipole absorption) or $j = \frac{3}{2}^+$ (magnetic dipole), and so it could be $P_{3/2}$ or $D_{3/2}$. The choice between the states may be made by measuring the polarisation of the recoil proton at 90^0 in the *c*-system as suggested by Sakurai (1958b; see also Peierls, 1960). The argument may be given briefly as follows. If the state is mainly $D_{3/2}$, it will interfere with the tail of the contribution from the $P_{3/2}$ resonance at 300 MeV and hence cause polarisation. If, on the other hand, it is also a $P_{3/2}$ state no interference and hence no polarisation can occur. The experimental results (Fig. 14.11) show strong polarisation and therefore indicate N_2^* is a $D_{3/2}$ state. The graph is taken from the paper of Mencuccini, Querzoli and Salvini (1961); the negative sign implies that the polarisation is opposite to the vector $\mathbf{q} \times \mathbf{k}$ where $\mathbf{q}$ and $\mathbf{k}$ are the momenta of the incoming photon and outgoing pion respectively.

The data on the photoproduction of π^+-mesons is complicated by the photoelectric term (§ 14.1 (d)) and so will not be discussed.

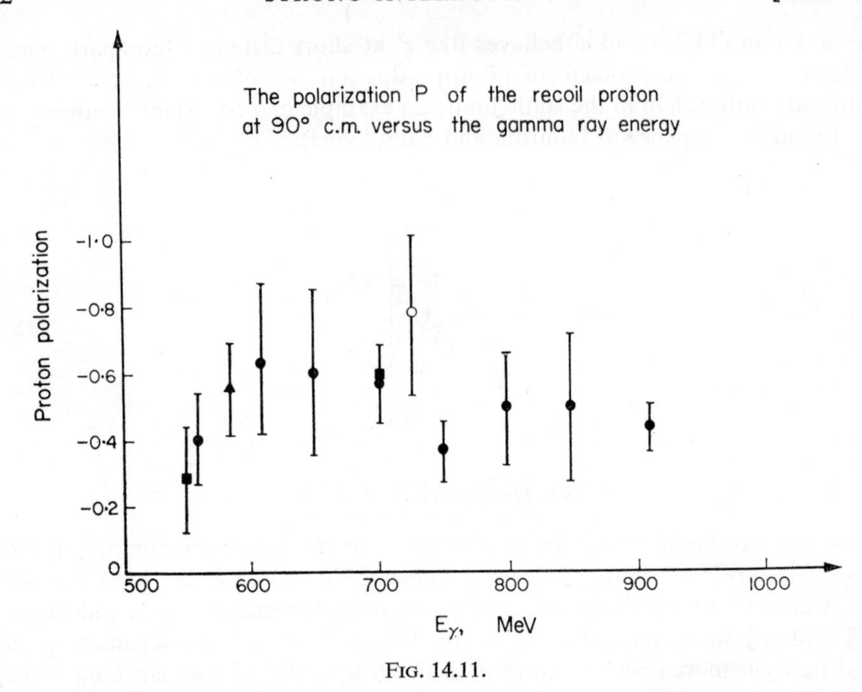

FIG. 14.11.

We next consider π–p scattering. The angular distributions in elastic π–p scattering have been measured by a number of groups (see, for example, Helland *et al.*, 1963). The data can be fitted by expressions of the following type:

$$\frac{d\sigma_{sc}}{d\Omega} = \sum_{n=0}^{n_{max}} a_n \cos^n \theta \tag{14.30}$$

where θ is the angle of scattering in the c-system. The behaviour of the coefficients a_n can be related to angular momentum states since

$$\frac{d\sigma_{sc}}{d\Omega} = |g(\theta)|^2 + |h(\theta)|^2 = \sum_{n=0}^{n_{max}} a_n \cos^n \theta$$

where g and h were given in (14.2)

$$g(\theta) = \sum_{l=0}^{\infty} [(l+1) f_{l+} + l f_{l-}] P_l^0 (\cos \theta)$$

$$h(\theta) = -\sum_{l=0}^{\infty} (f_{l+} - f_{l-}) P_l^1 (\cos \theta).$$

Thus

$$a_n = \operatorname{Re} \sum_{i \leq j} b_{ijn} f_i f_j^* \tag{14.31}$$

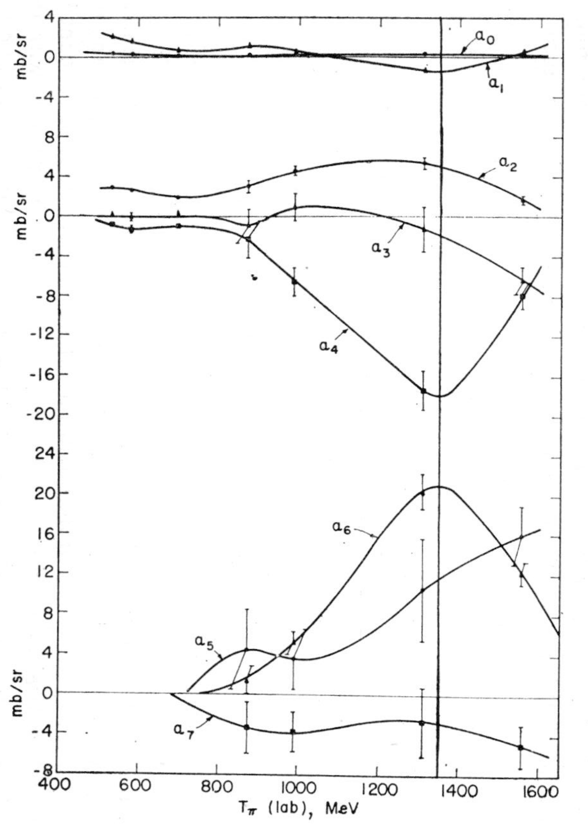

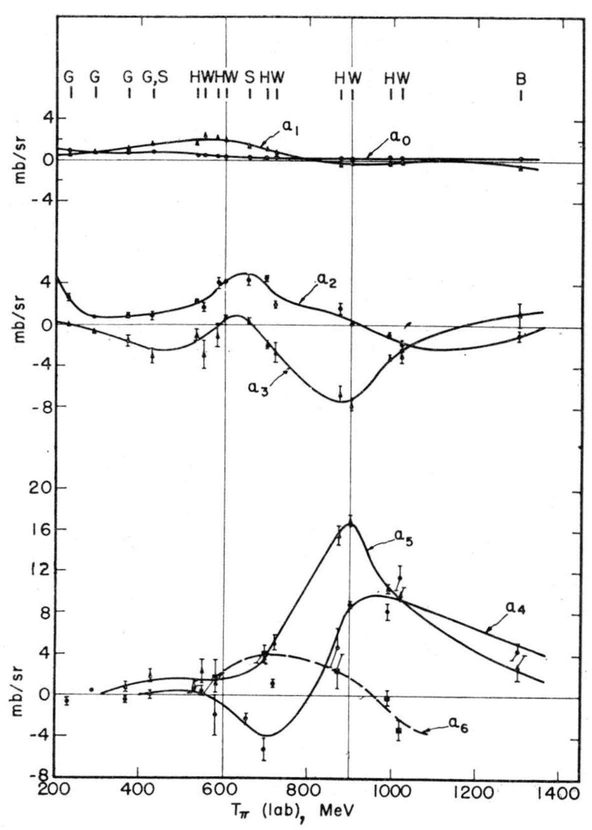

FIG. 14.12. Coefficients a_n for πp scattering versus pion kinetic energy in the laboratory; left- and right-hand diagrams represent π^+ and π^- data respectively (Helland *et al.*, 1963).

where the coefficients b_{ijn} can be obtained by straightforward (but tedious) expansion. It is apparent that $n_{max} = 2l_{max}$, where l_{max} represents the largest orbital angular momentum state which makes a significant contribution.

Let us now consider the N_2^* resonance. Helland *et al.* remark that the values of the coefficients a_n do not indicate the prominence of any single partial wave state (Fig. 14.12). Measurements of the polarisation of the recoil protons have yielded data which are consistent with assignments of either $\frac{3}{2}^+$ or $\frac{3}{2}^-$ to this resonance (Moyer, private communication).

The evidence from pion–nucleon scattering is slightly less ambiguous about the state N_3^*; Fig. 14.12 shows that at 900 MeV large coefficients a_3, a_4 and a_5 are required to fit the data, but no term involving $\cos^6 \theta$. Let us consider the implications arising from the fact that a_5 is large and $a_6 \sim 0$. The appropriate coefficients can be shown to be

$$a_5 = 105\,\mathrm{Re} f_{2-} f_{3+}^* + 112{\cdot}5\,\mathrm{Re} f_{2+} f_{3-}^* + 45\,\mathrm{Re} f_{2+} f_{3+}^*$$

$$a_6 = 43{\cdot}75\,|f_{3+}|^2 + 262{\cdot}5\,\mathrm{Re} f_{3-} f_{3+}^*$$

and so the conditions a_5 large, $a_6 \sim 0$ can be fulfilled either by f_{3-} large and $f_{3+} \sim 0$ or a fortuitous combination of signs between the real and imaginary parts of the two functions. The more obvious choice is that f_{3-} is large; that is, the isobar N_3^* is an $F_{5/2}$ state. Thus the parity of this state is opposite to that which is favoured for the N_2^* state ($D_{3/2}$). This conclusion is in agreement with the observation of large polarisation of the recoil proton in the process $\gamma + p \rightarrow p + \pi^0$ at 900 MeV (Fig. 14.11). A detailed analysis by Peierls (1960) of the photoproduction data at 1000 MeV also favours the assignment of $F_{5/2}$ to this resonance.

TABLE 14.6

Peak	Energy (MeV)	Width (MeV)	T	L_j^{parity}
N_1^*	1238	90	$\frac{3}{2}$	$P_{3/2}^+$
N_2^*	1510	60	$\frac{1}{2}$	$D_{3/2}^-$
N_3^*	1680	100	$\frac{1}{2}$	$F_{5/2}^+$
N_4^*	1900	200	$\frac{3}{2}$	$F_{7/2}^+$, $G_{7/2}^-$?

Let us, finally, consider the state N_4^*; the smallness of the coefficient a_7 indicates that N_4^* is unlikely to have angular momentum $j = \frac{9}{2}$ or larger, and the large value for a_6 suggests that the quantum states $F_{7/2}$ or $G_{7/2}$ may be prominent. We may therefore summarise the data on the peaks as in Table 14.6, with the understanding that only the quantum numbers for N_1^* are unambiguously known.

14.2. ELASTIC NUCLEON–NUCLEON SCATTERING

14.2(a). *Polarisation and related parameters*

The analysis of the experimental data on nucleon–nucleon scattering is a much more difficult task than for pion–nucleon scattering. From a field theoretical point of view the problem is complicated by virtual pion–nucleon and pion–pion processes. Even a phenomenological analysis is more difficult since both particles possess spin, and, since both particles are massive, many orbital angular momentum states can participate at comparatively low energies. The problem is further complicated since transitions involving changes in l can occur without violating total angular momentum and parity conservation, for example $^3P_2 \leftrightarrow {}^3F_2$.

Let us consider how many independent parameters are required to make a complete phenomenological analysis of proton–proton scattering at a given energy (Chamberlain *et al.*, 1957). Let us assume that orbital angular momentum states up to some value l_{max} are present. The Pauli exclusion principle restricts p–p interactions to singlet and triplet states for even and odd values of l respectively (§ 9.2(g)); thus there are $(2l_{max} + 1)$ states if l_{max} is even and $(2l_{max} + 2)$ if l_{max} is odd.† In addition mixing parameters must be introduced to allow for transitions of the type $^3P_2 \leftrightarrow {}^3F_2$ and there are $\frac{1}{2}(l_{max} - 2)$ or $\frac{1}{2}(l_{max} - 1)$ of these functions required respectively if l_{max} is even or odd. Thus a total of $\frac{5}{2}l_{max}$ parameters are required if l_{max} is even, and $\frac{1}{2}(5l_{max} + 3)$ if it is odd.

Now consider a measurement of the angular distribution in a p–p scattering experiment. Since the particles are identical the angular distribution must be symmetric about 90° in the c-system and can be represented as

$$\frac{d\sigma_{sc}}{d\Omega} = \sum_{n=0}^{n=l_{max}} a_n \cos^{2n} \theta \qquad (14.32)$$

and so there are only $(l_{max} + 1)$ coefficients a_n. This number is clearly insufficient to fix the $\frac{5}{2}l_{max} + \left(\begin{smallmatrix}0\\3/2\end{smallmatrix}\right)$ parameters mentioned above, and so more sophisticated measurements must be undertaken. The information which may be obtained from additional experiments has been given by a number of workers (see, for example, Chamberlain *et al.*, 1957; Puzikov, Ryndin and Smorodinsky, 1957); for example polarisation measurements yield an additional l_{max} independent coefficients, and it can be shown that it is possible to overdetermine the $\frac{5}{2}l_{max} + \left(\begin{smallmatrix}0\\3/2\end{smallmatrix}\right)$ parameters (this requirement is essential in any case, if only because of the large errors associated with some of the experiments). The phenomenological analysis of the n–p data is also capable (in principle) of yielding a unique solution.

Let us now consider the experimental data. Angular distributions for elastic scattering are shown in Figs. 14.13 and 14.14 (Macgregor, Moravcsik and Stapp,

† These numbers are easily understood by writing out the first few terms for the angular momentum states $^1S_0$, ^3P_j, $^1D_2$, ^3F_j,...

1960). Measurements using hydrogen and deuterium have shown that n–n scattering is essentially equivalent to p–p scattering (Fig. 14.13), and p–n to n–p scattering (Dzhelepov *et al.*, 1956). Thus nucleon–nucleon forces are charge symmetric.

One interesting feature is the remarkable flatness of the proton–proton differential cross-sections up to 400 MeV. Although this behaviour suggests s-wave

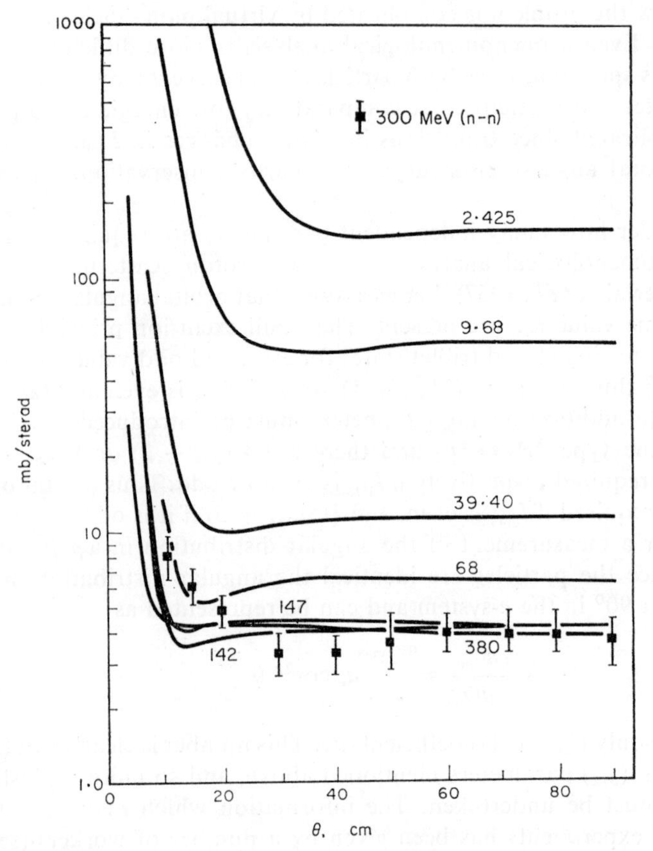

FIG. 14.13. Cross-sections for elastic scattering in proton–proton collisions (MacGregor, Moravcsik and Stapp, 1960).

scattering the explanation is not tenable above 200 MeV since the maximum permissible s-wave cross-section is smaller than that observed experimentally. In fact it is apparent from the n–p data that many partial waves contribute – the cross-section at 27 MeV is proportional to $1 + \frac{1}{3}\cos^2\theta$ thus indicating that a strong p-wave contribution has already set in.

Now let us examine the polarisation data. The degree of polarisation in nucleon–nucleon scattering is determined by a double scattering experiment,

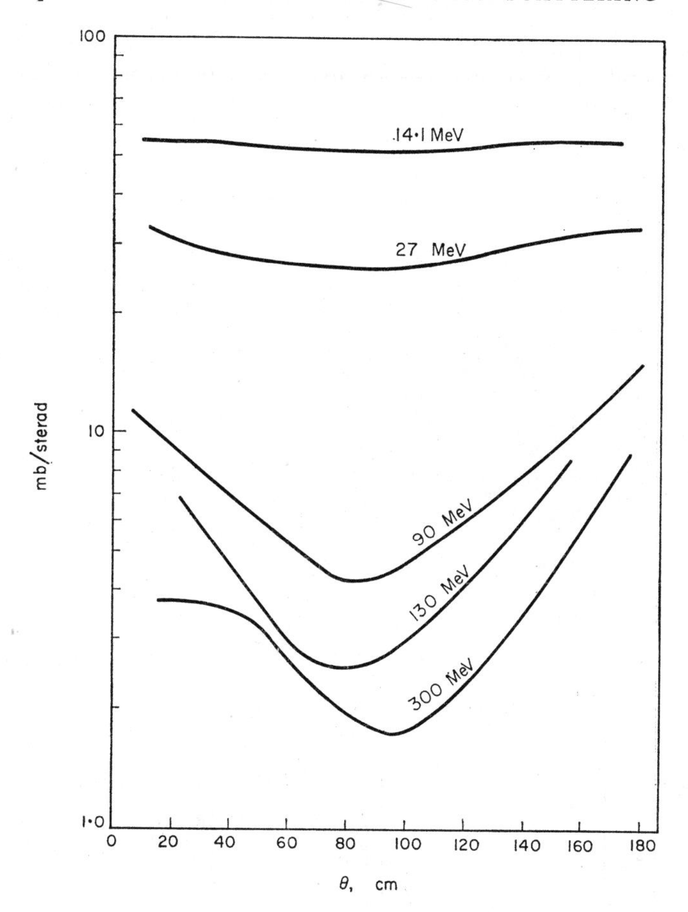

FIG. 14.14. Cross-sections for elastic scattering in neutron–proton collisions (MacGregor Moravcsik and Stapp, 1960).

whose principles we have discussed in § 6.2(f). If a beam of particles, which acquires polarisation P_1 in an initial collision, strikes an unpolarised target, the angular distribution of the particles following the second collision is given by (6.75)

$$I_2(\theta, \varphi) = I_{02}(1 + P_1 P_2(\theta)\, \mathbf{n}_1 \cdot \mathbf{n}_2) = I_{02}(1 + P_1 P_2(\theta) \cos \varphi_2) \quad (14.33)$$

where I_{02} represents the angular distribution if $P_1 = 0$, and $\mathbf{n}_1$ and $\mathbf{n}_2$ represent normals to the first and second scattering planes. This equation leads to a left–right asymmetry as given in (6.76)

$$e_2(\theta) = \frac{I_2(\theta, 0) - I_2(\theta, \pi)}{I_2(\theta, 0) + I_2(\theta, \pi)} = P_1 P_2(\theta).$$

Thus a knowledge of P_1 and e_2 would fix P_2. The principle of the measurement of polarisation is therefore as follows (see, for example, Chamberlain *et al.*, 1957). A proton beam is scattered from a target, say beryllium, giving polarisation P_1, and then from a hydrogen target (P_2) and the asymmetry e_2 is measured; the

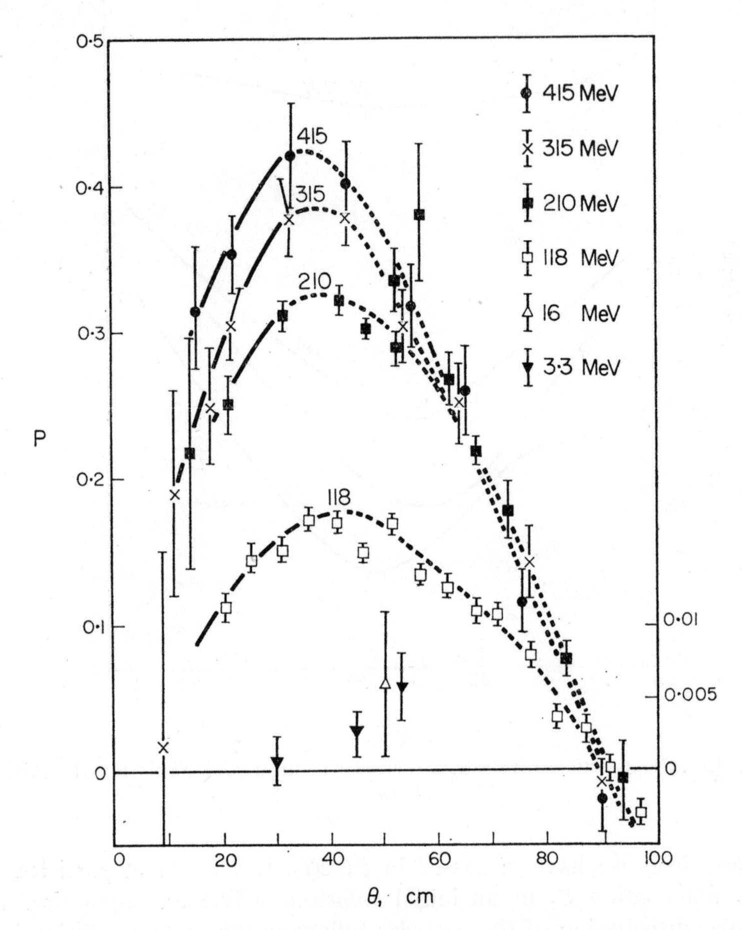

FIG. 14.15. Polarisation in *pp* scattering as a function of energy (MacGregor, Moravcsik and Stapp, 1960).

experiment is then repeated by substituting a second beryllium target for the hydrogen one, yielding an asymmetry $e_1 = P_1P_1$ and hence P_2 is obtained (the absolute sign of P_2 is a matter of convention). The results for the polarisation occurring in *p–p* scattering experiments are displayed in Fig. 14.15 (Macgregor, Moravcsik and Stapp, 1960). The polarisation vanishes at 90° in *p–p* scattering since the requirements of symmetry imply that $P(\theta) = -P(\pi - \theta)$ for identical particles.

Now let us consider triple scattering experiments; these examine relative spin orientations in an experiment. Let us assume we have a beam of particles with polarisation $\langle\bar{\sigma}_1\rangle = P_1\mathbf{n}_1$ striking a target. The polarisation of the scattered beam $\langle\bar{\sigma}_2\rangle$ may be described in terms of its components in three mutually perpendicular directions; these were chosen by Wolfenstein (1954) as

$$\mathbf{n}_2 = \frac{\hat{\mathbf{k}}_2 \times \hat{\mathbf{k}}_2'}{|\hat{\mathbf{k}}_2 \times \hat{\mathbf{k}}_2'|}, \quad \mathbf{s}_2 = \mathbf{n}_2 \times \hat{\mathbf{k}}_2', \quad \hat{\mathbf{k}}_2' \tag{14.34}$$

where $\hat{\mathbf{k}}_2$ and $\hat{\mathbf{k}}_2'$ respectively represent unit vectors along the directions of the particle before and after scattering at the second target. The behaviour of the scattered beam can then be represented as

$$\begin{aligned} I_2(\theta, \varphi)\langle\bar{\sigma}_2\rangle = I_{02}\{&[P_2 + D\langle\bar{\sigma}_1\rangle \cdot \mathbf{n}_2]\,\mathbf{n}_2 \\ &+ [A\langle\bar{\sigma}_1\rangle \cdot \hat{\mathbf{k}}_2 + R\langle\bar{\sigma}_1\rangle \cdot (\mathbf{n}_2 \times \hat{\mathbf{k}}_2)]\,\mathbf{s}_2 \\ &+ [A'\langle\bar{\sigma}_1\rangle \cdot \hat{\mathbf{k}}_2 + R'\langle\bar{\sigma}_1\rangle \cdot (\mathbf{n}_2 \times \hat{\mathbf{k}}_2)]\,\hat{\mathbf{k}}_2'\} \end{aligned} \tag{14.35}$$

where the coefficients P_2, D,... are scalar functions of angle and energy. This equation is constructed by remembering that $\langle\bar{\sigma}_2\rangle$ is an axial vector and then inserting on the right-hand side all the independent axial vector terms which can be constructed from $\langle\bar{\sigma}_1\rangle$, $\mathbf{n}_2$, $\hat{\mathbf{k}}_2$ and $\hat{\mathbf{k}}_2'$. If the initial beam is unpolarised ($\langle\bar{\sigma}_1\rangle = 0$) the equation reduces to $\langle\bar{\sigma}_2\rangle = P_2\mathbf{n}_2$. It should be noted that not all of the coefficients are independent. Using time reversal arguments Wolfenstein showed that $R' = -A$ for an infinitely heavy target and that

$$\frac{A + R'}{A' - R} = \tan\frac{\theta}{2} \tag{14.36}$$

for the nonrelativistic collision of two particles of equal mass.

In order to determine the coefficients D, A, R, A' and R' a third scattering is required; this scattering measures the projection of $\langle\bar{\sigma}_2\rangle$ in a direction $\mathbf{n}_3$ which is normal to the third scattering plane. The differential cross-section following this scattering may be taken from equations (6.74) and (6.75)

$$I_3(\theta_3, \varphi_3) = I_{03}(1 + \langle\bar{\sigma}_2\rangle \cdot P_3(\theta)\,\mathbf{n}_3) \tag{14.37}$$

where we have made appropriate alternations to the subscripts; P_3 may be regarded as the analysing power of the final scattering experiment.

Since polarisation along the direction of motion cannot be detected by a single scattering the third scattering is chosen to measure either $\langle\bar{\sigma}_2\rangle \cdot \mathbf{n}_2$ or $\langle\bar{\sigma}_2\rangle \cdot \mathbf{s}_2$ (that is $\mathbf{n}_3$ parallel to $\mathbf{n}_2$ or $\mathbf{s}_2$). The corresponding asymmetries are written as e_{3n} and e_{3s}, where

$$e_3 = \frac{I_3(+) - I_3(-)}{I_3(+) + I_3(-)} = \langle\bar{\sigma}_2\rangle \cdot P_3\mathbf{n}_3 \tag{14.38}$$

where (+) and (−) correspond to $\cos\varphi_3 = \pm 1$ ($\mathbf{n}_3$ parallel to $\pm\mathbf{n}_2$ or $\pm\mathbf{s}_2$). If we combine this equation with equations (14.35) and (14.33) we find for $\mathbf{n}_3$ parallel to $\mathbf{n}_2$ and $\mathbf{s}_2$

$$e_{3n} = P_3\langle\bar{\boldsymbol{\sigma}}_2\rangle\cdot\mathbf{n}_2 = P_3\frac{(P_2 + DP_1\mathbf{n}_1\cdot\mathbf{n}_2)}{(1 + P_1P_2\mathbf{n}_1\cdot\mathbf{n}_2)} \tag{14.39}$$

$$e_{3s} = P_3\langle\bar{\boldsymbol{\sigma}}_2\rangle\cdot\mathbf{s}_2 = P_3\frac{RP_1\mathbf{n}_1\cdot\mathbf{n}_2\times\hat{\mathbf{k}}_2}{(1 + P_1P_2\mathbf{n}_1\cdot\mathbf{n}_2)}.$$

The parameter A does not appear in the second equation since

$$\langle\bar{\boldsymbol{\sigma}}_1\rangle\cdot\hat{\mathbf{k}}_1' \equiv \langle\bar{\boldsymbol{\sigma}}_1\rangle\cdot\hat{\mathbf{k}}_2 = 0$$

after the first scattering (compare § 6.2(f)). The parameter D is normally measured for the condition $\mathbf{n}_1\cdot\mathbf{n}_2 = \cos\varphi_2 = +1$; that is all three scattering planes are parallel to each other, R is measured under the condition $\mathbf{n}_1\cdot\mathbf{n}_2\times\hat{\mathbf{k}}_2 = \sin\varphi_2 = 1$, that is the scattering planes are perpendicular. The polarisations P_1, P_2 and P_3 are determined in an auxiliary series of experiments which measure left–right asymmetries in double scatterings.

The parameter D has a simple physical interpretation. If we consider the situation where $\cos\varphi_2 = 1$ in (14.39) and assume that the beam is completely polarised in its first scattering ($P_1 = 1$) then

$$\langle\bar{\boldsymbol{\sigma}}_2\rangle = \mathbf{n}_2\frac{P_2 + D}{1 + P_2}.$$

Thus D represents the degree of depolarisation since $D = 1$ implies there is no depolarisation in the second scattering.

Since $\langle\bar{\boldsymbol{\sigma}}_1\rangle\cdot\hat{\mathbf{k}}_2 = 0$ after the first scattering, the parameter A is not detectable in a straightforward triple scattering experiment; it may be measured by inserting a magnetic field between the first and second scatterers. The direction of the field must be perpendicular to $\mathbf{n}_1$ and $\hat{\mathbf{k}}_1'(\equiv\hat{\mathbf{k}}_2)$ and the spin may then be given a component along its direction of motion. The final scattering plane is then set perpendicular to the second scattering plane as in the R experiment. In a similar manner the parameter R' may be measured by placing a magnetic field between second and third scatterers, whilst A' requires two fields (neither parameter has, as yet, been measured).

Further information on the scattering process can be obtained by studying the spin correlations of the two participating nucleons. This can be done by using polarised targets or by studying spin correlations of both outgoing particles. The problem has been studied by Oehme (1955) and by Stapp (1955, 1956). In the correlation experiment the dyadic $\langle\bar{\boldsymbol{\sigma}}\bar{\boldsymbol{\sigma}}_t\rangle$ is used where $\bar{\boldsymbol{\sigma}}_t$ refers to the spin of the target particle. If we work in the laboratory system and use nonrelativistic kinematics, we may use (14.34) to introduce the notation $\hat{\mathbf{k}}' = \mathbf{s}_t = \hat{\mathbf{P}}$ and

$\hat{\mathbf{k}}'_t = -\mathbf{s} = -\hat{\mathbf{K}}$ where the subscripts t refer to the target particle. The dyadic can then be specified in terms of its components $\mathbf{nn}$, $\hat{\mathbf{K}}\hat{\mathbf{P}}$, $\hat{\mathbf{K}}\mathbf{n}$ and $\mathbf{n}\hat{\mathbf{P}}$, and the most general expression is given by

$$\begin{aligned} I_2\langle\bar{\sigma}\bar{\sigma}_t\rangle_2 = I_{02}\{&(C_{nn} + C^P_{nn}\langle\bar{\sigma}_1\rangle\cdot\mathbf{n}_2)\,\mathbf{n}_2\mathbf{n}_2 \\ &+ (C_{KP} + C^P_{KP}\langle\bar{\sigma}_1\rangle\cdot\mathbf{n}_2)\,\hat{\mathbf{K}}_2\hat{\mathbf{P}}_2 \\ &+ C^P_{nK}\langle\bar{\boldsymbol{\sigma}}_1\rangle\cdot(\mathbf{n}_2\times\hat{\mathbf{k}}_2)\,\hat{\mathbf{K}}_2\mathbf{n}_2 \\ &+ C^P_{nP}\langle\bar{\boldsymbol{\sigma}}_1\rangle\cdot(\mathbf{n}_2\times\hat{\mathbf{k}}_2)\,\mathbf{n}_2\mathbf{P}_2\}. \end{aligned} \tag{14.40}$$

In the above expression the terms with superscript P represent the correlation parameters, which can be measured when the incident beam is polarised; in practice only the coefficients for an unpolarised beam have been measured (C_{nn} and C_{KP}). The term C_{nn} is measured by first scattering nucleons in a nucleon–nucleon collision and then observing the subsequent behaviour of both particles in further scatterings; all scatterings occur in the same plane and it may be shown that

$$C_{nn} = \frac{(L, L) + (R, R) - (L, R) - (R, L)}{(L, L) + (R, R) + (L, R) + (R, L)}\,\frac{1}{P_f P_{ft}}$$

where (L, L) implies that both second scatterings are to the left, and P_f and P_{ft} refer to the analysing powers of the final scattering systems. In the C_{KP} experiments the final scattering planes are perpendicular to the initial plane.

Systematic measurements of D, R and A for p–p scattering over wide angular ranges have been made at energies between 95 and 315 MeV, and also at 635 MeV; measurements on C_{nn} and C_{KP} at fixed angles (mainly 90° in the c-system) have also been made (references may be found in the excellent article of Macgregor, Moravcsik and Stapp 1960; later references may be found in the paper of Gotow, Lobkowicz and Heer, 1962). Data on n–p scattering is more scanty.

14.2(b). *Phase shifts for elastic nucleon–nucleon scattering*

The evaluation of phase shifts for nucleon–nucleon scattering from the observable parameters $d\sigma_{sc}/d\Omega$, P, D,... is a lengthy process.

As an intermediate step an invariant scattering operator is introduced. One possible form was introduced by Wolfenstein and Ashkin (1952) for p–p scattering

$$\begin{aligned} M(\mathbf{k}, \mathbf{k}') = BS &+ C(\boldsymbol{\sigma}_1 + \boldsymbol{\sigma}_2)\cdot\mathbf{n} \\ &+ \tfrac{1}{2}G(\boldsymbol{\sigma}_1\cdot\mathbf{q}\boldsymbol{\sigma}_2\cdot\mathbf{q} + \boldsymbol{\sigma}_1\cdot\mathbf{p}\boldsymbol{\sigma}_2\cdot\mathbf{p})\,T \\ &+ \tfrac{1}{2}H(\boldsymbol{\sigma}_1\cdot\mathbf{q}\boldsymbol{\sigma}_2\cdot\mathbf{q} - \boldsymbol{\sigma}_1\cdot\mathbf{p}\boldsymbol{\sigma}_2\cdot\mathbf{p})\,T \\ &+ N(\boldsymbol{\sigma}_1\cdot\mathbf{n}\boldsymbol{\sigma}_2\cdot\mathbf{n})\,T \end{aligned} \tag{14.41}$$

$$\mathbf{n} = \frac{\mathbf{k} \times \mathbf{k}'}{|\mathbf{k} \times \mathbf{k}'|}, \quad \mathbf{p} = \frac{\mathbf{k} + \mathbf{k}'}{|\mathbf{k} + \mathbf{k}'|}, \quad \mathbf{q} = \frac{\mathbf{k} - \mathbf{k}'}{\mathbf{k} + \mathbf{k}'}$$

where the subscripts 1 and 2 refer to the two particles and not to the scattering event. S and T represent projection operators for the singlet and triplet states and B, C, G, H and N are functions of θ and energy. Equivalent expressions have been given by Stapp (Macgregor, Moravcsik and Stapp, 1960) and by Puzikov, Ryndin and Smorodinsky (1957).

The M-operator of equation (14.41) is the equivalent of that introduced in (6.62) for the scattering of a particle of spin $\frac{1}{2}$ from a spin 0 target. It is apparent from equations (6.71) and (6.70) that

$$\frac{d\sigma_{sc}}{d\Omega} = \tfrac{1}{4}\,\mathrm{tr}\; MM^{\dagger}$$

$$P\frac{d\sigma_{sc}}{d\Omega} = \tfrac{1}{4}\,\mathrm{tr}\; MM^{\dagger}\; \boldsymbol{\sigma}\cdot\mathbf{n}$$

and the other observables may be also related to M through (6.58) – for example

$$D\frac{d\sigma_{sc}}{d\Omega} = \tfrac{1}{4}\,\mathrm{tr}\; M\boldsymbol{\sigma}\cdot\mathbf{n}M^{\dagger}\boldsymbol{\sigma}\cdot\mathbf{n}$$

and upon evaluation of the traces, relations between B, C, G, H and N and the observables are obtained.

The complete programme for the relation of the observables to phase shifts may be summarised as follows:

(1) observables $d\sigma_{sc}/d\Omega$, P, D,... to coefficients of M,

(2) coefficients to matrix elements in spin space,

(3) matrix elements to phase shifts.

The programme is thus a more sophisticated version of that discussed in § 14.1 for pion–nucleon scattering. Detailed discussion of it, together with appropriate tables, may be found in the article by Macgregor, Moravcsik and Stapp (1960); see also Stapp (1955).

The programme described above is most easily carried out in reverse, that is by starting with a set of phase shifts and calculating the observables. Despite the fact that the phase shifts are, in principle, overdetermined, several acceptable phase shifts can be found at the higher energies. All but two sets have been discarded by

(1) comparing with data for the process

$$p + p \rightarrow d + \pi^{+}$$

where the parity of the pion and spin of the deuteron limit the p–p system to certain angular momentum states,

(2) requiring a sensible behaviour of the phase shifts as a function of energy,

(3) using a one-pion exchange contribution to represent the large angular momentum states (Cziffra *et al.*, 1959).

The argument behind (3) may be summarised as follows. For a given incident momentum k, the particles with large impact parameter d (distance of closest approach) will encounter the long range part of the force. Since $kd \sim l$ these particles possess large angular momentum. If the long range interaction potential is of the Yukawa form e^{-mr}/r, where m is the mass of the particle exchanged, then the longest range force is due to the exchange of one pion. The scattering amplitude was therefore calculated in Born approximation for a one-pion exchange and low angular momentum states (below ${}^1G_4$) were removed by a projection operator. The high angular momentum states calculated in the one pion approximation plus a phase shift analysis of the lower states were then fitted to the experimental data.

Many analyses of the nucleon–nucleon scattering experiments have been made (see the *CERN Conference Report*, 1962, for recent details). In Table 14.7 we display phase shift solutions given in the paper of Macgregor, Moravcsik and Stapp (1960).

14.2(c). *Interpretation of the scattering data*

The problem of nucleon–nucleon scattering has been discussed under three chronological headings:

(1) potential models,

(2) meson and field theoretical methods,

(3) dispersion relations.

Each method has drawn upon information obtained by previous work and often mixtures of the methods have been used. An excellent discussion of the different approaches may be found in the article by Moravcsik and Noyes (1961).

We have already stated that the transition matrix in each state of isospin is determined by five independent terms (14.41), and so it might be expected that five functions are required to explain the behaviour of the nucleon–nucleon scattering amplitude in a potential theory. A suitable compilation (see, for example, Okubo and Marshak, 1958) is given by

$V_1(r)$	central potential	(14.42)
$V_2(r)\,(1 + \boldsymbol{\sigma}_1 \cdot \boldsymbol{\sigma}_2)$	spin potential	
$V_3(r)\left(\left(\boldsymbol{\sigma}_1 \cdot \dfrac{\mathbf{r}}{r}\right)\left(\boldsymbol{\sigma}_2 \cdot \dfrac{\mathbf{r}}{r}\right) - \dfrac{1}{3}\boldsymbol{\sigma}_1 \cdot \boldsymbol{\sigma}_2\right)$	tensor potential ($r = \lvert\mathbf{r}\rvert$)	
$V_4(r)\,\mathbf{L} \cdot (\boldsymbol{\sigma}_1 + \boldsymbol{\sigma}_2)$	spin–orbit potential	
$V_5(r)\,(\mathbf{L} \cdot \boldsymbol{\sigma}_1)\,(\mathbf{L} \cdot \boldsymbol{\sigma}_2)$	spin–orbit–tensor potential	

TABLE 14.7. *Phase Shifts for*

(As obtained by ordinary analysis at 310 MeV and by modified analyses at 310,

	$^1S_0$	$^3P_0$	$^3P_1$	$^3P_2$	$^1D_2$	ε_2
310 MeV ‡ 1	−10·1	−14·3	−26·6	16·1	12·9	−1·0
310 MeV ‡ 2	−19·5	−36·0	−11·7	18·8	4·4	−9·3
310 MeV ‡ 1	−8·92	−11·27	−27·49	16·65	11·87	−1·55
310 MeV ‡ 2	−28·99	−27·92	−8·71	21·05	4·78	−7·55
210 MeV ‡ 1	5·9	2·1	−20·9	17·0	7·5	−1·6
210 MeV ‡ 2	−15·7	−28·4	−2·1	17·9	2·8	−5·8
145 MeV	11·6	16·8	−16·7	14·2	7·4	−3·1
95 MeV	23·5	14·1	−12·9	10·1	4·0	
95 MeV	21·1	18·1	−11·1	11·0	4·2	
68 MeV	33·2	12·7	−10·2	7·7	2·7	

‡ OPEC = one-pion exchange contribution.

‡ 1 solution 1.

‡ 2 solution 2.

The necessity for potentials of this type can be argued on physical grounds. The first two potentials are necessary to explain the spin dependence of nuclear forces in the limits of *s*-wave interaction. A tensor interaction (that is an alignment of spin and position of the particles) is necessary to explain the quadrupole moment of the deuteron (Blatt and Weisskopf, 1952). Spin-orbital forces are required to explain the large polarisation at high energies and its behaviour as a function of energy.

The relation of the phase shifts to the potentials follows the principles indicated in § 6.3(c), but, of course, a more sophisticated treatment is required because of the presence of the spin terms. Nevertheless, rough behaviour of the potentials may be inferred from inspection of Table 14.7 and the use of the simple arguments $kd \sim l$ (where d is the impact parameter) and "positive phase shifts are equivalent to attractive potentials". Firstly, the behaviour of the $^1S_0$ and $^1D_2$ phase shifts indicates a repulsive potential core of radius ~0·5 fermi surrounded by an attractive region. Secondly, strong spin–orbit potentials are obviously required from the behaviour of the triplet P terms; the energy dependence of these phase shifts can be achieved by using an attractive long-range tensor force (which is energy independent) and a short range spin–orbit force which is repulsive in the $^3P_0$ state.

Proton–Proton Scattering

210, 145, 95 and 68 MeV.) The table gives nuclear bar phase shifts in degrees

$^3F_2$	$^3F_3$	$^3F_4$	$^1G_4$	ε_4	$^3H_4$	$^3H_5$	$^3H_6$
0·8	−4·4	3·2	2·0	−1·2	1·5	0·1	1·3
−0·5	0·3	2·5	2·6	−1·5	2·1	−1·4	1·4
1·21	−3·53	3·54			OPEC‡		
−0·49	−0·03	3·33			OPEC		
−0·4	−2·0	0·3			OPEC		
1·2	−2·3	1·0			OPEC		
−1·8	1·0	0·07			OPEC		
		OPEC					
		OPEC					
		OPEC					

We consider, finally, the application of field theoretical arguments to the nucleon–nucleon scattering problem.† Since the Yukawa potential contains the term e^{-mr}/r where m represents the mass of the particle exchanged, we may make rough qualitative arguments relating potentials to regions over which various meson contributions might be expected to dominate. They are normally divided into three zones (Taketani, Nakamura and Sasaki, 1951):

(1) for distances greater than $\sim 1{\cdot}5$ pion Compton wavelengths ($\sim m_\pi^{-1}$) the interaction should be dominated by the one-pion exchange (§ 10.4(e)),
(2) $0{\cdot}7\ m_\pi^{-1} < r < 1{\cdot}5\ m_\pi^{-1}$ two-pion exchange,
(3) $r < 0{\cdot}7\ m_\pi^{-1}$ more complex systems.

The one-pion region can be treated with some confidence. We have seen that it can be used successfully in the phase shift analysis (§ 14.2(b)), and that the pion–nucleon coupling strength of the expected magnitude can be extracted from the pole in n–p scattering (§ 10.4(e)).‡ In addition the central and tensor potentials appear naturally in the cut off form of meson theory (§ 14.1(c))

$$H_{\text{int}} = \sum_{i=1,2} \frac{f}{m_\pi} \int d\mathbf{x}\varrho(\mathbf{x}-\mathbf{x}_i)\, \boldsymbol{\sigma}_i \cdot \nabla(\boldsymbol{\tau}_i \cdot \boldsymbol{\varphi}) \tag{14.43}$$

† Whilst the principles of applying the dispersion relations to nucleon–nucleon scattering problems are straightforward, their detailed application is not. We shall therefore not consider them.

‡ The pion properties may be also extracted from the p–p data (Signell, 1960).

The inner regions are less easily understood, and in particular the part played by the pion resonant systems could be extremely important (Scotti and Wong, 1963). Sakurai (1960) and Matthews (1961) have shown that the exchange of isotopic vector and scalar particles with spin unity in nucleon–nucleon scattering could lead to a natural explanation of the repulsive core and to a spin–orbit term of the correct sign. Furthermore, the contributions to the scattering amplitude arising from the exchange of a vector particle are of opposite sign for nucleon–nucleon and nucleon–antinucleon interactions (the situation is analogous to the exchange of a photon in Møller and Bhabha scattering). This change in sign causes the repulsive core to become an attractive well in nucleon–antinucleon scattering, and is obviously an attractive facet of the theory.

14.3. THE INELASTIC SCATTERING OF NUCLEONS AND PIONS

14.3(a). *Nucleon–nucleon inelastic scattering*

The threshold for nucleon–nucleon inelastic scattering occurs at a kinetic energy of $\sim$ 280 MeV in the laboratory system; it leads to the processes

$$\begin{array}{ll} p + p \to p + p + \pi^0 & n + p \to n + p + \pi^0 \\ \qquad\quad\; p + n + \pi^+ & \qquad\quad\; d + \pi^0 \\ \qquad\quad\; d + \pi^+ & \qquad\quad\; n + n + \pi^+ \\ & \qquad\quad\; p + p + \pi^- . \end{array}$$

The reactions $p + p \to d + \pi^+$ and $n + p \to d + \pi^0$ are of interest since they produce a test of charge independence. Since the isospins of the deuteron and pion are 0 and 1 respectively, the reaction must proceed through a pure $T = 1$ channel. Now the $n + p$ system is an equal mixture of $T = 0$ and $T = 1$ states whilst the p–p system is a pure $T = 1$ state. Thus the cross-section for $p + p \to d + \pi^+$ should be twice that for $n + p \to d + \pi^0$. The experimental data is in accord with this prediction (see, for example, Flyagin *et al.*, 1959).

The processes of pion production in nucleon–nucleon collisions appear to be dominated by the production of the nucleon isobars as intermediate states, for example

$$\begin{array}{l} p + p \to N_1^* + n \\ \qquad\qquad\; \downarrow \\ \qquad\qquad\; p + \pi^+ \end{array}$$

(Lindenbaum and Sternheimer, 1957, 1960; Mandelstam, 1958b). At nucleon kinetic energies below 1 GeV the N_1^* $(\frac{3}{2}, \frac{3}{2}^+)$ isobar alone is effective because of energy considerations. In Fig. 14.16 the trends in the cross-sections for the production of single pions in proton–proton collisions as a function of energy are displayed. The data is taken mainly from the paper of Mandelstam (1958b).

The angular momentum states participating in single pion production between threshold and a proton energy of 400 MeV have been subjected to a phenomenological analysis by Rosenfeld (1954). He showed that the process could be un-

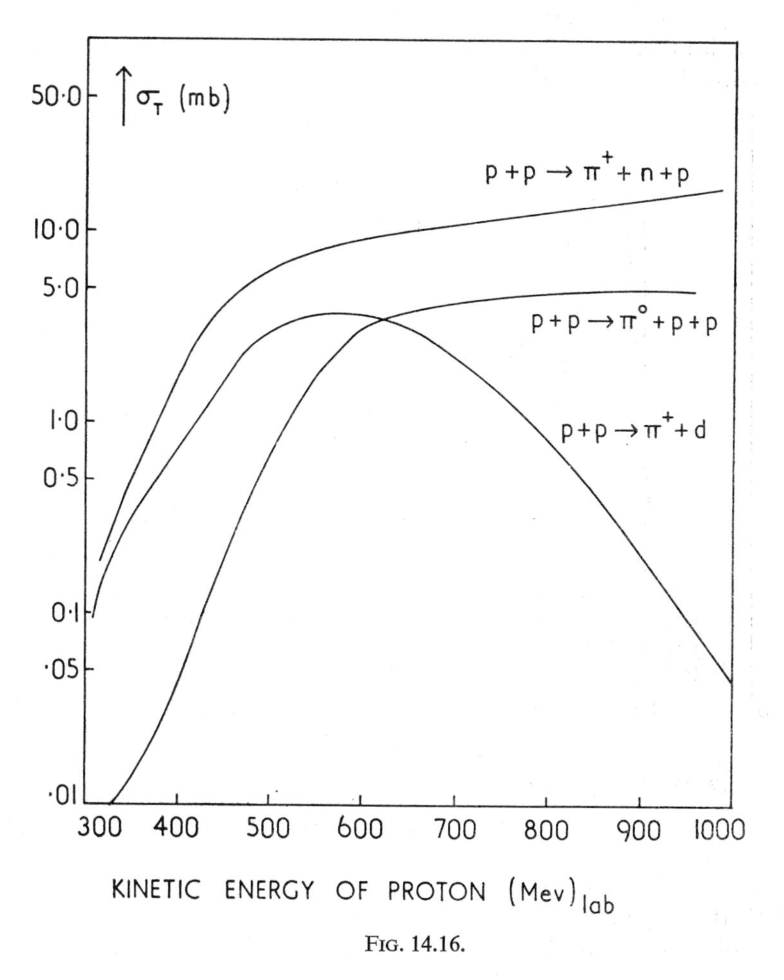

FIG. 14.16.

derstood if the reactions proceeded through a dominant mode ${}^1D_2 \rightarrow {}^3Sp_2$, where the capital letter indicates the orbital angular momentum of the two nucleons and the small letter that of the pion relative to the nucleon system. Now this transition corresponds to the production of the N_1^* $(\frac{3}{2}, \frac{3}{2}^+)$ isobar and a nucleon in an *s*-state relative to each other. An analysis of the experimental data up to 700 MeV by Mandelstam showed that isobar production in *p*-states was also necessary to explain the results at higher energies.

The dominance of the $\frac{3}{2}$, $\frac{3}{2}^+$ isobar also explains the behaviour of the ratio of the cross-sections for π^+ and π^0 production in p–p collisions. The transition ${}^1D_2 \rightarrow {}^3Sp_2$ is forbidden in π^0 production since the two protons cannot be in the ${}^3S$ state without violating the Pauli exclusion principle. Production must therefore take place through an angular momentum channel which is inhibited by centrifugal potential barrier effects, and therefore the cross-section is small. At higher energies this limitation is not so important and so the cross-section for π^0 production rises more rapidly than for π^+. This effect can be seen in Fig. 14.16.

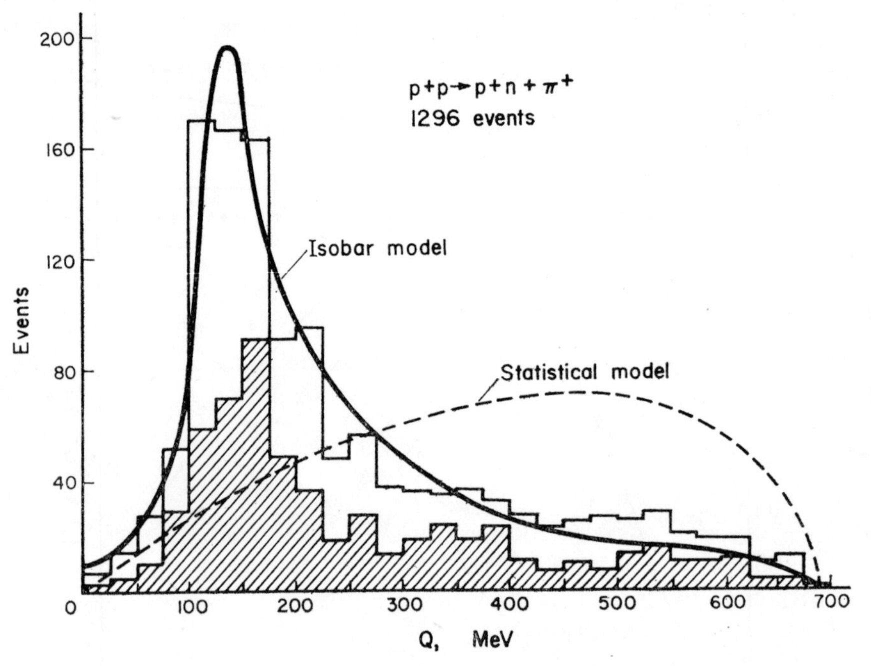

FIG. 14.17. Distribution of Q values in the pion–proton system for the reaction $p+p \rightarrow p+n+\pi^+$.

The effects of the isobars appear most clearly in the data above 1 GeV. In Fig. 14.17 the distribution of the Q values (the kinetic energy released in the break up of a hypothetical N^*) are plotted for the pion–proton system for the reaction

$$p + p \rightarrow p + n + \pi^+$$

at an incident proton energy of 2 GeV (reported by Detoeuf, 1961). The peak due to the N_1^* system is clearly visible. At higher energies the contributions from the states N_2^* and N_3^* start to appear. In Fig. 14.18 the momentum distributions for the recoil protons in the process

$$p + p \rightarrow p + X$$

are displayed (Duke *et al.*, 1961). The incident protons possessed a kinetic energy of 2·8 GeV. The arrows indicate the expected positions of the isobars.

The shapes of the curves appearing in Fig. 14.18 may be qualitatively explained by a one-pion exchange process, as illustrated in Fig. 14.19 (see, for example, Selleri, 1961). The exchange of a single light particle implies that the distributions should be strongly peaked in the forward direction, as can be seen in Fig. 14.18

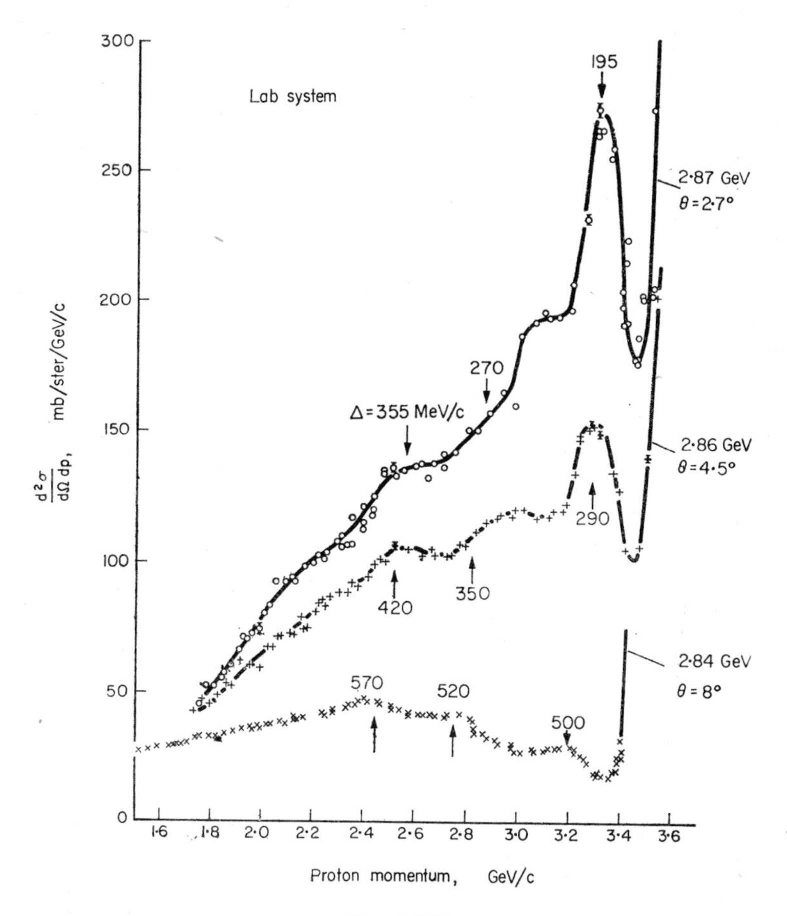

FIG. 14.18.

π N N* π° p p p p

FIG. 14.19.

14.3(b). *Pion–nucleon inelastic scattering*

Total cross-sections for the inelastic and elastic scattering of pions from protons are displayed in Fig. 14.20 and 14.21 (Helland *et al.*, 1962).

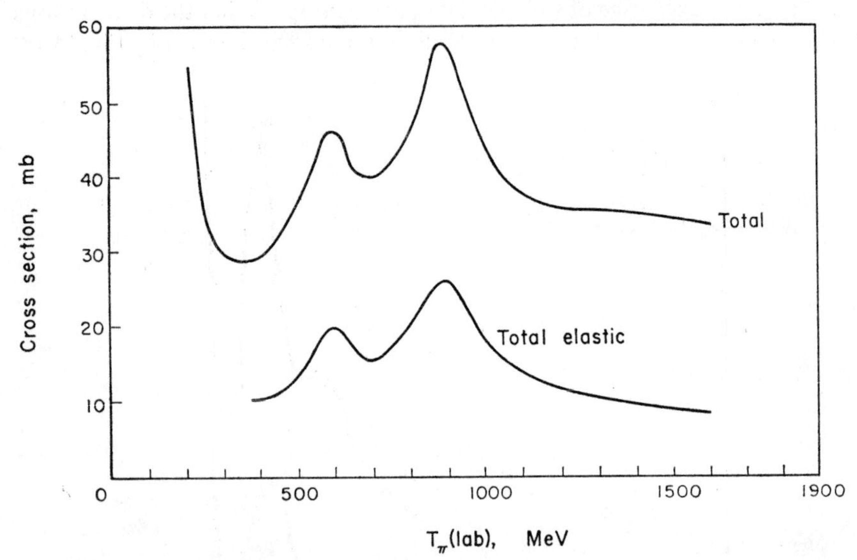

FIG. 14.20. Elastic and total cross-sections for $\pi^- p$ scattering. The difference represents the inelastic and charge exchange cross-sections.

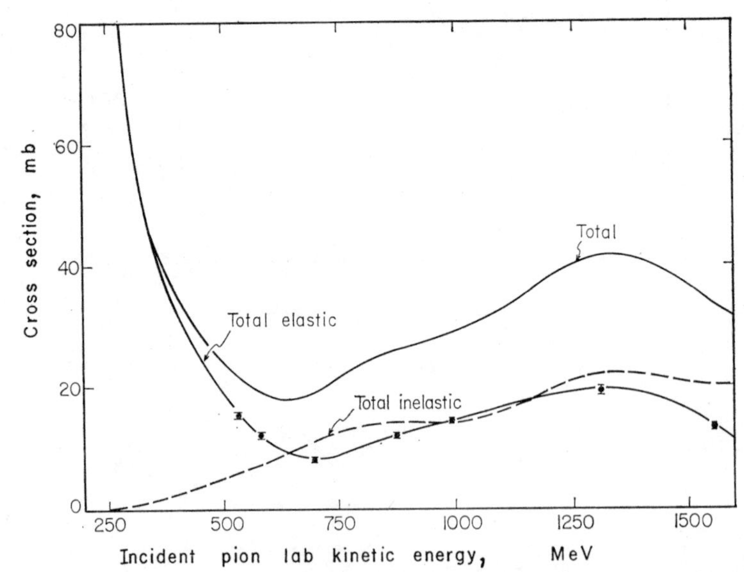

FIG. 14.21. Elastic, inelastic and total cross-sections for $\pi^+ p$ scattering (Moyer, private communication).

TABLE 14.8

T_π(MeV)	$\pi^- p$ (mb)					$\pi^+ p$ (mb)			
	$\pi^0\pi^0 n$	$\pi^-\pi^+ n$	$\pi^0\pi^- p$	$n\pi$ $n \geqq 3$	K, Λ, Σ	$\pi^+\pi^0 p$	$\pi^+\pi^+ n$	$n\pi$ $n \geqq 3$	K, Λ, Σ
300	~0·8								
500	1·0	0·5	2·0				0·1		
700	3·6	1·0	3·0			2·0	1·0		
900	3·0	10·0	8·0			9·0	2·0	0·8	0·03
1100	2·0	10·0	9·0	1·0	1·0	9·0	3·0	2·5	0·25
1300		8·0	11·0		1·0	10·0	4·6	4·8	0·4
1500		7·0	12·0			10·0	3·0	5·0	
1700									
1900					0·15				0·25

The data for the branchings into various inelastic channels is given in Table 14.8. The data is taken mainly from the Aix-en-Provence (1961) and the Geneva (1962) Conference reports, and interpolation has been made where necessary. Errors are of the order of ~1mb.

As in the case of nucleon–nucleon inelastic scattering the data on the pion–nucleon interaction shows evidence for the excitation of the $(\frac{3}{2}, \frac{3}{2}^+)$ nucleon isobar resonance. In Fig. 14.22 the momentum spectrum of π^0-mesons from the

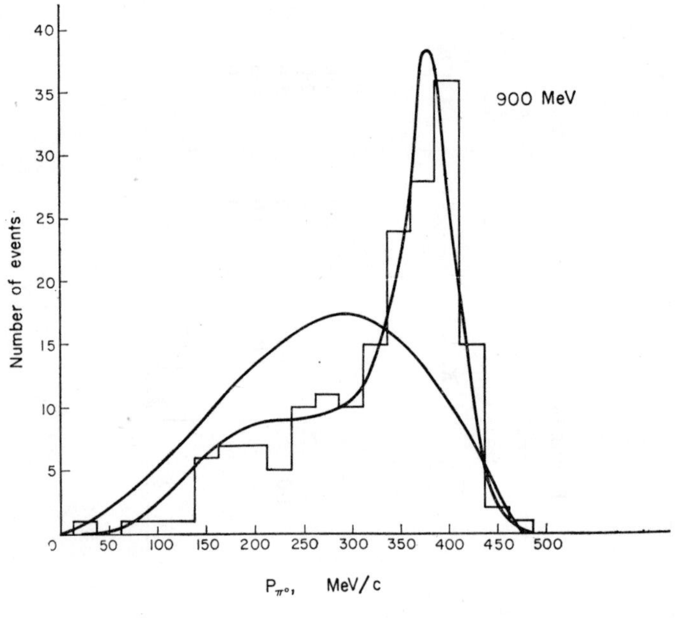

FIG. 14.22.

reaction $\pi^+ + p \to \pi^+ + \pi^0 + p$ at an incident pion energy of 900 MeV (Barloutaud *et al.*, 1961) is plotted, and the characteristic peak of the nucleon isobar is obvious. Nevertheless, the production of the isobar N_1^* cannot be the only process which participates. If this were true the appropriate Feynman diagram would look like Fig. 14.23, and assuming the separate existence of the isobar for a sufficiently long time it is a straightforward task to show that the ratio

$$\frac{\sigma(\pi^+ + p \to \pi^+ + \pi^0 + p)}{\sigma(\pi^+ + p \to \pi^+ + \pi^+ + n)} = 6{\cdot}5$$

(see, for example, Sternheimer and Lindenbaum, 1961). Table 14.8 shows that the experimental ratio decreases from ~ 4 at 900 MeV to ~ 2 at 1300 MeV. The difference may probably be accounted for by pion–pion interactions (Fig. 14.24), and in particular by the pion resonances. It is noteworthy that the $T = 1$ two-pion resonance corresponding to a mass of ~ 750 MeV (§ 13.3(a)) starts to appear in π^+p scattering at 900 MeV and is making significant contributions by 1090 MeV (Stonehill *et al.*, 1961). This is illustrated in Fig. 14.25. Techniques for extracting π–π scattering data from π–p cross-sections have been given by Chew and Low (1959) (compare with (13.7)).

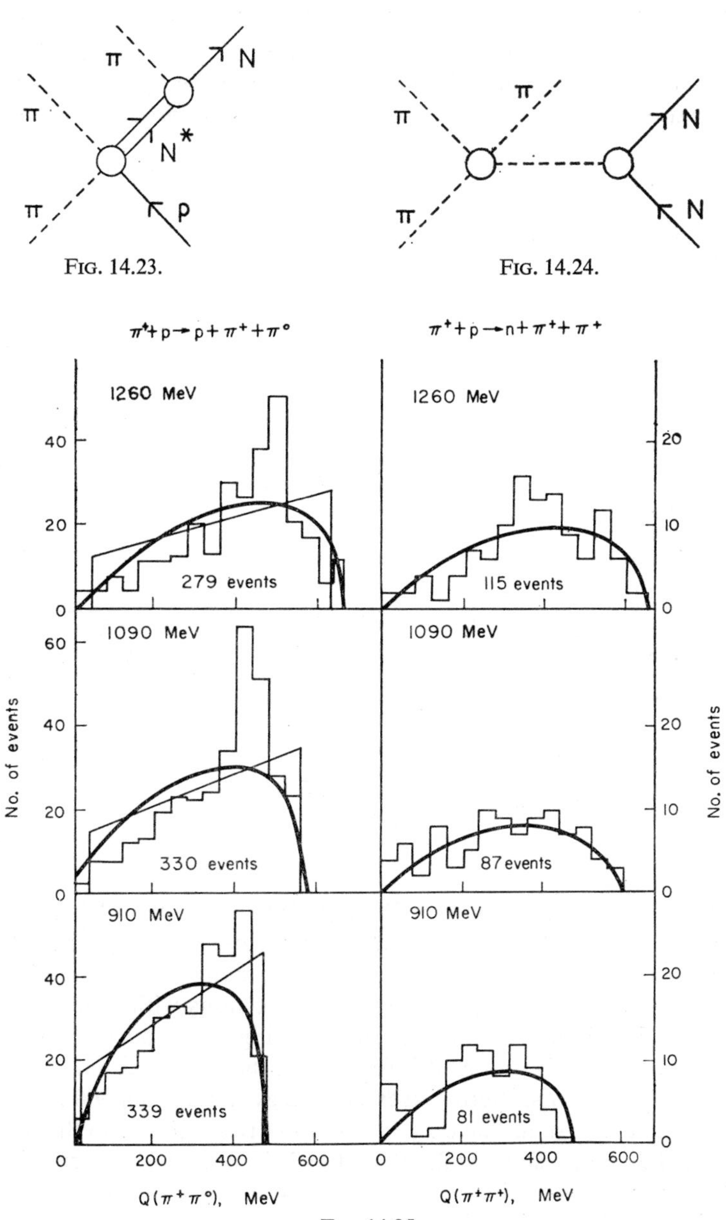

FIG. 14.23.

FIG. 14.24.

FIG. 14.25.

14.4. NUCLEON–ANTINUCLEON INTERACTIONS

14.4(a). *Scattering data*

The interaction of antinucleons with nucleons is considerably more complicated than nucleon–nucleon processes because many additional channels are open. Most of the available information concerns antiproton–proton interactions; that on the total cross-sections has been summarised in Fig. 14.26 which

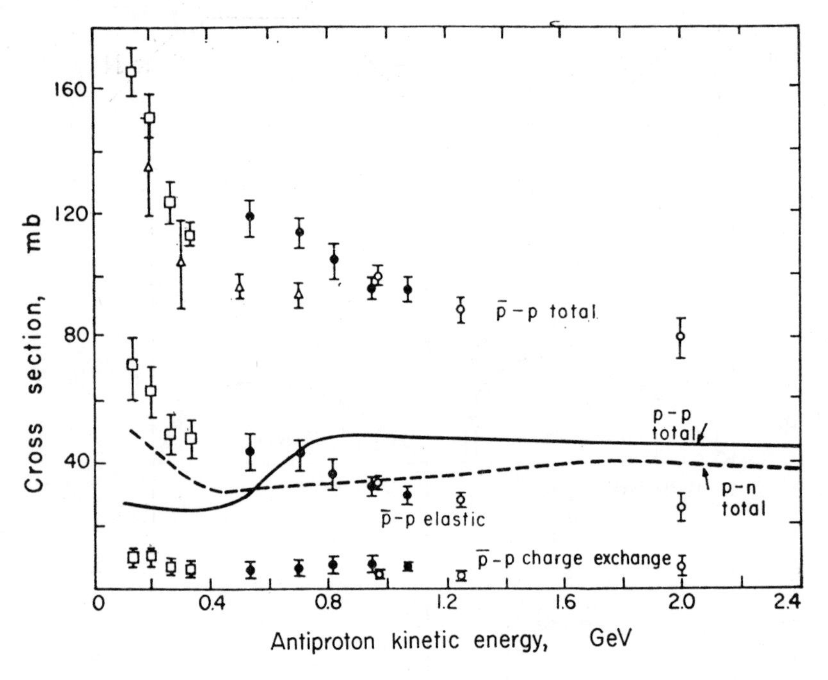

FIG. 14.26.

has been taken from the report of Wenzel (1960). The term 'charge exchange scattering' refers to $p + \bar{p} \rightarrow n + \bar{n}$; it remains close to $\sim$ 8 mb over the entire region. Measurements of the type $\bar{p}n$ have also been made by subtracting $\bar{p}p$ total cross-section from those for $\bar{p}d$ scattering. Corrections with some uncertainties must be made in this data to allow for the shielding of one nucleon by another in the deuteron. When this is done it is found that the cross-sections are effectively the same for $\bar{p}p$ and $\bar{p}n$ scattering over the energy range from 500 to 1100 MeV. Now the $\bar{p}p$ and $\bar{p}n$ systems have $T_3 = 0$ and -1 respectively, so that the former is in a mixed $T = 0, 1$ state whilst the latter is in a pure $T = 1$ state. Thus the equality of the cross-sections implies that nucleon–antinucleon scattering is virtually independent of isospin over the energy range 500–1100 MeV.

The cross-section for inelastic scattering is made up from annihilation processes, for example

$$p + \bar{p} \rightarrow n\pi$$

and genuine inelastic processes of the type

$$p + \bar{p} \rightarrow \pi^0 + p + \bar{p}.$$

A measurement of the production of single pions at 940 MeV (Solmitz, 1960) gave a figure of 5 ± 1 mb, in contrast with a value of ~ 20 mb for the processes $p + p \rightarrow \pi + N + N$ at this energy. Part of the difference may arise from the fact that only one nucleon can be excited to the $\frac{3}{2}, \frac{3}{2}^+$ state in $\bar{p}p$ scattering. The figure of 5 mb for the inelastic processes implies that the annihilation cross-

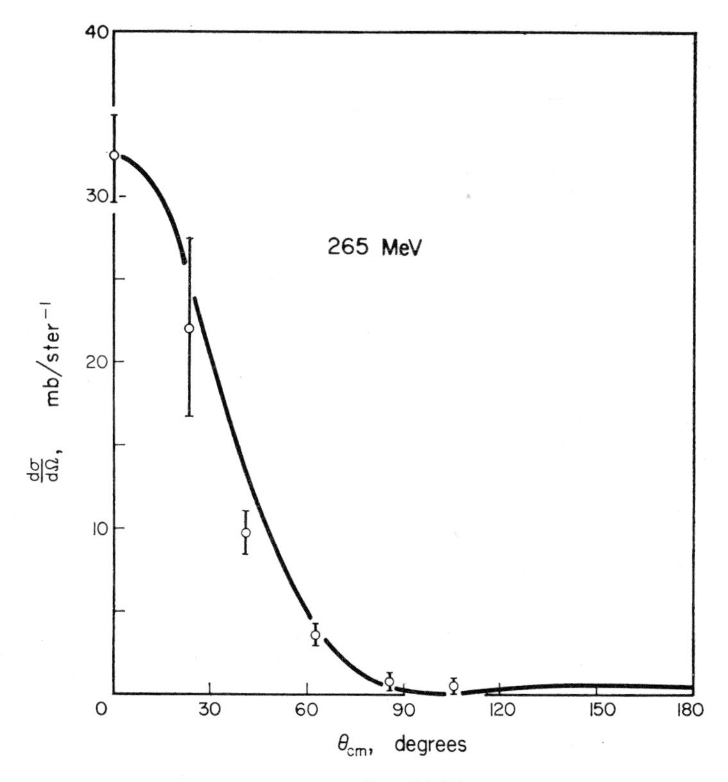

FIG. 14.27.

section is $\sim$ 50 mb at 940 MeV. Since the threshold for true inelastic processes starts at 280 MeV and their cross-sections are unlikely to fluctuate violently as a function of energy, then the above result taken in conjunction with Fig. 14.26 suggests that the annihilation cross-section falls away slowly with energy up to 1 GeV.

Measurements of the angular distributions occurring in elastic scattering have been made at a number of energies between 100 and 2000 MeV (Coombes *et al.*, 1958; Armenteros *et al.*, 1960). A typical curve is shown in Fig. 14.27. This curve can be seen to be considerably different to the angular distribution for elastic *p–p* scattering at comparable energies (Fig. 14.13). As the energy of the antiproton increases the angular distributions become more sharply peaked in the forward direction.

The large inelastic scattering cross-sections and the shape of the elastic scattering distributions strongly suggest that the proton–antiproton interaction corresponds to "black sphere" conditions (§ 6.1(c)). The total cross-sections are then given by (6.23)

$$\sigma_{sc} = \sigma_r = \pi R^2, \qquad \sigma_T = 2\pi R^2$$

and the differential cross-section by (6.24)

$$\frac{d\sigma_{sc}}{d\Omega} = R^2 \left[\frac{J_1(kR \sin \theta)}{\sin \theta} \right]^2$$

where R is a radius of interaction.

The solid line in Fig. 14.27 was calculated by Coombes *et al.*, using a value of R given by the measured total cross-section, $\sigma_T = 2\pi R^2$. The above equations also imply that the sum of the cross-sections for elastic plus charge exchange scattering should be half the total cross-section. An inspection of Fig. 14.26 shows that this relation is approximately satisfied, especially at the low energy end. The term R is not constant, however, but varies between $\sim 1{\cdot}6 \times 10^{-13}$ and $1{\cdot}1 \times 10^{-13}$ cm over the energy range of 150–2000 MeV. This is of the order of the pion Compton wavelength.

Not much quantitative work has been done on the $\bar{p}p$ process from a field-theoretical point of view. In principle information from proton–proton collisions could be used to examine $\bar{p}p$ interactions. However, it must be remembered that the *p–p* interaction occurs only in the $T = 1$ channel, whereas the $\bar{p}p$ reaction occurs in both $T = 1$ and $T = 0$ channels. Furthermore, we lack detailed knowledge about the proton–proton mechanisms inside distances of the order of a pion Compton wavelength. As we have already mentioned in § 14.2(c), the exchange of vector mesons could be important in this region.

14.4(b). *The annihilation of antiprotons at rest*

There are many channels open for proton–antiproton annihilation at rest, since the symmetry requirements are not very restrictive. The main two body decay modes are given in Table 14.9 (Armenteros *et al.*, 1962a; Chadwick *et al.*, 1962). A notable absentee from this list is the decay mode $p\bar{p} \to \pi^0\pi^0$; apart from the technical difficulty of detection, this mode is not expected if annihilation takes place in *s*-states. The spin and parity of the *s*-states for a proton–antiproton system are 0^- and 1^- since fermion and antifermion possess opposite intrinsic parity (§ 5.4(d)). However. In order to satisfy Bose–Einstein statistics two π° mesons must be in states of even orbital angular momentum and therefore parity. Thus conservation of parity and angular momentum cannot be simultaneously satisfied for the decay $p\bar{p} \rightarrow \pi^0\pi^0$ from s-states. Evidence is presented below which indicates that s-state capture is the normal mode of annihilation of the proton–antiproton pair.

TABLE 14.9

Mode	Branching ratio
$\pi^+\pi^-$	$(3{\cdot}95 \pm 0{\cdot}38)\ 10^{-3}$
K^+K^-	$(1{\cdot}31 \pm 0{\cdot}18)\ 10^{-3}$
$K^0\bar{K}^0$	$(0{\cdot}56 \pm 0{\cdot}8)\ 10^{-3}$
$\varrho\pi$	$(2{\cdot}7 \pm 0{\cdot}6)\ 10^{-2}$

As in the case of positronium, annihilation into photons is also possible, but as this is an electromagnetic interaction the branching ratio is negligible.

If the antiproton possesses kinetic energy, further channels open up, for example

$$p + \bar{p} \rightarrow \Lambda + \bar{\Lambda}.$$

The average number of pions produced in the annihilation at rest is $4{\cdot}78 \pm 0{\cdot}17$ (Chadwick *et al.*, 1961). Many of these arise because the $p\bar{p}$ system decays through pion resonant states (Sakurai, 1960).

The process $p + \bar{p} \rightarrow K^0 + \bar{K}^0$ has proved to be important, both as a means of examining the atomic states prior to $p\bar{p}$ annihilation (d'Espagnat, 1961), and as a method for determining the spin of the K^{0*}-meson (Schwartz, 1961). The process $p + \bar{p} \rightarrow K^0 + \bar{K}^0$ can proceed as

$$\begin{aligned} p + \bar{p} \rightarrow\ & K_1^0 + K_1^0 \quad \text{(a)} \\ & K_2^0 + K_2^0 \quad \text{(b)} \\ & K_1^0 + K_2^0 \quad \text{(c).} \end{aligned} \tag{14.44}$$

If the process is studied in a hydrogen-filled bubble chamber of conventional dimensions, mainly K_1^0 decays are recorded in the chamber because of the long lifetime of the K_2^0 particles. Therefore we shall concentrate on the processes (a) and (c). Let us examine the conservation laws for these two reactions; they are both strong and so we must conserve total angular momentum (j), spatial (ξ_P) and charge (ξ_C) parity.

The proton and antiproton are of opposite intrinsic parity (§ 5.4(d)), and K_1^0 and K_2^0-mesons are of opposite intrinsic charge parity (§ 12.9(d)). Therefore we may draw up Table 14.10 for the behaviour of the systems under the parity and charge conjugation operations (compare (11.103), (5.101) and (5.137).

TABLE 14.10

	Particles		
	$p\bar{p}$	$K_1^0K_1^0$	$K_1^0K_2^0$
ξ_P	$(-1)^{L+1}$	$(-1)^j$	$(-1)^j$
ξ_C	$(-1)^{L+s}$	$+1$	-1

In this table L and j refer to the orbital angular momentum of the initial system and the total angular momentum of the final system respectively.

The observed branching ratio for the processes (a) and (c) is

$$\frac{p + \bar{p} \to K_1^0 + K_1^0}{p + \bar{p} \to K_1^0 + K_2^0} = \frac{0}{54}$$

(Armenteros *et al.*, 1962a). An inspection of Table 14.10 shows that this result implies that capture from the ${}^3S_1$-state predominates in the process $p+\bar{p}\to K^0+\bar{K}^0$ (if we restrict our attention to $L = 0$ and 1). It then seems reasonable to assume that s-state capture dominates in all inelastic channels, as suggested theoretically (Day, Snow and Sucher, 1959).

Now consider capture in the s-state for the process

$$p + \bar{p} \to \bar{K}^0 + K^{0*}, \qquad K^{0*} \to K^0 + \pi^0.$$

If we assume zero spin for the K^{0*} state, then its intrinsic parity is odd relative to K^0 because of the parity of the pion, and so $\xi_P = (-1)^{j+1}$ for the $\bar{K}^0 + K^{0*}$ system. Conservation of j and P then occurs for $p\bar{p}$ annihilation in the ${}^1S_0$-state, and so $\xi_C = +1$ from Table 14.10. Thus the process of annihilation would proceed as follows:

$$p + \bar{p} \to \begin{Bmatrix} \bar{K}^0 + K^{0*} \\ K^0 + \bar{K}^{0*} \end{Bmatrix} \to \begin{Bmatrix} K_1^0 + K_1^0 + \pi^0 \\ K_2^0 + K_2^0 + \pi^0 \end{Bmatrix} \tag{14.45}$$

in order to preserve charge conjugation invariance. On the other hand, if the spin of the K^{0*} particle is one, capture from s and p-states can occur and mixtures of $K_1^0 + K_2^0$ mesons should be produced. The experimental data of Armenteros *et al.* (1962b) indicate the presence of these mixtures and so appear to be incompatible with zero spin for the K^* meson.

14.5. THE INTERACTIONS OF STRANGE PARTICLES

14.5(a). *The production of strange particles*

Total cross-sections for the associated production of strange particles in pion–nucleon collisions are displayed in Fig. 14.28. The data have been taken mainly from the paper of Wolf *et al.* (1961) and the conference reports for Aix-en-Provence (1961) and Rochester (1960). The peak in the cross-section for the production of ΛK^0 corresponds to the position of the peak N_3^* in the total π^-p cross-section (Fig. 13.1). This relationship is not unexpected since the Λ-hyperon has $T = 0$, and so the reaction $\pi^-p \rightarrow \Lambda K^0$ can only proceed through the $T = \frac{1}{2}$ channel. The mass of the Σ^- particle causes the threshold for the reaction $\pi^-p \rightarrow \Sigma^-K^+$ to be above the N_3^*-state.

The cross-sections appearing in Fig. 14.28 show that the production of strange particles represents a small fraction of the total inelastic cross-section for pion–

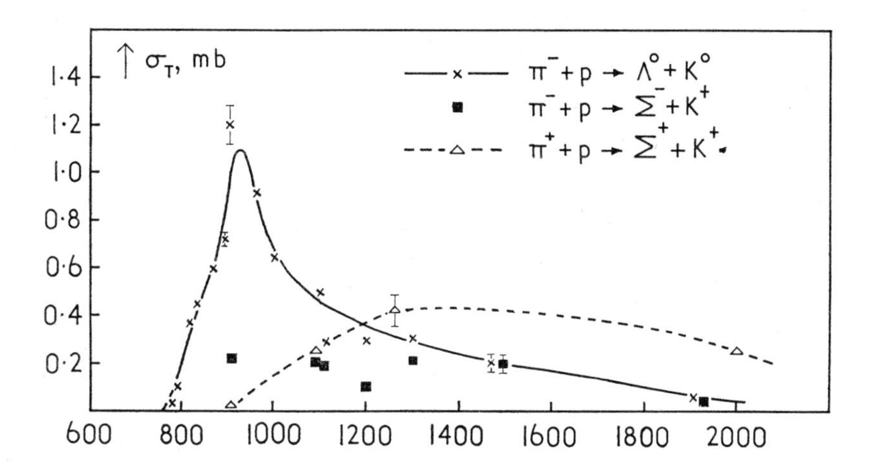

FIG. 14.28.

nucleon collisions. This behaviour persists up to high energies; for example cross-sections of ~ 1 mb for strange particle production are found in reactions initiated by pions and by protons with energies of ~ 20 GeV (Peyrou, 1961).

14.5(b). *Total cross-sections for kaon–proton scattering*

Cross-sections for the total and elastic scattering of $K^{\pm}$ mesons on protons are displayed in Figs. 14.29 and 14.30. At low energies only elastic scattering is possible in K^+p interactions. In contrast, the strangeness quantum number $S = -1$ for K^- mesons implies that inelastic channels for the production of Λ and Σ hyperons are open at all energies. This condition leads to large K^-p cross-sections at low energies. The peak in the K^-p scattering at 1·05 GeV/c corresponds to the Y_0^{***} resonance (§ 13.4(b)).

14.5(c). K^+-*nucleon scattering*

The interaction K^+p is in a pure $T = 1$ state. It is characterised by a remarkable simplicity in terms of partial waves, since up to a kaon kinetic energy of 450 MeV in the laboratory system (800 MeV/c in momentum) the interaction is predominantly in the *s*-state. An extensive series of measurements of the angular distribution in K^+p scattering have been made in the momentum range 140 to

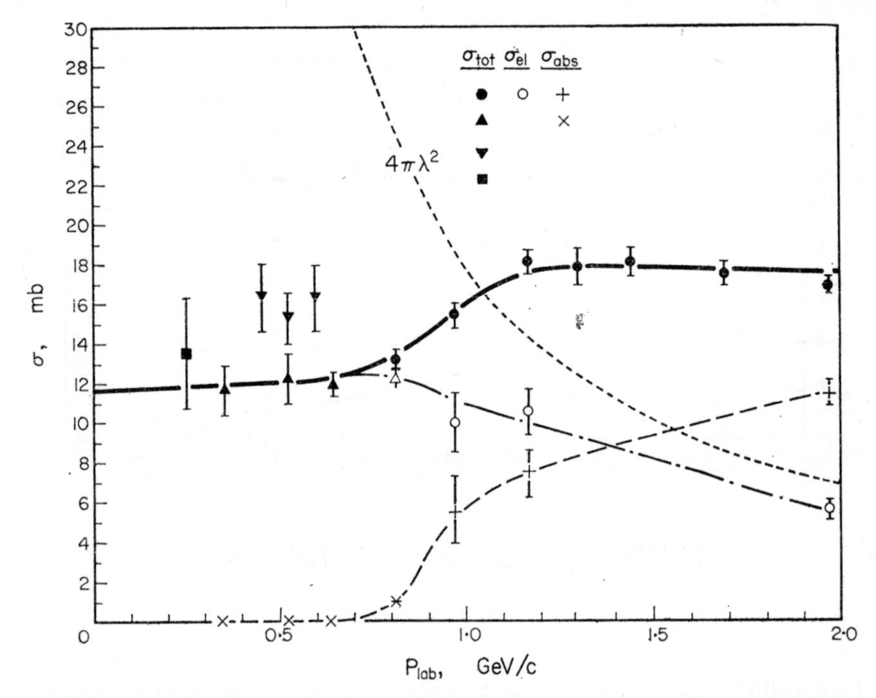

FIG. 14.29. Total and elastic scattering cross-sections for the K^+p interaction (Cook *et al.* 1962).

810 MeV/c. (Stubbs *et al.*, 1961; Goldhaber *et al.*, 1962). The isotropic character of the angular distributions and the linear dependence of the phase shift upon momentum (§ 6.3(d)) are consistent with *s*-wave scattering. Furthermore, constructive interference with the Coulomb scattering was observed, thus indicating negative phase shifts and a repulsive interaction (§ 6.3(c)). The data may be summarised with the aid of Table 14.11.

At the highest momentum an inelastic cross-section of $1{\cdot}0 \pm 0{\cdot}2$ mb was observed.

Measurements on K^+p scattering have also been made at 1·0, 1·2 and 2·0 GeV/c. (Cook *et al*,. 1962). In contrast to the data at lower energies, marked anisotropy is found.

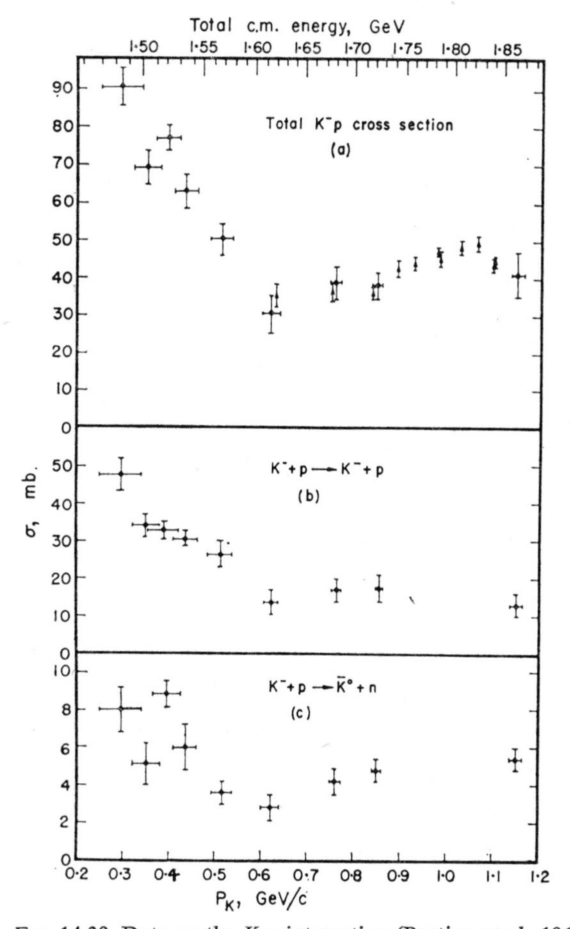

FIG. 14.30. Data on the K^-p interaction (Bastien *et al.*, 1962a).

TABLE 14.11

K^+ momentum (MeV/c)	$T = 1$ Total cross-section (mb)	s-wave fit Phase shift (degrees)
140	$9{\cdot}2 \pm 2{\cdot}1$	$-7{\cdot}2 \pm 0{\cdot}8$
175	$12{\cdot}5 \pm 2{\cdot}2$	$-10{\cdot}4 \pm 0{\cdot}9$
205	$11{\cdot}5 \pm 1{\cdot}7$	$-11{\cdot}7 \pm 0{\cdot}9$
235	$11{\cdot}2 \pm 1{\cdot}6$	$-13{\cdot}2 \pm 0{\cdot}9$
265	$10{\cdot}0 \pm 1{\cdot}6$	$-14{\cdot}0 \pm 1{\cdot}1$
355	$11{\cdot}7 \pm 1{\cdot}2$	$-20 \pm 1{\cdot}1$
520	$12{\cdot}2 \pm 1{\cdot}3$	$-29{\cdot}4 \pm 1{\cdot}7$
642	$12{\cdot}4 \pm 0{\cdot}9$	$-36{\cdot}2 \pm 1{\cdot}4$
810	$12{\cdot}2 \pm 0{\cdot}4$	$-47 \pm 1{\cdot}0$

The K^+n interaction involves both $T = 1$ and $T = 0$ states, and is therefore more complex than the K^+p case. An added complication is the fact that free neutrons are not available and so deuterons or more complex nuclei must be used as targets. Most of the work has been concentrated on the charge exchange reaction $K^+ + d \to K^0 + p + p$ (Chinowsky *et al.*, 1960). These workers have found evidence for the presence of *d*-waves in the $T = 0$ state at kinetic energies as low as 230 MeV (in the laboratory system).

14.5(d). *The K^-p interaction at low energies*

The K^-N reactions are unique in their complexity at low energy. We list below the open channels near threshold for the K^-p interaction

$$\begin{array}{lll} & T = 0 & T = 1 \\ K^- + p \to & K^- + p & K^- + p \\ & \bar{K}^0 + n & \bar{K}^0 + n \\ & & \Lambda + \pi^0 \\ & \overset{0}{\Sigma^\pm} + \overset{0}{\pi^\mp} & \overset{0}{\Sigma^\pm} + \overset{0}{\pi^\mp} \\ & \Lambda + 2\pi & \Lambda + 2\pi. \end{array} \tag{14.46}$$

Only the $T = 1$ channel is available for the K^-n interaction, but experimental problems then arise through the lack of free neutrons.

We shall concentrate on the two body final states since the probability of producing $\Lambda\, 2\pi$ is found to be small at low energies. If we consider the isospin channels, the tables of Clebsch–Gordan coefficients (Tables 9.4 and 9.6, pp. 391–2) yield the following base states:

$$|\psi_T(T_3 = 0)\rangle \equiv |\psi_0\rangle = \frac{1}{\sqrt{2}}(|K^-p\rangle - |\bar{K}^0 n\rangle) \tag{14.47}$$

$$|\psi_1\rangle = \frac{1}{\sqrt{2}}(|K^-p\rangle + |\bar{K}^0 n\rangle)$$

$$|\varphi_T(T_3 = 0)\rangle \equiv |\varphi_0\rangle = \frac{1}{\sqrt{3}}(|\Sigma^+\pi^-\rangle + |\Sigma^-\pi^+\rangle - |\Sigma^0\pi^0\rangle)$$

$$|\varphi_1\rangle = \frac{1}{\sqrt{2}}(|\Sigma^+\pi^-\rangle - |\Sigma^-\pi^+\rangle)$$

$$|\varphi_2\rangle = \frac{1}{\sqrt{6}}(|\Sigma^+\pi^-\rangle + |\Sigma^-\pi^+\rangle + 2\,|\Sigma^0\pi^0\rangle)$$

$$|\chi(T = 1)\rangle \equiv |\chi\rangle = |\Lambda\pi^0\rangle.$$

Thus the two body reactions lead to the following amplitudes in the isospin channels (the subscripts again refer to the isospin channel number, and we use the t-matrix notation of § 13.5(b))

(1) scattering

$$\langle K^- p|\, t\, |K^- p\rangle = \tfrac{1}{2}[t_1 + t_0] \tag{14.48}$$

$$\langle \overline{K}{}^0 n|\, t\, |K^- p\rangle = \tfrac{1}{2}[t_1 - t_0]$$

(2) absorption

$$\langle \Sigma^+\pi^-|\, t\, |K^- p\rangle = \frac{1}{\sqrt{6}}\, t_0(\Sigma) + \frac{1}{2}\, t_1(\Sigma)$$

$$\langle \Sigma^-\pi^+|\, t\, |K^- p\rangle = \frac{1}{\sqrt{6}}\, t_0(\Sigma) - \frac{1}{2}\, t_1(\Sigma)$$

$$\langle \Sigma^0\pi^0|\, t\, |K^- p\rangle = \frac{-1}{\sqrt{6}}\, t_0(\Sigma)$$

$$\langle \Lambda\pi^0|\, t\, |K^- p\rangle = \frac{1}{\sqrt{2}}\, t_1(\Lambda)$$

$$\sigma_r(0) = 6\sigma(\Sigma^0)$$

$$\sigma_r(1) = 2[\sigma(\Sigma^-) + \sigma(\Sigma^-) - 2\sigma(\Sigma^0) + \sigma(\Lambda)]$$

where the numbers in parentheses refer to the isospin channel number.

We next consider the angular momentum states. A careful analysis of bubble chamber data on K^-p interactions near threshold has been made by Humphrey and Ross (1962). The work covers interactions in the range 0 to 250 MeV/c (laboratory system). The branching ratios in the capture process at rest are listed in Table 14.12.

TABLE 14.12

Channel	Branching ratio
$\Sigma^-\pi^+$	0·45 ± 0·01
$\Sigma^+\pi^-$	0·21 ± 0·01
$\Sigma^0\pi^0$	0·28 ± 0·03
$\Lambda\pi^0$	0·06 ± 0·02
$\Lambda\pi^0\pi^0$	0·0013 ± 0·0006

The results of Humphrey and Ross for total cross-sections are displayed in Fig. 14.31; only data for the production of Σ^+ and Σ^- hyperons is shown, since it is difficult to make kinematic distinctions between the production of Λ^0 and Σ^0 particles when the kaon is not at rest. It can be seen that the K^-p interaction is highly absorptive, thus indicating that the imaginary part of the scattering amplitude is comparable with the real part. The angular distributions for the hyperon production processes are all consistent with isotropy, indicating that the K^-p interaction is predominantly s-wave. It was also found that the absorption cross-section approached the limit of π/k^2, which is set by unitarity for s-wave interactions (compare (6.37)), throughout the kaon momentum range (75 to 275 MeV/c in the laboratory system). The concept of s-wave interactions is further supported by theoretical arguments by Day, Snow and Sucher (1959; see also Snow, 1960) concerning K^-p absorption at rest.

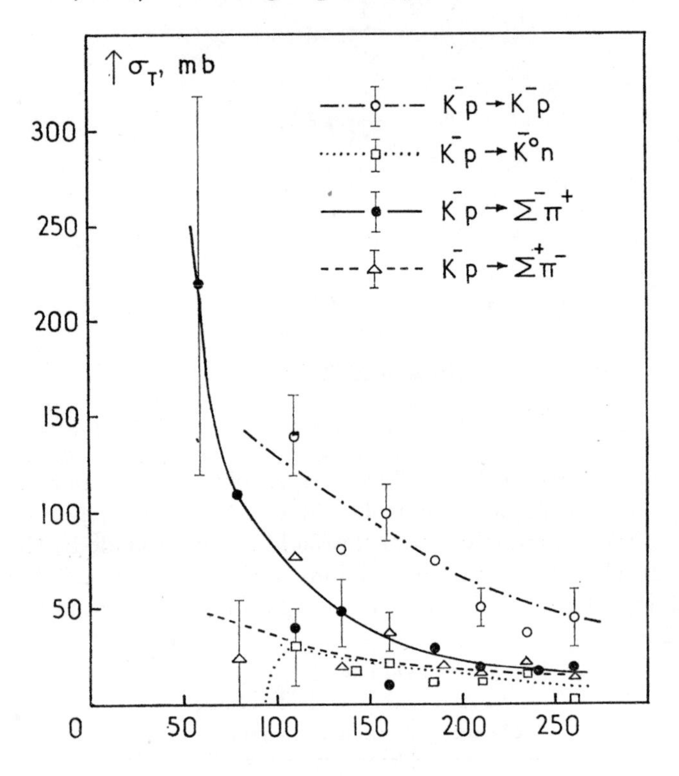

FIG. 14.31.

Thus at low energies the interaction probably proceeds through two channels, $j = \frac{1}{2}$, $l = 0$, $T = 0$ and $j = \frac{1}{2}$, $l = 0$, $T = 1$. A phenomenological analysis of the K^-p interaction at low energies has been made by Dalitz and Tuan (1960; see also Dalitz, 1961a). The method uses an effective range expansion (§ 6.3(d)) for the phase shifts and retains only the leading terms, that is

$$k \cot \delta = \frac{1}{A} + \tfrac{1}{2} k^2 r_0 + \cdots \sim \frac{1}{A}$$

where A represents the scattering length. Complex phase shifts are used (compare (6.13)), and so in the isospin channel T we may write

$$k \cot \delta_T = \frac{1}{A_T}, \qquad A_T = a_T + ib_T. \tag{14.49}$$

It is assumed that these parameters are independent of energy over the restricted range considered in the experiment. If we consider the formulae of § 6.1 (in particular (6.9), (6.10) and (6.17)) in conjunction with (6.101) and (14.48), it is then a straightforward task to show that

$$\sigma(K^-p \to K^-p) = \pi \left| \frac{A_0}{1 - ikA_0} + \frac{A_1}{1 - ikA_1} \right|^2 \tag{14.50}$$

$$\sigma(K^-p \to K^0n) = \pi \left| \frac{A_0}{1 - ikA_0} - \frac{A_1}{1 - ikA_1} \right|^2 \frac{k_0}{k}$$

$$\sigma_r(0) = \frac{4\pi}{k} b_0 \left| \frac{1}{1 - ikA_0} \right|^2$$

$$\sigma_r(1) = \frac{4\pi}{k} b_1 \left| \frac{1}{1 - ikA_1} \right|^2$$

where we have included a correction for the $K^-\bar{K}^0$ mass difference; in practice Coulomb corrections must also be included.

It can be seen that there are four parameters a_0, b_0, a_1 and b_1 and four cross-sections, and so a unique solution is in principle possible. Unfortunately, ambiguities with respect to signs arise in the analysis through reflections about the real and imaginary axes of A, furthermore an experimental problem arises in that Λ^0 and Σ^0 production cannot be distinguished except in K^-p capture at rest. Two further parameters are therefore introduced:

$$\varepsilon \equiv \frac{\sigma(\Lambda\pi^0)}{\sigma_r(1)} = \frac{|t_\Lambda(1)|^2}{|t_\Sigma(1)|^2 + |t_\Lambda(1)|^2}$$

and the ratio γ for $\Sigma^-\pi^+$ to $\Sigma^+\pi^-$ production at rest.

The analysis of Humphrey and Ross has led to the solutions shown in Table 14.13 where a_0, b_0, a_1 and b_1 are expressed in units of 10^{-13} cm.

TABLE 14.13

	a_0	b_0	a_1	b_1	ε	γ
Solution 1	−0·22	2·74	0·02	0·38	0·40	2·15
	±1·07	±0·31	±0·33	±0·08	±0·03	
Solution 2	−0·59	0·96	1·20	0·56	0·39	2·04
	±0·46	±0·17	±0·06	±0·15	±0·02	

Akiba and Capps (1962; see also Dalitz, 1962, 1963) have argued that solution 2 is preferable; the argument is based on a comparison with data for the processes $K^- p \to \Sigma^\mp \pi^\pm$ in the neighbourhood of the Y_0^{**} resonance.

14.5(e). *The hyperon–nucleon interaction*

Very little is known about the interaction of hyperons with free nucleons because of the short lifetime of these particles. The known data can be summarised in Table 14.14.

TABLE 14.14

Reaction	Hyperon momentum (MeV/c)	Cross section (mb)
$\Lambda + p \to \Lambda + p$	500 — 1000	40 ± 20 (a)
	400 — 1000	22 ± 6 (b)
$\Lambda + p \to \Sigma^+ + n$	700 — 800	30 ± 20 (c)
$\Lambda + p \to \Sigma^0 + p$	640 — 1000	9 ± 5 (b)
$\Sigma^+ + p \to \Sigma^+ + p$	500 — 1300	38^{+18}_{-14} (c)
$\Sigma^- + p \to \Sigma^- + p$	500 — 1300	10^{+6}_{-4} (c)

(a) Crawford *et al.* (1959b);
(b) Alexander *et al.* (1961);
(c) Stannard (1961).

Additional information has been obtained from the study of hypernuclei (complex nuclei containing a bound hyperon). It has been concluded that the

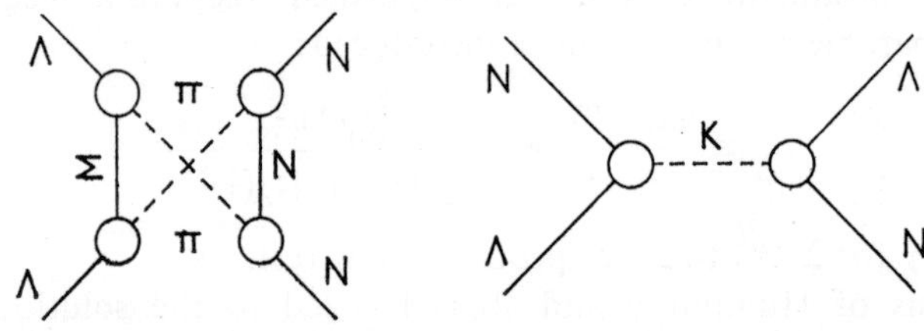

FIG. 14.32.

Λ-nucleon forces have a strong spin dependence, namely that the interaction in the ${}^1S_0$-state is strong whilst that in the ${}^3S_1$-state is relatively weak (Dalitz, 1961b).

The simplest diagrams for Λ–N interactions are shown in Fig. 14.32; they show that the Λ–N force is of shorter range than the N–N force, and that exchange forces can arise from K-meson exchange. The spin dependence of the Λ–N interaction arises naturally for even $\Sigma\Lambda$ parity, and Dalitz has shown that the strength of the singlet force indicates that $g_{NN\pi}^2 \sim g_{\Sigma\Lambda\pi}^2$.

14.6. STRONG INTERACTIONS AT GREAT ENERGIES

14.6(a) *Data on total cross-sections*

In Figs. 14.33, 14.34 and 14.35 curves are given which show the total cross-sections for the interactions of nucleons, antinucleons, pions and kaons with protons at great energies. It can be seen that at ~10 GeV the total nucleon–nucleon cross-section has settled down to a figure of ~40 mb and therefore

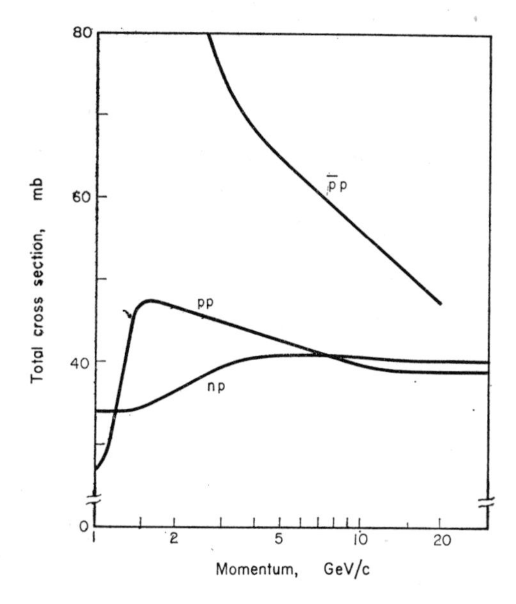

FIG. 14.33. Trends in nucleon–nucleon and antinucleon–nucleon scattering as a function of energy (Diddens *et al.*, 1962c).

remains effectively constant. Work with the cosmic radiation indicates that the total cross-sections have rougly the same value at 10^3 to 10^4 GeV.

An inspection of Fig. 14.33 to 14.35 suggests that the total cross-sections for the interaction of particles and antiparticles with nucleons tend to the same values as the energies increase. This behaviour was predicted by Pomeranchuk (1958). The general theorem may be written as

$$\sigma_{Tab} = \sigma_{T\bar{a}b} \quad (s \to \infty) \tag{14.51}$$

where a and $\bar{a}$ represent particle and antiparticle, and b the partner in the interaction. The symbol s represents the square of the total energy in the c-system.

We have obtained equation (14.51) by use of a Regge-type vacuum pole in § 10.5(f). Rather than discuss the general theorem we shall now show how it

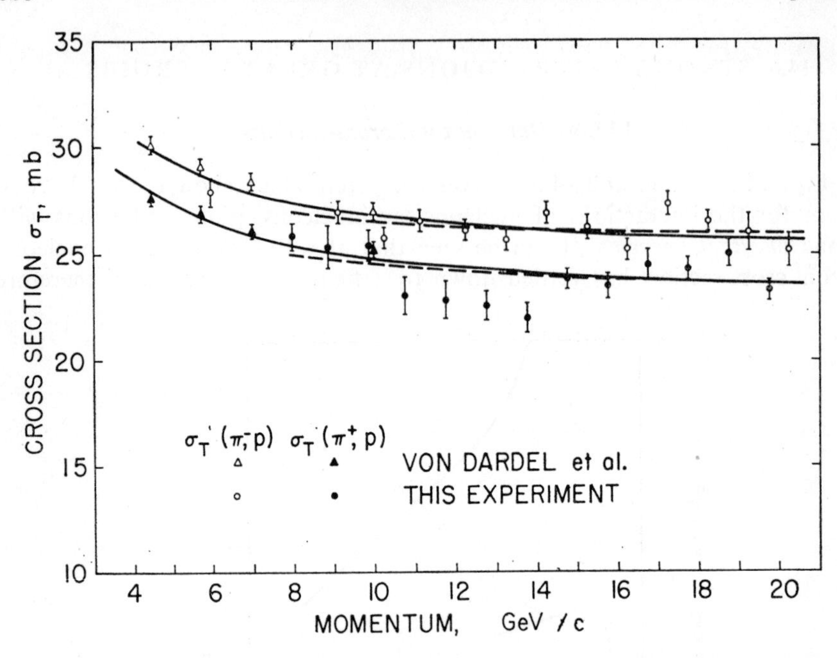

FIG. 14.34. Total cross-sections for $\pi^{\pm}$–p scattering (Lindenbaum *et al.*, 1961).

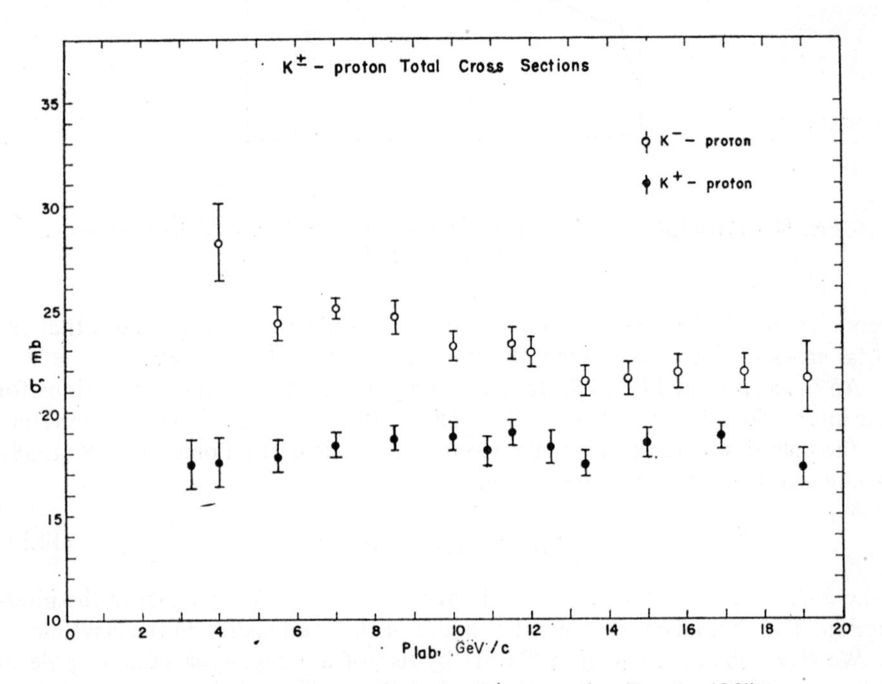

FIG. 14.35. Total cross-sections for $K^{\pm}p$ scattering (Peyrou, 1961).

may arise by the use of equation (10.129) for the forward scattering of pions

$$Re\, f_L^-(\omega) - Re\, f_L^+(\omega) = \frac{f^2}{4\pi} \frac{4\omega}{\omega^2 - \omega_R^2} + \frac{\omega}{2\pi^2} P \int_{m_\pi}^{\infty} d\omega' k_L' \frac{[\sigma_T^-(\omega') - \sigma_T^+(\omega')]}{\omega'^2 - \omega^2}$$

where the superscripts refer to π^- and π^+-mesons respectively. Now if we consider moderate values of ω, the quantities $\mathrm{Re}\, f_L(\omega)$ and $f^2[4\omega/(\omega^2 - \omega_R^2)]$ are nonzero but finite, but if the total cross-sections tend to constant values at high energies the integral then behaves like $[\sigma_T^-(\omega') - \sigma_T^+(\omega')] \log \omega'$, which diverges unless $[\sigma_T^-(\omega') - \sigma_T^+(\omega')] \to 0$ faster than $(\log \omega')^{-1}$ as $\omega' \to \infty$. We may therefore conclude that $\sigma_T^-(\omega') \to \sigma_T^+(\omega')$ as $\omega' \to \infty$. But π^- and π^+-mesons are charge conjugate particles and so equation (14.51) is satisfied for pion–nucleon scattering.

Equation (14.51) possesses an interesting corollary which we shall again demonstrate by pion–nucleon scattering. Consider σ_{Tab} and $\sigma_{T\bar{a}b}$ in isospin notation with the aid of equation (13.5)

$$\sigma_{Tab} \equiv \sigma_T(\pi^- p) = \tfrac{1}{3}\sigma_T(\tfrac{3}{2}) + \tfrac{2}{3}\sigma_T(\tfrac{1}{2})$$

$$\sigma_{T\bar{a}b} \equiv \sigma_T(\pi^+ p) = \sigma_T(\tfrac{3}{2})$$

where the numbers in brackets refer to the isospin channel. The operation of the Pomeranchuk theorem (14.51) then yields

$$\tfrac{1}{3}\sigma_T(\tfrac{3}{2}) + \tfrac{2}{3}\sigma_T(\tfrac{1}{2}) = \sigma_T(\tfrac{3}{2})$$

$$\sigma_T(\tfrac{1}{2}) = \sigma_T(\tfrac{3}{2})$$

The more general relation is

$$\sigma_{Tab}(T) = \sigma_{Tab}(T') \quad (s \to \infty) \tag{14.52}$$

where the T and T' in brackets refer to isospin. This relation implies that

$$\sigma_{Tnp} = \sigma_{Tpp} \quad (s \to \infty).$$

The trend of Fig. 14.33 appears to confirm this relation.

14.6(b). *Regge poles and total cross-sections*

The classical 'black sphere' treatment of scattering at high energies (§ 6.1(c)) predicts constant total cross-sections

$$\sigma_T = 2\pi R^2, \quad \sigma_{sc} = \pi R^2$$

and differential cross-sections which are functions of momentum transfer $(\sqrt{-t})$ alone

$$T(s, t) \to s f(t) \quad (s \to \infty) \tag{14.53}$$

(compare (7.95)). Gribov (1961) pointed out that this behaviour was inconsistent with the dual requirements of unitarity and analyticity in the Mandelstam representation. Furthermore, the predictions do not fit the experimental data since the diffraction peak and the total elastic cross-section for proton–proton scattering are observed to shrink with increasing energy (§ 10.5(d)).

If we wish to make the model compatible with the experimental data we must modify it to include an opacity factor a (§ 6.1(c)), and we then obtain the values

$$\sigma_T = a2\pi R^2, \qquad \sigma_{sc} = a^2\pi R^2$$

(compare (6.25)), so that observed constancy of σ_T and the falling ratio σ_{sc}/σ_T implies that the radius of interaction R and the transparency of the interaction system increase steadily with energy.

An alternative approach to the problems of high energy interactions is supplied by the Regge pole technique (§ 10.5). We have shown previously that the exchange of a Regge pole with the quantum numbers of the vacuum is capable of explaining the experimental features we have outlined above. We shall now examine the problem in a little more detail.

The conjectured behaviour of the Pomeranchuk pole, $\alpha_P(t) = 1$ at $t = 0$ (§ 10.5(d)), implies that the total cross-sections should be constant at high energies for all interactions. Whilst the experimental data tend towards this pattern at the highest energies at present available, they have not reached this condition completely – (see, for example, Fig. 14.33 for nucleon–nucleon scattering). Poles other than the vacuum pole can contribute, however, to the total cross-section. In order to estimate these cross-sections we employ the optical theorem

$$\text{Im } T_{ii} \propto s\sigma_T$$

which reduces to

$$\text{Im } T_{ii} = s\sigma_T$$

(compare (10.165)) if we make the usual assumption that spin effects play a minor role at great energies.

Let us consider nucleon–nucleon scattering in more detail with the aid of the optical theorem. For forward scattering only trajectories with baryon number zero can contribute to the scattering amplitude, that is P, π, η, ω, ϱ. However, π and η are excluded because they are pseudoscalar particles, and so the total cross-sections are given by

$$\sigma_{pp} = \sigma_P - G_\omega^2\left(\frac{s}{2M^2}\right)^{\alpha_\omega(0)-1} - G_\varrho^2\left(\frac{s}{2M^2}\right)^{\alpha_\varrho(0)-1} \tag{14.54}$$

$$\sigma_{\bar{p}p} = \sigma_P + G_\omega^2\left(\frac{s}{2M^2}\right)^{\alpha_\omega(0)-1} + G_\varrho^2\left(\frac{s}{2M^2}\right)^{\alpha_\varrho(0)-1}$$

$$\sigma_{pn} = \sigma_P - G_\omega^2\left(\frac{s}{2M^2}\right)^{\alpha_\omega(0)-1} + G_\varrho^2\left(\frac{s}{2M^2}\right)^{\alpha_\varrho(0)-1}$$

$$\sigma_{\bar{p}n} = \sigma_P + G_\omega^2\left(\frac{s}{2M^2}\right)^{\alpha_\omega(0)-1} - G_\varrho^2\left(\frac{s}{2M^2}\right)^{\alpha_\varrho(0)-1}$$

(Drell, 1962). The form of these equations may be readily understood with the aid of our discussion in § 10.5(d). The term σ_P represents the constant cross-section due to the Pomeranchuk pole, and the expression $2M^2$ has been factored out of the remaining terms in order to write the energy dependence in a convenient dimensionless form

$$\frac{s}{2M^2} \sim \frac{E_L}{M} \quad (s \to \infty)$$

(compare equation (10.139)). The functions G^2 then represent the remainder of the imaginary part of the forward scattering amplitudes (see (10.168) for example) multiplied by $(2M^2)^{\alpha(0)-1}$. The signs in σ_{pp} and $\sigma_{\bar{p}p}$ may be understood within the convention that the diffraction channel represents an attraction whereas the exchange of vector particles (ϱ and ω) gives repulsion between like particles and attraction between unlike particles (compare Coulomb scattering). The differences in signs between σ_{pp} and σ_{pn} and between σ_{pp} and $\sigma_{\bar{p}n}$ arise from consideration of isospin.

The difference in cross-section for pp and $\bar{p}p$ scattering is then due to the ϱ and ω trajectories whilst the pp–pn difference is due to the ϱ-trajectory only

$$\tfrac{1}{2}(\sigma_{\bar{p}p} - \sigma_{pp}) = G_\omega^2\left(\frac{s}{2M^2}\right)^{\alpha_\omega(0)-1} + G_\varrho^2\left(\frac{s}{2M^2}\right)^{\alpha_\varrho(0)-1}$$

$$\tfrac{1}{2}(\sigma_{pn} - \sigma_{pp}) = G_\varrho^2\left(\frac{s}{2M^2}\right)^{\alpha_\varrho(0)-1}.$$

The experimental data show that the former difference ($\sim$20 mb) is much larger than the latter ($\sim$2 mb) in the 10 GeV region (Fig. 14.33), and so the ϱ-trajectory can be neglected as a first approximation. However, it is clear from (14.54) and Fig. 14.33 that the vacuum and ω trajectories cannot explain the experimental data alone, since equation (14.54) implies that the pp and $\bar{p}p$ cross-sections should approach their limiting value σ_P in a symmetric manner. It has therefore been suggested by Drell (1962; see also Hadjioannou, Phillips and Rarita, 1962) that a third trajectory P' contributes to the scattering amplitude with an imaginary part at $t = 0$ which just cancels that from the ω-trajectory so that the pp cross-section is constant.† The amplitudes can therefore be written as

$$T\begin{pmatrix} pp \\ \bar{p}p \end{pmatrix} = T_P + T_{P'} \mp T_\omega$$

in an obvious notation. The trajectory P' has the same quantum numbers as the vacuum, and satisfactory agreement between experiment and theory can be reached if $\alpha_{P'}(0) = \alpha_\omega(0) \sim 0{\cdot}4$.

† Gatland and Moffat (1963) have shown that the effects attributed to this pole may be reproduced by using the Pomeranchuk pole and a cut in the angular momentum plane extending as far as Re $\alpha(t) = 1$ for all t values.

A similar conclusion can be reached concerning the total cross-sections for K^+p and K^-p scattering. Independent evidence for the existence of the P' trajectory has also been presented by Igi (1962) in a study of data from πp scattering. He showed that the measured energy dependence of the total cross-sections can be satisfied by an amplitude which represents the sum of two terms – one corresponding to the Pomeranchuk pole and the other with $\alpha(0) \sim 0{\cdot}5$. Since the ω-system has G parity -1 it cannot contribute to pion–nucleon scattering, and it therefore seems reasonable to conjecture that the second term corresponds to the P' amplitude of nucleon–nucleon scattering.

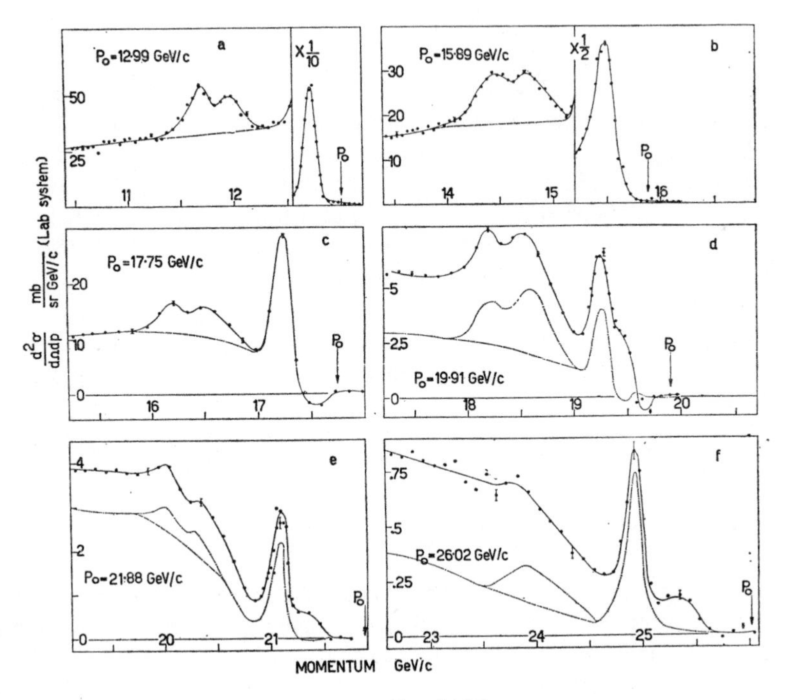

FIG. 14.36.

14.6(c). *Elastic and inelastic cross-sections*

The elastic scattering of protons in pp collisions has been discussed in § 10.5(d), and a trajectory given for $\alpha_P(t)$. This curve was based on the assumption that only the Pomeranchuk pole contributes to the scattering – a not unreasonable assumption since this pole dominates the other terms at high energies and small momentum transfers. As $\alpha_P(t) \to 0$, however, and then goes negative, the contribution from the other terms become increasingly important. For example the presence of the ω and P' terms could cause the width of the diffraction peak in $\bar{p}p$ scattering to be narrower than that for p–p scattering (Hadjioannu, Phillips and Rarita, 1962). An effect of this nature has been observed by Goldschmidt-Clermont *et al.* (1962).

In contrast to nucleon–nucleon scattering the work on pion–nucleon scattering has shown no evidence for a shrinkage in the diffraction pattern with increasing energy (§ 10.5(d)). This result cannot be explained either in terms of a single Pomeranchuk pole or a model involving three poles, as discussed in the previous section. It would therefore appear that if the Regge pole hypothesis is correct, its manipulation cannot be as simple as early results indicated.

We mention, finally, inelastic nucleon–nucleon scattering at high energies. Here the most notable feature is the appearance of two peaks in the momentum spectrum of the inelastically scattered particles (Cocconi *et al.*, 1961). The results of this experiment are illustrated in Fig. 14.36. The rest energies of the recoiling systems, corresponding to the peaks were found to correspond closely to the masses of the N_2^* and N_3^* nucleon isobars (1·51 and 1·69 GeV respectively). A notable feature is the absence of the N_1^* isobar, which appears in proton–proton inelastic scattering at lower energies (§ 14.3(a)). This isobar has isospin quantum number $T = \frac{3}{2}$ and so cannot be produced by the exchange of vacuum poles (Frautschi, Gell-Mann and Zachariasen, 1962).

APPENDIXES

A.1. NOTATION

In this book the metric

$$x = \mathbf{x},\ ict \equiv \mathbf{x},\ it \qquad (x_4 = it)$$

$$k = \mathbf{k},\ \frac{iE}{c} \equiv \mathbf{k},\ iE \qquad (k_4 = iE)$$

has been used. Vectors in three-dimensional spaces have been given in bold type and their components by Roman indices; vector operators have been written as

$$\operatorname{div} \mathbf{x} = \nabla \cdot \mathbf{x} \qquad \operatorname{grad} \varphi = \nabla \varphi \tag{A.1.1}$$

$$\operatorname{curl} \mathbf{x} = \nabla \times \mathbf{x} \qquad \nabla^2 \mathbf{x} = \left(\frac{\partial^2}{\partial x^2} + \frac{\partial^2}{\partial y^2} + \frac{\partial^2}{\partial y^2} \right) \mathbf{x}.$$

Four-vectors have been written in light type and Greek indices have been used to denote their components. The d'Alembertian operator has been written as

$$\begin{aligned} \Box^2 \varphi &= \left(\frac{\partial^2}{\partial x^2} + \frac{\partial^2}{\partial y^2} + \frac{\partial^2}{\partial z^2} - \frac{\partial^2}{\partial t^2} \right) \varphi \\ &= \left(\frac{\partial^2}{\partial x_1^2} + \frac{\partial^2}{\partial x_2^2} + \frac{\partial^2}{\partial x_3^2} + \frac{\partial^2}{\partial x_4^2} \right) \varphi. \end{aligned} \tag{A.1.2}$$

The Einstein convention for repeated suffices in summations has been normally followed

$$A_{\lambda\alpha} B_{\mu\alpha} = \sum_{\alpha} A_{\lambda\alpha} B_{\mu\alpha} \tag{A.1.3}$$

but when it has been especially important to draw attention to the summation this rule has been relaxed. In addition, summations involving the scalar products of the momentum four-vector have been written without indices in places in order to simplify notation

$$p_\lambda^2 = p_\lambda p_\lambda = p^2 \tag{A.1.4}$$

$$\gamma_\lambda p_\lambda = \gamma p.$$

A.2. UNITS

In the equations for quantum mechanical systems, the symbols $\hbar$ (Planck's constant/2π) and c (the velocity of light) occur repeatedly. It is therefore convenient to adopt a system of units which permits the elimination of these symbols from the quantum mechanical equations.

This result may be achieved by employing a system which reduces all quantities in the c.g.s. system to powers of $\hbar$ and c and one natural dimension – length. Thus consider a quantity A' with the following dimensions in the c.g.s. system:

$$[A'] = M^{\alpha}L^{\beta}T^{\gamma}.$$

In order to introduce the new quantity A with dimensions of length, L, we must combine $\hbar$ and c in powers such that the dimensions of M and T are zero. Now in the c.g.s. system $\hbar$ and c possess the following dimensions:

$$[\hbar] = ML^2T^{-1}, \qquad [c] = LT^{-1}.$$

Therefore

$$[A] = [A']f(\hbar, c) = [A'][\hbar]^{\delta}[c]^{\varepsilon}$$
$$= M^{\alpha}L^{\beta}T^{\gamma}(ML^2T^{-1})^{\delta}(LT^{-1})^{\varepsilon}$$

and to set M and T to zero powers we must have the following equations

$$[M] \qquad \alpha + \delta = 0 \qquad (\delta = -\alpha) \tag{A.2.1}$$

$$[T] \qquad \gamma - \delta - \varepsilon = 0 \qquad (\varepsilon = \alpha + \gamma)$$

and so

$$[A] = [A']\hbar^{-\alpha}c^{\alpha+\gamma} \tag{A.2.2}$$
$$= L^{\beta-\alpha+\gamma}.$$

Thus mass, for example, has dimensions (length)$^{-1}$ in the new system. Similarly, since energy has dimensions,

$$[E'] = ML^2T^{-2} \equiv M^{\alpha}L^{\beta}T^{\gamma}$$

in c.g.s. units, then

$$[E] = L^{2-1-2} = L^{-1}, \qquad E \equiv E'/\hbar c \tag{A.2.3}$$

so that

$$E'_{\text{ergs}} = E_{\text{natural units}} \times 1{\cdot}05 \times 10^{-27} \times 3 \times 10^{10}.$$

It is a straightforward matter to show that $\hbar = c = 1$ in natural units.

Now let us examine electronic charge, in electrostatic units

$$[e'] = M^{1/2}L^{3/2}T^{-1}$$

therefore, in natural units

$$[e] = L^0, \qquad e = \frac{e'}{\hbar^{1/2}c^{1/2}} \tag{A.2.4}$$

and

$$e'_{\text{esu}} = e_{\text{natural units}}\, \hbar^{1/2}c^{1/2}.$$

This gives us (approximately)

$$e = \sqrt{\frac{1}{137}}$$

in natural units. However, charge is defined by Coulomb's law through a proportionality factor, and a more convenient unit is provided by a rationalised system in which e has the approximate value

$$e \sim \sqrt{\left(\frac{4\pi}{137}\right)} \quad \text{or} \quad \frac{e^2}{4\pi} \sim \frac{1}{137}. \tag{A.2.5}$$

Numerical data on energy, momentum and mass (and sometimes length) are normally quoted in MeV units. The first three quantities are related by the equation

$$E^2 = p^2c^2 + m^2c^4$$

and momentum and mass have MeV/c and MeV/c^2 units respectively (the c is frequently suppressed). The MeV unit is related to other frequently encountered quantities in the following manner

$$\begin{aligned}
1\text{eV} &= \frac{e_{\text{esu}}}{\text{c}} \times 10^8 \text{ erg} \\
1 \text{ MeV} &= 1{\cdot}6021 \times 10^{-6} \text{ erg} \\
\hbar &= 1{\cdot}054 \times 10^{-27} \text{ erg sec} = 6{\cdot}582 \times 10^{-22} \text{ MeV sec} \\
\hbar c &= 1{\cdot}973 \times 10^{-11} \text{ MeV cm} \\
1 \text{ fermi} &= 10^{-13} \text{ cm} = (197 \text{ MeV/c})^{-1}
\end{aligned}$$

where the last relation arises from the de Broglie equation $\lambda\!\!\!^{-} = \hbar/p$.

A.3. MATRICES

A. 3(a). *Introduction*

A *matrix* is a square or rectangular set of numbers, which may be related to other matrices by a set of rules which we shall discuss below. We shall denote a matrix by, say, A and its *components or elements* by A_{ij}, where i and j denote respectively the row and column in which the element A_{ij} appears.

$$A = [A_{ij}] = \begin{matrix} A_{11}A_{12}A_{13} \dots A_{1m} \\ A_{21}A_{22}A_{23} \dots A_{2m} \\ A_{31}A_{32}A_{33} \dots A_{3m} \\ \dots\dots\dots\dots \\ A_{n1}A_{n2}A_{n3} \dots A_{nm} \end{matrix} \tag{A.3.1}$$

This matrix is said to be of order $(n \times m)$, since it has n rows and m columns. A matrix with a single row or column is called a *row* or *column vector* respectively.

A.3(b). *Matrix addition and multiplication*

Two matrices A and B may be added or subtracted if they are of the same order. The result is a third matrix

$$C = A \pm B \tag{A.3.2}$$

$$C_{ij} = A_{ij} \pm B_{ij}.$$

The matrices A and B may be multiplied in the order AB provided that the number of columns in A equals the number of rows in B. If this condition holds the matrices are said to be *conformable*. If A and B are of order $(n \times h)$ and $(h \times m)$ respectively, then the product C is of order $(n \times m)$. The matrix elements of C are given by

$$C_{ij} = \sum_{s=1}^{h} A_{is}B_{sj} \equiv A_{is}B_{sj} \tag{A.3.3}$$

where

$$i = 1, 2 \dots n, \qquad j = 1, 2 \dots m$$

The laws regarding distribution and association apply to matrix multiplication, that is

$$A(B + C)\,D = ABD + ACD$$

$$D = ABC = (AB)\,C = A(BC)$$

but in general $AB \neq BA$. When the equality holds the two matrices are said to *commute.*

The multiplication of conformable row (R) and column (C) vectors with each and with other matrices leads to the following results:

$$AC = C' \qquad RA = R' \tag{A.3.4}$$

$$RC = \text{scalar quantity} \qquad CR = B$$

where R' and C' are new row and column vectors respectively and A and B are matrices.

A.3(c). *Special matrices*

The additive and multiplication rules can be used to define a *null matrix* O, with the properties

$$O + A = A, \qquad OA = AO = O \tag{A.3.5}$$

that is, all of its elements are zero.

Another matrix which may be defined by the multiplication law is the unit *matrix* $\hat{1}$

$$\hat{1}A = A\hat{1} = A. \tag{A.3.6}$$

The elements of the unit matrix can be written as δ_{ij}, that is, they are all zero except along the diagonal.

Any matrix in which all the elements vanish except along a diagonal is known as a *diagonal matrix*; its elements can be represented as $A_i\delta_{ij}$. All diagonal matrices commute with each other, for example if A and B are diagonal

$$(AB)_{ik} = \sum_j A_i\delta_{ij}B_j\delta_{jk} = A_iB_i\delta_{ik} = (BA)_{ik}.$$

The *Hermitian adjoint* $A^\dagger$ of a matrix A may be obtained by transposing rows and columns and taking the complex conjugate of each element, so that

$$A^\dagger_{ij} = A^*_{ji}. \tag{A.3.7}$$

It is a straightforward matter to show that the Hermitian adjoint of a product of matrices is a product of the adjoints in reverse order

$$(ABC)^\dagger = C^\dagger B^\dagger A^\dagger. \tag{A.3.8}$$

A matrix is said to be *Hermitian* or *self-adjoint* if

$$A^\dagger = A \tag{A.3.9}$$

A matrix may possess an inverse which is defined by

$$AA^{-1} = \hat{1} = A^{-1}A. \tag{A.3.10}$$

Consider the product of two nonsingular† square matrices $AB = C$ and let us multiply the both sides by $B^{-1}A^{-1}$ and C^{-1} as indicated below, then

$$B^{-1}A^{-1}ABC^{-1} = B^{-1}A^{-1}CC^{-1} \tag{A.3.11}$$

$$C^{-1} = B^{-1}A^{-1}.$$

This result may extend to the product of any number of matrices.

† A singular matrix possesses the property $\det A = 0$.

If A is square and nonsingular the ij element of its inverse is given by the cofactor of A_{ij} divided by the determinant of A (see below)

$$A_{ij}^{-1} = \frac{\text{cofactor } A_{ij}}{\det A}. \tag{A.3.12}$$

A.3(d). *Evaluation of matrices*

The trace or spur of a square matrix is the sum of its diagonal elements

$$\operatorname{tr} A = \sum_i A_{ii}. \tag{A.3.13}$$

It is a simple matter to show that

$$\operatorname{tr} AB = \operatorname{tr} BA$$

The *determinant* of a square matrix may be obtained in the following manner. Let us assume that the matrix has n rows and columns; a total of $n!$ products may be obtained by taking one element from each row and column. Each product is then arranged so that the first subscripts of the elements are in their natural order. This operation leads to an even and odd set of $\frac{1}{2}n!$ terms; in the even set an even number of interchanges will bring the second subscripts to their natural order, and an odd number of changes is required for the second set. If plus and minus signs are given to the even and odd terms respectively, the algebraic sum of the $n!$ terms is the value of the determinant. Thus we find

$$\det A = \sum (-1)^h A_{1j_1} A_{2j_2} \ldots A_{nj_n} \tag{A.3.14}$$

where the summation is made over all permutations of $j_1, j_2, \ldots, j_n$ and h is the number of interchanges required to restore the natural order. As an example consider a matrix with $n = 3$

$$\begin{aligned}
\det A = \det &\begin{bmatrix} A_{11}A_{12}A_{13} \\ A_{21}A_{22}A_{23} \\ A_{31}A_{32}A_{33} \end{bmatrix} \\
&= A_{11}A_{22}A_{33} + A_{11}A_{23}A_{32} + A_{12}A_{23}A_{31} \\
&\quad + A_{12}A_{21}A_{33} + A_{13}A_{21}A_{32} + A_{13}A_{22}A_{31} \\
&= A_{11}A_{22}A_{33} - A_{11}A_{32}A_{23} + A_{31}A_{12}A_{23} \\
&\quad - A_{21}A_{12}A_{33} + A_{21}A_{32}A_{13} - A_{31}A_{22}A_{13}
\end{aligned} \tag{A.3.15}$$

The following rules for determinants arise from the definition (A.3.14):

(1) det $A = 0$ if all elements of a row or column are zero, or if all elements of a row (column) are identical with, or multiples of, the corresponding elements of another row (column),

(2) the value of a determinant does not change if rows and columns are interchanged,

(3) the value of a determinant changes sign if two rows (columns) are interchanged.

We mention two final definitions. The *complementary minor* of an element A_{ij} is the determinant obtained by striking out the row and column in which A_{ij} appears. The *cofactor* of A_{ij} is $(-1)^{i+j}$ times its complementary minor.

A.4. TENSORS

A tensor was defined in equation (3.41) as a system which behaved as

$$T'_{\lambda\mu\varrho\ldots} = a_{\lambda\alpha}a_{\mu\beta}a_{\varrho\gamma}\ldots T_{\alpha\beta\gamma\ldots}$$

during an orthogonal transformation of the coordinates. It is obvious that this definition does not have to be restricted to the coordinates of a space–time system. From the above definition one may show that

$$T_{\alpha\beta\gamma\ldots\eta}A_\alpha B_\beta C_{\gamma}\ldots X_\eta$$

is an invariant if A, B, C,... are vectors. This property gives an alternative definition of a tensor.

In addition systems are sometimes encountered which transform in the reverse manner to that defined above

$$T'^{\lambda\mu\varrho\ldots} = a^{-1}_{\alpha\lambda}a^{-1}_{\beta\mu}a^{-1}_{\gamma\varrho}\ldots T^{\alpha\beta\gamma\ldots}.$$

These tensors are called *contragradient tensors*; mixed tensors are also encountered (compare § 13.5(a)).

The rules for the manipulation of tensors may be summarised as follows:

(1) addition and subtraction – if these operations are performed with corresponding components of tensors of the same rank, a tensor of equal rank results

$$T_{\alpha\beta\gamma} \pm U_{\alpha\beta\gamma} = V_{\alpha\beta\gamma}. \qquad \text{(A.4.1)}$$

(2) a tensor of rank α and one of rank β may be multiplied to yield a tensor of rank $\alpha + \beta$, if all the components of the first tensor are multiplied by all of the components of the second tensor

$$T_{\alpha\beta\gamma\ldots\lambda\mu\varrho} = U_{\alpha\beta\gamma\ldots}V_{\lambda\mu\varrho\ldots} \qquad \text{(A.4.2)}$$

(3) contraction – a tensor of rank $(\alpha - 2)$ may be obtained from one of rank α by putting two indices equal and summing for this index

$$T_{\gamma\ldots} = T_{\alpha\alpha\gamma\ldots} \equiv \sum_\alpha T_{\alpha\alpha\gamma\ldots} \qquad \text{(A.4.3)}$$

We finally mention some special tensors.

(1) Symmetrical properties – if a tensor changes sign upon interchange of two of its indices it is called a *skew-symmetric* or *antisymmetric tensor*; if the original sign is retained it is called a *symmetric tensor*, for example

$$T_{\alpha\beta} = T_{\beta\alpha} \quad \text{symmetric} \tag{A.4.4}$$

$$T_{\alpha\beta} = -T_{\beta\alpha} \quad \text{antisymmetric.}$$

Symmetric and antisymmetric tensors retain their identities under transformations, for example if $T_{\alpha\beta} = T_{\beta\alpha}$ then

$$T'_{\lambda\mu} = a_{\lambda\alpha}a_{\mu\beta}T_{\alpha\beta} = a_{\mu\beta}a_{\lambda\alpha}T_{\beta\alpha} = T'_{\mu\lambda}$$

and if $T_{\alpha\beta} = -T_{\beta\alpha}$,

$$T'_{\lambda\mu} = a_{\lambda\alpha}a_{\mu\beta}T_{\alpha\beta} = -a_{\mu\beta}a_{\lambda\alpha}T_{\beta\alpha} = -T'_{\lambda\mu}.$$

(2) The Kronecker delta term represents the components of a tensor of rank two (the fundamental tensor) – proof, if we substitute $\delta_{\alpha\beta}$ for $T_{\alpha\beta}$ in

$$T'_{\lambda\mu} = a_{\lambda\alpha}a_{\mu\beta}T_{\alpha\beta}$$

we find

$$T'_{\lambda\mu} = a_{\lambda\alpha}a_{\mu\alpha} = \delta_{\lambda\mu}$$

by (3.37).

(3) There is a tensor with components $\varepsilon_{\alpha\beta\gamma\ldots}$ which is antisymmetric with respect to all pairs of indices, and whose components are

$$\begin{array}{rl} +1 & \text{if } \alpha\beta\gamma\ldots \text{ is an even permutation of } 1, 2, 3, \ldots \\ -1 & \text{if } \alpha\beta\gamma\ldots \text{ is an odd permutation of } 1, 2. 3, \ldots \\ 0 & \text{if an any two indices are identical.} \end{array} \tag{A.4.5}$$

As an example of the use of ε we may write equation (9.91) as

$$\begin{aligned} T_3 &= -\int d\mathbf{x}(\pi_1\varphi_2 - \pi_2\varphi_1) \\ &= -\varepsilon_{3ij}\int d\mathbf{x}\,\pi_i\varphi_j \end{aligned} \tag{A.4.6}$$

A.5. γ-MATRICES

The conditions for the α, β, and γ-matrices of the Dirac equation

$$\beta^\dagger = \beta, \qquad \alpha_k^\dagger = \alpha_k, \qquad \beta^2 = \alpha_k^2 = \hat{1} \tag{A.5.1}$$

$$(\alpha_k\alpha_l + \alpha_l\alpha_k) = 2\delta_{kl}, \qquad \alpha_k\beta + \beta\alpha_k = 0$$

$$\gamma_4 = \beta \qquad \gamma_k = -i\beta\alpha_k$$

$$\gamma_\mu\gamma_\lambda + \gamma_\lambda\gamma_\mu = 2\,\delta_{\mu\lambda}$$

discussed in §§ 3.3(b), (c) and (d), permit considerable latitude in assigning specific forms to the γ-matrices. We list below some sets which are often encountered in the literature:

$$\gamma_k = \begin{pmatrix} 0 & -i\sigma_k \\ i\sigma_k & 0 \end{pmatrix} \quad \gamma_4 = \begin{pmatrix} 1 & 0 \\ 0 & -1 \end{pmatrix} \quad \gamma_5 = \begin{pmatrix} 0 & -1 \\ -1 & 0 \end{pmatrix}$$

$$\gamma_k = \begin{pmatrix} 0 & i\sigma_k \\ -i\sigma_k & 0 \end{pmatrix} \quad \gamma_4 = \begin{pmatrix} 0 & 1 \\ 1 & 0 \end{pmatrix} \quad \gamma_5 = \begin{pmatrix} -1 & 0 \\ 0 & 1 \end{pmatrix}$$

$$\gamma_1 = \begin{pmatrix} 1 & 0 \\ 0 & -1 \end{pmatrix} \quad \gamma_2 = \begin{pmatrix} 0 & -i\sigma_2 \\ i\sigma_2 & 0 \end{pmatrix} \quad \gamma_3 = \begin{pmatrix} 0 & -1 \\ -1 & 0 \end{pmatrix}$$

$$\gamma_4 = \begin{pmatrix} 0 & -i\sigma_1 \\ i\sigma_1 & 0 \end{pmatrix} \quad \gamma_5 = \begin{pmatrix} 0 & -i\sigma_3 \\ i\sigma_3 & 0 \end{pmatrix}$$

where 1 and 0 represent 2×2 unit and null matrices respectively.

These sets are called the Dirac–Pauli, Kramers and Majorana representations respectively. The Dirac–Pauli set gives a simple interpretation of the spinor components of the Dirac equation, whereas the Kramers and Majorana representations have some advantages in discussing the transformation properties of the Dirac fields.

In most problems involving transition probabilities it is possible to avoid specific representations for the γ-matrices. The important quantity in such problems is the trace of the γ-matrix (§ 7.4(d)). By use of the properties

$$\gamma_\mu \gamma_\lambda + \gamma_\lambda \gamma_\mu = 2\delta_{\mu\lambda} \tag{A.5.2}$$

$$\mathrm{tr}\ \hat{1} = 4$$

where $\hat{1}$ represents a unit 4×4 matrix, the following properties of the traces may be proved (the terms A_λ, B_μ, C_ϱ and D_ν are four-vectors)

(1) $$\mathrm{tr}\ \gamma_\mu \gamma_\lambda = 4\,\delta_{\mu\lambda} \tag{A.5.3}$$

(2) $$\mathrm{tr}\ \gamma_\lambda \gamma_\mu \gamma_\varrho \gamma_\nu = 4(\delta_{\lambda\mu}\delta_{\varrho\nu} - \delta_{\lambda\varrho}\delta_{\mu\nu} + \delta_{\lambda\nu}\delta_{\mu\varrho}) \tag{A.5.4}$$

(3) If $(\gamma_\mu \gamma_\lambda \dots \gamma_\varrho)$ contains an odd number of γ-matrices

$$\mathrm{tr}\ (\gamma_\mu \gamma_\lambda \dots \gamma_\varrho) = 0 \tag{A.5.5}$$

(4) $$\mathrm{tr}\ \gamma_\lambda A_\lambda \gamma_\mu B_\mu = 4\,AB \tag{A.5.6}$$

(5) $$\mathrm{tr}\ \gamma_\lambda A_\lambda \gamma_\mu B_\mu \gamma_\varrho C_\varrho \gamma_\nu D_\nu$$

$$= 4[(AB)(CD) - (AC)(BD) + (AD)(BC)] \tag{A.5.7}$$

The proof of (A.5.5) is given in § 3.3(d).

We list, finally, some properties of the scalar products of γ-matrices (the summation convention is implied for repeated Greek indices)

(1) $$\gamma_\mu A_\mu \gamma_5 = -\gamma_5 \gamma_\mu A_\mu \qquad \text{(A.5.8)}$$

(2) $$\gamma_\mu A_\mu \gamma_\lambda B_\lambda = -\gamma_\lambda B_\lambda \gamma_\mu A_\mu + 2\, A_\lambda B_\lambda \qquad \text{(A.5.9)}$$

(3) $$\gamma_\mu a \gamma_\mu = 4a \qquad \text{(A.5.10)}$$

(4) $$\gamma_\mu \gamma_\lambda A_\lambda \gamma_\mu = -2\gamma_\lambda A_\lambda \qquad \text{(A.5.11)}$$

(5) $$\gamma_\mu \gamma_\lambda A_\lambda \gamma_\varrho B_\varrho \gamma_\mu = 4\, A_\lambda B_\lambda \qquad \text{(A.5.12)}$$

(6) $$\gamma_\mu \gamma_\lambda A_\lambda \gamma_\varrho B_\varrho \gamma_\nu C_\nu \gamma_\mu = -2\gamma_\nu C_\nu \gamma_\varrho B_\varrho \gamma_\lambda A_\lambda \qquad \text{(A.5.13)}$$

where a is a complex number. The above relations may be proved by applications (and expansions) of equations (A.5.2).

A.6. δ-FUNCTIONS

The Dirac δ-function may be defined by the following properties:

$$\delta(x-a) = 0 \qquad (x \neq a) \qquad \text{(A.6.1)}$$

$$\int_b^c dx\, \delta(x-a) = 1 \text{ if } a \text{ contained in interval } b \text{ to } c$$
$$= 0 \text{ if } a \text{ is not in this interval.}$$

The function can be conveniently expressed in mathematical form in the following manner:

$$\frac{1}{2\pi} \int_{-\infty}^{+\infty} dx\, e^{i(k_x - k'_x)x} = \delta(k_x - k'_x) \qquad \text{(A.6.2)}$$

where x is one-dimensional; the δ-function can also be given multidimensional form, for example if $(k-k')x$ is four-dimensional, then

$$\frac{1}{(2\pi)^4} \int_{-\infty}^{+\infty} d^4x\, e^{i(k-k')x} = \delta(k-k') \equiv \delta^4(k-k'). \qquad \text{(A.6.3)}$$

The following useful properties may be associated with the δ-function (Dirac, 1947):

$$\delta(x) = \delta(-x) \qquad \text{(A.6.4)}$$

$$\frac{\partial}{\partial x}\delta(x) = -\frac{\partial}{\partial x}\delta(-x)$$

$$x\delta(x) = 0$$

$$x \frac{\partial}{\partial x} \delta(x) = -\delta(x)$$

$$\delta(ax) = \frac{1}{a} \delta(x) \qquad a > 0$$

$$\delta(x^2 - a^2) = \frac{1}{2a} [\delta(x - a) + \delta(x + a)] \qquad a > 0$$

$$\int dx\, \delta(a - x)\, \delta(x - b) = \delta(a - b)$$

$$f(x)\, \delta(x - a) = f(a)\, \delta(x - a)$$

$$\int dx\, f(x)\, \delta[y(x)] = \sum_i \frac{f(x_i)}{|\partial y(x_i)/\partial x|}$$

where the points x_i are the real roots of $y(x) = 0$ in the interval of integration.

Several other singular functions may be associated with the δ-function; for example, if a system is discrete rather than continuous the Kroenecker δ-symbol may be used.

$$\frac{1}{V} \sum_{\mathbf{x}} e^{i(\mathbf{k}-\mathbf{k}')\cdot\mathbf{x}} = \delta_{\mathbf{kk}'} \qquad \text{(A.6.5)}$$

$$\delta_{\mathbf{kk}'} = \begin{cases} 1 & \mathbf{k} = \mathbf{k}' \\ 0 & \mathbf{k} \neq \mathbf{k}'. \end{cases}$$

Other useful functions are the step functions with the properties

$$\theta(t) = \frac{1}{2\pi i} \int_{-\infty}^{+\infty} da \frac{e^{iat}}{a - i\varepsilon} = \begin{cases} 1 & t > 0 \\ 0 & t < 0 \end{cases} \qquad \text{(A.6.6)}$$

$$\varepsilon(t) = \frac{1}{\pi i} P \int_{-\infty}^{+\infty} \frac{da}{a} e^{iat} = \theta(t) - \theta(-t) = \begin{cases} +1 & t > 0 \\ -1 & t < 0 \end{cases}$$

where P denotes the principal value.

A.7. SPHERICAL FUNCTIONS

A plane wave of wave number k (where $k = \hbar/p$), travelling in the positive z-direction, can be represented by the function e^{ikz}. This function can be expanded into a sum of spherical waves in the following manner:

$$e^{ikz} = \sum_{l=0}^{\infty} \sqrt{[4\pi(2l + 1)]}\, i^l j_l(kr)\, Y_l^0(\theta, \varphi) \qquad \text{(A.7.1)}$$

and if an arbitrary direction θ', φ' is chosen for the direction of the plane wave, then

$$e^{i\mathbf{k}\cdot\mathbf{r}} = 4\pi \sum_{l=0}^{\infty} \sum_{m=-l}^{m=+l} i^l j_l(kr)\, Y_l^{m*}(\theta', \varphi')\, Y_l^m(\theta, \varphi) \qquad \text{(A.7.2)}$$

(Blatt and Weisskopf, 1952). In these equations $j_l(kr)$ in a spherical Bessel function and $Y_l^0(\theta, \varphi)$ is a spherical harmonic. We list below some useful properties of these functions.

A.7(a). *Spherical Bessel functions*

The spherical Bessel function may be written as

$$j_l(\mu) = \left(\frac{\pi}{2\mu}\right)^{\frac{1}{2}} J_{l+1/2}(\mu) \qquad \text{(A.7.3)}$$

where J is an ordinary Bessel function and l is an integer (tables of Bessel functions may be found in Jahnke and Emde, 1945). The spherical function may be expressed as

$$j_l(\mu) = (-\mu)^l \left(\frac{1}{\mu}\frac{d}{d\mu}\right)^l \left(\frac{\sin\mu}{\mu}\right) \qquad \text{(A.7.4)}$$

and so

$$j_0(\mu) = \frac{\sin\mu}{\mu}, \qquad j_1(\mu) = \frac{\sin\mu}{\mu^2} - \frac{\cos\mu}{\mu}$$
$$j_2(\mu) = \left(\frac{3}{\mu^3} - \frac{1}{\mu}\right)\sin\mu - \frac{3}{\mu^2}\cos\mu. \qquad \text{(A.7.5)}$$

For small and large values of μ the function behaves like

$$j_l(\mu) = \frac{\mu^l}{(2l+1)!!} \qquad |\mu| \ll l \qquad \text{(A.7.6)}$$

$$j_l(\mu) = \frac{1}{\mu}\sin\left(\mu - l\frac{\pi}{2}\right) \qquad |\mu| \gg l$$

where

$$(2l+1)!! = 1 \times 3 \times 5, \ldots (2l+1).$$

A.7(b). *Spherical harmonics*

The spherical harmonics $Y_l^m(\theta, \varphi)$ satisfy the relations

$$Y_l^m(\theta, \varphi) = \left[\frac{2l+1}{4\pi}\frac{(l-m)!}{(l+m)!}\right]^{\frac{1}{2}} P_l^m(\cos\theta)\, e^{im\varphi} \qquad \text{(A.7.7)}$$

$$\mathbf{L}^2 Y_l^m = l(l+1)\, Y_l^m$$

$$L_z Y_l^m = m Y_l^m$$

$$L_\pm Y_l^m = (L_x \pm iL_y)\, Y_l^m = [l(l+1) - m(m \pm 1)]^{1/2}\, Y_l^{m \pm 1}$$

where we have followed the phase convention of § 5.3(b) in the last equation. In the above relations

$$l = 0, 1, 2, 3, \ldots$$

$$m = -l, (-l+1), \ldots (l-1), l$$

and P_l^m is an associated Legendre polynomial. For positive m

$$P_l^m(x) = (-1)^m (1 - x^2)^{m/2} \frac{d^m}{dx^m} P_l(x) \tag{A.7.8}$$

$$P_l(x) = \frac{1}{2^l l!} \frac{d^l}{dx^l} (x^2 - 1)^l.$$

In the first of the above equations we have followed the phase convention of Condon and Shortley (1935); most works use this convention.

The following relationships exist between the components $\pm m$ for a given value of l

$$P_l^{-m}(x) = (-1)^m \frac{(l-m)!}{(l+m)!} P_l^m(x) \tag{A.7.9}$$

$$Y_l^{-m}(\theta, \varphi) = (-1)^m\, Y_l^{m*}(\theta, \varphi).$$

Using (A.7.7) and (A.7.8) the lowest order spherical harmonics are given by

$$l = 0 \quad Y_0^0 = \frac{1}{\sqrt{(4\pi)}} \tag{A.7.10}$$

$$l = 1 \quad Y_1^0 = \sqrt{\left(\frac{3}{4\pi}\right)} \cos\theta \qquad Y_1^1 = -\sqrt{\left(\frac{3}{8\pi}\right)} \sin\theta\, e^{i\varphi}$$

$$l = 2 \quad Y_2^0 = \sqrt{\left(\frac{5}{16\pi}\right)} (3\cos^2\theta - 1)$$

$$Y_2^1 = -\sqrt{\left(\frac{15}{8\pi}\right)} \sin\theta \cos\theta\, e^{i\varphi}$$

$$Y_2^2 = \sqrt{\left(\frac{15}{32\pi}\right)} \sin^2\theta\, e^{2i\varphi}$$

$$l = 3 \qquad Y_3^0 = \sqrt{\left(\frac{7}{16\pi}\right)}(5\cos^3\theta - 3\cos\theta)$$

$$Y_3^1 = -\sqrt{\left(\frac{21}{64\pi}\right)}\sin\theta(5\cos^2\theta - 1)\,e^{i\varphi}$$

$$Y_3^2 = \sqrt{\left(\frac{105}{32\pi}\right)}\sin^2\theta\cos\theta\,e^{2i\varphi}$$

$$Y_3^3 = -\sqrt{\left(\frac{35}{64\pi}\right)}\sin^3\theta\,e^{3i\varphi}.$$

Values for $Y_l^{-m}(\theta, \varphi)$ can be obtained from (A.7.9).

The function Y_l^m satisfies the following conditions for orthonormality and completeness

$$\int_0^{2\pi} d\varphi \int_0^{\pi} d\theta \sin\theta Y_{l'}^{m'*}(\theta, \varphi)\, Y_l^m(\theta, \varphi) = \delta_{ll'}\delta_{mm'}$$

$$\sum_{l=0}^{\infty} \sum_{m=-l}^{m=+l} Y_l^{m*}(\theta', \varphi')\, Y_l^m(\theta, \varphi) = \delta(\varphi - \varphi')\,\delta(\cos\theta - \cos\theta'). \tag{A.7.11}$$

A.7(c). *The vector addition of angular momenta*

The addition of angular momenta has been discussed in § 5.3(e). The general relations are derived in Condon and Shortley (1935) and may be summarised as

$$\mathscr{Y}_J^M = \sum_{m_1=-j_1}^{m_1=+j_1} \sum_{m_2=M-m_1} C_J^{M}{}_{j_1}^{m_1}{}_{j_2}^{m_2}\, Y_{j_1}^{m_1}\, Y_{j_2}^{m_2} \tag{A.7.12}$$

$$Y_{j_1}^{m_1}\, Y_{j_2}^{m_2} = \sum_{J=|j_1-j_2|}^{J=j_1+j_2} C_J^{M}{}_{j_1}^{m_1}{}_{j_2}^{m_2}\, \mathscr{Y}_J^M \qquad (M = m_1 + m_2)$$

where the terms C are called vector addition or Clebsch–Gordan coefficients. They satisfy the following orthogonality relations (Blatt and Weisskopf, 1952):

$$\sum_{m_1=-j_1}^{m_1=+j_1} \sum_{m_2=-j_2}^{m_2=+j_2} C_J^{M}{}_{j_1}^{m_1}{}_{j_2}^{m_2}\, C_{J'}^{M'}{}_{j_1}^{m_1}{}_{j_2}^{m_2} = \delta_{JJ'}\,\delta_{MM'} \tag{A.7.13}$$

$$\sum_{J=|j_1-j_2|}^{J=j_1+j_2} \sum_{M=-J}^{M=+J} C_J^{M}{}_{j_1}^{m_1}{}_{j_2}^{m_2}\, C_J^{M}{}_{j_1}^{m_1'}{}_{j_2}^{m_2'} = \delta_{m_1m_1'}\,\delta_{m_2m_2'}$$

$$\sum_{M=-J}^{M=+J} \sum_{m_2=-j_2}^{m_2=+j_2} C_J^{M}{}_{j_1}^{m_1}{}_{j_2}^{m_2}\, C_J^{M}{}_{j_1'}^{m_1'}{}_{j_2}^{m_2} = \frac{2J+1}{2j_1+1}\,\delta_{j_1j_1'}\,\delta_{m_1m_1'}.$$

Tables of Clebsch–Gordan coefficients are given in Condon and Shortley (1935). We reproduce in Tables A.7.1 and A.7.2 the coefficients when j_2 is either $\frac{1}{2}$ or 1

TABLE A.7.1

	$C^{M\ m_1\ m_2}_{J\ j_1\ \frac{1}{2}}$	
	$m_2 = \frac{1}{2}$	$m_2 = -\frac{1}{2}$
$J = j_1 + \frac{1}{2}$	$\left[\frac{j_1 + M + \frac{1}{2}}{2j_1 + 1}\right]^{\frac{1}{2}}$	$\left[\frac{j_1 - M + \frac{1}{2}}{2j_1 + 1}\right]^{\frac{1}{2}}$
$J = j_1 - \frac{1}{2}$	$-\left[\frac{j_1 - M + \frac{1}{2}}{2j_1 + 1}\right]^{\frac{1}{2}}$	$\left[\frac{j_1 + M + \frac{1}{2}}{2j_1 + 1}\right]^{\frac{1}{2}}$

Numerical values of the coefficients associated with the products $Y^{m_1}_{1/2} Y^{m_2}_{1}$ and $Y^{m_1}_{1} Y^{m_2}_{1}$ are given in Tables 5.2 (p. 200) and 9.6 (p. 391) respectively.

The coefficients appear in problems associated with both isospin and angular momentum. As an example of the latter application, we may quote the relation

$$Y^m_l(\theta, \varphi)\, Y^{m'}_{l'}(\theta, \varphi) = \sum_L \sum_M \left[\frac{(2l + 1)\,(2l' + 1)}{4\pi(2L + 1)}\right]^{\frac{1}{2}} C^{0\ 0\ 0}_{L\ l\ l'}\, C^{M\ m\ m'}_{L\ l\ l'}\, Y^M_L(\theta, \varphi) \tag{A.7.14}$$

which yields the product of two spherical harmonics of the same angular variables as a linear combination of spherical harmonics.

A.7(d). *Projection operators*

It is often desirable to extract an expression relating to a specific angular momentum or isospin state from one which represents the effects of a combination of such states. This process can be achieved with the aid of projection operators.

The projection operator yields the value of a projection of an arbitrary vector in Hilbert space along a specified direction. It possesses the properties

$$P_\alpha |\alpha\rangle = |\alpha\rangle, \qquad P_\alpha |\beta\rangle = 0, \qquad \alpha \neq \beta \tag{A.7.15}$$

and satisfies the operator equations

$$\sum_\alpha P_\alpha = \hat{1}, \qquad P^2_\alpha = P_\alpha, \qquad P_\alpha P_\beta = 0, \qquad \alpha \neq \beta. \tag{A.7.16}$$

As an example of the use of projection operators, let us consider the problem of the scattering of spin 0 and spin $\frac{1}{2}$ particles (Lepore, 1950). The total angular momentum operator is

$$\mathbf{J} = (\mathbf{L} + \tfrac{1}{2}\boldsymbol{\sigma})$$

TABLE A.7.2

	$C^{M\ m_1\ m_2}_{J\ j_1\ 1}$		
	$m_2 = 1$	$m_2 = 0$	$m_2 = -1$
$J = j_1 + 1$	$\left[\dfrac{(j_1 + M)(j_1 + M + 1)}{(2j_1 + 1)(2j_1 + 2)}\right]^{\frac{1}{2}}$	$\left[\dfrac{(j_1 - M + 1)(j_1 + M + 1)}{(2j_1 + 1)(j_1 + 1)}\right]^{\frac{1}{2}}$	$\left[\dfrac{(j_1 - M)(j_1 - M + 1)}{(2j_1 + 1)(2j_1 + 2)}\right]^{\frac{1}{2}}$
$J = j_1$	$-\left[\dfrac{(j_1 + M)(j_1 - M + 1)}{2j_1(j_1 + 1)}\right]^{\frac{1}{2}}$	$\dfrac{M}{[j_1(j_1 + 1)]^{\frac{1}{2}}}$	$\left[\dfrac{(j_1 - M)(j_1 + M + 1)}{2j_1(j_1 + 1)}\right]^{\frac{1}{2}}$
$J = j_1 - 1$	$\left[\dfrac{(j_1 - M)(j_1 - M + 1)}{2j_1(2j_1 + 1)}\right]^{\frac{1}{2}}$	$-\left[\dfrac{(j_1 - M)(j_1 + M)}{j_1(2j_1 + 1)}\right]$	$\left[\dfrac{(j_1 + M + 1)(j_1 + M)}{2j_1(2j_1 + 1)}\right]^{\frac{1}{2}}$

and so

$$\mathbf{J}^2 = \mathbf{L}^2 + \boldsymbol{\sigma} \cdot \mathbf{L} + \tfrac{1}{4}\boldsymbol{\sigma}^2$$

$$\boldsymbol{\sigma} \cdot \mathbf{L} = \mathbf{J}^2 - \mathbf{L}^2 - \tfrac{1}{4}\boldsymbol{\sigma}^2$$

$$\boldsymbol{\sigma} \cdot \mathbf{L}\,|jls\rangle = [j(j+1) - l(l+1) - \tfrac{3}{4}]\,|jls\rangle.$$

Thus if we have two states corresponding to $j = l + \frac{1}{2}$ and $j = l - \frac{1}{2}$, which we shall designate as $|l\,+\rangle$ and $|l\,-\rangle$ respectively, then

$$\boldsymbol{\sigma} \cdot \mathbf{L}\,|l\,+\rangle = [(l + \tfrac{1}{2})(l + \tfrac{3}{2}) - l(l+1) - \tfrac{3}{4}]\,|l\,+\rangle = l\,|l_+\rangle$$

$$\boldsymbol{\sigma} \cdot \mathbf{L}\,|l\,-\rangle = -(l+1)\,|l\,-\rangle \tag{A.7.17}$$

Suitable projection operators for the $|l\,+\rangle$ and $|l\,-\rangle$ states are therefore given by

$$P_+ = \frac{l + 1 + \boldsymbol{\sigma} \cdot \mathbf{L}}{2l+1}, \qquad P_- = \frac{l - \boldsymbol{\sigma} \cdot \mathbf{L}}{2l+1} \tag{A.7.18}$$

since

$$P_+\,|l\,+\rangle = |l\,+\rangle, \qquad P_-\,|l\,-\rangle = |l\,-\rangle$$

$$P_+\,|l\,-\rangle = 0, \qquad P_-\,|l\,+\rangle = 0.$$

Now let us use these operators in a scattering problem. The scattering amplitude for a spinless system was given in (6.20) as

$$f(\theta) = \sum_{l=0}^{\infty} (2l+1) \frac{\sin \delta_l\, e^{i\delta_l}}{k} P_l^0(\theta) = \sum_l (2l+1) f_l P_l^0(\theta)$$

where we have used (6.18). If we introduce spin functions χ for initial and final states, we may modify the above relation to the following form for scattering between spin states i and f:

$$\begin{aligned} f_{fi}(\theta) &= \chi_f^\dagger \sum_l (2l+1)(P_+ f_{l+} + P_- f_{l-}) P_l^0 \chi_i \\ &= \chi_f^\dagger \sum_l (2l+1)\left(\frac{l+1+\boldsymbol{\sigma}\cdot\mathbf{L}}{2l+1} f_{l+} + \frac{l - \boldsymbol{\sigma}\cdot\mathbf{L}}{2l+1} f_{l-}\right) P_l^0 \chi_i \\ &= \chi_f^\dagger \sum_l \{[(l+1) f_{l+} + l f_{l-}]\, P_l^0 + (f_{l+} - f_{l-})\,\boldsymbol{\sigma}\cdot\mathbf{L} P_l^0\}\, \chi_i. \end{aligned} \tag{A.7.19}$$

Now spin 'up' and spin 'down' are orthogonal states and so $\chi_f^\dagger \chi_i = \delta_{fi}$ for the first expression on the right-hand side of the above equation; thus it represents an amplitude for no spin flip. Now consider the second term, we can write

$$\boldsymbol{\sigma} \cdot \mathbf{L} P_l^0 = -i \sin\theta \frac{\partial}{\partial(\cos\theta)} P_l^0\, \boldsymbol{\sigma} \cdot \mathbf{n}' = i P_l^1 \boldsymbol{\sigma} \cdot \mathbf{n}' \tag{A.7.20}$$

where $\mathbf{n}'$ is defined by the direction

$$\mathbf{k}' \times \mathbf{k} = \mathbf{n}' k^2 \sin \theta$$

that is opposite to the axial vector $\mathbf{n}$ of equation (6.68). Thus equation (A.7.19) becomes

$$\begin{aligned} f_{fi}(\theta) &= \chi_f^\dagger \sum_l \{[(l+1) f_{l+} + l f_l) P_l^0 + i(f_{l+} - f_{l-}) P_l^1 \boldsymbol{\sigma} \cdot \mathbf{n}'\} \chi_i \\ &= \chi_f^\dagger \sum_l \{[(l+1) f_{l+} + l f_{l-}] P_l^0 - i(f_{l+} - f_{l-}) P_l^1 \boldsymbol{\sigma} \cdot \mathbf{n}\} \chi_i . \end{aligned} \tag{A.7.21}$$

If this equation is compared with (6.62) the spin-flip and no spin-flip terms are easily recognisable.

$$g(\theta) = \sum_l [(l+1) f_{l+} + l f_{l-}] P_l^0 \quad \text{no flip}$$

$$h(\theta) = -\sum_l (f_{l+} - f_{l-}) P_l^1 \quad \text{spin flip}.$$

The most common application of projection operators is to spin and isospin systems. Using the methods associated with (A.7.17) and (A.7.18) we may obtain the projection operators listed below for isospin (or spin) systems $\frac{1}{2}$ and 1. We have used isospin operators in our illustrations.

(1) Isospin systems $\frac{1}{2}$, $\frac{1}{2}$:

$$\mathbf{I}^2 = \tfrac{1}{4}(\boldsymbol{\tau}_1 + \boldsymbol{\tau}_2)^2 \equiv i(i+1) \tag{A.7.22}$$

$$\begin{aligned} \boldsymbol{\tau}_1 \cdot \boldsymbol{\tau}_2 &\equiv +1 \quad \text{on} \quad i = 1 \\ &\equiv -3 \quad \text{on} \quad i = 0 \end{aligned}$$

$$P_0 = \frac{1 - \boldsymbol{\tau}_1 \cdot \boldsymbol{\tau}_2}{4}, \qquad P_1 = \frac{3 + \boldsymbol{\tau}_1 \cdot \boldsymbol{\tau}_2}{4}.$$

(2) Isospin systems $\frac{1}{2}$, 1 (compare (A.7.17) and (A.7.18)):

$$\mathbf{I}^2 = (\mathbf{T} + \tfrac{1}{2}\boldsymbol{\tau})^2 \equiv i(i+1) \tag{A.7.23}$$

$$\begin{aligned} \mathbf{T} \cdot \boldsymbol{\tau} &\equiv 1 \quad \text{on} \quad i = \tfrac{3}{2} \\ &\equiv -2 \quad \text{on} \quad i = \tfrac{1}{2} \end{aligned}$$

$$P_{1/2} = \frac{1 - \boldsymbol{\tau} \cdot \mathbf{T}}{3}, \qquad P_{3/2} = \frac{2 + \boldsymbol{\tau} \cdot \mathbf{T}}{3}.$$

(3) Isospin systems 1, 1:

$$\mathbf{I}^2 = (\mathbf{T}_1 + \mathbf{T}_2)^2 \equiv i(i+1) \tag{A.7.24}$$

$$\begin{aligned} \mathbf{T}_1 \cdot \mathbf{T}_2 &\equiv 1 \quad \text{on} \quad i = 2 \\ &\equiv -1 \quad \text{on} \quad i = 1 \\ &\equiv -2 \quad \text{on} \quad i = 0 \end{aligned}$$

$$P_0 = \tfrac{1}{3}(\mathbf{T}_1 \cdot \mathbf{T}_2 + 1)(\mathbf{T}_1 \cdot \mathbf{T}_2 - 1), \qquad P_1 = \tfrac{1}{2}(\mathbf{T}_1 \cdot \mathbf{T}_2 + 2)(1 - \mathbf{T}_1 \cdot \mathbf{T}_2)$$

$$P_2 = \tfrac{1}{6}(\mathbf{T}_1 \cdot \mathbf{T}_2 + 1)(\mathbf{T}_1 \cdot \mathbf{T}_2 + 2).$$

As an example of the application of a projection operator for isospin, consider pion–nucleon scattering. We then have isospin states of $\frac{1}{2}$ and $\frac{3}{2}$, and we may write the transition element between isospin bases β and α as

$$\begin{aligned} T_{\beta\alpha} &= \langle\beta|\, T \,|\alpha\rangle \\ &= \langle\beta|\, (T_{3/2}P_{3/2} + T_{1/2}P_{1/2}) \,|\alpha\rangle \\ &= T_{3/2}\,\langle\beta|\, P_{3/2} \,|\alpha\rangle + T_{1/2}\,\langle\beta|\, P_{1/2} \,|\alpha\rangle \\ &= T_{3/2}\,\langle\beta|\, \frac{2 + \boldsymbol{\tau}\cdot\mathbf{T}}{3} \,|\alpha\rangle + T_{1/2}\,\langle\beta|\, \frac{1 - \boldsymbol{\tau}\cdot\mathbf{T}}{3} \,|\alpha\rangle \\ &= \tfrac{1}{3}(2T_{3/2} + T_{1/2})\,\delta_{\beta\alpha} + \tfrac{1}{3}(T_{3/2} - T_{1/2})\,\langle\beta|\, \boldsymbol{\tau}\cdot\mathbf{T} \,|\alpha\rangle. \end{aligned} \tag{A.7.25}$$

Now consider $\langle\beta|\, \boldsymbol{\tau}\cdot T \,|\alpha\rangle$; if we use (9.93) we may write

$$\boldsymbol{\tau}\cdot\mathbf{T} = \begin{pmatrix} 0 & -i\tau_3 & i\tau_2 \\ i\tau_3 & 0 & -i\tau_1 \\ -i\tau_2 & i\tau_1 & 0 \end{pmatrix} \tag{A.7.26}$$

This matrix enables us to express $\boldsymbol{\tau}\cdot\mathbf{T}$ in a different manner. Since

$$\begin{aligned} \tau_i\tau_j + \tau_j\tau_i &= 2\delta_{ij} \\ \tau_i\tau_j - \tau_j\tau_i &= 2i\tau_k \end{aligned} \tag{A.7.27}$$

where i, j, k are written cyclically in the last equation, we can write $\tau_1\tau_2 = i\tau_3$ and so on. We may therefore construct a general matrix for a product $\tau_\beta\tau_\alpha$ as

$$\begin{array}{c|ccc} \beta\backslash\alpha & 1 & 2 & 3 \\ \hline 1 & 1 & i\tau_3 & -i\tau_2 \\ 2 & -i\tau_3 & 1 & i\tau_1 \\ 3 & i\tau_2 & -i\tau_1 & 1 \end{array} = \tau_\beta\tau_\alpha. \tag{A.7.28}$$

A comparison of the matrices (A.7.26) and (A.7.28) then shows that

$$\langle\beta|\ \boldsymbol{\tau}\cdot\mathbf{T}\ |\alpha\rangle = \delta_{\beta\alpha} - \tau_\beta\tau_\alpha = -\tfrac{1}{2}[\tau_\beta, \tau_\alpha] \tag{A.7.29}$$

where we have used the first of equations (A.7.27). Thus (A.7.25) becomes

$$T_{\beta\alpha} = \tfrac{1}{3}(2T_{3/2} + T_{1/2})\,\delta_{\beta\alpha} + \tfrac{1}{6}(T_{1/2} - T_{3/2})\,[\tau_\beta, \tau_\alpha]. \tag{A.7.30}$$

We note, finally, a projection operator for orbital angular momentum states. The Legendre polynominals satisfy the following normalisation condition:

$$\int_{-1}^{+1} d(\cos\theta)\, P_l^m(\cos\theta)\, P_{l'}^{m'}(\cos\theta) = \frac{2}{2l+1}\,\frac{(l+m)!}{(l-m)!}\,\delta_{ll'}. \tag{A.7.31}$$

Since we can normally expand scattering amplitudes as

$$f(\theta) = \sum_l a_l^m P_l^m\,(\cos\theta)$$

we can therefore write

$$a_l^m = \frac{2l+1}{2}\,\frac{(l-m)!}{(l+m)!}\int_{-1}^{+1} d(\cos\theta)\, P_l^m(\cos\theta)\, f(\theta). \tag{A.7.32}$$

A projection operator for scattering functions for a system involving particles of spin 0 and $\frac{1}{2}$ is quoted in § 14.1 (c).

A.8. COORDINATE SYSTEMS

Lorentz transformations between different reference frames were discussed in § 3.2. In most experiments only two are encountered – the laboratory and c-systems. The latter refers to the system in which the total linear momentum is zero.

The velocity and total energy of the c-system are given by

$$\boldsymbol{\beta}_c = \frac{\mathbf{P}}{E}, \qquad E_c = \sqrt{(E^2 - \mathbf{P}^2)} \tag{A.8.1}$$

where $\mathbf{P}$ and E represent the total momentum and energy in the laboratory system respectively (compare (3.30) and (3.31)). If only two particles are present it is a simple matter to show that their linear momentum in the c-system is given by

$$p_c = \frac{1}{2E_c}\{[E_c^2 - (m_1 + m_2)^2]\,[E_c^2 - (m_1 - m_2)^2]\}^{\frac{1}{2}} \tag{A.8.2}$$

where m_1 and m_2 represent the masses of the particles.

The relationship between energies and angles in the laboratory and c-systems can be obtained with the aid of the Lorentz transformation (compare (3.34) and § 3.2(e)).

$$\begin{pmatrix} \gamma & 0 & 0 & -i\beta_c\gamma \\ 0 & 1 & 0 & 0 \\ 0 & 0 & 1 & 0 \\ i\beta_c\gamma & 0 & 0 & \gamma \end{pmatrix} \begin{pmatrix} p\cos\theta \\ p\sin\theta \\ 0 \\ iE \end{pmatrix} = \begin{pmatrix} \gamma p\cos\theta + \beta_c\gamma E \\ p\sin\theta \\ 0 \\ i(\beta_c\gamma p\cos\theta + \gamma E) \end{pmatrix} = \begin{pmatrix} p_L\cos\theta_L \\ p_L\sin\theta_L \\ 0 \\ iE_L \end{pmatrix} \quad \text{(A.8.3)}$$

where

$$\gamma = \frac{1}{\sqrt{(1-\beta_c^2)}}$$

and the laboratory components are indicated by the subscript L. If the velocity of the particle in the c-system is less than β_c the maximum value of θ_L is given

$$\sin\theta_L = \frac{p}{m\gamma\beta_c}$$

where m is the mass and p the momentum of the particle in the c-system.

Angular distributions in the two systems are related by

$$\sigma(\theta, \varphi)\, d\Omega = \sigma_L(\theta_L, \varphi)\, d\Omega_L \quad \text{(A.8.4)}$$

where $\sigma(\theta, \varphi)$ is a cross-section per steradian. A useful relation is the fact that the double variable

$$\frac{1}{p}\,\frac{d^2\sigma}{dE d\Omega} \quad \text{(A.8.5)}$$

is a Lorentz invariant quantity.

A further useful expression is one which relates spin quantity directions in different coordinate systems (Henry, Schrank and Swanson, 1963). If a particle travels with a velocity β in the c-system at an angle θ to the direction of β_c, then in the laboratory frame the particle travels at an angle θ_L which is given by (A.8.3)

$$\tan\theta_L = \frac{p\sin\theta}{\gamma(p\cos\theta + \beta_c E)}.$$

If the rest mass of the particle is m and its spin points at an angle α to θ in the c-frame and α_L to θ_L in the laboratory frame, then the difference $\varepsilon = \alpha_L - \alpha$ is given by

$$\tan\varepsilon = \frac{m\beta_c\sin\theta}{p + \beta_c E\cos\theta}. \quad \text{(A.8.6)}$$

A.9. DATA†

TABLE A.9.1. *Atomic Constants*

Avogadro's number	$N = 6{\cdot}0249 \times 10^{23}$ mols/g-mole
Velocity of light	$c = 2{\cdot}99793 \times 10^{10}$ cm/sec
Electronic charge	$e = 4{\cdot}80286 \times 10^{-10}$ e.s.u.
	$= 1{\cdot}6021 \times 10^{-19}$ c
$1/2\pi$ (Planck's constant)	$\hbar = 1{\cdot}054 \times 10^{-27}$ erg sec
	$= 6{\cdot}5817 \times 10^{-22}$ MeV sec
	$\hbar c = 1{\cdot}9732 \times 10^{-11}$ MeV cm
Boltzmann's constant	$k = 8{\cdot}6167 \times 10^{-11}$ MeV/°C
Fine structure constant	$\alpha = \dfrac{e^2}{\hbar c} = \dfrac{1}{137{\cdot}037}$

Quantities associated with the electron mass

Rydberg	$R_\infty = \dfrac{me^4}{2\hbar^2} = 13{\cdot}605$ eV
Bohr radius	$a_\infty = \dfrac{\hbar^2}{me^2} = 0{\cdot}52917 \times 10^{-8}$ cm
Compton radius	$\lambda\!\!\!^{-} = \dfrac{\hbar}{mc} = \alpha a_\infty = 3{\cdot}8612 \times 10^{-11}$ cm
Thompson radius	$r_0 = \dfrac{e^2}{mc^2} = \alpha^2 a_\infty = 2{\cdot}81785 \times 10^{-13}$ cm
Bohr magneton	$\mu_B = \dfrac{e\hbar}{2mc} = 0{\cdot}57883 \times 10^{-14}$ MeV/G
$\frac{1}{2}$ Cyclotron frequency	$\dfrac{\omega}{2} = \dfrac{e}{2mc} = 8{\cdot}7945 \times 10^{6}$ rad/sec/G

† The information in these tables has been taken mainly from tables by Barkas and Rosenfeld (1961), Snow and Shapiro (1961) and Roos (1964). Charge has been quoted in e.s.u. in table A.9.1 ; α and related quantities have the same values in rationalised units.

TABLE A.9.2 *Particle Data*

Group	Particle	Mass (MeV)	Mean life (sec)	$j^{\xi P}$	B	T	S
Photon	γ	0	stable	1^-	0	–	–
Leptons	ν_e	$<0{\cdot}00025$	stable	$\frac{1}{2}$	0	–	–
	ν_μ	$<2{\cdot}5$	stable	$\frac{1}{2}$	0	–	–
	$e^\pm$	$0{\cdot}510976 \pm 0{\cdot}000007$	stable	$\frac{1}{2}$	0	–	–
	$\mu^\pm$	$105{\cdot}655 \pm 0{\cdot}010$	$(2{\cdot}210 \pm 0{\cdot}003)\, 10^{-6}$	$\frac{1}{2}$	0	–	–
Mesons	$\pi^\pm$	$139{\cdot}59 \pm 0{\cdot}05$	$(2{\cdot}55 \pm 0{\cdot}03)\, 10^{-8}$	0^-	0	1	0
	π^0	$135{\cdot}00 \pm 0{\cdot}05$	$(2{\cdot}2 \pm 0{\cdot}8)\, 10^{-16}$	0^-	0	1	0
	K^+ }	$493{\cdot}9 \pm 0{\cdot}2$	$(1{\cdot}224 \pm 0{\cdot}013)\, 10^{-8}$	0^-	0	$\frac{1}{2}$	$+1$
	K^- }			0^-	0	$\frac{1}{2}$	-1
	K^0 }	$497{\cdot}8 \pm 0{\cdot}6$	$K_1^0 (1{\cdot}00 \pm 0{\cdot}038)\, 10^{-10}$	0^-	0	$\frac{1}{2}$	$+1 (K^0)$
	K_1^0, K_2^0 }		$K_2^0 \left(6{\cdot}1 \begin{array}{l}+1{\cdot}6\\-1{\cdot}1\end{array}\right) 10^{-8}$				
	$\bar{K}^0$ }			0^-	0	$\frac{1}{2}$	$-1 (\bar{K}^0)$
Baryons	p } N	$938{\cdot}213 \pm 0{\cdot}01$	stable }	$\frac{1}{2}^+$	1	$\frac{1}{2}$	0
	n }	$939{\cdot}507 \pm 0{\cdot}01$	$(1{\cdot}013 \pm 0{\cdot}029)\, 10^{3}$ }				
	Λ	$1115{\cdot}36 \pm 0{\cdot}14$	$(2{\cdot}51 \pm 0{\cdot}09)\, 10^{-10}$	$\frac{1}{2}^+$	1	0	-1
	Σ^+ }	$1189{\cdot}4 \pm 0{\cdot}20$	$(0{\cdot}78 \pm 0{\cdot}03)\, 10^{-10}$ }				
	Σ^- } Σ	$1197{\cdot}6 \pm 0{\cdot}5$	$(1{\cdot}59 \pm 0{\cdot}05)\, 10^{-10}$ }	$\frac{1}{2}^+$	1	1	-1
	Σ^0 }	$1193{\cdot}2 \pm 0{\cdot}7$	$< 0{\cdot}1 \times 10^{-10}$† }				
	Ξ^- } Ξ	$1321{\cdot}2 \pm 0{\cdot}3$	$(1{\cdot}75 \pm 0{\cdot}05)\, 10^{-10}$ }	$\frac{1}{2}$	1	$\frac{1}{2}$	-2
	Ξ^0 }	$1315{\cdot}2 \pm 1{\cdot}0$	$(2{\cdot}80 \pm 0{\cdot}26)\, 10^{-10}$ }				
	Ω^-	1686 ± 12			1	0	-3

† Expected to be $\sim 10^{-19}$ sec on theoretical grounds.

TABLE A.9.3. *Resonant States*

Group	State	Mass (MeV)	Width (MeV)	$j^{\xi}P^{G}$†	B	T	S	Main decay modes
Pion	π	see Table A.9.2.		0^{--}	0	1	0	$\mu\nu$
	η	548·5	$\leqq 7$	0^{-+}	0	0	0	mixed
	ϱ	755	120 ± 10	1^{-+}	0	1	0	2π
	ω	783	$9{\cdot}5 \pm 2$	1^{--}	0	0	0	3π
	X^0	960	$\leqq 12$		0		0	$\eta\ 2\pi$
	φ	1019·5	$3{\cdot}1 \pm 0{\cdot}6$	1^{--}	0	0	0	$K\bar{K}$
	A_1	1090	125		0	1	0	$\varrho\pi$
	A_2	1310	80		0	1	0	$\varrho\pi$
	B	1220	~ 150		0	1	0	$\omega\pi$
	f	1260	~ 160	2^{++}	0	0	0	2π
	E	1410	~ 50		0		0	$\bar{K}K^*$
Kaon	K	see Table A.9.2.		0^-	0	$\frac{1}{2}$	± 1	mixed
	$\varkappa$	725	$\leqq 12$	$\geqq 1$	0	$\frac{1}{2}$	1	$K\pi$
	K^*	890	50	1^-	0	$\frac{1}{2}$	± 1	$K\pi$
	C	1230	~ 80					$K\varrho$
Nucleon	N	see Table A.9.2.		$\frac{1}{2}^+$	1	$\frac{1}{2}$	0	
	N_1^*	1238	90	$\frac{3}{2}^+$	1	$\frac{3}{2}$	0	$N\pi$
	N_2^*	1512	115	$\frac{3}{2}^-$(?)	1	$\frac{1}{2}$	0	
	N_3^*	1688	100	$\frac{5}{2}^+$(?)	1	$\frac{1}{2}$	0	
	N_4^*	1922	200	$\frac{7}{2}$(?)	1	$\frac{3}{2}$	0	
	N_5^*	2190	200	?	1	$\frac{1}{2}$	0	
	N_6^*	2360	200	?	1	$\frac{3}{2}$	0	
Hyperon	Λ	see Table A.9.2.		$\frac{1}{2}^+$	1	0	-1	
	Y_0^*	1405	50	$\frac{1}{2}^-$(?)	1	0	-1	$\Sigma\pi$
	Y_0^{**}	1520	15	$\frac{3}{2}^-$	1	0	-1	mixed
	Y_0^{***}	1815	70	$\frac{5}{2}^+$	1	0	-1	$\bar{K}N$
	Σ	see Table A.9.2.		$\frac{1}{2}^+$	1	1	-1	
	Y_1^*	1385	50	$\frac{3}{2}^+$	1	1	-1	$\Lambda\pi$
	Y_1^{**}	1660	40	$\frac{3}{2}$	1	1	-1	mixed
	Ξ	see Table A.9.2.		$\frac{1}{2}$	1	$\frac{1}{2}$	-2	
	Ξ^*	1532		$\frac{3}{2}^+$	1	$\frac{1}{2}$	-2	$\Xi\pi$

† The G assignments refer to pion systems only.

TABLE A.9.4. *Weak Decay Modes*

Particle	Decay rate (sec^{-1})	Decay mode	Branching ratio
μ	$4{\cdot}6\times10^5$	$e+\nu+\nu$†	~1
π	$4{\cdot}0\times10^7$	$\mu+\nu$	~1
		$e+\nu$	$(1{\cdot}21\pm0{\cdot}07)10^{-4}$
		$\pi^0+e+\nu$	$(1{\cdot}15\pm0{\cdot}22)10^{-8}$
$K^\pm$	$8{\cdot}2\times10^7$	$\mu+\nu$	$0{\cdot}642\pm0{\cdot}013$
		$\pi+\pi^0$	$0{\cdot}186\pm0{\cdot}009$
		$\pi^\pm+\pi^\mp+\pi^\mp$	$0{\cdot}057\pm0{\cdot}002$
		$\pi+\pi^0+\pi^0$	$0{\cdot}017\pm0{\cdot}003$
		$\pi^0+\mu+\nu$	$0{\cdot}048\pm0{\cdot}006$
		$\pi^0+e+\nu$	$0{\cdot}050\pm0{\cdot}005$
K_1^0	$1{\cdot}1\times10^{10}$	$\pi^++\pi^-$	$0{\cdot}665\pm0{\cdot}014$
		$\pi^0+\pi^0$	$0{\cdot}335\pm0{\cdot}014$
K_2^0	$1{\cdot}6\times10^7$	$\pi^0+\pi^0+\pi^0$	$0{\cdot}38\pm0{\cdot}07$
		$\pi^++\pi^-+\pi^0$	$0{\cdot}087\pm0{\cdot}023$
		$\pi^\pm+\mu^\mp+\nu$	$0{\cdot}250\pm0{\cdot}06$
		$\pi^\pm+e^\mp+\nu$	$0{\cdot}283\pm0{\cdot}06$
n	$1{\cdot}0\times10^3$	$p+e^-+\bar{\nu}$	1
Λ	4×10^9	$p+\pi^-$	$0{\cdot}685\pm0{\cdot}017$
		$n+\pi^0$	$0{\cdot}315\pm0{\cdot}017$
		$p+e^-+\nu$	$(0.85\pm0{\cdot}09)10^{-3}$
		$p+\mu^-+\nu$	$(0{\cdot}13\pm0{\cdot}06)10^{-3}$
Σ^+	$1{\cdot}3\times10^{10}$	$p+\pi^0$	$0{\cdot}51\pm0{\cdot}024$
		$n+\pi^+$	$0{\cdot}49\pm0{\cdot}024$
Σ^0	$>10^{11}$	$p+\gamma$	~1
Σ^-	$6{\cdot}3\times10^9$	$n+\pi^-$	~1
		$n+e^-+\nu$	$(1{\cdot}35\pm0{\cdot}4)10^{-3}$
		$n+\mu^-+\nu$	$(0{\cdot}88\pm0{\cdot}3)10^{-3}$
Ξ^-	$5{\cdot}7\times10^9$	$\Lambda+\pi^-$	~1
Ξ^0	$3{\cdot}2\times10^9$	$\Lambda+\pi^0$	~1

† No distinction is made between different types of neutrino in this table.

REFERENCES†

ABASHIAN, A. and HAFNER, E.M. (1958) *Phys. Rev. Lett.* **1,** 255.
ADAIR, R.K. (1955) *Phys. Rev.* **100,** 1540.
AHARONI, J. (1959) *The Special Theory of Relativity*, Oxford.
AKIBA, T. and CAPPS, R.H. (1962) *Phys. Rev. Lett.* **8,** 457.
AKIMOV, IU.K., SAVCHENKO, O.V. and SOROKO, L.M. (1960) *Proc. Int. Conf. on High Energy Physics at Rochester*, p. 49.
ALEXANDER, G., ANDERSON, J.A., CRAWFORD, F.S. LASKAR, W. and LLOYD, L.J. (1961) *Phys. Rev. Lett.* **7,** 348.
ALIKHANOV, A.I., GALAKTIONOV, YU.V., GORODKOV, YU.V., ELISEYEV, G.P. and LYUBIMOV, V.A. (1960) *JETP* **11,** 1380.
ALIKHANOV, A.I. *et al.* (1962) *Int. Conf. on High Energy Physics at CERN*, p. 423.
ALI-ZADE, C.A., GUREVICH, I.I., DOBRETSOV, YU.P., NIKOLSKII, B.A. and SURKOVA, L.V. (1959) *JETP* **9,** 940.
ALLEN, J.S. (1958) *The Neutrino*, Princeton.
ALLEN, J.S. (1959) *Rev. Mod. Phys.* **31,** 791.
ALSTON, M.H. *et al.* (1960) *Phys. Rev. Lett.* **5,** 520.
ALSTON, M.H. *et al.* (1961a) *Phys. Rev. Lett.* **6,** 300; (1961b) *ibid.* **6,** 698.
ALTMAN, A. and MACDONALD, W.M. (1962) *Nuclear Phys.* **35,** 593.
ALVAREZ, L.W. *et al.* (1957) *Nuovo Cim.* **5,** 1026.
ALVAREZ, L.W., EBERHARD, P., GOOD, M.L., GRAZIANO, W., TICHO, H.K. and WOJCICKI, S.G. (1959) *Phys. Rev. Lett.* **2,** 215.
AMATI, D. and FUBINI, S. (1962) *Ann. Rev. Nuclear Sci.*, **12,** 359.
AMATI, D., FUBINI, S. and STANGHELLINI, A. (1962) *Phys. Lett.* **1,** 29.
AMATI, D., FUBINI, S. and STANGHELLINI, A. (1962) *Int. Conf. on High Energy Physics at CERN*, p. 560.
ANDERSON, C.D. (1932) *Science* **76,** 238.
ANDERSON, H.L., FUJII, T., MILLER, R.H. and TAU, L. (1960) *Phys. Rev.* **119,** 2050.
ANDERSON, J.A., BANG, VO.X., BURKE, P.G., CARMONY, D.D. and SCHMITZ, N. (1961) *Phys. Rev. Lett.* **6,** 365.
ARMENTEROS, R., BARKER, K.H., BUTLER, C.C. and CACHON, A. (1951). *Phil. Mag.* **42,** 1113.
ARMENTEROS, R., COOMBES, C.A., CORK, B., LAMBERTSON, G.R. and WENZEL, W.A. (1960) *Phys. Rev.* **119,** 2068.
ARMENTEROS, R. *et al.* (1962a) *Int. Conf. on High Energy Physics at CERN*, p. 351; (1962b) *ibid.* 295.
ASHKIN, J., FAZZINI, T., FIDECARO, G., MERRISON, A.W., PAUL, H. and TOLLESTRUP, A.V. (1959) *Nuovo Cim.* **13,** 1240.
ASHKIN, A., PAGE, L.A. and WOODWARD, W.M. (1954) *Phys. Rev.* **94,** 357.
ASHMORE, A., RANGE, W.H., TAYLOR, R.T., TOWNES, B.M., CASTILLEJO, L. and PEIERLS, R.F. (1962) *Nuclear Phys.* **36,** 258.
ASTBURY, A., BARTLEY, J.H., BLAIR, I.M., KEMP, M.A.R., MUIRHEAD, H. and WOODHEAD, T. (1962) *Proc. Phys. Soc.* **79,** 1011.
ASTIER, A. *et. al.* (1961) *Aix-en-Provence Conf. Rep.*, CEN Saclay, **1,** 227.
BACASTOW, R., ELIOFF, T., LARSEN, R., WIEGAND, C. and YPSILANTIS, T. (1962) *Phys. Rev. Lett.* **9,** 400.

† References have been given to the translated versions of Soviet journals whenever possible. Reference is given to the first name only in papers with more than six authors.

BACKENSTOSS, G., HYAMS, B.D., KNOP, G., MARIA, P.C. and STIERLIN, U. (1961) *Phys. Rev. Lett.* **6,** 415.
BARBARO-GALTIERI, A., BARKAS, W.H., HECKMAN, H.H., PATRICK, J.W. and SMITH, F.M. (1962) *Phys. Rev. Lett.* **9,** 26.
BARDIN, R.K., BARNES, C.A., FOWLER, W.A. and SEEGER, P.A. (1962) *Phys. Rev.* **127,** 583.
BARDON, M., BERLEY, D. and LEDERMAN, L.M. (1959) *Phys. Rev. Lett.* **2,** 56.
BARDON, M., FRANZINI, P. and LEE, J. (1961) *Phys. Rev. Lett.* **7,** 23.
BARKAS, W.H., BIRNBAUM, W. and SMITH, F.M. (1956) *Phys. Rev.* **101,** 778.
BARKAS, W.H. and ROSENFELD, A.H. (1961) UCRL 8030.
BARLOUTAUD, R. (1961) *et al. Aix-en-Provence Conf. Rep.* CEN Saclay, **1,** 27.
BARNES, K.J. (1962) *Phys. Lett.* **1,** 166.
BARNES, S.W., ROSE, B., GIACOMELLI, G., RING, J., MIYAKE, K., and KINSEY, K. (1960) *Phys. Rev.* **117,** 226.
BARNES, V.E. *et al.* (1964) *Phys. Rev. Lett.* **12,** 204.
BARTLETT, D., DEVONS, S. and SACHS, A.M. (1962) *Phys. Rev. Lett.* **8,** 120.
BARTLETT, M.S. (1953) *Phil. Mag.* **44,** 249.
BASTIEN, P.L. *et al.* (1962a) *Int. Conf. on High Energy Physics at CERN*, p. 373.
BASTIEN, P.L. *et al.* (1962b) *Phys. Rev. Lett.* **8,** 114, 302 (E).
BEALL, E.F. *et al.* (1962) *Int. Conf. on High Energy Physics at CERN*, p. 368.
BEHRENDS, R.E., DREITLEIN, J., FRONSDAL, C. and LEE, W. (1962) *Rev. Mod. Phys.* **34,** 1.
BEHRENDS, R.E. and SIRLIN, A. (1961) *Phys. Rev.* **121,** 324.
BELINFANTE, F.J. and MØLLER, C. (1954) *Kgl. Danske Videnskab. Selskab. Mat.-Fys. Medd.* **28,** No. 6.
BERESTETSKII, V.B. (1949) *JETP* **19,** 673, 1130.
BERESTETSKII, V.B. (1961) *JETP* **12,** 993.
BERESTETSKII, V.B. KROKHIN, O.N. and KLEBNIKHOV, A.K. (1956) *JETP* **3,** 761.
BERINGER, R. and HEALD, M.A. (1954) *Phys. Rev.* **95,** 1474.
BERGIA, S., STANGHELLINI, A., FUBINI, S. and VILLI, C. (1961) *Phys. Rev. Lett.* **6,** 367.
BERKELMAN, K., FELDMAN, M., LITTAUER, R.M., ROUSE, G. and WILSON, R.R. (1963) *Phys. Rev.* **130, 2061.**
BERNARDINI, G. (1955) *Suppl. Nuovo Cim.* **11,** 104.
BERTANZA, L. *et al.* (1962) *Int. Conf. on High Energy Physics at CERN*, p. 279.
BETHE, H.A. and MARSHAK, R.E. (1947) *Phys. Rev.* **72,** 506.
BETHE, H.A. and MORRISON, P. (1956) *Elementary Nuclear Theory*, Wiley.
BERTOLINI, E. *et al.* (1962) *Int. Conf. on High Energy Physics at CERN*, p. 421.
BHABHA, H.J. (1936) *Proc. Roy. Soc. A*, **154,** 195.
BHOWMIK, B., GOYAL, D.P. and YAMDAGNI, N.K. (1961) *Nuovo Cim.* **22,** 296.
BINGHAM, H. (1964) CERN/TC/PHYSICS 64–13.
BISHOP, G.R. (1962) *Int. Conf. on High Energy Physics at CERN*, p. 753.
BJORKLAND, R., CRANDALL, W.E., MOYER, B.J. and YORK, H.F. (1950) *Phys. Rev.* **77,** 213.
BLACKETT, P.M.S. and OCCHIALINI, G.P.S. (1933) *Proc. Roc. Soc. A* **139,** 699.
BLANKENBECLER, R., COOK, L.F. and GOLDBERGER, M.L. (1962) *Phys. Rev. Lett.* **8,** 463.
BLATT, J.M. and JACKSON, J.D. (1949) *Phys. Rev.* **76,** 18.
BLATT, J.M. and WEISSKOPF, V.F. (1952) *Theoretical Nuclear Physics*, Wiley.
BLESER, E., LEDERMAN, L., ROSEN, J., ROTHBERG, J. and ZAVATTINI, E. (1962a) *Phys. Rev. Lett.* **8,** 128. (1962b) *ibid.* **8,** 288.
BLEULER, K. (1950) *Helv. Phys. Acta* **23,** 567.
BLIN-STOYLE, R.J. (1960) *Phys. Rev.* **118,** 1605; *ibid.* **120,** 181.
BLIN-STOYLE, R.J. (1961) *Rutherford Jubilee Int. Conf.* Heywood, p. 677.
BLIN-STOYLE, R.J. and LE TOURNEUX, J. (1961) *Phys. Rev.* **123,** 627.
BLOCH, F. (1946) *Phys. Rev.* **70,** 460
BLOCH, F., HANSEN, W.W. and PACKARD, M. (1946) *Phys. Rev.* **70,** 474.
BLOCH, F. and NORDSIECK, A. (1937) *Phys. Rev.* **52,** 54.

BLOCH, F., NICODEMUS, D. and STAUB, H. (1948) *Phys. Rev.* **74,** 1025.
BLOCK, M.M. *et al.* (1959) *Phys. Rev. Lett.* **3,** 291.
BLOCK, M. M., LENDINARA, L. and MONARI, L. (1962) *Int. Conf. on High Energy Physics at CERN*, p. 371.
BONETTI, A., LEVI-SETTI, R., PANETTI, M. and TOMASINI, G. (1953). *Nuovo Cim.* **10,** 1736.
BOWCOCK, J., COTTINGHAM, W.N., LURIE, D. (1960) *Nuovo Cim.* **16,** 918.
BROWN, R.H., CAMERINI, U., FOWLER, P.H., MUIRHEAD, H., POWELL, C.F. and RITSON, D.M. (1949), *Nature* **163,** 82.
BRUECKNER, K., SERBER, R. and WATSON, K. (1951) *Phys. Rev.* **81,** 575.
BURGY, M.T., KROHN, V.E., NOVEY, T.B., RINGO, G.R. and TELEGDI, V.L. (1958) *Phys. Rev.* **110,** 1214.
CABIBBO, N. (1963) *Phys. Rev. Lett.* **10,** 531,
CAMERINI, U., MUIRHEAD, H., POWELL, C.F. and RITSON, D.M. (1948) *Nature* **162,** 433.
CAMERINI, U. *et al.* (1962) *Phys. Rev.* **128,** 362.
CAPPS, R.H. and TAKEDA, G. (1956) *Phys. Rev.* **103,** 1877.
CARLSON, A.G., HOOPER, J.E. and KING, D.T. (1950) *Phil. Mag.* **41,** 701.
CARMONY, D., ROSENFELD, A.H. and VAN DE WALLE, R.T. (1962) *Phys. Rev. Lett.* **8,** 117.
CARMONY, D. and VAN DE WALLE, R.T. (1962) *Phys. Rev. Lett.* **8,** 73.
CARTWRIGHT, W.F., RICHMAN, C., WHITEHEAD, M.N. and WILCOX, H.A. (1953) *Phys. Rev.* **91,** 677.
CASSEN, B. and CONDON, E.U. (1936) *Phys. Rev.* **50,** 846.
CHADWICK, G.B. *et al.* (1961) *Aix-en-Provence Conf. Rep.* CEN Saclay, **1,** 269.
CHADWICK, G.B. *et al.* (1962) *Int. Conf. on High Energy Physics at CERN*, p. 69.
CHADWICK, J. (1932) *Proc. Roy. Soc. A* **136,** 692.
CHAMBERLAIN, O., SEGRÈ, E., WIEGAND, C. and YPSILANTIS, T. (1955) *Phys. Rev.* **100,** 947.
CHAMBERLAIN, O., SEGRÈ, E., TRIPP, R.D,. WIEGAND, C. and YPSILANTIS, T. (1957) *Phys. Rev.* **105,** 288.
CHAMBERLAIN, O. *et al.* (1962) *Phys. Rev.* **125,** 1696.
CHAN, C.H. (1962) *Proc. Phys. Soc.* **80,** 39.
CHARPAK, G. *et al.* (1961a) *Phys. Rev. Lett.* **6,** 128.
CHARPAK, G., FARLEY, F.J.M., GARWIN, R.L., MÜLLER, T., SENS, J.C. and ZICHICHI, A. (1961b) *Nuovo Cim.* **22,** 1043.
CHARPAK, G., FARLEY, F.J.M., GARWIN, R.L., MÜLLER, T., SENS, J.C. and ZICHICHI, A. (1962) *Phys. Lett.* **1,** 16.
CHEW, G.F. (1954) *Phys. Rev.* **95,** 1669.
CHEW, G.F. (1958) *Phys. Rev.* **112,** 1380.
CHEW, G.F. (1959) *Ann. Rev. Nuclear Sci.* **9,** 29.
CHEW, G.F. (1961) *Dispersion Relations* (Ed. Screaton), Oliver and Boyd.
CHEW, G.F. (1962) *Rev. Mod. Phys.* **34,** 394.
CHEW, G.F. and FRAUTSCHI, S. (1961) *Phys. Rev.* **123,** 1478.
CHEW, G.F. and FRAUTSCHI, S. (1962) *Phys. Rev. Lett.* **8,** 41.
CHEW, G.F., FRAUTSCHI, S. and MANDELSTAM, S. (1962) *Phys. Rev.* **126,** 1202.
CHEW, G.F., GOLDBERGER, M.L., LOW, F.E. and NAMBU, Y. (1957) *Phys. Rev.* **106,** 1337, 1345.
CHEW, G.F. and LOW, F.E. (1956) *Phys. Rev.* **101,** 1570, 1579.
CHEW, G.F. and LOW, F.E. (1959) *Phys. Rev.* **113,** 1640.
CHINOWSKY, W. *et al.* (1960) *Proc. Int. Conf. on High Energy Physics at Rochester*, p. 451.
CHINOWSKY, W., GOLDHABER, G., GOLDHABER, S., LEE, W. and O'HALLORAN, T. (1962) *Phys. Rev. Lett.* **9,** 330.
CHRÉTIEN, M. *et al.* (1962) *Phys. Rev. Lett.* **9,** 127.
CHRISTENSON, J.H., CRONIN, J.W., FITCH, V.L. and TURLAY, R. (1964) *Phys. Rev. Lett.* **13,** 138.
CLARK, D.L., ROBERTS, A. and WILSON, R. (1951) *Phys. Rev.* **83,** 649. (1952) *ibid.* **85,** 523.
COCCONI, G. (1962) *Int. Conf. on High Energy Physics at CERN*, p. 883.
COCCONI, G. *et al.* (1961) *Phys. Rev. Lett.* **7,** 450.
COESTER, F., HAMMERMESH, M. and TANAKA, K. (1954) *Phys. Rev.* **96,** 1142.

COFFIN, T., GARWIN, R.L., PENMAN, S., LEDERMAN, L.M. and SACHS, A.M. (1958) *Phys. Rev.* **109,** 973.
COHEN, E.R., CROWE, K.M. and DUMOND, J.M. (1957) *Fundamental Constants of Physics*, Interscience.
COHEN, E.R. and DUMOND, J.M. (1958) *Phys. Rev. Lett.* **1,** 291.
COLGATE, S.A. (1952) *Phys. Rev.* **87,** 592.
COLGATE, S.A. and GILBERT, F.C. (1952) *Phys. Rev.* **89,** 790.
COLEMAN, S. and GLASHOW, S.L. (1961) *Phys. Rev. Lett.* **6,** 423.
COLLINGTON, D.J., DELLIS, A.N., SANDERS, J.H. and TURBERFIELD, K.C. (1955) *Phys. Rev.* **99,** 1622.
COMPTON, A.H. (1923) *Phys. Rev.* **21,** 715.
CONDON, E.U. and SHORTLEY, G.H. (1935) *Theory of Atomic Spectra*, Cambridge.
CONVERSI, M., DI LELLA, L., TOLLER, M. and RUBBIA, C. (1962) *Phys. Rev. Lett.* **8,** 125.
CONVERSI, M., PANCINI, E. and PICCIONI, O. (1947) *Phys. Rev.* **71,** 209.
COOK, V. *et al.* (1961) UCRL 9386.
COOK, V., KEEFE, D., KERTH, L.T., MURPHY, P.L., WENZEL, W.A. and ZIPF, T.F. (1962) *Int. Conf. on High Energy Physics at CERN*, p. 364.
COOMBES, C.A., CORK, B., GALBRAITH, W., LAMBERTSON, G.R. and WENZEL, W.A. (1957) *Phys. Rev.* **108,** 1348.
COOMBES, C.A., CORK, B., GALBRAITH, W., LAMBERTSON, G.R. and WENZEL, W.A. (1958) *Phys. Rev.* **112,** 1303.
CORINALDESI, E. (1959) *Suppl. Nuovo Cim.* **14,** 369.
CORK, B., LAMBERTSON, G.R., PICCIONI, O. and WENZEL, W.A. (1956) *Phys. Rev.* **104,** 1193.
COWAN, C.L., REINES, F., HARRISON, F.B., KRUSE, H.W. and MCGUIRE, A.D. (1956) *Science* **124,** 103.
COWAN, E.W. (1954) *Phys. Rev.* **94,** 161.
CRAWFORD, F.S., CRESTI, M., GOOD, M.L., SOLMITZ, F.T., STEVENSON, M.L. and TICHO, H.K. (1959a) *Phys. Rev. Lett.* **2,** 114; (1959b) *ibid.* **2,** 174.
CRAWFORD, F.S. (1962) *Int. Conf. on High Energy Physics at CERN*, p. 827.
CRUSSARD, J., KAPLON, M.F., KLARMANN, J. and NOON, J.H. (1954) *Phys. Rev.* **93,** 253.
CULLIGAN, G., FRANK, S.G.F. and HOLT, J.R. (1959) *Proc. Phys. Soc.* **73,** 169.
CULLIGAN, G., LATHROP, J.F., TELEGDI, V.L., WINSTON, R. and LUNDY, R.A. (1961) *Phys. Rev. Lett.* **7,** 458.
CZIFFRA, P., MACGREGOR, M.H., MORAVCSIK, M.J. and STAPP, H.P. (1959) *Phys. Rev.* **114,** 880.
DALITZ, R.H. (1953) *Phil. Mag.* **44,** 1068.
DALITZ, R.H. (1954) *Phys. Rev.* **94,** 1046.
DALITZ, R.H. (1955) *Phys. Rev.* **99,** 915.
DALITZ, R.H. (1956) *Proc. Phys. Soc.* **69,** 527.
DALITZ, R.H. (1957) *Rep. Progr. Phys.* **20,** 163.
DALITZ, R.H. (1961a) *Aix-en-Provence Conf. Rep.*, CEN Saclay **2,** 151; (1961b) *Rutherford Jubilee Int. Conf.*, Heywood, p. 103.
DALITZ, R.H. (1962) *Int. Conf. on High Energy Physics at CERN*, p. 391.
DALITZ, R.H. (1963) *Ann. Rev. Nuclear Sci.* **13,** 339.
DALITZ, R.H. and LIU, L. (1959) *Phys. Rev.* **116,** 1312.
DALITZ, R.H. and TUAN, S. (1960) *Ann.of Phys.* **10,** 307.
DANBY, G. *et al.* (1962) *Phys. Rev. Lett.* **9,** 36.
DAVIDON, W.C. and GOLDBERGER, M.L. (1956) *Phys. Rev.* **104,** 1119.
DAVIS, R. (1955) *Phys. Rev.* **97,** 766.
DAY, T.B. and SNOW, G.A. (1959) *Phys. Rev. Lett.* **2,** 59.
DAY, T.B., SNOW, G.A. and SUCHER, J. (1959) *Phys. Rev. Lett.* **3,** 61.
DENNISON, D.M. (1927) *Proc. Roy. Soc.* **115,** 483.
DEPOMMIER, P., HEINTZE, J., RUBBIA, C. and SOERGEL, V. (1963) *Phys. Letters* **5,** 61.
DETOEUF, J.F. (1961), *Aix-en-Provence Conf. Rep.*, CEN Saclay, **2,** 57.

DEUTSCH, M. (1951) *Phys. Rev.* **83,** 866.
DEUTSCH, M. (1953) *Progr. Nuclear Phys.* **3,** 131.
DEVLIN, T.J., MOYER, B.J. and PEREZ-MENDEZ, V. (1962) *Phys. Rev.* **125,** 690.
DEVONS, S., GIDAL, G., LEDERMAN, L.M. and SHAPIRO, G. (1960) *Phys. Rev. Lett.* **5,** 330.
DIDDENS, A.N., LILLETHUN, E., MANNING, G., TAYLOR, A.E., WALKER, T.G. and WETHERELL, A.M. (1962a) *Phys. Rev. Lett.* **9,** 111; (1962b) ibid. **9,** 108; (1962c) ibid. **9,** 32.
DIRAC, P.A.M. (1928) *Proc. Roy. Soc.* A **117,** 610.
DIRAC, P.A.M. (1929) *Proc. Roy. Soc.* A 126, 360.
DIRAC, P.A.M. (1930) *Proc. Camb. Phil. Soc.* **26,** 361.
DIRAC, P.A.M. (1947) *The Principles of Quantum Mechanics*, 3rd ed., Oxford.
DONOVAN, P.F., ALBURGER, D.E. and WILKINSON, D.H. (1961) *Rutherford Jubilee Int. Conf.*, Heywood, p. 827.
DRELL, S.D. (1958) *Ann. of. Phys.* **4,** 75.
DRELL, S.D. (1962) *Int. Conf. on High Energy Physics at CERN*, p. 897.
DRELL S.D. and ZACHARIASEN, F. (1958) *Phys. Rev.* **111,** 1727.
DRISCOLL, R.L. and BENDER, P.L. (1958) *Phys. Rev. Lett.* **1,** 413.
DUDZIAK, W.F., SAGANE, R. and VEDDER, J. (1959) *Phys. Rev.* **114,** 336.
DUKE, P.J., CHADWICK, G.B., COLLINS, G.B., FUJII, T., HIEN, N.C. and TURKOT, F. (1961) *Aix-en-Provence Conf. Rep.*, CEN Saclay, **1,** 419.
DUNAITSEV, A., PETRUKHIN, V., PROKOSHKIN, IU. and RYKALIN, V. (1962) *Phys. Lett.* **1,** 138.
DURAND, L. (1961) *Phys. Rev. Lett.* **6,** 631.
DURBIN, R., LOAR, H. and STEINBERGER, J. (1951) *Phys. Rev.* **84,** 581.
DYSON, F.J. (1949) *Phys. Rev.* **75,** 486, 1736.
DYSON, F.J. (1951) *Advanced Quantum Mechanics*, lecture notes, Cornell University.
DYSON, F.J. (1952) *Phys. Rev.* **85,** 631.
DYSON, F.J. and MCVOY, K.W. (1957) *Phys. Rev.* **106,** 1360.
DZHELEPOV, V.P., KAZARINOV, YU.M., GOLOVIN, B.M., FLYAGIN, V.B. and SATAROV, V.I (1956) *Suppl. Nuovo Cim.* **3,** 61.
EBEL, M.E. and ERNST, F.J. (1960) *Nuovo Cim.* **15,** 173.
EGOROV, L.B., ZHURAVLEV, G.V., IGNATENKO, A.E., HSUANG-MING, LI., PETRASHKU, M.G. and CHULTEM, D. (1961) *JETP* **13,** 268.
EINSTEIN, A. (1905) *Ann. der Phys.* **17,** 132.
EISLER, F. *et al.* (1958) *Nuovo Cim.* **7,** 222.
EMIGH, C.R. (1952) *Phys. Rev.* **86,** 1028.
ERNST, F.J., SACHS, R.G. and WALI, K.C. (1960) *Phys. Rev.* **119,** 1105.
ERWIN, A.R., MARCH, R., WALKER, W.D. and WEST, E. (1961) *Phys. Rev. Lett.* **6,** 628.
D'ESPAGNAT, B. (1961) *Nuovo Cim.* **20,** 1217.
D'ESPAGNAT, B. and PRENTKI, J. (1956) *Nuclear Phys.* **1,** 33.
EVANS, R.D. (1955) *The Atomic Nucleus*, MCGRAW-HILL, New York.
EVANS, R.D. (1958) *Handbuch der Physik*, **34,** 218, Springer-Verlag, Berlin.
EVSEEV, V.S., KOMAROV, V.I., KUSH, V.Z., ROGANOV, V.S., CHERNOGOROVA, V.A. and SZYMCZAK, M.M. (1962) *JETP* **14,** 217.
FABRI, E. (1954) *Nuovo Cim.* **11,** 479.
FABIANI, F. *et al.* (1961) *Aix-en-Provence Conf. Rep.*, CEN Saclay, **1,** 151.
FEDERBUSH, P., GOLDBERGER, M.L. and TREIMAN, S.B. (1958) *Phys. Rev.* **112,** 642.
FELD, B.T. (1953) *Phys. Rev.* **89,** 330.
FELDMAN, G., MATTHEWS, P.T. and SALAM, A. (1960) *Nuovo Cim.* **16,** 549.
FEINBERG, G. (1958) *Phys. Rev.* **110,** 1482.
FEINBERG, G., and WEINBERG, S. (1961) *Phys. Rev. Lett.* **6,** 381.
FERMI, E. (1934), *Z. für Phys.* **88,** 161.
FERMI, E. (1950) *Nuclear Physics*, Chicago University Press.
FERMI, E. (1951) *Elementary Particles*, Yale University Press.
FERMI, E. (1955) *Suppl. Nuovo Cim.* **2,** 17.

FERRARI, G. *et al.* (1956) *CERN Symposium*, **2,** 230.
FERRELL, R.A. (1951) *Phys. Rev.* **84,** 858.
FERRO-LUZZI, M., TRIPP, R.D. and WATSON, M.B. (1962) *Phys. Rev. Lett.* **8,** 28.
FEYNMAN, R.P. (1949) *Phys. Rev.* **76,** 749, 769.
FEYNMAN, R.P. (1961) *Quantum Electrodynamics,* Benjamin.
FEYNMAN, R.P. and GELL-MANN, M. (1958) *Phys.Rev.* **109,** 193.
FLYAGIN, V.B., DZHELEPOV, V.P., KISELEV, V.S. and OGANESYAN, K.O. (1959) *JETP* **8,** 592.
FOELSCHE, H., FOWLER, E.L., KRAYBILL, H.L., SANFORD, J.R. and STONEHILL, D. (1962) *Phys. Rev. Lett.* **9,** 223.
FOLEY, K.J., LINDENBAUM, S.J., LOVE, W.A., OZAKI, S., RUSSELL, J.J. and YUAN, L.C.L. (1963) *Phys. Rev. Lett.* **10,** 376.
FOWLER, P.H., MENON, M.G.K., POWELL, C.F. and ROCHAT, O. (1951) *Phil. Mag.* **42,** 1040.
FOWLER, W.B., SHUTT, R.P., THORNDIKE, A.M. and WHITTEMORE, W.L. (1953) *Phys. Rev.* **91,** 1287.
FOWLER, W.B., SHUTT, R.P., THORNDIKE, A.M. and WHITTEMORE, W.L. (1954) *Phys. Rev.* **93,** 861.
FOWLER, W.B., SHUTT, R.P., THORNDIKE, A.M. and WHITTEMORE, W.L. (1955) *Phys. Rev.* **98,** 121.
FRANKEL, S. *et al.* (1962) *Phys. Rev. Lett.* **8,** 123.
FRAUENFELDER, H.R. *et al.* (1957) *Phys. Rev.* **106,** 386.
FRAUTSCHI, S.C., GELL-MANN, M. and ZACHARIASEN, F. (1962) *Phys. Rev.* **126,** 2204.
FRAZER, W. and FULCO, J. (1959) *Phys. Rev. Lett.* **2,** 365.
FRAZER, W. and FULCO, J. (1960) *Phys. Rev.* **117,** 1609.
FREEMAN, J.M., MONTAGUE, J.H., WEST, D. and WHITE, R.E. (1962) *Phys. Lett.* **3,** 136.
FREEMAN, J.M., MONTAGUE, J.H., MURRAY, G., WHITE, R.E. and BURCHAM, W.E. (1964) *Phys. Lett.* **8,** 115.
FRIEDRICH, W. and GOLDHABER, G. (1927) *Z. für Phys.* **44,** 700.
FRIEDMAN, J.I. and TELEGDI, V.L. (1957) *Phys. Rev.* **105,** 1681.
FRÖHLICH, H., HEITLER, H. and KEMMER, N. (1938) *Proc. Roy. Soc.* A **166,** 154.
FROISSART, M. (1961) *Phys. Rev.* **123,** 1053.
FUJII, A. and PRIMAKOFF, H. (1959) *Nuovo Cim.* **12,** 327.
FURRY, W.H. (1937) *Phys. Rev.* **51,** 125.
GAMMEL, J.L. and THALER, R.M. (1960) *Progr. in Elementary Particle and Cosmic Ray Phys.* North-Holland, **5,** 99.
GARDNER, J.H. and PURCELL, E.M. (1949) *Phys. Rev.* **76,** 1262.
GARDNER, J.H. (1951) *Phys. Rev.* **83,** 996.
GARWIN, R.L., LEDERMAN, L.M. and WEINRICH, M. (1957) *Phys. Rev.* **105,** 1415.
GASIOROWICZ, S. (1960) *Fortschr. der Phys.* **8,** 665.
GATLAND, I.R. and MOFFAT, J.W. (1963) *Phys. Rev.* **129,** 2812.
GELL-MANN, M. (1953) *Phys. Rev.* **92,** 833.
GELL-MANN, M. (1956) *Suppl. Nuovo Cim.* **4,** 848.
GELL-MANN, M. (1958) *Phys. Rev.* **111,** 362.
GELL-MANN, M. (1961) *CTSL Report* – 20.
GELL-MANN, M. (1962a) *Phys. Rev. Lett.* **8,** 263; (1962b) *Phys. Rev.* **125,** 1067.
GELL-MANN, M. (1964) *Phys. Lett.* **8,** 214.
GELL-MANN, M. and LEVY, M. (1960) *Nuovo Cim.* **16,** 705.
GELL-MANN, M. and PAIS, A. (1955a) *Glasgow Conf. on Nuclear and Meson Physics,* Pergamon p. 342; (1955b) *Phys. Rev.* **97,** 1387.
GELL-MANN, M. and ROSENFELD, A.H. (1957) *Ann. Rev. Nuclear Sci.* **7,** 407.
GELL-MANN, M., GOLDBERGER, M.L., NAMBU, Y. and OEHME, R. (unpublished).
GERSHTEIN, S. and ZELDOVITCH, J. (1956) *JETP* **2,** 576.
GIAMATI, G.C. and REINES, F. (1962) *Phys. Rev.* **126,** 2178.
GILBERT, W. and SCREATON, G.R. (1956) *Phys. Rev.* **104,** 1758.

GLASER, D. A. (1959) *Proc. Ninth Int. Conf. Rep. on High Energy Physics,, Kiev* Moscow, **2,** 242.
GLASSER, R. G., SEEMAN, N. and STILLER, B. (1961) *Phys. Rev.* **123,** 1014.
GLAUBER, R. J. (1958) *Lectures in Theoretical Physics,* Interscience, p. 315.
GOLDBERG, A. (1958) *Phys. Rev.* **112,** 618.
GOLDBERGER, M. L. (1960) *Dispersion Relations and Elementary Particles* (Ed.) De Witt and Omnes, Wiley.
GOLDBERGER, M. L., MIYAZAWA, H. and OEHME, R. (1955) *Phys. Rev.* **99,** 986.
GOLDBERGER, M. L., and TREIMAN, S. B. (1958a) *Phys. Rev.* **110,** 1178; (1958b) *ibid.* **111,** 354.
GOLDHABER, M., GRODZINS, L. and SUNYAR, A. W. (1958) *Phys. Rev.* **109,** 1015.
GOLDHABER, S. *et al.* (1962) *Phys. Rev. Lett.* **9,** 135.
GOLDSCHMIDT-CLEMONT, Y., KING, D. T., MUIRHEAD, H. and RITSON, D. M. (1948) **61,** 138.
GOLDSCHMIDT-CLEMONT, Y. *et al.* (1962) *Int. Conf. on High Energy Physics at CERN,* p. 84.
GOTOW, K., LOBKOWICZ, F. and HEER, E. (1962) *Phys. Rev.* **127,** 2206.
GOURDIN, M. and SALIN, PH. (1963) *Nuovo Cim.* **27,** 193.
GREGORY, B. P. (1962) *Int. Conf. on High Energy Physics at CERN,* p. 779.
GREGORY, B. P. LAGGARIGUE, A., LEPRINCE-RINGUET, L., MULLER, F. and PEYROU, Ch. (1954) *Nuovo Cim,* **11,** 292.
GRIBOV, V. N. (1961) *Nuclear Phys.* **22,** 249.
GRIBOV, V. N. and POMERANCHUK, I. YA. (1962) *Phys. Rev. Lett.* **8,** 343.
GUPTA, S. N. (1950) *Proc. Phys. Soc.* **A** 63, 681.
GUPTA, S. N. (1951) *Proc. Phys. Soc.* **A 64,** 426.
HAAG, R. (1955) *Kgl. Danske Videnskab. Selskab., Mat.- Fys. Medd.* **29,** No. 12.
HAAG, R. (1959) *Suppl. Nuovo Cim.* **14,** 131.
HABER-SCHAIM, U. (1956) *Phys. Rev.* **104,** 1113.
HADJIOANNOU, F., PHILLIPS, R. J. N. and RARITA, W. (1962) *Phys. Rev. Lett.* **9,** 183.
HAGEDORN, R. (1961) *Introduction to Field Theory and Dispersion Relations CERN,* 61–6.
HAMERMESH, M. (1962) *Group Theory and its Applications to Physical Problems,* Pergamon.
HAMILTON, J. (1956) *Proc. Camb. Phil. Soc.* **52,** 97.
HAMILTON, J. (1959) *The Theory of Elementary Particles,* Oxford.
HAMILTON, J. and WOOLCOCK, W. S. (1960) *Phys. Rev.* **118,** 291.
HAND, L. N., MILLER, D. G. and WILSON, R. (1963) *Rev. Mod. Phys.* **35,** 335.
HELLAND, J. A., DEVLIN, T. J., HAGGE, D. E., LONGO, M. J., MOYER, B. J. and WOOD, C. D. (1962) *Int. Conf. on High Energy Physics at CERN,* p. 3.
HELLAND, J. A. *et al.* (1963) *Phys. Rev. Lett.* **10,** 27.
HEISENBERG, W. (1932) *Z. für Phys.* **77,** 1.
HEISENBERG, W. (1934) *Z. für Phys.* **90,** 209.
HEISENBERG, W. (1943) *Z. für Phys.* **120,** 513, 673.
HEITLER, W. and HERZBERG, G. (1929) *Naturwiss.* **17,** 673.
HENRY, G. R., SCHRANK, G. and SWANSON, R. A, (1963) *PPAD report,* 506D.
HILDEBRAND, R. H. (1962) *Phys. Rev. Lett.* **8,** 34.
HODSON, A. L. *et al.* (1954) *Phys. Rev.* **96,** 1089.
HOFSTADTER, R. and MCINTYRE, J. A. (1949) *Phys. Rev.* **76,** 1269.
HOFSTADTER, R., DE VRIES, C. and HERMAN, R. (1961) *Phys. Rev. Lett.* **6,** 290.
HOOVER, J. I., FAUST, W. R. and DOHNE, C. F. (1952) *Phys. Rev.* **85,** 58.
HORI, T. (1927) *Z. für Phys.* **44,** 834.
HUGHES, V. W., MCCOLM, D. W., ZIOCK, K. and PREPOST, R. (1960) *Phys. Rev. Lett.* **5,** 63.
HUMPHREY, W. and ROSS, R. (1962) *Phys. Rev.* **127,** 1305.
HUTCHINSON, D. P., MENES, J., SHAPIRO, G., PATLACH, A. M. and PENMAN, S. (1961) *Phys. Rev. Lett.* **7,** 129.
IGI, K. (1962) *Phys. Rev. Lett.* **9,** 76.
IKEDA, M., OGAWA, S. and OHNUKI, Y. (1959) *Progr. Th. Phys.* **22,** 715.
JACKSON, J. D. (1961) *Dispersion Relations* (Ed. Screaton), Oliver and Boyd.
JACKSON, J. D., TREIMAN, S. B. and WYLD, H. W. (1957) *Phys. Rev.* **106,** 517.

JACOB, M. and MATHEWS, J. (1960) *Phys. Rev.* **117,** 854.
JACOB, M. and WICK, G.C. (1959) *Ann. of Phys.* **7,** 404.
JAHNKE, E. and EMDE, F. (1945) *Tables of Functions*, Dover.
JAUCH, J.M. and ROHRLICH, F. (1955) *The Theory of Electrons and Photons*, Addison-Wesley.
JONES, D.P., MURPHY, P.G. and O'NEILL, P.L. (1958) *Proc. Phys. Soc.* **72,** 429.
JOST, R. (1957) *Helv. Phys. Acta* **30,** 409.
KABIR, P.K. (1961) *Nuovo Cim.* **22,** 429.
KAPUSCINSKI, W. and EYMERS, J.G. (1929) *Proc. Roy. Soc.* 122, 58.
KEMMER, N., (1938) *Proc. Camb. Phil. Soc.* **34,** 354; *Proc. Roy. Soc.* A **166,** 127.
KENDALL, H.W. and DEUTSCH, M. (1956) *Phys. Rev.* **101,** 20.
KERTH, L.T. (1961) *Rev. Mod. Phys.* **33,** 389.
KIBBLE, T.W.B. (1960) *Phys. Rev.* **117,** 1159.
KINOSHITA, T. (1959) *Phys. Rev. Lett.* **2,** 477.
KINOSHITA, T. and SIRLIN, A. (1957) *Phys. Rev.* **108,** 844.
KLEIN, O. and NISHINA, Y. (1929) *Z. für Phys.* **52,** 853.
KLEPIKOV, N.P., MESCHERYAKOV, V.A. and SOKOLOV, S.N. (1960) Dubna preprint, D-584.
KONOPINSKI, E.J. (1959) *Ann. Rev. Nuclear Sci.* **9,** 99.
KONOPINSKI, E.J. and UHLENBECK, G.E. (1935) *Phys. Rev.* **48,** 7.
KRAMERS, H.A. (1927) *Atti. Congr. Int. Fisici, Como,* **2,** 545.
KRAMERS, H.A. (1937) *Proc. Amsterdam Acad.* **40,** 814.
KRASS, A.S. (1962) *Phys. Rev.* **125,** 2172.
KRETZSCHMAR, M.. (1961) *Ann. Rev. Nuclear Sci.* **11,** 1.
KROLL, N.M. and WADA, W. (1955) *Phys. Rev.* **98,** 1355.
KRONIG, R. (1926) *J. Opt. Soc. Amer.* **12,** 547.
KRUGER, H. (1961) UCRL 9322.
LAMB, W.E. (1951) *Rep. Progr. Phys.* **14,** 19.
LANDAU, L. (1957) *Nuclear Phys.* **3,** 127.
LANDAU, L. (1959) *Proc. Ninth Int. Conf. on High Energy Physics, Kiev* **2,** 95, Moscow.
LANDÉ, K., BOOTH, E.T., IMPEDUGLIA, J., LEDERMAN, L.M. and CHINOWSKY, W. (1956) *Phys. Rev.* **103,** 1901.
LAPORTE, O. (1924) *Z. für Phys.* **23,** 135.
LATTES, C.M.G., MUIRHEAD, H., POWELL, C.F. and OCCHIALINI, G.P.S. (1947) *Nature* **159,** 694.
LATTES, C.M.G., OCCHIALINI, G.P.S. and POWELL, C.F. (1947) *Nature* **160,** 453, 486.
LAYZER, A.J. (1960) *Phys. Rev. Lett.* **4,** 580.
LEE, T.D. (1961) CERN 61-30, 99.
LEE, T.D. (1962) *Phys. Rev. Lett.* **9,** 319.
LEE, T.D. and YANG, C.N. (1956a) *Phys. Rev.* **104,** 254; (1956b) *Nuovo Cim.* **3,** 749.
LEE, T.D. and YANG, C.N. (1957a) *Phys. Rev.* **105,** 1671; (1957b) BNL Rep. No. 443.
LEE, T.D. and YANG, C.N. (1958) *Phys. Rev.* **109,** 1755.
LEE, T.D. and YANG, C.N. (1960) *Phys. Rev. Lett.* **4,** 307.
LEE, T.D., YANG, C.N. and OEHME, R. (1957) *Phys. Rev.* **106,** 340.
LEE, Y.K., MO, L.W. and WU, C.S. (1963) *Phys. Rev. Lett.* **10,** 253.
LEHMANN, H. (1959) *Suppl. Nuovo Cim.* **14,** 153.
LEHMANN, H., SYMANZIK, K. and ZIMMERMAN, W. (1955) *Nuovo Cim.* **1,** 205.
LEHMANN, H., SYMANZIK, K. and ZIMMERMAN, W. (1957) ibid. **6,** 319.
LEPORE, J.V. (1950) *Phys. Rev.* **79,** 137.
LEVY, M. (1962) *Nuovo Cim.* **24,** 920.
LINDENBAUM, S.J. and STERNHEIMER, R.M. (1957) *Phys. Rev.* **105,** 1874.
LINDENBAUM, S.J. and STERNHEIMER, R.M. (1960) *Phys. Rev. Lett.* **5,** 24.
LINDENBAUM, S.J., LOVE, W.A., NIEDERER, J.A., OZAKI, S., RUSSELL, J.J. and YUAN, L.L. (1961) *Phys. Rev. Lett.* **7,** 352.
LONDON, G.W. (1963) *Rochester preprint* NYO-10274.
LOW, F.E. (1955) *Phys. Rev.* **97,** 1392.

LÜDERS, G. (1954) *Kgl. Danske Videnskab. Selskab*, Mat.-Fys. Medd. **28**, No. 5.
LUERS, D., MITTRA, I.S., WILLIS, W.J. and YAMAMOTO, S.S. (1964) *Phys. Rev.* **133**, B1276.
LUNDY, R.A. (1962) *Phys. Rev.* **125**, 1686.
LYNCH, G.R. (1962) *Proc. Phys. Soc.* **80**, 46.
LYNCH, G.R. OREAR, J. and ROSENDORFF, S. (1960) *Phys. Rev.* **118**, 284.
MCDONALD, W.S., PETERSON, V.Z. and CORSON, D.R. (1957) *Phys. Rev.* **107**, 577.
MACGREGOR, M.H., MORAVCSIK, M.J. and STAPP, H.P. (1960) *Ann. Rev. Nuclear Sci.* **10**, 291.
MCMASTER, W.H. (1961) *Rev. Mod. Phys.* **33**, 8.
MACQ, P.C., CROWE, K.M. and HADDOCK, R.P. (1958) *Phys. Rev.* **112**, 2061.
MAGLIĆ, B.C., ALVAREZ, L.W. ROSENFELD, A.H. and STEVENSON, M.L. (1961) *Phys. Rev. Lett.* **7**, 178.
MANDELSTAM, S. (1958a) *Phys. Rev.* **112**, 1344; (1958b) *Proc. Roy. Soc.* A **244**, 491.
MANDELSTAM, S. (1959) *Phys. Rev.* **115**, 1741, 1752.
MANDELSTAM, S. (1962) *Rep. Progr. Phys.* **25**, 99.
MARTIN, A. (1962) *Int. Conf. on High Energy Physics at CERN*, p. 566.
MATTHEWS, P.T. (1961) *Aix-en-Provence Conf. Rep.* CEN Saclay, **2**, 87.
MATTHEWS, P.T. and SALAM, A. (1951) *Rev. Mod. Phys.* **23**, 311.
MEITER, P.H.E. and BAUER, E. (1962) *Group Theory, The Application to Quantum Mechanics*, North-Holland.
MENCUCCINI, C., QUERZOLI, R. and SALVINI, G. (1961) *Aix-en-Provence Conf. Rep.*, CEN Saclay, **1**, 17.
MEYER, PH. and SALZMAN, G. (1959) *Nuovo Cim.* **14**, 1310.
MICHEL, L. (1950) *Proc. Phys. Soc.* **A 63**, 514.
MICHEL, L. (1952) *Phys. Rev.* **86**, 814.
MICHEL, L. (1957) *Rev. Mod. Phys.* **29**, 223.
MIYAKE, K., KINSEY, K.F. and KNAPP, D.E. (1962) *Phys. Rev.* **126**, 2188.
MØLLER, (1932) *Ann. der Phys.* **14**, 531.
MØLLER, C. (1945) *Kgl. Danske Videnskab Selskab., Mat.-Fys. Medd.* **23**, No. **1**.
MORAVCSIK, M.J. (1956) *Phys. Rev.* **104**, 1451.
MORAVCSIK, M.J. (1957) *Phys. Rev.* **105**, 267.
MORAVCSIK, M.J. and NOYES, H.P. (1961) *Ann. Rev. Nuclear Sci.* **11**, 95.
MORPURGO, G. and TOUSCHEK, B.F. (1955) *Nuovo Cim.* **1**, 1159.
MORSE, P.M. and FESHBACH, H. (1953) *Methods of Theoretical Physics*, MCGRAW-HILL, New York.
MOTT, N.F. and MASSEY, H.S.W. (1949) *Theory of Atomic Collisions*, Oxford.
NAKANO, T. and NISHIJIMA, K. (1953) *Progr. Th. Phys.* **10**, 581.
NAMBU, Y. *Phys. Rev. Lett.* **4**, 380.
NEDDERMEYER, S.H. and ANDERSON, C.D. (1938) *Phys. Rev.* **54**, 88.
NE'EMAN, Y. (1961) *Nuclear Phys.* **26**, 222.
NIGAM, B.P. (1963) *Rev. Mod. Phys.* **35**, 117.
NISHIJIMA, K. (1955) *Progr. Th. Phys.* **13**, 285.
O'CEALLAIGH, C. (1950) *Phil. Mag.* **41**, 838.
O'CEALLAIGH, C. (1951) *Phil. Mag.* **42** 1032.
OEHME, R. (1955) *Phys. Rev.* **98**, 147, 216.
OKUBO, S. (1962) *Prog. Th. Phys.* **27**, 949.
OKUBO, S. and MARSHAK, R.E. (1958) *Ann. Phys.* **4**, 166.
OKUBO, S., MARSHAK, R.E. and SUDARSHAN, E.C.G. (1959) *Phys. Rev.* **113**, 944.
OKUN, L.B. (1958) *JETP* **7**, 322.
OKUN, L.B. and PONTECORVO, B. (1957) *JETP* **5**, 1297.
OKUN, L.B. and SEHTER, Y.M. (1958) *Nuovo Cim.* **10**, 359.
OLSON, D.N., SCHOPPER, H.F. and WILSON, R.R. (1961) *Phys. Rev. Lett.* **6**, 286.
OPPENHEIMER, J.R. (1930) *Phys. Rev.* **35**, 562.
ORE, A. and POWELL, J.L. (1949) *Phys. Rev.* **75**, 1696.

OREAR, J. (1954) *Phys. Rev.* **96,** 1417.
PAIS, A. (1952) *Phys. Rev.* **86,** 663.
PAIS, A. (1962) *Phys, Rev. Lett.* **9,** 117.
PAIS, A. and TREIMAN, S.B. (1957) *Phys. Rev.* **105,** 1616.
PANOFSKY, W.K.H., AAMODT, R.L. and HADLEY, J. (1951) *Phys. Rev.* **81,** 565.
PAULI, W. (1933a) *Proc. Solvay Conf.*; (1933b) *Handbuch der Physik* **24,** 1, 233.
PAULI, W. (1941) *Rev. Mod. Phys.* **13,** 203.
PAULI, W. (1955) *Niels Bohr and the Development of Physics*, Pergamon, p. 30.
PAULI, W. (1958) UCRL 8213.
PAULI, W. and WEISSKOPF, V.F. (1934) *Helv. Phys. Acta* **7,** 709.
PEIERLS, R.F. (1960) *Phys. Rev.* **118,** 325.
PETERMANN, A. (1958) *Fortschr. der Phys.* **6,** 505.
PEVSNER, A. *et al.* (1961) *Phys. Rev. Lett.* **7,** 421.
PEYROU, C. (1961) *Aix-en-Provence Conf. Rep*, CEN Saclay, **2,** 103.
PICKUP, E., ROBINSON, D.K. and SALANT, E.O. (1961) *Phys. Rev. Lett.* **7,** 192.
PJERROU, G.M., PROWSE, D.J., SCHLEIN, P., SLATER, W.E., STORK, D.M. and TICHO, H.K. (1962) *Int. Conf. on High Energy Physics at CERN*, p. 289.
PLANO, R.J. (1960) *Phys. Rev.* **119,** 1400.
PLANO, R., PRODELL, A., SAMIOS, N., SCHWARTZ, M. and STEINBERGER, J. (1959) *Phys. Rev. Lett.* **3,** 525.
PLANO, R., SAMIOS, N., SCHWARTZ, M., and STEINBERGER, J. (1957) *Nuovo Cim.* **5,** 1700.
POIRIER, J.A. and PRIPSTEIN, M. (1963) *Phys. Rev.* **130,** 1171.
POMERANCHUK, I. YA. (1958) *JETP* **7,** 499.
PONTECORVO, B. (1959) *JETP* **10,** 1236.
PRIMAKOFF, H. (1959) *Rev. Mod. Phys.* **31,** 802.
PROWSE, D.J. and BALDO-CEOLIN, M. (1958) *Nuovo Cim.* **10,** 635.
PUPPI, G. (1962) *Int. Conf. on High Energy physics at CERN*, p. 713.
PURCELL, E.M., TORREY, H.C. and POUND, R.V. (1946) *Phys. Rev.* **69,** 37.
PUZIKOV, L., RYNDIN, R. and SMORODINSKY, J. (1957) *Nuclear Phys.* **3,** 436.
RASETTI, F., (1930) *Z. für Phys.* **61,** 598.
RASETTI, F., (1941) *Phys. Rev.* **60,** 198.
REGGE, T. (1959) *Nuovo Cim.* **14,** 951.
REGGE, T. (1960) *Nuovo Cim.* **18,** 947.
REINES, F. and COWAN, C.L. (1959) *Phys. Rev.* **113,** 273.
RICHARDSON, J.R. (1948) *Phys. Rev.* **74,** 1720.
RITSON, D.M. (Ed.) (1961) *Techniques of High Energy Physics*, Interscience.
ROBSON, J.M. (1955) *Phys. Rev.* **100,** 933.
ROCHESTER, G.D. and BUTLER, C.C. (1947) *Nature* **160,** 855.
ROOS, M. (1964) *Phys. Lett.* **8,** 1.
ROSEN, S.P. (1962) *Phys. Rev. Lett.* **9,** 186.
ROSENBLUTH, M.N. (1950) *Phys. Rev.* **79,** 615.
ROSENFELD, A.H. (1954) *Phys. Rev.* **96,** 139.
ROSENFELD, A.H. (1963) *Selected Topics on Elementary Particle Physics*, Academic Press.
RUBBIA, C. (1961) *Rutherford Jubilee Int. Conf.*, Heywood, p. 707.
RUDERMAN, M. and FINKELSTEIN, R. (1949) *Phys. Rev.* **76,** 1458.
SACHS, R.G. (1962) *Phys. Rev.* **126,** 2256.
SAKATA, S. (1956) *Progr. Th. Phys.* **16,** 686.
SAKURAI, J.J. (1958a) *Phys. Rev. Lett.* **1,** 40; (1958b) *ibid.*, 258.
SAKURAI, J.J. (1960) *Ann. Phys.* **11,** 1, 1960.
SAKURAI, J.J. (1963) *Phys. Rev.* **132,** 434.
SALAM, A. (1956) *Nuovo Cim.* **3,** 424.
SALAM, A. (1957) *Nuovo Cim.* **5,** 299.
SALAM, A. and WARD, J.C. (1961) *Nuovo Cim.* **20,** 419.

SAMIOS, N. P., BACHMAN, A. H., LEA, R. M., KALOGEROPOULOS, T. E. and SHEPHARD, W. D. (1962) *Phys. Rev. Lett.* **9**, 139.
SANDERS, J. H. (1961) *The Fundamental Atomic Constants*, Oxford.
SCHIFF, L. I. (1955) *Quantum Mechanics*, MCGRAW-HILL.
SCHUPP, A. A., PIDD, R. W. and CRANE, H. R. (1961) *Phys. Rev.* **121**, 1.
SCHWARTZ, M. (1961) *Phys. Rev. Lett.* **6**, 556.
SCHWEBER, S. S., BETHE, H. A. and DE HOFFMAN, F. (1955) *Mesons and Fields*, **1**, Row, Peterson.
SCHWINGER, J. (1948a) *Phys. Rev.* **73**, 416; (1948b) *ibid.* **74**, 1439.
SCHWINGER, J. (1949) *Phys. Rev..* **76**, 790.
SCHWINGER, J. (1951) *Phys. Rev..* **82**, 914.
SCHWINGER, J. (1953) *Phys. Rev.* **91**, 712; **94**, 1362.
SCOTTI, A. and WONG, D. Y. (1963) *Phys. Rev. Lett.* **10**, 142.
SELLERI, F. (1961) *Phys. Rev. Lett.* **6**, 64.
SEWARD, F. D., HATCHER, C. R. and FULTZ, S. C. (1961) *Phys. Rev.* **121**, 605.
SHWE, H., SMITH, F. M. and BARKAS, W. H. (1962) *Phys. Rev.* **125**, 1024.
SIGNELL, P. S. (1960) *Phys. Rev. Lett.* **5**, 474.
SIMONS, L.(1958) *Handbuch der Physik*, Springer Verlag, **34**, 139.
SMITH, J. H., PURCELL, E. M. and RAMSEY, N. F. (1957) *Phys. Rev.* **108**, 120.
SMORODINSKII, YA. (1959a) *Soviet Phys. Usp.* **2**, 1; (1959b) *Proc. Ninth Int. Conf. on High Energy Physics, Kiev.* **1**, 271, Moscow.
SNOW, G. A. (1960) *Proc. Int. Conf. on High Energy Physics at Rochester*, p. 407.
SNOW, G. A. and SHAPIRO, M. M. (1961) *Rev. Mod. Phys.* **33**, 231.
SOLMITZ, F. T. (1960) *Proc. Int. Conf. on High Energy Physics at Rochester*, p. 164.
SOMMER, H., THOMAS, H. A. and HIPPLE, J. A. (1951) *Phys. Rev.* **82**, 697.
SOSNOVSKII, A. N., SPIVAK, P. E., PROKOFIEV, IU. A., KUTIKOV, I. E. and DOBRININ, IU. P. (1959) *JETP* **8**, 739.
SRIVASTAVA, P. P. and SUDARSHAN, G. (1958) *Phys. Rev.* **110**, 765.
STANNARD, F. R. (1961) *Phys. Rev.* **121**, 1513.
STAPP, H. P. (1955) UCRL 3098.
STAPP, H. P. (1956) *Phys. Rev.* **103**, 425.
STAPP, H. P., YPSILANTIS, T. J. and METROPOLIS, N. (1957) *Phys. Rev.* **105**, 302.
STERNHEIMER, R. M. and LINDENBAUM, S. J. (1961) *Phys. Rev.* **123**, 333.
STEVENSON, M. L., ALVAREZ, L. W., MAGLIĆ, B. C. and ROSENFELD, A. H. (1962) *Phys. Rev.* **125**, 687.
STIENING, R. F., LOH, E. and DEUTSCH, M. (1963) *Phys. Rev. Lett.* **10**, 536.
STONEHILL, D. *et al.* (1961) *Phys. Rev. Lett.* **6**, 624.
STUBBS, T. F. *et al.* (1961) *Phys. Rev. Lett.* **7**, 188.
STÜCKELBERG, E. C. G. (1934) *Ann. der Phys.* **21**, 367.
de SWART, J. J. (1963) *Rev. Mod. Phys.* **35**, 916.
TAKETANI, M., NAKAMURA, S. and SASAKI, M. (1951) *Progr. Th. Phys.* **6**, 581.
TAMM, I. (1930) *Z. für Phys.* **62**, 545.
TANNER, N. (1957) *Phys. Rev.* **107**, 1203.
TELEGDI, V. L. (1960) *Proc. Int. Conf. on High Energy Physics at Rochester*, p. 713.
THIBAUD, J. (1933) *Compt. Rend.* **197**, 915.
TITCHMARSH, E. C. (1948) *Introduction to the Theory of Fourier Integrals*, Oxford.
TOMONAGA, S. and ARAKI, G. (1940) *Phys. Rev.* **58**, 90.
TREIMAN, S. B. (1959) *Phys. Rev.* **113**, 355.
TRIPP, R. D., WATSON, M. B. and FERRO-LUZZI, M. (1962a) *Phys. Rev. Lett.* **8**, 175; (1962b) *ibid.* **9**, 66.
TSAI, Y. S. (1961) *Phys. Rev.* **122**, 1898.
UHLENBECK, G. E. and GOUDSMIT, S. (1925) *Naturwiss.* **13**, 953.
WALKER, W. D. (1955) *Phys. Rev.* **98**, 1407.
WALKER, J. K. and BURG, J. P. (1962) *Phys. Rev. Lett.* **8**, 37.

WATSON, G. N. (1944) *Theory of Bessel Functions*, Cambridge.
WATSON, K. M. (1954) *Phys. Rev.* **95,** 228.
WATSON, M. B., FERRO-LUZZI, M. and TRIPP, R. D. (1963) *Phys. Rev.* **131,** 2248.
WEINBERG, S. (1958) *Phys. Rev.* **112,** 1375.
WEINBERG, S. (1960) *Phys. Rev. Lett.* **4,** 575.
WEINBERG, S. (1964) *Phys. Rev.* **133,** B1318.
WELTON, T. A. (1948) *Phys. Rev.* **74,** 1157.
WENTZEL, G. (1956) *Proc. Sixth Rochester Conf.* VIII-15.
WENZEL, W. A. (1960) *Proc. Conf. on High Energy Physics at Rochester*, p. 151.
WEYL, H. (1929) *Z. für Phys.* **56,** 330.
WEYL, H. (1931) *Theory of Groups and Quantum Mechanics*, Methuen.
WHEELER, J. A. (1937) *Phys. Rev.* **52,** 1107.
WICK, G. C. (1950) *Phys. Rev.* **80,** 268.
WICK, G. C. (1955) *Rev. Mod. Phys.* **27,** 339.
WIGHTMAN, A. S. (1956) *Phys. Rev.* **101,** 860.
WIGNER, E. P. (1932) *Nach. Gess. Wiss.*, Göttingen, p. 546,
WIGNER, E. P. (1949) *Proc. Amer. Phil. Soc.* **93,** 521.
WILLIAMS, E. J. and ROBERTS, G. E., (1940) *Nature* **145,** 102.
WILSON, R. R. (1958) *Phys. Rev.* **110,** 1212.
WILSON, R. R. (1961) *Aix-en-Provence Conf. Rep.*, CEN Saclay, **2,** 21.
WOLF, S. E. *et al.* (1961) *Rev. Mod. Phys.* **33,** 439.
WOLFENSTEIN, L. (1954) *Phys. Rev.* **96,** 1654.
WOLFENSTEIN, L. and ASHKIN, J. (1952) *Phys. Rev.* **85,** 947.
WOOLCOCK, W. S. (1961) *Aix-en-Provence Conf. Rep.*, CEN Saclay, **1,** 459.
WU, C. S., AMBLER, E., HAYWARD, R. W., HOPPES, D. D. and HUDSON, R. P. (1957) *Phys. Rev.* **105,** 1413.
XUONG, N. H. and LYNCH, G. R. (1961) *Phys. Rev. Lett.* **7,** 327.
YANG, C. N. (1957) *Rev. Mod. Phys.* **29,** 231.
YANG, C. N. and TIOMNO, J. (1950) *Phys. Rev.* **79,** 495.
YORK, C. M., LEIGHTON, R. B. and BJORNERUND, E. K. (1953) *Phys. Rev.* **90,** 167.
YORK, C. M., KIM, C. O. and KERNAN, W. (1959) *Phys. Rev. Lett.* **3,** 288.
YUAN, L. C. L. (1956) *CERN Symposium*, **2,** 195.
YUKAWA, H. (1935) *Proc. Phys. Math. Soc. Japan* **17,** 48.
ZEMACH, C. (1964) *Phys. Rev.* **133,** B1201.
ZWEIG, G. (1964) *CERN preprint* 8419/Th 412.

INDEX